AQA GCSE MATHS

FOUNDATION

Steve Fearnley
June Haighton
Steve Lomax
Peter Mullarkey
James Nicholson
Matt Nixon

Powered by **MyMaths**.co.uk

OXFORD
UNIVERSITY PRESS

OXFORD
UNIVERSITY PRESS

Great Clarendon Street, Oxford, OX2 6DP, United Kingdom

Oxford University Press is a department of the University of Oxford. It furthers the University's objective of excellence in research, scholarship, and education by publishing worldwide. Oxford is a registered trade mark of Oxford University Press in the UK and in certain other countries

British Library Cataloguing in Publication Data
Data available

978-0-19-835165-8

3 5 7 9 10 8 6 4 2

Paper used in the production of this book is a natural, recyclable product made from wood grown in sustainable forests. The manufacturing process conforms to the environmental regulations of the country of origin.

Printed by Bell and Bain Ltd, Glasgow

Acknowledgements
The publisher would like to thank David Bowles, John Guilfoyle and Katie Wood for their contributions to this book.

Although we have made every effort to trace and contact all copyright holders before publication this has not been possible in all cases. If notified, the publisher will rectify any errors or omissions at the earliest opportunity.

Cover: David Ashley/Shutterstock; **p2–3:** 1000 Words/Getty Images; **p10:** Yulia Nikulyasha Nikitina/Shutterstock; **p18:** Tania Sohlman/Shutterstock; **p19t:** Coprid/Shutterstock; **p19b:** BonD80/Shutterstock; **p24–25:** oatjo/Shutterstock; **p46–47:** jsorde/Shutterstock; **p28:** Pavol Kmeto/Shutterstock; **p57:** Robyn Mackenzie/Shutterstock; **p58:** gremlin/iStock; **p59:** Pinwheels/Dreamstime; **p62:** Ilya D. Gridnev/Shutterstock; **p63:** AridOcean/Shutterstock; **p68–69:** Carol Gauthier/Shutterstock; **p66:** Dmitry Naumov/Shutterstock; **p90–91:** STILLFX/Shutterstock; **p94:** Georgi Roshkov/Shutterstock; **p95:** Sergiy Kuzmin/Shutterstock; **p99:** Jultud/Shutterstock; **p103l:** All For You/Shutterstock; **p103r:** Seregam/Shutterstock; **p112–113:** Kondor83/Shutterstock; **p114–115:** Adam Gregor/Shutterstock; **p138–139:** Henrik Lehnerer/Shutterstock; **p141:** Valerio Pardi/Shutterstock; **p147:** Rawpixel/Shutterstock; **p148:** javarman/Shutterstock; **p160–161:** kesterhu/Shutterstock; **p182–185:** akg–images/Interfoto; **p187t:** tkemot/Shutterstock; **p187c:** Steve Allen/Shutterstock; **p187b:** James Clarke/Shutterstock; **p191t:** mikecphoto/Shutterstock; **p191b:** Offscreen/Shutterstock; **p194:** Difydave/iStock; **p200–201:** Rtimages/Shutterstock; **p226–227:** Franck Boston/Shutterstock; **p228–229:** BMCL/Shutterstock; **p235:** Kochneva Tetyana/Shutterstock; **p238t:** Zhukov Oleg/Shutterstock; **p238b:** Mikio Oba/Shutterstock; **p241:** hohl/iStock; **p250–251:** kosmos111/Shutterstock; **p270–271:** watchara/Shutterstock; **p280:** Brendan Howard/Shutterstock; **p283:** Ieva Geneviciene/Dreamstime; **p288–289:** Stephen Coburn/Shutterstock; **p306–307:** Carsten Peter/Speleoresearch & Films/National Geographic/Getty Images; **p311:** Nikonaft/Shutterstock; **p315l:** duckycards/iStock; **p315r:** bluestocking/iStock; **p315b:** Firmafotografen/iStock; **p319b:** duckycards/iStock; **p319t:** pawel.gaul/iStock; **p324–325:** zygomaticus/Shutterstock; **p326–327:** American Spirit/Shutterstock; **p348–349:** Science Photo/Shutterstock; **p353:** hakandogu/Shutterstock; **p356:** OUP; **p357:** duckycards/iStock; **p366–367:** katatonia82/Shutterstock; **p371:** HES Photography/Shutterstock; **p375:** Cheryl Casey/Dreamstime; **p386–387:** Pal Teravagimov/Shutterstock; **p394:** Ozgur Guvenc; **p395t:** Ozgur Guvenc; **p395b:** budgetstockphoto/BigStock; **p397:** northallertonman/Shutterstock; **p401:** vuk8691/iStock; **p408–409:** northallertonman/Shutterstock; **p426–427:** Shebeko/Shutterstock; **p428–429:** Robert Zp/Shutterstock; **p443:** Malll Themd/Shutterstock; **p457:** Natalia Dobryanskaya/Shutterstock

Approval message from AQA

This textbook has been approved by AQA for use with our qualification. This means that we have checked that it broadly covers the specification and we are satisfied with the overall quality. We have not however reviewed the InvisiPen links, and have therefore not approved this content. Full details of our approval process can be found on our website.

We approve textbooks because we know how important it is for teachers and students to have the right resources to support their teaching and learning. However, the publisher is ultimately responsible for the editorial control and quality of this book.

Please note that when teaching the GCSE Mathematics (8300) course, you must refer to AQA's specification as your definitive source of information. While this book has been written to match the specification, it cannot provide complete coverage of every aspect of the course.

A wide range of other useful resources can be found on the relevant subject pages of our website: www.aqa.org.uk.

Contents

About this book

This book has been specially created for the new AQA GCSE Mathematics examination (8300).

It has been written by an experienced team of teachers, consultants and examiners and is designed to help you obtain the best possible grade in your maths GCSE.

As well as mathematical fluency, Assessment Objective 1 (AO1), the new course places an increased emphasis on your ability to reason, AO2, and your ability to apply mathematical knowledge to problem solving, AO3. This change of emphasis is built into the way topics are covered in this book.

In each chapter the lesson are organised in pairs. The first lesson is focussed on helping you to master the basic skills required (AO1) whilst the second lesson applies these skills in questions that develop your reasoning and problem solving abilities (AO2 & 3).

Throughout the book four-digit MyMaths codes are provided allowing you to link directly, using the search bar, to related lessons on the MyMaths website: so you can see the topic from a different perspective, work independently and revise.

At the end of a chapter you will find a summary of what you should have learnt together with a review section that allows you to test your fluency with the basic skills (AO1). Depending on how well you do a *What next?* box provides suggestions on how you could improve even further. This includes links to InvisiPen worked solution videos contained on the accompanying online Kerboodle. Finally there is an Assessment section which allows you to practise exam-style questions (AO1 – 3).

At the end of the book you will find a guide to understanding key phrases and terms and a full set of answers to all the exercises.

The GCSE maths specification identifies two types of content at foundation level. A coloured band in the top-right corner indicates what type of content is included in a lesson.

	Standard	All students should develop confidence and competence with this content.
	Underlined	All students will be assessed on this content; more highly attaining students should develop confidence and competence with this content.

We wish you well with your studies and hope that you enjoy this course and achieve exam success.

1 Calculations 1

Introduction

When you go shopping in a supermarket, you are presented with hundreds of products, often looking similar, as well as lots of different offers. You need to be able to do arithmetic in your head to ensure that you are keeping within your budget and also that you choose the best value offer.

What's the point?

Being able to add, subtract, multiply and divide doesn't just mean that you're good at maths at school – it means that you can confidently look after your own finances in the real world.

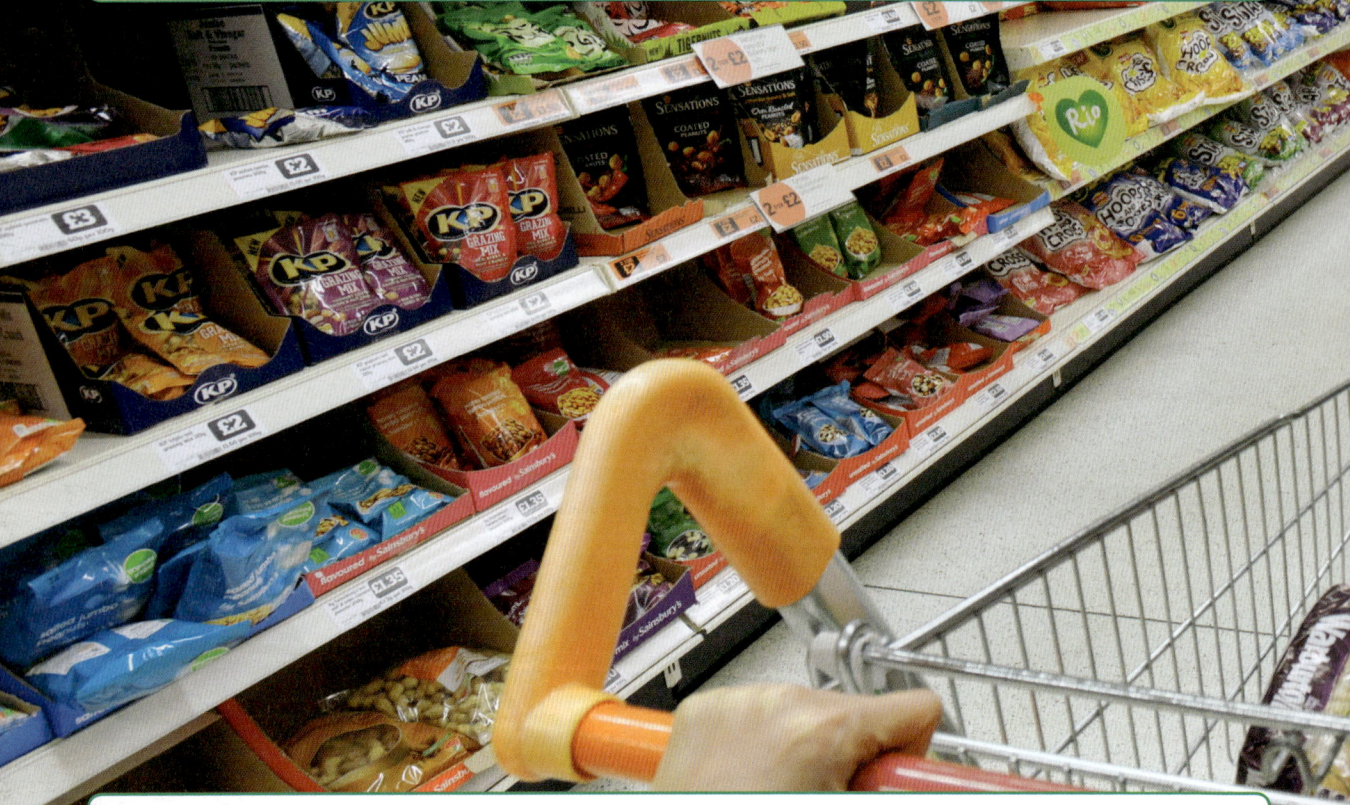

Objectives

By the end of this chapter you will have learned how to ...

- Use place value when calculating with decimals.
- Order positive and negative integers and decimals using the symbols $=$, $\neq$, $<$, $>$, $\leq$, $\geq$.
- Round to a number of decimal places or significant figures.
- Add and subtract positive and negative integers and decimals.
- Multiply and divide positive and negative integers and decimals.
- Use BIDMAS in multi-stage calculations.

Check in

1 Write the number three hundred and four in figures.

2 Calculate

    a 7×10 **b** $5 - 17$ **c** $40 \div 8$

3 Put these numbers in order starting with the smallest.

    $-8, -1, 2, -5, -3, 4$

Chapter investigation

Computers do not use the decimal system that we are familiar with, that is the digits 0 to 9. They use the binary system, which is composed only of the digits 1 and 0.

Investigate the binary system. How would you write the decimal number 37 in binary?

1.1 Place value

● In the **decimal system**, the value of each **digit** in a number depends upon its **place value**.

In the number 4237.65

Th	H	T	U	•	tenths	hundredths
4	2	3	7	•	6	5

> In words it is four thousand two hundred and thirty-seven point six five.

p.218

● Inequality and equality signs show the relationship between two numbers.

   $<$ means less than $\leqslant$ means less than or equal to $=$ means equal to

   $>$ means greater than $\geqslant$ means greater than or equal to $\neq$ means not equal to

EXAMPLE

Place the correct symbol $<$, $>$ or $=$ between the numbers in each pair.

  a 5.07 5.7   **b** 397 379   **c** -10 5   **d** -19 -24 **e** $\dfrac{3}{2}$ 1.5

  a $5.07 < 5.7$ **b** $397 > 379$ **c** $-10 < 5$ **d** $-19 > -24$ **e** $\dfrac{3}{2} = 1.5$

You can use a place value table to multiply and divide by a power of 10, such as 10 or 100.

● To multiply a number by 10 move all the digits one place to the left.

● To divide a number by 100 move all the digits two places to the right.

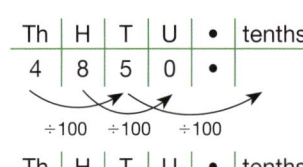

$37 \times 10 = 370$

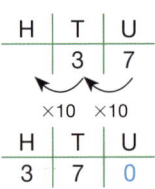

H	T	U
	3	7

×10 ×10

H	T	U
3	7	0

▲ The 0 holds the digits in place.

$4850 \div 100 = 48.5$

Th	H	T	U	•	tenths
4	8	5	0	•	

÷100 ÷100 ÷100

Th	H	T	U	•	tenths
		4	8	•	5

● **Negative numbers** are numbers below zero.

> −14 is further away from zero than −13, so it is smaller.

EXAMPLE

Place these numbers in **order**, starting with the smallest.

$-13, -14, 2, -5, -3, 4$

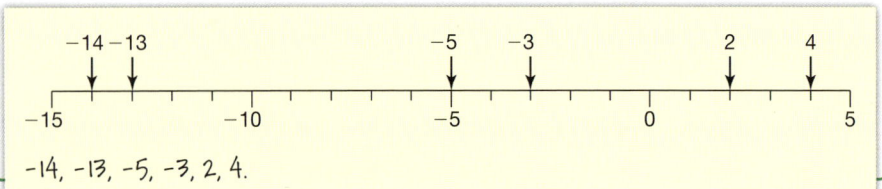

$-14, -13, -5, -3, 2, 4.$

Exercise 1.1S

1 Write each of these numbers in figures.

a Eighty seven

b One hundred and forty-three

c Four hundred and six

d Four hundred and sixty

e Two thousand and fifty-three

f Eight thousand, five hundred and three

g Eight thousand, five hundred and thirty

h Thirty-four thousand, six hundred and forty

i Thirty thousand, four hundred and sixty four

j Two hundred and six thousand, five hundred and three

2 Write each of these numbers in words.

a 456 **b** 13 200

c 115 020 **d** 460 340

e 4 325 400 **f** 55 670 345

g 45.8 **h** 367.03

i 4503.34 **j** 2700.02

3 Put each list of numbers in order, starting with the smallest.

a 56, 34, 9, 112, 178, 89, 139

b 2372, 1784, 2386, 1990, 3233, 3022

c 40 500, 45 045, 4555, 4005, 40 545

d 240 440, 204 044, 24 445, 42 024

4 Put each list of numbers in order, starting with the largest.

a 5.103, 5.099, 5.2, 5.12, 5.007

b 0.545, 0.55, 0.525, 0.5, 0.509

c 7.302, 7.403, 7.35, 7.387, 7.058

d 0.4, 4.2, 0.42, 42, 2.4

e 27.6, 26.9, 27.06, 26.97, 27.1

f 13.3, 14.15, 13.43, 13.19, 14.03

5 Place the correct symbol $=$ or $\neq$ between the numbers in each pair.

a 3.6   3.60

b 450   Four hundred and five

c 80.71   Eighty seven point one

d 9.50   Nine point five

6 Place the correct symbol $<$ or $>$ between the numbers in each pair.

a 15   16 **b** 21   12

c 3.7   7.3 **d** 6.9   7

e 3.01   3.002 **f** 14.9   14.99

7 Calculate these multiplications and divisions.

a 12×10 **b** 4×100

c $320 \div 10$ **d** $4600 \div 100$

e 30×10 **f** 4.6×10

g $230 \div 100$ **h** $659 \div 10$

i 34×1000 **j** 3.56×100

k $23.6 \div 10$ **l** 0.345×100

8 Put these lists of numbers in order from lowest to highest.

a $-13, -6, 0, 17, -12, 15$

b $0, -5, -6, -8, -3, -7$

c $2, 1, -2, 4, 3, -5$

d $-1.5, 3, 9, -3, 2, -8$

e $-3, 2, -5, -4.5, 3, -2$

f $3, 8, 6, -9, -1, 2$

g $-1, -3, 0, -4.5, 5.5, -2.5$

h $-5, -5.1, -6, -5.8, -5.7, -5.4$

9 Calculate these additions and subtractions.

a $4 + 12$ **b** $5 - 12$

c $7 - 3$ **d** $14 + 23$

e $34 - 17$ **f** $8 - 15$

g $-3 + 12$ **h** $-23 + 12$

i $-15 + 7$ **j** $-13 + 34$

k $-5 - 3$ **l** $-5 + 3$

m $-12 - 6$ **n** $21 - 17$

o $-8 + 3$ **p** $-4 + 8 - 2$

q $-12 - 3 - 5$ **r** $13 - 8 + 5$

s $-5 + 4 - 7$ **t** $-12 - 4 - 12$

Q 1069, 1072, 1103, 1392   SEARCH

1.1 Place value

- The value of each digit in a number depends on its place value.
- To multiply or divide by a power of 10, move the digits to the left when multiplying and to the right when dividing.
- Adding a negative number is the same as subtracting a positive number. Subtracting a negative number is the same as adding a positive number.

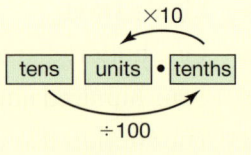

		×10	
tens	units	•	tenths

÷100

<	less than
≤	less than or equal to
>	greater than
≥	greater than or equal to
=	equal to
≠	not equal to

HOW TO

To solve place value questions
1. RTQ – Read the Question, and think what to do.
2. Apply your knowledge of place value.
3. ATQ – Answer the Question.

EXAMPLE

John pays a bill for £356.42, but has mixed up the place value of the digits 6 and 4.

a Write the correct amount. **b** How much has his mistake cost?

1. Swap 6 and 4.

a £354.62 ③

1. Find the difference.

b 356.42 − 354.62 = 1.8
£1.80 ③

2. Always put two digits after the decimal point when writing amounts of money.

EXAMPLE

Gemma's score on her mobile phone game is 3645. She needs to score 4000 to reach the next level. She can hit the following targets: 1, 5, 10, 15, 40, 50, 60, 300, 400, 500.

a Which three targets must she hit to score exactly 4000?

b She reaches 4000 and gets a bonus '100 × score!'
What is her new score?

1. Find the difference and partition by place value.

a 4000 − 3645 = 355 ② 355 = 300 + 50 + 5

3. Targets 300, 50 and 5. Or 300, 40 and 15.

b 400 000 ③

2. Digits move two places to the left; add two placeholder zeros.

EXAMPLE

An engineer measures the thickness of four sheets of metal.

2.05 mm   2.033 mm   2.4 mm   2.303 mm

If she piled up 100 of the thickest sheets, how high would the pile be?

1. Order the numbers to identify the thickest. It can help to add placeholder zeros.

2.050   2.033   2.400   2.303

2. Then rearrange into order, smallest first.

2.033 mm   2.05 mm   2.303 mm   2.4 mm

Multiply 2.4 by 100 by moving digits two places left.

3. 240 mm

> If it helps, read the digits after the point as fifty, thirty three, four hundred, and three hundred and three.

Exercise 1.1A

1 Dan has given a customer a bill for £356.28. He realises he has mixed up the 6 and the 2. How much does he have to pay back to the customer?

2 Veneer is a thin sheet of attractive wood. A joiner has a pile of 10 sheets of oak veneer. Each sheet is 0.1 cm thick.

  a How thick is the pile in mm?

  b A sheet of veneer is glued on to the top of different blocks of wood. What is the new overall thickness in mm of

   i a 5 cm block

   ii a 3.5 cm block

   iii a 12.25 cm block?

  c What would be the new thicknesses of the blocks above if veneer was glued to the top and the bottom?

3 Bev is training for a 100 metre race. Her practice times, in seconds, are

  12   11.06   11.5   11.2   11.12

  What is

  a Bev's fastest time

  b her slowest time?

  c In the race she improves her fastest time by three hundredths of a second. What time did she run?

4

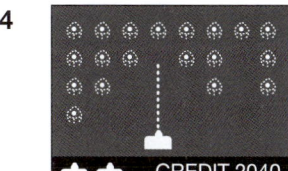

  a Leia is playing an old space invader game. Her score is 2040.
  What is her new score after she has the following score increases in turn?

  5, 70, 200, 1100?

  b She hits a bonus target scoring her 25 more. Then she gets hit. Her new score is 140. Explain what the hit did to decrease her score.

5 Six students raise money for charity. A benefactor gives money to each student using a rule to make it fair. Dipak records the information in a table.

Money raised	Benefactor donation
Abi – £32.40	£324
Bobby – £56	£560
Charles – £8.05	£80.50
Dipak – £40.53	**i**
Ellen – **ii**	£1020
Fergus – 98 p	**iii**

  a Describe the rule that the benefactor uses.

  b Fill in the gaps in Dipak's table.

  c How much is raised altogether?

*6 Sarah is investing some money by buying shares. She wants some mining shares and looks up prices on the Internet on Monday.

Share Prices	
ABC Gold	72.00 p
Diamond Holdings	2.73 p
Potash Mining Ltd	0.92 p
Tungsten Mines Inc.	1883 p

  a What is the cost, in pounds, of

   i 1000 shares in Diamond Holdings

   ii 100 shares in Potash Mining Ltd

   iii 10 shares in Tungsten Mines Inc.

   iv 1000 shares in ABC Gold?

  b Sarah paid £9.20 for shares in Potash Mining Ltd.
  How many shares did she buy?

  c Sarah bought 1000 shares in Diamond Holdings on Monday. On Tuesday, the price went up by 0.1 p.
  If Sarah had bought the shares on Tuesday, how much extra would she have paid for them?

1069, 1072, 1103, 1392   SEARCH

1.2 Rounding

p.180

You can **round** a number to the nearest whole number, 10, 100, 1000, and so on; or to a given number of **decimal places** (dp) or **significant figures** (sf).

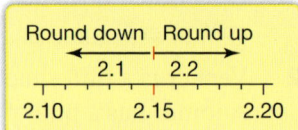

- To round a number, look at the next (smaller) digit
  - next digit = 0, 1, 2, 3 or 4 ⇒ round down
  - next digit = 5, 6, 7, 8 or 9 ⇒ round up.

EXAMPLE

Round 16.473

   a to the nearest whole number   **b** to 1 dp   **c** to 2 dp   **d** to the nearest 10.

   a 16 Look at the tenths digit.   **b** 16.5 Look at the hundredths digit.   **c** 16.47 Look at the thousandths digit.   **d** 20 Look at the units digit.

- The first digit that is not zero in a number is called the **first significant figure**. It has the highest value in the number.
  The next, smaller, digit is the second significant figure, and so on.

> Always look at the next digit to the right, to decide which way to round.

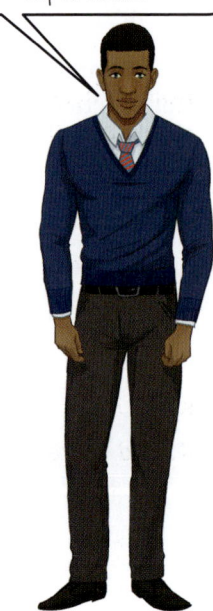

You can round a number to a given number of significant figures in the same way.

EXAMPLE

Round these numbers to 1 sf.

   a 7560   **b** 52.3   **c** 1.5

You might find a place value table helpful.

	Th	H	T	U	•	tenths
a	7	5	6	0	•	
b			5	2	•	3
c				1	•	5

   a 8000 First significant figure is 7, the next digit is 5, so round up.

   b 50 First significant figure is 5, the next digit is 2, so round down.

   c 2 First significant figure is 1, the next digit is 5, so round up.

EXAMPLE

Round 54.76 to 2 sf.

Find the 2nd significant figure and look at the next digit.

T	U	•	tenths	hundredths
5	4	•	7	6

The next digit is 7, so the 2nd significant figure is rounded up.

54.76 = 55 (2 sf).

Exercise 1.2S

1 Round these numbers to the nearest 10.

    a 48   **b** 89

    c 483   **d** 792

    e 2638   **f** 6193

2 Round these numbers to the nearest 100.

    a 343   **b** 484

    c 882   **d** 2732

    e 5678   **f** 16491

> You could sketch a number line to help you.

3 Round these numbers to the nearest 1000.

    a 3448   **b** 2895

    c 4683   **d** 36927

    e 62532   **f** 261932

4 Round these numbers to the nearest
    i 1000   **ii** 100   **iii** 10.

    a 3472   **b** 81382

    c 1236.4   **d** 283.4

    e 13998   **f** 9999

5 Round these numbers to the nearest whole number.

    a 4.8   **b** 3.9

    c 11.6   **d** 25.074

    e 16.286   **f** 435.972

    g 3.7   **h** 8.7

    i 18.63   **j** 69.49

    k 109.9   **l** 6.899

6 Round these numbers to 1 dp.

    a 0.27   **b** 2.89

    c 3.82   **d** 12.48

7 Round these numbers to
    i 2 dp   **ii** 1 sf.

    a 0.327   **b** 2.869

    c 3.802   **d** 14.458

8 Round these numbers to the
    i nearest 10   **ii** nearest 100
    iii nearest 1000.

    a 3487   **b** 3389

    c 14853 m   **d** £57792

    e 92638 kg   **f** £86193

    g 3438.9   **h** 74899.36

9 Round these numbers to the nearest whole number.

    a 3.738   **b** 28.77

    c 468.63   **d** 369.29

    e 19.93   **f** 26.9992

    g 100.501   **h** 0.001

10 Round these numbers to
    i 3 dp   **ii** 2 dp   **iii** 1 dp.

    a 3.4472   **b** 8.9482

    c 0.1284   **d** 28.3872

    e 17.9989   **f** 9.9999

    g 0.003987   **h** 2785.5555

11 Round these numbers to
    i 3 sf   **ii** 2 sf   **iii** 1 sf.

    a 8.3728   **b** 18.82

    c 35.84   **d** 278.72

    e 1.3949   **f** 3894.79

    g 0.008372   **h** 2399.9

    i 8.9858   **j** 14.0306

    k 1403.06   **l** 140306

12 Round these numbers to the accuracy given.

    a 17.034 (1 dp)   **b** 2307 (2 sf)

    c 7.0169 (2 sf)   **d** 10.608 (2 dp)

    e 14624 (3 dp)   **f** 0.0500 (3 dp)

    g 101.101 (2 sf)   **h** 9.999 (2 sf)

    i 94952.6 (3 sf)   **j** 9999.55 (1 dp)

1.2 Rounding

- To **round** to the nearest whole number, 10, 100 or 1000, or to a number of **decimal places** or **significant figures** look at the next digit: five or more rounds up, four or less rounds down.
- For decimal places count from the decimal point.
- For significant figures count from the first **non-zero** digit.

HOW TO

To solve a problem involving rounding
① RTQ – Read the Question, and decide what to do.
② Apply your knowledge of rounding.
③ ATQ – Answer the Question.

EXAMPLE

A school buys 12 laptops at £234.49 each.
The headteacher need to know the cost to the nearest 100.
She calculates it to be £2400.

    a   How did she get £2400?

    b   Provide a better approximation.
        Explain why your approximation is better.

   a   2400 ÷ 12 = £200   ① Work out what price she used for each laptop.
        She rounded £234.49 to the nearest 100, then multiplied by 12. ② ③

   b   234.49 x 12 = 2813.88   ② Calculate the total cost using the exact values
                                and then round the answer.

       2813.88 ≈ £2800   ③

       £2800 is a better approximation because by only rounding at the end, the calculation
       is more accurate.

EXAMPLE

Samit uses scientific scales to find the mass of four common creatures.
Mouse 22.158 g   Wren 8.4325 g   Spider 0.006 25 g   Fly 0.000 012 5 g

He wants to compare their masses; should he round to 2 dp or 2 sf?
Give your reasons.

   a   Round each mass to 2 dp and 2 sf. ① ②
       Mouse 22.16 g (2 dp)   Wren 8.43 g (2 dp)   Spider 0.01 g (2 dp)   Fly 0.00 g (2 dp)
       Mouse 22 g (2 sf)   Wren 8.4 g (2 sf)   Spider 0.0063 g (2 sf)   Fly 0.000 0013 g (2 sf)
       He should round to 2 sf.
       Rounding to 2 dp would suggest the fly has no mass. ③

> The mass of 0.00 g (2 dp) is mathematically correct but not very useful!

Exercise 1.2A

1 A supplier sells 7 tons of gravel at £124.99 per ton. He wants to give his customer a rough idea of the cost. Which rounded value is more accurate? Why?

    a $7 \times £100 = £700$ **b** $874.93 \approx £900$

2 Votes for four politicians were declared.

    CON   25 958   LIB   2705

    LAB   26 057   UKIP   5651

    The local newspaper decides to round these to the nearest 1000 in its report.

    a   What would each result be reported as?

    b   Why would there be a problem if the election officials also decided to round to the nearest 1000?

    c   Which power of 10 could they round to, without having the same problem?

3 Nicki decides to give some money to a charity each week for a month. She finds 10% of her earnings, then rounds this off to the nearest pound and sends it to the charity. She works different hours each week.

    Her weekly wages for February are

    £598.30   £645   £658.46   £720.32

    a   How much has she given to the charity by the end of the month?

    b   Would she give more or less if she had just given 10% of her month's earnings, rounded off at the end of the month? By how much?

4 Scientific calculators often show long decimal numbers on the display screen.

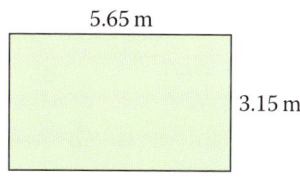

    Here are four special numbers

    $\pi = 3.141\,592\,653\ldots$

    $e = 2.718\,281\,828$

    $\dfrac{1}{6} = 0.166\,666\,666\ldots$

    $\sqrt{200} = 14.142\,13$

4 Litha noticed that $\pi \approx 3.14$ to 2 dp and to 3 sf; and that $e \approx 2.72$ to 2 dp and 3 sf. She said this proves that rounding to 2 dp and 3 sf is the same. Use the other special numbers to show she is not correct.

5 The diagram shows a plan of Phil's living room.

$$5.65\,\text{m}$$

    3.15 m

    Find the area of Phil's living room to

    a   2 dp   **b**   2 sf

    c   Phil wants to estimate the cost of a new carpet. Should he use the answer from part **a** or **b**? Give your reasons.

    d   A carpet fitter wants to know how much carpet to order. Should she use the answer from part **a** or **b**? Give your reasons.

*6 A student has to work out $\sqrt{3} \times 25^3$ and leave her answer correct to 3 sf. She makes a mistake and rounds each number to 3 sf before multiplying them.

    a   What answer does she get?

    b   She then tries again, this time rounding at the end. What is the difference between her two answers?

*7 HM revenue rounds your earnings down to the nearest pound and charges you tax as a percentage of your earnings.

    If your earnings for January, February and March are

    £121.28, £75.80, £151.60

    would you pay more or less tax if you rounded using maths rules rather than the HM revenue rules?

p.96

p.280

1.3 Adding and subtracting

There are two rules for adding and subtracting negative numbers.

- Adding a negative number is the same as subtracting a positive number.

- Subtracting a negative number is the same as adding a positive number.

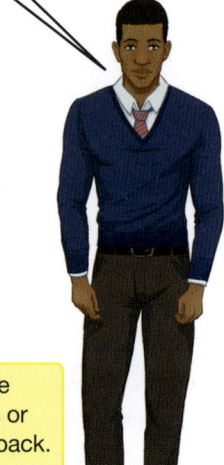

Subtraction is the inverse of addition, it undoes it. So, you can think of $-(-2)$ as undoing the inverse, that is, adding a positive!

EXAMPLE

Calculate **a** $-5 + -6$ **b** $+4 - -2$

a $-5 + -6 = -5 - 6 = -11$ **b** $+4 - -2 = 4 + 2 = 6$

You can use tricks to add and subtract numbers in your head.

- Use **partitioning** to split a number into useful smaller parts.

- Use **compensation** when a number is close to a whole number, a multiple of 10 or a multiple of 100.
  You round the number and make a correction.

You could use number pairs or counting on/back.

EXAMPLE

Use a mental method to calculate these.

a $19.5 - 7.2$ **b** $5.8 + 4.8$

a $19.5 - 7.2 = 19.5 - 7 - 0.2$ Split smaller number into parts according to place value.
          $= 12.3$ Subtract the units and then the tenths.

b $5.8 + 4.8 = 5.8 + 5 - 0.2$ Write 4.8 as $5.0 - 0.2$.
        $= 10.6$

To use a written method

- Always estimate the answer before you start. It will help you check your answer.

- Set out the calculation in columns and line up the decimal points.

EXAMPLE

Use a written method to calculate these.

a $102.773 + 28.47$ **b** $26.44 - 1.105$

a Estimate: $100 + 30 = 130$ **b** Estimate: $30 - 1 = 29$

$$\begin{array}{r} 102.773 \\ +\ 28.470 \\ \hline 131.243 \\ \hline {\scriptstyle 1\ 1\ 1} \end{array}$$

Remember to include the 'carry' digits.

$$\begin{array}{r} 26.44\overset{3\ 10}{\cancel{4}}0 \\ -\ 1.105 \\ \hline 25.335 \end{array}$$

'Borrow' from the next column.

Add placeholder zeros to ensure your numbers are correctly aligned.

Exercise 1.3S

1 Calculate

a	$150 + 120$	**b**	$170 - 90$
c	$160 + 170$	**d**	$130 - 90$
e	$1900 + 1900$	**f**	$210 - 140$
g	$320 + 110$	**h**	$510 - 120$

2 Calculate

a	$13 + -5$	**b**	$6 + -8$
c	$12 + -3$	**d**	$4 + -4$
e	$-5 + -8$	**f**	$-3 + -11$
g	$-11 + -3$	**h**	$15 - -5$
i	$4 - -8$	**j**	$-2 - -5$
k	$-12 - -7$	**l**	$-14 - -8$
m	$-16 - -20$	**n**	$-13 + -12$
o	$-13 - -12$	**p**	$13 + -12$
q	$-12 + 7 - 4$	**r**	$-12 + -7 - 4$
s	$-20 + -5 - -2$	**t**	$-12 - 7 + -4$

3 Use a mental method for each of these calculations.
Use jottings to show your method.

a	$257 + 98$	**b**	$448 + 112$
c	$427 + 523$	**d**	$256 + 552$
e	$354 + 213$	**f**	$561 + 328$
g	$16.2 - 1.9$	**h**	$5.8 + 14.9$

4 Use an appropriate written or mental method for each of these calculations.

a	$62 - 47$	**b**	$83 - 68$
c	$487 - 356$	**d**	$852 - 728$
e	$548 - 387$	**f**	$589 - 387$
g	$33.4 + 15.2$	**h**	$34.6 + 13.7$
i	$19.8 + 8.8$	**j**	$18.7 + 26.5$
k	$8.7 - 2.5$	**l**	$15.8 - 8.4$
m	$26.3 - 7.9$	**n**	$53.6 - 27.8$
o	$25 + 38 + 68$	**p**	$123 + 76 - 58$
q	$173 - 27 + 56$	**r**	$327 + 176 - 255$

5 Calculate these using a written method.

a	$3.52 + 4.6$	**b**	$13.62 + 2.9$
c	$8.5 + 14.81$	**d**	$75.8 + 28.39$
e	$17.3 - 4.22$	**f**	$16.6 - 3.47$
g	$37.7 - 18.86$	**h**	$57.28 - 38.4$
i	$16.4 + 9.87$	**j**	$49.2 + 7.72$
k	$9.42 - 5.9$	**l**	$26.9 + 9.82$
m	$36.57 - 8.59$	**n**	$36.28 - 17.4$

6 Use a mental or written method to work out these calculations.

a	$23.4 + 13.4$	**b**	$24.6 + 53.7$
c	$19.7 + 7.4$	**d**	$27.8 + 14.3$
e	$9.6 - 3.4$	**f**	$16.7 - 9.6$
g	$16.3 - 7.8$	**h**	$61.7 - 33.8$

7 Calculate these using a written method.

a	$4.32 + 6.4$	**b**	$16.32 + 3.4$
c	$4.5 + 13.61$	**d**	$73.2 + 68.79$
e	$16.3 - 8.25$	**f**	$12.6 - 7.87$
g	$67.3 - 28.56$	**h**	$47.38 - 28.7$
i	$25.3 + 8.76$	**j**	$38.1 + 6.61$
k	$8.31 - 4.8$	**l**	$15.8 + 8.79$
m	$25.46 - 7.48$	**n**	$47.39 - 18.5$

8 Calculate these using a mental or written method.

a $12.3 + 2.7 + 7.08$

b $38.76 + 16.9 - 8.32$

c $61.3 + 14.85 + 7.02$

*9 Harry's grandad uses a different method for subtractions. He adds the carried digit to the subtracted number.

$$\begin{array}{r} 5\ {}^1 2\ .\ 8\ {}^1 0 \\ -{}^2 \not{1}\ 9\ .\ {}^6\not{8}\ 4 \\ \hline 3\ 3\ .\ 2\ 6 \end{array}$$

Use this method to calculate
$67.362 - 21.454$.

1.3 Adding and subtracting

RECAP

- Partitioning and compensation can be used to help you do calculations in your head.
- Always estimate your answer first, so you can check your answer later.
- Written methods often require you to line up numbers in a column according to their place value. You might need to 'carry' digits between columns.

It can help to find number pairs to multiples of 10.

HOW TO

To solve a problem using addition and subtraction

(1) RTQ – Read the Question and decide what you need to do.

(2) First estimate your answer and then use an appropriate written or mental method.

(3) ATQ – Answer the Question; use your estimate to check your answer.

EXAMPLE

Katie obtained these marks on Paper 1 of her maths exam.

$$3, 5, 2, 7, 5, 8, 11, 4, 5, 0, 7$$

a   Paper 1 is out of 100. How many marks did she lose?

She scored 74 on Paper 2.

b   She needs to get a total of 140 marks to pass Papers 1 and 2. Did she pass? Give your reasons?

a (1) You need to find her total mark, then subtract that from 100.

$(3 + 5 + 2) + (7 + 5 + 8) + (11 + 4 + 5) + 7$

$= 10 + 20 + 20 + 7$ (2)

$= 57$

$100 - 57 = 43.$ Count on from 57

$57 + 3 + 40 = 100$

She lost 43. (3)

b (1) Add 74 to the total from part **a**.

$57 + 74 = 131$ (2) $50 + 70 = 120$

$7 + 4 = 11$

$120 + 11 = 131$

No, beacuse 131 < 140. (3)

EXAMPLE

Dennis receives two payments of £568.42 and £76 for work done. He also has to pay a bill for £265.16

What does he have left after paying his bill?

(1) Add the payments and then subtract the bill.

Estimate 570 + 80 = 650

$\begin{array}{r} 568.42 \\ +76.00 \\ \hline 644.42 \end{array}$ (2) Use zeros to line up columns.

Estimate 650 − 270 = 380

$\begin{array}{r} 644.42 \\ -265.16 \\ \hline 379.26 \end{array}$ (2) Take care when carrying.

(3) Money left = £379.26

Exercise 1.3A

1 Jamie has scored the following runs so far this cricket season.

7, 15, 23, 5, 8, 1, 12, 9, 0, 42 and 5.

He wants to score at least 150 runs.
How many more runs does he need to score?

2 Find the missing digits.

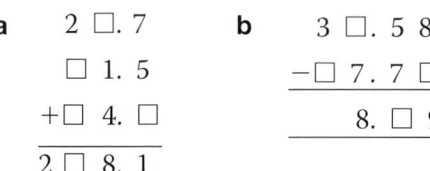

a
```
  2 □ . 7
    □ 1 . 5
  +□ 4 . □
  ─────────
  2 □ 8 . 1
```

b
```
  3 □ . 5 8
 −□ 7 . 7 □
 ──────────
    8 . □ 9
```

3 Copy and complete the pyramid.
Each number is the sum of the two numbers below it.

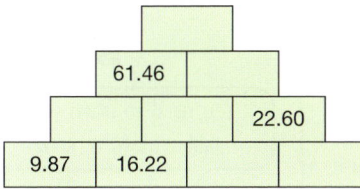

4 Three mothers look at some toddler books.
These are the prices.

'Ants' £6.78   'Bees' £5.42

'Cows' £2.99   'Ducks' £1.28

Each mother pays with a £20 note and gets the following change.

a Rebekah   £4.81

b Suzie   £10.23

c Tarah   £8.60

Which books did each mother buy?

5 Here are the populations of four villages

Thornton   4675   Norton   12 250

Brawton   562   Lockton   853

a What is the range of population (the difference between biggest and smallest)?

b What is the total population?

c New houses are built in Brawton. It is no longer the smallest of the four villages. At least how many people have moved in?

6 Five friends run a league table of points scored on a computer game they all play.

Name	Score
Asha	6523
Bryony	12 653
Callum	7350
Dora	964
Edward	7000
Total	

a Copy and complete the table, but put the friends in order, with the player with the most points at the top.

b Dora's score is recalculated following an error and she now has the most-points by 5 points.
By how much was her original score wrong?

*7

1026	432
742	342
1484	522

a Find a pair of numbers in this table where their total is twice their difference.

b Can you find a pair where the total is three times the difference?

c Can you find four numbers where one pair adds up to twice the total of the other pair?

8 Lucia is making a bean casserole.
Here is her recipe.

Bean casserole	
Kidney beans	1.6 kg
Red onions	0.375 kg
Celery	0.15 kg
French beans	
Tomatoes	1.2 kg

The total mass of ingredients is 3.525 kg.
Find the mass of French beans.

1.4 Multiplying and dividing

You need to remember these rules for negative numbers.

- Negative × positive = negative
- Negative × negative = positive
- Positive ÷ negative = negative
- Negative ÷ positive = negative
- Negative ÷ negative = positive

Multiplication is repeated addition, and division is repeated subtraction. So multiplying two negatives makes a positive because you are repeatedly subtracting a negative, that is, you are inversing an inverse.

If the signs are different the answer will be negative. If the signs are the same the answer will be positive.

Two useful mental methods to multiply and divide are partitioning and compensation.
$29 \times 2 = 30 \times 2 - 2 = 60 - 2 = 58$

To multiply or divide decimals using a written method, first **estimate** the answer. Then ignoring the decimal point set up the column multiplication or division as normal. Finally insert the decimal in the answer using your estimate.

You need to learn your tables up to 12 × 12.

EXAMPLE

Calculate
a 18.5×7.9 **b** $47.592 \div 1.8$

a   Estimate 20 × 8 = 160

```
    1 8 5
  ×   7 9
  1 6 6 5
+ 1 2 9 5 0
  1 4 6 1 5
```

b   Estimate 50 ÷ 2 = 25

```
        2 6 4 4
    18)4 7 5 9 2
      - 3 6        2 × 18 = 36
        1 1 5
      - 1 0 8      6 × 18 = 108
          7 9
        - 7 2      4 × 18 = 72
          7 2
        - 7 2
            0
```

Use your estimate to check your answer and place the decimal point.

146.15   26.44

- When there is more than one operation, the order in which you apply them is
BIDMAS   (**B**rackets, **I**ndices or powers, **D**ivision, **M**ultiplication, **A**ddition, **S**ubtraction)

EXAMPLE

Calculate
a $\dfrac{34 \times 8}{5 + 11}$

b $30 \div (15 - (12 - 7))$

Always write the calculation a line at a time, so you can see each operation clearly.

a $= \dfrac{(34 \times 8)}{(5 + 11)}$ This is the same as $(34 \times 8) \div (5 + 11)$
   $= 272 \div 16$ First do the brackets.
   $= 17$ Then do the division.

b $= 30 \div (15 - 5)$ Start with the innermost brackets.
   $= 30 \div 10$ Then do the next set of brackets.
   $= 3$ Finally work out the division.

Exercise 1.4S

For questions **1–4**, work out the answer mentally; use jottings to show your method.

1 Calculate

   a $+5 \times -3$ **b** $+2 \times -9$

   c $+7 \times -3$ **d** $-8 \times +7$

   e $-4 \times +9$ **f** $-6 \times +2$

2 Calculate

   a $+5 \times -5$ **b** $+4 \times -8$

   c $-8 \times +9$ **d** $-4 \times +5$

   e -3×-10 **f** -7×-7

   g $+8 \times +2$ **h** $+5 \times -4$

3 Calculate

   a $-18 \div +9$ **b** $-20 \div +4$

   c $-30 \div -6$ **d** $-12 \div -3$

   e $-66 \div +3$ **f** $+47 \div -47$

   g $-80 \div -2$ **h** $+24 \div +6$

4 Calculate

   a 14×7 **b** 19×8

   c 21×13 **d** 17×19

5 Check your answers to question **4** using a written method.

6 Use a written method to work out

   a 4.7×5.3 **b** 1.53×2.8

   c 21.6×4.9 **d** 33.65×3.89

   e 21.58×1.99 **f** 42.77×8.64

7 Use a written method to work out

   a $34.83 \div 9$ **b** $5.425 \div 7$

   c $7.328 \div 8$ **d** $451.8 \div 60$

   e $54.39 \div 3$ **f** $58.65 \div 17$

   g $66.4 \div 16$ **h** $185.76 \div 24$

   i $7.752 \div 1.9$ **j** $3.055 \div 1.3$

8 Use a written method to work out these correct to two decimal places.

   a $14.73 \div 2.8$ **b** $51.99 \div 1.8$

   c $193.8 \div 0.14$ **d** $1013 \div 5.77$

   e $23.78 \div 0.83$ **f** $65.79 \div 0.59$

9 Use a calculator to check your answers to questions **6–8**.

10 Use your knowledge of place value to find these given that $43 \times 67 = 2881$

   a 4.3×6.7 **b** 430×0.067

   c $2881 \div 670$ **d** $28.81 \div 430$

   e $2.881 \div (0.43 \times 0.67)$

11 Calculate these using the order of operations.

   a $2 + 8 \times 3$ **b** $4 \times 11 - 7$

   c $4 \times 3 + 5 \times 8$ **d** $5 + 12 \div 6 + 3$

   e $(2 + 9) \times 3$ **f** $(1.5 + 18.5) \div 4$

   g $(12 + 3) \times (14 - 2)$

12 Calculate these using the order of operations.

   a $(4 + 3) \times 2^2$ **b** $3^2 \times (15 - 7)$

   c $(6^2 - 16) \div 4$ **d** $2^4 \times (3^2 - 2 \times 4)$

   e $128 \div (2 + 2 \times 3)^2$

13 Calculate

   a $\dfrac{7^2 - 9}{5 \times 8}$ **b** $\dfrac{4 \times 8}{4^2}$

   c $\dfrac{15 \times 4}{6 \times 5}$ **d** $\dfrac{2 \times (3 + 4)^2}{7}$

   e $\dfrac{(6 + 4)^2}{20} + 7 \times 5$ **f** $\dfrac{6 + (2 \times 4)^2 + 7}{11}$

14 Calculate

   a $12 + 6 - 4$ **b** $5 \times 4 \div 2$

   c $40 \div 10 \div 2$ **d** $28 - 12 - 4$

   e $13(2 + 5)$ **f** $14(2 + 6)$

   g $5^2 + 9(8 - 3)$ **h** $4^2 + 3(16 - 9)$

   i $4 + (12 - (3 + 12))$

   j $120 \div (8 \times (7 - 2))$

15 Calculate each of these. Where appropriate, give your answer to 2 dp.

   a $(15.7 + 1.3) \times (8.7 + 1.3)$

   b $\dfrac{7^2}{(2.3 \times 4)^2}$ **c** $\dfrac{(7 + 5)^2}{\sqrt{25} + 7 \times 8}$

1.4 Multiplying and dividing

- Partitioning and compensation can be used for mental calculations.
- Exact calculations can be done using the written column methods used for integers, with an estimate used to position the decimal place.

×	+ve	−ve
+ve	+ve	−ve
−ve	−ve	+ve

Brackets
Indices
Division/Multiplication
Addition/Subtraction

HOW TO

To solve problems involving multiplication or division

① RTQ and decide what you need to do.

② Calculation an estimate.

③ Use an appropriate method to do the exact calculation.

④ ATQ.

EXAMPLE

Miss Smith is organises a school trip for 176 students and 12 staff.
Coaches can carry 56 people.
If Miss Smith hires enough coaches for everybody,
how many spare seats are there?

① Find the total number of passengers divided by 56 and *round up*.

176 + 12 = 188

188 ÷ 56 = 3 + remainder   ②   An exact answer is not needed, so use a mental method.

③ 4 × 56 = 4 × 50 + 4 × 6

$\qquad$ = 200 + 24 = 224

224 − 188 = 36.

④ There will be 36 spare places.

EXAMPLE

A good approximation to convert miles to kilometres is to say 1 mile is 1.6 kilometres.

a   How far is 127 miles in km?   **b**   How far is 480 km in miles?

① Part **a** requires a multiplication and part **b** a division.

a   Estimate ≈ 130 × 2 = 260 ②

```
    1 2 7     ③ An exact
  ×   1 6        answer is
  ─────────     needed, so
  1 2 7 0       use a written
+   7 6 2       method.
    1 4
  ─────────
  2 0 3 2
```

203.2 km ④

b   480 ÷ 1.6

$\qquad$ = 300 miles

You could use a written method, or you might spot that

3 × 16 = 48

48 ÷ 16 = 3

480 ÷ 1.6 = 300

Exercise 1.4A

1 Rugby league teams have 13 in the starting line-up and four substitutes. Eight teams are in a tournament.

a How many players are there altogether?

b Each team has six forwards, and three substitute forwards. How many forwards are in the tournament?

2 Nancy receives £6 pocket money per week.

a How much does she receive in February?

b How much in a year? (52 weeks)

c She multiplies the February amount by 12, but gets the wrong answer for a year. Why?

3 Valerie buys some decorative gravel for her garden. She decides on Scottish pebbles.

Scottish pebbles	
1 bag	£5.75
5 bags	£22.50
3 bags cover	1 m²

She measures her garden. It is 8 m².

a How many bags should she order?

b What does her order cost?

c Her friend suggests she orders an extra bag. Why is this a good idea?

4 The maths department orders some exercise books for the new term. There are 30 graph books in a pack, and 25 lined books in a pack. There are 1200 students in the school. Graph packs cost £12.80, lined packs £11.25

a The department decides to buy enough to give each student 1 graph book and 1 exercise book. How many packs of each type do they buy?

b What is the cost of the order?

5 Mick is tiling his floor with a mixed tile pattern.

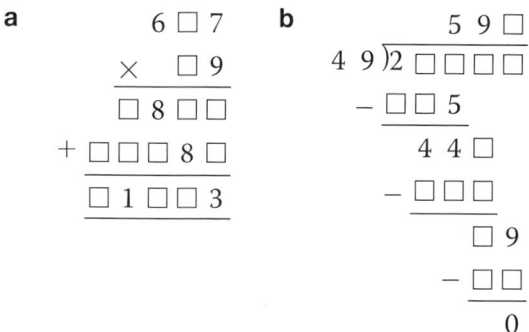

Packs of 12	
Small square	£12.95
Packs of 6	
Medium	£11.65
Large	£20.95
Rectangle	£18.95

The tile company says that for 100 square feet he will need 24 small squares, 25 medium squares, 25 large squares and 26 rectangles. His room is 12 feet wide by 25 feet long.

a How many packs of each size does Mick need?

b How much will the tiles cost?

6 Copy and complete these calculations.

a
```
      6 □ 7
   ×    □ 9
   ─────────
    □ 8 □ □
 + □ □ 8 □
 ───────────
   □ 1 □□ 3
```

b
```
          5 9 □
   4 9 )2 □ □ □ □
      − □ □ 5
        ───────
          4 4 □
        − □ □ □
        ───────
            □ 9
          − □ □
          ───────
              0
```

7 Steve's car manual says his car does '42 mpg'. Mpg is miles per gallon.

A gallon is 4.55 litres. A litre of diesel costs £1.38.

Steve drives from Leeds to Edinburgh, a distance of 210 miles. How much should diesel for the trip cost?

p.144

Summary

Checkout
You should now be able to...

	Test it Questions
✔ Use place value when calculating with decimals	**1, 2**
✔ Order positive and negative integers and decimals using the symbols $=, \neq, <, >, \leqslant, \geqslant$.	**3, 4**
✔ Round to a number of decimal places or significant figures.	**5, 6**
✔ Add and subtract positive and negative integers and decimals.	**7 – 9**
✔ Multiply and divide positive and negative integers and decimals.	**10 – 13**
✔ Use BIDMAS in multi-stage calculations.	**14, 15**

Language — Meaning — Example

Language	Meaning	Example
Decimal system	A number system using a base of 10.	**Thousands Hundreds Tens Units • tenths hundredths** 1000 100 10 1 $\frac{1}{10}$ $\frac{1}{100}$ 5 6 3 8 • 2 7
Digit	The individual symbols 0, 1, 2, 3, 4, 5, 6, 7, 8, 9 that are used on their own or put together to make numbers.	5, 6, 3, 8, 2 and 7 are digits. In the number 5638.27 the 6 has a value of 6 hundreds.
Place value	The value of a digit depends on its position in the number.	
Negative numbers	A negative number is less than zero.	$-3, -2, -1$
Rounding	Making a number less accurate but easier to estimate with.	$552.631 = 553$ (nearest whole number)
Decimal places (dp)	The number of digits after the decimal point. A number can be rounded to a given number of decimal places.	$552.631 = 552.63$ (2 dp)
Significant figures (sf)	Describe the relative importance of digits in a number. A number can be rounded to a given number of significant figures.	In the number 0.00487 the 4 is the 1st significant figure, 8 the 2nd and 7 the 3rd.
First significant figure	The first digit from the left that is not zero.	
Partitioning	Splitting a number into smaller numbers which add up to the original number.	$127 = 100 + 20 + 7$ $152 = 80 + 40 + 32$ $5.1 + 12.7 = 5.1 + 10 + 2 + 0.7$
Compensation	One number is rounded to simplify the calculation, then the answer is adjusted to compensate for the original change.	$142 - 39 = (142 - 40) + 1 = 102 + 1 = 103$ $158 - 18.9 = (158 - 20) + 1.1 = 139.1$

Significant figures table:

4 sf	3 sf	2 sf	1 sf
45920	45900	46000	50000
78.02	78.0	78	80
0.003256	0.00326	0.0033	0.003

Review

1   Work out the value of these expressions.

    a 67×100 **b** 8.52×10

    c 0.24×1000 **d** 0.05×100

2   Work out the value of these expressions.

    a $450 \div 10$ **b** $6210 \div 1000$

    c $7.9 \div 100$ **d** $0.06 \div 10$

3   Copy the numbers and write $>$ or $<$ between them to show which is larger.

    a $905 \cdot 961$ **b** $14.7 \cdot 14.9$

    c $0.7 \cdot 0.09$ **d** $0.214 \cdot 0.22$

4   Put each list of numbers in order, starting with the smallest.

    a   53909 503099 530909 503909 53099

    b   4.3   4.289   4.32   4.09   4.29

    c -8 9 -4 0 -14

5   Round

    a   845   to the nearest 10

    b   25.3   to the nearest whole number

    c   0.846   to 1 decimal place

    d   62.938   to 2 decimal places.

6   Round to the stated level of accuracy

    a   351   1 sf   **b**   5070   2 sf

    c   45.72   3 sf   **d**   0.0845 1 sf

    e   0.0902 2 sf   **f**   0.99   1 sf.

7   Calculate.

    a $4 + -5$ **b** $-3 + -2$

    c $5 - -1$ **d** $-3 - -5$

    e $-7 - -4 + 3$ **f** $6 - -3 + -12$

8   Calculate using mental or written methods.

    a $467 + 891$ **b** $14.9 + 23.5$

    c $905.4 + 8.67$ **d** $0.58 + 6.821$

9   Calculate using mental or written methods.

    a $965 - 45$ **b** $657 - 389$

    c $257.4 - 38.2$ **d** $9.57 - 5.9$

10 Calculate

    a $-3 \times +7$ **b** -8×-4

    c $-35 \div +7$ **d** $-52 \div -4.$

11 Calculate using mental methods.

    a 12×20 **b** 25×16

    c $240 \div 6$ **d** $960 \div 120$

12 Calculate using written methods.

    a 47×63 **b** 192×78

    c $224 \div 8$ **d** $312 \div 12$

13 Calculate using written methods.

    a 3.2×5.6 **b** 4.31×2.7

    c $19.6 \div 8$ **d** $118.65 \div 21$

14 Evaluate these without using a calculator.

    a $13 + 5 \times 4$ **b** $20 - 12 \div 4$

    c $5 \times (2 + 9)$ **d** 5×3^2

    e $12 + \sqrt{36}$ **f** $7(9 - 1) \div 2^2$

15 Use your calculator to work these out.

    a $4.8 + 5.2 \times 6$

    b $\dfrac{39 - \sqrt{25 + 24}}{4^2}$

What next?

<table>
<tr><td rowspan="6">Score</td><td>0 – 5</td><td></td><td>Your knowledge of this topic is still developing.
To improve look at MyMaths: 1001, 1004, 1005, 1007, 1020, 1028, 1068, 1069, 1072, 1103, 1167, 1392, 1393, 1916, 1917</td></tr>
<tr><td>6 – 12</td><td></td><td>You are gaining a secure knowledge of this topic.
To improve your fluency look at InvisiPens: 01Sa – q</td></tr>
<tr><td>13 – 15</td><td></td><td>You have mastered these skills. Well done you are ready to progress!
To develop your problem solving skills look at InvisiPens: 01Aa – f</td></tr>
</table>

Assessment 1

1   Carli tries to order each set of numbers from smallest to largest.
    Carli has made some mistakes.
    Explain where Carli has made a mistake and put each list of numbers in order,
    starting with the smallest.

   a   8, 19, −33, 44, 303, 576   [3]   **b**   −19, −576, 8, 33, 44, 303   [2]

2   Ben tries to put these numbers in ascending order.

    $42 \div 100, 0.3 \times 10, 4236 \div 1000, 516 \div 10, 42 \times 100, 216 \times 1000$

    Has he ordered the numbers correctly? Give reasons for your answer.   [4]

3   In the number grids shown the number in each cell is the sum of the two adjacent
    cells above it. Copy and complete the grids shown.

a
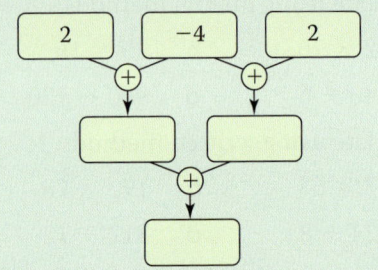
[3]

b
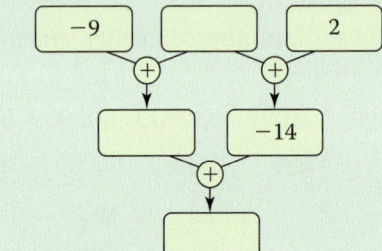
[3]

4   A magic square is a square grid of numbers where each number is different.
    The sum of the numbers in each row, each column and each diagonal is the same.

   a   What number must be the sum of each row, column and diagonal?   [1]

   b   Fill in the missing values in the magic square.   [5]

7		
6	8	10

5   The world's tallest man, Robert Wadlow, was 271.78 (2 dp) cm tall.
    The world's tallest woman, Yao Defen, is 233.34 cm (2 dp) tall.

   a   A challenger to the world's tallest man record measured his height as 271.8 cm
           to 1 dp. Has the challenger definitely beaten the world record?
           Give reasons for your answer.   [1]

   b   A challenger to the world's tallest woman record measured her height as
           233.341 cm. Has the challenger definitely beaten the world record?
           Give reasons for your answer.   [1]

6   Dave is 36 and Jane is 44. Jane says that she and Dave are the same age to 1 sf.

   a   Is Jane correct?   [1]

   b   Will Dave and Jane be the same age to 1 sf in one years time?
           Give reasons for your answer.   [2]

   c   How old will Dave and Jane be the next time their ages are the same to 1 sf?   [2]

7   As the Earth spins on its axis, everything on the Earth's surface moves with it.
    The distance travelled in one day due to the Earth's rotation is $3.142x$, where x is the diameter of the circular path.
    Abena lives on the equator and Edward lives in the UK.

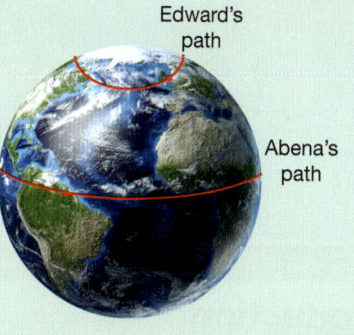

Edward's path

Abena's path

$x = 12\,756$ for Abena and $x = 8134$ for Edward

a   Write both values of x to 2 sf.   [2]

b   Use your answer to part **a** to estimate how much further Abena travels than Edward in one day.   [3]

c   Explain how using values of x to correct to 1 sf would affect the estimate in part **b**.   [3]

8   Maria writes some calculations.
    By working out the answers, find the 'odd one out' in each group.

a   A 3.45×8 B 4.6×6 C 3.95×7 [1]

b   A 3.68×9 B 4.74×7 C 4.14×8 [1]

c   A $36.6 \div 8$ B $55.2 \div 12$ C $45.75 \div 10$ [1]

d   A $222.2 \div 20$ B $88.8 \div 8$ C $199.8 \div 18$ [1]

9   Amelia has eight cards labelled **A** to **H**.
    The bottom half of each card has a question and the top half has an answer.
    Choose any card and answer the question.
    Find the card with the matching answer and answer the question at the bottom.
    Now find the card with the answer. Keep going until you have answered all the questions and write down the sequence of your answers.
    The question on your final card should match the answer on your first card.   [10]

A	B	C	D
20.16	6.8	29.6	17.71
$7.4 \times 4 = ?$	$88.55 \div 5 = ?$	$4.88 \times 2 = ?$	$134.2 \div 11$

E	F	G	H
46.27	9.76	12.2	5.4
$48.6 \div 9 = ?$	$6.61 \times 7 = ?$	$6.72 \times 3 = ?$	$27.2 \div 4 = ?$

10 Jasmine's bike has wheels of circumference 2.5 m.
    When Jasmine cycles to school, the wheels go round 850 times.
    How far does Jasmine cycle to school?   [3]

11 Work out the following without using a calculator. Include units in your answer.

a   A bag of sweets has mass 113 g includes wrappings of 0.5 g. Each sweet has mass 4.5 g. How many sweets are in the bag?   [2]

b   A football stadium has 135 400 m² of seating for its fans. Each fan is allowed 5.4 m² of space. How many fans, to the nearest 1000, is the stadium capable of holding?   [2]

12 Write one pair of brackets in each calculation to make the answer correct.

a $3 + 4 \times 5 + 2 = 37$ [1]   b $60 \div 5 + 7 + 5 = 10$ [1]

2 Expressions

Introduction

The real world is messy and complicated, and always in motion. Algebra is a vital part of maths because it attempts to describe aspects of the world, such as fluid flow or the forces acting on a suspension bridge. Through equations and formulae, algebra provides a mathematical model to describe the real-world situation, from which understanding can be gleaned and predictions made. It is only able to do this if it makes assumptions that simplify the situation, meaning that the model is only ever an inaccurate reflection of the real world. However simplifying a situation helps us to understand the forces that lie behind it.

What's the point?

Without algebra, we would not be able to work with large mechanical forces – so there would be no skyscrapers or suspension bridges; we would also not be able to understand electronics, so there would be no tablets or mobile phones.

Objectives

By the end of this chapter you will have learned how to …

- Use algebraic notation.
- Substitute numbers into formulae and expressions.
- Use and understand the words expressions, equations, formulae, terms and factors.
- Collect like terms and simplify expressions involving sums, products, powers and surds.
- Use the laws of indices.
- Multiply a single term over a bracket.
- Take out common factors in an expression.

Check in

1 Work out these powers.

    a 3^2 **b** 2^2 **c** 4^2

2 There are 5 CDs in a packet.
How many CDs are there in 3 packets?

3 Work out

    a $4 - {-3}$ **b** $2 + {-3}$ **c** $-3 + 5$ **d** $-4 - {-1}$

    e -3×2 **f** $4 \times {-2}$ **g** $6 \div {-3}$ **h** $-8 \div {-2}$

4 **a** Write all the factors of these numbers.

        i 18   **ii** 12   **iii** 24

    b Write all the common factors of 18, 12 and 24.

    c What is the highest common factor of 18, 12 and 24?

Chapter investigation

Think of a number between 1 and 10.

- Double it.
- Add 4.
- Halve your answer.
- Take away the number you first thought of.

What do you notice? Investigate why this is the case.

Can you invent different instructions that give similar results?

2.1 Terms and expressions

A **variable** is a letter used to represent something unknown. Variables can be used to make **expressions**, **equations** and **formulae**.

p.124

- An **expression** is a meaningful collection of mathematical symbols without an = sign.
- **Terms** are groups of symbols in an expression separated by + and − signs.

In the expression $6x - 5y + 2$, '$+6x$', '$-5y$' and 2 are all terms; x and y are variables that can take any value.

- An **equation** is an expression containing *one* = sign and at least one unknown to be found.

In the equation $x + 5 = 7$, the variable x can take only the value $x = 2$.

- A **formula** is an equation involving several variables. It is often related to a real-life application.

In the formula $E = mc^2$ the variables E and m can take many values but are linked by the formula.

There are several conventions used in algebra that you must learn.

$3m$ ✓	$m + m + m$ ✗	
$4y$ ✓	$4 \times y$ ✗	
ab ✓	ba ✗	
$4ab$ ✓	$ab4$ ✗	
$4(x + 2)$ ✓	$(x + 2)4$ ✗	
x^2 ✓	xx ✗	

- Multiplication signs aren't written.
- Terms involving letters are written in alphabetical order.
- Terms involving letters and numbers are written in alphabetical order with the number first.
- If there is a bracket write the term outside the bracket first.
- x squared is written using the power two.

EXAMPLE

Write an algebraic expression for these descriptions.

a   A number, n, add 4.

b   A number, n, take away 6.

c   A number, n, multiplied by 7.

d   A number, n, divided by 2.

e   A number, n, multiplied by itself.

a $n + 4$

b $n - 6$

c $7n$

d $\dfrac{n}{2}$ or $\dfrac{1}{2}n$

e n^2

The conventions save you writing and make it easier to tell when two expressions are equal.

EXAMPLE

Find the value of these expressions when $b = 8$.

a $2b$ b $b - 6$ c $\dfrac{b}{2}$ d b^2

a $2b = 2 \times 8$ b $b - 6 = 8 - 6$ c $\dfrac{b}{2} = 8 \div 2$ d $b^2 = 8^2$

     $= 16$ $= 2$ $= 4$ $= 8 \times 8 = 64$

EXAMPLE

Find the value of these expressions when $a = 4$ and $b = -2$.

a $a + b$ b $a - b$ c ab d $\dfrac{a}{b}$

a $a + b = 4 + -2$ b $a - b = 4 - -2$ c $ab = 4 \times -2$ d $\dfrac{a}{b} = \dfrac{4}{-2}$

    $= 2$ $= 6$ $= -8$ $= 4 \div -2 = -2$

Algebra   Expressions

Exercise 2.1S

1 Match the algebraic expression with the correct description.

$n + 2$	A number subtract 2
$n - 2$	A number divided by 2
$2n$	2 take away a number
$\dfrac{n}{2}$	A number multiplied by 2
$2 - n$	A number add 2

2 Write algebraic expressions for these descriptions.

 a x add 4 **b** y add 6

 c p subtract 5 **d** d take away 7

 e 7 take away x **f** x subtracted from 7

3 Write algebraic expressions for these descriptions.

 a x times 4 **b** y multiplied by 6

 c 5 times p **d** d multiplied by 7

 e double f **f** treble y

4 Write algebraic expressions for these descriptions.

 a x divided by 4 **b** y divided by 6

 c p divided by 5 **d** x divided by 2

 e one half of x **f** three quarters of z

5 Match the algebraic expression with the correct description.

$4x + 2$	x add 4 all multiplied by 2
$2x + 4$	x add 2 all multiplied by 4
$2(x + 4)$	x multiplied by 4 add 2
$4(x + 2)$	x multiplied by 2 add 4

6 Do you agree with these statements? Give your reasons.

 a m^2 is the same as $2m$.

 b The value of $5b$ is 10 because $a = 1$, $b = 2$, $c = 3$, $d = 4$, ..., etc.

 c When $y = 3$ the value of $7y$ is 73.

 d When $p = 3$ the value of p^2 is 6.

7 Work out these expressions when $b = 10$.

 a $b + 4$ **b** $b + 7$

 c $b + 5$ **d** $b + 3$

 e $b - 2$ **f** $b - 8$

 g $b - 8$ **h** $b - 12$

8 Work out these expressions when $y = 12$.

 a $2y$ **b** $6y$

 c $4y$ **d** $8y$

 e $10y$ **f** $20y$

 g $\dfrac{y}{2}$ **h** $\dfrac{y}{3}$

 i $\dfrac{24}{y}$ **j** $\dfrac{120}{y}$

9 Work out these expressions when

 i $x = 3$ and $y = 9$

 ii $x = 5$ and $y = -1$.

 a $x + y$ **b** $x - y$

 c $y - x$ **d** xy

 e $\dfrac{x}{y}$ **f** $\dfrac{y}{x}$

10 Work out these expressions.

 a d^2 when $d = 4$ **b** p^2 when $p = 5$

 c m^2 when $m = 10$ **d** c^2 when $c = 12$

 e p^2 when $p = 0$ **f** x^2 when $x = -2$

 g y^2 when $y = -5$ **h** x^2 when $x = -10$

11 Match the algebraic expression with the correct description.

$2x^2$	x squared plus 2 squared
$(2x)^2$	x plus 2 squared
$(x + 2)^2$	x multiplied by 2 squared
$x^2 + 2^2$	x squared multiplied by 2

12 The sides of a rectangle are represented by a and b. Give your reasons why the perimeter can be found using the formula $P = 2a + 2b$.

1158, 1186, 1187   SEARCH

2.1 Terms and expressions

- In algebra, you use letters to represent unknown numbers.
- You can replace letters with number values. This is called **substituting**.
- An expression is a collection of letters and numbers with no = sign.

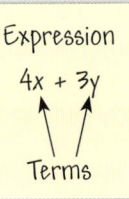

Expression

$4x + 3y$

Terms

HOW TO

To write expressions and equations in algebra
1. Give every **unknown** a letter and write it down.
2. Translate the words into letters and symbols.
3. Test your expression using real values... or ...substitute in the value given.

EXAMPLE

a   Write an expression for the number of pens in 3 boxes plus 5 extra pens.

b   If there are 10 pens in each box how many pens are there in total?

1. Give every unknown a letter.   number of pens in one box = m
2. Translate into letters and symbols. $3m + 5$ 3 boxes + 5 pens.
3. Substitute in the value given. $m = 10$ so $3 \times 10 + 5 = 35$ pens.

EXAMPLE

In a fruit shop, apples cost 20 p each and oranges cost 15 p each.
Write an expression for the cost of x apples and y oranges.

1. number of apples = x
   number of oranges = y
2. cost of x apples at 20 p + cost of y oranges at 15 p
   $20x + 15y$
3. Test your expression using actual values.
   3 apples and 2 oranges would cost $3 \times 20 + 2 \times 15$
   $= 60 + 30 = 90$ p

EXAMPLE

Paul, Rashid and Sara count the number of books in their school bags.
Paul has p books, Rashid has r books and Sara has s books.
These two equations are true.
$r = 2p$ and $s = r + 2$

a   Paul says that he has half the number of books as Rashid. Is he correct?

b   Rashid says that he has two more books than Sarah. Is he correct?

c   Show that $p + r + s = 12$ when $p = 2$.

a   This is the same as saying Rashid has twice as many books as Paul.
   2. Translate into letters and symbols.
   $r = 2p$. This is true, Paul is correct.

b $r = s + 2$ this is the same as saying   2
   $r - 2 = s$
   $s = r - 2$ is false, Rashid is wrong. $s = r + 2$

c   3. Substitute $p = 2$ to find r.
   $r = 2p = 2 \times 2 = 4$
   Substitute $r = 4$ to find s.
   $s = r + 2 = 4 + 2 = 6$
   Substitute $p = 2$, $r = 4$ and $s = 6$.
   $p + r + s = 2 + 4 + 6 = 12$

Exercise 2.1A

1 Match each expression in box **A** with an expression in box **B**.

A

n oranges at 5p each	$4m$
$3 \times n$ $n + n$	$7 \times m$
m toys at £3 each	$6n$

B

3 chews at n pence each $m \times 4$

$7m$ $5n$ $3m$ $2n$

6 stamps at n pence each

2 Tony's Taxis calculates its fares using the formula

$$F = 2 + 4n$$

where F is the fare in pounds and n is the number of miles.

Carla's Cabs calculates its fares using the formula

$$F = 6 + 3n$$

Jessica wants to travel 7 miles. Which company should she use?

3 A rent-a-car company charges for its cars using the formula

$$C = 30 + 20n$$

where C is the cost in £s and n is the number of days hired.

a Bobbi says that it costs an extra £30 a day for each extra day that you rent a car. Do you agree with Bobbi? Give your reasons.

b What is the charge for 10 days' hire?

c The company also charges a cleaning fee when the car is returned. Bobbi is charged a total of £260 for car rental and cleaning.

What is the maximum number of days that Bobbi hired the car for?

4 Work out the value of each capital letter, then read the coded word.

$m = 3$ $n = 8$ $p = 5$ $q = -2$

$H = 4n + 2q$ $L = 6m + 3p$
$O = m^2 + 7$ $A = 2q + 10$
$T = pq$ $C = 2n - 3p$ $E = np + 2m$

I	28	16	I	16	33	6	-10	46

5 Daniel has x DVDs.
Lisa has twice as many DVDs as Daniel.
Sareeta has 3 fewer DVDs than Lisa.
Write an expression in terms of x for the number of DVDs Sareeta has.

6 In one month, Dan sends x texts.
Alix sends 4 times as many texts as Dan.
Kris sends 8 more texts than Alix.
How many texts does Kris send?

7 In a pizza takeaway
- a medium pizza has 6 slices of tomato
- a large pizza has 10 slices of tomato.

Write an expression for the total number of slices of tomato needed for c medium and d large pizzas.

8 Write algebraic expressions for the cost of

a f teas and g scones

b j fruit juices and k flapjacks

c x teas, y milks and z scones

d p milks, q fruit juices and r flapjacks.

Café price list

Tea	50p
Fruit juice	80p
Milk	60p
Scone	30p
Flapjack	40p

9 Gemma and Paul both worked out $2x^2$ when $x = 6$.
Who was right? Give your reasons.

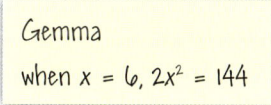

Gemma

when $x = 6$, $2x^2 = 144$

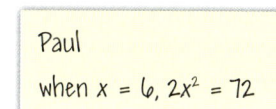

Paul

when $x = 6$, $2x^2 = 72$

10 Audrey, Billie and Cerys count the amount of money they each have. Audrey has £a, Billie has £b and Cerys has £c.
These two equations are true.
$$b = 3a \qquad c = a + b$$

a Audrey says that she has three times as much money as Billie. Cerys says that she has four times as much money as Audrey. Are they both correct? Give reasons for your answers.

b Audrey has £5. How much money do the three friends have in total?

2.2 Simplifying expressions

Expressions are made up of a collection of terms.
Terms can be made up of coefficients and variables.

Expressions can be **simplified** by **collecting like terms**.

● Like terms have the same letter or symbol.

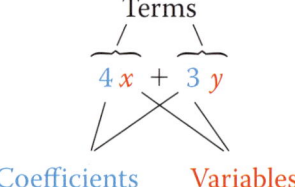

Terms

$$4x + 3y$$

Coefficients   Variables

$4a + 3b + 2a + b$ can be simplified by collecting the 'like terms'
$4a$ and $+2a$ and, $+3b$ and $+b$.

$4a + 3b + 2a + b = 6a + 4b$

Expressions can be simplified using the rules of algebraic notation.

$$b \times 4 \times a = 4ab \qquad p \div 2 = \frac{p}{2}$$

By convention you write the
coefficient first and then the
variables in alphabetical order.

EXAMPLE

Simplify

a $4p + p + 2p$

b $4a - 3b + 2a + 5b$

c $5x - 2y - 3 + 3x$

d $3a + 4 + 2y$

---

a $7p$ □□□□ + □ + □□ = □□□□□□□ = 7□

b $(4a) - (3b) + (2a) + 5b = 6a + 2b$ Collect a and b terms separately
$4a + 2a = 6a$ $-3b + 5b = 2b$

c $(5x) - (2y) - 3 (+ 3x) = 8x - 2y - 3$ Collect x, y and numbers separately
$5x + 3x = 8x$

d This cannot be simplified.

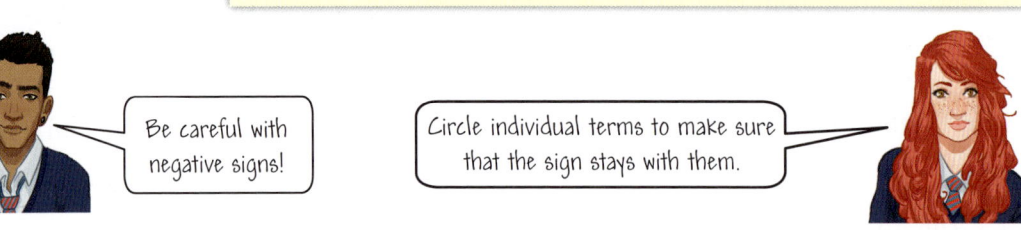

Be careful with
negative signs!

Circle individual terms to make sure
that the sign stays with them.

EXAMPLE

Simplify

a $y^2 + 3y^2$

b $2ab + 5a - 3b$

c $ab^2 + 6ab^2$

d $5ab^2 + 2a^2b - 3ab^2$

---

a $4y^2$

b This cannot be simplified.

c $7ab^2$

d $(5ab^2) + (2a^2b) - (3ab^2) = 2ab^2 + 2a^2b$ Collect the ab^2 and a^2b terms
separately.

EXAMPLE

Simplify

a $2 \times y \times w \times 3$ **b** $\frac{1}{2}x + \frac{1}{4}x$ **c** $2p \div 10$ **d** $4 \div p$ **e** $3r + 3 \times 2r$

---

a $6wy$ **b** $\frac{3}{4}x$ **c** $\frac{2p}{10} = \frac{p}{5}$ **d** $\frac{4}{p}$ **e** $3r + 6r = 9r$

Leave the coefficient as a fraction.

Exercise 2.2S

1 Simplify these expressions.

a $d + d + d + d$ **b** $q + q + q$

c $4a + 2a + a + a$ **d** $y + 3y + y + 7y$

e $4t + t + t + 3t$ **f** $e + 2e + 3e + 4e$

g $x + x + 2$ **h** $y + 4 + 3y + 5$

2 Harry thinks that $k + k + k = k^3$.

Will thinks that $k + k + k = k3$.

Do you think either Harry or Will are correct?
Give reasons for your answer.

3 Simplify these expressions.

a $d + 5d - 2d$ **b** $4q - 2q + q$

c $4a - 2a - a + 6a$ **d** $y - 3y - y + 7y$

e $4t + t - t - 3t$ **f** $e - 2e - 3e + 4e$

g $5k + 4 - 2k$ **h** $8y - 5 - 2y + 11$

i $8p - 4 - 12p + 11$

j $8b + 4d - 12 + 4bd$

4 Find five expressions that simplify to $12m$.

5 Zara thinks that $3d + 2e + 4d + 5e = 14de$.
Do you agree with Zara?

Give reasons for your answer.

6 Simplify these expressions.

a $a + 2b + a$ **b** $4t + 3s + 2t$

c $d + 2e + e + 3e$ **d** $4y + 2y + w + 7w$

e $2p + f + 3p + 3f$ **f** $e + 2f + 3g + 4h$

g $5k + 4 + 2d + 2k$ **h** $8y + 5 + 2y + 11$

i $3p + 4 + 2p + 2q$

j $8b + 5 - 2a + 6ab$

7 Simplify these expressions.

a $a - 2b + 6a$

b $4t - 3s + 2t$

c $d - 2e + 3d - 3e$

d $5y - 3y - w + 6w$

e $6p - 4f - 9p + 6f$

f $2e - 2f - 3g + 4f$

g $5k + 4 + 2d - 8k$

h $8b - 5a - 2b + 11$

8 Simplify these expressions.

a $2ab + 6ab$ **b** $4st - 3st + 2st$

c $de - 2de + 5d - 3e$

d $5wy - 3y - wy + 6wy$

e $4fp + fp - 9p + 6f$

f $2e + 2ef + 3fg + 4gh$

9 Liz thinks that $a^2 + a^2 + a^2 = 6a$

Do you agree with Liz?
Give reasons for your answer

10 Simplify these expressions.

a $d^2 + d^2$ **b** $b^2 + 3b^2$

c $4a^3 + 5a^3$ **d** $2y^2 + 4y + y^2$

e $4t^3 + 5t^3 - 6t^3$ **f** $5k^3 + 4 - 2k^3$

11 Simplify these expressions.

a $ad^2 + ad^2$ **b** $3ab^2 + 10ab^2$

c $4xy^3 + 5xy^3$ **d** $2xy^2 + 4xy + xy^2$

e $4st^3 + 5s^3t - 6st^3$ **f** $8y^2 - 5y - 2x + 11x^2$

12 Find five expressions involving indices that
can be simplified to $2a^2 + 3b^2$

13 Simplify these expressions.

a $\frac{1}{3}x + \frac{1}{3}x$ **b** $\frac{3}{4}y - \frac{1}{4}y$

c $\frac{2}{3}a + \frac{1}{2}b - \frac{1}{3}a$ **d** $\frac{1}{2}x + \frac{1}{4}y - \frac{1}{4}x$

14 Simplify these expressions.

a $3 \times a$ **b** $a \times 3$

c $a \times b \times 3$ **d** $b \times 3 \times a$

e $3 \times p \times 2$ **f** $a \times 4 \times a$

g $3 \div a$ **h** $a \div 3$

i $10 \div 2p$ **j** $6d \div 12$

k $3q + 4 \times 2q$ **l** $r - 5 \times 3r$

m $2 \times 4s - 3 \times 3s + 5 \times 4t$

*15 Simplify these expressions.

a $\dfrac{3 \times b \times 8}{6 \times a}$

b $\dfrac{4ab + 6ab}{5a}$

c $\dfrac{7a + 9ab - 4a}{3a}$

Q 1178, 1179   SEARCH

2.2 Simplifying expressions

- Terms with the same letter are called like terms.
- You can simplify expressions by collecting like terms.

> $2a$ and $6a$ are like terms
> $2a$ and $4b$ are not like terms
> a and a^2 are not like terms

1. Read the question. You may need to use your knowledge from other areas of mathematics.
2. Simplify your expression using the rules of algebra.
3. Remember to answer the question.

> $2a + 6a = 8a$
> $7a + 2b - 3a - 3b = 4a - b$

Which two triangles have the same perimeter?

A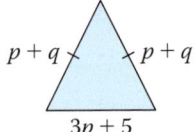
$p + q$ $p + q$
$3p + 5$

B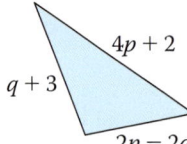
$4p + 2$
$q + 3$
$2p - 2q$

C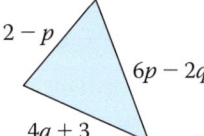
$2 - p$
$6p - 2q$
$4q + 3$

1. Perimeter is the distance around the edge of a 2D shape.

   A $p + q + p + q + 3p + 5$ **B** $q + 3 + 4p + 2 + 2p - 2q$ **C** $2 - p + 6p - 2q + 4q + 3$

2. Collect like terms to simplify your expressions.

   $p + q + p + q + 3p + 5$ $q + 3 + 4p + 2 + 2p - 2q$ $2 - p + 6p - 2q + 4q + 3$
   $= 5p + 2q + 5$ $= 6p - q + 5$ $= 5p + 2q + 5$

3. Say which two triangles have the same perimeter.

   A and C

> p.144

Joe has written expressions for the area and perimeter of this shape.
$A = 46xy$ cm²
$P = 21x + 16y$ cm
Is he correct? Give your reasons.

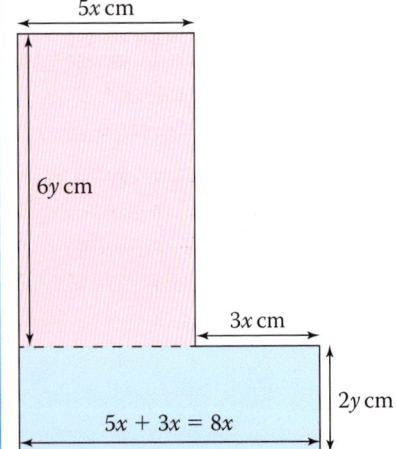

$5x$ cm
$6y$ cm
$3x$ cm
$2y$ cm
$5x + 3x = 8x$

1. Work out the area of each rectangle separately.
   Area of a rectangle = length × width
   Area of pink rectangle = $5x \times 6y = 30xy$
   Area of blue rectangle = $8x \times 2y = 16xy$

2. Collect like terms to find the total area
   Total area = $30xy + 16xy = 46xy$ cm²

3. Yes, he is correct.

1. Work out the perimeter of the whole shape.
   Don't include the dotted line!
   Total perimeter = $5x + 6y + 3x + 2y + 8x + 2y + 6y$
   $= 16x + 16y$ cm   2. Collect like terms.

3. No, he is not correct.
   His expression is $5x$ too big.

Exercise 2.2A

1 Bags of peanuts come in two sizes.
There are x peanuts in a small bag.
There are y peanuts in a large bag.
Sebastian buys 1 small bag and 1 large bag.
Gabi buys 2 small bags and 1 large bag.
Kofi buys 3 small and 2 large bags.

   a Sebastian and Gabi combine their bags of peanuts.
Do they have more peanuts than Kofi?

   b Write an expression for the total number of peanuts that all three friends buy.

2 Make sets of three matching expressions, using one expression from each box in each set.

A
$3x + 5y - x + 2y$
$2x - 4y + 3x + 2y$
$2x + 4y - x$
$2y + 3x - x + 3y$

B
$2x + 5y$
$3y + 7x + 4y - 5x$
$5x - 2y$
$3x + 6y - 2x - 2y$

C
$7y - 3x + 5x - 2y$
$4x + 4y - 3x$
$7y + 2x$
$2x - 4y + 2y + 3x$

3 Three students tried to simplify $3m + 5$.
Which of them did it correctly?

Sara	Paul	Abdul
$3m + 5 = 8m$	$3m + 5 = 15m$	$3m + 5 = 3m + 5$

4 Rearrange each set of cards to make a correct statement.

a
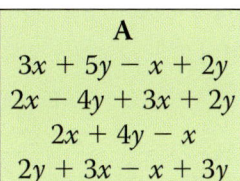

b
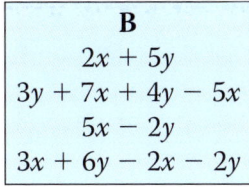

c
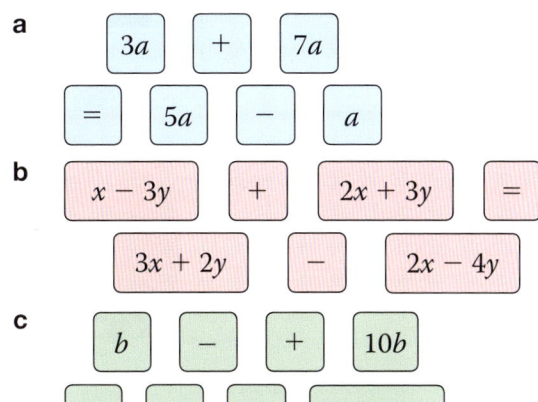

5 a Write a simplified expression for the
   i perimeter
   ii area of this rectangle.

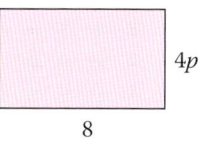

   b What are the measurements of a rectangle with perimeter $6x + 4y$ and area $6xy$?

   c Another rectangle has area $4ab$.
Write down two possible perimeters.

6 Find the area of this shape.

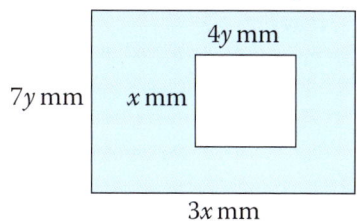

7

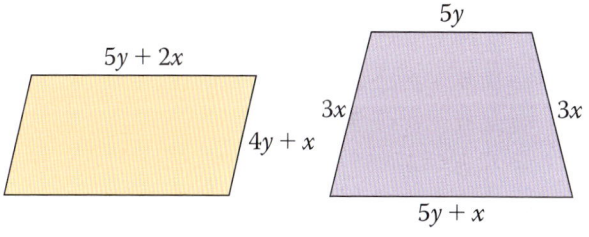

   a The triangle and the parallelogram have the same perimeter. Find the length of each side of the triangle.

   b Does the rectangle or the trapezium have the greater perimeter?
Give reasons for your decision.

   c A regular pentagon has a perimeter that is equal to the sum of the rectangle's perimeter and the trapezium's perimeter. Find the length of each side of the pentagon.

1178, 1179 SEARCH

2.3 Indices

Repeated multiplications of the same number or letter can be written using **index form**.

$$2^4 = 2 \times 2 \times 2 \times 2$$

2 is the **base**

4 is the **index** or **power**

There is a set of rules known as the index laws.

- $y^a \times y^b = y^{a+b}$ To multiply powers add the indices.
- $y^a \div y^b = y^{a-b}$ To divide powers subtract the indices.
- $(y^a)^b = y^{ab}$ To raise a power to a power multiply the indices.

p.280

EXAMPLE

Simplify these expressions, leaving your answer in index form.

a $2^6 \times 2^3$

b $2^6 \div 2^3$

c $(2^3)^2$

'Index form' means leave the answer in the form a^n.

a $2^6 \times 2^3 = 2^{6+3}$ $2^6 \times 2^3 = (2 \times 2 \times 2 \times 2 \times 2 \times 2) \times (2 \times 2 \times 2)$
      $= 2^9$ $= 2^{6+3} = 2^9$

b $2^6 \div 2^3 = 2^{6-3}$ $2^6 \div 2^3 = \dfrac{2^6}{2^3} = \dfrac{2 \times 2 \times 2 \times 2 \times 2 \times 2}{2 \times 2 \times 2}$
      $= 2^3$ $= 2^{6-3} = 2^3$

c $(2^3)^2 = 2^{3 \times 2}$ $(2^3)^2 = (2 \times 2 \times 2)^2 = (2 \times 2 \times 2) \times (2 \times 2 \times 2)$
      $= 2^6$ $= 2^{3 \times 2} = 2^6$

You can apply the index laws to letters or combinations of letters and numbers.

EXAMPLE

Simplify these expressions.

a $p^3 \times p^5$

b $b^6 \div b^3$

c $4p^3 \times 3p^5$

d $12p^6 \div 4p^2$

e $(5a^3)^2$

f $p^9 \times p^{-7}$

g $\dfrac{q^{-3}}{q^{-5}}$

h $(w^6)^{-4}$

a $p^3 \times p^5 = p^{3+5} = p^8$

b $b^6 \div b^3 = b^{6-3} = b^3$

c $4p^3 \times 3p^5 = 4 \times 3 \times p^3 \times p^5$
     $= 12 \times p^{3+5} = 12p^8$

Multiply and divide any numbers as normal.

d $12p^6 \div 4p^2 = \dfrac{12 \times p^6}{4 \times p^2} = \dfrac{12^3 \times p \times p \times p \times p \times p \times p}{4 \times p \times p}$
     $= 3 \times p^{6-2} = 3p^4$

e $(5a^3)^2 = (5a^3) \times (5a^3) = 5^2 \times a^{3 \times 2} = 25a^6$

f $p^9 \times p^{-7} = p^{9+-7} = p^2$

The index laws work just the same for negative indices.

g $\dfrac{q^{-3}}{q^{-5}} = q^{-3} \div q^{-5} = q^{-3--5} = q^2$

h $(w^6)^{-4} = w^{6 \times -4} = w^{-24}$

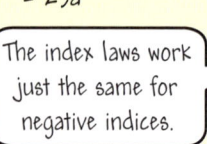

EXAMPLE

Simplify

$\dfrac{4a^6 \times 6a^4}{8a^2}$

Treat the numbers and letters separately.

$\dfrac{4a^6 \times 6a^4}{8a^2} = \dfrac{4 \times 6}{8} \times \dfrac{a^6 \times a^4}{a^2} = \dfrac{24}{8} \times \dfrac{a^{6+4}}{a^2}$ Add indices for multiplication.

$= 3 \times a^{10-2} = 3a^8$ Subtract indices for division.

Algebra Expressions

Exercise 2.3S

1 Write these expressions using index notation.

a $3 \times 3 \times 3 \times 3$

b $2 \times 2 \times 2 \times 2 \times 2 \times 2$

c $4 \times 4 \times 5 \times 5 \times 5$

d $3 \times 3 \times 2 \times 3 \times 2 \times 2$

e $p \times p \times p$

f $p \times q \times q \times p \times p$

g $2 \times r \times 2 \times 2 \times r \times r \times r$

h $3 \times s \times s \times 3 \times t \times s \times t$

2 Work out the value of these expressions.

a 5^3 **b** 2^3

c 1^5 **d** 6^2

e $(-1)^3$ **f** $\left(\dfrac{1}{2}\right)^2$

3 Simplify these expressions leaving your answers in index form.

a $3^2 \times 3^4$ **b** $2^2 \times 2^7$

c $5^2 \times 5^5$ **d** $6^5 \times 6^2$

e $7^4 \times 7^5$ **f** $11^3 \times 11^7$

4 Simplify these expressions leaving your answers in index form.

a $5^6 \div 5^3$ **b** $3^8 \div 3^4$

c $8^9 \div 8^4$ **d** $2^{14} \div 2^7$

e $6^4 \div 6^8$ **f** $4^3 \div 2^4$

5 Simplify these expressions leaving your answers in index form.

a $(2^3)^2$ **b** $(5^4)^6$

c $(3^3)^7$ **d** $(8^2)^8$

e $(7^4)^{-2}$ **f** $\left(\left(\dfrac{1}{4}\right)^2\right)^3$

6 Use your calculator to check your answers to questions **2** to **5**. For example, **3a**, $3^2 \times 3^4 = 9 \times 81 = 729$ and $3^6 = 729$.

7 Kyle thinks that $a^5 \times a^2 = a^{10}$.
Do you agree with Kyle? Give your reasons.

8 Simplify these expressions.

a $a^2 \times a^4$ **b** $y^2 \times y^8$

c $b^2 \times b^6$ **d** $p^5 \times p^9$

e $h^4 \times h^8$ **f** $s^3 \times t^9$

9 Tracey thinks that $4y^5 \times 2y^2 = 6y^7$ because *'the index rules say that you add the powers when two terms are multiplying each other'.*

Do you agree with Tracey?
Give reasons for your answer.

10 Simplify these expressions.

a $3x^5 \times x^2$ **b** $5y^2 \times y^5$

c $4b^2 \times 3b^6$ **d** $2p^4 \times 5p^7$

e $5h^5 \times 6h^6$ **f** $4s^3 \times 3t^4$

11 Andy thinks that $12p^{12} \div 3p^4 = 9p^8$ because *'the index rules say that you subtract the powers when two terms are dividing each other'.*

Do you agree with Andy?
Give reasons for your answer.

12 Simplify these expressions.

a $10y^6 \div 5y^2$ **b** $6a^9 \div 3a^3$

c $20k^7 \div 4k^3$ **d** $18p^8 \div 6p^3$

e $35x^{10} \div 7x^4$ **f** $4x^8 \div 8y^4$

13 Simplify these expressions.

a $(a^3)^2$ **b** $(y^2)^6$ **c** $(k^3)^5$

d $(p^7)^8$ **e** $(a^3)^7$ **f** $(a^3)^7$

14 Simplify these expressions.

a $(2a^3)^2$ **b** $(3y^2)^6$ **c** $(5k^3)^2$

d $(6p^7)^3$ **e** $(2a^3)^7$ **f** $(4a^4)^4$

15 Simplify these expressions.

a $y^{-5} \times y^7$ **b** $x^2 \times x^{-4}$

c $a^{-1} \times a^{-5}$ **d** $h^{-2} \div h^4$

e $\dfrac{p^3}{p^{-1}}$ **f** $\dfrac{p^{-4}}{p^{-3}}$

16 Simplify these expressions.

a $(b^{-2})^3$ **b** $(k^4)^{-3}$ **c** $(q^{-3})^{-9}$

17 Simplify these expressions.

a $4y^{-2} \times 3y^3$ **b** $3a^{-5} \times 2a^{-2}$

c $45b^{-3} \div 5b^2$ **d** $7k^{-2} \div 14k^{-3}$

18 Simplify these expressions.

a $2a^3 \times 4a^4 \times a^5$ **b** $\dfrac{m^{11}}{m^4}$

c $\dfrac{12y^9}{3y^3}$ **d** $\dfrac{3y^3 \times 6y^5}{2y^4}$

2.3 Indices

RECAP

- An **index** is a power. The **base** is the number which is raised to this power.
- You can simplify expressions with same base using the three index laws.
  - When multiplying, add the indices.
  - When dividing, subtract the indices.
  - To find the 'power of a power' multiply the indices.

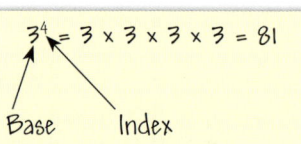

$3^4 = 3 \times 3 \times 3 \times 3 = 81$

Base   Index

HOW TO

① Read the question carefully. You may need to apply your knowledge from other areas of maths.

② Use the index laws to calculate or simplify.

③ Answer the question. Don't forget units.

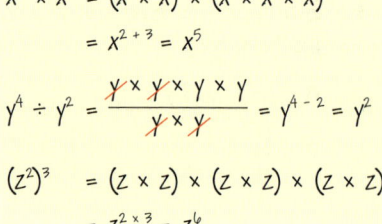

$x^2 \times x^3 = (x \times x) \times (x \times x \times x)$
$= x^{2+3} = x^5$

$y^4 \div y^2 = \dfrac{y \times y \times y \times y}{y \times y} = y^{4-2} = y^2$

$(z^2)^3 = (z \times z) \times (z \times z) \times (z \times z)$
$= z^{2 \times 3} = z^6$

EXAMPLE

These two rectangles have the same area.
What is the length of the missing side?

A

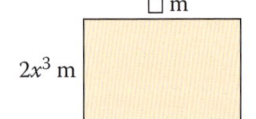

$4x^2$ m

$3x^2$ m

B

☐ m

$2x^3$ m

① Area of a rectangle = length × width

Area of **A** = $3x^2 \times 4x^2$

② Apply the index laws.

$= 3 \times 4 \times x^2 \times x^2 = 12 \times x^{2+2} = 12x^4$ m^2

③ Set area of **A** equal to the area of **B**.

$2x^3 \times \square \, x^{\square} = 12x^4$

$2 \times 6 = 12$

$3 + 1 = 4$

Missing length = $6x$ m

$2 \times \square = 12$

$x^3 \times x^{\square} = x^4$

Don't forget your units!

EXAMPLE

Which two of these expressions are the same?

A $(2p^2q)^3$

B $3p^2 \times 2q^3$

C $\dfrac{6p^7 \times 4q^4}{3pq}$

① You need to simplify all three expressions to see which is the odd one out.

②

A Everything inside the bracket is cubed.

$(2p^2q)^3 = 2^3 \times p^{2 \times 3} \times q^3$
$= 8p^6q^3$

B $3p^2 \times 2q^3$
$= 3 \times 2 \times p^2 \times q^3$
$= 6p^2q^3$

C Treat numbers and letters separately.

$\dfrac{6p^7 \times 4q^4}{3pq} = \dfrac{6 \times 4}{3} \times \dfrac{p^7}{p} \times \dfrac{q^4}{q}$

$= \dfrac{24}{3} \times p^6 \times q^3 = 8p^6q^3$

③ A and C are the same.

Exercise 2.3A

1 Match each of the pairs.

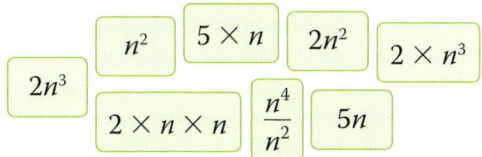

2 A cube has sides of length $2xy^2$ cm. Find the volume of the cube. `p.312`

3 Find an expression for the area of each shape.

a

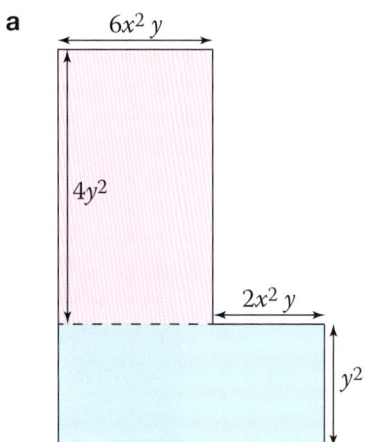

$6x^2y$

$4y^2$

$2x^2y$

y^2

b

$4a^2b$ ab^2 ab^2

$5b^3$

4 Which two expressions are the same?

A $(pq)^2 \times (p^3q)^2 \div p^2q$

B $\dfrac{(pq)^3 \times p^2q}{pq}$

C $(pq)^2 \times p^3q^2 \div pq$

5 Which two expressions are the same?

A $(3x - x)^3 \times y^2$

B $(y + 3y)^2 \times x^3$

C $(xy + xy)^2 \times 2x$

6 A rectangle has area $20p^4q^2$.
The width of the rectangle is $2p^2q$.
Find the perimeter of the rectangle.

7 A square has area $16a^2b^6$.
Find the length of the sides of the square.

8 Rearrange each set of cards to make a correct statement.

a

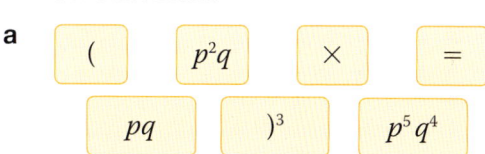

$($ p^2q $\times$ $=$

pq $)^3$ p^5q^4

b

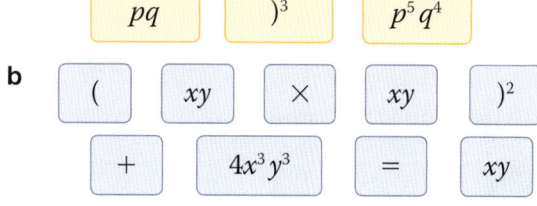

$($ xy $\times$ xy $)^2$

$+$ $4x^3y^3$ $=$ xy

9 Write a simplified expression for the area of this triangle. `p.144`

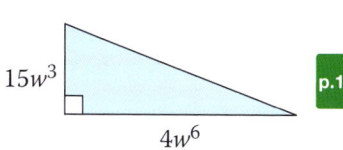

$15w^3$ $4w^6$

10 x^4 expanded is $x \times x \times x \times x$

a Simplify $x^6 \div x^4$ by expanding the powers.

b Simplify $x^4 \div x^6$ by

   i expanding the powers

   ii using the index laws.

c Comment on your answer to part **b**.

11 Find the missing powers.

a $5^x = 25$ **b** $4^x = 64$

c $2^x \times 2^x = 2^6$ **d** $3^x \times 3^3 = 3^{10}$

e $10^x \div 10^2 = 10^3$ **f** $7^8 \div 7^x = 7^3$

g $(2^x)^4 = 2^{16}$ **h** $4^{2x} \times 4 = 4^7$

*12 Simplify $((x^4)^2)^5$

*13 Are these statements true or false?

Give reasons for your answers.

a $(xy^2)^3 = (xy^3)^2$

b $(2xy^2)^3 = 6x^3y^6$

c $\dfrac{x^{-2} \times x^{-3}}{x^{-5}} = 1$

d $(x^2 + x \times x)^2 = 2x^4$

Q 1033   SEARCH

2.4 Expanding and factorising 1

Expressions can be **expanded** by multiplying a single term over a bracket or **factorised** by finding common factors.

- To expand an expression multiply *every* term inside the brackets by the term outside the bracket.
- To factorise an expression find the **highest common factor** (HCF) which divides all the terms and take this outside using brackets.

$$3(x + 2) = 3x + 6$$

$$4x + 2xy = 2x(2 + y)$$

Expanding is the opposite of factorising.

p.272

EXAMPLE

Expand these expressions.

a $4(y + 5)$ **b** $p(p - 3)$ **c** $4b(b - 3)$

a $4(y + 5)$
$= 4 \times y + 4 \times 5$
$= 4y + 20$

b $p(p - 3)$
$= p \times p - p \times 3$
$= p^2 - 3p$

c $4b(b - 3)$
$= 4b \times b - 4b \times 3$
$= 4b^2 - 12b$

Expand means 'multiply out'.

Brackets can be expanded using the 'grid' method'.

$\times$	y	$+5$
4	$4y$	$+20$

$\times$	p	-3
p	p^2	$-3p$

$\times$	b	-3
$4b$	$4b^2$	$-12b$

EXAMPLE

Expand and simplify $6(t + 2) - 5(t - 2)$

$6(t + 2) - 5(t - 2) = 6t + 12 - 5t + 10$
$= t + 22$

$-5(t - 2) = -5 \times t - 5 \times -2 = -5t + 10$

Collect like terms $6t - 5t = t$ and $+12 + 10 = +22$

Factorising is the opposite of expanding.

EXAMPLE

Factorise completely these expressions.

a $3x + 15$ **b** $p^2 - 7p$ **c** $12a^2bc - 30ab^2c$

a $3x + 15$
$= 3 \times x + 3 \times 5$
$= 3(x + 5)$

3 is a factor of $3x$ and 15

b $p^2 - 7p$
$= p \times p - 7 \times p$
$= p(p - 7)$

p is a factor of p^2 and $7p$

c $12a^2bc - 30ab^2c$
$= 2 \times 6 \times a \times a \times b \times c - 5 \times 6 \times a \times b \times b \times c$
$= 6abc(2a - 5b)$

6, a, b and c are all factors $12a^2bc$ and $30ab^2c$

You can always check your factorisation by multiplying out the bracket.

Algebra Expressions

Exercise 2.4S

1 Expand these expressions.

    a $4(y + 2)$ **b** $6(b + 7)$

    c $7(y + 3)$ **d** $12(d + 5)$

    e $3(t + 8)$ **f** $\frac{1}{2}(w + 10)$

2 Sandie thinks that $5(x + 4) = 5x + 4$.
Do you agree with Sandie?
Give reasons for your answer.

3 Expand these expressions.

    a $-4(x + 5)$ **b** $-6(b + 3)$

    c $-7(t + 2)$ **d** $-3(d + 8)$

    e $-10(t + 8)$ **f** $-8(w + 9)$

4 Expand these expressions.

    a $-3(x - 5)$ **b** $-2(b - 8)$

    c $-7(t - 8)$ **d** $-7(d - 10)$

    e $-9(9 - t)$ **f** $-8(6 - w)$

5 Expand these expressions.

    a $y(y + 2)$ **b** $b(b + 7)$

    c $y(y + 3)$ **d** $d(d + 5)$

    e $t(t + 8)$ **f** $w(w + 9)$

6 Zac thinks that $x(x + 4) = x^2 + 4$.
Do you agree with Zac?
Give reasons for your answer.

7 Expand these expressions.

    a $y(y^2 - 2)$ **b** $b(b - 6)$

    c $3y(y + 3)$ **d** $2d(d - 5)$

    e $7t(t - 8)$ **f** $9w(7 - w)$

    g $ab(a + 5b)$ **h** $\frac{st}{4}(2s + 3t)$

8 Expand and simplify these expressions.

    a $4(x + 3) + 5(x + 6)$

    b $8(y + 3) + 5(y - 3)$

    c $5(t - 2) + 8(t + 2)$

    d $9(p - 3) + 4(p - 4)$

    e $6(b - 1) - 7(b - 2)$

    f $2(m - 3) - 7(4 - m)$

9 Find all the common factors of

    a $2x$ and 6   **b** $4y$ and 12

    c   10 and $20j$ **d** m^2 and m

    e $2s^2$ and $2s$ **f** $12z$ and $8z^2$

10 The answer is $20x + 35$.
Write down a possible question of the form
$a(bx + c)$ where a, b and c are integers.

11 Factorise completely $6x + 12$

12 Factorise completely these expressions.

    a $4p + 8$ **b** $5y + 10$

    c $3d + 21$ **d** $9k + 72$

    e $6b + 24$ **f** $6w + 54$

13 Find five expressions that cannot be
factorised.

14 Ahmed thinks that $12p + 20pq$ factorised
completely is $2(6p + 10pq)$.
Do you agree with Ahmed?
Give reasons for your answer.

15 Factorise completely these expressions.

    a $p + 8pt$ **b** $y + 6xy$

    c $d + 7bd$ **d** $t - 12st$

    e $6b + 24bc$ **f** $6w + 54wy$

    g $16ab - 40b$ **h** $15q - 45p$

16 Factorise completely these expressions.

    a $p^2 + 8p$ **b** $y + 6y^2$

    c $w + 4w^3$ **d** $ab - 4ab^2$

    e $6b^2 + 24bc$ **f** $12x^2yz + 36xy^2z$

    g $mufc + rfc$ **h** $mickey + mouse$

17 The cards show expansions and
factorisations. Match the cards in pairs.

     $4(x + 3)$ $4x^2 - 3x$ $3(x - 4)$

     $4x + 3x^2$ $3x - 12$

     $x(4 + 3x)$ $4x + 12$ $x(4x - 3)$

Q 1155, 1247   SEARCH

Summary

Checkout

You should now be able to...

Test it

Questions

✔ Use algebraic notation.	1
✔ Substitute numbers into formulae and expressions.	2 – 5
✔ Use and understand the words expressions, equations, formulae, terms and factors.	6, 7
✔ Collect like terms and simplify expressions involving sums, products, powers and surds.	8, 9
✔ Use the laws of indices.	10
✔ Multiply a single term over a bracket.	11
✔ Take out common factors in an expression.	12

Language   Meaning

Example

Language	Meaning	Example
Expression	A collection of letters and numbers without an = sign.	$5x - 2$
Equation	Contains an = sign and an unknown letter to be solved.	$6x + 2 = 14$ $\quad x = 2$ $x^2 = 5 + 4x$ $\quad x = 5$ or -1
Formula **Formulae**	Contains an = sign and describes a relationship between two or more letters.	$V = IR$
Term	One of the quantities in an expression. Terms are linked with addition or subtraction signs.	In the expression $4x^3 + 3x^2 - 7y + 9$ $4x^3$, $3x^2$, $7y$ and 9 are all terms.
Substituting	Replacing a letter with a number and working out the value.	Substituting $x = 2$ in $4x^2 + 3x$ gives $4 \times 2^2 + 3 \times 2 = 22$
Unknown	An unknown quantity represented by a letter.	$3x + 4 = 16$ The unknown value of x can be found by solving the equation.
Index **Base** **Power**	In index notation, the index or power shows how many times the base has to be multiplied. The plural of index is **indices**.	Power or index $5^3 = 5 \times 5 \times 5$ Base
Index laws	A set of rules for calculating with numbers written in index notation.	$3^2 \times 3^5 = 3^7$ $\quad a^m \times a^n = a^{m+n}$ $5^6 \div 5^2 = 5^4$ $\quad a^m \div a^n = a^{m-n}$ $(2^3)^4 = 2^{12}$ $\quad (a^m)^n = a^{mn}$
Coefficient	A number in front of a letter that shows how many of that letter are required.	In $6x + 1$, 6 is a coefficient.
Brackets	Used to show part of an expression that has to be evaluated before the rest of the expression.	In $3(x + 9)$, 9 is added to x before multiplying by 3.
Simplify	Expand brackets, collect like terms or factorise to make an expression easier to use.	$2(6x + 3) - 3x + 2y = 12x + 6 - 3x + 2y$ $= 9x + 2y + 6$
Highest common factor	The largest number or algebraic expression that divides exactly into two or more expressions.	HCF of 15 and 35 is 5. HCF of $15x$ and $3xy$ is $3x$.

Review

1 Simplify these expressions.

    a $y \times 13$ **b** $y \times 7 \times x$

    c $x + x + x$ **d** $y \times y \times y$

    e $2 \times x \div 4$ **f** $4yx + yx$

2 Pens cost 15 p each.

    a How much do 9 pens cost?

    b How much do y pens cost?

3 Substitute $t = 7$ to find the value of these expressions.

    a $5t$ **b** $t + 8$

    c $10 - 2t$ **d** t^2

4 Substitute $x = 3$ and $y = -4$ to find the value of these expressions.

    a $7y$ **b** xy **c** $2y^2$

    d $2x - 3y$ **e** $x^2 + 2y$ **f** $2x^2 + y^2$

5 Calculate the value of these expressions when $m = 4$, $n = 2$ and $p = -1$.

    a $\dfrac{3m}{n}$ **b** $\dfrac{m}{p}$ **c** $\dfrac{2p - 8}{5}$

    d $\dfrac{5m - 3n}{7}$ **e** $\dfrac{mp}{n}$ **f** $\dfrac{(5m + 2p)}{(n + 4)}$

6

$$5x - y \qquad S = \frac{D}{T} \qquad 3x - 2 = 7$$

Give an example of each of these from the box.

    a equation   **b** formula

    c expression

7 List the terms in each of these expressions.

    a $5f + 6g - 2h$ **b** $5p - 6q + q^2$

8 Write each of these expressions in the simplest way possible.

    a $5a \times b$ **b** $c + c + c + c$

    c $d \times d \times d$ **d** $2f \times 5g$

    e $4 \times 2e - 5 \times 3$ **f** $a^2 \times a + a^2 + 3$

9 Simplify these expressions.

    a $7r - 5r$

    b $2a + 5b + 3a + b$

    c $7d - 4e - 3e - 2d$

    d $2x + 3x^2 - x$

10 Simplify these expressions involving indices.

    a $c^2 \times c^4$ **b** $d^8 \div d^3$

    c $(r^3)^4$ **d** $t^9 \times t^{-3}$

    e $2u^3 \times 3u^5$ **f** $\dfrac{12v^7}{4v^6}$

11 Expand the brackets in these expressions.

    a $2(a + 1)$ **b** $8(4b - 2c)$

    c $-5(3d - 4)$ **d** $h(h + 2)$

12 Fully factorise these expressions.

    a $14b + 7$ **b** $8 - 4c$

    c $3x^2 + 6x$ **d** $4ab - 12b$

What next?

Score			
	0 – 5		Your knowledge of this topic is still developing. To improve look at MyMaths: 1033, 1155, 1158, 1178, 1179, 1186, 1187, 1247
	6 – 10		You are gaining a secure knowledge of this topic. To improve your fluency look at InvisiPens: 02Sa – f
	11 – 12		You have mastered these skills. Well done you are ready to progress! To develop your problem solving skills look at InvisiPens: 02Aa – f

Assessment 2

1   Debs has 4 packets of crisps, each worth c pence. She says it cost her $c + c + c + c$ pence. Her friend Gill says the crisps cost Debs $4c$ pence.
    Who is right? Give reasons for your answer.   [2]

2   Vic is V years old.
    a   How old was Vic 4 years ago?   [1]
    b   How old is Vic's daughter if she is half of Vic's age?   [1]
    c   How old will Vic's daughter be in five years time?   [1]
    d   How old will Vic be when his daughter is twice her current age?   [2]

3   The book of a film costs £b, a DVD of the same film costs £d and the Blu-ray version costs £r.
    What do the following statements mean?
    a $d = 8$ [1]   b $r - d = 7$ [1]   c $b = \frac{1}{2}d$ [1]
    d $r + d = 23$ [1]   e $b + d + r = 27$ [1]

4   A rectangle has a length of l m and a width of w m.
    a   What expression represents the perimeter of the rectangle?   [1]
    A second rectangle is 5 m longer and 5 m narrower.
    b   What expression represents the perimeter of this rectangle?   [1]
    c   Do the two rectangles have the same perimeter? Give reasons for your answer.   [2]

5   The formula for the curved surface area of a cone is $A = \pi r l$, where r is the radius of the base and l is the slant height.
    a   Find A for a cone with base radius 5 cm and slant height 10 cm.   [2]
    b   Find r when $A = 45\,\text{m}^2$ and $l = 4$ m.   [3]
    c   Find l when $A = 126.4\,\text{in}^2$ and $r = 12.3$ in.   [3]

6   a   Amanda, Bengt, Colin, Davide, Erik and Fiona try to put the expression $9x - 4x - 2$ into its simplest form. Here are their results

        Amanda $13x + 2$ Bengt $5(x - 2)$ Colin $5x + 2$

        Davide $9 - 6x$ Erik $3x$ Fiona $5x - 2$.

        Who is correct? Give reasons for your answer.   [1]

7   There are two different charges to post letters of two different sizes. Small letters must be smaller than 24 cm in length and 16.5 cm wide. Large letters have maximum dimensions of 35.3 cm by 25 cm. Small letters cost 62 p to post and large letters 93 p.
    a   Using the letter S to represent the number of small letters posted and L the number of large letters, write down a formula to represent the cost C of sending some letters of both sizes.   [2]
    b   Navinder posts 4 small letters and 2 large letters.
        Use your formula to confirm that Navinder paid £4.34.   [2]
    c   Julie posts 5 small letters and 3 large letters.
        Use your formula to calculate how much they cost to post.   [2]
    d   Amelia gets 31 birthday cards through the post. 26 are small letter size and the rest are large letter size. How much did her friends and family spend to send them?   [2]
    e   Eilidh has to post 7 items. 3 are small letter size, two measure 25 cm by 15 cm and the other two are large letter size. How much does she have to pay to post the items? [3]

Algebra Expressions

8 The area of a trapezium is given by the formula $A = \dfrac{(a + b) \times h}{2}$ where a and b are the parallel sides and h is the perpendicular height.
Find the area of the trapezium when $a = 2z^2$, $b = 3z^2$ and $h = 4z$. [3]

9 Romeo buys Juliet a present. It is a cuboid with a square base of side y cm and height 20 cm.

   a Calculate, in terms of y, the total surface area of the cuboid in its simplest form. [4]

   b The volume is 200 cm³. Romeo says that the exact value of $y = 10$.
     Is he correct? Give reasons for your answer. [4]

10 Wanda is W years old.

   a Wanda has twin brothers who are 5 years less than twice Wanda's age.
     The total of the ages of the three children can be written as $\square(W - \square)$.
     Find the missing values. [4]

   b Wanda's dad's age is three years more than four times Wanda's.
     Her mum is 2 years younger than her dad.
     The sum of Wanda's parents' ages can be written as $\square(\square W + \square)$.
     Find the missing values. [5]

11 A room is L-shaped with a width w.
Greta draws a plan of the room with these dimensions:
Can Greta's diagram be correct?
Give reasons for your answer. [6]

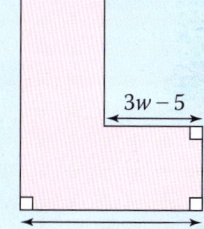

12 A rectangular pyramid has a base with width $2p$ and length $5p$.
Two of the four triangular sides have an area of $2p(3p + 2)$ each.
The other two have an area of $5p(p - 3)$ each.
Find, simplify and factorise the expression for the total surface area of the pyramid. [5]

13 Mark draws a rectangle $ABCD$ as shown. He joins AB and CD with the line EF which is parallel to BC and AD.

$AD = a$ cm
$AE = (a + 4)$ cm
$FC = 9$ cm

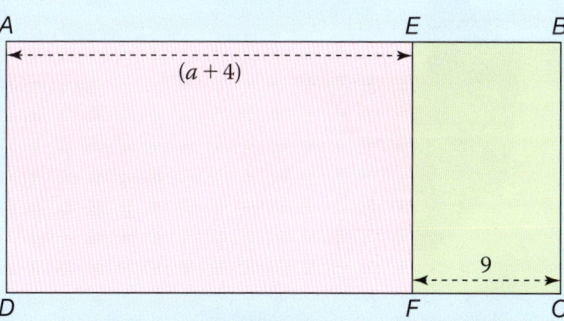

   a He says that in its simplest form the length of $DC = 9a + 4$.
     Is he correct? If not, give the correct expression. [2]

   b Find, simplify and factorise the expression for the perimeter of the rectangle $ABCD$. [3]

   c Write down the expressions for the areas of $AEFD$ and $EBCF$.
     Give a geometric reason for your answers. [3]

3 Angles and polygons

Introduction

Tiling is a fascinating topic that is highly mathematical, involving angles and shapes. There are some wonderful tiling patterns to be found in architecture, particularly in Islamic floors to be found in palaces and mosques.

The tiling shown in this picture is from the Alhambra Palace in Granada, Spain.

What's the point?

An understanding of angles and shapes allows us to create beautiful things.

Objectives

By the end of this chapter you will have learned how to ...

- Describe and apply the properties of angles at a point, on a line and at intersecting and parallel lines.
- Derive and use the sum of angles in a triangle.
- Derive and apply the properties and definitions of special types of quadrilaterals.
- Solve geometrical problems on coordinate axes.
- Identify and use congruence and similarity.
- Deduce and use the angle sum in any polygon and derive properties of regular polygons.

Check in

1 Measure this line in

    a millimetres   **b** centimetres.

2 *Estimate* the size of these angles.

    a   **b**

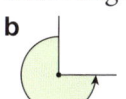

3 Give the value of these angles.

    a   **b**   **c**   **d**

Chapter investigation

A quadrilateral has been drawn on a 3 × 3 square dotty grid.

How many different quadrilaterals can you find?

3.1 Angles and lines

$90°$ is a **right angle**.

An **acute** angle is less than $90°$.

An **obtuse** angle is more than $90°$ but less than $180°$.

A **reflex** angle is more than $180°$ but less than $360°$.

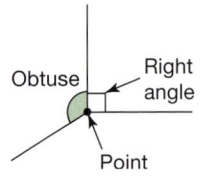 Obtuse — Right angle — Point

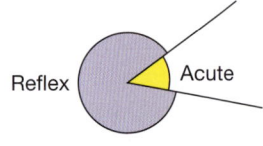 Reflex — Acute

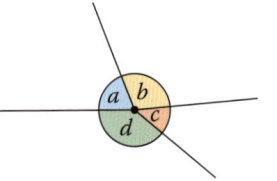 

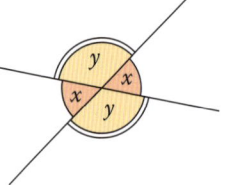

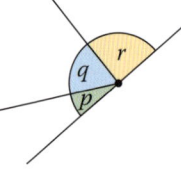

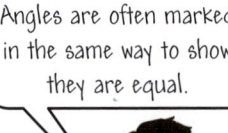 Angles are often marked in the same way to show they are equal.

▲ $a + b + c + d = 360°$ ▲ $x + y = 180°$ ▲ $p + q + r = 180°$

- Angles at a point add up to $360°$.
- Angles at a point on a straight line add up to $180°$.
- Vertically opposite angles are equal.

EXAMPLE

Work out the values of x, y and z.

Give reasons for your answers.

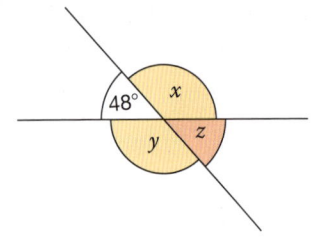

$x = 180° - 48° = 132°$ Angles on a straight line add to $180°$.

$y = x = 132°$ Vertically opposite angles are equal.

$z = 48°$ Vertically opposite angles are equal.

 Other reasons are often possible. You can also use angles on a straight line to find y and z.

Perpendicular lines meet at a right angle.

Parallel lines are the same distance apart everywhere along their length.

A line that crosses parallel lines makes two types of angle.

- **Alternate** angles are equal.
- **Corresponding** angles are equal.

◄ Alternate angles
$d = f$ and $c = e$
Look for a Z or Σ shape.

◄ Corresponding angles
$a = e$, $b = f$, $c = g$ and $d = h$
Look for a F or $⅂$ shape.

Use arrows to show that lines are parallel.

EXAMPLE

Find the angles marked by letters. State whether each answer is acute, obtuse or reflex.

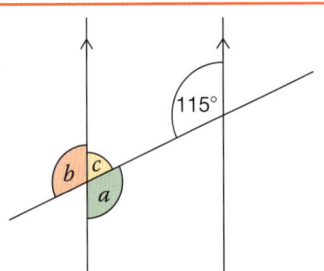

$a = 115°$ Alternate angles are equal.

$b = 115°$ Corresponding angles are equal.

$c = 180° - 115° = 65°$

Angles on a straight line add to $180°$.

$115°$ is an obtuse angle.

$65°$ is an acute angle.

Exercise 3.1S

1 Work out the unknown angles.
State whether each answer is acute,
obtuse or reflex.

a

80° *a*

b

b
200°

c

c 106°
156°

d

110°
d 85°

2 **a** Work out the size of each angle labelled
by a letter.

i

49° *p*

ii

r
52°

iii

155°
q

b Which answer is obtuse?

3 Find the unknown angles.
Give a reason for each answer.

a

u
v
w
108°

b

z
34°
x
y

4 Write down the size of lettered angle.
Give a reason for each answer.

a

a
50°

b

114°
b

4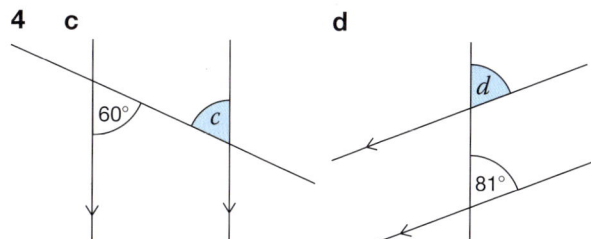

c

60° *c*

d

d
81°

5 Find the unknown angles.
Give a reason for
each answer.

q
p *r*
54°

6 Find each angle
marked by a letter.
In each case give a
reason.

78°

t *s*
u *v*

7 **a** Write down the letter that marks
i an acute angle
ii an obtuse angle
iii a reflex angle.

b Find the size of
each of these angles.
Give reasons.

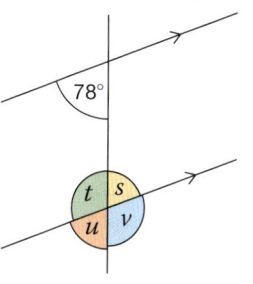

x
y
z
70°

8 Find the size of each angle
marked with a letter. Give your reasons.

a

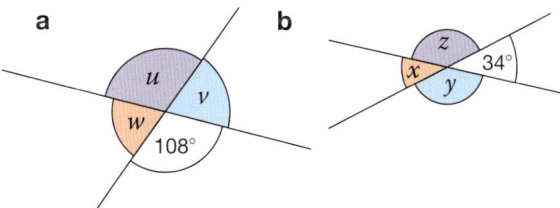

a
b *c*
d
57°

b

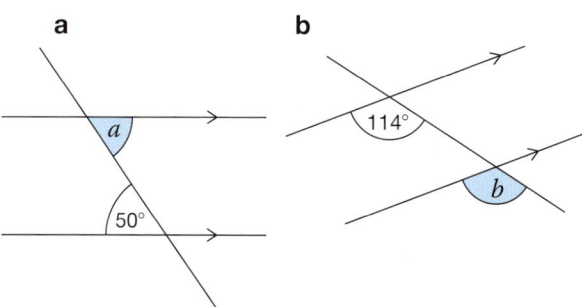

s 98°
t
u
73° *q*
p *r*

3.1 Angles and lines

RECAP

- Angles at a point add up to 360°.
- Angles at a point on a straight line add up to 180°.
- Vertically opposite angles are equal.
- When two lines are parallel, alternate angles are equal and corresponding angles are equal.

Angles can be used to give directions or describe turns.

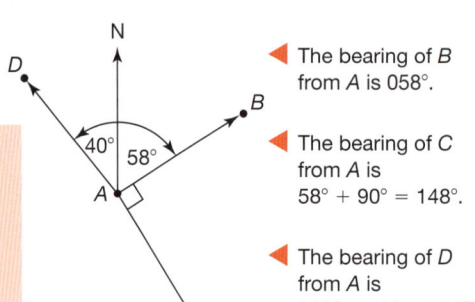

- A **three-figure bearing** is a
  - three-figure angle
  - measured *clockwise*
  - from *north*.

◀ The bearing of *B* from *A* is 058°.

◀ The bearing of *C* from *A* is 58° + 90° = 148°.

◀ The bearing of *D* from *A* is 360° − 40° = 320°.

The north line is at the point where the bearing is *from*.

HOW TO

To solve problems involving angles and parallel lines

① Draw a sketch. Include any angles that you know or are given.

② Look for parallel lines or places where angles meet at a point or on a line.

③ Use rules from the Skills section to find the angle(s) you need.

EXAMPLE

Find the angles between the hands of a clock at 5 o'clock.

② The 12 equal angles at the centre add up to 360°.

③ $x = 360° \div 12$ (angles at a point add to 360°)

$\quad = 30°$

Angle between the hands = 5 × 30° = 150°.

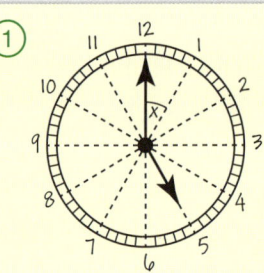

EXAMPLE

The bearing of a ship, *S*, from a lighthouse, *L*, is 245°.

Find the bearing of the lighthouse from the ship.

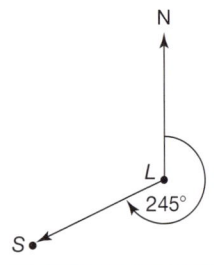

① For a bearing *from* the ship, draw a north line at *S*.

② $x = 245° - 180° = 65°$

The angle on a straight line is 180°.

③ $y = x = 65°$

Alternate angles are equal.
The bearing of the lighthouse from the ship is 065°.

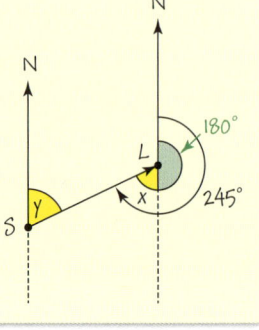

Exercise 3.1A

1 **a** Draw a sketch to show the eight compass points: N, NE, E, SE, S, SW, W and NW.

   b Write each compass point as a three-figure bearing.

2 **a** Tanya is facing north-west. She turns anti-clockwise through 90°. Which direction is she now facing?

   b Ashad is facing west. He turns to face north-east. Describe the angle he turns through.

3 **a** Find the angle between the hands of a clock at each of these times.

   i 3 o'clock   **ii** 4 o'clock

   iii 10 o'clock   **iv** half past one.

   b Find the angle turned through in 10 minutes by

   i the minute hand

   *ii the hour hand.

4 Four swimmers set out from a boat as shown. Write each direction as a three-figure bearing.

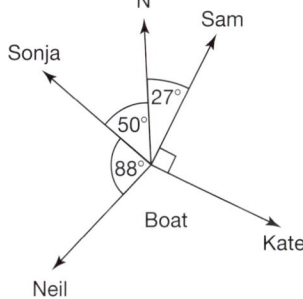

5 **a** A boat leaves a port on a bearing of 198°, then turns clockwise through 68°. Find the bearing of the new course.

   b A boat leaves a port on a bearing of 024°, then turns anti-clockwise through 96°. Find the bearing of the new course.

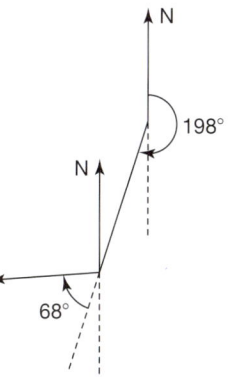

6 **a** The bearing of Aton from Barnum is 054°. Find the bearing of Barnum from Aton.

   b The bearing of Padsey from Mulfield is 296°. Find the bearing of Mulfield from Padsey.

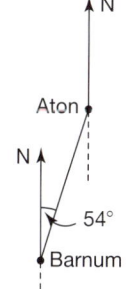

7 The bearing of a plane from an airfield is 167°. On what bearing should the plane fly to go to the airfield?

8 A ship leaves a port on a bearing of 195°. The captain is told to return to the port. On what bearing must the ship travel?

*9 A plane takes off from a runway in a NW direction. It then turns through an angle of 75° to its right. A helicopter, flying on a bearing of 232°, needs to turn to fly in the same direction. What turn must the helicopter make?

*10 The diagram shows three ports *P*, *L* and *Y* and the capital *C* of an island.

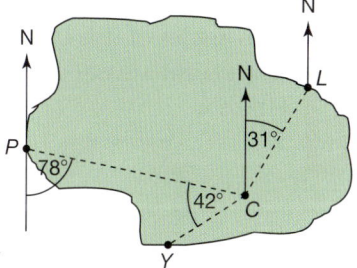

   a Find the bearing of each port from the capital.

   b Find the bearing of the capital from each port.

11 The diagram shows a triangle with angles *a*, *b* and *c*. One side of the triangle is extended and a parallel line added.

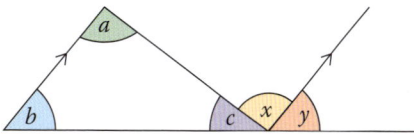

   a Copy and complete the following.

   $x = a$ because ....

   $y = b$ because...........................

   b Explain how this shows that the angles of a triangle add up to 180°.

Triangles and quadrilaterals

- The sum of the angles of a triangle = 180°.
- The sum of the angles of a quadrilateral = 360°.

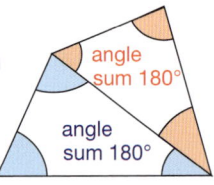

▶ A quadrilateral split into two triangles.

angle sum 180°
angle sum 180°

Triangles
In a **scalene** triangle, the sides and angles are all different.
An **isosceles** triangle has 2 equal sides and 2 equal 'base' angles.
An **equilateral** triangle has 3 equal sides. Each angle is 60°.

Triangles can also be

acute-angled,   **right-angled**   or **obtuse-angled.**

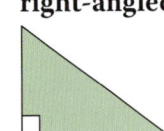

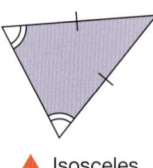

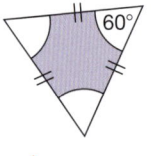

▲ Isosceles   ▲ Equilateral

Quadrilaterals
In a **parallelogram** both pairs of opposite sides are parallel.
A **trapezium** has only 1 pair of parallel sides.
A **rhombus** is a parallelogram with 4 equal sides.
A **kite** has 2 pairs of equal adjacent sides.
A **rectangle** is a parallelogram whose angles are all right angles.
A **square** is a rectangle with 4 equal sides.

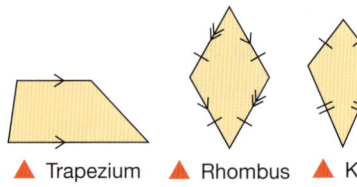

▲ Trapezium   ▲ Rhombus   ▲ Kite

EXAMPLE

Two angles of this **isosceles trapezium** are equal to 48°.
Find x.

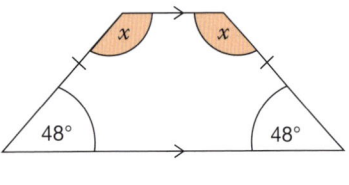

$2x = 360° - 48° - 48°$ Angle sum of quadrilateral = 360°.

$= 264°$

$x = 264° \div 2$

$= 132°$

The other angles are both 132°.

An isosceles trapezium has two equal sides.
It is symmetrical, so the unknown angles are equal.

EXAMPLE

Find the angles marked by letters.

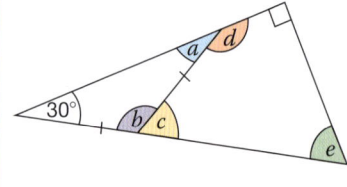

$a = 30°$ Equal base angles of isosceles triangle.

$b = 180° - 30° - 30°$ Angle sum of triangle = 180°.

$= 120°$

$c = 180° - b = 60°$ Angles on a straight line add up to 180°.

$d = 180° - a = 150°$ Angles on a straight line add up to 180°.

$e = 360° - 90° - d - c$ Angle sum of quadrilateral = 360°.

$e = 360° - 90° - 150° - 60°$

$= 60°$

Or using the angle sum of the large right-angled triangle
$e = 180° - 30° - 90° = 60°$.

Exercise 3.2S

1 Answer Yes or No.

    a Is an equilateral triangle an acute-angled triangle?

    b Can an obtuse-angled triangle contain two obtuse angles?

2 **a** Find the lettered angles.

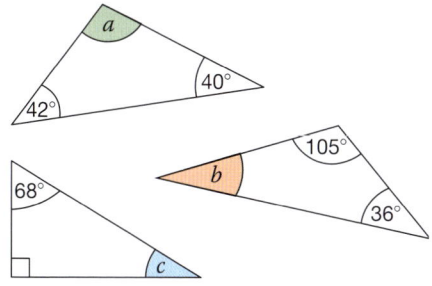

    b Which letter marks an obtuse angle?

3 **a** Find the angles marked by letters.

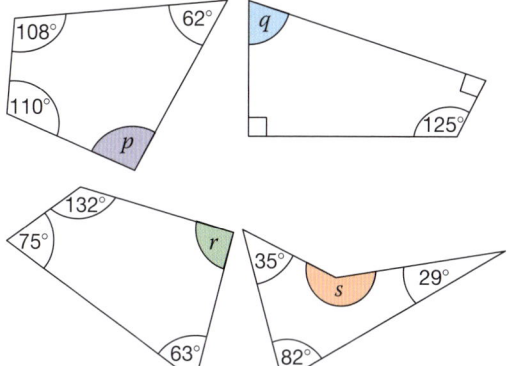

    b Write down the letter that marks
      i a right angle
      ii a reflex angle.

4 Work out the angles marked by letters.

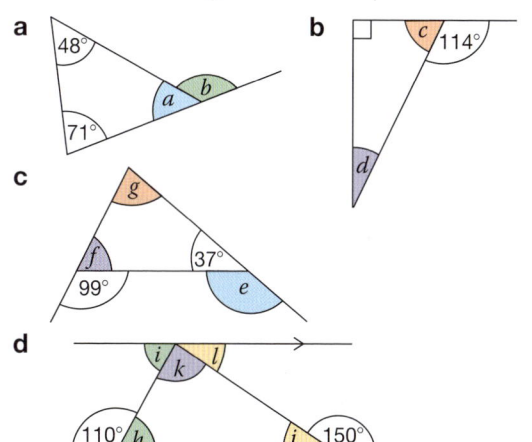

5 Find the lettered angles.

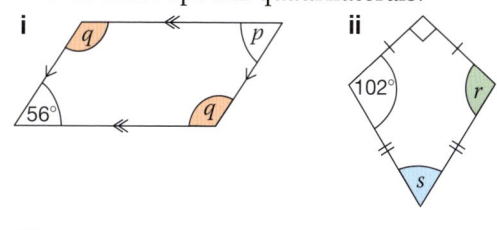

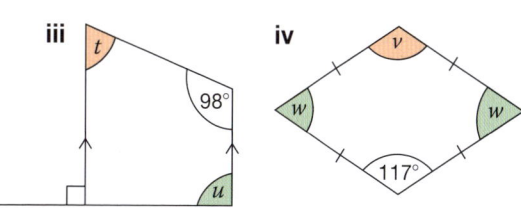

6 **a** Name these special quadrilaterals.

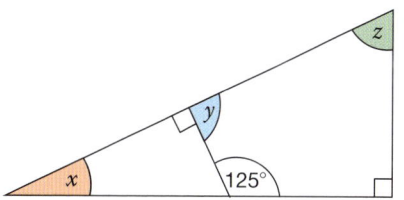

    b Find the angles marked by letters.

7 Work out the angles marked by letters.

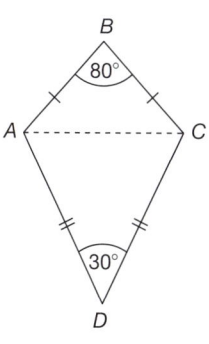

8 *ABCD* is a kite.
  Angle *ABC* = 80°
  Angle *ADC* = 30°

    a Find
      i angle *BAC*
      ii angle *BCA*
      iii angle *BAD*
      iv angle *BCD*

    *b For any kite *ABCD*, does angle *BAD* = angle *BCD*? Explain your answer.

1082, 1102, 1130, 1141   SEARCH

3.2 Triangles and quadrilaterals

- The sum of the angles of a triangle = 180°.
- The sum of the angles of a quadrilateral = 360°.

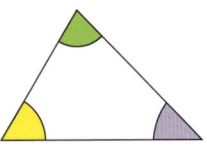

▲ If you cut out a triangle and tear off its corners you can fit them exactly onto a straight line.

HOW TO

To solve an angles problem involving triangles and quadrilaterals

① Sketch a diagram (unless one is given).
Mark (or look for) known angles, equal or parallel sides.

② Apply geometrical properties from **3.1** and **3.2** to fill in missing angles until you find what you need. Say which results you are using.

③ Use geometrical properties to answer the questions.

EXAMPLE

In quadrilateral $ABCD$, angle A is 56°, angle B is 94° and the other angles are equal.

Find the other angles.

② Angle sum of a quadrilateral is 360°.

$x + x = 360° - 56° - 94°$

$2x = 210°$

$x = 210° ÷ 2$

$\quad = 105°$

③ Angle C and angle D are both 105°.

① Mark the equal unknown angles x.

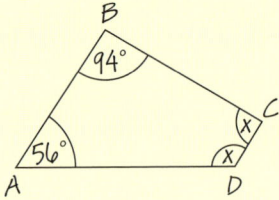

EXAMPLE

Tao writes

'$x + y = 98°$ Corresponding angles'

a Find the correct value of $x + y$.
Give reasons.

b What mistake has Tao has made?

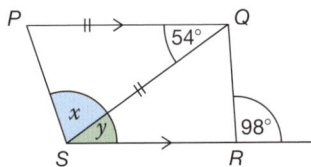

a ② PQ is parallel to SR, so y and 54° are alternate angles.

③ $y = 54°$ Alternate angles, PQ parallel to SR.

② In isosceles triangle PQS, both base angles must be x.

③ $2x = 180° - 54° = 126°$ Equal base angles in isosceles triangle.

$x = 126° ÷ 2 = 63°$

$x + y = 63° + 54° = 117°$

b ② Tao thought $x + y$ and 98° were corresponding angles.

This would mean that PS and QR would be parallel.

③ Tao thought that PS was parallel to QR.

This is the **converse** rule.
If corresponding angles are equal, then the lines are parallel.

Exercise 3.2A

1 a Two angles in a triangle are 23° and 84°. What is the third angle?

b Two angles in an isosceles triangle are 70° and 40°. What is the third angle?

c Three angles in a quadrilateral are 54°, 67° and 123°. What is the fourth angle?

d Two angles in a parallelogram are 60° and 60°. What are the other two angles?

2 Four squares will fit around a point without leaving any gaps or overlaps. What combinations of squares and equilateral triangles will likewise fit around a point?

3 Find the angles marked by letters.

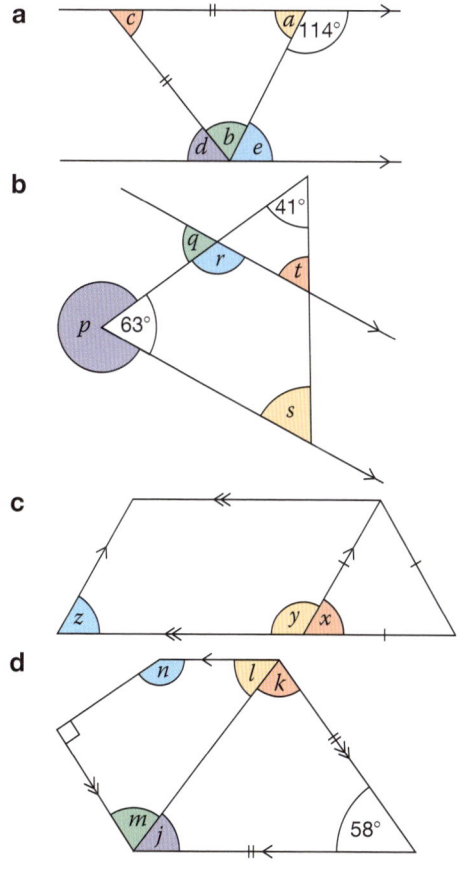

a

b

c

d

4 a Say why a triangle cannot have more than one obtuse angle.

b What type of angles are the others?

5 Sally says that a parallelogram with 4 equal angles must be a rectangle. Is Sally correct? Give your reasons.

6 On x and y axes from 0 to 8, draw and describe each shape.

p.290

Shape	Vertices (corners) at
a	(0, 1), (2, 1), (3, 2), (1, 2)
b	(2, 3), (3, 5), (1, 5)
c	(4, 5), (6, 6), (8, 5), (6, 4)
d	(2, 7), (4, 6), (5, 7), (4, 8)
e	(6, 1), (7, 1), (6, 3)
f	(5, 0), (5, 4), (4, 3), (4, 2)

7 The diagram shows a kite *KLMN*. Cutting the kite along *diagonal KM* gives two *identical obtuse-angled triangles.*

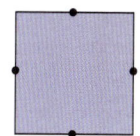

a i Describe the triangles that you get by cutting the kite along the diagonal *NL*.

ii Describe the triangles that you get by cutting the kite along both diagonals.

b Describe the triangles you get by cutting these quadrilaterals along their diagonals.

    i Rhombus   **ii** Rectangle

    iii Square

8 a Draw a square and join the mid-points of the sides. What shape is the result?

b Repeat part **a**, starting with

    i a rectangle   **ii** a rhombus

    iii a kite   **iv** parallelogram

    v a quadrilateral in which all the angles are different.

*9 In quadrilateral *ABCD*: *BC* and *AD* are parallel, angle *BCD* = 130°, angle *BDC* = 20°, and angle *BAD* = 120°. Show that triangle *BAD* is isosceles.

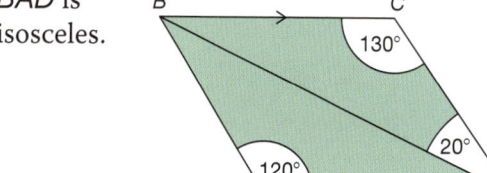

3.3 Congruence and similarity

Congruent shapes are exactly equal in size and shape: equal sides and equal angles.

You can prove that two triangles are congruent by showing that they have

● Three equal sides (SSS)

● Two sides and the angle *between* them equal (SAS)

● Two angles and a **corresponding** side equal (ASA)

● A right angle, the hypotenuse and another side equal (RHS)

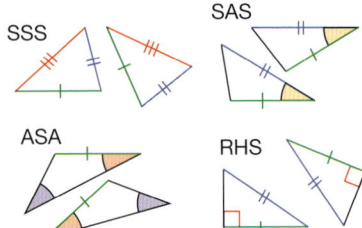

EXAMPLE

a Is triangle A congruent to triangle B?

b Is triangle X congruent to triangle Y?

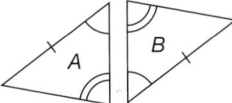
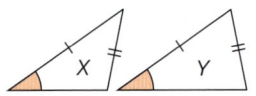

a Yes   ASA   The equal sides are corresponding because they are both opposite the double-marked angles.

b No   The equal angle is not *between* the two marked sides.

Similar shapes are the same shape but different in size.
Their angles are equal in size.
The sides are multiplied or divided by the same **scale factor**.
Triangles PQR and ABC are similar because they have equal angles.

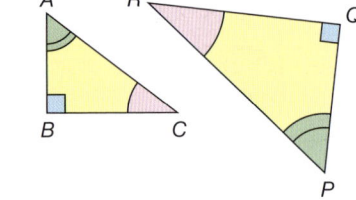

● The scale factor $= \dfrac{PQ}{AB} = \dfrac{QR}{BC} = \dfrac{RP}{CA}$

EXAMPLE

a Show that triangle ADE is similar to triangle ABC.

b Calculate the length of
 i AC
 ii CE.

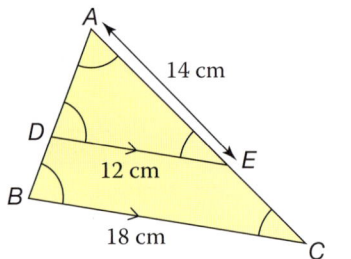

a $\angle ADE = \angle ABC$ Corresponding angles.

   $\angle AED = \angle ACB$ Corresponding angles.

   $\angle DAE = \angle BAC$ The same angle.

   Triangles ADE and ABC are similar.   Equal angles.

b i The scale factor $= \dfrac{BC}{DE} = \dfrac{18}{12} = \dfrac{3}{2} = 1.5$

     $AC = 1.5 \times 14 = 21\,\text{cm}$

ii $CE = AC - AE$

     $= 21\,\text{cm} - 14\,\text{cm} = 7\,\text{cm}.$

BC and DE are corresponding sides.
AC and AE are corresponding sides.

Exercise 3.3S

1

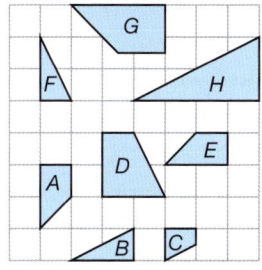

Write down the letters of the shapes that are

a congruent to triangle *B*

b similar to triangle *B*

c congruent to trapezium *A*

d similar to trapezium *A*.

2 In each part state whether the pair of triangles is congruent. Give your reasons.

a 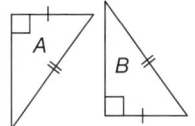 **b**

c 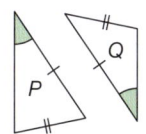 **d**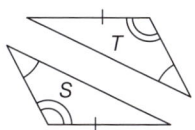

3 State which two triangles are congruent. Give your reasons.

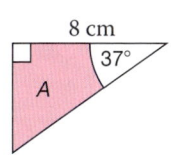 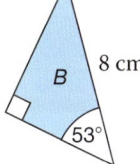 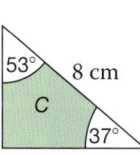

4 Which two shapes are similar? Give your reasons.

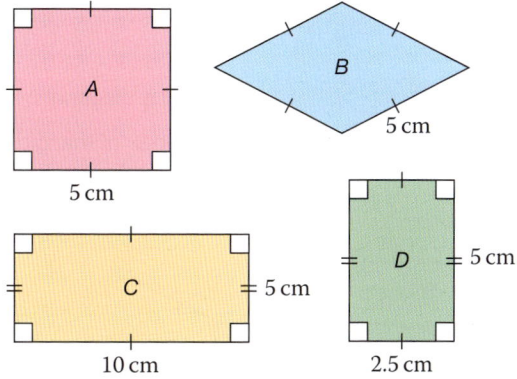

5 **a** Show that triangle *STR* is similar to triangle *PQR*.

  b Calculate the length of

   i *SR*   **ii** *PS*.

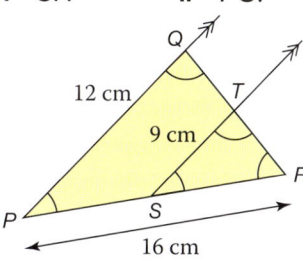

6 **a** Show that triangle *ABC* is similar to triangle *EDC*.

  b Calculate the length of

   i *CD*   **ii** *BD*.

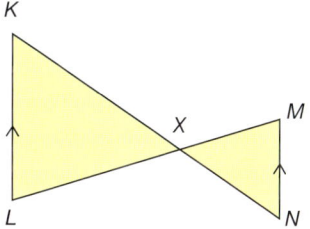

7 **a** Show that triangle *KLX* is similar to triangle *NMX*.

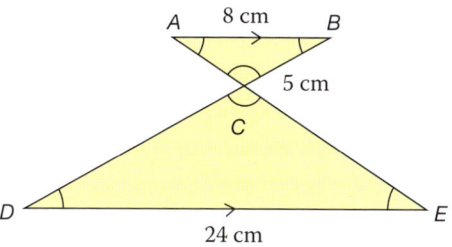

  b $KX = 60\,cm$, $LX = 50\,cm$, $KL = 45\,cm$ and $XM = 30\,cm$. Calculate

   i *MN*   **ii** *XN*.

Did you know...

Plants often use similar shapes. The leaves of ferns are similar as are the smaller leaflets.

3.3 Congruence and similarity

RECAP

- Congruent shapes are exactly the same shape and size. Triangles are congruent if the following are equal

    SSS   3 sides

    SAS   2 sides and the angle between them

    ASA   2 angles and a corresponding side

    RHS   a right angle, the hypotenuse and 1 other side.

- Similar figures are the same shape but different in size. All angles are the same but lengths are multiplied or divided by the same scale factor.

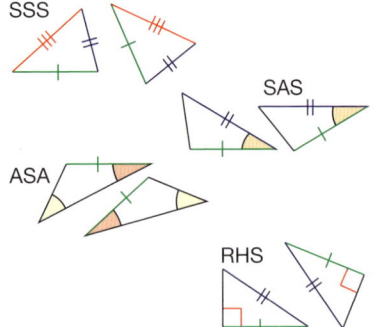

HOW TO

To solve a similarity/congruence problem

(1) Sketch a diagram (unless one is given). Label known angles, equal or parallel sides.

(2) Look for congruent or similar shapes.

(3) Use congruence to prove facts about shapes.

Use similarity to find lengths of objects that have been enlarged or reduced using the scale factor.

EXAMPLE

Kay has a photograph that is 4 inches wide and 6 inches high.

She wants the photo enlarged to make a poster 30 inches wide.

What will be the height of the poster?

(2) The poster will be similar to the photo. Corresponding sides will be in the same ratio.

Scale factor = $\frac{30}{4}$ = 7.5

(3) Height of poster = 7.5 × 6 = 45
The height of the poster will be 45 inches.

Or as the height of the photo is $1\frac{1}{2}$ times its width, the height of the poster will be $1\frac{1}{2}$ × 30 = 45 inches.

(1)   4 inches   30 inches

6 inches

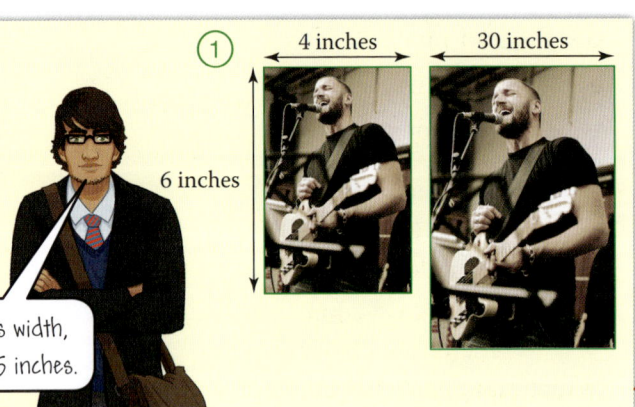

EXAMPLE

Show that the opposite sides of a parallelogram are equal in length.

(3) Give reasons why triangles *PQR* and *RSP are* congruent.

∠QRP = ∠SPR   Alternate angles.

∠QPR = ∠SRP   Alternate angles.

Side PR is in both triangles.

Triangles PQR and RSP are congruent.   ASA

So PQ = RS and QR = PS by congruence.

(1) Drawing a diagonal

(2) gives a pair of triangles.

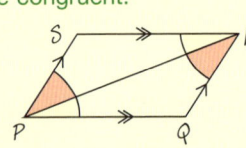

This is a proof.
Just measuring the sides of a parallelogram is *not* a proof.

Exercise 3.3A

1 A photograph is 7 inches wide by 5 inches high.

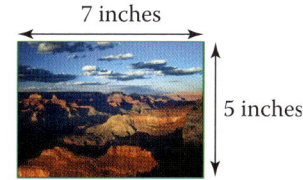

7 inches

5 inches

Jack wants a similar poster 24 inches tall. What will be the width of the poster?

2 A diagram is 10 cm wide and 12 cm high. A photocopier is used to reduce the size of the diagram.

    a Find the new height of the diagram when the new width is 8 cm.

    b Find the new width of the diagram when the new height is 8 cm.

3 Identify *all* the congruent triangles in this kite. Say how you know that each pair of triangles is congruent.

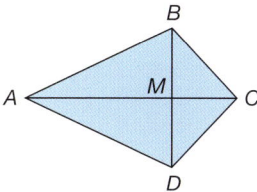

4 *PQR* is an isosceles triangle.
S is the mid-point of *QR*.

    a Show that triangle *PQS* is congruent to triangle *PRS*.

    b **i** Write down the angle that is equal to angle *PSQ*.

       ii Explain why angle *PSQ* is a right angle.

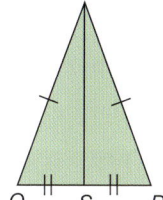

5 In rectangle *ABCD*, *AB* is parallel to *DC*, *AD* is parallel to *BC* and all the angles are 90°.

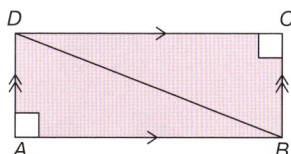

    a Show that triangle *ABD* is congruent to triangle *CDB*.

    b Explain how this proves that the opposite sides of a rectangle are equal in length.

6 In the kite *ABCD*, *AB* = *AD* and *BC* = *CD*. Show that angle *B* = angle *D*.

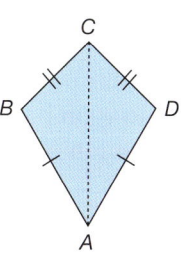

7 A shop sells picture frames in these sizes.

Size	Width	Height
Mini	24 cm	30 cm
Small	30 cm	40 cm
Medium	40 cm	50 cm
Large	50 cm	70 cm
Extra Large	60 cm	80 cm

Which of these sizes are similar in shape? Explain your answer.

8 Are these triangles congruent, similar or neither? Give your reasons.

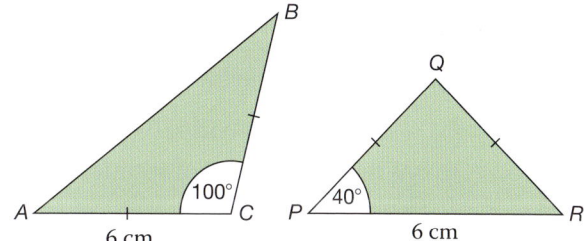

9 Find all the missing sides and angles in this diagram.

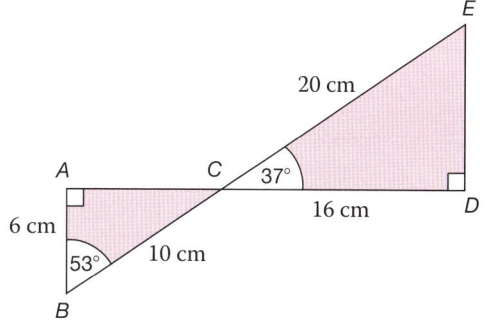

10 Find out the dimensions of A1, A2, A3, A4, A5 and A6 sheets of paper.
Are all the sizes similar in shape?
Describe how the sizes are related.

1119, 1148   SEARCH

3.4 Polygon angles

A **polygon** is a two-dimensional shape with straight sides.
A **regular** polygon has all its sides equal *and* all its angles equal.
A **pentagon** has 5 sides, a **hexagon** has 6 sides,
an **octagon** has 8 sides and a **decagon** has 10 sides.

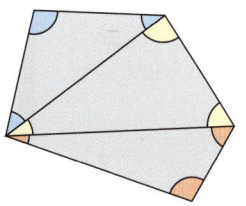

- The sum of the **interior angles** of *any* polygon
  = (number of sides − 2) × 180°.

- The sum of the **exterior angles** of *any* polygon = 360°.

- At each vertex: interior angle + exterior angle = 180°.

▲ Pentagon (5 sides)
Angle sum = 3 × 180° = 540°

EXAMPLE

A heptagon has 7 sides,
3 angles of 135° and
3 angles of 125°.
Find the other angle.

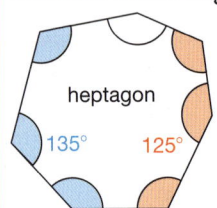

heptagon

135°   125°

A heptagon can be split into 5 triangles.

Angle sum of a heptagon = 5 × 180°
                        = 900°

The other angle = 900° − 3 × 135° − 3 × 125°
                = 900° − 405° − 375°
                = 120°

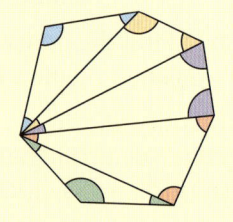

EXAMPLE

a   Find the size of the interior angles of a regular hexagon.

b   **i**   How many lines of symmetry does a regular hexagon have?

   ii   What is the order of rotational symmetry of a regular hexagon?.

a   Sum of 6 equal exterior angles = 360°
   Each exterior angle = 360° ÷ 6 = 60°
   Interior angle = 180° − 60° = 120°
   Or
   Sum of interior angles = (6 − 2) × 180° = 720°
   Each interior angle = 720° ÷ 6 = 120°

b   **i**   6 lines of symmetry.
   ii   Order 6.

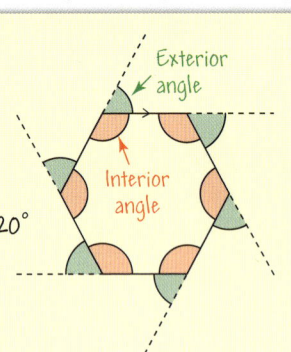

Exterior
angle

Interior
angle

Going round a polygon, the
total angle turned
through = 360°.

Imagine shrinking the
polygon to a point.

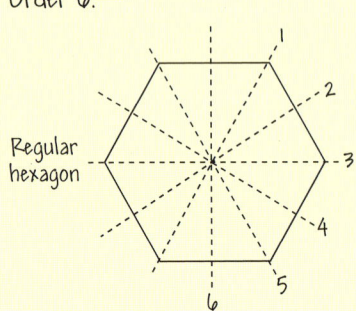

Regular
hexagon

1
2
3
4
5
6

You can use tracing
paper to check line and
rotational symmetry.

Exercise 3.4S

1 Explain how to find the sum of all the interior angles in an octagon.

2 Find the angle x.

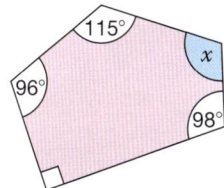

3 Copy and complete this table.

Regular polygon	No. of sides	Exterior angle	Interior angle
Triangle	3		
Quadrilateral	4		
Pentagon	5		
Hexagon	6		
Heptagon	7		
Octagon	8		
Nonagon	9		
Decagon	10		

4 Two of the angles of a hexagon are right angles and three angles are equal to 132°. Find the other angle.

5 A dodecagon is a polygon with 12 sides. Eleven of the angles of a dodecagon are equal to 154°.
Calculate the other angle.

6 A regular polygon has 15 sides.

  a Use the angle sum of this polygon to work out the size of an interior angle.

  b Use the sum of the exterior angles to check your answer to part **a**.

7 A regular polygon has 18 sides.

  a Calculate the size of an interior angle of this polygon.

  b Use another method to check your answer to part **a**.

8 The diagram shows a regular heptagon.

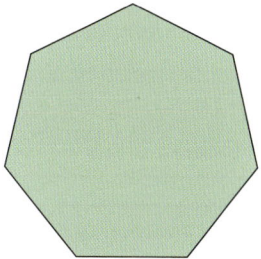

  a Use tracing paper to copy the heptagon.

  b What is the order of rotational symmetry of this shape?

  c Draw all the lines of symmetry on your copy.

9 **a** Write down the order of rotational symmetry of

  i the regular octagon

  ii the regular decagon.

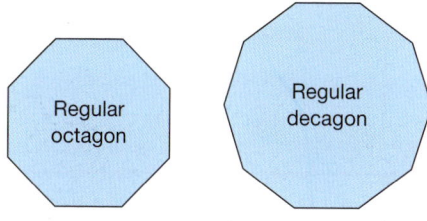

  b How many lines of symmetry has

  i the regular octagon

  ii the regular decagon ?

*10 The diagram shows a regular nonagon divided into congruent triangles.

  a Find the angles marked x and y.

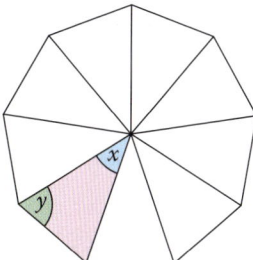

  b Use the value of y to check your answers in question **3** for a regular nonagon.

3.4 Polygon angles

RECAP

- The sum of the exterior angles of any polygon is 360°.
- The sum of the interior angles of an n-sided polygon is $(n - 2) \times 180°$.
- Interior angle + exterior angle = 180°

▲ Designers can create elaborate patterns using simple polygons.

HOW TO

To solve an angle problem involving polygons

1. Draw a sketch (if needed).
2. Decide whether to use the sum of the interior or exterior angles.
3. Find the answer.

EXAMPLE

A regular polygon has interior angles of 156°.

How many sides does the polygon have?

② Find the exterior angle.

Exterior angle = 180° – 156° = 24°   Angles on a straight line.

③ The sum of the exterior angles is 360°, so divide to find out how many there are.

Number of exterior angles = 360° ÷ 24° = 15

The polygon has 15 sides.   The number of sides = number of angles.

① Sides of polygon

156°

Exterior angle

EXAMPLE

a   Find the interior angle of a regular 12-sided polygon.

b   Explain why regular 12-sided polygons

   i   will not fit together exactly at a point.

> A tessellation is a tiling of the plane in which there are no gaps and no overlaps.

   ii   will fit together exactly at a point with equilateral triangles.

a   ② Find the exterior angle then the interior angle.

   ③ Exterior angle = 360° ÷ 12   Sum of exterior angles
             = 30°   = 360°.

   Interior angle = 180° – 30°
              = 150°   Angles on a straight line.

① Interior angle

Exterior angle

> Sum of interior angles = 10 × 180° = 1800°
> Each interior angle = 1800° ÷ 12 = 150°

b   i   ② Shape will fit together exactly if the sum of the angles at a point is 360°.

   ③ Sum of 2 interior angles = 300°   <360°. Gaps

      Sum of 3 interior angles = 450°   >360°. Overlaps

      The polygons do not fit together at a point.

   ii   ③ Each angle in an equilateral triangle = 60°

      300° + 60° = 360°

      Two regular 12-sided polygons and an equilateral triangle will fit together at a point.

Exercise 3.4A

1 Find the number of sides of a regular polygon that has exterior angles of 12°.

2 The diagram shows part of a regular polygon.

Calculate the number of sides of the polygon.

3 A regular polygon has interior angles of 177°. How many sides does this polygon have?

4 a Copy and complete the following table for regular polygons.

Number of sides	Exterior angles	Interior angles
10		
	20°	
		165°
40		
	5°	
		176°

b As the number of sides increases describe what happens to the size of

    i the exterior angles

    ii the interior angles.

5 A regular octagon, a regular hexagon and a regular pentagon are brought together at a point.

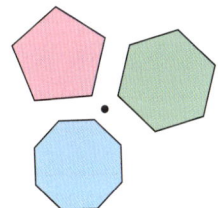

Do they fit together

    A with overlaps

    B with gaps

    C without overlaps or gaps?

Give reasons for your answer.

6 Find angles x and y in this regular octagon.

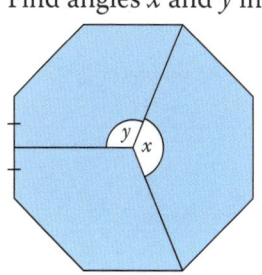

*7 Two sides of a regular pentagon are extended to form angle x. Find the size of angle x.

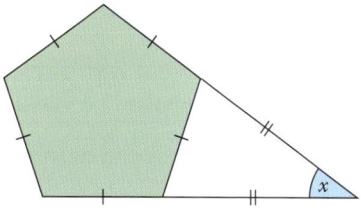

*8 Calculate angle y.

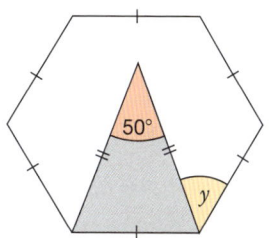

*9 The diagram shows a regular hexagon attached to a square.

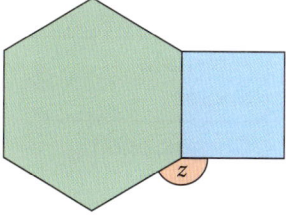

    a Calculate angle z.

    b Name one or more regular polygons that would fit exactly into this space.

10 A regular polygon can be made by fitting together squares and equilateral triangles. How many sides do the inside and outside polygons have?

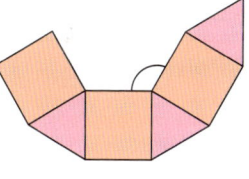

Did you know…

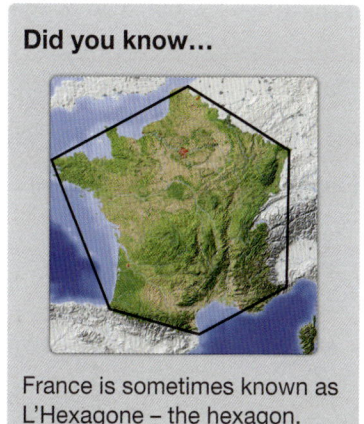

France is sometimes known as L'Hexagone – the hexagon.

 1100, 1320   SEARCH

Summary

Checkout
You should now be able to...

		Test it Questions
✔	Describe and apply the properties of angles at a point, on a line and at intersecting and parallel lines.	1, 2
✔	Derive and use the sum of angles in a triangle.	3, 4
✔	Derive and apply the properties and definitions of special types of quadrilaterals.	5 – 7
✔	Solve geometrical problems on coordinate axes.	8
✔	Identify and use congruence and similarity.	9, 10
✔	Deduce and use the angle sum in any polygon and derive properties of regular polygons.	11

Language	Meaning	Example
Acute angle	An angle smaller than a right angle	
Right angle	90° or one-quarter turn.	
Obtuse angle	Greater than 90° but smaller than 180°.	
Reflex angle	Greater than 180° but smaller than 360°.	
Alternate angles	When referring to parallel lines: angles in the corners of a Z shape.	Alternate angles Corresponding angles
Corresponding angles	When referring to parallel lines: angles under the arms of an F shape.	
Three-figure bearing	A direction defined by a three-figure angle measured clockwise from north.	East is 090°. South-west is 225°.
Polygon	A 2D shape with straight edges.	Pentagon (5), Hexagon (6), Octagon (8), Decagon (10).
Regular	All sides are equal and all angles are equal.	A regular quadrilateral is a square.
Triangle	A three sided polygon.	Right-angled, equilateral, isosceles, scalene.
Quadrilateral	A four sided polygon.	Square, rectangle, rhombus, trapezium, parallelogram, kite.
Congruent	Exactly the same shape and size.	
Similar	The same shape but different in size.	
Scale Factor	The ratio of corresponding lengths in two similar shapes.	A and B are similar; the scale factor is 2. A and C are congruent.
Interior angle	The angle between two sides inside a polygon.	
Exterior angle	The angle between one side of a polygon and the next side extended.	

Review

1 Calculate the size of angles *a*, *b*, *c* and *d* and give a reason for each answer.

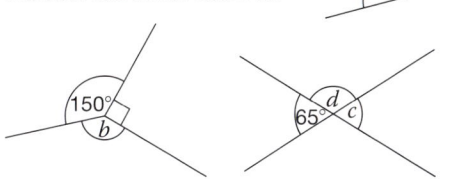

2 Calculate the size of angles *a* and *b* and give a reason for each answer.

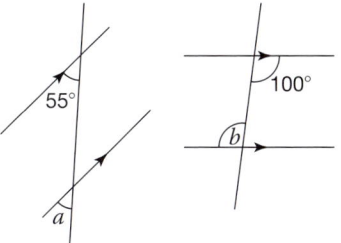

3 Calculate the size of angle *ABC* in this triangle.

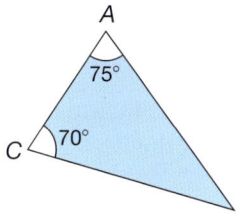

4 The triangle *DEF* is not drawn accurately. Which side is the same length as side *DE*?

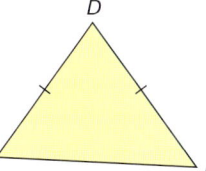

5 Use your knowledge about triangles to prove that the angles in a quadrilateral add up to 360°.

6 Calculate the size of angles *a* and *b* in the rhombus.

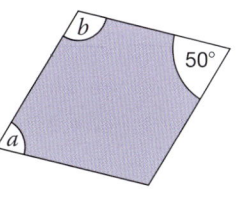

7 Which quadrilateral is being described below?

   a Two pairs of parallel and equal sides.

   b One pair of parallel sides and no equal sides.

   c Two pairs of equal sides but no parallel sides.

8 Draw coordinate axes with *x* and *y* from 0 to 6. Now plot these points A (3, 5), B (6, 2) and C (1, 0).

   Join up the dots to form a triangle, what type of triangle is this?

9 Are these two triangles congruent?

   Give a reason for your answer.

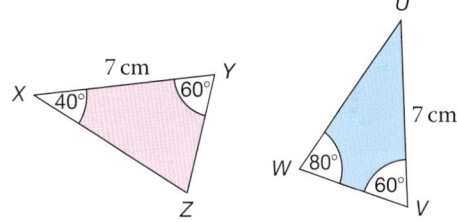

10 These triangles are similar, what is the length of *DE*?

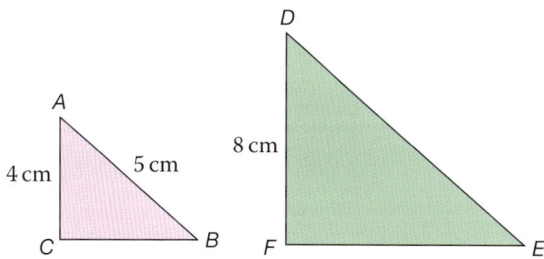

11 What do the interior angles of a pentagon add up to?

What next?

Score			
0 – 4			Your knowledge of this topic is still developing.
			To improve look at MyMaths: 1082, 1086, 1100, 1102, 1109, 1119, 1130, 1320, 1141, 1148
5 – 9			You are gaining a secure knowledge of this topic.
			To improve your fluency look at InvisiPens: 03Sa – j
10 – 11			You have mastered these skills. Well done you are ready to progress!
			To develop your problem solving skills look at InvisiPens: 03Aa – h

Assessment 3

1 How many turns do these actions require?

    a The hour hand of a clock moving for 8 hours. [2]

    b The Earth rotating about the Sun for 9 months. Ignore the Earth spinning on its own axis. [2]

    c An athlete running 3 laps of a track. [1]

2 What is the angle between the hour hand and the minute hand of a clock at

    a 5 o'clock [1]   **b** half past 8 [2]

3 The cruise ship 'Black Watch' sails on a bearing of 070°. To avoid a storm it changes course to a bearing of 130°. What angle has it turned through? [2]

4 Three villages, East Anywhere, Middle Anywhere and West Anywhere, are the vertices of the triangle shown. The bearing of East Anywhere from West Anywhere is 075°.

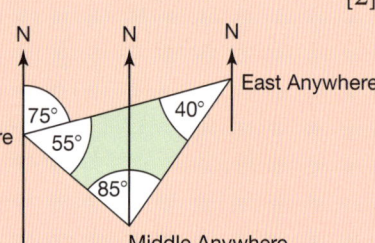

    Write down the bearing of

    a Middle Anywhere from West Anywhere [1]

    b West Anywhere from East Anywhere [2]

    c Middle Anywhere from East Anywhere. [2]

5 Rafa and Sunita were standing at the top of a hill. Rafa was facing due north and Sunita was facing due south. Explain why they could see each other. [1]

6 Siobhan fits three triangles together at one point.

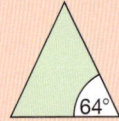 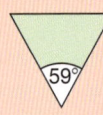 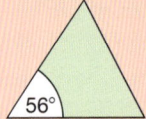

    Will they make a straight line? Give reasons for your answer. [2]

7 Sienna says that the missing angles have these values

$a = 56°$ $b = 75°$ $c = 46°$ $d = 148°$ $e = 121°$ $f = 121°$.

    a 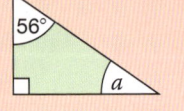 [2]   **b** 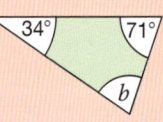 [2]   **c** 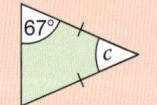 [2]

    d 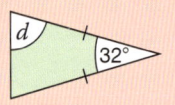 [2]   **e**  [4]

    Decide if her value for each angle is correct or incorrect.
Give reasons for your answers.

8 Manuel made the following statements about triangles. Write down if each of his statements is true or false. If the statement is true draw an example. If the statement is false explain why it is false.

    a Some triangles have two obtuse angles. [2]

    b Some triangles have one obtuse angle and two acute angles. [2]

    c Some triangles have one right angle, one acute angle and one obtuse angle. [2]

    d Some triangles have two acute angles and one right angle. [2]

9 Mark draws triangle *XZY* so that it is congruent to triangle *ABC*.
Write down the sizes of all the angles in triangle *XZY*. [2]

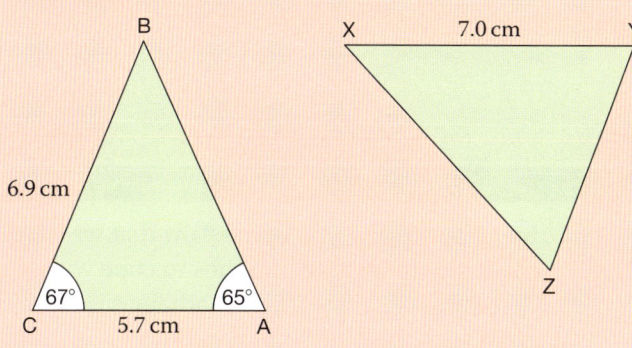

10 Sunita is standing in the shade of a tree, as shown.

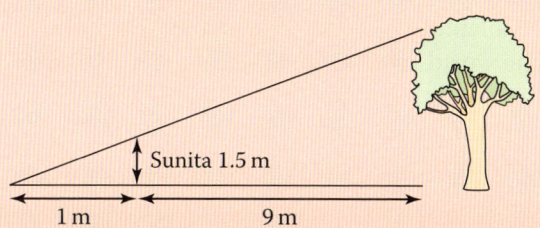

How high is the tree? [2]

11 a Natasha draws a pentagon that has angles of 125°, 155° and 74°.
She wants to draw the other two angles so that they are equal.
What size should Natasha draw these two angles? [3]

b Shivani draws an octagon that has five angles each of 114°.
She wants to draw the other three angles so that they are equal.
What size should Shivani draw these three angles? [3]

c Tyler draws a hexagon that has three angles of 137° each. He wants to draw
the remaining three angles so that they have the ratio $w : w + 120 : w - 30$.
What value should Tyler use for w? [4]

d Dean draws a regular polygon with 60 sides.
What size should he use for the interior angles? [2]

12 Stewart wants to make a patio with seven identical
hexagonal paving slabs, as shown.

a He says that UVWY is a trapezium.
Is he correct?
Give reasons for your answer. [2]

b What angle does he need to use for
i the external angle [2]
ii the internal angle [2]
of the regular hexagons?

c What size does he need to make angle VXY? [4]

d He says that triangle UWY is isosceles.
Is he correct?
If not, what type of triangle is it? [1]

4 Handling data 1

Introduction

In the modern world, there is an ever-increasing volume of data being continually collected, analysed, interpreted and stored. It was estimated that in 2007, there was 295 exabytes (or 295 billion gigabytes) of data being stored around the world. It has increased significantly since then. To put this into perspective, if all that data were recorded in books, it would cover the area of China in 13 layers of books. With all this information being generated, it is important that we have the mathematical techniques to cope with it. Statistics is the branch of mathematics that deals with the handling of data.

What's the point?

Availability of data helps us to understand the world around us. Whether it is scientists are looking for trends in global warming, or consumers looking at cost comparison data to help inform purchasing decisions.

Objectives

By the end of this chapter you will have learned how to …

- Construct and interpret frequency tables and two-way tables.
- Construct and interpret pictograms, bar-line charts and bar charts.
- Interpret and construct pie charts and know their appropriate use.
- Compare distributions using median, mean, mode and range and identify outliers.

Check in

1   Put these numbers in order of size, smallest first.

   a 37, 42, 17, 6, 30, 19, 26, 29

   b 118, 135, 106, 121, 130, 115

   c 156, 145, 154, 165, 166, 155, 144

2   Calculate

   a $63 + 58$ **b** $48 + 69$ **c** $73 + 95 + 84$ **d** $138 + 275$ **e** $63 + 5$

   f $38 - 15$ **g** $96 - 47$ **h** $136 - 54$ **i** $258 - 69$ **j** $432 - 166$

3   Calculate the value of the angle x.

   a

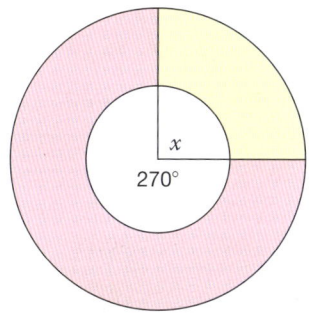

   b

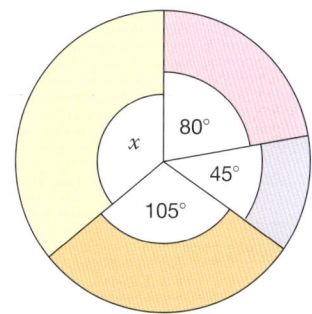

4   Calculate mentally:

   a $360 \div 3$ **b** $360 \div 8$ **c** $360 \div 6$

   d $360 \div 5$ **e** $360 \div 60$ **f** $360 \div 18$

   g $360 \div 12$ **h** $360 \div 20$ **i** $360 \div 36$

Chapter investigation

What is the average age of students in your class?

4.1 Organising data

When you collect data you first have to organise it. Using a **tally chart** is a good way to do this.

EXAMPLE

A class has 30 students. Lisa asks each student whether they walk to school (Y) or not (N).

Y	N	N	Y	N	Y	N	N	Y	N
Y	N	Y	Y	N	N	N	Y	N	N
N	N	N	N	N	Y	Y	Y	N	Y

a Complete a data collection sheet (a **tally chart**).

b Find the proportion of students in the class who walk to school.

a

Do you walk to school?	Tally	Number
Yes	ЖЖ II	12
No	ЖЖ ЖЖ ЖЖ III	18

b The proportion of students who walk to school is 12 out of 30.

Did you know…

The National Census is carries out every 10 years in the UK. The information is used to plan services like schools and hospitals. The next Census will be in 2021.

● You can organise data in a table, such as a **frequency table**, using **rows** and **columns**.

Method of travel	Number of students
Car	14
Bus	11
Walk	5

● A **two-way table** shows more detail and links two types of information, for example, method of travel and gender.

	Boys	Girls
Car	4	10
Bus	6	5
Walk	2	3

14 students travelled by car.

5 girls travelled by bus.

Exercise 4.1S

1 Chris counts the number of passengers in passing cars.

1 0 2 0 1 0 1 3 1 0 3 2
0 1 0 0 2 1 0 0 0 1 2

Number of passengers	Tally	Number of cars
0		
1		
2		
3		

a Copy and complete the data collection sheet to show this information.

b Calculate the total number of cars that passed.

2 James uses a data collection sheet for a survey to find his class' favourite soup. He limits the choice to Tomato (T), Vegetable (V), Fish (F) or Other (O).

T T T V F V V T T V
F O T V O O T V T F
O T V O T V O O O O

a Copy and complete the tally chart to show the data.

Type of soup	Tally	Number of students

b Calculate the number of students in James's class.

c State the class' favourite soup.

3 Felicity counts the number of occupants in passing cars.

1 2 1 4 3 1 2 1 5 4
1 1 1 2 1 3 1 4 1 1
1 2 1 1 1 2 1 1 4 2
2 2 1 2 1 3 2 1 2 1

a Construct a data collection sheet to show this data.

b Calculate the total number of cars that passed.

c Calculate the total number of occupants of the cars that passed.

4 Tickets to see an Irish band cost £5, £10, £15 or £20. Tickets with the following prices are sold one morning.

> These values are all in pounds (£).

15 10 10 5 5 5 5 5
10 20 20 5 5 5 5 10
5 10 5 15 15 20 20 5
5 10 10 10 10 5 5 5
10 5 10 15 20 20 20 20

Price	Tally	Number of tickets
£5		
£10		
£15		
£20		

a Copy and complete the tally chart.

b How many £5 tickets were sold?

c How many tickets were sold altogether?

5 A tetrahedron dice is numbered 1, 2, 3, 4.

a Make a data collection sheet to show these scores.

4 2 1 3 1 3 4 2
3 3 2 2 2 3 4 4
2 4 1 1 2 4 1 2
1 2 1 3 2 1 4 3
2 3 1 4 4 2 3 1

b State the number of times a 3 was rolled.

6 There are two cinemas in a Cinecomplex. The number of people in each cinema is shown in the two-way table.

	Cinema	
	1	2
Adult	31	47
Child	12	8

Calculate the number of

a adults in the Cinecomplex

b children in the Cinecomplex

c people in Cinema 1

d people in Cinema 2

e people in the whole Cinecomplex.

4.1 Organising data

RECAP

RECAP

- You can organise data in a frequency table.
- A two-way table shows more detail and links two types of information.

Frequency tables and two-way tables make it easy to spot trends within the data set. They are also useful for organising large data sets and non-numerical data.

HOW TO

① Construct or complete the table or diagram.

② Interpret information from the table.

③ Add entries in columns or rows if necessary to find the answer.

EXAMPLE

A class of 30 students study either History or Geography.

There are 13 girls in the class, with 8 girls and 9 boys studying Geography.

How many boys study History?

① Draw a two-way table to show the information that you know.

② Use the totals to fill in the other entries.

	History	Geography	Total
Boys		9	30 − 13 = 17
Girls	13 − 8 = 5	8	13
Total		9 + 8 = 17	30

↓

	History	Geography	Total
Boys	17 − 9 = 8	9	17
Girls	5	8	13
Total	30 − 17 = 13	17	30

③ Read the answer from the table.

8 boys study History.

EXAMPLE

The number of televisions in each house in Fern's street is shown in the frequency table.

a Calculate the number of houses in Fern's street.

b Calculate the total number of televisions in Fern's street.

① Complete the table to show the total number of televisions.

a ② Total number of houses
= 1 + 5 + 12 + 9 + 1 = 28

b ③ Multiply the number of TVs by the number of houses for each row in the table. Find the total.

No. of TVs	No. of houses	No. of TVs × no. of houses
0	1	0 × 1 = 0
1	5	1 × 5 = 5
2	12	2 × 12 = 24
3	9	3 × 9 = 27
4	1	4 × 1 = 4

Total number of televisions = 0 + 5 + 24 + 27 + 4 = 60

Exercise 4.1A

1 The table shows some information about the eye colour of children in a nursery.

	Brown	Blue	Other	Total
Boys	5		6	36
Girls				
Total	23	28		64

Are there more blue-eyed girls or brown-eyed boys?

2 A group of 40 students study either French or Spanish.
There are 17 boys in the group, with 15 girls and 10 boys studying French.
How many girls study Spanish?

3 The table shows the number of students in a school who take part in the Duke of Edinburgh scheme.

Year 8	13
Year 9	12
Year 10	11

There are

● seven year 8 boys

● five Year 10 girls

● equal numbers of boys and girls in year 9.

Emily says that there are the same number of boys and girls taking part.
Do you agree with Emily? Say why.

4 Sophie did a survey to find the number of DVDs owned by the students in her class.
The results are shown in the frequency table.

Number of DVDs	Number of students
0	1
1	8
2	6
3	8
4	2
5	3

a Calculate the number of students in Sophie's class.

b Calculate the total number of DVDs owned by the whole class.

5 In a traffic survey, the colour and speed of 100 cars are recorded. The results are summarised in the two-way table.

	Not speeding	Over the speed limit
Red	5	55
Not red	10	30

Maria claims that drivers of red cars tend to break speed limits. Use the two-way table to decide whether you agree with Maria.

6 Kyra and Ravi collect data on the number of minutes per day students in their year play computer games.

Kyra collects this data from a sample of 14 students.

40   26   75   55   33   39   28

66   67   71   64   37   52   47

Kyra starts to make a frequency table to display the data.

Minutes	Frequency
26	1
27	0
28	1
29	0

Ravi uses this grouped frequency table instead.

Minutes	Frequency
20 to 29	2
30 to 39	3
40 to 49	2
50 to 59	2
60 to 69	3
70 to 79	2

Give one advantage and one disadvantage of using a grouped frequency table to display the data.

Representing data 1

You can use a **pictogram** to display data.

● Pictograms use symbols to show the size of each category.

You may have to use part of a symbol to **represent** some quantities.

EXAMPLE

The number of cars that a car salesman sells is given in the table.

Week	Cars sold
1	4
2	6
3	12
4	10

Draw a pictogram to illustrate this information.

Use 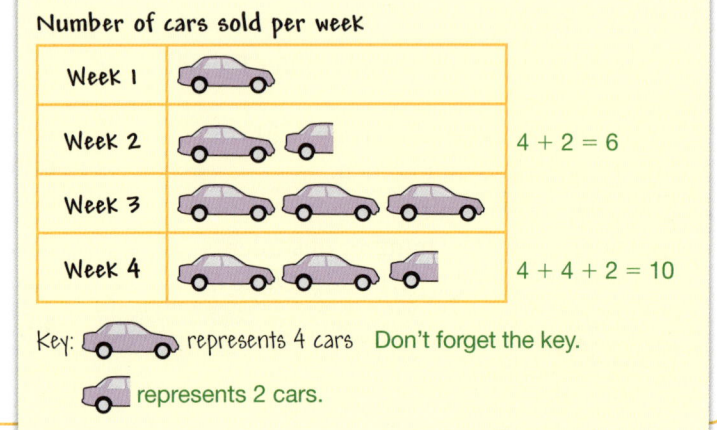 to represent 4 cars.

Number of cars sold per week

Week 1	
Week 2	4 + 2 = 6
Week 3	
Week 4	4 + 4 + 2 = 10

Key: represents 4 cars   Don't forget the key.

represents 2 cars.

You can use a **bar chart** to display data.

Bar charts give a visual picture of the size of each category.

● A bar chart should have
  ● labels on each axis
  ● bars with equal width
  ● equal gaps between the bars
  ● values on the vertical axis evenly spaced starting at zero.

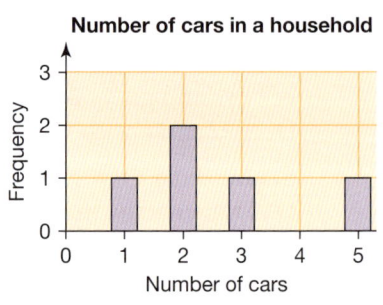

Number of cars in a household

The bars can be horizontal or vertical.

Bar-line charts are a good way to display (discrete) numerical data.

EXAMPLE

A class are asked to name one favourite pet. The results are shown.

Pet	Dog	Cat	Guinea pig	Other
Number of students	16	9	12	3

Draw a bar chart to show this information.

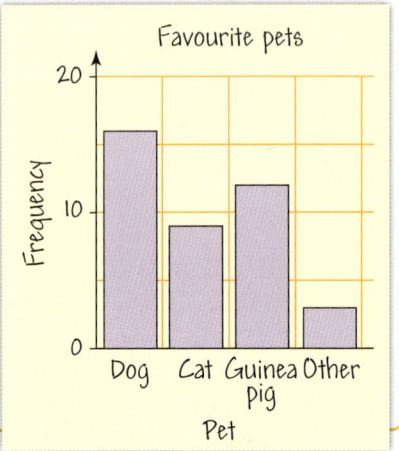

Favourite pets

Use a sensible scale on the vertical axis. Try to make the graph as large as possible.

Statistics   Handling data 1

Exercise 4.2S

1 The pictogram shows where people use the internet the most.
Copy the pictogram and represent this information on your diagram.
Library 6   School 9   Work 17

Internet usage	
Home	🖥🖥🖥🖥
Internet cafe	🖥🖥◳
Library	
School	
Work	

Key:
🖥 represents 2 people

2 The results of a survey about people's favourite hot drink are given in the table.

Letting 🥤 represent 20 people, draw a pictogram for this data.

Tea	80
Coffee	60
Hot chocolate	50
Soup	30
Other	20

3 The number of foreign language teachers at a school is shown in the pictogram.

a Which subject has only one teacher?

b How many French teachers teach at the school?

c How many more German teachers are there than Russian teachers?

d Calculate the total number of foreign language teachers at the school.

Number of language teachers	
French	🧍🧍🧍𝄪
Spanish	🧍🧍🧍
Russian	🧍
German	🧍🧍🧍𝄪
Italian	𝄪

Key: 🧍 represents 2 teachers

4 The number of concerts held at various venues is given.

Venue	Number of concerts
NEC	10
Arena	15
MEN	20
NIA	18
Wembley	9

Draw a bar chart to show this information.

5 The number of Bank Holidays in different countries is shown.

Draw a bar chart to show this information.

Country	Number of Bank Holidays
UK	8
Italy	16
Iceland	15
Spain	14

6 The size of donations to a charity are shown in the frequency table.
Draw a bar-line chart to show this information.

Resort	Cost (£)
£1	90
£2	100
£5	80
£10	70
£20	110

7 The numbers of buses that stop at a village through the week are shown on the bar chart.

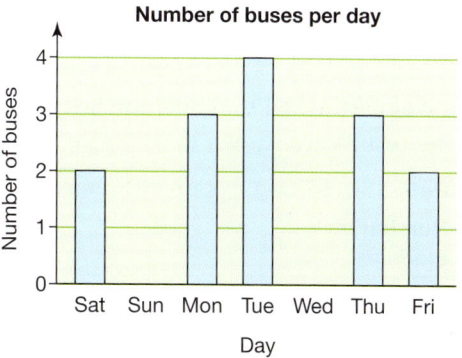

Number of buses per day

a How many buses stop at the village on Thursday?

b On which days is there no bus service?

c Calculate the total number of buses that stop at the village throughout the week.

Q 1193, 1205   SEARCH

4.2 Representing data 1

- A pictogram uses symbols to show the size of each category.

- A bar chart uses a bar to show the number of entries in each category.

- Pictograms and bar charts show all the data, but in categories. They are good for comparing the frequencies of different categories.

It can be time consuming to draw lots of symbols when using pictograms for large data sets.

HOW TO

To solve a problem involving a statistical graph

① Read information from the diagram.
Make sure that you read the scale or key carefully.

② Use the information from the diagram to answer the question.

EXAMPLE

The pictogram shows the number of students attending school in a week.

Four students were absent on Monday.

Calculate the number of students that were absent on Thursday.

Number of students at school	
Monday	◯◯◯◯
Tuesday	◯◯◖
Wednesday	◯◯◯
Thursday	◯◕
Friday	◯◯◿

Key: ◯ represents 4 students

① Each circle represents 4 students, so three-quarters of a circle represents 3 students.

Number of students attending on Thursday = 4 + 3 = 7

② Number of students attending on Monday = 4 + 4 + 4 + 4 = 16

Total number of students = 16 + 4 = 20   There were 4 students absent on Monday.

Subtract the number of students who attended on Thursday from the total.

Number of students absent on Thursday = 20 − 7 = 13

EXAMPLE

A class are asked to name one favourite pet. The results are shown in the table.

Pet	Dog	Cat	Guinea pig	Other
Number of students	16	9	12	3

Jess uses her computer to create this graph.

Make two criticisms of the graph.

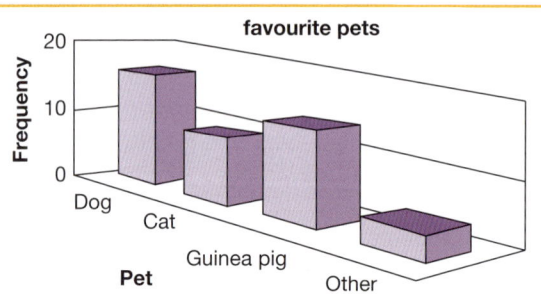

① Compare the graph with the frequency table and graph in 4.3 Skills. Look at the scale of the graph.

The scale chosen makes it is difficult to read information from the graph.

② Bar charts should let you compare the different categories.

The perspective makes it difficult to compare the heights of the bars.

Exercise 4.2A

1 The pictogram shows the number of people who eat different take-away food. 240 people were surveyed in total.

Take-away food choices

Italian	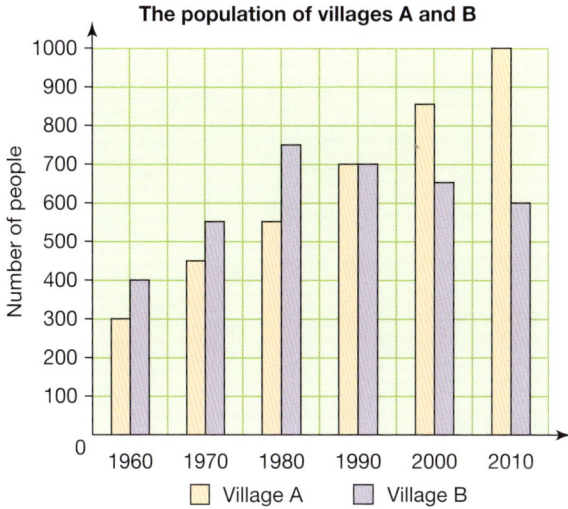
English	
Chinese	
Indian	

How many people ate Chinese food?

2 The dual bar chart shows the population of two villages.

The population of villages A and B

a During what year were the populations of the two villages the same?

b Compare the changes in population of the two villages over time.

3 A newspaper headline says

28 athletes from the United States won medals at the 2014 Winter Olympics.

● Team USA won more bronze medals than gold medals.

● A quarter of the medals were silver medals.

Draw a bar chart to show the number of bronze, silver and gold medals that could have been won by the team.

4 The number of students in different year groups is given in the frequency table.

Department	Frequency
Year 7	120
Year 8	126
Year 9	130
Year 10	128
Year 11	125

a Give one disadvantage of using a pictogram to display the data.
Marnie draws this bar chart for the data.

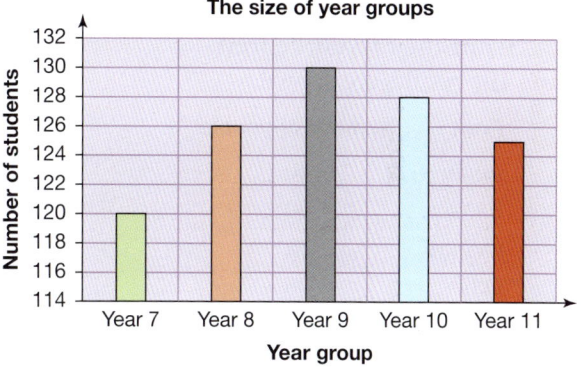
The size of year groups

b Give one reason why Marnie's graph is misleading.

5 Travel insurance prices are shown on the bar chart.

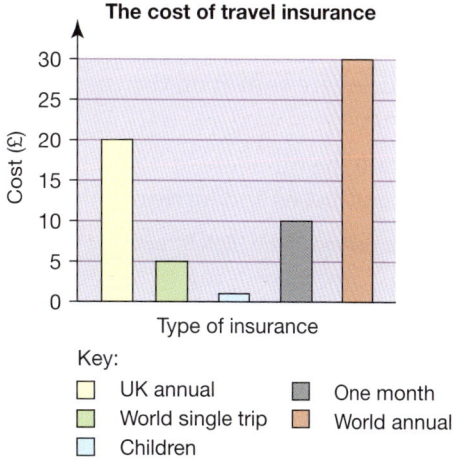
The cost of travel insurance

Key:
- ☐ UK annual
- ☐ World single trip
- ☐ Children
- ☐ One month
- ☐ World annual

Suzie plans to make four trips abroad during the year. She can either buy World Single Trip insurance each time or World Annual insurance.
Which is her cheaper option? Give reasons.

🔍 1193, 1205   SEARCH

4.3 Representing data 2

You can use a **pie chart** to display data.

Pie charts use a circle to give a quick visual picture of all the data.

The size of each **angle** shows the size of each **category**.

- A pie chart shows
  - the **proportion** or fraction of each category compared to the whole circle
  - all the data, but in categories.

Favourite fruit

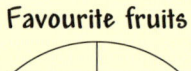

Apple is a quarter of all the data.

Banana is the biggest **sector**.

All the data must be included.

EXAMPLE

240 people are asked to name their favourite fruit. The results are shown.

Fruit	Number of people
Apple	50
Banana	80
Orange	72
Other	38

Draw a pie chart to illustrate the information.

Calculate the angle for one person.

$$360° ÷ 240 = 1.5°$$

Calculate the angles for each category.

Apple	$50 × 1.5° = 75°$
Banana	$80 × 1.5° = 120°$
Orange	$72 × 1.5° = 108°$
Other	$38 × 1.5° = 57°$

Add to check. $\underline{360°}$

Measure, colour and label the sectors.

Favourite fruits

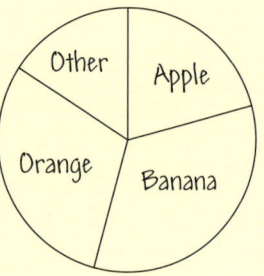

- You can interpret information from a pie chart by using the angles for each category to find the frequencies.

EXAMPLE

60 vehicles are shown on the pie chart.

Calculate the numbers of cars, vans, buses and lorries.

Vehicles parked in the High Street

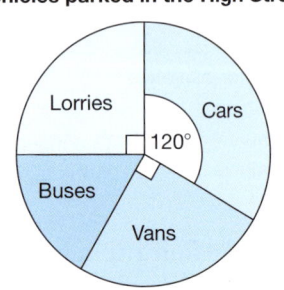

There are 360° in a circle.

60 vehicles represents 360°.

1 vehicle represents $360° ÷ 60 = 6°$

The angle for buses $= 360° − (90° + 90° + 120°) = 60°$

Number of cars $= 120 ÷ 6 = 20$

Number of vans $= 90 ÷ 6 = 15$

Number of buses $= 60 ÷ 6 = 10$

Number of lorries $= 90 ÷ 6 = \underline{15}$

Total number of vehicles $= \quad 60$

Use a protractor to measure the angles in the pie chart.

Exercise 4.3S

1 Alvin makes eight sandwiches.

1 tuna

2 cheese and tomato

3 chicken

2 corned beef

Draw a pie chart to show this information.

2 Seven boys and five girls attend an after-school homework club.

   a Calculate the total number of students.

   b Calculate the angle one student represents in a pie chart.

   c Calculate the angles to represent boys and girls.

   d Draw a pie chart to show the information.

3 The weather record for 60 days is shown in the frequency table. This gives the predominant weather for that particular day.

   a Calculate the angle that one day represents in a pie chart.

   b Calculate the angle of each category in the pie chart.

Weather	Number of days
Sunny	15
Cloudy	18
Rainy	14
Snowy	3
Windy	10

   c Draw a pie chart to show the data.

4 A school fete is open from 10 am to 4 pm.

   a Calculate the number of minutes the school fete is open.

A teacher has offered to help.
She spends these times on each stall.

   b Draw a pie chart to show this information.

Stall	Time
Bat the Rat	30 mins
Hook a Duck	25 mins
Smash a Plate	35 mins
Roll a Coin	80 mins
Tombola	70 mins
Break 1	60 mins
Break 2	60 mins

5 The pie chart shows the survey results for the favourite band of 100 people.

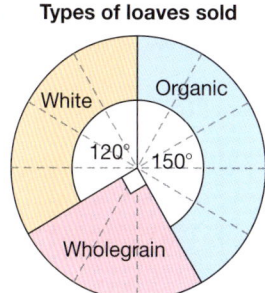

Favourite bands

   a What fraction of the pie chart represents

     i Kasabian

     ii The Prodigy

     iii Glasvegas?

   b Calculate the number of people who voted for

     i Kasabian   **ii** The Prodigy

     iii Glasvegas.

6 A shop sells 12 loaves of three different types: organic, wholegrain and white.

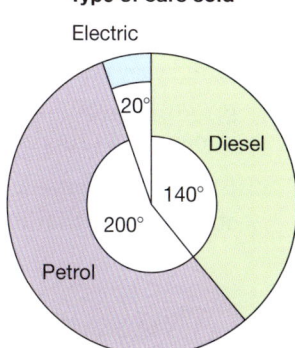

Types of loaves sold

   a Copy and complete   12 loaves = 360°

                                   1 loaf = ___°

   b Calculate the number of loaves that are

     i organic   **ii** wholegrain

     iii white.

7 A car dealer sells 18 cars in one week of three different types: diesel, petrol and electric.

Type of cars sold

   a Calculate the angle that represents one car.

   b Calculate the number of cars sold that are

     i diesel   **ii** petrol

     iii electric.

1206, 1207   SEARCH

4.3 Representing data 2

RECAP

- **Pie charts** use sectors of a circle to represent the size of each category.
- The size of the sector angle is proportional to the frequency.
- The angles in a pie chart must add up to 360°.
- Draw a key for each sector or label each sector clearly.

Woodley FC 2005 season

(Pie chart showing Win, Lose, Draw)

HOW TO

① Read information from the diagram. Make sure that you read the scale or key carefully.

② Use the information from the diagram to answer the question.

Use a protractor to measure the angles accurately.

EXAMPLE

Josh records the number of jars of sauce sold in his deli.

He draws a pie chart to show the information.

Josh wants to compare the sales of each flavour he sells.

Give three disadvantages of Josh's chart.

Jars of sauce sold

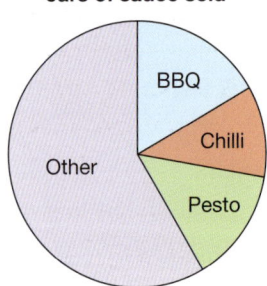

① Pie charts show the proportion of the total, but don't show the frequency of each category.

② The pie chart doesn't have any information about the number of jars.

Look again at the pie chart – is it easy to compare the sales of the flavours?

It is difficult to compare the BBQ, chilli and pesto sauces using the pie chart as the angles are similar.

Does the pie chart have information about all of the flavours in the shop?

There is no information about which flavours are 'other', which is the largest category.

EXAMPLE

Janelle asks people in Easton and Westville to name their favorite fruit.

The results are shown in the pie chart. Janelle says that 'There *are more people* in Easton whose favorite fruit is banana than there are in Westville.'

Explain why Janelle is wrong and rewrite her statement so that it is true.

Favourite fruits   **Favourite fruits**

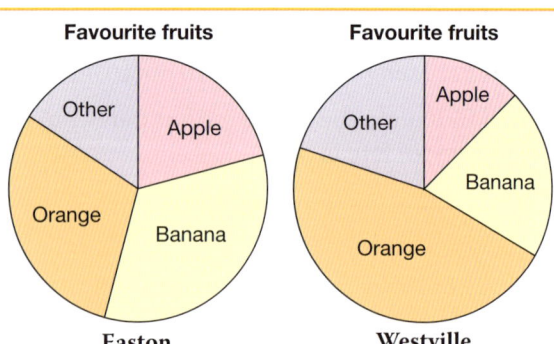

Easton   Westville

① Pie charts don't show the frequency of each category.

② Janelle doesn't know how many people in Westville are represented in the pie chart. Although the sector is larger for Easton, it could represent fewer people. Pie charts show the proportion of each category.

There <u>is a higher proportion of people</u> in Easton whose favorite fruit is banana than there are in Westville.

Exercise 4.3A

1 This pie chart shows the number of tourists visiting some regions in England over a certain period.

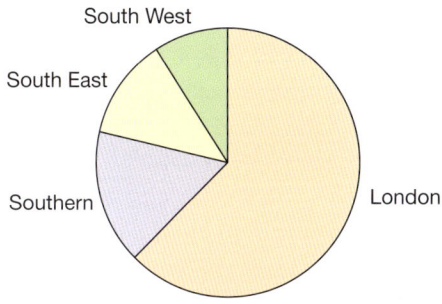

Tourist destinations

The pie chart is not drawn to scale.

Complete this table.

Region	Number of tourists	Angle
London	225000	
Southern		60°
South East	35000	
South West		40°
	360000	

2 The number of drinks sold during lunchtime in a shop are shown in the table.

Drink	No. sold	Drink	No. sold
Apple	6	Lemonade	3
Blackcurrant	5	Mocha	2
Cappuccino	8	Orange	7
Cola	12	Pineapple	4
Espresso	3	Tea	9
Latte	5	Water	9

Explain one disadvantages of using a pie chart to display the data.

3 The test scores of five students are shown.

Name	Mark
Abi	55
Carlos	52
Daniel	58
Emily	51
Parminder	57

Explain one disadvantages of using a pie chart to display the data.

4 The pie charts show the number of year 11 students studying different languages at two nearby schools.

Language students   **Language students**

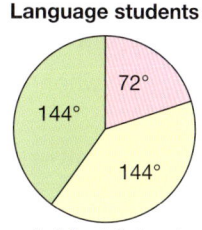   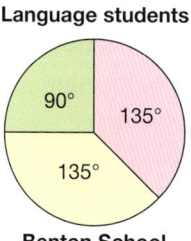

Ashford School   Benton School

☐ French ☐ Spanish ☐ German

a Jermaine says "twice as many students at Ashford school study German compared to those studying Spanish". Do you agree with Jermaine? Give a reason for your answer.

b Lydia says it must be the case that more students at Ashford schools study French than at Benton school. Is she correct? You must give reasons for your answer.

5 The bar chart shows the number of letters delivered in one week to No. 10 and No. 12.

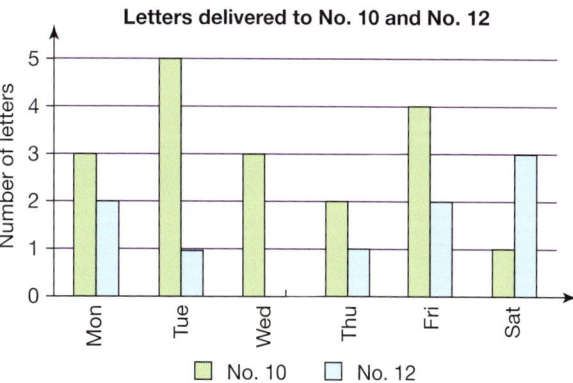

Letters delivered to No. 10 and No. 12

☐ No. 10   ☐ No. 12

a Draw an accurate pie chart to show the number of letters delivered each day to No. 10.

b Draw an accurate pie chart for No. 12.

c Mark wants to compare the proportion of letters received on each day for the two houses. Should he use the bar chart or the pie charts? Give your reasons.

d Peter want to compare the number of letters received each day for the two houses. Should he use the bar chart or the pie charts? Give your reasons.

4.4 Averages and spread 1

- The **mean** of a set of data is the total of all the values divided by the number of values.

- The **mode** is the value that occurs most often.

- The **median** is the middle value when the data is arranged in order.

- The **range** is the highest value – lowest value.

Outliers are values that 'lie outside' most of the other values of a set of data.

In this data set:

1, 1, 2, 2, 3, 4, 4, 4, 16

16 is an outlier.

The mean and range are both affected by outliers.

EXAMPLE

Ten people took part in a golf competition. Their scores are shown in the frequency table.

Calculate the mean, mode and median of the scores.

Score	Frequency
67	1
68	4
69	3
70	1
71	1

4 people scored 68.

1 person scored 71.

The results can be written in numerical order.

67, 68, 68, 68, (68, 69), 69, 69, 70, 71   Mean = 687 ÷ 10 = 68.7   Mode = 68 (occurs 4 times)

Median = (68 + 69) ÷ 2 = 68.5

There are 10 values so the median is the mean of the 5th and 6th value.

Alternatively, you can calculate the mean directly from the frequency table.

Score	Frequency	Score × Frequency
67	1	67
68	4	272
69	3	207
70	1	70
71	1	71
	10	687

68 + 68 + 68 + 68 or 68 × 4.

The total of all the scores of the 10 golfers.

Mean = 687 ÷ 10 = 68.7

Median = (5th value + 6th value) ÷ 2 = (68 + 69) ÷ 2 = 68.5

Mode = 68 as 68 occurs the most often.

The mode is 68, not 4.

There are advantages and disadvantages of the three types of averages.

	Mode	Median	Mean
Advantages	• Can be used with all types of data (numbers, colours …) • Not affected by outliers • Mode is an actual value in the data set	• Not affected by outliers • Easy to calculate	• Uses all of the values
Disadvantages	• Not always a mode • Can be more than one mode	• Not always an actual value in the data set	• Not always an actual value in the data set • Can be affected by outliers

Exercise 4.4S

1 Calculate the mean for each set of numbers.

    a   4, 9, 7, 12   **b**   8, 11, 8

    c   3, 2, 2, 3, 0   **d**   1, 9, 8, 6

    e   2, 2, 4, 1, 0, 3   **f**   23, 22, 25, 26

    g   17, 19, 19, 20, 18, 17, 16

    h   103, 104, 105   **i**   14, 10, 24, 12

    j   4, 6, 7, 6, 4, 4, 3, 7, 3, 6

2 **a**   Write these numbers in order, smallest first.

       i   7, 16, 8, 5, 4, 3, 3

      ii   11, 12, 10, 9, 9

     iii   38, 35, 24, 37, 34

     iv   101, 98, 103, 97, 99, 97, 95

      v   3, 2, 0, 0, 1, 2, 1, 2, 3

    b   Use your answers to find the median of each set of numbers.

    c   Which data sets contain outliers? Have the outliers affected the median?

3 Calculate the mode and range of each set of numbers.

    a   −6, 0, 1, 1, 1, 1, 2, 2, 2, 3, 3, 4, 4

    b   5, 5, 6, 6, 6, 7, 7, 7, 7, 8, 8, 8, 8, 8

    c   10, 11, 11, 11, 12, 12, 13, 14

    d   21, 22, 23, 24, 24, 25, 25, 25, 36

    e   8, 8, 8, 9, 9, 10, 11

    f   4, 3, 5, 5, 6, 6, 4, 3, 4, 5, 6, 5, 3

    Which data sets contain outliers? Have the outliers affected the mode or range?

4 The numbers of flowers on eight rose plants are shown in the frequency table.

Number of flowers	Tally	Frequency
3	||||	4
4	||	2
5	||	2

    a   List the eight numbers in order of size, smallest first.

    b   Calculate the mean, mode, median and range of the eight numbers.

5 The number of days that 25 students were present at school in a week are shown in the frequency table.

    a   List the 25 numbers in order of size, smallest first.

    b   How many students were present for 5 days of the week?

    c   Calculate the mean, mode, median and range of the 25 numbers.

Number of days	Tally	Frequency
0		0
1	||||	4
2	|||| |	6
3	||	2
4	||||	5
5	|||| |||	8

6 For these sets of numbers work out the

     i   range   **ii**   mode

     iii   mean   **iv**   median.

    a   5, 9, 7, 8, 2, 3, 6, 6, 7, 6, 5

    b   45, 63, 72, 63, 63, 24, 54, 73, 99, 65, 63, 72, 39, 44, 63

    c   97, 95, 96, 98, 92, 95, 96, 97, 99, 91, 96

    d   13, 76, 22, 54, 37, 22, 21, 19, 59, 37, 84

    e   89, 87, 64, 88, 82, 88, 85, 83, 81, 89, 90

    f   53, 74, 29, 32, 67, 53, 99, 62, 34, 28, 27, 27, 27, 64, 27

    g   101, 106, 108, 102, 108, 105, 106, 109, 103, 105, 107, 104, 104, 105, 105

7 **a**   Subtract 100 from each of the numbers in question **6 g** and write down the set of numbers you get.

    b   For your set of numbers in **a**, work out the

       i   range   **ii**   mode

      iii   mean   **iv**   median

    c   Compare your answer for the range with **6 g**.

    What do you notice?

4.4 Averages and spread 1

- To calculate the mean of a set of data, add all the values and divide by the number of values.

- To find the median, arrange the data in order and choose the middle value.

- To find the mode, choose the value that occurs most often.

- To find the range, subtract the smallest value from the largest value.

> The mean, median and mode are averages. The range is a measure of spread.

HOW TO

To compare data sets

(1) Calculate the mean, median, mode, range or interquartile range for each set of data.

(2) To compare the averages, look at the mean, median or mode.

(3) To compare the spread look at the range.

EXAMPLE

A team of 7 girls and a team of 8 boys did a sponsored run for charity.
The distances the girls and boys ran are shown.

Girls

Distance (km)	Frequency
1	3
2	2
3	1
4	1
5	0

Boys

3, 5, 5, 3, 4, 4, 5, 5
all distances in kilometres

By calculating the mean, median and range, compare each set of data.

(1) Calculate the mean, median and range of each set of data.

Girls

Mean = (3 + 4 + 3 + 4 + 0) ÷ 7 = 14 ÷ 7 = 2

Median = 2   The 4th distance is the middle value.

Range = 4 - 1 = 3   Highest value – lowest value.

Boys

Mean = 34 ÷ 8 = 4.25

Median = (4 + 5) ÷ 2 = 9 ÷ 2 = 4.5   The mean of the 4th and 5th distances.

Range = 5 - 3 = 2   Highest value – lowest value.

(2) Compare the mean and median.

The mean and median show the boys ran further on average than the girls.

(3) Compare the range.

The range shows that the girls' distances were more spread out than the boys' distances.

Exercise 4.4A

1 The number of bottles of milk delivered to two houses is shown in the table.

	Sat	Sun	Mon	Tues	Wed	Thur	Fri
Number 45	2	0	1	1	1	1	1
Number 47	4	0	2	2	2	2	2

  a Calculate the range for Number 45 and Number 47.

  b Use your answers for the range to compare the number of bottles of milk delivered to each house.

2 The number of cars at each house on Ullswater Drive are

2 4 1 0 1 2 1 2 3 2

  a Copy and complete the frequency table.

Number of cars	Tally	Number of houses
0		
1		
2		
3		
4		

  b Calculate the mean, mode and median number of cars for Ullswater Drive.
The mean, mode and median number of cars at each house on Ambleside Close are

Mean	Mode	Median
0.7	0	1

  c Use the mean, mode and median to compare the number of cars on Ullswater Drive and Ambleside Close.

3 The range of these numbers is 17.

    34   40   25   ?

Find two possible values for the unknown number.

4 Monica has five numbered cards.
One of the cards is numbered −2.6.
Monica's cards have

  ● range = 7.2

  ● median = 3.5

  ● mode = 4.1

Write down the five numbers on Monica's cards.

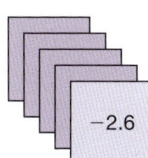

5 The bar chart shows the test results for a class of 20 students.

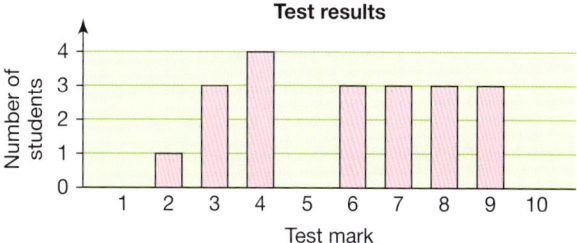

Test results

  a Copy and complete the frequency table to illustrate these results.

Mark	1	2	3	4	5	6	7	8	9	10
Number of students	0	1								

  b Calculate the mean, mode, median and range for the 20 students.

  c If a new student joined the class and got a mark of 10, how would this affect your ~~three~~ four answers to part **b**?

6 There are nine passengers on a bus.
The mean age of the passengers is 44 years old.
Jasmine gets on the bus.
Jasmine is 14 years old.
Find the new mean age of the passengers on the bus.

7 Reuben counted the raisins in 21 boxes.
The mean number of raisins per box was 14.1 (1 dp).
Reuben records the information for 20 boxes in the table.

Number of raisins	Number of boxes
13	5
14	8
15	7

Find number of raisins in the last box.

8 The mean mark in a statistics test for a class was 84%.
There are 32 students in the class, 12 of whom are girls.
The mean mark in the test for these girls was 93%.
Work out the mean mark in the statistics test for the boys.

Summary

Checkout

You should now be able to...

✓ Construct and interpret frequency tables and two-way tables.	**1, 2**
✓ Construct and interpret pictograms, bar-line charts and bar charts.	**3, 4**
✓ Interpret and construct pie charts and know their appropriate use.	**5, 6**
✓ Compare distributions using median, mean, mode and range and identify outliers.	**7 – 9**

Language | Meaning | Example

Language	Meaning	Example
Population	The whole group of people or items that are to be investigated.	To find out how long students spend each week doing homework at one school with a population of 1000, survey a group of 50 students from that school.
Sample	A set chosen to represent a population.	
Survey	Gather data from a sample to find out the characteristics of a population.	
Data collection sheet	A sheet on which data can be recorded and organised.	
Tally chart	A data collection sheet for data collected by counting.	
Frequency table	A table that records the frequency of each piece of data.	
Frequency	The number of times each piece of data occurs.	2 cars are red
Pictogram	A frequency diagram using a symbol to represent a number of units of data.	See lesson 4.3.
Bar-chart	The height of each bar represents the frequency.	See lesson 4.3.
Bar-line chart	The length of each line represents the frequency.	See lesson 4.3.
Pie chart	A circular chart divided into sectors. The size of the angle of each sector is proportional to the frequency.	See lesson 4.4.
Mean	An average found by adding all the values together and dividing by the number of values.	Data: 1, 1, 2, 2, 4, 5, 7, 8, 8, 8, 9, 22 Mean $= 77 \div 11$ $\quad = 7$
Mode	The value that occurs most often.	Mode $= 8$
Median	The middle value when the data is arranged in order of size. If there is an even number of data, the median is the mean of the middle two values.	1, 1, 2, 2, 4, 5, 7, 8, 8, 8, 9, 22 Median $= 6$
Range	The difference between the largest value and the smallest value.	Range $= 22 - 1 = 21$
Outlier	A value that lies outside most of the other values in a set of data.	22 is an outlier.

Example frequency table (from Example column):

Colour of car	Tally	Frequency
Red	II	2
Silver	IIII III	8
Black	IIII IIII	10
Other	IIII	4

Review

1 Organise this data into a frequency table.

3, 4, 1, 3, 4, 3, 4, 3, 2, 3, 1, 2, 3, 3, 4, 4, 2, 2, 3

2 The results of a handedness survey are shown. Use the two-way table to find the number of

a people surveyed

b boys

c left-handed girls

d right-handed people.

	Left	Right
Girls	2	15
Boys	1	12

3 The pictogram shows the number of children playing football each lunchtime at a school.

Number of football players

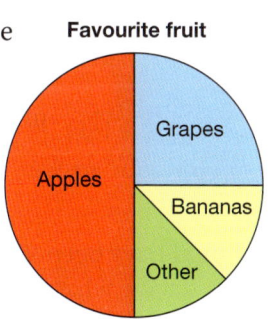

represents 4 children

a How many children played football on a

    i Monday   **ii** Wednesday?

b Ten children played on a Thursday and 15 on a Friday, complete a copy of the pictogram to show this information.

4 Draw a bar-line chart to represent the data in question **1**.

5 The pie chart shows the favourite fruit of 80 people.

How many people prefer

a apples

b bananas?

Favourite fruit

Grapes, Apples, Bananas, Other

6 Suzy spends 60p of her pocket money on a snack, £1.80 on a magazine and saves £3.60.

Show this information in a pie chart.

7 Jacob counts the number of sweets in a selection of packets, his results are

5, 8, 4, 3, 7, 9, 3, 6, 7, 8, 7, 5

Calculate.

a the mean   **b** the median

c the mode   **d** the range.

8 A sample is taken of people visiting a garden. This is done by recording all the people that arrive between 10:00 am and 10:15 am.

Their ages are recorded as

45, 37, 24, 8, 38, 40, 59, 68, 41, 80

a Use this sample to estimate the mean age of all people visiting the garden.

b What are the possible issues with estimating the mean in this way?

9 A zookeeper shows the number of birds in different aviaries at a zoo.

13, 11, 2, 5, 24, 34, 12, 11, 24, 30 , 1, 35, 11

Write down the

a highest number

b range

c modal number

d median number of birds in an aviary.

In a different zoo the median number of birds per aviary is 15 and the range is 20.

e Compare the two zoos.

What next?

Score	0 – 3		Your knowledge of this topic is still developing. To improve look at MyMaths: 1192, 1193, 1202, 1205, 1206, 1207, 1214, 1215, 1254
	4 – 7		You are gaining a secure knowledge of this topic. To improve your fluency look at InvisiPens: 04Sa – n
	8 – 9		You have mastered these skills. Well done you are ready to progress! To develop your problem solving skills look at InvisiPens: 05Aa – e

Assessment 4

1 These figures show the number of one pint cartons of milk bought in a small corner shop over 50 days.

5	2	1	1	2	3	4	1	2	4
2	2	2	3	1	4	0	1	4	2
3	2	4	6	1	2	0	2	3	3
0	1	3	2	1	1	3	4	1	5
4	0	2	2	0	1	1	1	1	0

    a Draw a frequency table for this data. [3]

    b Draw a bar chart for this data. [3]

2 A teacher collected information about the way that the 35 students in her class access information electronically.

	Smart phone	Tablet/Laptop	Broadband on a home computer	Internet café
Girls	15	9	12	0
Boys	20	16	10	3

    a How many students use a smart phone? [1]

    b How many boys have a tablet/laptop? [1]

    c How many more boys than girls use an internet café? [1]

    d Explain why the sum of the entries in the table does not add up to the total number of students in the class. [1]

3 In a group of 35 business executives, 15 wear glasses and 5 are left-handed.
4 executives are left-handed but don't wear glasses.
Display this information in a two-way table. [4]

4 A lorry driver recorded the distances driven in 10 journeys (in miles).

56   113   88   67   163   90   88   109   135   121

    a Calculate the

        i mean   [2]   **ii** median   [2]   **iii** mode. [1]

    b Does he average 100 miles per day? Give reason for your answer. [2]

    c Calculate the range. [2]

    d He realises that one distance recorded as 88 should have been 98.
Without performing any additional calculations, decide what effect this will have on the mean, median, mode and range of the distances.
Give reason for your answers. [4]

5   The bar chart shows the different types of coins
in Harry's till.

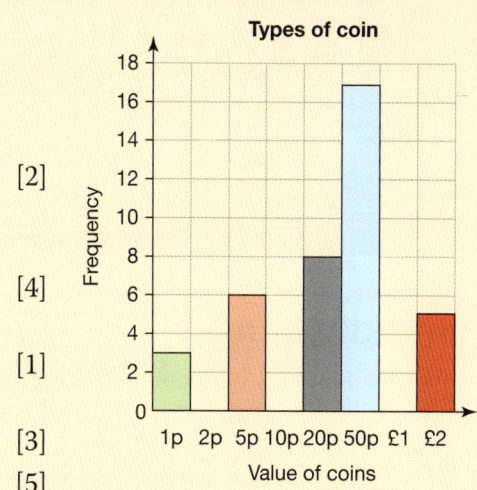

Types of coin

a   There are also seven 2 p coins, seventeen
10 p coins and six £1 coins in his till.
Add this data to the bar chart.   [2]

b   Draw a pictogram representing the
data shown. Use one circle to represent
two coins.   [4]

c   Which coin is found most often in
the till?   [1]

d   What is the value of the median coin in
the till?   [3]

e   Find the mean value of the coins in the till.   [5]

6   Mia records all the tickets she sells on the train service from Paris to Milan on one day.
Draw a pie chart to show this information.

Ticket class	Business	Standard	First	Child	Senior
Frequency	17	18	14	6	25

[5]

7   This pie chart shows how 1200 hospital staff members get to work in summer.

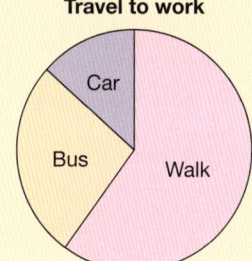

Travel to work

a   What is the modal form of travel?   [1]

b   Which is the least favourite form of travel?   [1]

c   What percentage of staff take the bus?   [2]

d   How many staff drive to work?   [3]

e   How many staff cycle to work?   [1]

8   The table shows information about the number of children
and pets in different families.

a   How many families have 1 pet?   [2]

b   How many 2 child families have 1 pet?   [1]

c   How many families are there where the
number of pets is the same as the number
of children?   [2]

d   What is the modal number of pets?   [1]

e   Calculate the mean number of pets per family.   [4]

		Number of pets					
		0	1	2	3	4	5
	1	3	5	4	1	0	1
Number of children in family	2	4	2	2	0	0	0
	3	1	0	2	1	0	0
	4	0	1	0	0	0	0
	5	2	1	0	0	0	0

5 Fractions, decimals and percentages

Introduction

Many food and drink products that you buy come with nutrition information clearly displayed on the labelling. This is usually in the form of percentages, for example 'saturated fat 5%, total carbohydrate 12%, calcium 9%, etc'. Product manufacturers are expected by law to display this information and the labels often tend to have a similar format.

What's the point?

Awareness of what you are eating and drinking is important in achieving a healthy, balanced diet and percentages allow you to monitor this.

Fat Total	1g
Fat Saturated	2g
Fat Trans	2g
Carbohydrate	0g
Sugar	23g
Dietary Fibre	15g
Sodium	0g
	45mg

Recommended Daily Allowances

Vitamin A	0%	Vitamin C	0%
Calcium	15%	Iron	2%

	per Serving	Intake (per Serving)	100mL
Energy	607 kJ (145Cal)	7%	243kJ (58Cal)
Protein	2.5g	5%	1.0g
Fat, Total	0.6g	0.9%	0.2g
- saturated	0.2g	0.9%	0.1g
Carbohydrate	31.2g	10%	12.5g
- sugars	8.0g	9%	3.2g
Sodium	815mg	36%	325mg

Objectives

By the end of this chapter you will have learned how to ...

- Convert between terminating decimals and their corresponding fractions.
- Compare decimals and fractions using the symbols $>$ and $<$.
- Find fractions and percentages of amounts.
- Add and subtract simple fractions and mixed numbers.
- Multiply and divide simple fractions and mixed numbers.
- Convert between fractions, decimals and percentages.

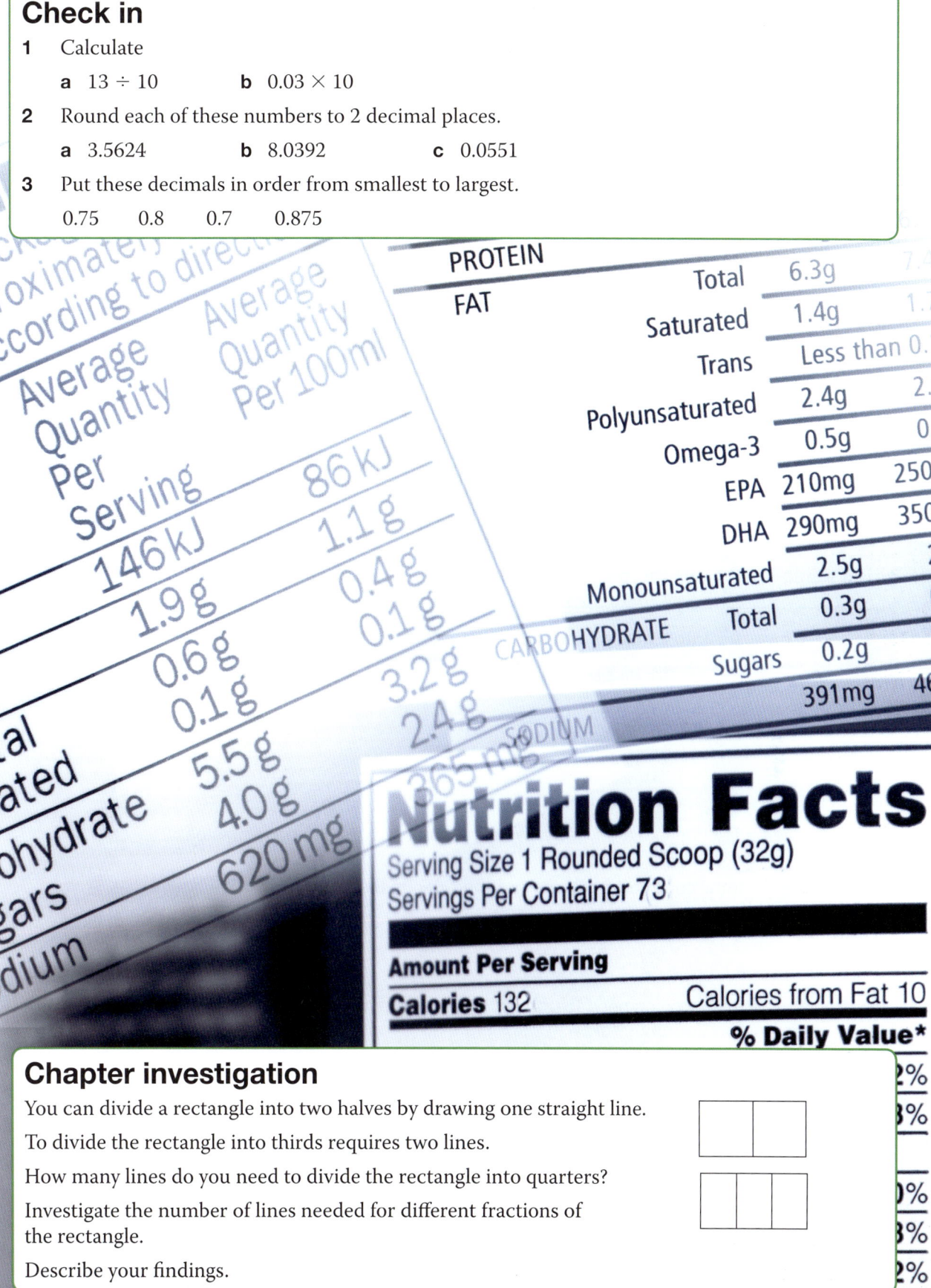

Check in

1 Calculate

    a $13 \div 10$ **b** 0.03×10

2 Round each of these numbers to 2 decimal places.

    a 3.5624   **b** 8.0392   **c** 0.0551

3 Put these decimals in order from smallest to largest.

    0.75   0.8   0.7   0.875

Chapter investigation

You can divide a rectangle into two halves by drawing one straight line.

To divide the rectangle into thirds requires two lines.

How many lines do you need to divide the rectangle into quarters?

Investigate the number of lines needed for different fractions of the rectangle.

Describe your findings.

5.1 Decimals and fractions

- A fraction describes part of a whole, where the whole is divided into *equal* parts.

EXAMPLE

Here is a fuel gauge. What fraction of the fuel tank is filled?

E [======================   ] F

$\frac{3}{5}$ full The fuel gauge is divided into five equal sections.
Three of the sections are coloured, which indicates fuel.

- You find equivalent fractions by multiplying or dividing the numerator and denominator by the same number.

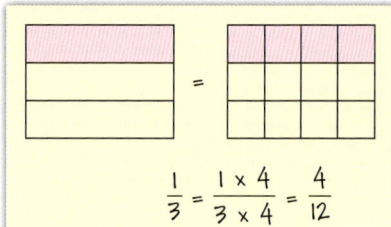

$$\frac{1}{3} = \frac{1 \times 4}{3 \times 4} = \frac{4}{12}$$

- You write a fraction in its **simplest form** by dividing its numerator and denominator by *all* common factors.

Fractions greater than 1, such as $\frac{5}{3}$, can be written as

mixed numbers, $1\frac{2}{3}$,   or   **improper fractions**, $\frac{5}{3}$.

EXAMPLE

a   Convert $\frac{13}{8}$ into a mixed number.   **b**   Convert $1\frac{3}{5}$ into an improper fraction.

Partition into proper fractions.   Partition

a $\frac{13}{8} = \frac{8}{8} + \frac{5}{8} = 1 + \frac{5}{8} = 1\frac{5}{8}$ **b** $1\frac{3}{5} = 1 + \frac{3}{5} = \frac{5}{5} + \frac{3}{5} = \frac{8}{5}$

- A **decimal** is another way of writing a fraction.
- To convert a fraction into a decimal, divide the numerator by the denominator.

You should learn some common fractions and their decimal equivalents.

$\frac{1}{4} = 0.25$	$\frac{5}{10} = \frac{1}{2} = 0.5$	$\frac{3}{4} = 0.75$	$\frac{7}{2} = 3.5$	$\frac{3}{8} = 0.375$	$\frac{1}{5} = 0.2$

- To compare fractions you need to convert them to decimals, or write each fraction with a **common denominator**.

EXAMPLE

Order these fractions from smallest to largest.
$\frac{3}{8}$, $\frac{4}{7}$, $\frac{5}{14}$

Find equivalent fractions with the same **common denominator**.

$\overset{\times 7}{\frac{3}{8} = \frac{21}{56}}$ $\overset{\times 8}{\frac{4}{7} = \frac{32}{56}}$ $\overset{\times 4}{\frac{5}{14} = \frac{20}{56}}$ $\frac{20}{56} < \frac{21}{56} < \frac{32}{56}$

$\underset{\times 7}{}$ $\underset{\times 8}{}$ $\underset{\times 4}{}$

$\frac{5}{14} < \frac{3}{8} < \frac{4}{7}$ Use the original fractions in your answer.

p.4

Exercise 5.1S

1 Write the fraction of each of these shapes that is shaded.

a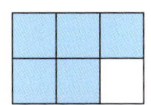
b

c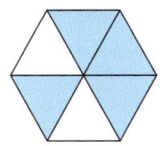
d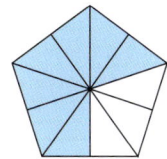

2 Write the fraction indicated by each letter.

a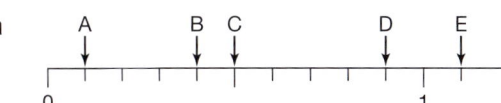

b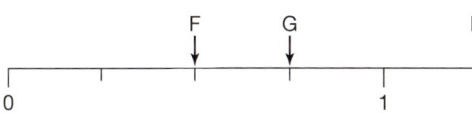

c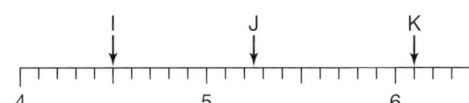

3 i What fraction of each shape is shaded?

ii Write each fraction in its simplest form.

a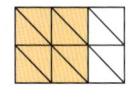
b

c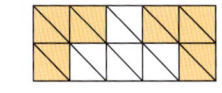
d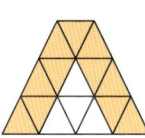

4 Write each fraction in its simplest form.

a $\frac{4}{12}$ b $\frac{21}{28}$ c $\frac{24}{40}$ d $\frac{28}{63}$

e $\frac{45}{72}$ f $\frac{42}{126}$ g $\frac{64}{144}$ h $\frac{23}{93}$

5 Convert each of these fractions to an improper fraction.

a $1\frac{1}{2}$ b $3\frac{2}{3}$ c $4\frac{3}{8}$ d $2\frac{2}{9}$

e $5\frac{6}{7}$ f $7\frac{4}{5}$ g $8\frac{8}{11}$ h $12\frac{4}{7}$

6 Convert each fraction to a mixed number.

a $\frac{5}{4}$ b $\frac{8}{5}$ c $\frac{11}{7}$ d $\frac{9}{4}$

e $\frac{11}{5}$ f $\frac{20}{7}$ g $\frac{23}{5}$ h $\frac{28}{9}$

7 Find the missing number in each pair of equivalent fractions.

a $\frac{2}{3} = \frac{\square}{12}$ b $\frac{3}{4} = \frac{\square}{36}$ c $\frac{5}{7} = \frac{40}{\square}$

d $\frac{7}{8} = \frac{\square}{64}$ e $\frac{12}{30} = \frac{\square}{5}$ f $\frac{6}{7} = \frac{\square}{105}$

g $\frac{5}{4} = \frac{\square}{68}$ h $\frac{\square}{10} = \frac{154}{220}$ i $\frac{3}{\square} = \frac{18}{48}$

8 Convert each decimal to a fraction in its simplest form.

a 0.1 b 0.6 c 0.75

d 1.25 e 1.5 f 1.2

9 Convert these fractions to decimals.

a $\frac{3}{10}$ b $\frac{4}{5}$ c $\frac{1}{4}$

d $1\frac{1}{10}$ e $\frac{14}{10}$ f $\frac{8}{5}$

10 Write the number each of the arrows is pointing to as a fraction and a decimal.

a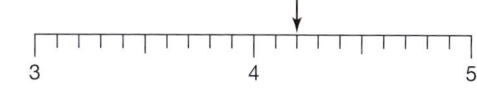

b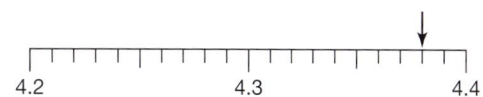

c

11 a Copy this decimal number line.

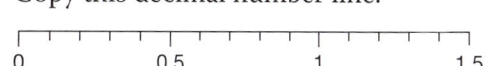

b Mark these fractions on your number line.

i $\frac{8}{10}$ ii $\frac{3}{5}$ iii $\frac{13}{10}$

12 Put these lists of numbers in order, starting with the smallest.

a $\frac{11}{5}$, 2.13, 2.09, $\frac{53}{25}$, 2.07

b 0.345, $\frac{7}{20}$, 0.325, $\frac{3}{10}$, 0.309

c $\frac{33}{25}$, 1.4, 11.35, 1.387, $\frac{529}{500}$

d 5.306, $\frac{661}{125}$, 5.308, 5.29, $\frac{53}{10}$

🔍 1016, 1019, 1042, 1075 SEARCH

5.1 Decimals and fractions

- Fractions describe parts of a whole.
  - You can find equivalent fractions by multiplying the numerator and denominator by the same number.
  - You can convert a fraction to a decimal and vice versa.
  - You can order fractions by converting them to a decimal, or writing them as equivalent fractions with the same common denominator.

$$\frac{\text{Numerator}}{\text{Denominator}}$$

$$\text{Numerator}$$

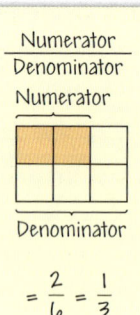

$$\text{Denominator}$$

$$= \frac{2}{6} = \frac{1}{3}$$

HOW TO

To solve problems involving fractions and decimals
1. RTQ and decide what to do.
2. If appropriate convert to decimals/fractions or equivalent fractions.
3. ATQ using the correct form of fraction or decimal, depending on the question. Don't forget units if appropriate.

EXAMPLE

Arthur is sorting out old drill bits, sized in fractions of an inch: $\frac{3}{8}, \frac{1}{4}, \frac{5}{16}, \frac{1}{8}, \frac{1}{2}, \frac{3}{16}$
He thinks one is missing, so he lines them up in order of size.

Which size do you think is missing? Give your reasons.

1. 2. Convert to equivalent fractions with a common denominator.

$$\frac{6}{16} \quad \frac{4}{16} \quad \frac{5}{16} \quad \frac{2}{16} \quad \frac{8}{16} \quad \frac{3}{16}$$

2, 3, 4, 5, 6 and 8 sixteenths are all in the list.
So $\frac{7}{16}$ is likely to be the missing size.   3.

EXAMPLE

Each of these shapes is partly shaded.

Shape A   Shape B   Shape C

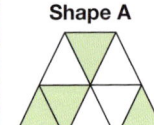   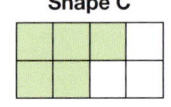

Which shape has the greatest fraction shaded? Give reasons for your answer.

1. Calculate the fraction of each shape that is shaded, then order the fractions by size.

Shape A $= \frac{4}{8} = \frac{1}{2} = 0.5$ 2. Compare equivalent decimals.

Shape B $= \frac{6}{10} = 0.6$

Shape C $= \frac{5}{8} = 0.625$

3. Shape C because 0.625 > 0.6 > 0.5

EXAMPLE

Dave says that $\frac{3}{7}$ is closer to $\frac{1}{2}$ than $\frac{2}{5}$.
Is he correct?
Give your reasons.

1. You need to compare each fraction.
2. $\frac{3}{7} = 0.423$ (3 sf) $\frac{2}{5} = 0.4$ $\frac{1}{2} = 0.5$
   $0.423 - 0.4 = 0.023$ $0.5 - 0.423 = 0.077$
3. No, because 0.023 < 0.077   Remember to include your reasons.

Exercise 5.1A

1 Wesley earns £300 a week. He pays £91 each week in tax. He also saves £60 per week. What fraction of his weekly wage does Wesley

  a pay in tax   **b** save?

2 A mechanic likes to hang her spanners in order of size.

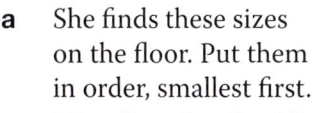

  a She finds these sizes on the floor. Put them in order, smallest first.

$$\frac{17}{32} \quad \frac{9}{16} \quad \frac{7}{16} \quad \frac{3}{8} \quad \frac{15}{32} \quad \frac{1}{2}$$

  b A spanner is missing from her set. What size do you think it is? Why?

3 A weather forecast lists the chances of different weather tomorrow.

The probability of rain is 0.5   **p.162**

The probability of sunny spells is $\frac{3}{10}$.

The probability of wind is 0.25

The probability of snow is 0.01

The probability of light cloud is $\frac{4}{5}$.

  a Andreas says there is more chance of wind than sunny spells. Is he correct?

  b Order the probabilities from highest to lowest.

  c Do you think tomorrow is a good day for a picnic? Why?

4 Eric is trying to decide which maths topics he needs to do more work on following a recent test. He has the following scores.

    Algebra   9 out of 10

    Fractions   3 out of 5

    Geometry   5 out of 8

    Graphs   16 out of 20

    Statistics   6 out of 12

  a He decides to revise any topic where he got less than three quarters correct. Which topics are on his revision list?

  b Eric looks at his answers for the topic on fractions. Find which two are wrong and correct them for him.

    i $\frac{1}{2} = 0.5$

4 b   **ii** $\frac{3}{4} = 0.7$

    iii $\frac{7}{25} = 0.28$

    iv $1\frac{4}{5} = 1.8$

    v $2\frac{4}{5} = 2.4$

5 Each of these shapes is partly shaded.

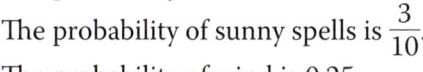

Shape 1   Shape 2   Shape 3

  a Put the shapes in order from least to most shaded.

  b Put the shapes in order from most to least shaded.

6 Rearrange these cards so that each row has four cards of the same value.

$1\frac{1}{2}$	2.20	0.625	$\frac{9}{6}$
$2\frac{1}{5}$	$\frac{5}{8}$	1.5	$\frac{11}{5}$
$\frac{20}{32}$	$\frac{30}{20}$	$\frac{22}{10}$	$\frac{10}{16}$

7 Explain which fraction is larger, $\frac{1}{3}$ or $\frac{2}{5}$.

8 Leonore argues that $\frac{69}{40} > 1\frac{58}{80}$, and Peter argues that $\frac{69}{40} < 1\frac{58}{80}$. Who is correct? Why?

*9 Rory has 13 CDs, 15 DVDs and some computer games in his collection. If his CDs represent $\frac{13}{37}$ of his collection, what fraction are his computer games?

10 Match the equivalent pair of cards.

$\frac{1}{2}a + 0.4b - 0.1a + \frac{1}{4}b$	$0.4a + 0.65b$
$1.8a + 0.4b - 1\frac{2}{5}a + \frac{1}{5}b$	$0.4a + 0.6b$
$\frac{1}{10}a + \frac{7}{4}b + \frac{2}{5}a + 1.1b$	$0.5a + 0.65b$

 Q 1016, 1019, 1042, 1075   SEARCH

5.2 Fractions and percentages

- You find fractions of a quantity by multiplying.

For example, two thirds of 5 is $\frac{2}{3} \times 5 = \frac{2 \times 5}{3} = \frac{10}{3} = 3\frac{1}{3}$

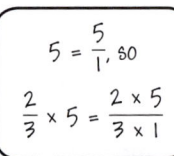

$5 = \frac{5}{1}$, so

$\frac{2}{3} \times 5 = \frac{2 \times 5}{3 \times 1}$

- When the quantity and the denominator of the fraction have a **common factor**, you can **cancel** this factor before multiplying.

For example, $\frac{2}{\underset{1}{3}} \times 24^{8} = \frac{2}{1} \times 8 = 16$

> You can also do this calculation mentally: $24 \div 3 = 8$, $2 \times 8 = 16$.

EXAMPLE

Calculate **a** $\frac{3}{4}$ of 28   **b** $\frac{5}{8}$ of 6   **c** $\frac{4}{9}$ of 12   **d** $\frac{5}{9}$ of 25

a $\frac{3}{\underset{1}{4}} \times 28^{7} = 3 \times 7 = 21$ Cancel by common factor 4 before multiplying.

b $\frac{5}{\underset{4}{8}} \times 6^{3} = \frac{5 \times 3}{4} = \frac{15}{4} = 3\frac{3}{4}$

c $\frac{4}{\underset{3}{9}} \times 12^{4} = \frac{4 \times 4}{3} = \frac{16}{3} = 5\frac{1}{3}$ You can write 5.$\dot{3}$, but it is simpler to leave it as a fraction.

d $\frac{5}{9} \times 25 = \frac{125}{9} = 13\frac{8}{9}$

- A percentage (%) is a fraction with a denominator of 100.
  It tells you how many parts there are per 100.

- To convert a percentage to a fraction, write it as a fraction with a denominator of 100. To convert to a decimal, divide the percentage by 100.

$50\% = \frac{50}{100} = 0.5$ $60\% = \frac{60}{100} = 0.6$ $25\% = \frac{25}{100} = 0.25$

- You can use mental methods to find percentages of amounts.

To find 50%, divide by 2.   To find 25%, divide by 4.
To find 10%, divide by 10.   To find 1%, divide by 100.

- To find a percentage of an amount multiply the quantity by an equivalent decimal or fraction.

For example, to find 47% of an amount, multiply the amount by $\frac{47}{100}$ or 0.47.

EXAMPLE

Calculate   **a** 45% of 60 cm   **b** 34% of 85 kg   **c** 16% of £25

a 50% of 60 = 30   50% is a half.
    5% of 60 = 3   Divided by 10.
    45% of 60 cm = 30 cm − 3 cm
            = 27 cm   Include units.

b $34\% = \frac{34}{100} = 0.34$ Use decimal equivalent.
    34% of 85 kg = 0.34 × 85
            = 28.9 kg   By calculator.
            = 29 kg (2 sf)

c 16% of £25 = $\frac{\overset{4}{16}}{\underset{41}{100}} \times 25^{1}$ Cancel the 25 and 100.
                Then cancel 16 and 4.
    = £4

Exercise 5.2S

1 Use a mental method to calculate each of these amounts.

  a $\frac{1}{2}$ of 40 sheep   **b** $\frac{1}{3}$ of 15 apples

  c $\frac{1}{5}$ of 25 shops   **d** $\frac{1}{4}$ of 48 marks

2 Calculate each of these, leaving your answer in its simplest form.

  a $5 \times \frac{1}{2}$ **b** $8 \times \frac{1}{4}$

  c $8 \times \frac{1}{3}$ **d** $13 \times \frac{1}{7}$

  e $\frac{1}{12} \times 24$ **f** $\frac{1}{3} \times 4$

3 Calculate each of these, leaving your answer in its simplest form.

  a $6 \times \frac{2}{3}$ **b** $5 \times \frac{3}{4}$

  c $2 \times \frac{2}{5}$ **d** $4 \times \frac{7}{6}$

  e $5 \times \frac{9}{20}$ **f** $\frac{4}{5} \times 28$

  g $\frac{4}{9} \times 30$ **h** $\frac{11}{18} \times 14$

4 **a** What fraction of this shape is shaded?

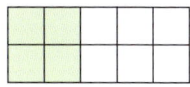

  b What percentage is shaded?

  c Copy the shape and shade in more squares so that 70% of the shape is shaded.

5 Calculate these percentages without using a calculator.

  a 50% of £60   **b** 50% of 40

  c 50% of 272 p   **d** 10% of 40

  e 10% of 370 p   **f** 1% of £700

  g 50% of 12 kg   **h** 50% of £31

  i 1% of 200   **j** 1% of 420 m

  k 5% of £80   **l** 10% of £340

  m 5% of £120   **n** 20% of 210 km

  o 20% of $530   **p** 25% of £300

  q 20% of £32   **r** 5% of 28 m

6 Calculate these fractions of an amount.

Leave your answers as fractions in their simplest form where appropriate.

  a $\frac{3}{10}$ of €40   **b** $\frac{2}{5}$ of £70

  c $\frac{3}{4}$ of 50 m   **d** $\frac{4}{7}$ of 64 km

  e $\frac{3}{8}$ of £1000   **f** $\frac{5}{6}$ of 70 mm

  g $\frac{11}{12}$ of 1500 m   **h** $\frac{4}{13}$ of 60 g

7 Calculate these fractions of an amount.

Where appropriate round your answer to 2 dp.

  a $\frac{8}{15}$ of 495 kg   **b** $\frac{9}{10}$ of $5000

  c $\frac{5}{9}$ of 8 kg   **d** $\frac{7}{9}$ of 1224 cups

  e $\frac{13}{18}$ of 30 tonnes   **f** $\frac{4}{15}$ of 360°

  g $\frac{12}{31}$ of 360°   **h** $\frac{13}{15}$ of 1 hour

  i $\frac{17}{15}$ of £230   **j** $\frac{22}{7}$ of 120 cm

8 Calculate these percentages of an amount.

Show all the steps of your working.

  a 11% of £18   **b** 60% of 7300 km

  c 8% of £30   **d** 2% of €3000

  e 7% of 60 m   **f** 13% of 40 cm

  g 75% of 48 m   **h** 3% of £70

  i 12% of £17   **j** 16% of 87 km

  k 8% of £38   **l** 32% of €340

  m 17% of 65 m   **n** 73% of 46 cm

  o 85% of 148 m   **p** 2% of £76.40

  q 25% of £85   **r** 35% of 325p

9 Maurice says,

'To find 10%, you divide by 10. So to find 20%, you just divide by 20.'

Explain why Maurice is wrong and suggest a better rule for him.

5.2 Fractions and percentages

- To find a fraction of a quantity, multiply the quantity and the fraction.
- To find a percentage of a quantity, multiply it by an equivalent decimal or fraction.

Multiplying by a fraction is the same as dividing by the denominator, then multiplying by the numerator.

HOW TO

To solve problems involving fractions and percentages
1. RTQ and decide what you need to do.
2. Decide which method to use: written, mental or with a calculator.
3. ATQ and don't forget to include any units.

EXAMPLE

VAT (Value Added Tax) is added on to the cost of items. VAT is 20% of the price. Donna bought a camera for £360 and a case for £17.65. She was not charged VAT. How much money did she save by buying the items VAT-free?

① Calculate 20% of each item's price, then add the two amounts together.

10% = £36   ② 10% is the same as dividing by 10.   10% = £1.765   ② Divide by 10, don't round.

20% = £72   Double it.   20% = £3.53   Double it.

£72 + £3.53 = £75.53. She saved £75.53 ③

EXAMPLE

Shauna budgets for some of her weekly spending. Whatever she earns, she puts 50% aside for general spending and the rest she splits into four shares.

  i $\frac{1}{4}$ for food   ii $\frac{1}{5}$ for travel   iii $\frac{3}{10}$ for electricity   iv $\frac{3}{8}$ for rent

a She earns £840. Work out how much she puts away for each item.

b Shauna has made a mistake. What is the problem?

a ① Multiply £840 by each fraction.   ② You can do these without a calculator.

50% of £840 = £420

  i $\frac{1}{4}$ of £420 = £105   Divide by 4.   ii $\frac{1}{5}$ of £420 = £84   Divide by 5.

  iii $\frac{1}{10}$ of £420 = £42   Divide by 10.   iv $\frac{1}{8}$ of £420 = £52.50   Divide by 8.

  $\frac{3}{10}$ = £126   Then multiply by 3. $\frac{3}{8}$ = £157.50   Then multiply by 3.

b £105 + £84 + £126 + £157.50 = £472.50   Add up the total.

£472.50 > £420   The fractions she chose add up to more than one whole. ③

EXAMPLE

Tom has £42. He spends 30% of it on Monday. On Tuesday he spends $\frac{3}{4}$ of the remainder. How much does he spend on Tuesday?

① Work out how much is left after Monday.   Then work out $\frac{3}{4}$ of what is left.

30% of £42 is 0.3 × 42

② 0.1 × 42 = 4.2   Calculate 10%.   42 − 12.6 = 29.4

3 × 4.2 = £12.6   Multiply by 3.   29.4 × 0.75 = £22.05 $\frac{3}{4}$ = 0.75

③ He spends £22.05 on Tuesday.

Exercise 5.2A

1 Leave your answer in its simplest form.

a What is the mass of 4 packets each with a mass of $\frac{1}{5}$ kg?

b A cake has mass $\frac{7}{20}$ of a kg. What is the mass of 10 cakes?

2 Jim works overtime for two weeks. He earns £190 in the first week and £224.50 in week 2. He pays 20% income tax on each amount.

How much tax has he paid at the end of the two weeks?

3 Rearrange these cards to make three correct statements.

25% of	£450	= £128
30% of	£640	= £130
20% of	£520	= £135

4 Jack and Jill fetched 24 pints of water in their bucket. Jill drank a tenth of this, Jack a quarter. They gave three eighths to Dame Dob, and a fifth to Jill's mother.

a How much water does each person get?

b How much was left over?

5 Phil works out that yesterday he spent a third of the day asleep, $\frac{3}{8}$ of the day watching TV, a tenth eating, and a quarter at school.

Is he correct?
How can you tell?

6 Julia earns a salary of $47 800 per year. She is awarded a 27% pay rise. What is her new salary?

*7 The label on a chocolate bar gives information about the nutritional value of the bar. A 220 g bar of chocolate contains 19% saturated fat.

If one bar is 32 squares of chocolate, how many grams of saturated fat are in 4 squares?

*8 Rajin earns £18 500 per year; £6550 is untaxed, the rest is taxed 20%.

a How much tax does he pay per month?

b What percentage of his salary is taxed?

*9 Gareth is drawing a pie chart to display information on students' favourite lessons. He uses the fractions liking each subject to work out the angles for the pie chart.

p.78

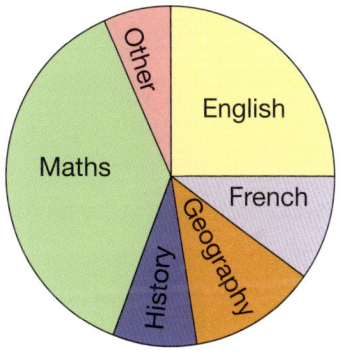

English	$\frac{1}{4}$	History	$\frac{1}{12}$
French	$\frac{1}{10}$	Maths	$\frac{3}{8}$
Geography	$\frac{1}{8}$	Others	$\square$

What angle is the 'Others' sector?

10 Farmer Macdonald has died leaving his 17 horses to his three sons. The only condition is that the eldest son gets *exactly* half the horses, the middle son *exactly* $\frac{1}{3}$ and the youngest *exactly* $\frac{1}{9}$. Seeing them struggle to do this, their neighbour lends them her horse. Suddenly they can divide up the horses and still be able to give the neighbour's horse back to her. Is this possible? Say how.

1018, 1030, 1031   SEARCH

5.3 Calculations with fractions

● You can only add and subtract fractions if they have a common denominator.

$$\frac{2}{8} + \frac{3}{8} = \frac{2+3}{8} = \frac{5}{8}$$

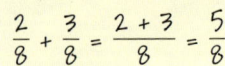

● To add or subtract fractions with different denominators, first change them to equivalent fractions that share a common denominator.

$$\frac{11}{12} + \frac{1}{3} = \frac{11}{12} + \frac{4}{12} = \frac{\cancel{15}^{5}}{\cancel{12}^{4}} = \frac{5}{4} = 1\frac{1}{4}$$

p.272 ● To multiply fractions, multiply the numerators and then the denominators and cancel any common factors.

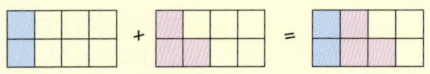

You should always give your answers in their simplest forms and change any improper fractions to mixed numbers.

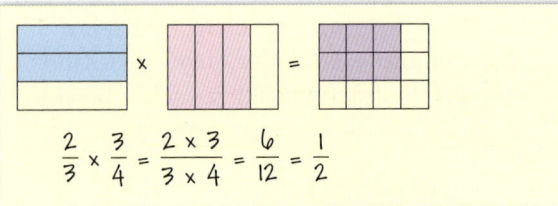

$$\frac{2}{3} \times \frac{3}{4} = \frac{2 \times 3}{3 \times 4} = \frac{6}{12} = \frac{1}{2}$$

To multiply by a mixed number you must first convert it to an improper fraction.

● To divide by a fraction multiply by its **reciprocal**.

The reciprocal of a fraction is the fraction 'turned upside down'.

$$\div 4 \rightarrow \times \frac{1}{4} \qquad \div \frac{1}{3} \rightarrow \times 3 \qquad \div \frac{5}{6} \rightarrow \times \frac{6}{5}$$

A number multiplied by its reciprocal is 1.

EXAMPLE

Work out each of these calculations.

a $\frac{3}{5} + \frac{11}{16}$ **b** $\frac{2}{7} - \frac{1}{4}$ **c** $\frac{3}{8} \times \frac{5}{9}$ **d** $\frac{3}{5} \div 8$ **e** $\frac{3}{8} \div \frac{6}{7}$ **f** $2\frac{4}{5} \div 3\frac{1}{2}$

a $\frac{3}{5} + \frac{11}{16} = \frac{3 \times 16 + 5 \times 11}{5 \times 16} = \frac{48 + 55}{80} = \frac{103}{80} = 1\frac{23}{80}$ **b** $\frac{2}{7} - \frac{1}{4} = \frac{2 \times 4 - 1 \times 7}{7 \times 4} = \frac{8-7}{28} = \frac{1}{28}$

The LCM of 5 and 16 is 80.   The LCM of 7 and 4 is 28.

c $\frac{\cancel{3}^{1}}{8} \times \frac{5}{\cancel{9}_{3}} = \frac{1}{8} \times \frac{5}{3} = \frac{1 \times 5}{8 \times 3} = \frac{5}{24}$ **d** $\frac{3}{5} \div 8 = \frac{3}{5} \times \frac{1}{8} = \frac{3 \times 1}{5 \times 8} = \frac{3}{40}$

Cancel as soon as you can.   The reciprocal of 8 is $\frac{1}{8}$.

e $\frac{3}{8} \div \frac{6}{7} = \frac{\cancel{3}^{1}}{8} \times \frac{7}{\cancel{6}_{2}} = \frac{7}{16}$ **f** $2\frac{4}{5} \div 3\frac{1}{2} = \frac{14}{5} \div \frac{7}{2} = \frac{\cancel{14}^{2}}{5} \times \frac{2}{\cancel{7}_{1}} = \frac{4}{5}$

The reciprocal of $\frac{6}{7}$ is $\frac{7}{6}$.   Change to mixed numbers
$2 \times 5 + 4 = 14$ and $3 \times 2 + 1 = 7$.

Exercise 5.3S

> Always give your answer in its simplest form.

1 Find an equivalent fraction for each fraction. Both of your new fractions should have the same denominator.

  a $\frac{1}{2}$ and $\frac{1}{3}$ **b** $\frac{1}{5}$ and $\frac{1}{3}$ **c** $\frac{1}{2}$ and $\frac{1}{5}$

  d $\frac{3}{10}$ and $\frac{1}{3}$ **e** $\frac{1}{3}$ and $\frac{3}{7}$ **f** $\frac{5}{6}$ and $\frac{3}{4}$

2 Calculate these additions and subtractions.

  a $\frac{1}{3}+\frac{1}{3}$ **b** $\frac{2}{5}+\frac{1}{5}$ **c** $\frac{3}{10}+\frac{7}{10}$

  d $\frac{7}{8}+\frac{5}{8}$ **e** $\frac{13}{7}-\frac{6}{7}$ **f** $\frac{5}{16}-\frac{1}{16}$

  g $\frac{12}{15}+\frac{8}{15}$ **h** $\frac{19}{16}-\frac{7}{16}$ **i** $\frac{12}{35}+\frac{8}{35}$

3 Change these improper fractions into mixed numbers.

  a $\frac{5}{4}$ **b** $\frac{9}{5}$ **c** $\frac{13}{8}$

  d $\frac{17}{4}$ **e** $\frac{22}{7}$ **f** $\frac{100}{11}$

4 Change these mixed numbers to improper fractions.

  a $1\frac{3}{4}$ **b** $1\frac{7}{16}$ **c** $1\frac{5}{9}$

  d $2\frac{4}{7}$ **e** $5\frac{1}{5}$ **f** $4\frac{3}{4}$

5 Calculate these additions and subtractions.

  a $1\frac{2}{3}+\frac{2}{3}$ **b** $4\frac{2}{7}-\frac{5}{7}$ **c** $3\frac{2}{6}-4\frac{3}{6}$

6 Calculate these additions and subtractions.

  a $\frac{1}{3}+\frac{1}{2}$ **b** $\frac{1}{4}+\frac{3}{5}$ **c** $\frac{3}{5}-\frac{1}{3}$

  d $\frac{4}{5}-\frac{2}{7}$ **e** $\frac{5}{8}+\frac{1}{3}$ **f** $\frac{4}{9}+\frac{2}{5}$

  g $\frac{7}{9}-\frac{2}{11}$ **h** $\frac{7}{15}+\frac{3}{7}$ **i** $\frac{8}{16}-\frac{3}{28}$

7 Calculate these multiplications.

  a $3\times\frac{1}{2}$ **b** $6\times\frac{1}{3}$ **c** $10\times\frac{1}{3}$

  d $15\times\frac{1}{7}$ **e** $\frac{1}{10}\times25$ **f** $\frac{1}{3}\times13$

8 Calculate these multiplications.

  a $3\times\frac{2}{3}$ **b** $6\times\frac{2}{3}$ **c** $5\times\frac{2}{3}$

  d $4\times\frac{3}{20}$ **e** $\frac{3}{5}\times10$ **f** $\frac{11}{8}\times7$

9 Calculate these divisions.

  a $4\div\frac{1}{2}$ **b** $2\div\frac{1}{5}$ **c** $2\div\frac{1}{7}$

  d $10\div\frac{1}{2}$ **e** $12\div\frac{1}{4}$ **f** $22\div\frac{1}{10}$

10 Calculate these additions and subtractions.

  a $\frac{4}{5}+\frac{2}{3}$ **b** $1\frac{1}{2}+\frac{3}{5}$ **c** $1\frac{1}{3}+1\frac{1}{4}$

  d $1\frac{2}{7}+\frac{3}{5}$ **e** $2\frac{2}{5}-\frac{1}{3}$ **f** $3\frac{3}{8}-1\frac{1}{2}$

  g $4\frac{1}{3}-2\frac{3}{4}$ **h** $3\frac{4}{7}-2\frac{8}{9}$ **i** $5\frac{2}{8}+3\frac{7}{2}$

11 Calculate these divisions.

  a $4\div\frac{2}{3}$ **b** $7\div\frac{2}{5}$ **c** $2\div\frac{5}{6}$

  d $12\div\frac{6}{7}$ **e** $20\div\frac{5}{12}$ **f** $5\div\frac{7}{9}$

  g $3\div1\frac{1}{2}$ **h** $3\div1\frac{2}{5}$ **i** $11\div\frac{15}{4}$

12 Calculate these multiplications.

  a $\frac{2}{5}\times\frac{3}{4}$ **b** $\frac{3}{5}\times\frac{3}{4}$ **c** $\frac{5}{7}\times\frac{3}{4}$

  d $\frac{4}{7}\times\frac{3}{5}$ **e** $\frac{5}{6}\times\frac{4}{5}$ **f** $\frac{3}{8}\times\frac{7}{9}$

  g $\frac{3}{5}\times\frac{10}{9}$ **h** $\frac{15}{16}\times\frac{12}{5}$ **i** $\left(\frac{3}{7}\right)^2$

  j $1\frac{3}{4}\times\frac{2}{7}$ **k** $3\frac{2}{3}\times\frac{7}{11}$ **l** $1\frac{3}{8}\times1\frac{2}{5}$

13 Calculate these divisions.

  a $4\div\frac{2}{5}$ **b** $\frac{2}{3}\div\frac{4}{5}$ **c** $\frac{4}{5}\div\frac{3}{4}$

  d $\frac{4}{7}\div\frac{2}{3}$ **e** $\frac{3}{7}\div\frac{4}{9}$ **f** $\frac{3}{5}\div\frac{1}{3}$

  g $\frac{3}{4}\div3$ **h** $\frac{4}{7}\div5$ **i** $\frac{4}{11}\div5$

  j $\frac{7}{4}\div\frac{2}{3}$ **k** $\frac{7}{4}\div\frac{3}{2}$ **l** $\frac{9}{5}\div\frac{5}{3}$

  m $1\frac{1}{2}\div\frac{3}{4}$ **n** $2\frac{1}{4}\div\frac{2}{3}$ **o** $2\frac{2}{5}\div\frac{9}{7}$

14 Work out these multiplications.

  a $\frac{2}{3}\times\frac{9}{10}\times\frac{5}{7}\times\frac{14}{6}$ **b** $\frac{3}{7}\times\frac{15}{27}\times\frac{21}{25}$

5.3 Calculations with fractions

- To add and subtract fractions, convert to equivalent fractions with a common denominator, then add or subtract the numerators.

- To multiply fractions, convert any mixed numbers to improper fractions, multiply the numerators and the denominators and cancel any common factors.

- Dividing by a fraction is the same as multiplying by its reciprocal.

$$6\frac{1}{4} - \frac{5}{6} = \frac{25}{4} - \frac{5}{6}$$
$$= \frac{75 - 10}{12}$$
$$= \frac{65}{12} = 5\frac{5}{12}$$

$$6\frac{1}{4} \div \frac{5}{6} = \frac{\overset{5}{\cancel{25}}}{\overset{}{\cancel{4}}_2} \times \frac{\overset{3}{\cancel{6}}}{\cancel{5}_1}$$
$$= \frac{15}{2} = 7\frac{1}{2}$$

HOW TO

To solve problems involving calculations with fractions
1. RTQ and decide what you need to do ($+$, $-$, $\times$ or $\div$).
2. Use the appropriate written method or, if allowed, your calculator.
3. ATQ, give your answer in its simplest form and include any units.

> There is a fraction key on your calculator, but you should learn to be confident in fraction arithmetic.

EXAMPLE

Tim works in a cafe. At the end of the night he has three measuring jugs with milk left in them. One has half a litre in it, the second $1\frac{1}{4}$ litres, the third has $\frac{1}{3}$ of a litre. He tries to pour all the milk into a 2 litre bottle. Will it fit? You must show your working.

1. Add the three fractions together to see if the total is more or less than 2.

$$\frac{1}{2} + 1\frac{1}{4} + \frac{1}{3} = \frac{6}{12} + \frac{15}{12} + \frac{4}{12}$$ 　2. A common denominator is 12.

$$= \frac{25}{12} = 2\frac{1}{12}$$ 　Convert to a mixed number.

It does not fit in the bottle because $2\frac{1}{12} > 2$. 　3.

EXAMPLE

Rachael sews together hexagonal patches to make a quilt. Each triangle is equilateral.

What fraction of the patches do the red triangles represent?

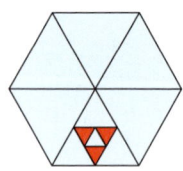

1. The red triangles are a fraction of a fraction of a fraction of the patch: multiply.

2. $\frac{1}{4} \times \frac{1}{6} = \frac{1}{24}$ 　　Triangle 2 is $\frac{1}{4}$ of triangle 1. 　　Triangle 1 $= \frac{1}{6}$

$\frac{3}{4} \times \frac{1}{24} = \frac{3}{96} = \frac{1}{32}$ 　　Red triangles are $\frac{3}{4}$ of triangle 2.

3. The red triangles are $\frac{1}{32}$ of the tile.

Exercise 5.3A

1 Graham cuts three pieces of piping from a pipe six inches long. The pieces measure $1\frac{1}{2}$ inches, $2\frac{1}{4}$ inches and $1\frac{5}{8}$ inches.

   a What do the three pieces add up to?

   b How much pipe is spare?

2 Asif says that $\frac{1}{4} + \frac{1}{3} = \frac{2}{7}$.

   a Explain what he has done wrong.

   b What is the correct answer?

3 A paint pot can hold $2\frac{3}{4}$ litres of paint. Hector buys $11\frac{1}{5}$ litres of emulsion paint. How many times can he *fill* the paint pot with emulsion paint?

4 Beatrice works overtime one weekend. She is paid at *time and a quarter*, that is, she is paid $1\frac{1}{4}$ hours for every hour she works. She works for 4 hours on Saturday, and $5\frac{1}{2}$ hours on Sunday. She gets paid £15 per hour.

   a How many hours overtime does she work?

   b How much money does she earn from the overtime?

5 A computer chip is a rectangle measuring $\frac{7}{8}$ cm by $\frac{3}{4}$ cm.

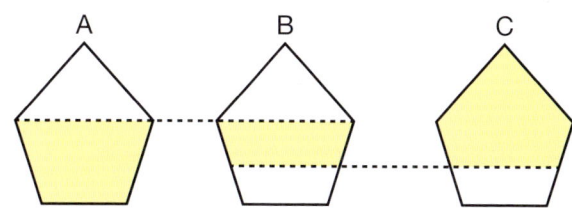

   a What is the perimeter of the rectangle?

   b What is the area of the rectangle?

6 Here are eight fractions.

$\frac{1}{2}$	$\frac{3}{4}$	$\frac{3}{5}$	$\frac{1}{8}$
$\frac{4}{3}$	$\frac{5}{2}$	$\frac{3}{8}$	$2\frac{1}{2}$

   a Which pairs add up to a whole number?

   b Which pair adds up to a half?

   c Which pair multiplies to make 1?

   d Which pair multiplies to make $1\frac{1}{2}$?

   e Which pair have the same value?

6 **f** Find a pair where one is twice the other.

   g Find a pair where one is a third of the other.

   h How many times does the smallest fraction go into the largest fraction?

   i What is the largest take away the smallest?

   j Which two fractions are reciprocals of each other?

7 Here are three identical shapes.
$\frac{4}{7}$ of A is shaded.
$\frac{3}{5}$ of C is shaded.

A   B   C

How much of B is shaded?

8 Fill the empty squares to make each statement true, horizontally and vertically. Enter each fraction in its simplest form.

$\frac{1}{2}$	$+$	$\frac{1}{6}$	$=$	
$\times$		$+$		$\star$
	$\times$		$=$	
$=$		$=$		$=$
$\frac{3}{20}$	$\times$		$=$	$\frac{1}{10}$

Which operation must replace the '$\star$'?

*9 Dave piles up five sheets of card. The thicknesses of the sheets, in fractions of an inch, are

$$\frac{1}{32}, \quad \frac{1}{16}, \quad \frac{3}{32}, \quad \frac{1}{8} \quad \text{and} \quad \frac{1}{4}$$

   a Which card does he take away from the pile so that the pile is half an inch tall?

   b How many of the thinnest card would he need to pile up to equal the thickest?

1017, 1040, 1046, 1047   SEARCH

5.4 Fractions, decimals and percentages

- To convert a **terminating** decimal to a fraction, write the decimal as a fraction with **denominator** 10, 100, 1000, ... according to the number of decimal places. Then simplify.

- To convert a percentage to a fraction, write it as a fraction with a denominator of 100 and simplify.

- To convert a percentage to a decimal, divide by 100.

> A terminating decimal stops, for example, 0.625

EXAMPLE

Convert these to fractions. **a** 0.306 **b** 45% **c** 32.5%

a $0.306 = \frac{306}{1000} = \frac{153}{500}$ 3 decimal places so denominator is 1000.

b $45\% = \frac{45}{100} = \frac{9}{20}$ **c** $32.5\% = \frac{32.5}{100} = \frac{325}{1000} = \frac{65}{200} = \frac{13}{40}$

- To convert a fraction to a decimal divide the numerator by the denominator.
- To convert a fraction to a percentage
  - write as an equivalent fraction with denominator 100, take the numerator.
  - convert to a decimal, multiply by 100%.

> $\frac{1}{4} = \frac{25}{100} = 0.25$
>
> $\frac{25}{100} \times 100\% = 25\%$

EXAMPLE

Write these fractions as **i** decimals **ii** percentages **a** $\frac{17}{20}$ **b** $\frac{3}{8}$

a **i** $\frac{17}{20} = \frac{85}{100} = 0.85$ **ii** 85%

b **i** $\frac{3}{8} = 3 \div 8 = 8)\overline{3.000}^{0.375}$ **ii** 37.5%

p.4 When comparing fractions, decimals an percentages, it is a good idea to first convert them to the same representation.

EXAMPLE

Order these from lowest to highest: 45% $\frac{5}{8}$ 0.52

You can use any method and convert all to any representation.

$\frac{5}{8} = 0.625 = 62.5\%$ $0.52 = 52\%$

$45\%, 0.52, \frac{5}{8}$ Write the answer using their original forms.

- A **recurring decimal** has digits that keep repeating.

> The dots show which digits repeat.

$\frac{1}{3} = 1 \div 3 = 0.333333 \ldots = 0.\dot{3}$ $\frac{4}{33} = 4 \div 33 = 0.121212 \ldots = 0.\dot{1}\dot{2}$

$2\frac{2}{3} = 2.666666\ldots = 2.\dot{6}$ $\frac{1}{7} = 0.142857142857 \ldots = 0.\dot{1}4285\dot{7}$

Exercise 5.4S

1 Write these percentages as decimals.

  a 67%   **b** 78%   **c** 99%

  d 70%   **e** 39%   **f** 88%

  g 150%   **h** 125%   **i** 99.9%

  j 110%   **k** 75%   **l** 37.6%

2 Write these decimals as percentages.

  a 0.32   **b** 0.22   **c** 0.85

  d 0.03   **e** 0.54   **f** 0.63

  g 0.38   **h** 0.375   **i** 0.333

  j 1.25   **k** 0.0015   **l** 0.995

3 Write these decimals as fractions in their simplest forms.

  a 0.8   **b** 0.28   **c** 0.325

  d 0.05   **e** 0.12   **f** 0.375

4 Change these fractions to decimals. Give your answers to 2 decimal places as appropriate.

  a $\dfrac{3}{10}$ **b** $\dfrac{7}{25}$ **c** $\dfrac{7}{12}$

  d $\dfrac{9}{15}$ **e** $\dfrac{15}{7}$ **f** $\dfrac{7}{2}$

5 Write these percentages as fractions in their simplest forms.

  a 25%   **b** 40%   **c** 65%

  d 15%   **e** 145%   **f** 55%

6 Write each of these decimals as a fraction in its simplest form.

  a 0.3   **b** 0.6   **c** 0.64

  d 0.45   **e** 1.375   **f** 1.08

  g 3.2375   **h** 3.0625   **i** 4.25

7 Change these fractions to decimals without using a calculator.

  a $\dfrac{3}{10}$ **b** $\dfrac{11}{25}$ **c** $\dfrac{26}{25}$

  d $\dfrac{124}{200}$ **e** $\dfrac{27}{60}$ **f** $\dfrac{39}{75}$

  g $\dfrac{42}{150}$ **h** $3\dfrac{21}{60}$ **i** $\dfrac{89}{25}$

8 Change these fractions into decimals using an appropriate method. Give your answers to two decimal places where necessary.

  a $\dfrac{22}{50}$ **b** $\dfrac{2}{3}$ **c** $\dfrac{27}{20}$

  d $\dfrac{11}{15}$ **e** $\dfrac{8}{7}$ **f** $1\dfrac{2}{5}$

  g $2\dfrac{11}{66}$ **h** $\dfrac{11}{13}$ **i** $\dfrac{22}{7}$

9 Write each of these percentages as a fraction in its simplest form.

  a 60%   **b** 90%   **c** 35%

  d 35%   **e** 1%   **f** 362%

  g 15.25%   **h** 2.125%   **i** 0.145%

10 Write each of these fractions as a percentage without using a calculator.

  a $\dfrac{27}{50}$ **b** $\dfrac{2}{5}$ **c** $\dfrac{17}{20}$

  d $\dfrac{13}{25}$ **e** $\dfrac{2}{3}$ **f** $\dfrac{48}{200}$

  g $1\dfrac{3}{15}$ **h** $\dfrac{33}{75}$ **i** $\dfrac{3}{8}$

11 Write these percentages as decimals.

  a 37%   **b** 7%   **c** 189%

  d 45%   **e** 200%   **f** 2.5%

12 Write these decimals as percentages.

  a 0.72   **b** 0.2   **c** 2

  d 0.0003   **e** 3.0   **f** 0.01

13 Write these fractions as percentages. Give your answers to 1 decimal place as appropriate.

  a $\dfrac{48}{70}$ **b** $\dfrac{16}{25}$ **c** $\dfrac{17}{19}$

  d $1\dfrac{11}{12}$ **e** $\dfrac{5}{19}$ **f** $1\dfrac{1}{8}$

14 Order these numbers from lowest to highest.

  a $\dfrac{5}{11}$ 42% $\dfrac{43}{100}$ 0.423

  b $\dfrac{2}{3}$ 0.67   60% $\dfrac{13}{25}$

15 Order these numbers from highest to lowest.

  a 0.375 $\dfrac{7}{2}$ 33% $\dfrac{13}{40}$

  b 1.33 $\dfrac{7}{5}$ 45% $1\dfrac{4}{9}$

*16 Write these fractions as recurring decimals, using the correct notation.

  a $\dfrac{2}{3}$ **b** $\dfrac{3}{11}$ **c** $\dfrac{2}{9}$

  d $\dfrac{1}{6}$ **e** $\dfrac{2}{7}$ **f** $\dfrac{3}{7}$

Q 1015, 1016, 1029, 1074   SEARCH

5.4 Fractions, decimals and percentages

RECAP

- You can convert between fractions, decimals and percentages.
- When comparing it is best to first convert to all fractions, all decimals, or all percentages.

> Try drawing a number line and marking an it all the equivalent fractions, decimals and percentages that you know.

HOW TO

To solve problems involving equivalence between fractions, decimals and percentages
① RTQ and decide what you need to do.
② Apply the appropriate conversion technique, if needed.
③ ATQ and include units if appropriate.

EXAMPLE

Vita is comparing her marks in four exams. Each mark is given in a different way.

A 15 out of 20   **B** 92%   **C** $\dfrac{47}{50}$ **D** 0.758

a What is the range of her marks?

b Vita says 'the difference between my best and second best marks is greater than the difference between my worst two marks'. Is Vita correct? Give your reasons.

① Order the marks by converting them all to percentages.

② **A** 15 out of 20 = $\dfrac{15}{20}$ = $\dfrac{75}{100}$ = 75%   **B** 92%   Leave as it is.

C $\dfrac{47}{50}$ = $\dfrac{94}{100}$ = 94%   **D** 0.758 = 75.8%   Multiply by 100.

C > B > D > A

a ① Range = largest − smallest   **b** C − B = 94% − 92% = 2%

Range = 94% − 75% = 19%   D − A = 75.8% − 75% = 0.8%

2% > 0.8%   So Vita is correct. ③

EXAMPLE

A tennis club increased its membership from 32 to 44. What percentage of the 44 members are new?

44 − 32 = 12   ① Work out the number of new members as a fraction.

$\dfrac{12}{44}$ = $\dfrac{3}{11}$ = 0.272727... = 0.2$\dot{7}$ ② By calculator.

27% (2 sf)   ③ Convert to a percentage.

EXAMPLE

A regular pentagon is divided as shown.

① Convert to percentages and add together.

② 0.13 = $\dfrac{13}{100}$ = 13% $\dfrac{1}{8}$ = $\dfrac{0.125}{8)1.000}$ = 12.5%

48 + 13 + 12.5 = 73.5

③ 100% − 73.5% = 26.5%

What percentage of the shape is not coloured?

Exercise 5.4A

1 A supermarket surveys its customers about waiting times. 46 out of 800 customer said they queued for more than 5 minutes on average when at the store.

Can the supermarket manager claim that '95% of our customers say they only ever queue for less than 5 minutes'? Give your reasons.

2 Violet is doing a survey to find the percentage of girls in classes in her year group. She lists the proportion of each class that are girls.

10A 15 out of 30   10B 48%

10C $\frac{3}{5}$ 10D 0.72

Which class has the highest fraction of

  a girls   **b** boys?

3 George's store is changing the price of some goods for summer.

		Old price	New price
A	Sandals £12.		£10.80
B	Umbrellas £7.65		£6.12
C	Torches £6		£5.70
D	Scarves £12.50		25% off
E	Coats £87		$\frac{1}{3}$ off.
F	Gloves £7.25		£1.50 off

  a Work out the decrease in price as a percentage of the old price for items **A**, **B** and **C**.

  b Which item, **A** – **F**, has had the biggest price reduction in terms of percentage of the original price?

4 A £1400 television drops to £1330 in a shop sale. Online it is on offer and costs 90% of the normal price. Which is the better deal? Show your working.

5 The population of a village in 2001 was 200. By 2011 it was 232. Assuming no one left the village between 2001 and 2011, what fraction of the population in 2011 was not living in the village in 2001?

6 Jessica sees bargain signs in two bookshops. She wants to buy two books, each normally £6.90.

6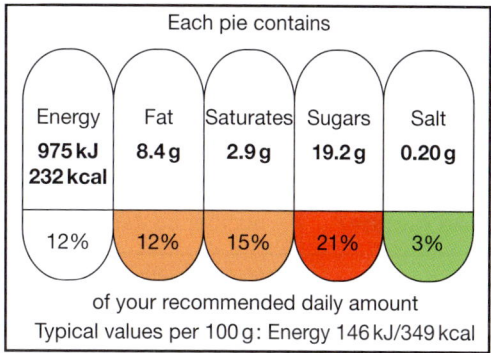

  a Her friend Shannon says it doesn't matter which shop she goes to. Is she correct? Show your working to justify your answer.

  b Chloe wants to buy two books, one costing £5.60, the other £4.68. Stonewaters says the half price offer applies to the cheaper book. Which shop should she go to? Show your working.

*7 A food label shows nutritional information for an individual apple pie.

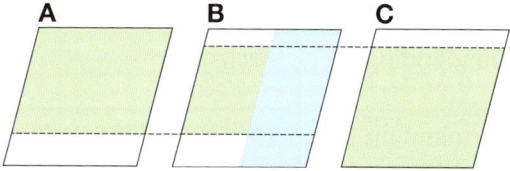

  a The bottom figures show that 19.2 g of sugar is 21% of a recommended daily amount. How many grams is a recommended daily amount?

  b What are the recommended daily amounts of fat, saturates and salt?

  c The label says that 100 g of pie would provide 349 kcal. Angela has lost the bit of the label with the total mass of the pie. What is the mass, in grams, of one pie?

*8 Three identical parallelograms are shaded. $\frac{4}{5}$ of **A** is shaded. 85% of **C** is shaded.

A   B   C

50% of B is coloured blue.

What percentage of B is shaded altogether?

   1015, 1016, 1029, 1074   SEARCH

Summary

Checkout
You should now be able to...

	Test it Questions
✔ Convert between terminating decimals and their corresponding fractions.	1 – 4
✔ Compare decimals and fractions using the symbols $>$ and $<$.	5
✔ Find fractions and percentages of amounts.	6, 7
✔ Add and subtract simple fractions and mixed numbers.	8
✔ Multiply and divide simple fractions and mixed numbers.	9, 10
✔ Convert between fractions, decimals and percentages.	11

Language	Meaning	Example
Fraction	A fraction compares the size of a part with the size of a whole. All the parts that make up the whole have to be the same size.	$\frac{5}{8}$ is shaded
Equal	Exactly the same quantity or size.	$\frac{1}{2} = \frac{2}{4}$
Numerator	The part of a fraction above the line.	$\dfrac{2}{5}$ ← Numerator
Denominator	The part of a fraction below the line.	← Denominator
Mixed number	A number containing a whole number and a fraction.	$1\frac{1}{3}$ and $3\frac{5}{8}$ are mixed numbers.
Improper fraction	A fraction with the numerator larger than the denominator.	$\frac{5}{2}$ and $\frac{102}{51}$ are improper fractions.
Decimal	A way of expressing values of fractions less than 1.	$0.376 = \frac{3}{10} + \frac{7}{100} + \frac{6}{1000} = \frac{376}{1000}$
Decimal equivalent	A number written as a decimal that has the same value as a fraction or percentage.	$\frac{5}{8} = 0.625$
Common factor	A factor that is shared by two or more numbers.	$15 = 3 \times 5 \qquad 35 = 5 \times 7$ 5 is a common factor of 15 and 35.
Cancel	Common factors in the numerator and denominator of a fraction can be 'cancelled'.	$\frac{2}{3}$ of $12 = \frac{2}{3} \times 12^4 = 2 \times 4 = 8$ 3 is a common factor of 12 and 3.
Lowest common denominator	The smallest number into which the denominators of two or more fractions will divide.	The lowest common denominator of $\frac{1}{6}$ and $\frac{1}{8}$ is 24. $\quad \frac{1}{6} = \frac{4}{24}, \; \frac{1}{8} = \frac{3}{24}$
Ascending	Going up.	1, 2, 3, 4, 5, 6 are in ascending order.
Descending	Going down.	6, 5, 4, 3, 2, 1 are in descending order.
Terminating	A decimal with a definite number of digits.	$\frac{1}{8} = 0.125$
Recurring decimal	A decimal with a repeating pattern that goes on forever.	$\frac{1}{3} = 0.333... = 0.\dot{3}$ $\frac{9}{11} = 0.818181... = 0.\dot{8}\dot{1}$

Review

1 Convert these mixed numbers to improper fractions.

    a $1\frac{2}{5}$ **b** $3\frac{4}{7}$

2 Convert these improper fractions to mixed numbers.

    a $\frac{9}{4}$ **b** $\frac{11}{6}$

3 Write these decimal numbers as fractions in their simplest form.

    a 0.3   **b** 0.25

    c 0.88   **d** 0.05

4 Convert these fractions to decimals.

    a $\frac{1}{2}$ **b** $\frac{7}{10}$

    c $\frac{2}{100}$ **d** $\frac{25}{40}$

5 For each pair of fractions, write which is the larger fraction.

    a $\frac{2}{7}$ and $\frac{1}{5}$ **b** $\frac{8}{3}$ and $\frac{21}{8}$

6 Calculate the following fractions of amounts.

    a $\frac{1}{5}$ of 45   **b** $\frac{3}{7}$ of 28

    c $1\frac{1}{3}$ of 6   **d** $\frac{9}{8}$ of 32

7 Calculate the following percentages of amounts.

    a 25% of 64   **b** 40% of 120

    c 15% of 80   **d** 110% of 90

8 Calculate

    a $\frac{3}{11} + \frac{4}{11}$ **b** $\frac{4}{5} - \frac{1}{10}$

    c $\frac{3}{8} + \frac{2}{3}$ **d** $1\frac{2}{5} - \frac{1}{4}$

    e $5\frac{2}{3} - 2\frac{3}{4}$ **f** $2\frac{5}{6} + 1\frac{1}{4}$

9 Calculate

    a $4 \times \frac{5}{8}$ **b** $\frac{2}{7} \times \frac{1}{5}$

    c $\frac{4}{5} \times \frac{3}{8}$ **d** $1\frac{2}{3} \times \frac{5}{6}$

10 Calculate

    a $\frac{4}{5} \div 10$ **b** $4\frac{2}{3} \div 7$

    c $5 \div \frac{1}{4}$ **d** $4 \div 1\frac{3}{7}$

    e $\frac{5}{6} \div \frac{2}{3}$ **f** $5\frac{1}{4} \div \frac{3}{7}$

11 Copy and complete the table to show the conversions between fractions, decimals and percentages. Ensure your fractions are fully simplified.

Fraction	Decimal	Percentage
$\frac{3}{5}$		
	0.01	
		65%
	1.2	

What next?

Assessment 5

Do not use a calculator, with the exception of questions **14** and **16**.

1 Jasmin ordered the following fractions as shown. Reorder them correctly. [4]

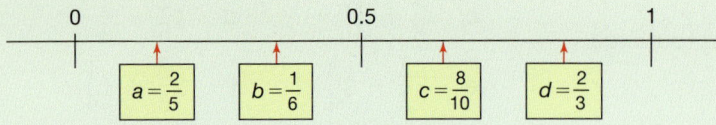

2 **a** Kylie says that $\frac{4}{5} = \frac{10}{15}$. Is Kylie right? [2]

  b Christie says that any fraction that has the same numerator as denominator
  is equivalent to 1. Is Christie right? Give reasons for your answer. [2]

3 **a** Write $\frac{16}{64}$ in its lowest terms. [1]

  b Ben answers the question like this.

  Cancel the 6s $\frac{16}{64}$ which leaves $\frac{1}{4}$. Therefore $\frac{16}{64} = \frac{1}{4}$.

  Describe the mistake that Ben has made. [2]

  c Give a similar example to the one in part **b** that shows the same mistake. [1]

4 **a** Ahmed tries to write the following numbers in order from biggest to smallest.

  558.0   55.80   5 580   5.580   0.055 80   0.5580

  Is his order correct? If not, write the numbers in the correct order. [2]

  b Darren says that the number that is halfway between

  i   6.4 and 6.6 is 6.5   **ii**   9.5 and 9.6 is 9.59

  iii 2.2 and 2.21 is 2.25   **iv** 6.99 and 7 is 6.95.

  For each pair of numbers, decide if Darren is correct or not.

  In the cases where he is not correct, give the correct halfway number. [4]

5 A fruit-bowl contains apples, bananas and pears. Two-fifths of the fruit in the bowl
are apples. 45% of the fruit in the bowl are bananas. $\frac{3}{20}$ of the fruits are pears.

  a What percentage of the fruit in the bowl are apples? [1]

  b What percentage of the fruit in the bowl are pears? [1]

  c What fraction of the fruit in the bowl are bananas? [1]

  d Which fruit has the biggest percentage of the bowl's contents?
  Explain your answer. [2]

6 'Mr Slim' put two identical pairs of trousers in his summer sale.
One pair said 'One quarter off'. The other pair said '20% discount'.
Which pair of trousers would you buy? Give reasons for your answer. [2]

7 A shop has a sale with the following discounts. What is the value, in £s, of each discount?

  a 30% off £400   [2]   **b** 70% off £54   [2]   **c** 35% off £90   [2]

  d 22% off £65   [3]   **e** 36% off £1   [1]   **f** $47\frac{1}{2}$% off £800   [3]

8 The number of students in a class last year was 50.
This year the number of students has increased by 20%.
How many students are there this year? [2]

9 Branson asked some friends which e-mail provider they use.

Provider	Number of boys	Number of girls
Air	5	0
Minus-net	1	2
Chat-Chat	3	6
PT	8	8
Redyonder	8	4
	Total 25	Total 20

   a Which provider do 20% of the girls use? [1]

   b Which provider do 20% of the boys use? [1]

   c Which provider do 20% of the total number of pupils use? [2]

   d Branson said 'In my survey, PT was equally popular with both boys and girls.'
      Was Branson correct or incorrect?
      Give your reasons. [2]

10 a What percentage of the word MISSISSIPPI is made up by the letter S? [2]

   b Old MacDougal had a farm.

      He had 21 pigs, 8 lambs, 150 sheep, 8 calves, 27 cows, 35 chickens and 1 bull.

      What percentage of his livestock were chickens? [2]

11 In the 4×100 m relay, the first three runners took the following proportion of the total race time.

   First runner 25%   Second runner $\frac{7}{25}$ Third runner 0.35

   a What decimal of the time was taken by the 4th member? [2]

   b Which team member ran the fastest leg? [1]

   c Which team member ran the slowest leg? [1]

12 G. Russet has an orchard. The orchard contains $3\frac{3}{4}$ hectares of apple trees.
   He needs to treat $\frac{2}{5}$ of the area for disease prevention.
   What area does he need to treat? [3]

13 In a football match the goalkeeper kicked the ball from the goal line for $\frac{5}{9}$ of the length of the pitch. Another player then kicked it a further $\frac{7}{20}$ of the way.
   If the length of the pitch is 90 yards, how many yards further is the opposite goal line? [4]

14 In her garden, busy Lizzie planted flowers in 27% of the total area.

   a The total area was $540 \, \text{m}^2$. What area did she use for flowers? [2]

   b 36% of the *remainder* of her garden was taken up by the patio.
      Find the area of the patio, to the nearest m^2. [3]

15 In 'Topmarks College', $\frac{8}{11}$ of the students are girls.

   Of these girls $\frac{3}{4}$ are brunette and of these brunettes $\frac{5}{9}$ wear earrings.
   What fraction of the school students are brunette girls who wear earrings? [2]

16 'FALSEPRINT' film laboratories sell prints in sizes 12.5 cm by 7.5 cm and 15 cm by 10 cm. Their adverts say that their 15 cm by 10 cm prints are more than 50% bigger than the 12.5 cm by 7.5 cm size. Are they correct? [4]

17 Convert the following fractions into recurring decimals.

   a $\frac{1}{3}$ [2]   b $\frac{5}{9}$ [2]   c $\frac{6}{7}$ [2]   d $\frac{7}{11}$ [2]

18 Put the following list in ascending order of size.

   0.34 $\frac{3}{8}$ $33\frac{1}{3}\%$ $\frac{5}{14}$ 0.334   33.3% [5]

Life skills 1: The business plan

Four friends – Abigail, Mike, Juliet and Raheem – are planning to open a new restaurant in their home town of Newton-Maxwell. They have a lot to think about and organise!

They start by creating a business plan. This plan needs to include: market research to understand their potential customers, what their costs and revenues are expected be, how big a loan they could afford to borrow and how any profits should be shared.

Task 1 – Market research

The friends decide to investigate how much people are prepared to pay for a three course meal. They carry out a small pilot survey. This involves stopping people in the street and asking them a few questions.

a Draw a grouped dual bar chart to show the ages of the women and men interviewed. Use the groups 10–19, 20–29, 30–39, 40–49, 50–59, 60–69. Describe what this shows.

b Abigail and Mike think that men will be prepared to pay more for a good meal than women. Do the results back up this theory? Calculate averages to justify your conclusions.

Pilot survey results (15 men and 15 women)

M 24, £33	F 20, £23	F 22, £25
M 37, £36	M 62, £33	F 47, £36
M 47, £35	M 42, £32	F 19, £16
F 52, £32	M 31, £22	M 66, £25
M 26, £24	M 55, £40	F 38, £35
F 18, £20	M 39, £35	M 40, £30
M 21, £21	F 23, £21	M 20, £30
F 58, £40	F 35, £32	F 32, £28
F 22, £30	F 61, £37	F 28, £20
M 23, £27	M 51, £27	F 44, £34

Key   Gender (M/F) Age (years), amount prepared to spend

Task 2 – Projected revenue

The friends are estimating the revenue (money coming in) for their restaurant for the first year. To do this they make some assumptions. Use their assumptions to answer the following questions.

a What would the mean amount paid for a meal be after VAT is added?

b Use the amount paid for a meal excluding VAT to estimate the revenue for the first year.

c Write down a formula for the profit (money left after costs), P, in terms of the other variables listed to the right.

d Find P if G = £40 000, S = £80 000 and C = £50 000.

e Make S the subject of the formula you found in part c.

> The cost of a meal includes Value Added Tax (VAT) charged at 20%.

Assumptions

- The mean amount paid for a meal, excluding VAT, is £24.42.
- The restaurant is open 364 days a year.
- The mean number of meals sold a day is 25.

Variables

G = cost of food bought by restaurant

S = salary costs C = other costs

R = revenue P = profit

Broad age band	Percentage
16–24	19
25–49	47
50–64	18
65 and over	16

 Percentage of the population in Newton-Maxwell in different age bands.

Task 3 – The survey

Following from the pilot survey, the friends do a larger survey of 200 people. They decide their sample should have the same proportion of people from each age group in their sample as the percentage in the total population of Newton-Maxwell.

How many people from each of the age groups should they include in their sample?

Ownership shares

Abigail $\frac{2}{5}$ Raheem $\frac{1}{4}$

Mike and Juliet both own an equal share of the remainder.

Task 4 – Shares in the business

The friends invested different capital (initial amounts of money) into the business. Based on this, they each own shares that determine the fraction of profit they are entitled to.

a What fraction of the business do Mike and Juliet each have?

b Draw a pie chart to show how much of the business each person owns.

c How much profit would each owner get from a yearly profit of £50 000?

Three year repayment formula

C = amount borrowed A = amount repaid each year

i = annual interest rate (AIR), expressed as a decimal

$$C = A\left(\frac{1}{1 + i} + \frac{1}{(1 + i)^2} + \frac{1}{(1 + i)^3}\right)$$

Task 5 – The business loan

The friends decide to take out a business loan in order to equip the restaurant.

They borrow £C at an annual interest rate (AIR) of i (expressed as a decimal), and repay an amount £A at the end of each year for three years, as shown by the three year repayment formula.

a What is the value of i if the annual interest rate is 6%?

b If they decide that the maximum they can afford to repay each year is £5000, how much can they borrow at an AIR of

   i   6%   **ii**   8%?

c If they decide that they need to borrow £30 000, how much will each yearly repayment be at an AIR of

   i   6%   **ii**   8%?

6 Formulae and functions

Introduction

Nurses often use mathematical formulae when they are administering drugs, for example, in converting from one unit to another, calculating the amount of a drug based on somebody's mass, or working out concentrations from solutions.

Working with formulae is a topic within algebra, and is a good example of the practical use of mathematics.

What's the point?

The ability to apply a mathematical formula accurately when calculating a patient's dose of a particular medicine is vitally important.

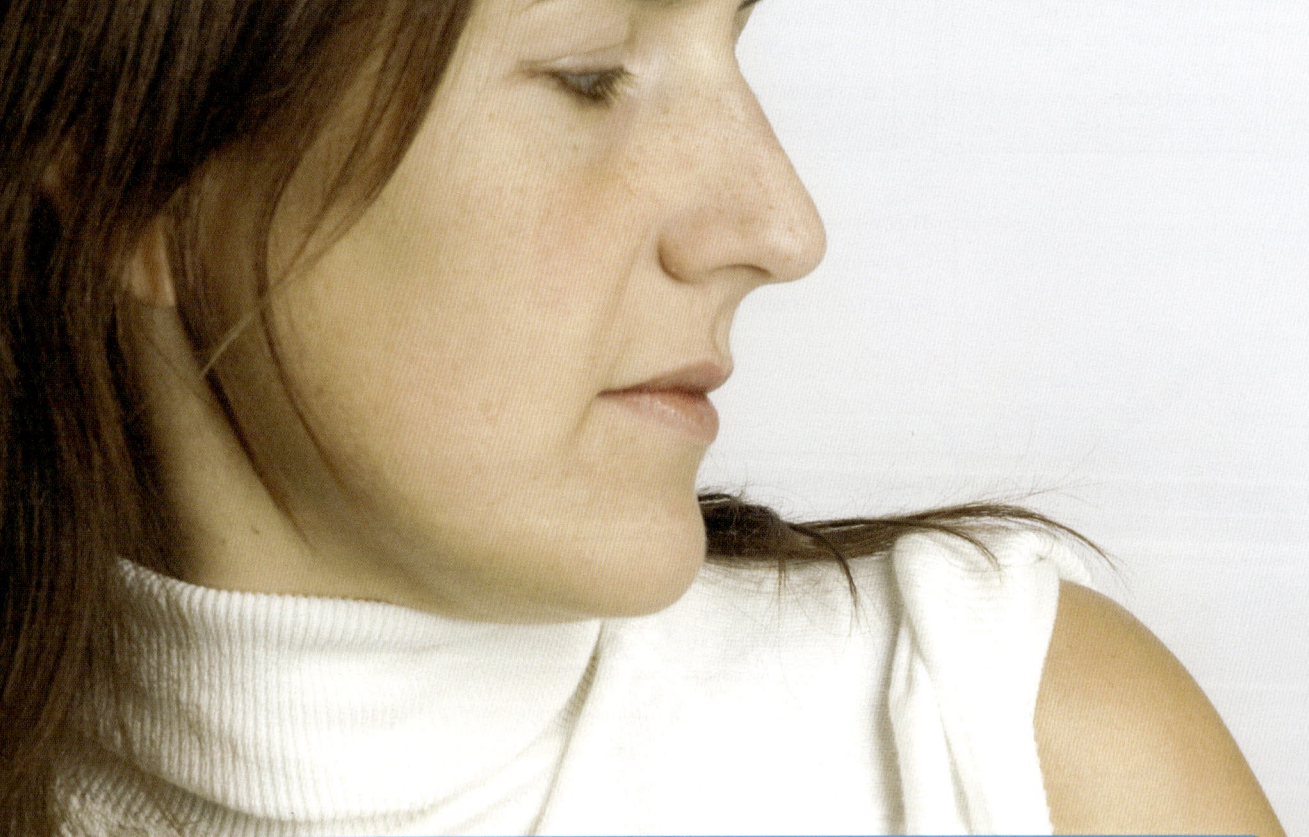

Objectives

By the end of this chapter you will have learned how to ...

- Substitute numerical values into formulae and expressions.
- Rearrange formulae to change the subject.
- Identify inequalities, equations, formulae and identities.
- Expand double brackets.
- Factorise quadratic expressions of the form $x^2 + bx + c$ and the difference of two squares.

Check in

1 Follow the order of operations to work out these calculations.

    a $3 \times 5 - 12 + 2$ **b** $4 + 2 \times 3 - 1$

    c $6 \div 3 + 2$ **d** $3 \times 4 - 4 \div 2$

2 Multiply out

    a $3(x + 1)$ **b** $2(x - 1)$

    c $4(2x + 3)$ **d** $3(4x - 2)$

3 Factorise

    a $3x + 9$ **b** $6x + 2$

    c $3y - 9$ **d** $4y + 12$

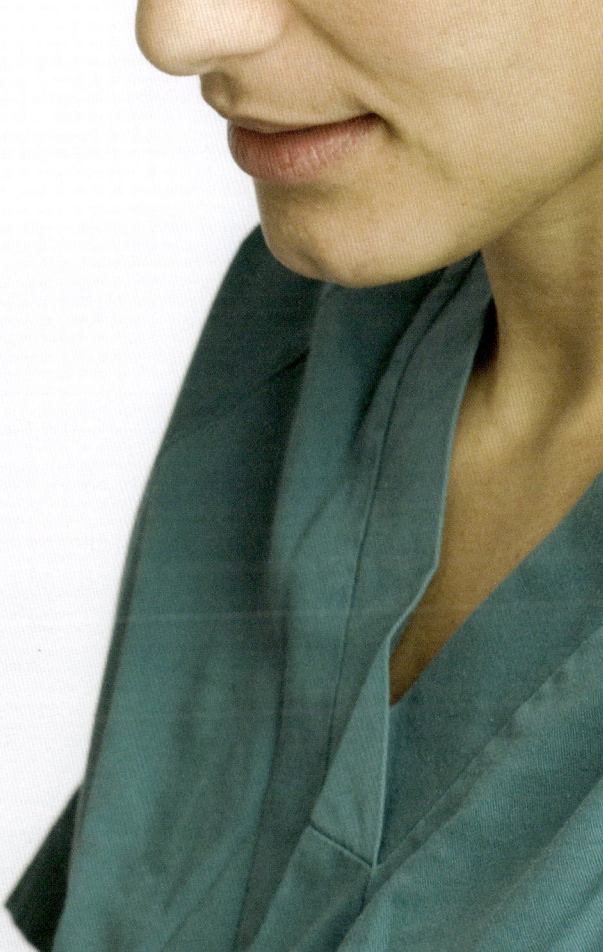

Chapter investigation

This grid of numbers uses each of the numbers from 1 to 9.

4	9	2
3	5	7
8	1	6

Every row, column and diagonal adds up to 15. It is called a magic square.

Create your own magic square.

6.1 Substituting into formulae

● A **formula** is an equation representing the relationship between two or more **variables**.

p.26

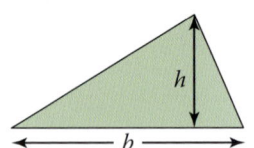

$$F = ma \quad v = u + at$$
$$E = mc^2 \quad P = 2(a + b)$$

Formulae are used to calculate the value of a variable.

The area of a triangle can be calculated using this formula.

$$A = \frac{1}{2}bh$$

Here b represents the length of the base of the triangle and h its perpendicular height.

● To evaluate a formula **substitute** numerical values for the variables and use the BIDMAS rules to work out the value of the resulting expression.

p.16

EXAMPLE

Using the formulae given, calculate the value of H.

a $H = d + 5$ when $d = 4$ **b** $H = d - 5$ when $d = 10$ **c** $H = 5d$ when $d = 5$

a $H = 4 + 5$

     $= 9$

b $H = 10 - 5$

     $= 5$

c $H = 5 \times 5$ | $5d$ means $5 \times d$ |

     $= 25$

EXAMPLE

Using the formula, $W = 2x + y^2$, calculate

a W when $x = 4, y = 5$ **b** W when $x = 0, y = -3$ **c** x when $W = 21$ and $y = 3$

The values need to be substituted into the correct variables in the formula.

a $W = 2x + y^2$

     $W = (2 \times 4) + 5^2$

     $W = 8 + 25$

     $= 33$

b $W = 2x + y^2$

     $W = (2 \times 0) + (-3)^2$

     $W = 0 + 9$

     $= 9$

c $W = 2x + y^2$

     $21 = 2x + 3^2$

     $12 = 2x$

     $x = 6$

Be careful when substituting negative numbers.

You can create a formula.

EXAMPLE

The cost of a taxi journey is £2 per mile plus £3 standard charge.

a   Derive a formula for the cost, £C, of a taxi journey in terms of the number of miles, m, travelled.

b   Use the formula to calculate the cost of a taxi journey of 4.5 miles.

c   A taxi journey costs £10. How long was the journey?

a $C = 3 + 2m$

           A 1 mile journey costs £3 + 1 × £2

           A 2 mile journey costs £3 + 2 × £2

           A journey costs £3 plus the number of miles multiplied by 2.

b $C = 3 + 2 \times 4.5 = 2 + 9 = £11$

c $10 = 3 + 2m$ -3

     $7 = 2m$ $\div 2$

     $m = \dfrac{7}{2} = 3.5$ miles

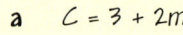

Algebra   Formulae and functions

Exercise 6.1S

1 Using the formula, $A = y + 4$, calculate the value of A when $y = 5$.

2 Using the formula, $P = k - 6$, calculate the value of P for these values of k.

    a $k = 10$ **b** $k = 20$

    c $k = 6$ **d** $k = 4$

3 Ian is using the formula $B = 5w$.

    When $w = 4$, Ian calculates $B = 54$.

    Do you agree with Ian? Explain your answer.

4 Using the formula, $D = 5p$, calculate the value of D for these values of p.

    a $p = 2$ **b** $p = 5$

    c $p = 7$ **d** $p = 0$

    e $p = -2$ **f** $p = -5$

5 Using the formula, $T = \dfrac{m}{4}$ calculate the value of T for these values of m.

    a $m = 12$ **b** $m = 20$

    c $m = 16$ **d** $m = 40$

    e $m = 0$ **f** $m = 2$

6 Using these formulae, calculate the value of F when $a = 9$.

    a $F = 2a + 5$ **b** $F = 4a - 6$

    c $F = 2(a + 5)$ **d** $F = a^2$

    e $F = 2a^2$ **f** $F = \sqrt{a}$

7 Using the formula, $P = 2a + 2b$ calculate the value of P for these values of a and b.

    a $a = 2, b = 4$ **b** $a = 3, b = 5$

    c $a = 0, b = 10$ **d** $a = -2, b = 0$

    e $a = b = 5$ **f** $a = -1, b = -2$

8 Using the formula $v = u + at$, complete this table.

	u	a	t	v
a	4	5	2	
b	0	2	4	
c	5	−2	3	
d		2	3	8
e	−2		6	10

9 The formula to convert temperatures in °C to temperatures in °F is $F = \dfrac{9}{5}C + 32$.

    Using the formula, convert a temperature of 25 °C to °F.

10 Derive a formula for A for each of these situations.

	Relationship
a	A is equal to 6 more than p
b	A is equal to 5 less than m
c	A is equal to double k
d	A is equal to 6 add 5 lots of t
e	A is equal to 3 times d add 7
f	A is equal to t add 7, multiplied by 3
g	A is equal to the square of y

11 The approximate circumference, C, of a circle can be calculated by multiplying the diameter, d, of the circle by 3. p.230

    a Derive a formula for the circumference, C, of a circle.

    b Use the formula to find the approximate circumference of a circle with diameter 10 cm.

*12 The time for pendulum to complete a cycle can be calculated using this formula.

$$T = 2\pi\sqrt{\dfrac{L}{g}}$$

    where L is the length of the pendulum, and g is the acceleration due to gravity.

    Calculate the value for T when $L = 25$ cm and $g = 9.8$ m/s².

13 a Calculate the perimeter of this rectangle.

        4 cm

        10 cm

    b The perimeter of a rectangle can be calculated using the formula, $P = 2(a + b)$, where a and b are the lengths of the longer and shorter sides. Using the formula, calculate the perimeter of the rectangle.

    c If the perimeter of a rectangle is 40 cm, find at least five sets of values for a and b.

6.1 Substituting into formulae

- A formula is an equation representing the relationship between two or more variables.
- Formulae are used to calculate the value of a variable

Use the BIDMAS rules when evaluating a formula.

HOW TO

To evaluate a formula
1. Write down the formula that you are using.
2. Write it out again with the variables substituted by numbers.
3. Carry out the calculation and round if appropriate.
4. Check if any units need to be included in the final answer.

EXAMPLE

p.312

The formula for the volume of a cylinder is $V = \pi r^2 h$ where r is the radius of the circular cross-section and h is the height of the cylinder. The diameter of a tin of soup is 7 cm. The height of the tin is 11 cm. Find the volume of soup in the tin.

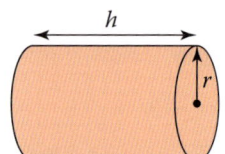

1. $V = \pi r^2 h$
2. $V = \pi \times 3.5^2 \times 11$ Radius = diameter ÷ 2.
   $= 423.329...$ Using a calculator.
   $= 423$ (nearest whole number)
3. The volume of soup in the tin is 423 cm³.

EXAMPLE

Here are two facts about motion

Facts	Formulae
Acceleration = Change in velocity ÷ Time	$a = \dfrac{v - u}{t}$
Change in momentum = Force × Time	$mv - mu = Ft$

a   Work out the value of a when $u = 2$ m/s, $v = 14$ m/s and $t = 3$ seconds

b   Work out the value of u when $m = 5$, $v = 7$, $F = 2$ and $t = 10$.

a $a = \dfrac{v - u}{t} = \dfrac{14 - 2}{3} = \dfrac{12}{3} = 4$ ①②③

Acceleration = 4 m/s²   ④

b $5 \times 7 - 5 \times u = 2 \times 10$ ①②

$35 - 5u = 20$ ③ −20

$15 - 5u = 0$ +5u

$15 = 5u$ ÷5

$3 = u$

Exercise 6.1A

1 The formula for the surface area of a sphere is $S = 4\pi r^2$ where r is the radius of the sphere.

    a The radius of a table-tennis ball is 20 mm. Show that the surface area of the ball is about 5027 mm²

    b The radius of the Moon is 1737 km. Find the surface area of the Moon.

2 Here are some facts about energy

Fact	Formula
Gravitational potential energy = mass × gravity × height	$G = mgh$
Kinetic energy = $\frac{1}{2}$ × mass × velocity squared	$K = \frac{1}{2}mv^2$
Work done = Force × distance	$W = Fd$
Work done = Energy transferred	$W = E$
Power = Energy ÷ time	$P = \dfrac{E}{t}$

    a Calculate W when $F = 20$ and $d = 10$

    b Calculate the value of Power when Energy = 1200 and time = 240

    c Work out the value of E if $F = 35$ and $d = 6$

    d Calculate gravitational potential energy when $m = 0.24$, $g = 9.8$ and $h = 10.5$

    e Calculate kinetic energy when mass is 0.24 kg and velocity is −3.5 m/s²

3 Speed is defined as distance divided by time.

    a Write a formula connecting speed, s, distance, d, and time, t.

    b Find s when $d = 150$ miles, $t = 3$ hours.

    c Find s when $d = 60$ km, $t = 0.5$ hours.

    d Find d when $s = 30$ km/hr, $t = 20$ mins.

    e Find t when $d = 64$ m, $s = 8$ m/min.

4 Here are some facts about electricity.

Fact	Formula
Power = Voltage × Current	$P = VI$
Voltage = Current × Resistance	$V = IR$
Charge = Current × Time	$Q = It$
Energy = Voltage × Charge	$E = VQ$

    a Calculate V when $I = 5$ and $R = 0.2$

    b Calculate E when $V = 50$ and $Q = 4$

    c Work out the value of Power when Voltage = 240 and Current = 25

    d Work out the value of Current when Time = 5 and Charge = 60

5 Ellie uses the formula
$$s = ut + \frac{1}{2}at^2$$
to calculate s when $u = 0$, $t = 2$ and $a = 3$. This is her working. Ellie has made three mistakes. Find her mistakes and the correct answer for s.

$s = 0 \times 3 + \dfrac{1}{2} \times 2 \times 3^2$

$\quad = 3 + \dfrac{1}{2} \times 6^2$

$\quad = 3 + \dfrac{1}{2} \times 36$

$\quad = 3 + 18$

$s = 21$

6 An amount of money, A, is invested in a bank. The rate of compound interest per year is $x\%$. The formula shows the value of the investment, T, after n years.

$$T = A\left(1 + \frac{x}{100}\right)^n$$

    a Melvin invests £12 000 at an interest rate of 3%. What is the value of his investment after

        i 5 years   **ii** 10 years   **iii** 25 years?

    b If the bank had offered simple interest, the formula for T would have been

$$T = A\left(1 + \frac{nx}{100}\right)$$

    What is the value of the investment at 25 years if simple interest is offered?

6.2 Using standard formulae

- A **formula** is an equation showing the relationship between two or more **variables**.

The perimeter of a rectangle is given by the formula

$$P = 2(a + b)$$

P is the **subject** of the formula for the perimeter of a rectangle. You can rearrange a formula to make other variables the subject.

An equivalent formula with a as the subject is $a = \dfrac{P}{2} - b$

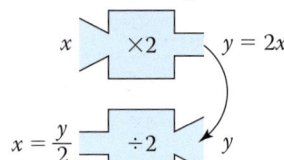

EXAMPLE

Rearrange these formulae to make x the subject.

a $A = x + 2$ **b** $T = x - 3$ **c** $M = 4x$ **d** $P = \dfrac{x}{5}$

> You must do the same to both sides of an equation to keep it balanced.
>
> **a** $A - 2 = x$ **b** $T + 3 = x$ **c** $\dfrac{M}{4} = x$ **d** $5P = x$
>
>   Subtract 2   Add 3   Divide by 4   Multiply by 5
>
> $x = A - 2$ $x = T + 3$ $x = \dfrac{M}{4}$ $x = 5P$

You can think of a function as a machine for converting an input value to an output value for the subject.

Inverse operations come in pairs.
addition ↔ subtraction
multiplication ↔ division

$x \rightarrow \boxed{\times 2} \rightarrow y = 2x$

$x \rightarrow \boxed{\div 3} \rightarrow \boxed{+5} \rightarrow y = \dfrac{x}{3} + 5$

$x = \dfrac{y}{2} \leftarrow \boxed{\div 2} \leftarrow y$

$x = 3 \times (y - 5) \leftarrow \boxed{\times 3} \leftarrow \boxed{-5} \leftarrow y$

To make the input the subject of the formula, run the function machine in reverse and use inverse operations.

- To make an input the subject of a formula, apply inverse operations in reverse order.

EXAMPLE

Rearrange these formulae to make p the subject.

a $A = 2p + d$ **b** $T = mp - c$ **c** $Q = \dfrac{p}{4} + 2$ **d** $M = \dfrac{p + b}{a}$

> $p \rightarrow \times 2 \rightarrow + d \rightarrow A$ $p \rightarrow \times m \rightarrow - c \rightarrow T$ $p \rightarrow \div 4 \rightarrow + 2 \rightarrow Q$ $p \rightarrow + b \rightarrow \div a \rightarrow M$
>
> **a** $A - d = 2p$ $-d$ **b** $T + c = mp$ $+c$ **c** $Q - 2 = \dfrac{p}{4}$ -2 **d** $aM = p + b$ $\times a$
>
> $\dfrac{A - d}{2} = p$ $\div 2$ $\dfrac{T + c}{m} = p$ $\div m$ $4(Q - 2) = p$ $\times 4$ $aM - b = p$ $-b$
>
> $p = \dfrac{A - d}{2}$ $p = \dfrac{T + c}{m}$ $p = 4(Q - 2)$ $p = aM - b$

Exercise 6.2S

1 Write down the inverse operation to these operations.

  a $+2$ **b** -3 **c** $\div 7$ **d** $\times 6$

  e $\times -2$ **f** $+6$ **g** -1 **h** $\div 5$

2 Rearrange these formulae to make a the subject.

  a $H = a + 2$ **b** $H = a - 2$

  c $H = 2a$ **d** $H = \dfrac{a}{2}$

3 Rearrange each of these formulae to make x the subject.

  a $y = x + 3$ **b** $y = x + c$

  c $y = x - 4$ **d** $y = x - c$

  e $y = 5x$ **f** $y = xc$

  g $y = \dfrac{x}{6}$ **h** $y = \dfrac{x}{c}$

4 Write down the sequence of inverse operations that reverse these sequences of operations.

  a $\times 2 \rightarrow +1$ **b** $+3 \rightarrow \div 4$

  c $-1 \rightarrow \times 4$ **d** $\div 6 \rightarrow +2$

  e $\times 5 \rightarrow -3$ **f** $+1 \rightarrow \times \dfrac{1}{2}$

  g $-2 \rightarrow \div \dfrac{1}{3}$ **h** $\div 4 \rightarrow -3$

5 Rearrange each of the formulae to make p the subject.

  a $A = 2p + 5$ **b** $A = 2p + d$

  c $A = 3p - 6$ **d** $A = 3p - d$

  e $H = 4pd$ **f** $H = apd$

  g $H = \dfrac{p}{4} + t$ **h** $H = \dfrac{p + t}{4}$

  i $W = \dfrac{p}{b} - t$ **j** $M = \dfrac{p - t}{b}$

6 Rearrange each of these formulae to make b the subject.

  a $H = b + a^2$ **b** $A = b - a^2$

  c $Q = 2b + a^2$ **d** $F = 2b - a^2$

  e $M = p^2 + b$ **f** $L = bp^2$

  g $T = a^2 b$ **h** $W = \dfrac{b}{a^2}$

7 **i** Draw function machines to represent these formulae.

  ii Hence or otherwise make x the subject of the formulae.

7 **a** $y = 2 - x$ **b** $y = \dfrac{3}{x}$

  c $y = c - mx$ **d** $y = \dfrac{k}{x}$

8 Mo is rearranging the formula $y = 10 - x$ to make x the subject of the formula.

  He thinks the answer is $x = y - 10$.

  Do you agree with Mo? Explain your answer.

9 Rearrange each of these formulae to make x the subject.

  a $y = 5 - x$ **b** $y = 20 - x$

  c $y = m - x$ **d** $y = 2b - x$

  e $y = s^2 - x$ **f** $y = \sqrt{p} - x$

10 Rearrange this formula to make T the subject.

  $$S = \dfrac{D}{T}$$

11 Rearrange each of these formulae to make m the subject.

  a $A = \dfrac{5}{m}$ **b** $y = \dfrac{20}{m}$

  c $A = \dfrac{x}{m}$ **d** $H = \dfrac{2b}{m}$

  e $L = \dfrac{t}{2m}$ **f** $y = \dfrac{\sqrt{p}}{m}$

12 Rearrange this formula to make y the subject.

  $$A = \dfrac{2}{y} + 3$$

13 Rearrange each of these formulae to make x the subject.

  a $y = \sqrt{x}$ **b** $y = \sqrt{x} + 2$

  c $y = \dfrac{\sqrt{x}}{3}$ **d** $y = 2b\sqrt{x}$

  e $A = x^2$ **f** $P = x^2 + 2$

  g $H = \dfrac{x^2}{3} + 2$ **h** $T = \dfrac{x^2 + 2}{5}$

*14 Make x the subject of this formula.
  $px + q = rx + s$

15 The time taken for a pendulum to complete one cycle can be calculated using the formula

  $$T = 2\pi \sqrt{\dfrac{L}{g}}$$

  where L is the length of the pendulum and g is the acceleration due to gravity.

  Make L the subject of the formula.

Q 1159, 1171 SEARCH

6.2 Using standard formulae

- The variable on the left of the equals sign is called the **subject** of the formula.
- Formulae can be rearranged to make other variables the subject of the formula.
  - Addition and subtraction are inverses of each other.
  - Multiplication and division are inverses of each other.
  - 'Square' and 'square root' are inverses of each other.

Inverse operations

$+ \leftrightarrow -$

$\times \leftrightarrow \div$

$^2 \leftrightarrow \sqrt{}$

$y = 2x + 3$

$y - 3 = 2x$

$\dfrac{y - 3}{2} = x$

To change the subject of a formula

① Identify the operations that are applied to the new subject and their order.

② Apply the inverse operations in reverse order.

③ Stop when only the new subject appears on one side.

A car mechanic has a £45 callout charge plus £40 for every hour he works.

a Write a formula for the number of hours worked as a function of the total fee.

b How long did a job take if the fee was £245?

a Start by creating a formula for the fee given the hours worked.

Let F = fee   and h = hours worked

F = 'callout charge' + 40 × 'hours worked'

$= 45 + 40h$ ① Order of operations: $h \rightarrow \times 40 \rightarrow + 45$

Change the subject to h.

$F - 45 = 40h$ -45 ② Reversed, inverse operations: $F \rightarrow -45 \rightarrow \div 40$

$\dfrac{F - 45}{40} = h$ $\div 45$

$h = \dfrac{F - 45}{40}$ ③ Write with subject on the left.

b $h = \dfrac{245 - 45}{40} = \dfrac{200}{40} = 5$ hours

The kinetic energy, K, of an object with mass, m, moving at speed, v, is given by $K = \dfrac{1}{2}mv^2$.

Write a formula for v given K and m.

$K = \dfrac{1}{2}mv^2$ ① $v \rightarrow ^2 \rightarrow \times m \rightarrow \times \dfrac{1}{2}$

$2K = mv^2$ ② $\div \dfrac{1}{2}$, this is the same as $\times 2$.

$\dfrac{2K}{m} = v^2$ $\div m$

$\sqrt{\dfrac{2K}{m}} = v$ $\sqrt{}$

$v = \sqrt{\dfrac{2K}{m}}$ ③

Use BIDMAS to help you decide in what order the operations are applied.

Exercise 6.2A

1 The cost, C, of an electricity bill is calculated from the number of units used, u, using this formula.

$$C = 17.5 + 0.1u$$

How many units were used if the cost is

a £37.50 **b** £32.50 **c** £74.50

2 Here are some facts about forces and motion.

Facts	Formulae
Acceleration = Change in velocity ÷ Time	$a = \dfrac{v - u}{t}$
Change in momentum = Force × Time	$mv - mu = Ft$
Force = Mass × Acceleration	$F = ma$
Momentum = Mass × Velocity	$P = mv$
Pressure = Force ÷ Area	$P = \dfrac{F}{A}$

a Rearrange $mv - mu = Ft$ to make t the subject.

b Write a formula for Acceleration in terms of Force and Mass.

c Write a formula for Force in terms of Pressure and Area.

d Write a formula for Area in terms of Pressure and Force.

e Rearrange $a = \dfrac{v - u}{t}$ to make v the subject.

*f Rearrange $mv - mu = Ft$ to make m the subject.

3 This is the formula for finding the density of an object is.

$$\text{Density} = \frac{\text{Mass}}{\text{Volume}}$$

a Write a formula for finding mass when volume and density are known.

b Write a formula for volume in terms of density and mass.

4 The equation of a straight line is $ax + by = c$. Rearrange the formula to make y the subject.

5 The volume of a cylinder is given by the formula $V = \pi r^2 h$. Rearrange the formula to change the subject to

a h **b** r.

6 The formula for the surface area of a sphere is $S = 4\pi r^2$ where r is the radius of the sphere. Rearrange the formula to find an expression for π.

7 Pythagoras' theorem states that $c^2 = a^2 + b^2$. Rearrange this formula to make b the subject.

8 Freddie and Asha have both rearranged the formula $c = a(x - b)$ to make x the subject. They have both found correct answers.

Freddie
$x = \dfrac{c}{a} + b$

Asha
$x = \dfrac{c + ab}{a}$

Describe how they got their answers and why they are equal.

9 The formula $F = \dfrac{9}{5}C + 32$ to convert a temperature in degrees Celsius, C, to a temperature in degrees Fahrenheit, F.

a Create a spreadsheet and use the formula functionality to return the value of F in cell B1 when you enter the value of C in cell A1.

b Rearrange the formula to make C the subject.

c Modify the spreadsheet to return the value of C in cell C1 based on the value of F in cell B1.

d Choose values to enter for C in cell A1. Check that cells A1 and C1 always display the same value.

6.3 Equations, identities and functions

You need to remember some key vocabulary used in algebra.

Equation	An expression containing *one* = sign and at least one unknown. It is only true for *particular* values of the unknown.	$5x - 7 = 13$ $x = 4$
Identity	An equation that is true for *every* value of the unknown(s). The = sign is usually replaced by $\equiv$ for 'identically equal to'.	$2x + 6 + 3x - 13$ $\equiv 5x - 7$
Formula	An equation involving several variables; often associated with a real application.	$E = mc^2$
Inequality	A statement relating two expressions that are *not* equal.	$5x - 7 > 13$

p.26

> You use the symbols $>, \geqslant, \leqslant$ and $\leqslant$ in inequalities.

EXAMPLE

Use one of these words to describe these expressions.

Equation	Identity
Formula	Inequality

a $\quad 3x - 2 = 5$

b $\quad 4x - 2 \leqslant 5$

c $\quad x(x + 2) \equiv x^2 + 2x$

d $\quad F = \dfrac{9}{5}C + 32$

a Contains an = sign. How many values of x is it true for?

$3x - 2 = 5 \implies 3x = 7 \implies x = 2\dfrac{1}{3}$

Equation — Only true for one value of x.

b Inequality — Contains one of the inequality symbols: $<, \leqslant, \geqslant, >$.

c Contains an $\equiv$ sign. How many values of x is it true for?

$x(x + 2) = x \times x + x \times 2$

$\qquad\qquad = x^2 + 2x$

Identity — True for all values of x.

d Formula — It is the formula for converting temperatures in Celsius to temperatures in Fahrenheit.

EXAMPLE

Show that these identities are true.

a $\quad 4p + 6p \equiv 10p$

b $\quad 4x + 5y - 2x \equiv 2x + 5y$

a $\quad 4p + 6p \equiv (p + p + p + p) + (p + p + p + p + p + p)$

$\qquad\qquad\qquad \equiv 10p$

b $\quad 4x + 5y - 2x \equiv (4 - 2)x + 5y$ Collect like terms.

$\qquad\qquad\qquad = 2x + 5y$

EXAMPLE

Show that these identities are true.

a $\quad 4(x + 2) \equiv 4x + 8$

b $\quad 2(3 + 4x) + 3(2 - 5x) \equiv 12 - 7x$

a $\quad 4(x + 2) \equiv 4 \times x + 4 \times 2$

$\qquad\qquad \equiv 4x + 8$

$\times$	x	$+2$
4	$4x$	$+8$

b $\quad 2(3 + 4x) + 3(2 - 5x)$

$\qquad \equiv 6 + 8x + 6 - 15x$

$\qquad \equiv 12 - 7x$ Collect like terms.

Exercise 6.3S

1 Match the correct entries.

$x + 3 < 10$
$2x + 3x \equiv 5x$
$2x + 1 = 6$
$v = u + at$

Equation
Formula
Identity
Inequality

2 Write an example of

    a an equation   **b** a formula

    c an inequality   **d** an identity.

3 Are these identities true or false? Give your reasons.

    a $4a + 5b \equiv 9ab$

    b $3p + 3p \equiv 6p^2$

    c $4a + 5a \equiv 9a$

    d $10p - 5 \equiv 5p$

    e $4a + 2b + 3a \equiv 7a + 2b$

    f $3d + 2e + 3e \equiv 8de$

    g $6a + 2b - 4a \equiv 2a + 2b$

4 Complete these identities by collecting like terms.

    a $2a + 7a \equiv \ldots\ldots\ldots$

    b $2p + 3p \equiv \ldots\ldots\ldots$

    c $10a - 4a \equiv \ldots\ldots\ldots$

    d $10p + 2p - 3p \equiv \ldots\ldots\ldots$

    e $4a^2 + 3a^2 \equiv \ldots\ldots\ldots$

    f $a + 5b + 3a \equiv \ldots\ldots\ldots$

    g $3d - 2e + 3e + 5d \equiv \ldots\ldots\ldots$

    h $\square + \square - \square \equiv 2a + 2b$

5 Are these identities true or false? Give your reasons.

    a $4(a + 2) \equiv 4a + 2$

    b $3(x + 2) \equiv 3x + 6$

    c $5(y - 2) \equiv 5y - 10$

    d $y(y + 3) \equiv 2y + 3$

    e $x(x - 4) \equiv x^2 - 4x$

6 Show that these are all identities.

    a $5a + 10 \equiv 5(a + 2)$

    b $3x + 12 \equiv 3(x + 4)$

    c $5y - 15 \equiv 5(y - 3)$

    d $y^2 + 3y \equiv y(y + 3)$

    e $x^2 - 4 \equiv (x + 2)(x - 2)$

7 Show that these are identities.

    a $4(a + 2) + 2(a + 1) \equiv 6a + 10$

    b $3(x + 2) + 4(x - 1) \equiv 7x + 2$

    c $5(y - 2) + 3(y - 3) \equiv 8y - 19$

    d $y(y + 3) + 2(y + 3) \equiv y^2 + 5y + 6$

    e $x(x - 4) + x(x + 2) \equiv 2x^2 - 2x$

8 Find the values of a and b such that these identities are true.

    a $2(x + 2) + 5(x + 1) \equiv ax + b$

    b $3(x - 2) + 4(x + 3) \equiv ax + b$

    c $5(y + a) + 3(y - b) \equiv 8y - 19$

    d $y(y + a) + 2(y + b) \equiv y^2 + 3y + 4$

    e $x(x - 4) + 2x(x - 3) \equiv ax^2 - bx$

9 Classify these expressions as equations, identities or formulae.

    a $V = \dfrac{4}{3} \pi r^3$

    b $x + yz = yz + x$

    c $3w = 4(w - 2) + 8$

    d $a + bc = d$

    e $y = x - 1$

    f $x^n x^m = x^{n+m}$

    g $20 - p = -(p - 20)$

    h $a^2 + b^2 = c^2$

    i $s = ut + \dfrac{1}{2} at^2$

    j $3(x - 2)^2 = 27$

    k $4a + 2(a - 3) = 6(a - 1)$

    l $3p + 2(4 - p) = 8$

10 The perimeter of a rectangle can be found using the formula $P = 2a + 2b$ or $P = 2(a + b)$. Show that these formulae are identical.

6.3 Equations, identities and functions

- An equation is a statement that the left hand side equals the right hand side.
- A formula is an equation conecting two or more variables.
- An inequality is a statement about two quantities that are different.
- An identity is a statement that is true for all values of the variables used.

Equation
$2(x + 3) = 10$ $(x = 2)$
Identity
$2(x + 3) \equiv 2x + 6$
Inequality
$2(x + 3) > 0$
Formula
$PV = nRT$

HOW TO

To solve problems involving equations and identities
① Consider the information given in the problem.
② Use algebra to create statements about the problem.
③ Make a final conclusion using correct symbols.

EXAMPLE

p.144

This rectangle is divided into two smaller rectangles.

a Use the diagram to create an identity.

b The area of the whole rectangle is 88 square units. Use this information to create an equation.

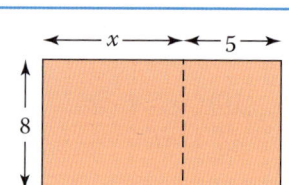

① Calculate the area as a whole and as a sum of parts.

a Area of whole rectangle = $(x + 5) \times 8$
$= 8(x + 5)$

② Area = length × width.

Areas of small rectangles = $x \times 8 + 5 \times 8$
$= 8x + 40$

$8(x + 5) = 8x + 40$ ③

b $8x + 40 = 88$ ÷ 8
$x + 5 = 11$

EXAMPLE

Complete these expressions to create identities.

a $\boxed{}(x + \boxed{}) \equiv 7x + 21$

b $6(5x - \boxed{}) + 4x(\boxed{}x + 5) \equiv 8x^2 + \boxed{}x - 42$

① Label each missing number. Expand each bracket and compare terms.

a $\boxed{1}(x + \boxed{2}) \equiv \boxed{1}x + \boxed{1} \times \boxed{2}$ ②
$\equiv 7x + 21$
$\boxed{1} = 7$ Coefficient of x.
$\boxed{1} \times \boxed{2} = 21$ Constant term.
$7 \times \boxed{2} = 21$
$\boxed{2} = 3$
③ $7(x + 3) \equiv 7x + 21$

b $6(5x - \boxed{1}) + 4x (\boxed{2}x + 5) \equiv 8x^2 + \boxed{3}x - 42$
$30x - 6 \times \boxed{1} + 4 \times \boxed{2} \times x^2 + 20x \equiv 8x^2 + \boxed{3}x - 42$ ②
$4 \times \boxed{2} = 8$ Coefficient of x^2.
$\boxed{2} = 2$
$30 + 20 = \boxed{3}$ Coefficient of x.
$\boxed{3} = 50$
$-6 \times \boxed{1} = -42$ Constant term.
$\boxed{1} = 7$
③ $6(5x - 7) + 4x(2x + 5) \equiv 8x^2 + 50x - 42$

Exercise 6.3A

1 For each of the following diagrams

   i create an identity

   ii write an equation.

a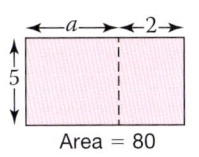

b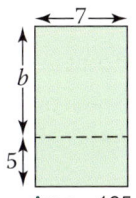

c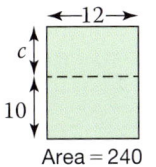

d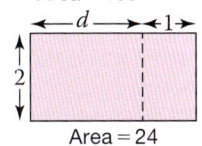

e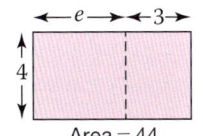

f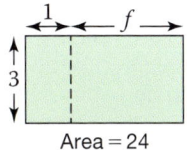

2 Create an identity using these rectangles.

a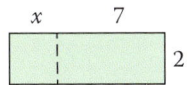

b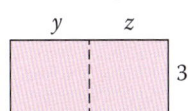

c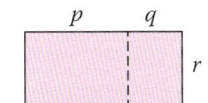

d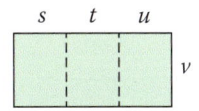

3 Create an identity using each of these diagrams.

a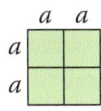

b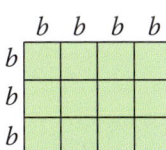

4 Copy and complete these expressions to create identities.

   a $3(x + 4) \equiv 3x + \square$

   b $9(x - 6) \equiv \square x - 54$

   c $6(x - 12) \equiv \square x - \square$

   d $\square(x + 4) \equiv \square x + 16$

   e $3(x + \square) \equiv \square x + 33$

   f $\square(x + \square) \equiv 5x + 35$

5 Copy and complete these expressions to create identities.

   a $x(2x + \square) \equiv \square x^2 + 6x$

   b $\square x(3x + 2) \equiv 12x^2 + \square x$

   c $\square x(7x - \square) \equiv 14x^2 - 16x$

   d $\square x(\square x - 5) \equiv 16x^2 - 20x$

   e $6(2x - \square) + 3(\square x - 4)$
       $\equiv 15x - 24$

   f $7(7x - \square) + 2x(\square x + 9)$
       $\equiv 6x^2 + \square x - 56$

6 Decide whether each of the following statements is always true, sometimes true or never true.

   a An equation is a formula.

   b A formula is an equation.

   c A formula is a function.

   d A function is a formula.

   e An inequality is an equation.

   f An inequality is a function.

   g An equation uses the letter x.

   h A formula is written using algebra.

*7 For each of the following diagrams

   i write an inequality.

   ii create an identity.

a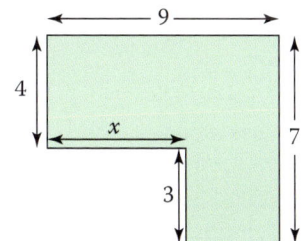

b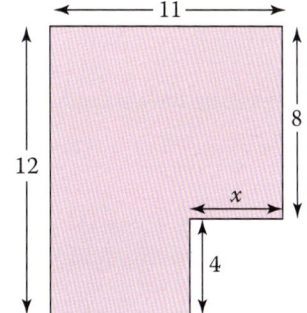

6.4 Expanding and factorising 2

Quadratic expressions can be **expanded** by multiplying two **binomials** or **factorised** by finding common factors.

- Expanding quadratics involves removing brackets to create an equivalent expression.
- Factorising quadratics involves finding common factors and adding brackets.

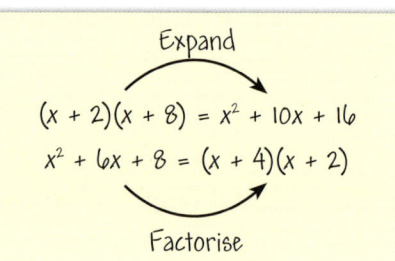

Expand
$(x + 2)(x + 8) = x^2 + 10x + 16$
$x^2 + 6x + 8 = (x + 4)(x + 2)$
Factorise

Expanding is the opposite of factorising.

EXAMPLE

Expand and simplify these double brackets.

a $(x + 5)(x + 2)$ **b** $(x + 5)(x - 2)$ **c** $(x - 5)(x - 2)$

Each term in the first bracket is multiplied by each term in the second bracket.

a $(x + 5)(x + 2)$ **b** $(x + 5)(x - 2)$ **c** $(x - 5)(x - 2)$

$= x^2 \boxed{+ 5x} \boxed{+ 2x} + 10$ $= x^2 \boxed{+ 5x} \boxed{- 2x} - 10$ $= x^2 \boxed{- 5x} \boxed{- 2x} + 10$

$= x^2 + 7x + 10$ $= x^2 + 3x - 10$ $= x^2 - 7x + 10$

Collect like terms to simplify the answer.

There are several ways to help you remember what to do.

a Use a grid.

×	x	$+5$
x	x^2	$+5x$
$+2$	$+2x$	$+10$

b Use FOIL.

First $x \times x$
Outer $x \times -2$
Inner $5 \times x$
Last 5×-2

c Use a smiley face.

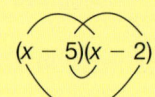

$(x - 5)(x - 2)$

Be careful with negative numbers.

Factorising is the opposite of expanding.

EXAMPLE

Factorise completely these quadratic expressions.

a $x^2 + 11x + 18$ **b** $y^2 - 3y - 10$ **c** $a^2 - 9a + 20$

a $x^2 + 11x + 18$ **b** $y^2 - 3y - 10$ **c** $a^2 - 9a + 20$

$= (x + 9)(x + 2)$ $= (y - 5)(y + 2)$ $= (a - 5)(a - 4)$

You can always check your factorisation by expanding out the brackets.

a You can use a partially filled grid to help you find the factors.

×	x	?
x	x^2	
?		$+18$

Test a list of factors.

1, 18	$\Rightarrow$ 19	✗
$-1, -18$	-19	✗
2, 9	11	✓
$-2, -9$	-11	✗
3, 6	9	✗
$-3, -6$	-9	✗

Include negative numbers in the pairs of factors.

What two numbers have a product of 18 and a sum of 11?

EXAMPLE

Factorise $x^2 - 4$

$x^2 - 4 = (x + 2)(x - 2)$

Check by expanding
$x^2 + 2x - 2x - 4$
$= x^2 - 4$

This is an example of a special type of factorisation known as the 'difference of two squares'.

Algebra   Formulae and functions

Exercise 6.4S

1 Copy and complete these statements.

  a $(x + 6)(x + 2) = x^2 + \square x + \square$

  b $(x + 5)(x + 3) = x^2 + \square x + \square$

  c $(x + 8)(x + 2) = x^2 + \square x + \square$

2 Expand and simplify these expressions.

  a $(x + 3)(x + 6)$ **b** $(x + 4)(x + 3)$

  c $(y + 2)(y + 7)$ **d** $(y + 6)(y + 4)$

  e $(a + 3)(a + 2)$ **f** $(a + 7)(a + 6)$

3 Reece is expanding $(2x + 3)(x + 5)$ using a grid method.

×	x	$+5$
$2x$	$2x^2$	$+5x$
$+3$	$+3x$	$+8$

$(2x + 3)(x + 5) =$
$2x^2 + 3x + 5x + 8$
$= 2x^2 + 11x + 8$

Do you agree with Reece's answer? Give your reasons.

4 Copy and complete these statements.

  a $(x + 6)(x - 2) = x^2 + \square x - \square$

  b $(x + 5)(x - 3) = x^2 + \square x - \square$

  c $(x - 2)(x + 8) = x^2 + \square x - \square$

  d $(x - 6)(x + 2) = x^2 - \square x - \square$

  e $(x - 5)(x + 3) = x^2 - \square x - \square$

  f $(x - 8)(x + 2) = x^2 - \square x - \square$

5 Expand and simplify these expressions.

  a $(x + 5)(x - 2)$ **b** $(x + 4)(x - 3)$

  c $(y + 2)(y - 7)$ **d** $(y - 6)(y + 4)$

  e $(a - 3)(a + 2)$ **f** $(a - 7)(a + 6)$

6 Copy and complete these statements.

  a $(x - 6)(x - 2) = x^2 - \square x + \square$

  b $(x - 5)(x - 3) = x^2 - \square x + \square$

  c $(x - 8)(x - 2) = x^2 - \square x + \square$

7 Expand and simplify these expressions.

  a $(x - 5)(x - 2)$ **b** $(x - 4)(x - 3)$

  c $(y - 2)(y - 7)$ **d** $(y - 6)(y - 4)$

  e $(a - 3)(a - 2)$ **f** $(a - 7)(a - 6)$

8 Expand and simplify these expressions.

  a $(x + 3)^2$ **b** $(y + 2)^2$

  c $(y - 4)^2$ **d** $(x - 1)^2$

9 Copy and complete this statement.

  $\square(x + \square) = x^2 + 12x$

10 Factorise completely these quadratic expressions.

  a $x^2 + 5x$ **b** $x^2 + 7x$

  c $2x^2 + 12x$ **d** $12x^2 + 6x$

11 Suki is expanding two brackets. The answer is $x^2 + 12x + 20$.

  What is the question Suki is working on?

12 Factorise completely these quadratic expressions.

  a $x^2 + 5x + 6$ **b** $x^2 + 7x + 10$

  c $y^2 + 7y + 12$ **d** $y^2 + 8y + 15$

  e $a^2 + 11a + 24$ **f** $a^2 + 13a + 12$

  g $x^2 + 20x + 91$ **h** $x^2 + 20x + 9$

13 Factorise completely these expressions.

  a $x^2 + x - 6$ **b** $x^2 + 9x - 22$

  c $y^2 + y - 12$ **d** $y^2 + 2y - 15$

  e $x^2 - 10x - 24$ **f** $x^2 + 4x - 10$

14 Factorise completely these expressions.

  a $x^2 - 6x + 8$ **b** $x^2 - 8x + 12$

  c $y^2 - 9y + 18$ **d** $y^2 - 12y + 27$

  e $p^2 - 10p + 24$ **f** $p^2 - 14a + 13$

  g $x^2 - 24x + 23$ **h** $x^2 - x + 10$

15 Factorise completely these expressions.

  a $x^2 + 16x + 48$ **b** $y^2 - 2y - 48$

  c $b^2 + 2b - 48$ **d** $a^2 - 26a + 48$

  e $x^2 + 28x - 60$ **f** $x^2 - 19x + 60$

16 Factorise completely these expressions.

  a $x^2 + 6x + 9$ **b** $y^2 - 6y + 9$

  c $y^2 - 4y + 4$ **d** $x^2 + 8x + 16$

17 Factorise these quadratic expressions.

  a $x^2 - 9$ **b** $y^2 - 25$

  c $b^2 - 100$ **d** $h^2 - 81$

  e $y^2 - 64$ **f** $a^2 - 225$

  g $x^2 - 60$ **h** $x^2 + 36$

*18 Expand and simplify these expressions.

  a $(2x + 1)(x + 3)$ **b** $(x - 2)(3x + 1)$

  c $(4x + 1)(3x + 1)$ **d** $(2x - 3)(3x + 2)$

1150, 1151, 1157   SEARCH

6.4 Expanding and factorising 2

RECAP

- To expand double brackets, multiply each term in the first bracket by each term in the second bracket.
- To factorise an expression you reverse the process and rewrite it using brackets.

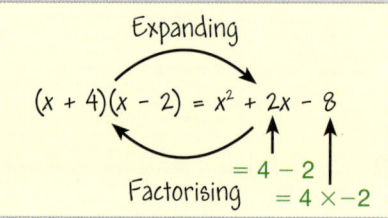

HOW TO

To solve problem involving quadratic expressions

1. RTQ and decide whether you need to expand or factorise.
2. If expanding, remember to collect like terms.
   If factorising, check your answer by expanding.
3. ATQ

Since quadratic expressions involve squared terms they can be related to area problems.

EXAMPLE

This rectangle has been split into four smaller rectangles.

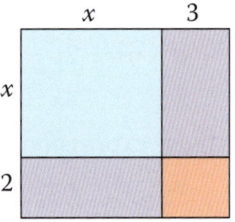

p.144

Use the diagram to justify this algebraic identity.
$(x + 3)(x + 2) \equiv x^2 + 5x + 6$

1. Calculate the total area two ways: as a whole or as a sum of its parts.
   Area of rectangle = length × width
2. Whole area $= (x + 3) \times (x + 2)$
   $= (x + 3)(x + 2)$

   Blue area $= x \times x = x^2$

   Purple area $= 3 \times x + x \times 2 = 3x + 2x = 5x$

   Red area $= 3 \times 2 = 6$

   Total area $= x^2 + 5x + 6$
3. Both calculations must give the same area.
   $(x + 3)(x + 2) = x^2 + 5x + 6$

EXAMPLE

a Prove this identity. $(a + b)(a - b) \equiv a^2 - b^2$

b Hence or otherwise evaluate these expressions *without* using a calculator.

i 107×93 **ii** $78^2 - 22^2$

a $(a + b)(a - b) \equiv a \times a - \cancel{a \times b} + \cancel{b \times a} - b \times b$ 2 Using FOIL.
$\equiv a^2 - b^2$

b **i** $107 \times 93 = (100 + 7)(100 - 7)$ **ii** $78^2 - 22^2 = (78 + 22)(78 - 22)$
$= 100^2 - 7^2$ $= 100 \times 56$
$= 10\,000 - 49$ $= 5600$ 3
$= 9951$ 3

Exercise 6.4A

1 Use these diagrams to justify the given algebraic identity.

a
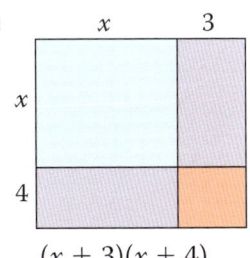
$(x + 3)(x + 4)$
$= x^2 + 7x + 12$

b
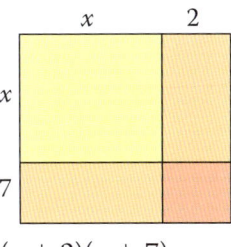
$(x + 2)(x + 7)$
$= x^2 + 9x + 14$

c
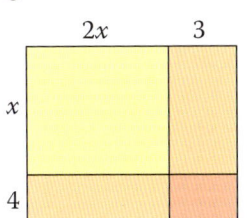

d
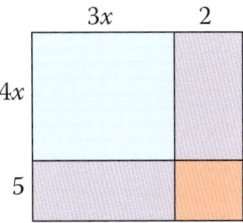

2 Derive an algebraic identity using these diagrams.

a
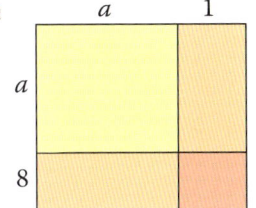

b
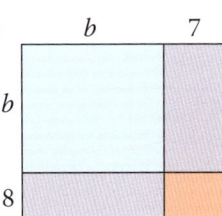

c
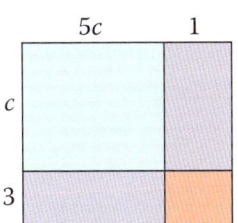

d
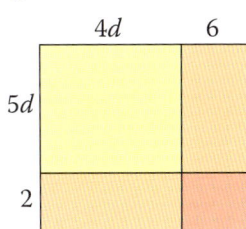

3 *Without* using a calculator, work out the value of these expressions.

a $67^2 - 33^2$ **b** $34^2 - 26^2$

c $61^2 - 59^2$ **d** $55^2 - 45^2$

e 62×58 **f** 47×33

g 47×53 **h** 54×46

4 **a** By expanding $(a + b)^2$ create an identity.

    b Hence or otherwise, work out the value of these expressions *without* using a calculator.

       i $132^2 + 2 \times 132 \times 268 + 268^2$

       ii $2.1^2 + 2 \times 2.1 \times 1.1 + 1.1^2$

       iii $79^2 - 2 \times 79 \times 19 + 19^2$

5 Jason is trying to factorise $x^2 + 5x + 6$. He says, '1 and 5 multiply to give 5 and 1 and 5 add to give 6 so $x^2 + 5x + 6 = (x + 1)(x + 5)$.'

    a Jason is wrong, say why.

    b Give the correct factorisation.

6 Copy these expressions, filling in the missing values.

    a $(x + \square)(x + 6) = x^2 + 9x + \square$

    b $(x + 4)(x - \square) = x^2 + \square x - 8$

    c $(x + 1)(x + 2) + (x + 4)(x - \square)$
       $= \square x^2 + 3x - \square$

7 This rectangle has area 17.

Show that $(x - 9)(x + 9) = 0$

8 Katie arranges a set of square tiles to make a rectangle that is $x + 2$ tiles long and $x + 4$ tiles wide.
Show that adding one more tile allows Katie to make a square.

9 Copy and complete these pyramids. Each entry is the product of the two entries immediately below it.

a
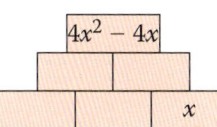
*b

*10 Simplify these algebraic fractions.

    a $\dfrac{x^2 + 3x}{x}$ **b** $\dfrac{y^2 - 7y}{y - 7}$

    c $\dfrac{y^2 - 5y + 6}{y - 3}$ **d** $\dfrac{x - 4}{x^2 + 3x - 28}$

1150, 1151, 1157   SEARCH

Summary

Checkout
You should now be able to...

Test it
Questions

	Test it Questions
✔ Substitute numerical values into formulae and expressions.	1 – 3
✔ Rearrange formulae to change the subject.	4
✔ Identify inequalities, equations, formulae and identities.	5, 6
✔ Expand double brackets.	7
✔ Factorise quadratic expressions of the form $x^2 + bx + c$ and the difference of two squares.	8 – 10

Language	Meaning	Example
Variable	A letter used to represent a number.	$3x + 4$ x is the variable.
Like terms	Terms that contain exactly the same variables and exactly the same powers.	$3x^2$ and $5x^2$ are like terms. $3x^2$ and $5x$ are not like terms.
Function machine	A diagram that indicates the order in which operations have to be done.	
Input	The value put in to a function.	$3 \rightarrow \boxed{\times 2} \rightarrow 6$ Input = 3 Operation = ×2 Output = 6
Output	The end value of a function after the operations have been applied.	
Operation	An operation is a rule for processing numbers.	
Inverse	The inverse operation reverses the effect of the original operation.	The inverse of $+4$ is -4. The inverse of $\times 2$ is $\div 2$.
Subject	The variable before the equals sign in a formula.	$V = IR$ V is the subject of the formula.
Rearrange	Rewrite an equation or formula as an equivalent version with a different variable as the subject.	$V = IR$ rearranged to make I the subject of the formula is $I = \dfrac{V}{R}$
Identity	An equation that is true for every possible value.	$\dfrac{a}{4} \equiv 0.25 \times a$
Function	A function is a rule that links each input value with one output value.	$f(x) = 2x$
Expand	Remove the brackets in an expression by multiplying.	$(3x + 1)(2x - 3) = 6x^2 - 9x + 2x - 3$ $= 6x^2 - 7x - 3$
Factorise	Find common factors in an expression and write it using brackets; the reverse of expanding.	$x^2 + 5x = x(x + 5)$ $x^2 - 6x + 5 = (x - 1)(x - 5)$
Quadratic	A quadratic expression contains a square term such as x^2 as the highest power.	$6x^2 - 7x - 3$

Review

1 This formula gives the relationship between density, D, mass, m, and volume, v.

$$D = \frac{m}{v}$$

Calculate the density when

  a $m = 15$ g and $v = 5$ cm^3

  b $m = 25$ g and $v = 2$ cm^3.

2 Use the formula $A = \frac{1}{2}bh$ to calculate A when

  a $b = 8$ and $h = 4$

  b $b = 5$ and $h = 9$.

3 Devi earns £x per hour, Jo earns twice as much as Devi and Alice earns £y less than Devi.

  a   Write an expression for the amount earned by

      i   Jo   **ii**   Alice.

  b   If Jo earns £30 per hour, how much does Devi earn per hour?

4 Rearrange each formula to make A the subject.

  a $3 + A = b$ **b** $2A = d$

  c $5A - c = F$ **d** $\frac{A + h}{2} = J$

  e $A^2 - 2K = L$ **f** $b = \frac{2}{A}$

5

> $4z + 2$ $3b \times 4b \equiv 12b^2$
>
> $F = ma$ $2y + 3 = 7$
>
> $y = 3x + 4$

Give an example of each of these from the box.

  a   Equation   **b**   Identity

  c   Expression   **d**   Formula

  e   Function

6 Use algebra to show that this identity is true.

$$2(3x + 1) - 4 \equiv 6x - 2$$

7 Expand the brackets and simplify these expressions.

  a $(x + 3)(x + 5)$ **b** $(x - 6)(x - 2)$

  c $(x - 7)(x + 4)$ **d** $(2x + 5)(3x - 1)$

8 Factorise these quadratic expressions.

  a $x^2 + 5x$ **b** $12x^2 - 3x$

9 Factorise these quadratic expressions.

  a $x^2 + 5x + 4$ **b** $x^2 - 7x + 6$

  c $x^2 - 2x - 8$ **d** $x^2 + 3x - 10$

10 Factorise these quadratic expressions.

  a $x^2 - 36$ **b** $4x^2 - 25$

What next?

Score			
	0 – 4		Your knowledge of this topic is still developing. To improve look at MyMaths: 1150, 1151, 1155, 1157, 1158, 1159, 1167, 1171, 1186, 1187, 1247
	5 – 8		You are gaining a secure knowledge of this topic. To improve your fluency look at InvisiPens: 06Sa – m
	9 – 10		You have mastered these skills. Well done you are ready to progress! To develop your problem solving skills look at InvisiPens: 06Aa – d

Assessment 6

1 a Amy uses this rule to work out how far away a thunderstorm is.
Count the number of seconds, t, between the time you see the lightning and the time you hear the thunder. Divide the number by 5 to work out the distance, m, in miles.
Write down a formula connecting m and t. [1]

 b Use your formula to find

 i how far away a thunderstorm is when $t = 15$ seconds [1]

 ii how far away a thunderstorm is when $t = 2\frac{1}{2}$ seconds [1]

 iii how long the time would be for a storm $2\frac{1}{2}$ miles away. [1]

2 Jamie uses 450 g raspberries and 550 g sugar to make 1 kg raspberry jam.

 a Write a formula connecting J, the mass of jam, R, the mass of raspberries and S, the mass of sugar. [1]

 b How much fruit does Jamie need to make 4 kg of jam? [2]

 c How much sugar does Jamie need to make 6 kg of jam? [2]

 d Jamie made 9 kg jam. How much fruit and sugar did he use? [2]

 e Jamie picked 2.25 kg raspberries from his garden and had enough sugar.
How much jam could he make? [2]

3 Mr and Mrs Perfectparents use this formula to work out how many hours of sleep, s their children need depending on their age a (years): $s = 15 - 0.75a$.

 a Find s when **i** $a = 2$ [2]   **ii** $a = 10$. [1]

 b Find a when **i** $s = 6$ [3]   **ii** $s = 9$. [1]

 c Use the formula to find out how much sleep an 18 year old needs.
Is this a sensible formula? Give reasons for your answer. [2]

4 David uses the formula $a^2 = b$. He works out

 a the value of b when $a = 7$. He says the answer is $b = 49$. [1]

 b the value of a when $b = 121$. He says the answer is $a = 11$. [2]

 Is David correct? If he is incorrect, work out the correct answer.

5 a Carlo tried to expand and simplify these expressions.
Is Carlo correct? Show your working.

 i $(p - 4)(p - 7) = p^2 + 28$ [2]   **ii** $(v + 9)(v - 7) + (4 - 5v)^2 = 6v^2 - 47$ [5]

 b Carlo then tried to factorise these expressions.
Is Carlo correct? Show your working.

 i $z^2 + 13z + 36 = (z + 4)(z + 9)$ [2]   **ii** $v^2 - 100 = (v - 10)^2$ [2]

6 a Write down a formula to find the number of days, D, there are in W weeks. [1]

 b Robert the builder's digger is broken and he needs to hire a new one.
The hire firm charges a fixed charge of £50 and £250 for each day Bob uses it.
Write a formula to show how much it costs Bob to hire a digger for D days. [1]

7 'TextUnending' has a pay as you go phone that charges
- 15p per minute between peak times of 09:00 and 20:00
- 12p per minute at other times, called off-peak
- 14p per text at all times.

'TextUnending' rounds all extra seconds up to the next minute.

a Using the letters P for the cost (in pence) and m for the time in minutes, write down a formula to work out the cost of

    i a call at peak times [1]   **ii** a call at off-peak times [1]

    iii making n texts. [1]

b **i** Hugo made a call for 3 min 20 s at 13:15 and sent 8 text messages.
How much was he charged? [4]

    ii Albert made one call for 6 min 10 s at 08:01 and another for 14 min 1 s at 09:01.
He sent 17 text messages in total. How much was he charged? [5]

c Cheng made 4 off-peak calls, lasting 2 min 43 s, 7 min 17 s, 23 min 42 s and 4 min 55 s.
She also sent 25 text messages. Work out how much she paid. [5]

8 Selina makes the following statements. For each statement either show that it is always true or find an example to show that it is false.

a An odd number times an even number is always odd. [1]

b Prime numbers have one factor. [1]

c The sum of two even numbers is always even. [2]

d Any number squared is more than 0. [1]

e Two prime numbers multiplied together are always odd. [1]

f The sum of three consecutive even numbers is always divisible by 6. [2]

g The square of any number is never a prime number. [2]

9 Work out the rule that turns each input in these function machines into its corresponding output.

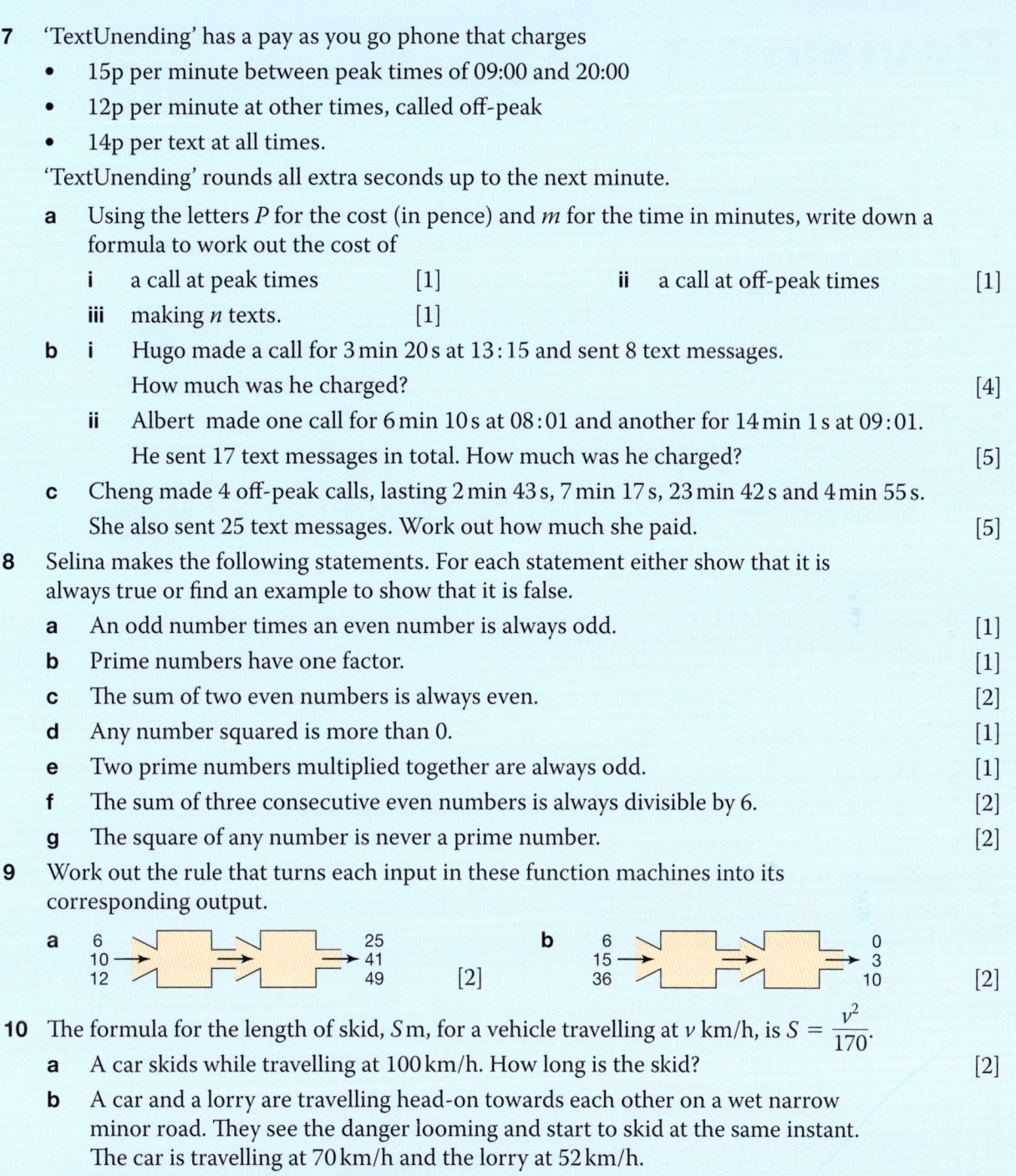

a
6
10 →
12

→ 25
→ 41
→ 49 [2]

b
6
15 →
36

→ 0
→ 3
→ 10 [2]

10 The formula for the length of skid, S m, for a vehicle travelling at v km/h, is $S = \dfrac{v^2}{170}$.

a A car skids while travelling at 100 km/h. How long is the skid? [2]

b A car and a lorry are travelling head-on towards each other on a wet narrow minor road. They see the danger looming and start to skid at the same instant.
The car is travelling at 70 km/h and the lorry at 52 km/h.
Calculate the minimum distance they were apart if they just stop in time. [4]

c Rearrange the formula to make v the subject. Hence find the speed of a vehicle which skidded for 60 m. Give your answer to the nearest km/h. [2]

Revision 1

1 The mass of 80 identical boxes is 10 kg. What is the mass of 45 of these boxes? Give your answer in kilograms. [3]

2 Jenni gets the bus to work. The fare is £3.50 for a return journey. She works Monday to Wednesday, Friday and Saturday. Would Jenni save money if she bought a weekly ticket costing £15? Give reasons for your answer. [4]

3 Oliver is finding the 'COOL' value of numbers. To find the 'COOL' value of a number

  1 square the number

  2 if the squared number has more than 1 digit, add the digits together

  3 repeat this step until you get a single digit. This is the 'COOL' value of the number.

  a Find the 'COOL' values for the numbers 9 and 28. [3]

  b There are only 4 'COOL' values for all the numbers. Find all of them. [3]

  c Which numbers are the same as their 'COOL' values? [2]

  d There are two consecutive numbers between 10 and 15 with the same 'COOL' value. What are they? [3]

  e 16 has a 'COOL' value of 4. What is the next number after 16 to have a 'COOL' value of 4? [2]

4 The numbers 0 to 99 are in this grid.

0	1	2	3	4	5	6	7	8	9
10	11	12	13	14	15	16	17	18	19
20	21	22	23	24	25	26	....	....	....
30	31	32	33	34	....	....	....	....	....
40	41	42	43	....	....	....	....	....	....

Five cells are coloured to form a T shape. The one shown is T_2.

4 $T_2 = 1 + 2 + 3 + 12 + 22 = 40$.

  a Work out T_{13}. [2]

  b Work out T_{65}. [1]

  c Write down an expression for T_x. Simplify your answer. [3]

  d Find x when $T_x = 105$. [2]

  e Is the value of T_{20} possible? Give reasons for your answer. [1]

  f Find the value of x that gives the largest value of T_x. Show your working. [3]

5 PQ and PR are a pair of stepladders standing on a horizontal floor. $PQ = PR$.

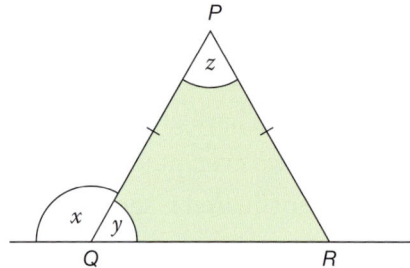

  a Angle x is 130°. Find the values of y and z. Give reasons for your answers. [4]

  b The ladders slip on the floor until angle z becomes 116°. Find the new values of x and y. [3]

  c What sort of triangle would PQR be if x was 120°? Give reasons for your answer. [2]

6 The data shows the number of nights and number of guests staying in a hotel.

Number of nights (x)	1	2	3	4	5	6	7	8	9
Number of guests (f)	4	9	3	6	8	11	8	4	2

  a Calculate the total number of guests that stayed in the hotel. [1]

  b Calculate the mean and median number of nights that guests stayed. [5]

  c Find the range. [1]

7 a A rectangle has sides $3y - 2$ cm and $2y + 6$ cm. Work out its perimeter. [3]

b A cuboid has edges $3y - 2$ cm, 6 cm and 8 cm. Work out its surface area. [4]

c Two sides of an equilateral triangle are labelled $3x + 2$. The other side is 35 cm. Write down an equation in x and solve it to find x. [4]

8 *ABCDEFGH* is a regular octagon.

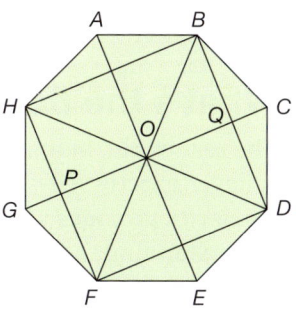

a Sara makes the following statements. For each statement decide if Sara is correct or not. Fully describe the shape in each case.

  i *HBD* is an isosceles triangle. [2]

  ii *HBDF* is a trapezium. [2]

  iii *HBCG* is a parallelogram. [3]

b Name any triangle

  i congruent to *HOP* [1]

  ii similar to *HOP* [1]

  iii with 1 line of symmetry [1]

  iv with 1 line parallel to *AC*. [1]

c Find the value of

  i angle *ABC* [3]

  ii angle *BDF* [1]

  iii angle *OAB* [1]

  iv angle *FHD* [1]

  v angle *POF* [1]

  vi angle *HOC*. [1]

9 Sunita has £7.50. She spends $\frac{1}{5}$ on a coffee and $\frac{5}{12}$ of what she has left on a sandwich.

a How much did Sunita spend on her coffee? [1]

b How much did she spend on her sandwich? [3]

c How much did she have left at the end? [1]

10 There are 450 campers on a campsite, 24% are teenagers, 81 campers are over 65.

a How many are teenagers? [2]

b What percentage of the total number of campers are over 65? [2]

c How many campers were neither teenagers nor over 65? [2]

11 Ally the chemist is diluting sulphuric acid. Flask A contains 80 ml of sulphuric acid and flask B contains 100 ml of water. Ally transfers 20 ml of water from flask B into flask A.

a What fraction of flask A is water? [1]

b What fraction of flask A is acid? [1]

The contents of flask A is mixed together and a 20 ml spoonful is transferred to flask B.

c How many ml of acid and water does the spoon hold? [2]

d Fill in the table of volumes for each flask. [4]

e Use the table to write the ratio of acid to water in

Flask A	Acid	____ ml
	Water	____ ml
Flask B	Acid	____ ml
	Water	____ ml

  i flask A [1]

  ii flask B. [1]

12 The formula for the radius, r, of a sphere given its surface area, A, is given by

$$r = \sqrt{\frac{A}{4\pi}}.$$

a Find r when $A = 65 \text{ cm}^2$. [2]

b Write down the value of A which gives a radius of 1. Show your working. [2]

c Rearrange the formula to make A the subject. [2]

d Use this formula to find the value of A when $r = 15.4$ cm [2]

13 The bar and pie chart show the same information. Complete both charts.

[4]

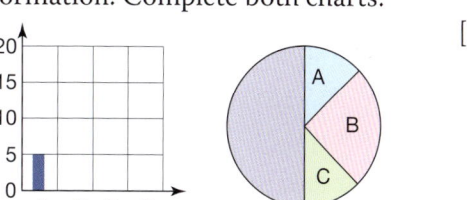

7 Working in 2D

Introduction

Self-similarity is the property whereby an entire shape is mathematically similar to a part of itself. What this means is that if you 'zoom in' on a small corner of the shape, you get an exact replica of the original shape itself. Self-similarity is used in fractal images, like the one you can see here, and it has real-world use in describing the structure of coastlines, as well as the natural growth of plants such as ferns, and the formation of crystals and snowflakes.

What's the point?

The real world, being mathematically untidy, is never exactly self-similar. However self-similarity provides a highly useful model in understanding the complex geometries seen in nature, which can't usually be reduced to simple rectangles and circles.

Objectives

By the end of this chapter you will have learned how to …

● Accurately measure and draw line segments and angles.

● Use standard units for lengths and areas

● Use bearings.

● Interpret maps and scale drawings.

● Know and apply formulae to calculate the area of triangles, parallelograms and trapezia.

● Identify, describe and construct reflections, rotations, translations and enlargements.

Check in

1 Evaluate

   a $6 \times \frac{8}{5}$ **b** $9 \times 2\frac{1}{2}$ **c** $6 \times 1\frac{3}{4}$

2 Evaluate

   a $5.8 + 2$ **b** $14.8 + 0.7$ **c** $6.4 + 2.6$

3 Measure this line

   a in millimetres   **b** in centimetres.

Chapter investigation

Create a snowflake!

Step 1   Draw an equilateral triangle.

Step 2   Draw equilateral triangles on each of the three sides
           (carefully – you'll have to divide each side into three equal parts).

Step 3   You now have 12 sides.
           Draw equilateral triangles on each of these sides.

If it's still not snowflaky enough, try once more – but you will find it starts getting very fiddly!

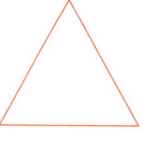

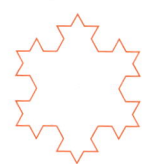

7.1 Measuring lengths and angles

p.256

p.50

RECAP

- The ratio scale 1 : *a* means real length = *a* × length on the map.
- A bearing is a three-figure angle measured *clockwise from north*.

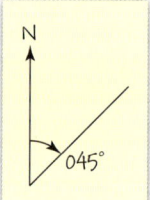

N

045°

HOW TO

① Choose a scale, unless one is given.
② Work out or measure the lengths or angles required.
③ Draw a diagram, unless one is given.
④ Give the answers including units.

EXAMPLE

A boat travels for 26 miles on a bearing of 115°, then 34 miles on a bearing of 263°.

a **i** Find the direction in which the boat must travel to return directly to the start.

  ii Find the length of the return journey.

b Explain how the accuracy of the answers depends on the scale you have used.

① Use a scale that gives a reasonably sized diagram.

a Using a scale of 1cm to represent 5 miles.

② Work out the distances you need.

Real (miles)		Diagram (cm)
5 miles		1cm
26 miles	÷ 5	5.2cm
34 miles	÷ 5	6.8cm

④ Find the bearing and distance.

  i Bearing for return journey = 034°

  ii Distance on map = 3.6cm

    Length of journey = 3.6 × 5 = 18 miles

b Errors in measuring are less important on a big diagram.

The larger the scale, the more accurate the answers will be.

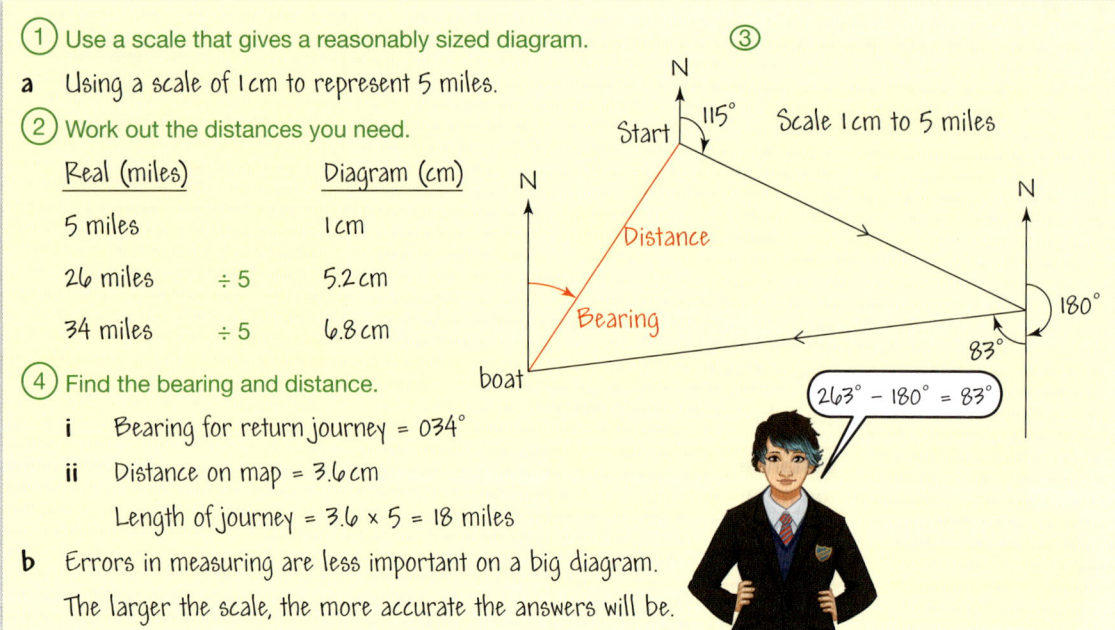

③

N

Start 115° Scale 1cm to 5 miles

N

Distance

Bearing

boat

N

180°

83°

263° − 180° = 83°

EXAMPLE

This is part of a plan of Seth's kitchen.

Seth says a 1000 mm wide cooker will fit in the space.

Is Seth correct? Explain your answer.

scale 1 : 50		wall	
kitchen units		space for cooker	

① ② ③ Measure the space on the diagram.

Width of space = 19 mm

④ Compare the real size of the space with the cooker width.

Real width of space = 19 × 50 = 950 mm

Seth is wrong because 1000 mm is more than the width of the space.

Measure in mm so that you don't need to change units.

Exercise 7.1A

1 A plane leaves airport *A* and flies 84 miles on a bearing of 067°.

The plane then changes course to 312° and travels a further 96 miles to land at airport *B*.

   a Find the distance and bearing for a direct return journey from airport *B* to airport *A*.

   b Describe one way you could improve the accuracy of your answers.

2 Liam sees a lifeboat on a bearing of 138° from Mevagissey. Kim sees the same lifeboat on a bearing of 215° from the Rame Head chapel.

Kim says the lifeboat is over half a kilometre nearer to her than to Liam.

Use tracing paper to copy the coastline and work out if Kim is correct.

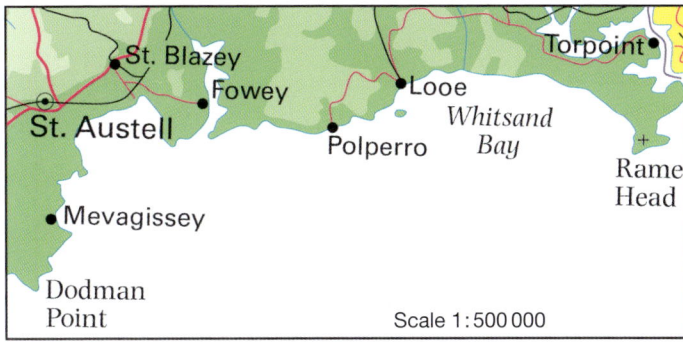

3 The sketch shows the dimensions of Sunita's bedroom.

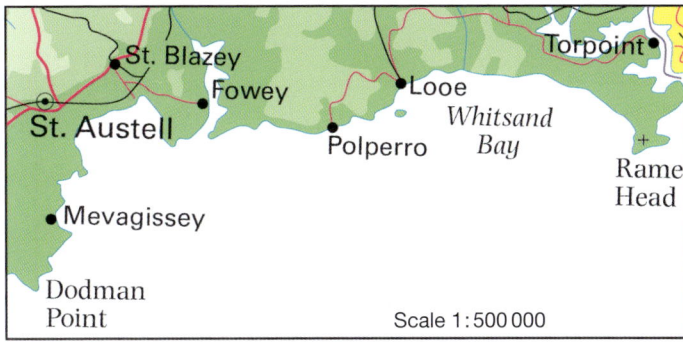

Not to scale

Sunita has written down the dimensions of the furniture she needs to fit in.

Item	length	width
Bed	190 cm	90 cm
Computer desk	120 cm	80 cm
Wardrobe	100 cm	58 cm
Drawers	87 cm	43 cm
Bookshelves	120 cm	20 cm

Draw an accurate scale drawing of Sunita's bedroom.

Show how she could realistically arrange the items.

4 Estimate the length of the truck.

1.6 m

*5 Sam is a salesman based in Douglas on the Isle of Man.

The table gives the towns that Sam must visit this week.

Each day Sam always starts from Douglas and returns to Douglas.

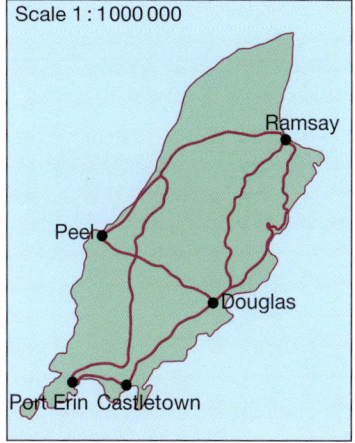

Scale 1 : 1 000 000

Monday	Peel
Tuesday	Ramsay
Wednesday	Castletown
Thursday	Port Erin
Friday	Peel and Ramsay

Sam has enough fuel in his car to travel 250 miles.

Will Sam need any more fuel for these journeys?

You must show your working.

1086, 1103, 1117, 1146   SEARCH

7.2 Area of a 2D shape

Area is the amount of space inside a 2D shape.
In the metric system, area is measured in
mm², **cm²**, **m²** or **km²**.

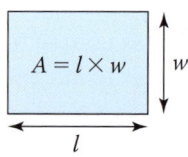
▲ Rectangle

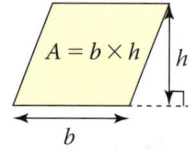
▲ Parallelogram

- Area of a rectangle = length × width

p.230
- Area of a parallelogram
  = base × perpendicular height

- Area of a triangle
  = $\frac{1}{2}$ base × perpendicular height

p.316
- Area of a trapezium
  = $\frac{1}{2}$ sum of the parallel sides × perpendicular height

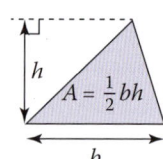
▲ Triangle   ▲ Trapezium

EXAMPLE

Find the area of these shapes.

a
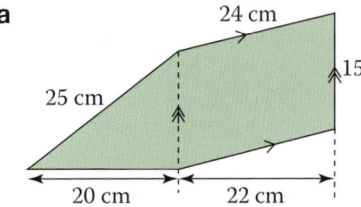

b
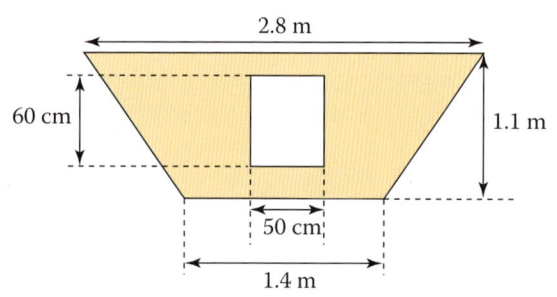

a Take care to use the perpendicular heights.

Area of triangle = $\frac{1}{2}$ × 20 × 15

= 150

Area of parallelogram = 15 × 22

= 330

Area of shape = 150 + 330

= 480 cm²

b Area of trapezium = $\frac{1}{2}$ (2.8 + 1.4) × 1.1

= $\frac{1}{2}$ × 4.2 × 1.1

= 2.31 m²

Area of hole = 0.5 × 0.6

= 0.3 m²

Area of shape = 2.31 − 0.3

= 2.01 m²

Units must be
consistent.

50 cm = 0.5 m

60 cm = 0.6 m

EXAMPLE

The area of a triangle is 350 cm². The length of its base is 28 cm.

Calculate the height of the triangle.

Substitute the known values into the formula, then rearrange.

$A = \frac{1}{2} × b × h$

$350 = \frac{1}{2} × 28 × h$

$350 = 14h$

$h = 350 ÷ 14 = 25$

The height of the triangle = 25 cm.

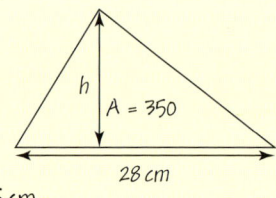

Sketch the triangle
if it helps.

Exercise 7.2S

1 Find the area of these shapes.

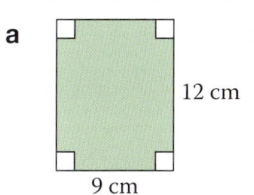

a

b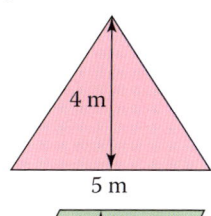

c d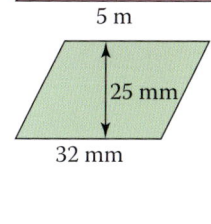

2 Find the area of these shapes.

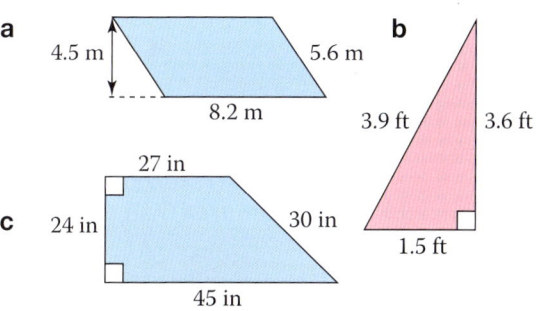

a

b

c

In the old, imperial system area is measured in square-inches (in²), square-feet (ft²), etc.

3 Copy and complete the table.

	Name of shape	Area of shape
a		
b		
c		
d		

a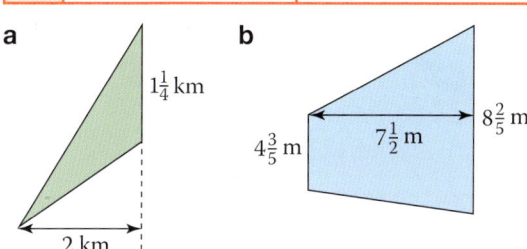

b

c d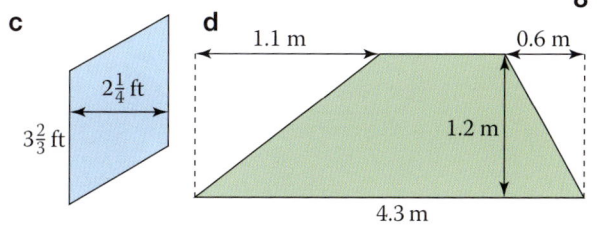

4 Copy the table. Find the missing values.

Shape	Base	Height	Area
Rectangle	16 mm		240 mm²
Parallelogram		45 cm	1800 cm²
Triangle	4 m		5 m²

5 Find the area of these shapes.

a b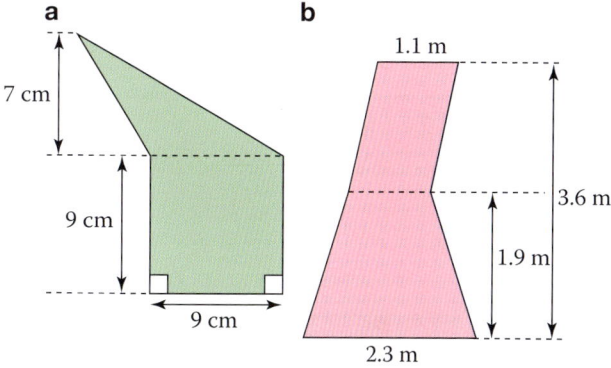

6 The area of a trapezium is 60 cm².

The height of the trapezium is 8 cm and the length of one parallel side is 10 cm.

Find the length of the other parallel side.

*7 Find the area of these shapes.

a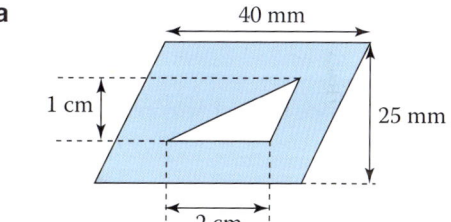

b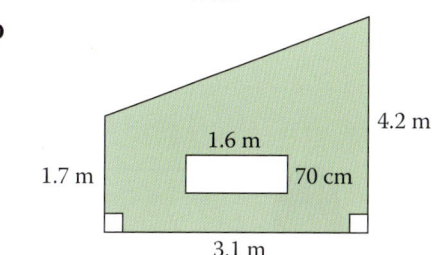

8 Find the area of this side of a house.

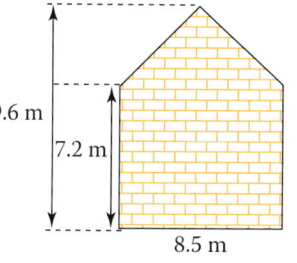

7.2 Area of a 2D shape

RECAP

- Area of a rectangle = length × width
- Area of a parallelogram
  = base × perpendicular height
- Area of a triangle = $\frac{1}{2}$ base × perpendicular height
- Area of a trapezium
  = $\frac{1}{2}$ sum of the parallel sides × perpendicular
  height

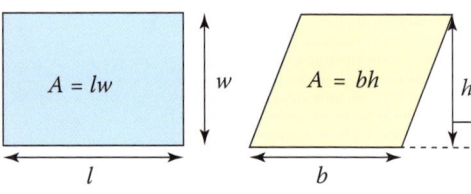

$A = lw$ w $A = bh$ h

l b

▲ Rectangle   ▲ Parallelogram

HOW TO

p.230

p.316

To find the area of a shape

① Draw a diagram (if needed) and identify simple shapes in it.

② Decide which formula(e) to use.

③ Find the areas of the simple shapes, including the units.

④ State your conclusion and, when asked to, your reasons.

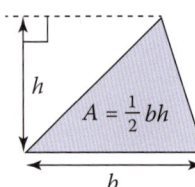

h $A = \frac{1}{2}bh$

b

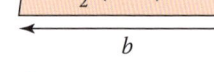

a

$A = \frac{1}{2}(a+b)h$ h

b

▲ Triangle   ▲ Trapezium

EXAMPLE

A farmer measured two fields. He says that the second field has a larger area than the first.

a Is the farmer correct?

b What assumptions have you made that could affect your answer?

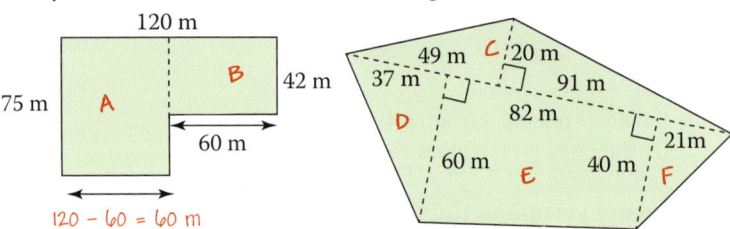

120 m

B   42 m

75 m   A

60 m

120 − 60 = 60 m

49 m   C   20 m
37 m   91 m
D   82 m
    21m
60 m   E   40 m   F

a ① Divide the first field into rectangles.

② For a rectangle $A = lw$
Length of A = 120 − 60 = 60 m

③ Area of A = 75 × 60 = 4500
Area of B = 60 × 42 = 2520
Total area = 4500 + 2520 = 7020 m²

The second field is divided into triangles and a trapezium. The 2 triangles on top can be worked out as 1 triangle, C.

For a triangle $A = \frac{1}{2}bh$ and for a trapezium $A = \frac{1}{2}(a+b)h$

Area of C = $\frac{1}{2}$ × 140 × 20   = 1400

Area of D = $\frac{1}{2}$ × 60 × 37   = 1110

Area of E = $\frac{1}{2}$ × (60 + 40) × 82 = 4100

Area of F = $\frac{1}{2}$ × 40 × 21   = 420

Total area = 1400 + 1110 + 4100 + 420 = 7030 m²

You can check this by dividing the shape with a horizontal line or subtracting areas.

④ The farmer is correct because 7030 is more than 7020.

b The measurements are assumed to be accurate.

If they are not, then the farmer may be wrong.

Remember to answer the question!

Geometry   Working in 2D

Exercise 7.2A

1 Annabel says the area of this plot is over $\frac{1}{4}$ hectare. (1 hectare = 10 000 m²)

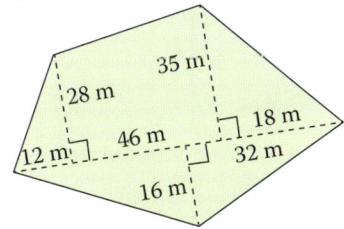

a Is Annabel correct?

b What assumptions could affect the answer to **a**?

2

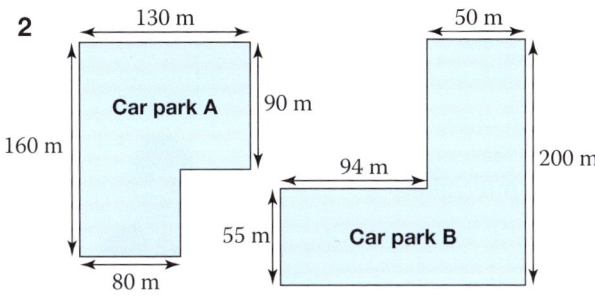

a Which car park has the greater area?

b What else will affect which car park can hold most cars?

3 Pete says six times more purple material than yellow material is used in this kite.

a Is Pete correct? Show how you decide.

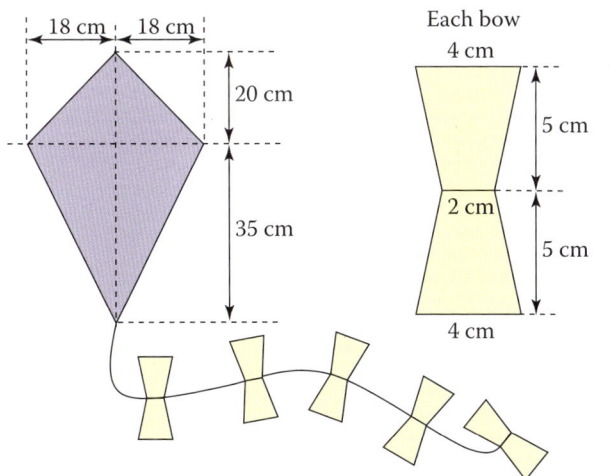

b If a yellow shape is cut out of a rectangular piece of material what is the smallest amount of material that is wasted?

4 Use x and y axes from 0 to 6 on centimetre square paper.

a Plot and join these points to make an arrow.
(0, 3), (3, 6), (5, 6), (3, 4), (6, 4)
(6, 2), (3, 2), (5, 0), (3, 0), (0, 3)

b Find the area of the arrow.

c Use a different method to check your answer to part **b**.

*5 The diagram shows the dimensions of Jim's bathroom floor.

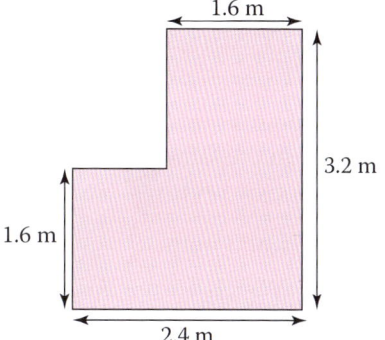

Jim wants to cover the floor with square tiles with sides of length 400 mm.

The tiles cost £29.50 for each pack of 5 tiles.

Jim says it will cost less than £200 to tile the floor.

Is Jim correct? Show your working.

*6 Amy has a rectangular map 30 cm long and 20 cm wide. Amy says 'Over half of this map lies within 4 cm of the edge'.

Is Amy correct? Show your workings.

Did you know…

A traditional unit of area is the acre, about 4050 m². It was defined as the area a pair of oxen could plough in a day.

1108, 1128, 1129   SEARCH

7.3 Transformations 1

A **transformation maps** points in the **object** to points in the **image** and causes the position and, for enlargements, the size of shapes to change.

Points that do not move in a transformation are called **invariant**.

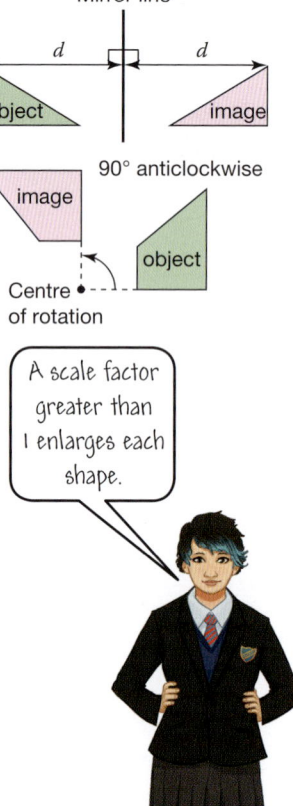

Mirror line

90° anticlockwise

Centre of rotation

- All points on the **mirrror line** are invariant in a **reflection**.
- The **centre of rotation** is invariant in a **rotation**.
- The **centre of enlargement** is invariant in an **enlargement**.

A scale factor greater than 1 enlarges each shape.

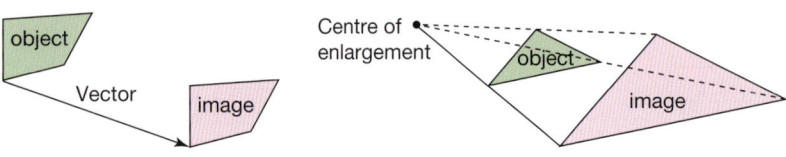

In a **translation**, *all* points move by the same distance in the same direction. A translation by **vector** $\begin{pmatrix} 3 \\ -1 \end{pmatrix}$ moves every point 3 units to the right and 1 unit down.

In an enlargement the distance from the centre of enlargement to every other point is multiplied by a **scale factor**.

p.56

- In a reflection, rotation or translation the image and object shapes are **congruent**.
- In an enlargement the image and object shapes are **similar**.

p.290
p.400

EXAMPLE

Quadrilateral Q has vertices at (2, 1), (3, 1), (3, 3) and (1, 2). Draw the image of Q after

a reflection in the line $y = -2$

b rotation of 90° anti-clockwise about $(-2, 1)$

c translation by vector $\begin{pmatrix} -8 \\ 2 \end{pmatrix}$

d enlargement, centre (2, 2) scale factor 3.

Transform each vertex of Q and join them to find the image.

a Each vertex has an image on the opposite side of the mirror line, an equal distance from it.

b Use tracing paper to copy Q, then rotate it to find the image.

c Each point moves 8 units left and 2 units upwards.

d Each image point is 3 times as far from the centre as the corresponding object point.

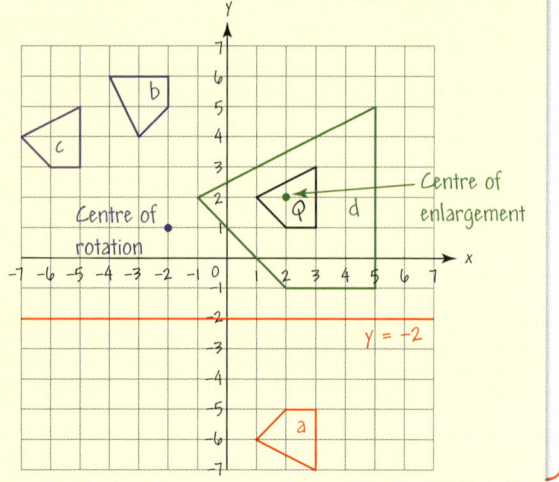

To find the line $y = -2$ join points like (1, -2), (2, -2) and (3, -2).

Exercise 7.3S

1 Copy the diagram onto squared paper.

Draw the reflection of each shape in the mirror line.

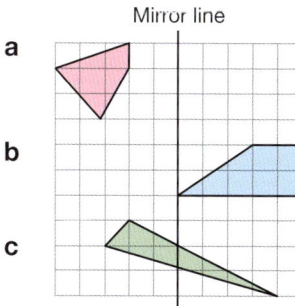

a

b

c

2 Copy triangle T and point C onto squared paper.

Draw the image of T after each of the following rotations.

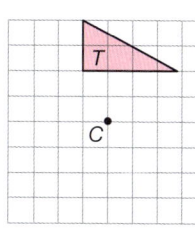

a Rotation of 90° clockwise about C.

b Rotation of 180° clockwise about C.

c Rotation of 90° anti-clockwise about C.

3 Copy triangle T from question **2** again.

Draw the image of T after translation by

a $\begin{pmatrix} 2 \\ 3 \end{pmatrix}$ **b** $\begin{pmatrix} -3 \\ 4 \end{pmatrix}$ **c** $\begin{pmatrix} 4 \\ -5 \end{pmatrix}$ **d** $\begin{pmatrix} -4 \\ -2 \end{pmatrix}$

4 Copy trapezium A and points C and D onto the middle of a sheet of squared paper.

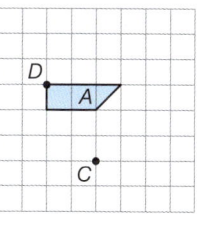

Draw the image of A after each of the following enlargements.

a Enlargement, centre C, scale factor 3.

b Enlargement, centre D, scale factor 4.

> For questions **5** and **6** use x and y axes from −4 to 4.

5 Triangle T has vertices at (1, 1), (4, 0), and (2, 3). Draw T and its image after

a reflection in the x axis

b reflection in the y axis.

6 Parallelogram P has vertices (1, 2), (4, 1), (4, 3) and (1, 4). Draw P and its image after

a rotation of 90° clockwise about (0, 0)

b rotation of 180° clockwise about (0, 0)

c rotation of 90° anti-clockwise about (0, 0).

> For questions **7** and **8** use x and y axes from −6 to 6.

7 a Draw and name the shape A with vertices at (−2, 1), (−2, 4), (−1, 4), (−1, 3) and (0, 1).

b Draw the image of A after each of these translations

i $\begin{pmatrix} 4 \\ 2 \end{pmatrix}$ **ii** $\begin{pmatrix} -1 \\ -7 \end{pmatrix}$ **iii** $\begin{pmatrix} 3 \\ -6 \end{pmatrix}$ **iv** $\begin{pmatrix} -3 \\ 1 \end{pmatrix}$

8 a Draw and name the shape X with vertices at (1, 0), (2, 2), (3, 0) and (2, −2).

b Draw the image of X after

i enlargement, centre (0, 0), scale factor 2

ii enlargement, centre (4, 0), scale factor 3.

*9 Use x and y axes from −7 to 7.

a Join points $A(-4, 2)$, $B(-3, 4)$, $C(-4, 5)$ and $D(-5, 4)$ to form a quadrilateral.

What special type of quadrilateral is $ABCD$?

b Reflect $ABCD$ in the line $y = -1$. Label the image $A_1B_1C_1D_1$.

c Rotate $ABCD$ 90° clockwise about (0, 1). Label the image $A_2B_2C_2D_2$.

d Translate $ABCD$ by vector $\begin{pmatrix} 9 \\ -7 \end{pmatrix}$. Label the image $A_3B_3C_3D_3$.

e Enlarge $ABCD$ using centre (−4, 4), scale factor 2. Label the image $A_4B_4C_4D_4$.

10 a Describe a transformation in which the point (1, 1) is invariant.

b Describe a transformation in which the points (1, 1) (2, 1) and (3, 1) are invariant.

Q 1099, 1113, 1115, 1127 SEARCH

7.3 Transformations 1

To describe a	give
● Reflection	The position of the mirror line (invariant).
● Rotation	The angle of rotation
	The direction, clockwise or anti-clockwise
	The centre of rotation (invariant).
● Translation	The vector or the distance and direction
● Enlargement	The scale factor
	The centre of enlargement (invariant).

● A point or line is invariant if it does not move in a transformation.

p.400

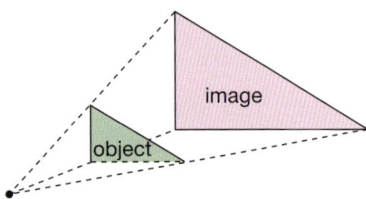

▲ The mirror line is halfway between the object and image. The mirror line is invariant.

HOW TO

To identify a transformation

① Decide which type of transformation is involved and find the information needed. Use tracing paper if you wish.

② Give a full description of the transformation.

▲ To find the centre of enlargement, draw lines between corresponding points. The centre of enlargement is invariant.

EXAMPLE

Write a full description of the transformation that maps the flag *F* onto

a *A*   **b** *B*   **c** *C*   **d** *D*.

a ① *A* has changed size, it is twice as tall and twice as wide as *F*: enlargement. Join corresponding points to find the centre of the enlargement.

② Enlargement, scale factor 2, centre (7, 1)

b ① *B* is in the same orientation but a different position: translation. *F* moves left by 9 squares and down by 5 squares.

② Translation by vector $\begin{pmatrix} -9 \\ -5 \end{pmatrix}$

c ① *C* is 'flipped': reflection. Points like (–1, 1), (–1, 3) and (–1, 4) lie halfway between *F* and *C*.

② Reflection in mirror line x = –1

d ① *D* is turned, 90° clockwise: rotation. Use tracing paper to find the centre of rotation.

② Rotation 90° clockwise about (3, –2).

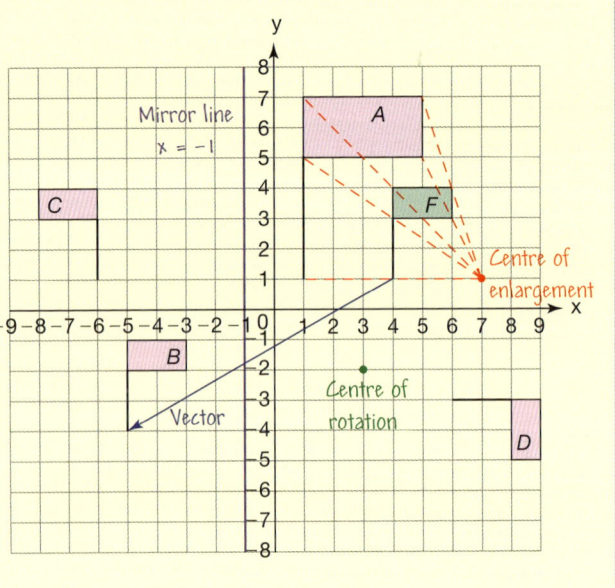

Exercise 7.3A

1 Write a full description of the transformation that maps

a *A* onto *B*

b *B* onto *A*

c *A* onto *C*

d *B* onto *D*

e *D* onto *B*.

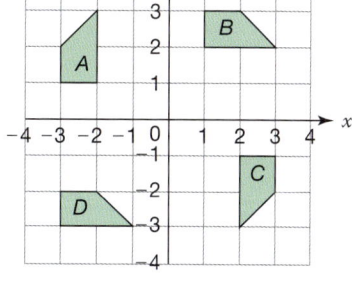

2 Describe fully the transformation that maps

a *P* onto *Q*

b *Q* onto *P*

c *P* onto *S*

d *R* onto *Q*

e *P* onto *R*.

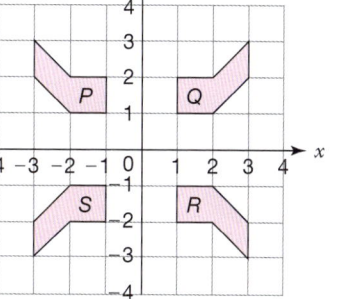

3 **a** Describe fully the transformation that maps

i *A* onto *B* **ii** *A* onto *C* **iii** *C* onto *B*

iv *D* onto *E* **v** *E* onto *F* **vi** *D* onto *F*

vii *G* onto *H* **viii** *I* onto triangle *J*.

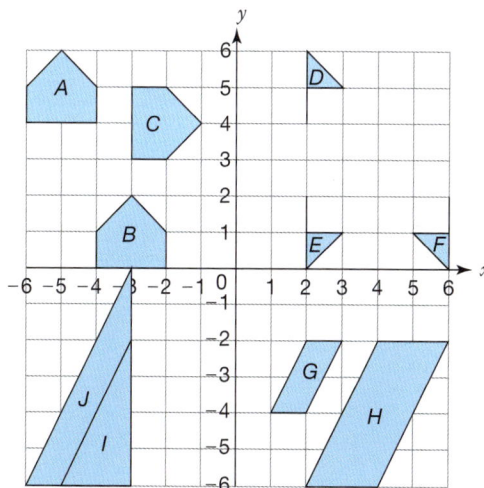

b For each transformation in part **a** list any points or lines which are invariant under that transformation.

4 The table gives the vertices of a kite and its image after a transformation.

Kite	(2, 0)	(3, 2)	(2, 3)	(1, 2)
Image	(6, 0)	(9, 6)	(6, 9)	(3, 6)

a What happens to the co-ordinates?

b Describe fully the transformation.

5 **a** Draw triangle *T* with vertices at (0, −1), (3, −1) and (2, 1) and its image, *S*, with vertices at (−6, −4), (6, −4) and (2, 4).

b Give a full description of the transformation that maps *T* onto *S*.

6 On a clock, describe the transformation of

a the minute hand

*b the hour hand during

i 30 minutes **ii** 10 minutes.

*7 Sue says a rotation of 90° clockwise about (1, 4) maps *A* onto *B*.

Is Sue correct? Give your reasons.

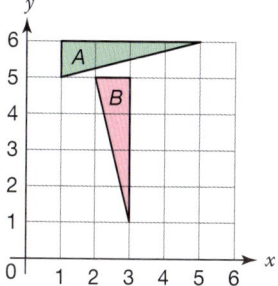

*8 Describe 3 *different* transformations that map *A* onto *B*.

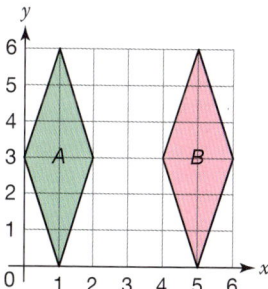

9 Investigate what happens to co-ordinates when a shape is

a reflected in

i the *x*-axis **ii** the *y*-axis

*iii $y = x$ ***iv** $y = -x$

b rotated through

i 90° clockwise about (0, 0)

ii 180° clockwise about (0, 0)

iii 90° anti-clockwise about (0, 0)

1099, 1113, 1115, 1127 SEARCH

7.4 Transformations 2

Two or more transformations may be combined.

The result may be equivalent to a single transformation.

To describe

- a reflection, give the mirror line.
- a rotation, give the centre and angle and say whether the rotation is clockwise or anti-clockwise.
- a translation, give the distance and direction or a vector.
- an enlargement, give the scale factor and the centre of enlargement.

— A scale factor greater than 1 makes the shape bigger.
— A scale factor between 0 and 1 makes the shape smaller.

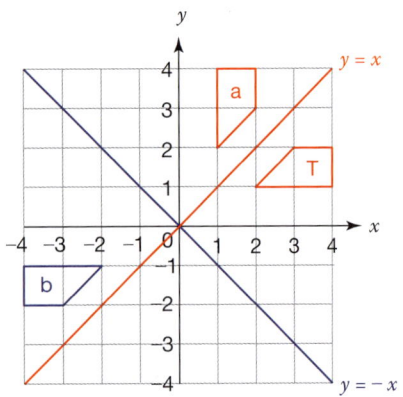

▲ Reflection of shape T in $y = x$ followed by reflection in $y = -x$

p.288

The line $y = x$ goes through points like $(-2, -2)$, $(-1, -1)$, $(2, 2)$ and $(3, 3)$.

The line $y = -x$ goes through points like $(-2, 2)$, $(-1, 1)$, $(2, -2)$ and $(3, -3)$.

EXAMPLE

a   Draw the triangle with vertices A (1, 1), B (7, 1) and C (1, 5).

b   Draw the image of ABC after an enlargement, centre $(-7, 3)$, scale factor $\frac{1}{2}$. Label the image $A_1B_1C_1$.

c   Draw the image of $A_1B_1C_1$ after translation by vector $\begin{pmatrix} 4 \\ -4 \end{pmatrix}$. Label the image $A_2B_2C_2$.

d   Describe the single transformation that maps ABC onto $A_2B_2C_2$.

a   See diagram.

b   Each vertex of $A_1B_1C_1$ is half as far from the centre as the corresponding vertex of ABC.

c   Each vertex of $A_1B_1C_1$ moves 4 units right and 4 units downwards.

d   The sides of triangle $A_2B_2C_2$ are half as long as those of ABC: enlargement. To find the centre of this enlargement join and extend AA_2, BB_2 and CC_2.

Enlargement, centre (1, −5),

scale factor $\frac{1}{2}$.

Remember to give all three pieces of information.

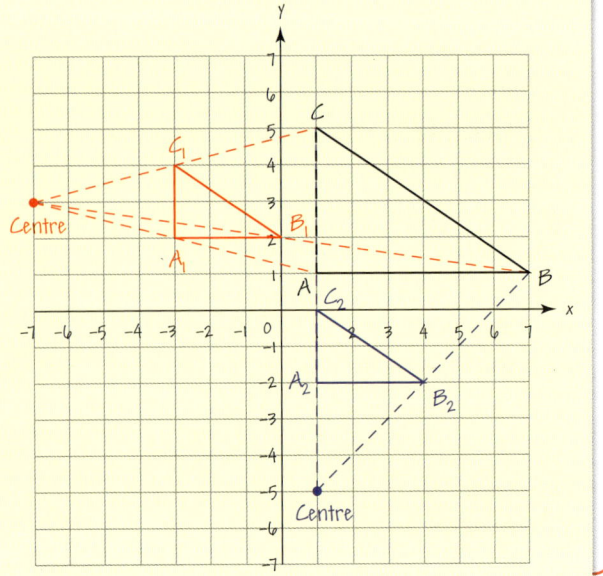

Exercise 7.4S

1 Copy the diagram onto squared paper.

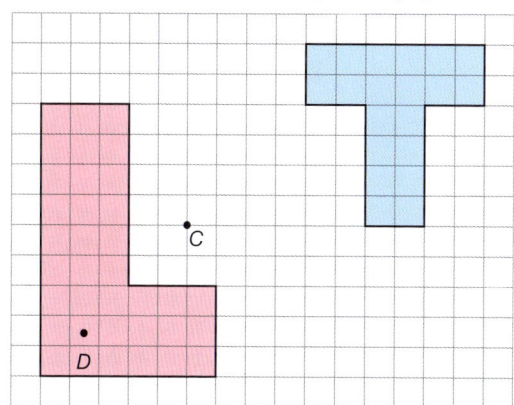

a Draw the enlargement of the 'T-shape' with centre C and scale factor $\frac{1}{2}$.

b Draw the enlargement of the 'L-shape' with centre D and scale factor $\frac{1}{3}$.

2 Copy the diagram.

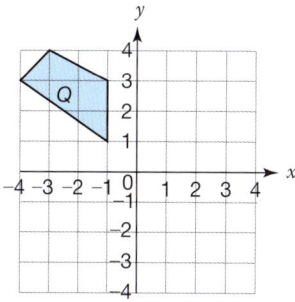

a Reflect Q in the x axis. Label the image R.

b Reflect R in the y axis. Label the image S.

c Describe the single transformation that maps Q onto S.

3 a On x and y axes from −4 to 4, draw triangle A with vertices at (−2, 1), (−3, 1) and (−4, 3).

b Rotate A through 90° clockwise about the origin, (0, 0). Label the image B.

c Rotate B through 180° clockwise about the point (2, 0). Label the image C.

d Describe the single transformation that maps A onto C.

For questions **4** and **7** use x and y axes from −6 to 6.

4 a Draw and name the quadrilateral, A, with vertices at (−6, 1), (0, 1), (−2, 5) and (−6, 5).

b Draw the image of A after an enlargement, centre (0, −5), scale factor $\frac{1}{2}$. Label the image B.

c Draw the image of B after translation by vector $\begin{pmatrix} 3 \\ 5 \end{pmatrix}$. Label the image C.

d Describe the single transformation that maps A onto C.

5 a Draw the parallelogram with vertices $A(-3, 4)$, $B(2, 4)$, $C(3, 6)$ and $D(-2, 6)$.

b Draw the image of ABCD after reflection in the line $y = 2$. Label the image $A_1B_1C_1D_1$.

c Draw the image of $A_1B_1C_1D_1$ after reflection in the line $y = -3$. Label the image $A_2B_2C_2D_2$.

d Describe the single transformation that maps ABCD onto $A_2B_2C_2D_2$.

6 a Draw pentagon P with vertices at (3, 1), (6, 1), (6, 2), (4, 3) and (3, 2).

b Draw the image of P after reflection in the line $y = x$. Label the image Q.

c Draw the image of Q after reflection in the line $y = -x$. Label the image R.

d Describe the single transformation that maps P onto R.

*7 a Draw the quadrilateral with vertices $A(-5, -3)$, $B(-2, -6)$, $C(-2, -2)$ and $D(-4, -2)$.

b Draw the image of ABCD after rotation of 90° clockwise about B. Label the image $A_1B_1C_1D_1$.

c Draw the image of $A_1B_1C_1D_1$ after translation by vector $\begin{pmatrix} 1 \\ 9 \end{pmatrix}$. Label the image $A_2B_2C_2D_2$.

d Describe the single transformation that maps ABCD onto $A_2B_2C_2D_2$.

e Describe the single transformation that maps $A_2B_2C_2D_2$ onto ABCD.

Q 1125   SEARCH

7.4 Transformations 2

p.400

RECAP

To identify the type of transformation compare the object and the image.

- Congruent shapes, same orientation: **translation**
  Give distance and direction.
- Congruent shapes, image 'flipped': **reflection**
  Give mirror line.
- Congruent shapes, image turned: **rotation**
  Give centre, angle and direction.
- Similar shapes: **enlargement**
  — image enlarged,   scale factor > 1
  — image reduced,   scale factor < 1.
  Give centre and scale factor.

▲ Geometric patterns on tiles, wallpaper and fabric are often made by combining or repeating transformations.

HOW TO

To identify a transformation

1. Draw a diagram (unless one is given)
2. Decide which type of transformation is involved and find the information needed.
3. Give a full description of the transformation.

EXAMPLE

Lines *M* and *N* cross at 45°.

a **i** Reflect *T* in line *M*, then reflect the resulting image in line *N*.

ii Describe the single transformation that maps *T* onto the image after the two reflections.

b What happens if the order of reflections is reversed?

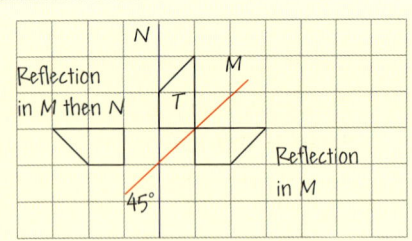

a **i** ① In a reflection, vertices that touch the mirror line do not move. For other vertices, the *perpendicular* distance from the mirror line is the same after the reflection.
The first diagram shows the images.

ii ② The trapezium has rotated – use tracing paper to find the centre.

③ Rotation of 90° anti-clockwise about the point of intersection of the mirrors.

b ① The second diagram shows the images when the order of reflections is reversed.

② The trapezium has rotated but in the other direction – use tracing paper to find the centre.

③ The overall transformation is now a rotation of 90° clockwise about the point of intersection of the mirrors.
The direction of rotation has reversed.

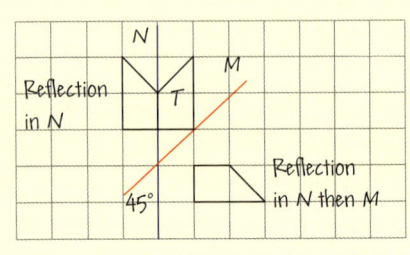

Exercise 7.4A

1 Repeating a transformation on each coloured L-shape gives this pattern.

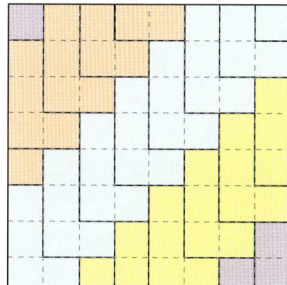

Give a full description of the transformation.

2 Describe fully a transformation that gives this pattern when repeatedly applied to the flag F.

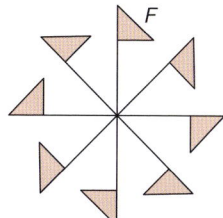

3 a i On a copy of this diagram, reflect T in M, then reflect the image in N.

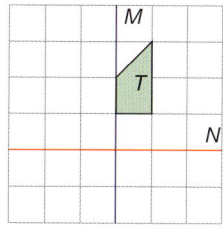

    ii Describe the single transformation that maps T onto the final image.

  b Does the final image change if the order of reflections is reversed?

4 a i On a copy of this diagram, reflect T in M, then reflect the image in N.

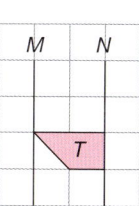

    ii Describe the single transformation that maps T onto the final image.

  b What happens if the order of reflections is reversed?

5 Describe a single transformation that is equivalent to translation by vector $\begin{pmatrix} 4 \\ -2 \end{pmatrix}$ followed by a translation by vector $\begin{pmatrix} 3 \\ 5 \end{pmatrix}$.

6 Mark says that reflection in $x = 1$ followed by reflection in $y = 2$ has the same effect as a clockwise rotation of 180° about (2, 1). Is this true? Illustrate your answer with a diagram.

7 A transformation is applied twice to a hexagon.

The result is as shown.

Describe two possible transformations.

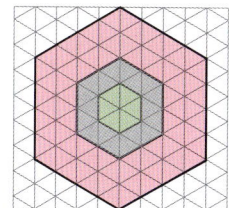

*8 a Jan says that a rotation followed by another rotation is always equivalent to a third rotation.

    Is this true? Give your reasons.

  b Ahmed says that a reflection followed by another reflection is always equivalent to a third reflection.

    Is this true? Give your reasons.

9 Triangle T is enlarged by scale factor 2 with centre O to give shape T_1. Shape T_1 is enlarged by scale factor 2 with centre O to give shape T_2.

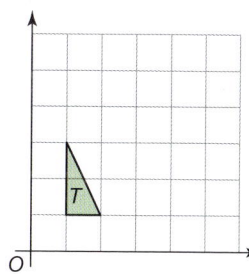

  a What single transformation returns shape T_2 to triangle T?

  b Draw another triangle and test if it is invariant under the two enlargements described in the question and the transformation you found in part **a**.

10 Investigate how transformations are used in patterns on wallpapers, fabric and tiles. Make your own geometric patterns by combining or repeating transformations.

Summary

Checkout

You should now be able to...

	Test it Questions
✔ Accurately measure and draw line segments and angles.	1, 2
✔ Use standard units for lengths and areas	1 – 3
✔ Use bearings.	2
✔ Interpret maps and scale drawings.	2
✔ Know and apply formulae to calculate the area of triangles, parallelograms and trapezia.	3
✔ Identify, describe and construct reflections, rotations, translations and enlargements.	4 – 6

Language	Meaning	Example
Length	Length is a measure of distance.	Millimetres, centimetres, metres and kilometres are all measures of length. Length can be measured with a ruler.
Angle	The amount that one straight line is turned relative to another that it meets or crosses.	Angles are measured in degrees. One degree is $\frac{1}{360}$ th of a complete turn. Use a protractor to measure an angle.
Area	The amount of space occupied by a 2D shape.	Area = 12 units^2 Perimeter = 14 units
Perimeter	The total distance around the edges that outline a shape.	
Transformation	A geometric mapping that takes the points in an **object** to points in an **image**.	Rotation, reflection, translation, enlargement.
Translation	A transformation in which all the points in the object are moved the same distance and in the same direction.	
Reflection / Mirror line	A transformation that moves points to an equal distance on the opposite side of a mirror line.	
Rotation / Centre of rotation	A transformation that turns points through a fixed angle whilst keeping their distance from the centre of rotation fixed.	anticlockwise
Enlargement / Scale factor / Centre of Enlargement	A transformation that moves points a fixed multiple, the scale factor, of their distance from the centre of enlargement.	$\frac{1}{3}$
Invariant	Does not change under a transformation.	A mirror line under reflection.

Review

1 Measure the size of

   a angle *RTS*

   b side *RT*.

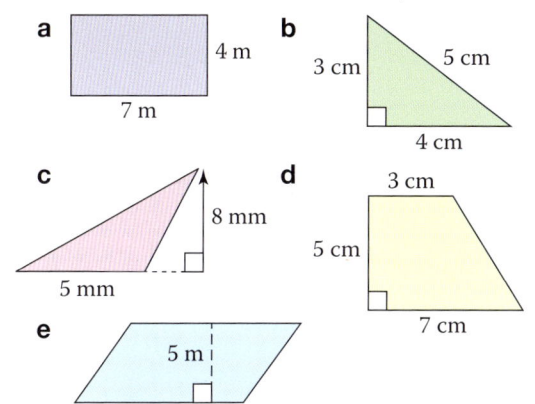

2 *A* is 2 km north of *B*. *C* is 3 km on a bearing of 065° from *B*. Make a scale drawing of *A*, *B* and *C* using the scale 2 cm to 1 km.

3 Work out the area of these shapes.
Remember to state the units of your answers.

   a 4 m, 7 m

   b 3 cm, 5 cm, 4 cm

   c 8 mm, 5 mm

   d 3 cm, 5 cm, 7 cm

   e 5 m, 11 m

4 Copy this diagram.

   a Reflect triangle *A* in the *x*-axis and label the image *B*.

   b Rotate triangle *A* 180° about (0, 0) and label the image *C*.

   c Describe the transformation of the object *C* to the image *B*.

5 On a copy of this diagram

   a translate triangle *A* by the vector $\begin{pmatrix} -4 \\ -6 \end{pmatrix}$ and label the image *C*,

   b describe the transformation that maps *A* to *B*.

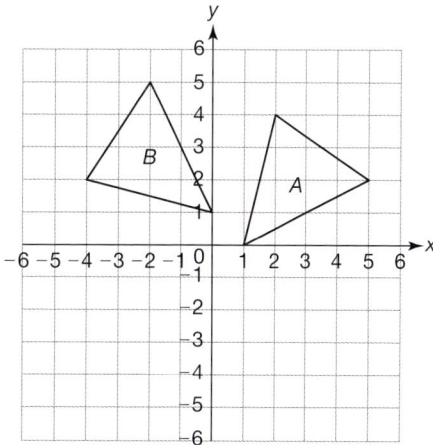

6 On a copy of this diagram

   a enlarge *A* by scale factor 3 from centre of enlargement (1, 9),

   b describe the transformation that maps *B* back onto *A*.

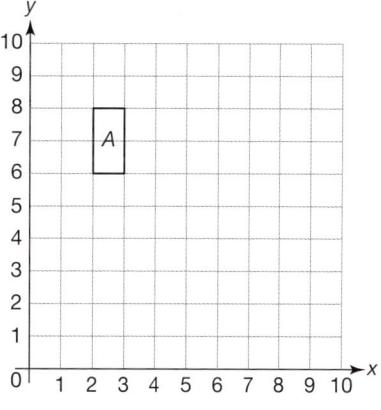

What next?

Score			
	0 – 2		Your knowledge of this topic is still developing. To improve look at MyMaths: 1086, 1099, 1103, 1108, 1113, 1115, 1117, 1125, 1127, 1128, 1129, 1146
	3 – 5		You are gaining a secure knowledge of this topic. To improve your fluency look at InvisiPens: 07Sa – m
	6		You have mastered these skills. Well done you are ready to progress! To develop your problem solving skills look at InvisiPens: 07Aa – e

Assessment 7

1 a Karl says that to convert from cm to m you divide by 100.
Marta says that you multiply by 100. Who is correct? [1]

  b Karl says that to convert from km to mm you divide by 1000 000.
Marta says you divide by 10 000. Who is correct? [1]

2 a Draw a quadrilateral, *ABCD*, with sides *BC* = 6.5 cm, *AD* = *DC* = 7.5 cm and angles
∠*ADC* = 105° and ∠*BCD* = 60°. [2]

  b Measure **i** *AB* **ii** ∠*DAB* **iii** ∠*ABC*. [3]

  c What type of quadrilateral is *ABCD*? [1]

  d Use a ruler to find the midpoint of each side. Join the midpoints to form another
quadrilateral. What type of quadrilateral is this? [2]

3 Briony looks at these angles. She says that angle

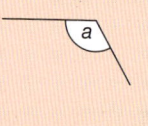 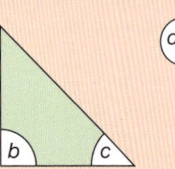 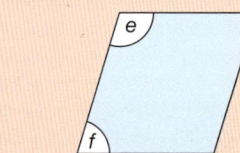 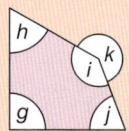

  a is obtuse **b** is acute **c** is acute **d** is reflex **e** is obtuse **f** is reflex

  g is obtuse **h** is reflex **i** is obtuse **j** is obtuse **k** is acute

Do you agree with Briony? Give reasons for your answers. [11]

4 Three churches, *A*, *B* and *C*, are the vertices of the
triangle shown. The bearing of *B* from *C* is 065°.

Work out and write down the bearings of

  a *C* from *B* **b** *C* from *A*

  c *B* from *C* **d** *A* from *C*

  e *A* from *B*. [10]

5 Layla is given these shapes all with measurements in cm.

  a 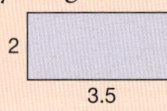 **b** 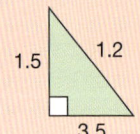 **c**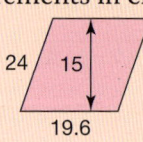

  d 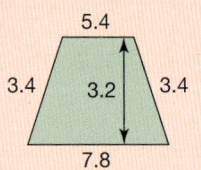 **e** 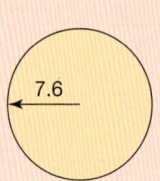 **f**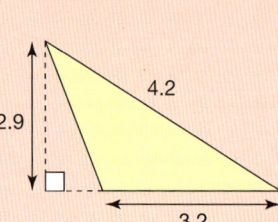

She works out that the areas of shapes **a** to **f** are

  a 7 cm² **b** 2.1 cm² **c** 294 cm² **d** 42.24 cm² **e** 23.9 cm² **f** 6.09 cm²

She then works out that the perimeters of shapes **g** to **j** are

  g 18 cm **h** 32 cm **i** 20 cm **j** 12 cm

For each shape, decide if Layla is correct or not.
If she is not correct, work out the correct area or perimeter. [18]

6 Square patio slabs come in 3 sizes, 1 m × 1 m, 2 m × 2 m and 3 m × 3 m.
Farakh wants to build a square patio of side 7 m.

   a How many of the 1 m × 1 m slabs would Farakh need?   [1]

   b Can Farakh build a square patio using just 2 m × 2 m or just 3 m × 3 m slabs?   [3]
Give reasons for your answer.

   c Can Farakh build a square patio using a mix of 2 m × 2 m and 3 m × 3 m slabs?   [4]
Draw a diagram to explain your answer.

7 Mark translates the red triangle. Match each translation to
the correct triangle on the diagram.

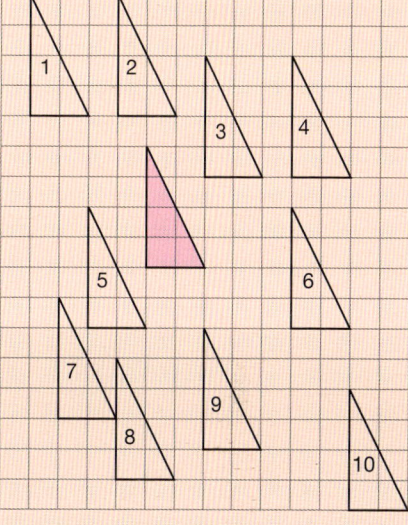

   a 5 right and 3 upwards

   b 4 left and 5 upwards

   c 3 left and 5 downwards

   d 5 right and −2 upwards

   e −2 to the right and −2 upwards

   f $\begin{pmatrix} 2 \\ 3 \end{pmatrix}$ **g** $\begin{pmatrix} -1 \\ 5 \end{pmatrix}$

   h $\begin{pmatrix} 7 \\ -8 \end{pmatrix}$ **i** $\begin{pmatrix} -1 \\ -7 \end{pmatrix}$ [9]

8 A rocket takes off from its launching pad and travels
25 miles horizontally and 45 miles vertically.

   a Show the path of the rocket on a squared grid. Take 1 square width/length
as representing 10 miles.   [2]

   b Write the vector which represents this translation.   [1]

At this position the first stage is ejected and the next stage takes it a further
15 miles horizontally and 20 miles vertically.

   c Draw this new translation at the end of your first diagram.   [2]

   d Write this new translation as a vector.   [1]

   e Using your grid, find the **single** vector that represents the whole journey.   [2]

   f How are the vectors in parts **b**, **d** and **e** related?   [1]

9 a Peter transforms the triangle, A, into
the triangle B.
Describe this transformation fully.   [3]

   b Rotate triangle A 90° clockwise, about the origin.
Label your triangle C.   [3]

   c Reflect triangle C in the line shown.
Label your triangle, D.   [3]

   d Fully describe the transformation
that maps A onto D.   [2]

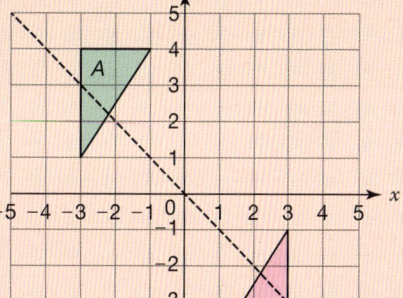

8 Probability

Introduction

When did you last look up at the night sky and see a shooting star? It is a rare event, although if you know when and where to look for meteor showers you will greatly increase your chances of seeing one. The world is full of uncertainty, from the unpredictable appearance of shooting stars or earthquakes, to manmade events like the result of a hockey match. Probability is the branch of mathematics that deals with the study of uncertainty and chance.

What's the point?

You apply probability whenever you weigh up everyday risks. For example, 'should I take an umbrella today?' A basic understanding of probability allows you to be more prepared for whatever life throws at you!

Objectives

By the end of this chapter you will have learned how to ...

- Use experimental data to estimate probabilities and expected frequencies.
- Calculate theoretical probabilities and expected frequencies using the idea of equally likely events.
- Compare theoretical probabilities with experimental probabilities.
- Recognise mutually exclusive events and exhaustive events and know that the probabilities of mutually exclusive exhaustive events sum to 1.

Check in

1 Cancel these fractions to their simplest form.

   a $\frac{12}{15}$ **b** $\frac{8}{10}$ **c** $\frac{5}{20}$ **d** $\frac{15}{25}$ **e** $\frac{10}{10}$

2 Order these decimals in size, smallest first.

   a 0.25   0.2   0.3

   b 0.7   0.8   0.75

   c 0.85   0.8   1

3 Work out each of these fraction calculations.

   a $\frac{1}{3} + \frac{2}{3}$ **b** $\frac{3}{10} + \frac{7}{10}$ **c** $1 - \frac{9}{10}$ **d** $1 - \frac{4}{5}$ **e** $1 - \frac{3}{4}$

4 Work out each of these decimal calculations.

   a $1 - 0.2$ **b** $1 - 0.7$ **c** $1 - 0.9$

Chapter investigation

Two people can play an old game called 'Rock, Paper, Scissors'.

In the game you make a shape with your hand.

● Rock is a closed fist.

● Paper is an open palm with closed fingers.

● Scissors is two fingers held like scissor blades.

The players reveal their 'hand' simultaneously.

Rock beats scissors, paper beats rock, and scissors beats paper.

Is there a best strategy for playing this game?

8.1 Probability experiments

- Probability measures how likely an **event** is to happen.
- All probabilities have a value from 0 (impossible) to 1 (certain).

For a regular dice with faces numbered one to six

The probability of obtaining seven = 0

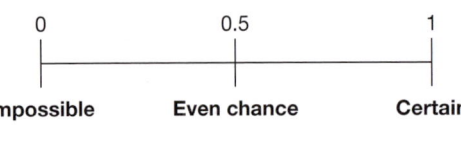

Impossible   Even chance   Certain

The probability of obtaining a number between one and six = 1

The probability of obtaining an even number = 0.5

▲ Dice have been used for thousands of years. This one is from ancient Egypt.

If the weather is sunny for 3 out of the 4 days just before a barbecue. You can predict that there is a 75% probability that the weather will be sunny for the barbecue.

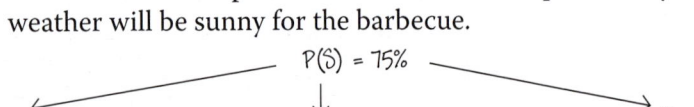

P(S) = 75%

P for probability   S for sunny   75% or 0.75 or $\frac{3}{4}$

This is an **estimated probability**, which is also called a **relative frequency**.

This can be calculated from an experiment by performing **trials**.

- Estimated probability is called the **relative frequency**.
- Relative frequency = $\dfrac{\text{Number of favourable trials}}{\text{Total number of trials}}$

You can give probabilities as fractions, decimals or percentages.

EXAMPLE

A drawing pin is thrown 10 times and lands point up in 7 cases.

a Estimate the probability that a drawing pin lands point up

The drawing pin is now thrown another 90 times and lands point up another 67 times.

b Calculate the relative frequency of the drawing pin landing point up using all the observations made so far.

a P(Point up) = $\frac{7}{10}$ = 0.7 = 70%

b P(Point up) = $\frac{7 + 67}{10 + 90}$ = $\frac{74}{100}$ = 0.74 = 74%   There were a total of 74 out of 100.

0.74 is a more reliable estimate because it is based on more information.

Exercise 8.1S

1 a Rafael and Sebastian have played 12 tennis matches against each other in the last year. Rafael won 6 out of 12 matches.
Estimate the probability that Sebastian wins the next match between the two players.

b Jo drives down the same road every morning on the way to work. On 24 journeys she had to stop at traffic lights 16 times. What is the probability that Jo will have to stop at traffic lights on her next journey to work?

c A piece of toast is dropped twenty times. It lands butter side down 14 times. Estimate the probability that toast will land butter side up when dropped again.

d A bag contains red and green balls. A ball is taken out at random, its colour noted down and then put back in the bag. This is repeated several times: red occurs 12 times and green occurs 8 times. What is the probability that the next ball pick is
i red   **ii** green?

2 Hari works at Quick-Fix garage.
Over the last month, 40 customers have come to the garage with a flat tyre caused by a puncture.
Hari records which tyre was punctured.

Tyre	Frequency
Front left	8
Front right	7
Back left	13
Back right	12

Estimate the probability that the next customer with a flat tyre has a puncture on their

a front left tyre   **b** front right tyre
c back left tyre   **d** back right tyre.

Give your answer as a

i fraction **ii** decimal **iii** percentage.

3 Rory wants to pick a red ball and can choose bag A or bag B.

Bag A	Colour	Red	Blue
	Frequency	1	2

Bag B	Colour	Red	Blue
	Frequency	2	8

a Which bag should Rory choose? Give reasons for your answer.

b Rory takes out two balls without replacing them in the bag. Rory wants to pick two reds. Which bag should he choose? Explain your answer.

4 There are 10 coloured balls in a bag. One ball is taken out and then replaced in the bag. The colours of the ball are shown in the table.

Colour	Red	Green	Blue
Frequency	9	14	27

a How many times was a ball taken out of the bag?

b Estimate the probability of taking out
i a red ball   **ii** a green ball
iii a blue ball.

c How many balls of each colour do you think are in the bag?

d How could you improve your estimate?

5 Toss a coin until you get a head, counting how many tosses you make, including the one on which the head appears.

a Do this 20 times and calculate the relative frequency from your observations for each of the values you have recorded.

b Comment on your result.

6 Throw an ordinary dice until you get a six, counting how many throws you make, including the one on which the six appears.

a Do this twenty times and calculate the relative frequency from your observations for each of the values you have recorded.

b Comment on your result.

8.1 Probability experiments

RECAP

- You can estimate the probability of an event by conducting an experiment.
- Estimated probability is called the **relative frequency**.

$$\text{Relative frequency} = \frac{\text{Number of favourable trials}}{\text{Total number of trials}}$$

HOW TO

To estimate the probability of an event
1. Think first in terms of words – likely / evens / unlikely.
2. Calculate the relative frequency of the event.
3. Answer the question. Remember that the estimated probability becomes more reliable as you increase the number of trials.

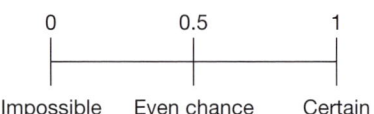

0   0.5   1

Impossible   Even chance   Certain

EXAMPLE

John and Kerri are trying to decide if a coin is fair.
When John throws the coin, 7 out of 10 are heads. He says, that heads are more likely than tails. When Kerri throws the coin, 19 out of 50 are heads. She says, that heads are less likely than tails. Who is correct?

1. If the coin is fair then there is an even chance, 0.5, of seeing heads.

2. John   Relative frequency $= \frac{7}{10} = 0.7$ Kerri   Relative frequency $= \frac{19}{50} = 0.38$

   Suggests that heads are more likely than tails.   Suggests that heads are less likely than tails.

3. Combining the results gives a more accurate estimate.

   Combined   Relative frequency $= \frac{7 + 19}{10 + 50} = \frac{26}{60} = 0.43 \, (2\,dp) < 0.5$

   This suggests that Jake is correct, but they should do more trials to be sure.

EXAMPLE

Aleesha takes a ball out of a bag, records the colour and places the ball back in the bag. She repeats this 40 times and records her results in a table.

Colour	Yellow	Green	Blue
Frequency	16	4	20
Relative frequency			

a   Complete Aleesha's table.

b   Which of these bags *could be* Aleesha's bag? Give reasons for your answer.

A    B    C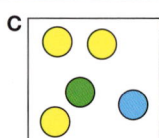

2. Use the formula for relative frequency.

Total number of trials $= 16 + 4 + 20 = 40$
Yellow $16 \div 40 = 0.4$
Green $4 \div 40 = 0.1$
Blue $20 \div 40 = 0.5$

Colour	Yellow	Green	Blue
Frequency	16	4	20
Relative frequency	0.4	0.1	0.5

b 1. Bag A has an even chance of yellow and even chance of blue. Green is impossible.

3. A is not her bag because bag 1 has no green balls.

1. Bag B has an even chance of yellow, and blue is more likely than green.

3. B could be her bag. The probabilities are close to the relative frequencies from the experiment.

3. Bag C yellow is likely, blue and green are unlikely.

3. C could be her bag. Aleesha should do more trials to get more reliable results.

Exercise 8.1A

1 Elise spins this spinner 4 times. She says that if the spinner is fair then the spinner will land on each colour once. Is Elise correct?

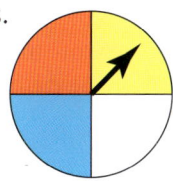

2 A bag contains 100 coloured balls. Xavier, Yvonne and Zoe each select a ball from the bag, record if the ball is red and replace the ball. The table shows their results

	Number of trials	Number of red balls
Xavier	5	4
Yvonne	20	16
Zoe	100	95

a Xavier says that the probability of choosing a red ball is $\frac{4}{5}$. Criticise his statement.

b Zoe says that the bag must contain 95 red balls. Is she correct?

c **i** Explain why the most accurate estimate of the relative frequency of choosing a red ball is 0.92.

    ii Does the bag contains 92 red balls?

3 Alik spins a spinner and records the result in a table. He adds each column when he lands on a new colour.

Colour	Red	White	Blue
Frequency	10	5	
Relative freq.		0.2	0.4

a Copy and complete Alik's table. What do you notice about the sum of the relative frequencies?

b How many of these spinners could be Alik's?

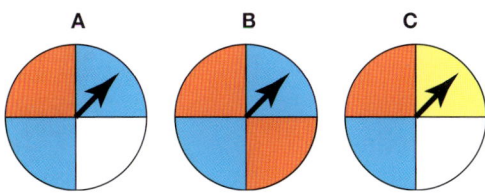

c Draw three possible spinners that could be Alik's spinner. You can divide your spinners into as many sectors as you like.

4 Marcie rolls a six-sided dice 60 times. Her results are shown in the tally chart.

Score	Frequency
1	卌 卌 卌 II
2	卌 卌 III
3	卌 卌 卌 卌 II
4	卌 III

a Which of these nets could be the net of Marcie's dice? Give reasons for your answer.

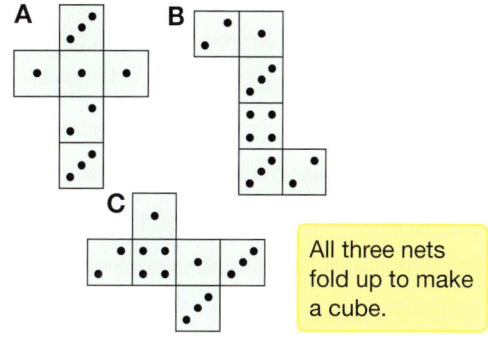

All three nets fold up to make a cube.

b Draw another net that *could* be Marcie's dice.

c Which net is most likely to be the net of Marcie's dice?

5 Alice and Aubrey are trying to decide if a spinner is fair.
The spinner is divided into three equal sections: black, white and green.
Alice spins the spinner 20 times. She lands on black 10 times and green 6 times.
Aubrey spins the spinner 30 times. She lands on green 11 times and white 12 times.
Is the spinner biased? Give your reasons.

*6 Keith thinks that fewer people are born on a Saturday or Sunday than on a weekday because any planned births will be scheduled during the week.

Keith collects information from 100 friends of different ages and finds that Saturday and Sunday seem to be as common as other days.
Suggest why the information he has collected does not mean that his idea about the days on which people are born is wrong.

Q 1209, 1210, 1211   SEARCH

8.2 Expected outcomes

If you know the probability of an event, you can calculate how many times you **expect** an outcome to happen.

- Expected frequency = number of trials × probability of the event

EXAMPLE

Calculate the expected number of heads if you spin a fair coin

a 6 times   **b** 30 times

c Sam spins a coin 30 times. She get a head 21 times.
Sam decides that the coin is **biased** (unfair). Do you agree?

a $6 \times 0.5 = 3$ There is an even chance of spinning a head.

b $30 \times 0.5 = 15$

c The coin could be biased as 21 is larger than 15. However, Sam needs to spin the coin a lot more times before she can decide.

Each spin of the coin is a **trial**.

- If you increase the number of trials, the outcome of an experiment becomes closer to the **theoretical** (expected) outcome.

EXAMPLE

Estimate the number of fours you would expect see if you throw a fair dice 300 times.

There are 6 possible outcomes which are equally likely.

$300 \times \dfrac{1}{6} = 50$

'On average' there will be 50 of each of the numbers 1, 2, 3, 4, 5 and 6 in the 300 trials

EXAMPLE

Calculate the expected number of twos if you throw a fair dice 20 times.

$20 \times \dfrac{1}{6} = 50 = 3.33.....$

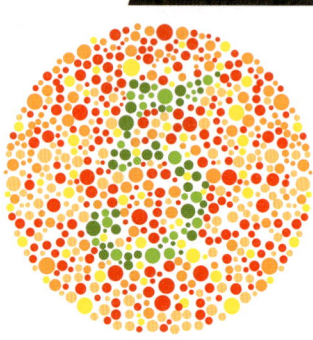

The expected number does not have to be a whole number. You could say that the answer is 'about 3'.

You can give probabilities as decimals, fractions and percentages.

EXAMPLE

8% of men have red-green colour-blindness.
Women are rarely colour-blind.
A group of 65 men and 40 women are attending a meeting.
Estimate how many people in the group suffer from red-green colour-blindness.

$65 \times 0.08 = 5.2$ About 5 are likely to be colour-blind.

Only the men are likely to be colour-blind.

Exercise 8.2S

1 A spinner is made from a circle divided into 4 equal sectors, coloured red, blue, green and yellow. How many times would you expect to see the colour blue if the spinner is spun

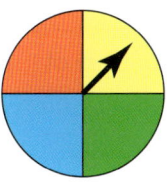

    a 20 times   **b** 9 times?

2 The probability of a drawing pin landing point up when dropped is $\frac{3}{4}$.
Estimate how many times it will land point up if it is dropped

    a 20 times   **b** 7 times.

3 All the counters in a bag are either green or black. The probability of choosing a green counter is $\frac{1}{3}$.

A counter is taken from the bag and then replaced.
The experiment is carried out 20 times.
Calculate the expected number of

    a green counters

    b black counters.

4 A spinner has 12 equal sections as shown in the diagram.

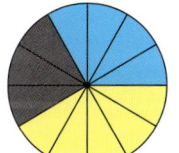

    a Estimate how many times the arrow will land on a yellow section if it is spun 60 times.

    b Estimate how many times the arrow will land on a blue section if it is spun 40 times.

5 Approximately 11% of people are left-handed. A group of 40 people are at a meeting.
What is the expected number of left-handed people attending the meeting?

6 When two fair coins are tossed the probability of getting two tails is $\frac{1}{4}$.

If 15 people each toss two coins, estimate the number of people who see two tails.

7 Three coins are thrown. The probability of the coins not being all heads or all tails is $\frac{3}{4}$.

If 25 people each throw three coins, estimate the number of people who do not see all heads or all tails.

8 A biased dice has the probabilities shown in the table. The dice is thrown 250 times and the outcomes recorded.
Calculate the expected numbers of each score that will be seen.

Score	1	2	3	4	5	6
Probability	0.1	0.2	0.25	0.15	0.1	0.2

9 Weather records show that approximately 30% of April days in a certain village have some rain. Estimate the number of days the village will have some rain in April next year.

10 A financial advisor claims that, on average, 70% of his recommended shares have increased in value. He has recommended 20 shares in the current year.
How many shares would you expect to increase in value over this year.

11 A set of cards have the numbers 1 to 9 on them. They are shuffled and the top card turned over.

The number is recorded and the process repeated until 100 observations have been recorded.
Estimate how many times you will see

    a an even number

    b a square number

    c a prime number

    d a factor of 36.

1211, 1264   SEARCH

8.2 Expected outcomes

RECAP
- Expected frequency = number of trials × probability of the event
- This is an estimate of how many times 'on average' you will see the outcome if you repeat the experiment.

HOW TO

To estimate the number of times an event will happen

① Find the total number of trials.

② Find the probability the event will happen – this may be given to you, or you may need to work it out from the information provided.

③ Calculate the expected frequency.

> What is actually seen can vary considerably when an experiment is repeated.

EXAMPLE

Katrina plays a game at a fair. Each try costs £1.
If the spinner lands on the WIN zone you win £2.
If the spinner lands on the SAFE zone you get your £1 back.
Katrina plays the game 10 times.
How much prize money should she expect to win?

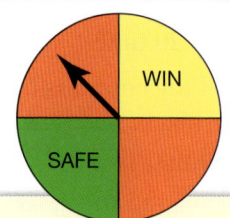

① There are 10 trials.

② Probability of winning £2 = $\frac{1}{4}$

② Probability of winning £1 back = $\frac{1}{4}$

③ Times she wins £2 = $10 \times \frac{1}{4} = 2.5$

Times she wins £1 = $10 \times \frac{1}{4} = 2.5$

Expected prize money = $(2.5 \times £2) + (2.5 \times £1) = £5 + £2.50 = £7.50$

Playing 10 games costs £10. Katrina should expect to lose £2.50.

EXAMPLE

Julia spins the spinner a number of times in an experiment.
She expects the spinner to land on red 10 times.

a How many times does Julia spin the spinner?

b When Julia carries out the experiment, the spinner lands on white 23 times. Is the spinner biased?

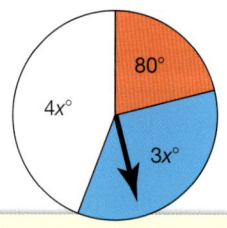

a ① number of trials × probability = expected frequency

There are 360° around a point.

$\frac{80}{360} \times$ number of spins = 10

$\frac{2}{9} \times$ number of spins = 10

Simplify the fraction

Number of spins = $10 \div \frac{2}{9}$

$= 10 \times \frac{9}{2} = 45$

b ② Find x and the angle of the white section.

$4x + 3x + 80 = 360$

$7x + 80 = 360$ Collect like terms.

$7x = 280$ Subtract 80 from both sides.

$x = 40$ Divide both sides by 7.

White has an angle of $4x° = 160°$.

③ White is twice as likely as red.

Julia should expect the spinner to land on white 20 times. 23 is close to 20, so the spinner is not biased.

Exercise 8.2A

1 A biased coin is flipped 5 times.

The ratio of heads to tails is 2:3.

   a The coin is flipped 30 times. Estimate the number of times you will see a head.

   b The coin is flipped 40 times. Estimate the numbers of times you will see tails.

2 *ABCD* is a square and *E* and *F* are the midpoints of sides *AD* and *BC*.

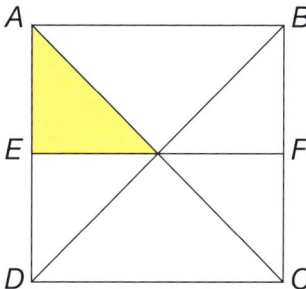

   a If 40 points are selected at random in the square how many would you expect to see in the shaded area?

   b If this shape was on the bottom of an open box and 40 small counters were dropped into the box, do you think you would see more, less, or about the same in the shaded area as in part **a**?

3 A spreadsheet generates a series of random digits. In a set of 150 random digits, estimate how many of them will be less than 4.

4 From looking at past results, Sanjiv knows that his favourite team have won 55% of their home league games and drawn 25%.

If the team plays 10 games, how many games would you expect the team to lose?

5 The table below shows the proportions of the hair colour of male and female students in Newcastle.

	Fair	Dark
Male	50%	50%
Female	60%	40%

A class in Newcastle has 18 boys and 15 girls. Estimate the number of dark haired students in the class.

6 The diagram shows a spinner with 5 coloured sides, but the spinner is not symmetrical.

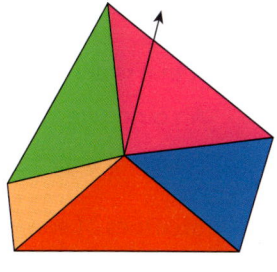

Aisha thinks the probability of the spinner landing on a particular colour is proportional to the length of the side, and draws up a probability table.

Colour	Purple	Blue	Red	Brown	Green
Probability	$\frac{1}{4}$	$\frac{1}{6}$	$\frac{1}{4}$	$\frac{1}{12}$	$\frac{1}{4}$

If Aisha spins the spinner 60 times, estimate the number of times she will land on each colour (if her probabilities are right).

*7 A red bag has 5 white and 10 black balls in it, and a blue bag has 5 white and 5 black balls in it. All the balls are identical except for their colour.

   a You choose a ball from the red bag at random and note its colour before returning it to the bag. If you do this 30 times, estimate how many times you will see a white ball.

   b If you have used the blue bag instead, would you expect to see a white more often, less often or about the same?

*8 The coloured sections on this spinner are all equal. Shawn is playing a game at a school fete which charges 50p a go. She wins a prize if her spinner lands on a blue sector.

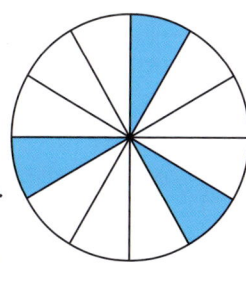

   a If Shawn has 24 goes, how often would she be likely to win?

   b If the prize is £1, how much profit does the fete make on average?

1211, 1264   SEARCH

8.3 Theoretical probability

- A trial is an activity or experiement.
- An outcome is the result of a trial.
- An event is one or more outcomes of a trial.

> Rolling a dice is a trial.
> The outcomes of rolling a dice are 1, 2, 3, 4, 5, 6.
> Rolling an even number is an event.

The outcomes of some experiments are equally likely, for example, throwing a fair coin or rolling a fair dice.

- The **theoretical probability** of an event is

$$\text{Probability of event happening} = \frac{\text{Number of outcomes in which event happens}}{\text{Total number of all possible outcomes}}$$

EXAMPLE

A fair dice is rolled and the number showing on the top is scored. Find the probability of these events.

a 3 is scored.   **b** A factor of 12 is scored.

c A prime number is scored.

> There are 6 possible outcomes which are equally likely.
>
> **a** $P(3) = \frac{1}{6}$ There is one 3.
>
> **b** $P(\text{factor of 12}) = \frac{5}{6}$ Five outcomes are factors of 12: 1, 2, 3, 4 and 6.
>
> **c** $P(3) = \frac{3}{6} = \frac{1}{2}$ Three outcomes are prime: 2, 3 and 5.

The probability of an event can be written as P(event).

Theoretical probability is based on equally likely outcomes.

- You use experimental probability when the event is unfair or biased or when the theoretical outcome is unknown.

The closer experimental probability is to theoretical probability the less likely it is that there is bias.

EXAMPLE

Sami threw a drawing pin 10 times and it landed point up 7 times. Cristiano threw a similar drawing pin 43 times and it landed point up 31 times.

a Calculate the relative frequency of the pin landing point up for each person.

b Which of these is the better estimate of the probability?

> **a** Sami $\frac{7}{10} = 0.7$ Cristiano $= \frac{31}{43} = 0.72$ (2 dp)
> **b** Cristiano used the largest sample so he has the better estimate.

- The experimental probability becomes a better estimate for the theoretical probability as you increase the number of trials.

Exercise 8.3S

1 A fair dice is rolled and the number showing on the top is scored.
Find these probabilities.

    a   P(6)   **b**   P(less than 3)

    c   P(factor of 10)   **d**   P(square number)

2 Are the outcomes in each case equally likely? If they are, give the probability of each outcome. If they are not, explain why.

    a   The score showing when a fair die is thrown.

    b   Whether a drawing pin lands point up or down when it is thrown in the air.

    c   The day of the week a baby is born on.

    d   The number of times I have to toss a fair coin until I see a head.

    e   What the letter is, if a letter is chosen at random from a book.

3 All the counters in a bag are either green or black. At the start the probability a counter chosen at random from the bag is green is $\frac{1}{4}$. Counters are not replaced when they are taken out.

The first two counters taken out of the bag are green.

What is the least number of black counters that must still be in the bag?

4 A class of 24 students are each given a similar biased dice and asked to record how many times they saw a six when they threw their dice 20 times.

7	3	6	6	8	9	5	7
6	8	7	8	6	7	4	9
7	7	6	7	8	6	9	5

    a   Give the highest and lowest relative frequencies found by any of the students.

    b   Find the best possible estimate of the probability of a six showing on the biased dice from the results given.

5 For the following events give a value between 0 and 1 for the probability of the event described happening.

If you have to estimate the probability then write (est) after your answer.

    a   You score a 4 when you throw a fair dice

    b   A set of cards has the numbers 1-9 written on one of each of the cards. You pick a card at random and it is an even number.

    c   You score less than 7 when you throw a fair dice.

    d   In their 11th match of a league season the top team is playing at home to the bottom team. The match is a draw,

    e   Your score is a factor of 60 when you throw a fair dice.

6 Michelle wonders how often there is a difference of more than 2 when you throw a pair of fair dice. She does an experiment, noting down how often it happens after each block of 10 throws.

5	4	5	3	2	4	6
3	3	3	5	4	2	4

    a   Give the relative frequencies at the end of each block of 10 throws.

The first two are $\frac{5}{10} = 0.5$, and $\frac{9}{20} = 0.45$.

    b   What is the best estimate Michelle has of the probability of getting a difference of more than 2?

7 Three friends were investigating how often a biased dice showed a six.
Sergio got 10 sixes from 40 throws.
Annalise got 8 sixes from 40 throws
Dominika got 21 sixes from 80 throws.
Calculate the best estimate of throwing a six.

8.3 Theoretical probability

- You use theoretical probability for fair activities, for example rolling a fair dice.
- For unfair or biased activities, or where you cannot predict the outcome, you use experimental probability (relative frequency).
- Relative frequency is the proportion of successful trials in an experiment.

Theoretical probability
$$= \frac{\text{Number of favourable outcomes}}{\text{Total number of outcomes}}$$
Relative frequency
$$= \frac{\text{Number of successful trials}}{\text{Total number of trials}}$$

HOW TO

To compare theoretical probability with relative frequency

(1) Start by assuming that the activity is fair and calculate the theoretical probability of the event.

(2) Use the results of the experiment to calculate the relative frequency.

(3) If the relative frequency and theoretical probability are very different, then the activity could be biased.

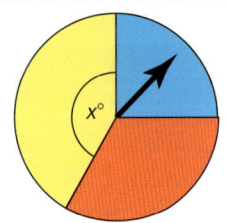

The more trials you carry out, the more reliable the relative frequency will be.

EXAMPLE

Kaseem rolled a dice 50 times and in 14 of those he scored a 2. Is the dice biased towards 2? Explain your answer.

(1) Theoretical probability $2 = \frac{1}{6} = 0.166\ldots$ If the dice is fair, then every outcome is equally likely.

(2) Relative frequency of rolling a $2 = \frac{14}{50} = 0.28$ Converting to decimals makes it easier to compare.

(3) The relative frequency is higher than the theoretical probability.
The dice appears to be biased towards 2.

EXAMPLE

A fair spinner is divided into blue, red and yellow sectors.
Valerie knows that the blue sector is one quarter of the spinner.
Valerie can't remember the size of angle x, but she knows that it is either 150° or 160°.
She spins the spinner 100 times and records her results in a table.

Colour	Blue	Yellow	Red
Frequency	23	42	35

Find the size of angle x. Explain your answer.

(1) Calculate the theoretical probabilities for each angle. There are 360° in a circle.

$x = 150°$ $P(\text{Yellow}) = \frac{150}{360} = 0.416..$ $x = 160°$ $P(\text{Yellow}) = \frac{160}{360} = 0.444\ldots$

Angle for red $= 360 - (150 + 90) = 120$ Angle for red $= 360 - (160 + 90) = 110$

If $x = 150°$, $P(\text{Red}) = \frac{120}{360} = 0.333\ldots$ $P(\text{Red}) = \frac{110}{360} = 0.305\ldots$

(2) Use the information in the table to find the relative frequency.

Relative frequency of yellow $= \frac{42}{100} = 0.42$ Relative frequency of red $= \frac{35}{100} = 0.35$

(3) As the spinner is fair, the relative frequency and theoretical probabilities should be similar.

The relative frequencies are closest to the probabilities for $x = 150°$.

Exercise 8.3A

1 The table below shows the gender and hair colour of 30 students in a class.

	Fair	Dark
Male	7	9
Female	8	6

A student is chosen at random from the class. What is the probability that the student is

a a girl   **b** a dark-haired boy.

2 Amber rolled a dice 100 times and in 71 of those rolls her number was a factor of 12.

Is the dice biased towards 5?

Give your reasons.

3 *ABCD* is a square and *E* and *F* are the midpoints
of sides *AD* and *BC*.

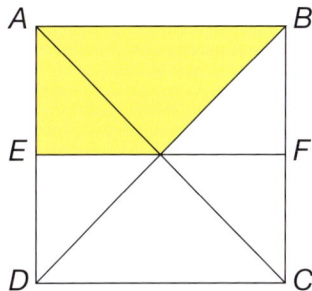

If a point is selected at random in the square what is the probability it is in the shaded area?

4 The square with sides 4 cm shown has four quarter circles shaded.
A point is chosen at random in the square. Calculate the probability the point is in the shaded area.

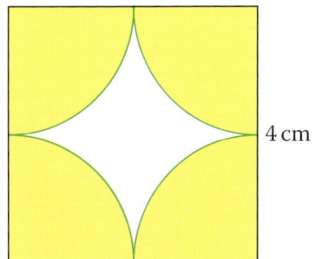

4 cm

5 Bag 1 contains 8 red balls and 12 green balls. Bag 2 contains 5 red balls and 7 green balls. The balls are identical apart from colour.

a You choose a ball from bag 1 at random. What is the probability you choose a green ball?

All the balls are now put into one bag and you choose a ball at random.

b What is the probability you choose a red ball?

6 The coloured sections on the spinner are all equal.

If the spinner lands on a green section, the score is doubled.

If the spinner lands on a blue section, the score is halved.

Shawn is playing a game with her friend Andrea who goes first, with the result shown.

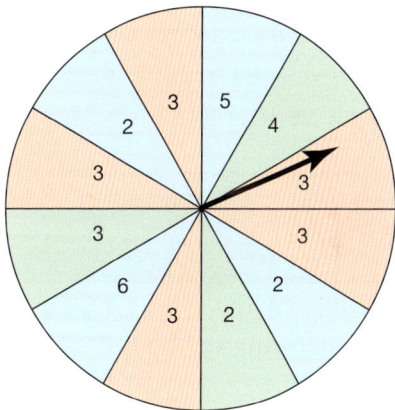

What is the probability that Shawn scores more than Andrea?

*7 Iveta has a spinner divided equally into 19 sectors.

The sectors are white, black and red.

There are the same number of black and red sectors.

Iveta spins the spinner 80 times and records her results in a table.

Colour	White	Black	Red
Frequency	17	32	31

Explain why the spinner could be biased.

Q 1210, 1264   SEARCH

8.4 Mutually exclusive events

● Two events are **mutually exclusive** if they cannot both happen at the same time.

A set of events is exhaustive if they include all possible outcomes.

EXAMPLE

You throw an ordinary dice. Here are three possible events.

A A prime number **B** An odd number **C** A square number

Which of these pairs of events are mutually exclusive?

a A and B   **b** A and C   **c** B and C

A: 2, 3 and 5   B: 1, 3 and 5   C: 1 and 4

a A and B both contain 3 and 5, so not mutually exclusive

b A and C have nothing in common, so they are mutually exclusive

c B and C both contain 1, so not.

● If a set of mutually exclusive events are also **exhaustive** then their probabilities sum to 1.

EXAMPLE

The probability that a hockey team wins a match is 0.65. The probability of a draw is 0.2. What is the probability that they lose the match?

The outcomes win, draw and lose are mutually exclusive.

$0.65 + 0.2 + P(Lose) = 1$

$P(Lose) = 1 - (0.65 + 0.2) = 0.15$

● Probability of an event happening = 1 – probability of event not happening

$P(A) = 1 - P(\text{not } A)$

EXAMPLE

The probability that you are selected for a random body search by an electronic security monitoring system is 0.06. What is the probability you will not be selected?

$1 - 0.06 = 0.94$ The probabilities add to 1.

EXAMPLE

The probability that a child in a nursery class has a cat is 0.3

    a Find the probability that the child does not have any pets.

    b Find the probability that the child has a hamster.

These events may not be mutually exclusive.

    a Not known, they can have other animals as pets.

    b Not known, a child may have either a cat or a hamster, both or neither!

Exercise 8.4S

1 For each of the following pairs of events say whether or not they are mutually exclusive.

a P: a girl has red hair
Q: she has brown eyes

b F: a boy gets the same score on 2 dice
G: his total score is odd

c M: a girl gets the same score on 3 dice
N: her total score is odd

d X: a girl gets the same score on 3 dice
Y: her total score is a multiples of 6.

2 A pack of cards contains 60 cards in four colours: black, red, green and blue. There are 15 of each colour.
The black cards carry the numbers 1 to 15.
The red cards are multiples of 2.
The green cards are multiples of 3.
The blue cards are multiples of 6.
The top card is turned over.
For each pair of events, say whether or not they are mutually exclusive.

a A: the card is black, and
B: it is an even number

b C: the card is red, and
D: it is an odd number

c E: the card is green, and
F: it is a factor of 20

d G: the card is blue, and
H: it is a prime number

3 All the counters in a bag are either green or black. The probability a counter chosen at random from the bag is green is $\frac{1}{3}$.

a Find the probability that a counter chosen at random is blue.

b Find the probability that a counter chosen at random is black.

4 All the counters in a bag are either green, white or black.
The probability a counter chosen at random from the bag is green is $\frac{1}{3}$, and for white it is $\frac{1}{2}$.
What is the probability a counter chosen is

a black **b** not white.

5 Seamons has two packs of 12 cards, each labelled 1 to 12. He removes all the multiples of 4 from one of the packs, combines the two packs and shuffles them. He turns over the top card. Find the probability that the card is

a a multiple of 6

b an even number

c not a prime number.

6 A biased dice has the probabilities shown in the table.

Score	1	2	3	4	5	6
Probability	0.1	0.2	x	0.3	0.1	0.1

Find the probability of getting
a a three

b an even number

c not a prime number

7 A fair dice has a triangle, square, circle, rectangle, sphere and cylinder showing on the six sides. The dice is rolled.
Find the probability that

a it shows a 3D shape

b it does not show a 2D shape with straight edges.

8 Olympic, United and City are the names of 3 teams playing in a 5-a-side league. The probability Olympic will win the league is 0.2, which is the same as City.
United are twice as likely as Olympic or City to win the league. Find the probability that

a United win the league

b City do not win the league

c the league is won by a team other than one of these three teams.

8.4 Mutually exclusive events

- Events are mutually exclusive if they cannot happen at the same time.
- If the events are also exhaustive, then the probabilities sum to 1.
- $P(A) = 1 - P(\text{not } A)$.

> You could list the outcomes in a table.

HOW TO

① Read the question carefully and think about the possible outcomes. It often helps to list them.

② Use what you know about mutually exclusive and exhaustive events to calculate probabilities.

EXAMPLE

A red and a blue dice are thrown together and the highest score is recorded.
Find the probability that the score recorded is 5 or less.

① There are 36 possible outcomes, and these can be shown in a table.

	1	2	3	4	5	6
1	1	2	3	4	5	6
2	2	2	3	4	5	6
3	3	3	3	4	5	6
4	4	4	4	4	5	6
5	5	5	5	5	5	6
6	6	6	6	6	6	6

> The complementary event is just another way of saying 1 minus the event.

② It is easier to find the **complementary** event

$$(5 \text{ or less}) = 1 - P(6)$$
$$= 1 - \frac{11}{36}$$
$$= \frac{25}{36}$$

EXAMPLE

There are a number of balls in a bag, identical except for their colour.

A ball is taken from the bag at random.

The probability of taking

- green is $\frac{1}{6}$
- blue is $\frac{1}{4}$
- black is $\frac{1}{2}$.

Are any other colours in the bag? Give your reasons.

② If the events are exhaustive then their probabilities add up to 1.

$$\frac{1}{6} + \frac{1}{4} + \frac{1}{2} = \frac{2}{12} + \frac{3}{12} + \frac{6}{12} = \frac{11}{12}$$

The probabilities add up to a number less than 1. There must be at least one other colour.

Exercise 8.4A

1 A red and a blue dice are thrown together and the difference between the scores is recorded.

  a Draw a table to show the possible outcomes.

  b What is the probability of a difference of
    i exactly one
    ii zero
    iii more than 1?

2 There are a number of balls in a bag, identical except for their colour.
A ball is taken from the bag at random.
Serena says that the probability of taking a green is $\frac{1}{4}$, taking a blue is $\frac{1}{3}$ and taking a black is $\frac{1}{2}$.
Can Serena be correct? Give reasons for your answer.

3 There are a number of balls in a bag, identical except for their colour.
A ball is taken from the bag at random.
Arinda says that the probability of taking a green is $\frac{1}{6}$, taking a blue is $\frac{1}{3}$, taking a white is $\frac{1}{8}$ and taking a black is $\frac{3}{8}$.

  a Can you tell if Arinda is correct? Give reasons for your answer

  b Can you tell if there are any other colours? Give reasons for your answer.

4 A bag contains at least green, blue and white balls. The probability a ball chosen at random is green is $\frac{1}{7}$. Blue is twice as likely to be chosen as green and white is twice as likely to be chosen as blue. Say why you know there are no other colours in the bag.

5 A bag of sweets contains 4 vanilla fudge pieces, 3 white chocolate caramels, 3 chocolate covered fudge pieces and 2 caramel fudge pieces.
What is the probability that a sweet chosen at random contains

  a chocolate

  b fudge

  c caramel?

Why do these probabilities add to more than 1?

6 A set of cards with the numbers 1 to 10 is shuffled and a card chosen at random.
Here are four possible events.
A A prime number   **B** A factor of 36
C An even number   **D** An odd number

  a List any pairs of mutually exclusive events.

  b List any pairs of exhaustive events.

  c Explain why A and C are not mutually exclusive.

7 The table below shows the gender and hair colour of the 28 students in a class.

	Fair	Dark	Red
Male	6	9	3
Female	7	6	0

A student is chosen at random from the class, and here are four possible events.
M the student is male
F the student is female
R the student has red hair
D the student has dark hair

Giving a reason, explain whether the following statements are true or false.

  a M and F are mutually exclusive.

  b M and F are exhaustive.

  c M and D are mutually exclusive.

  d F and R are mutually exclusive.

8 A red and a blue dice are thrown together and the blue score is subtracted from the red score.

  a Draw up a table to show the possible outcomes.

  b What is the probability the score is
    i −2 **ii** 5 **iii** less than 5?

9 The probabilities for a biased dice are shown in the table.

Score	1	2	3	4	5	6
Probability	0.1	0.2			0.2	0.1

Getting a 4 is three times as likely as the probability getting a 3.
Find the probability of getting a

  a 3 **b** 4 **c** prime number.

Summary

Checkout

You should now be able to...

✔	Use experimental data to estimate probabilities and expected frequencies.	1 – 3
✔	Calculate theoretical probabilities and expected frequencies using the idea of equally likely events.	1 – 3
✔	Compare theoretical probabilities with experimental probabilities.	3
✔	Recognise mutually exclusive events and exhaustive events and know that the probabilities of mutually exclusive exhaustive events sum to 1.	4 – 7

Language	Meaning	Example
Trial	An activity or experiment.	Rolling a single dice.
Outcome	The result of a trial.	Rolling a single dice and getting a 2.
Event	One or more outcomes of a trial.	When you roll a single dice you can get a 1, 2, 3, 4, 5 or 6.
Impossible	It cannot happen. The probability of it happening is 0.	You roll a single dice and the number is even and odd.
Certain	It must happen. The probability of it happening is 1.	You roll a single dice and the number is less than 7.
Likely	It has a better chance of happening than not happening. The probability is more than 0.5 and less than 1.	You roll a dice and the number is greater than 1.
Unlikely	It has a worse chance of happening than not happening. The probability is more than 0 and less than 0.5.	You roll a dice and the number is greater than 4.
Even chance	It has exactly the same chance of happening as not happening. The probability of it happening is 0.5.	You flip a coin and get a head.
Relative frequency	The **experimental probability** of an outcome after several trials is $$\frac{\text{Number of times the outcome happened}}{\text{Number of times the activity was done}}$$	A drawing pin is thrown 100 times. It lands point down 72 times. The relative frequency of the pin landing point down is $\frac{72}{100} = 0.72$
Expected frequency	How many times you expect the outcome to happen. Number of trials × probability of the event	The expected frequency of getting a 5 when you roll a fair dice 40 times is $\frac{1}{6} \times 40 = 6.666... = 6.\dot{6}$ about 7 times.
Theoretical probability	The predicted value found by $$\frac{\text{Number of ways the outcome could happened}}{\text{Number of possible outcomes}}$$	There is 1 way of rolling a two on a fair dice. There are 6 different possible outcomes. The theoretical probability of rolling a two is $\frac{1}{6}$.
Bias/ Biased	All outcomes are not equally likely.	A dice has the numbers 1, 1, 2, 3, 4, 5 on its faces. It is more likely that you will roll a 1 than any of the other outcomes. The dice is biased.
Equally likely	All outcomes have the same probability of happening.	When you roll a fair dice all the outcomes {1, 2, 3, 4, 5, 6} are equally likely. They all have a probability of $\frac{1}{6}$.

Review

1 Jake counts the spots on a number of ladybirds he finds in a park.

Number of Spots	Frequency
2	8
7	20
10	12
other	10

   a What is the relative frequency of ladybirds with 7 spots?

   b Assuming there are 500 ladybirds in the park, estimate how many will have exactly 10 spots.

2 A spinner with four colours on it is repeatedly spun and the results recorded.

Outcome	Relative frequency
Blue	0.6
Red	0.16
Green	
Yellow	0.04

   a What is the relative frequency of green?

   b Use these results to estimate how many times the spinner would land on red if it were spun 150 times.

3 An unbiased dice is thrown 20 times and the number 5 occurred 2 times.

   a What is the relative frequency of a 5?

   b What is the theoretical probability of a 5?

   The dice is now thrown an additional 100 times.

   c What would you expect to happen to the relative frequency of a 5?

4 A bag contains 7 white and 3 black counters. A counter is selected at random.

   What is the probability the counter is

   a black

   b red

   c white or black?

5 A fair coin is flipped twice. What is the probability of getting 'heads' twice?

6 A cupboard contains lots of textbooks. The probability that a randomly selected book is a maths textbook is $\frac{3}{11}$. What is the probability that it is not a maths textbook?

7 Sam always takes either a cheese, tuna or chicken sandwich for lunch each day. The probability Sam takes a cheese sandwich is $\frac{1}{7}$. He is twice as likely to take a tuna sandwich as a cheese sandwich.

   Calculate the probability of taking

   a a tuna sandwich

   b a chicken sandwich.

What next?

Score			
	0 – 3		Your knowledge of this topic is still developing. To improve look at MyMaths: 1209, 1210, 1211, 1262, 1263, 1264
	4 – 6		You are gaining a secure knowledge of this topic. To improve your fluency look at InvisiPens: 08Sa – f
	7		You have mastered these skills. Well done you are ready to progress! To develop your problem solving skills look at InvisiPens: 08Aa – e

Assessment 8

1 It is impossible to thrown a coin five times and see tails every time.
Is this statement correct? Give a reason for your answer. [1]

2 A fair coin is thrown and lands on heads twice. What is the probability of a head on the third throw? Give reasons for your answer. [2]

3 A dice is thrown 100 times and the following results are obtained.

Score	1	2	3	4	5	6
Frequency	14	17	22	18	15	14

Calculate the relative frequency of scoring an odd number. [2]

4 One in every eight children in a town has asthma.
How many children would you expect to have asthma

a in the school of 1600 children [1]

b in a road where 14 children live [1]

c in a class of 34 children? [1]

5 The sides of a biased spinner are labelled 1, 2, 3, 4 and 5.
The probability that the spinner will land on each of the numbers 1, 3 and 4 is given in the table.

Number	1	2	3	4	5
Probability	x	0.14	0.26	0.22	x

The probability that the spinner lands on either of the numbers 1 or 5 is the same.

a It is claimed that $x = 0.2$. Say why this is wrong and find the correct value of x. [3]

The spinner is spun 200 times.

b How many times should you expect the spinner to land on 3? [1]

c How many times should you expect the spinner to land on an odd number? [2]

d How many times should you expect the spinner to land on a prime number? [2]

6 Adam and Ben buy identical packets of mixed vegetable seeds. They pick out the pea seeds from their packets. Their results are shown in the tables.

Adam's results

Number of packets	Number of seeds in each packet	Number of pea seeds in each packet
5	30	5, 5, 8, 3, 6

Ben's results

Number of packets	Number of seeds in each packet	Total number of pea seeds
10	30	51

a Write down an estimate of the probability of selecting a pea seed at random from a packet using

   i Adam's results   ii Ben's results. [3]

b Whose results are likely to give the better estimate of the probability? Give reasons for your answer. [2]

7 A fridge contains 20 bottles of juice.
There were four bottles each of apple, blueberry, grape, orange and pineapple.

a A bottle is taken at random from the fridge. What is the probability that it is apple juice? [1]

b All of the orange juice is taken from the fridge. A bottle is taken at random from those remaining in the fridge, what is the probability that it is either grape of pineapple juice? Write your answer in its simplest form. [2]

8 A box of cat food contains 30 individual packets.
The probability that a tuna packet is chosen from the box is 0.3
How many packets of tuna packets are in the box? [2]

9 The probability that the 8:45 train into Manchester Piccadilly station arrives early is 0.18
The probability that the train arrives on time is 0.31

   a A frustrated passenger complains that the train is late more often than it is on
time or early. Is the passenger correct? [2]

   b A commuter takes 225 journeys on this train during one year.
How many days can the commuter expect to be on time? [2]

10 The table below shows the probabilities of selecting raffle tickets from a drum.
The tickets are coloured red, white or blue and numbered 1, 2, 3 or 4.
A Raffle Ticket is taken at random from the drum.

   a Find the probability that

      i it is white and numbered 3 [1]

      ii it is numbered 1 [2]

      iii it is red [2]

      iv it is either blue or numbered 2. [3]

   b What raffle ticket is impossible to draw?
Give reasons for your answer. [1]

		Number			
		1	**2**	**3**	**4**
Colour	Red	$\frac{2}{25}$	$\frac{1}{25}$	$\frac{3}{50}$	$\frac{1}{10}$
	White	$\frac{9}{50}$	$\frac{7}{50}$	$\frac{3}{25}$	$\frac{1}{50}$
	Blue	$\frac{3}{50}$	$\frac{4}{25}$	$\frac{1}{25}$	0

11 At a busy road junction, the traffic lights operate as follows

red: 40 seconds   red and amber: 10 seconds   green: 40 seconds   amber: 10 seconds

Work out the probability of each event. Give your answer as a simplified fraction.

   a the lights show green [1]   **b** there is an amber light showing [2]

   c there is not a red light showing [2]   **d** either green or amber lights are showing [2]

12 a Are these pairs of events mutually exclusive?
Give reasons for your answers.

      i Red section and multiples of 3. [2]

      ii Blue section and prime numbers. [2]

      iii White section and multiples of 7. [2]

   b Every number on the spinner is increased by 1.
How does this change your answers to part **a**? [6]

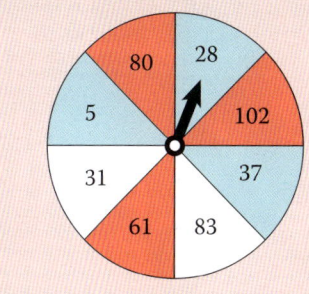

13 During December the probability that it is snowing is 0.1. The probability that it is sunny is 0.3
Is the probability that it is snowing or sunny in December is 0.4?
You must give a reason for your answer. [2]

14 A regular pentagon, two squares and a triangle are joined
to create a spinner.
Assume that the spinner is unbiased.

   a Find the probability of landing on the blue region of
the spinner. [4]

   b How does the assumption that the spinner is unbiased
affect your answer to part **a**? [1]

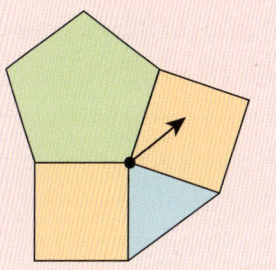

9 Measures and accuracy

Introduction

During the 1936 Olympic Games held in Berlin in Nazi Germany, the US athlete Jesse Owens broke the 100 m men's world record, with a time of 10.2 s, a record that remained unbroken for 20 years afterwards. He also took the gold medal in the long jump in the same games, and he is widely believed to have been the greatest athlete ever. It is said that Hitler was not impressed by Owens's victories, as he had hoped that the games would be a validation of Aryan supremacy.

The current 100 m world record is held by Usain Bolt of Jamaica, with a time of 9.58 s in 2009. Note the increased accuracy in the reporting of the time in 2009 compared with 1936.

What's the point?

Being able to rely on accurate decimal measurements is vital in athletics. In a 'photo-finish' with two competitors crossing the finishing line apparently together, 0.01 seconds can be the difference between gold and silver!

Objectives

By the end of this chapter you will have learned how to …

- Round numbers and measures to an appropriate degree of accuracy.
- Use approximation to make estimates.
- Check calculations using approximation and estimation.
- Use standard units of length, mass, volume, capacity, time and area.
- Use inequality notation to state error intervals and interpret limits of accuracy.

Check in

1 Work out these multiplications and divisions.

    a 40×10 **b** 14×1000 **c** 3.1×10 **d** 13.4×100

    e 6.3×1000 **f** $400 \div 10$ **g** $6000 \div 100$ **h** $430 \div 100$

    i $640 \div 1000$ **j** $3.1 \div 10$ **k** 0.034×100 **l** $0.78 \div 10$

2 What is the temperature on this thermometer?

    $-10°C$ $0°C$ $10°C$

3 Gavin is running at 5 mph.
He runs for 3 hours.
How far does he travel?

Chapter investigation

Most of the quantities you can measure, such as length and mass, are based on the decimal number system; however time is not. During the French Revolution, it was proposed to replace the standard system of hours and minutes with French Revolutionary Time, in which time would be measured using powers of 10.

Investigate French Revolutionary Time and write a short report.

At what times would your school day start and finish using this system?

9.1 Estimation and approximation

p.8

● An **approximation** of a value can be found by rounding.

A useful first approximation is to one **significant figure** but other approximations may be better.

EXAMPLE

Use approximations to answer these calculations.

a $76.5 + 184.2$ **b** $12.3 - 8.9$ **c** $20 - 14.53$

a $\approx 80 + 200 = 280$

b $\approx 12 - 9 = 3$

c $\approx 20 - 15 = 5$

Round each number appropriately before doing the calculation.

Remember
BIDMAS
Brackets
Indices (or powers)
Division or
Multiplication
Addition or
Subtraction

● **Approximations** can be used to **estimate** an answer before starting a calculation. Estimates are useful for checking your answer or adjusting place value.

EXAMPLE

Estimate the answer to this calculation. $\dfrac{4.23 \times 5.89}{9.7}$

$\approx \dfrac{4 \times 6}{10} = \dfrac{24}{10} = 2.4$ Round each number to the nearest whole number.

The symbol $\approx$ means 'approximately equal to'.

EXAMPLE

Estimate the answers to these calculations. **a** $\dfrac{8.93 \times 28.69}{0.48 \times 6.12}$ **b** $\dfrac{17.4 \times 4.89^2}{0.385}$

Round each number to one significant figure.

a $\approx \dfrac{9 \times 30}{0.5 \times 6}$

$= \dfrac{270}{3} = 90$

b $\approx \dfrac{20 \times 5^2}{0.4}$

$= \dfrac{20 \times 25}{0.4} = \dfrac{500}{0.4} = \dfrac{5000}{4} = 1250$

EXAMPLE

Estimate $\sqrt{5}$ to 1 dp. Do *not* use a calculator.

$\sqrt{4} = 2$, $\sqrt{9} = 3$. So $2 < \sqrt{5} < 3$ $\sqrt{5}$ is between 2 and 3.

Try 2.5

$\begin{array}{r} 2.5 \\ \times\ 2.5 \\ \hline 500 \\ +125 \\ \hline 6.25 \end{array}$ $6.25 > 5$
$2 < \sqrt{5} < 2.5$

Try 2.2

$\begin{array}{r} 2.2 \\ \times\ 2.2 \\ \hline 440 \\ +\ 44 \\ \hline 4.84 \end{array}$ $4.84 < 5$
$2.2 < \sqrt{5} < 2.5$

Try 2.25

$\begin{array}{r} 2.25 \\ \times\ 2.25 \\ \hline 45000 \\ 4500 \\ +\ 1125 \\ \hline 5.0625 \end{array}$ $5.0625 > 5$
$2.2 < \sqrt{5} < 2.25$

$\sqrt{5} = 2.2$ (1 dp)

Number   Measures and accuracy

Exercise 9.1S

1 Round each of these numbers to
 i 1 sf   **ii** the nearest integer.

 a 8.3728   **b** 18.82

 c 35.84   **d** 278.72

 e 1.3949   **f** 3894.79

 g 0.008 372   **h** 2399.9

 i 8.9858   **j** 14.0306

 k 1403.06   **l** 140 306

2 Round each of these numbers to
 i 3 sf   **ii** the nearest 100.

 a 3492   **b** 611 129

 c 0.003721   **d** 859

 e 9999   **f** 859 237

> When asked to estimate an answer you should
> show how you obtained your estimate.

3 Estimate answers to these calculations by
 rounding to the accuracy given.

 a $37.43 \div 3.52$ (1 sf)

 b 2.497×1.99 (1 dp)

 c $6342 \div 897$ (2 sf)

 d $50.73 \div 17.41$ (2 sf)

 e 0.0875×123.4 (1 sf)

 f $0.64 \div 1.60$ (2 dp)

4 Estimate answers to these calculations.

 a $4.88 + 3.07$ **b** $216 + 339$

 c $0.0049 + 0.00302$ **d** $43.89 - 28.83$

 e 3.77×0.85 **f** $44.66 \div 0.89$

5 Estimate the answers to these calculations.

 a 3.76×4.22 **b** 17.39×22.98

 c $\dfrac{4.59 \times 7.9}{19.86}$ **d** $54.31 \div 8.8$

6 Estimate the answers to these calculations.

 a $\dfrac{29.91 \times 38.3}{3.1 \times 3.9}$ **b** $\dfrac{16.2 \times 0.48}{0.23 \times 31.88}$

 c $(5.2^2 - 4.1^2)^2$ **d** $\dfrac{63.8 \times 1.7^2}{1.78^2}$

 e $\sqrt{(2.037 \div 0.041)}$

 f $\sqrt{(27.6 \div 0.57)}$

7 Estimate the answers to these calculations.

 a 4.98×6.12 **b** $17.89 + 21.91$

 c $\dfrac{5.799 \times 3.1}{8.86}$ **d** $34.8183 - 9.8$

 e $\dfrac{32.91 \times 4.8}{3.1}$ **f** $(2.45 \times 4.1)^2$

 g $7.8 + \dfrac{4.79}{\sqrt{47.7} + 7.56}$

 h $(9.8^2 + (9.2 - 0.438))^2$

8 Estimate these square roots to 1 dp.
 Do *not* use a calculator.

 a $\sqrt{2}$ **b** $\sqrt{8}$

 c $\sqrt{10}$ **d** $\sqrt{15}$

 e $\sqrt{20}$ **f** $\sqrt{26}$

 g $\sqrt{32}$ **h** $\sqrt{45}$

 i $\sqrt{70}$ **j** $\sqrt{85}$

 Use a calculator to check your estimates.

9 Estimate the answers to these calculations.

 a $2.6 \times 2.5 \times 1.9$ **b** 7.7×8.1

 c $\sqrt{26} + \sqrt{7}$ **d** $\sqrt{13} \times \sqrt{16}$

10 Estimate the answers to these calculations.

 a $\dfrac{9.7^2 - 3.9^2}{2 \times 19}$ **b** $\dfrac{35.9 \times 1.9}{0.49}$

 c $\dfrac{5.8^2 - 1.5 \times 1.8^2}{7.59 + 6.89}$

 d $\dfrac{(\sqrt{80} - \sqrt{24})^2 - \sqrt{63}}{3.01 \times 2.45 + 0.499}$

11 Estimate the answers to these calculations.

 a $\dfrac{23.4 - 1.85}{1.9 - 1.85}$ **b** $(24.9 - 24.5) \times 200$

 Say how you should decide what accuracy
 to approximate the numbers.

12 If Peeta buys a drink for £1.25, some crisps
 for 80p and a banana for 47p, roughly how
 much has he spent?

13 Irwin estimates $\dfrac{47.3 \times 18.9}{8.72} \approx 100$.

 What numbers could he have used to get his
 estimate?

9.1 Estimation and approximation

RECAP
- Rounding can be used to approximate numbers.
- Approximations can be used to estimate answers to complex calculations.
- You can use estimates to check answers and for adjusting place value.

> Rounding to 1sf is usually helpful!

HOW TO

To solve problems using approximations
1. RTQ and decide what you need to do.
2. Estimate using sensible approximations.
3. ATQ and round you answer to a suitable degree of accuracy.

EXAMPLE

Philip says 22 feet is 70 m to 1 sf.

a Use an estimate to show he is wrong.
   Suggest what he might have done wrong.
b Is your estimate reasonable? Give your reasons.

> 1 inch = 2.5400 cm
> 1 foot = 12 inches

1. Using 1 dp should give a good estimate.
a 1 inch ≈ 2.5 cm   2. Round to 1 dp.
   1 foot ≈ 30 cm 12×2.5
   22 feet ≈ $22 \times 0.3 = 6.6$ m
3. Philip was wrong by a power of 10.
   He may have wrongly thought 30 cm = 3 m.
   Make sure you justify your reasons mathematically.

1. Compare to an accurate value.
b $2.54 \times 12 \times 22 = 670.56$ cm   2. Calculator
   $= 6.71$ m (2 dp)
3. My estimate is 11 cm (6.71 – 6.6) below the real value. This difference is much less than the overall length, 7 m. The estimate is reasonable.

EXAMPLE

Monty is planting a new rectangular lawn. He is told that he needs between 45 and 60 grams of seed per square metre of lawn. The space he has prepared is 4.8 m × 5.1 m. Seed can be bought in 800 g bags. Find

a an estimate for the area of the space   b an estimate for how much seed he needs.
c How many bags should he buy? Why?

1. It is better to have too much seed than too little, so round values up where appropriate.
a $5 \times 5 = 25$ m²   2. Round to 1 sf.   b 25×60 g = 1500 g
                                                     25×45 g = 1125 g   2
                                                     He needs between 1125 g and 1500 g.   3
c He should buy 2 bags.   3
   Two bags contain 1600 g which is more than the maximum he needs.

EXAMPLE

Bo hired a car to travel from Leeds to Birmingham, a distance of 118 miles. He returned the car to Leeds. A small car costs £39.98 to hire plus 15p per mile to run. A large car costs £68.92 to hire and 12p per mile to run. Which car did Bo use if he chose the cheapest one.

1. Estimate the cost of fuel there and back and add it to the hire fee.
Large $120 \times 0.12 \times 2 + 70 ≈ 14 \times 2 + 70 ≈ 30 + 70 ≈ £100$ 2. Use $12 \times 12 = 144 ≈ 140$ and $14 \times 2 = 28 ≈ 30$.
Small $120 \times 0.15 \times 2 + 40 ≈ 120 \times 0.3 + 40 ≈ 40 + 40 ≈ £80$ 2. Use $15 \times 2 = 30$ and $12 \times 3 = 36 ≈ 40$.
3. The small car. It is about £20 cheaper.

Exercise 9.1A

1 Phil wants to know how many centimetres are in a yard. He knows an inch is 2.5400 cm and there are 36 inches in a yard.

a Estimate how many cm are in a yard by rounding to 1 sf first.

b How far off the exact answer is this? Was it a good estimate? Why?

2 Ron is staining a fence. The tin says one litre covers 6 m². Stain comes in 5 litre pots. Ron's fence is 1.8 m high, and 48 m long.

a Estimate the area of his fence.

b How many pots of stain should he buy? Show your working.

3 a Estimate these answers in your head.

    i $258 + 362$ **ii** $64 \div 27$

    iii 62.7×211.8 **iv** $96.7 - 64.8$

b Explain if each answer in **a** would be an overestimate or an underestimate.

4 Mark says that $\dfrac{19.3 \times 204}{3.8}$ is about 100.

Is he correct? Give your reasons.

5 Gosia is making jam to fill some 4 fluid ounce jars.
She knows that 1 fluid ounce = 28.3495 ml.

a Estimate how many of her jars she can *fill* if she makes 550 ml of jam.

b About how much jam would she need to make to fill one more jar.

6 Chen makes these estimates.

> **i** $1.86 \times 5.432 \approx 10$
>
> **ii** $17.543 \times 543.25 \approx 10\,000$
>
> **iii** $12.15 \div 17.55 \approx \dfrac{2}{3}$

For each calculation

a say what rounded values Chen might have used

b suggest a way to make each estimate more accurate.

7 Caroline has 490 MB of space left on her camera's SD card. She is taking photographs which are 6.448 MB in size.

a Approximately how many can she take?

b If 1 GB = 1000 MB and she has a 1.5 GB memory card, about how many photos has she taken when the card is full?

8 Paul's mobile phone gives a summary of his bank account. On 28 February he had about £2000 in his account.

06/03	ATM TOWN	−£100.00
06/03	DC DIRECT DEBIT	−£156.45
06/03	DDEBIT	−£99.30
06/03	CASH TRM MAR 06	−£80.00
04/03	CHQ IN	+£84.37
02/03	ABC & G	+£59.21
01/03	SALARY	+£1758.64
28/02	BALANCE	£2058.63

a During the first week of March, approximately how much money was

    i paid in (+)   **ii** paid out (−)?

b Approximately how much money did he have in the bank on 6 March?

*9 Greta is driving from London to Newcastle, a distance of 278 miles. Her car does 43 miles per gallon.
1 gallon is about 5 litres.

a Estimate how many litres she will use on her journey to Newcastle.

b In fact 1 gallon = 4.54609 litres. Will she use more or less diesel than you estimated? Give your reasons.

*10 Estimate the number of jelly beans in a 1 litre jar.

9.2 Calculator methods

You can use a scientific calculator to carry out more complex calculations that involve decimals.

EXAMPLE

Use your calculator to work out $3.46 + 2.9 \times 4.8$
Give your answer to one decimal place.

Estimate = 3 + 3 × 5 = 3 + 15 = 18   BIDMAS

You type [3] [.] [4] [6] [+] [2] [.] [9] [×] [4] [.] [8] [=]

The calculator should display
```
3.46+2.9×4.8
        17.38
```

17.38 = 17.4 (1 dp)   Check ≈ 18 ✓

Modern calculators have the BIDMAS rules built into them. Typing 2 + 3 × 4 gives 14 not 20!

A scientific calculator has **bracket** keys.

EXAMPLE

Use your calculator to work out $(3.9 + 2.2)^2 \times 2.17$
Write all the figures on your calculator display.

Estimate = $(4 + 2)^2 \times 2 = 6^2 \times 2 = 36 \times 2 = 72$ BIDMAS

You type [(] [3] [.] [9] [+] [2] [.] [2] [)] [x²] [×] [2] [.] [1] [7] [=]

The calculator should display
```
(3.9+2.2)2×2.17
       80.7457
```

80.7457   Check ≈ 72 ✓

You can use the bracket keys to do calculations where the **order of operations** is not obvious.

EXAMPLE

Use a calculator to work out the value of this expression. $\dfrac{21.42 \times (12.4 - 6.35)}{(63.4 + 18.9) \times 2.83}$
Write all the figures on the calculator display.

Estimate = $\dfrac{20 \times (12 - 6)}{(60 + 20) \times 3} = \dfrac{120}{240} = 0.5$

Rewrite the calculation as
$(21.42 \times (12.4 - 6.35)) \div ((63.4 + 18.9) \times 2.83)$
Type this into the calculator.

```
(21.42×(12.4-6.35))÷((63.4+18.9)×2.83)
```
➡
```
(21.42×(12.4-
 0.55640 1856
```

0.556 401 856   Check ≈ 0.5 ✓

EXAMPLE

Write 10 000 seconds in hours, minutes and seconds.

Do the calculation in stages.

Convert 10 000s to minutes, then convert the remainder back to seconds.
```
10000÷60
166.6666667
```
```
Ans-166
0.666666667
```
```
Ans×60
        40
```

10 000 s = 166.6 mins = 166 mins, 40 s

Convert 166 mins to hours, then convert the remainder back to minutes.
```
166÷60
2.766666667
```
```
Ans-2
0.766666666
```
```
Ans×60
        46
```

166 mins = 2.76 hrs = 2 hrs, 46 mins

10 000 s = 2 hrs, 46 mins, 40 s   Check 40 + 60 × (46 + 60 × 2) = 10 000 ✓

Exercise 9.2S

For each question first estimate the answer and then use your calculator to find an accurate answer.

1 Calculate these giving your answer to one decimal place.

   a $3.4 + 6.2 \times 2.7$

> Estimate $\approx 3 + 6 \times 3$
> $= 3 + 18$
> $=$ _____
> Exact $=$ _____

   b $1.98 \times 11.7 - 4.6$

   c $7.8 + 19.3 \div 4.12$

   d $2.09 \times 2.87 + 3.25 \times 1.17$

   e $13.67 \div 1.75 + 3.24$

   f $1.2 + 3.7 \times 0.5$

   g $802.6 \times 2.014 \div 3.92$

2 Calculate these giving all the figures on your calculator display.

   a $(2.3 + 5.6) \times 3^2$

   b $2.3^2 \times (12.3 - 6.7)$

   c $(2.8^2 - 2.04) \div 2.79$

   d $7.2 \times (4.3^2 + 7.4)$

   e $11.33 \div (6.2 + 8.3^2)$

   f $(2.5^2 + 1.37) \times 2.5$

3 Calculate these giving your answers to one decimal place.

   a $\dfrac{5.4 + 3.8}{4.5 - 2.9}$

   b $\dfrac{3.8 - 1.67}{4.3 - 2.68}$

   c $\dfrac{12.4 + 5.8}{14.5 - 3.9}$

   d $\dfrac{13.08 - 2.67}{2.13 + 2.68}$

4 Convert these times into weeks, days, hours, minutes and seconds.

   a $50\,000\,\text{s}$ **b** $100\,000\,\text{s}$

   c $500\,000\,\text{s}$ **d** $1\ \text{million}\,\text{s}$

   e $10\ \text{million}\,\text{s}$ **f** $30\ \text{million}\,\text{s}$

5 Calculate these giving your answers to two significant figures.

   a $3.2 \times (2.8 - 1.05)$

   b $2.8^2 \times (9.4 - 0.083)$

   c $16 \div (5.1^2 \times 7.2)$

   d $(3.8 + 8.9) \times (2.2^2 - 7.6)$

   e $1.8^3 + 4.7^3$

   f $52 \div (4.6 - 1.8^2)$

6 Calculate these giving all the figures on your calculator display.

   a $\dfrac{462.3 \times 30.4}{(0.7 + 4.8)^2}$

   b $\dfrac{13.58 \times (18.4 - 9.73)}{(37.2 + 24.6) \times 4.2}$

7 Calculate these giving all the figures on your calculator.

   a $\dfrac{165.4 \times 27.4}{(0.72 + 4.32)^2}$

   b $\dfrac{(32.6 + 43.1) \times 2.3^2}{173.7 \times (13.5 - 1.78)}$

   c $\dfrac{24.67 \times (35.3 - 8.29)}{(28.2 + 34.7) \times 3.3}$

   d $\dfrac{1.45^2 \times 3.64 + 2.9}{3.47 - 0.32}$

   e $\dfrac{12.93 \times (33.2 - 8.34)}{(61.3 + 34.5) \times 2.9}$

   f $\dfrac{24.7 - (3.2 + 1.09)^2}{2.78^2 + 12.9 \times 3}$

*8 Find the [Ans] and [x^{-1}] reciprocal buttons on your calculator. Enter a number and press [=], make a note of the number. Now repeat this sequence of buttons and note the number you get.

Repeat this sequence of buttons ten times. Now start again with a new first number. Describe what you find.

1043, 1932, 1933   SEARCH

9.2 Calculator methods

- Know how to use *your* calculator to work out complex calculations.
- Know how to apply the order of operations with your calculator.

HOW TO

To solve problems using your calculator

(1) RTQ and decide what to do. Estimate the answer.

(2) Use your calculator and check the answer agrees with your estimate.

(3) ATQ – interpret your calculator display in the context of the question.

EXAMPLE

Put brackets into each of these calculations to make them correct.

a $11.3 + 7.2 \times 2.5 = 46.25$ **b** $21.6 - 3 \times 4 + 3.2 = 0$

(1) If you cannot see where the brackets should go, try different positions until one works. Estimating helps.

a $(11 + 7) \times 3 = 54$ (1) Estimate. Using   **b** $22 - (3 \times 4) + 3 = 13$

$11 + (7 \times 3) = 32$ '× 3' for '× 2.5' will $(22 - 3) \times 4 + 3 = 79$

give bigger values. $22 - 3 \times (4 + 3) = 1$

$(11.3 + 7.2) \times 2.5 = 46.25$ (2) (3) $21.6 - 3 \times (4 + 3.2) = 0$ (2) (3)

EXAMPLE

Joseph books 4 tickets online for a show, at £25.90 each. He has to pay a £2.90 booking fee per ticket and a 'single transaction fee' of £2.85. Tickets 'on the door' cost £30 each.

Did he save money by booking online? How much?

(1) Calculate the 'online' and 'on door' costs and take the difference.

Online cost = 4 × (25.90 + 2.90) + 2.85   On door cost = 4 × 30

Estimate = 4 × 30 + 3 ≈ 120   = £120

Exact = £118.05   (2) ≈ 120 ✓

Difference = 120 − 118.05 = £1.95

Yes, he saved £1.95   (3)

EXAMPLE

A daytime taxi rate is set at £2.60 standing charge, plus £1.80 per mile, plus 20p for 40 seconds waiting time. 30p is added for every extra adult.

Harry and Denise hire a taxi for a 4 mile journey, with 2 minutes waiting time.
Harry gives the driver £15 and says 'keep the change.' What tip did he give?

(1) Total cost = standing charge + mileage cost + waiting time + extra adult.

Total cost = 2.60 + 4 × 1.80 + (2 × 60 ÷ 40) × 0.20 + 0.30   2 mins = 2 × 60 secs

Estimate = 3 + 4 × 2 + 3 × 0.2 + 0.3 $\dfrac{2 \times 60}{40} = 3$

= 3 + 8 + 0.6 + 0.3

= 12

Exact = £10.70   (2) ≈ 12 ✓

Harry's tip = 15.00 − 10.70 = £4.30   (3)

Number Measures and accuracy

Exercise 9.2A

1 **i** Estimate, than use your calculator to decide which of these calculations are correct.

   a $7.2 + (8.3 \times 3.4) = 35.42$

   b $36 \div (2.5 + 5.5) = 19.9$

   c $(36 \div 2.5) + 5.5 = 4.5$

   d $(21 - 3) \times (4 + 3) = 126$

  ii Move the brackets where necessary to make all the answers correct.

2 The night rate for taxi fares is set at £3.40 standing charge, plus £2.20 per mile, plus 20p for every 30 seconds. 40p is added for every additional adult.

   a Charlotte takes a taxi for 8 miles, which involves 5 minutes waiting. How much is the fare for this journey?

   b Robyn and her two sisters share a taxi for a 3 mile journey, with 3 minutes of waiting time. She gives the driver a tip which is 10% of the fare. How much does she pay?

3 Carpet tiles are sold in boxes of 12. Style A is £2.25 per tile, Style B is £3.95, and Style C is £4.49. They are only sold in full boxes.

   Ashley creates a pattern of tiles that requires $\frac{1}{3}$ of a box of Style C, $\frac{3}{4}$ of a box of style A and $\frac{1}{2}$ a box of Style B.

   a Estimate the cost of her pattern.

   b Use your calculator to find the exact cost by doing just one calculation.

   c She sells the leftover tiles for a profit of £5. How much did she sell the leftover tiles for?

   d If her pattern was twice the size, could she still buy only three boxes of tiles?

4 Estimate which calculations, **A**–**D** gives an answer that is

   a zero   **b** a whole number

   c more than 100   **d** a negative.

   Use you calculator to check if you were correct.

   A $97.31 \times 2 - 26.3 \times 7.5$

   B $(15.2^2 + 7) \div 2$

   C $5.7 \times 16 - 11 \times 3.2$

   D $\dfrac{2.3^2 - (3.2 + 2.09)}{0.7 \times 5 - 1.4 \times 3.8}$

5 Barry sees the following offer.

> **Vericheep Fone OFFER**
> Monthly Fee £12.99
>
> FREE - 200 texts every month
> FREE - 200 voice minutes every month
> Extra text messages 3.2p each
> Extra minutes 5.5p each

   His current mobile phone costs him £22.99 per month but he gets unlimited free texts and voice minutes.

   a What would the new offer cost Barry

   i in February, when he used 189 texts and 348 voice minutes?

   ii in March, when he used 273 texts and 219 voice minutes?

   b Explain if the new offer is good for Barry.

*6 The speed of sound depends, amongst other things, on the temperature of the air it travels through. If T stands for temperature in degrees Celsius, then the speed of sound, in feet per second, is

$$\frac{\sqrt{(273 + T)} \times 1087}{16.52}$$

   a What is the difference in the speed of sound at 0 °C compared to 18 °C?

   b To convert feet per second to miles per hour, you multiply by 15, then divide by 22. Calculate the speed of sound in miles per hour at 11.5 °C using a single calculation.

9.3 Measures and accuracy

You can measure **length**, **mass** and **capacity** using metric units.

- Length is a measure of distance.

  Standard units are mm, cm, m, km.

- Mass is a measure of the amount of matter in an object.
  Standard units are g, kg, tonnes.

- Capacity measures the amount of fluid that a container will hold.
  Standard units are ml, cl, l.

- Volume is the amount of space a solid shape takes up.
  Standard units are cm³, m³.

> The weight of an object is actually a measure of the force of gravity on that object.

Compound measures such as speed and density describe how one quantity changes in proportion to another.

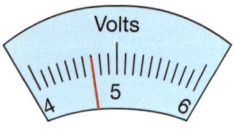

$D = S \times T$ $M = D \times V$
$S = D \div T$ $D = M \div V$
$T = D \div S$ $V = M \div D$

- **Speed** = $\dfrac{\text{Distance}}{\text{Time}}$

  Units such as m/s, km/h

- **Density** = $\dfrac{\text{Mass}}{\text{Volume}}$

  Units such as g/cm³, kg/m³

Density is a measure of the amount of mass per unit of volume.

p.312

EXAMPLE

Find the density of a piece of wood with cross-section area 42 cm², length 12 cm and mass 693 g.

Volume of a prism = cross-section area × length

Volume = 42 × 12 = 504 cm³

Density = mass ÷ volume

Density = 693 ÷ 504 = 1.375 g/cm³

Measurements are not exact. Their accuracy depends on the precision of the measuring instrument, the skill of the person making the measurement and whether they have been rounded.

Volts

◀ To the nearest volt the meter reads 5 volts.
$4.5 \leqslant x < 5.5$

- If a quantity is measured to within a given **error interval** then

  measurement $-\frac{1}{2}$ error interval $\leqslant$ true value $<$ measurement $+\frac{1}{2}$ error interval

EXAMPLE

The mass of a meteorite is given as 235.6 g. Find the limits of accuracy of the mass.

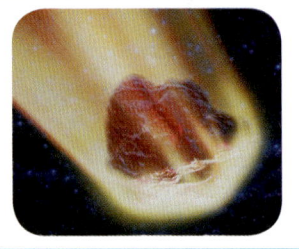

The **implied accuracy** is one decimal place, that is, the mass is given correct to the nearest 0.1 g.

$235.55 \leqslant \text{mass} < 235.65$ $\frac{1}{2} \times 0.1 = 0.05$

In calculations, give your answers to an appropriate degree of accuracy.

A measurement of 2.3 cm has an **implied accuracy** of 1 decimal place.

A measurement of 3.50 m has an implied accuracy of 2 decimal places.

If all measurements in a question are to 1 dp, give your answers to 1 dp.

Exercise 9.3S

1 Choose one of these metric units to measure each of these items. Write the appropriate abbreviation next to your answers.

millimetre	gram	millilitre	centimetre
kilogram	centilitre	metre	tonne
litre	kilometre		

   a your height

   b amount of tea in a mug

   c your weight

   d length of a suitcase

   e weight of a suitcase

   f distance from Paris to Madrid

   g quantity of drink in a can

   h amount of petrol in a car

   i weight of an elephant

   j weight of an apple

> You should learn these equivalences.
> Length: 10 mm = 1 cm, 100 cm = 1 m, 1000 m = 1 km.
> Mass: 1000 g = kg, 1000 kg = 1 tonne.
> Capacity: 1000 ml = 1 litre 100 cL = 1 litre, 1000 cm³ = 1 litre

2 Convert these measurements to the units shown.

   a 20 mm = ___ cm **b** 400 cm = ___ m

   c 450 cm = ___ m **d** 4000 m = ___ km

   e 0.5 cm = ___ mm **f** 4.5 kg = ___ g

   g 6000 g = ___ kg **h** 6500 g = ___ kg

   i 2500 kg = ___ t **j** 3 litres = ___ ml

3 Calculate the average speed for these journeys. Give the units in your answers.

   a 10 miles in 4 hours

   b 16 km in 30 minutes

   c 1 hour to travel 80 km

   d 2 km in quarter of an hour

   e 6 m in 20 minutes

   f 120 miles in half a day

4 Calculate the time taken if the average speed is 60 mph and the distance travelled is 150 miles.

5 Calculate the distance travelled if the average speed is 15 mph and the journey takes $2\frac{1}{2}$ hours.

6 Calculate the density.

   a Mass = 480 kg Volume = 1 m³

   b Mass = 250 g Volume = 6 cm³

7 A cube of side 2 cm weighs 40 grams.

   a Find the density of the material from which the cube is made, giving your answer in g/cm³.

   b A cube of side length 2.6 cm is made from the same material. Find the mass of this cube, in grams.

8 Emulsion paint has a density of 1.95 kg/litre. Find

   a the mass of 4.85 litres of the paint

   b the number of litres of paint that would have a mass of 12 kg.

9 Each of these measurements was made correct to one decimal place. Find the error interval for each measurement.

   a 5.8 m   **b** 16.5 litres

   c 0.9 kg   **d** 6.3 N

   e 10.1 s   **f** 104.7 cm

   g 16.0 km   **h** 9.3 m/s

10 Find the error intervals for these measurements, which were made to varying degrees of accuracy.

   a 6.7 m   **b** 7.74 litres

   c 0.813 kg   **d** 6 N

   e 0.001 s   **f** 2.54 cm

   g 1.162 km   **h** 15 m/s

11 Find the limits of accuracy of these measurements, which are correct to the nearest 5 mm.

   a 35 mm   **b** 40 mm

   c 110 mm   **d** 4.5 cm

9.3 Measures and accuracy

- You can measure length, mass and capacity using metric units.
- A compound measure, such as speed or density, connects two different measurements.
- Most measures are approximate due to rounding and their true value lies within an error interval. The error is half a unit above and below the degree of accuracy given for the measurement.

$$\text{Speed} = \frac{\text{Distance}}{\text{Time}}$$

$$\text{Density} = \frac{\text{Mass}}{\text{Volume}}$$

HOW TO

To solve problems involving measure and accuracy

1. RTQ and decide what to do.
2. Convert everything to the same units if appropriate, and use your knowledge of measures and accuracy.
3. ATQ, use suitable units and give your answer to an appropriate degree of accuracy.

EXAMPLE

Stella left home at 11:25 to go for a bicycle ride. She travelled 42 km at an average speed of 7 m/s. If she left home at 11:25 when did Stella get back?

1. Use time = distance ÷ speed and add this to 11:25.
2. Time = $\frac{42\,km}{7 m/s}$ = $\frac{42\,000}{7}$ s = 6000 s

    6000 s = 100 mins   1 min = 60 secs
          = 1 hr 40 mins   1 hr = 60 mins
3. Finish time = 11:25 + 1 hr + 40 mins = 13:05

EXAMPLE

Jack wants to send four copies of a book in a parcel. According to his scales, which are accurate to the nearest 5 g, a book weighs 375 g. Would you advise Jack to only put £4.35 worth of stamps on his parcel? Give your reasons.

Parcel rates	
Weight	Price
< 1.5 kg	£4.35
≥ 1.5 kg	£7.30

1. Find the maximum weight of 4 books including measurement error.
2. Max weight 1 book = 375 + 2.5 = 377.5 g $\frac{1}{2}$ × 5 g = 2.5 g

    Max weight 4 books = 4 × 377.5
                     = 1510 g

    1510 g is over 1.5 kg   1.5 kg = 1500 g
3. No, the 4 books could weigh over 1.5 kg. He should pay £7.30 to be safe.

EXAMPLE

Stuart puts his honey into 1lb jars. 1lb ≈ 454 g.

a How many jars can he fill with 10 kg of honey?

b He sells the jars at £4.50 each. How much money does he make?

a 1 How many lots of 1lb in 10 kg.
    10 kg = 10 000 g   2
    10000 ÷ 454 = 22.02...
    He can fill 22 jars.   3 Need full jars.
b 22 × 4.50 = £99   2 3

Exercise 9.3A

1 Sunita is travelling. It takes her 25 minutes to walk to the station, 8 minutes wait for her train, 1 hour 42 minutes on the train, 5 minutes wait for a taxi, and 25 minutes in the taxi.

If she leaves at 11 am, what time does she arrive?

2 Talal walks 5.4 km in an hour. What is his speed in

   a m/hr   **b** m/min   **c** m/s?

3 Bradley is competing in a cycle race over four stages in Britain and France.

   Stage 1   100 miles   Stage 2   75 miles
   Stage 3   232 km   Stage 4   84 km

   He knows that 5 miles is roughly 8 km

   He covers about 40 km every hour.

   a How long does the race take him?

   b What is his average speed in miles per hour (mph)?

4 Ella gets 6 pints of milk each week. One pint is approximately 568 ml.

   a How many millilitres does Ella get?

   b Ella's milkman stops delivering milk in pint bottles.

   How many 1-litre bottles must she order to have at least 6 pints?

5 A bag contains 80 g of sugar before 35 g of sugar is removed. Both measurements are given to the nearest gram. What is the largest and smallest amount of sugar that could be left in the bag?

6 A rectangular plot of land is 35 m by 28 m. An estate agent advertises the plot as being over 1000 m². Can this be true? Give your reasons.

7 A van has a safe load of 450 kg. It is being used to move boxes that weigh 30 kg to the nearest kg. How many boxes can the van safely carry? You must show your working.

8 Jamie is given a metal block in the shape of a cuboid measuring 25 cm by 30 cm by 1.2 m. The block weighs 655 kg. He is also given this information about the densities of five metals.

Metal	Density (kg/m³)
Titanium	4500
Tin	7280
Steel	7850
Brass	8525
Copper	8940

   a Jamie says the block is made of steel. Is he correct? Give your reasons.

   b If the block was made of copper, what would it weigh?

*9 A car travels 315 miles in 7 hours 30 mins. It uses 45 litres of petrol.

   a What is the car's average speed?

   b An advert for this type of car says it does 35 mpg (miles per gallon). A gallon is approximately 4.5 litres. Is the advert fair? Give your reasons.

Summary

Checkout

You should now be able to...

	Test it Questions
✔ Round numbers and measures to an appropriate degree of accuracy.	1
✔ Use approximation to make estimates.	2, 3
✔ Check calculations using approximation and estimation.	3
✔ Use standard units of length, mass, volume, capacity, time and area.	4 – 9
✔ Use inequality notation to state error intervals and interpret limits of accuracy.	10, 11

Language   Meaning

Language	Meaning	Example
Approximation	A less accurate value of a number, usually obtained by rounding, which is easier to work with.	$\pi = 3.141592654...$ $= 3.1$ (1 dp) $= 3$ (1 sf)
Estimate	A calculation made using approximate values or a judgement. Use the symbol $\approx$ to show approximations have been used.	$\pi \times 1.9^2 \approx 3 \times 2^2 = 12$ Exact $= 11.34114...$
Significant figures (sf)	Describes the relative importance of digits in a number. A number can be rounded to a given number of significant figures.	$256.46 = 260$ (2 sf)
Mass	A measure of the amount of matter in an object. **Weight** measures the force of gravity on a mass.	The mass of a bag of sugar is 1 kg.
Capacity	A measure of the amount of fluid that a 3D shape will hold.	Volume $= 8000\,cm^3$ Capacity $= 8$ litres
Volume	A measure of the amount of 3D space occupied by an object.	
Speed	A measure of the distance travelled by an object in a certain time.	The speed limit on UK motorways is 70 mph.
Density	A measure of the amount of matter in a certain volume.	The density of iron is $7.87\,g/cm^3$.
Accuracy	How close a measured or calculated quantity is to the exact value.	$1.239648 = 1.2$ is accurate to 1 dp $= 1.24$ is accurate to 2 dp
Implied accuracy	The accuracy of a value implied from the number of significant figures or decimal places given.	1.3 has implied accuracy of 1 dp. This means that the actual value could be between 1.25 and 1.35.

Review

1 Round each of these numbers to the nearest

   i 3 sf   **ii** 2 sf   **iii** 1 sf

   a 8752   **b** 14.981

   c 0.0682   **d** 0.5092

2 Estimate the answers to these calculations.

   a 18.5×11.9 **b** $\dfrac{105^2 + 265}{95}$

   c $\dfrac{3.4^2 + 5.9 \times 1.9^2}{15.33 - 8.91}$

   d $\sqrt{62.9 \div 4.21}$

3 Use your calculator to work these out.
   In each case first write an estimate for
   your answer. Write all the figures on your
   calculator display.

   a $9.6 - 3.7 \div 2.49$ **b** $4.8 \times (3.2^2 + 6.7)$

   c $\dfrac{5.92 + 3.78}{12.51 - 3.29}$ **d** $\dfrac{326.2 \times 4.15}{(5.2 - 2.9)^2}$

4

| m^2 | cm | litres | cm^3 | kg |

   Choose from the box a unit of

   a length   **b** volume

   c capacity   **d** mass

   e area.

5 Convert these lengths to metres.

   a 350 cm   **b** 145 mm

   c 0.2 km   **d** 932 cm

6 Simon starts running at 09:50 and stops at
   11:42, how long did he run for?

7 Isaac is paid £14.50 per hour.

   a How much will he be paid if he works
         300 minutes?

   b If Isaac is paid £137.75, how many hours
         has he worked for?

8 A car travels at 20 m/s. How long does it
   take to travel

   a 100 m   **b** 0.5 km?

9 A 30 cm^3 cylinder weighs 150 grams.

   a Find the density of the cylinder giving
         your answer in g/cm^3.

   b A cube is made of the same material as
         the cylinder. It weighs 135 grams. Find

      i   the volume of the cube

      ii its side length.

10 Helen cuts wood of length, x, which is
   90 cm measured correctly to the nearest
   centimetre. Write an inequality to describe
   the range of values of x.

11 Amaan's mass is measured and lies
   in the interval $59.5\,\text{kg} < m \leqslant 60.5\,\text{kg}$.
   How accurate was the measurement?

What next?

<table>
<tr><td rowspan="3">Score</td><td>0 – 4</td><td></td><td>Your knowledge of this topic is still developing.
To improve look at MyMaths: 1002, 1004, 1005, 1006, 1043, 1067, 1121, 1246, 1932, 1933</td></tr>
<tr><td>5 – 9</td><td></td><td>You are gaining a secure knowledge of this topic.
To improve your fluency look at InvisiPens: 09Sa – f</td></tr>
<tr><td>10 – 11</td><td></td><td>You have mastered these skills. Well done you are ready to progress!
To develop your problem solving skills look at InvisiPens: 09Aa – d</td></tr>
</table>

Assessment 9

1 David makes the following incorrect statements. Rewrite the statements correctly.

a 32.34 to 3 sf is 32 [1] **b** 5678 to the nearest 100 is 5600 [1]

c 0.203762 to 1 sf is 0 [1] **d** 312 to the nearest 10 is 31 [1]

e 294.82 to 2 sf is 294.82 [1] **f** 256 133 to the nearest 1000 is 256 [1]

2 Anna, Bart and Christian make these estimates for the following calculations. For each calculation, decide who makes the best estimate.

		Anna	Bart	Christian	
a	5.8 × 6.5	30	36	40	[1]
b	20.2 × 5.6	100	110	120	[1]
c	7.9 × 8.8	56	66	72	[1]
d	6.6 ÷ 1.3	5	6	9	[1]
e	7.9 ÷ 0.91	7	8	9	[1]

3 The rate at which a person blinks, on average, is about 6 times per minute.
Mia says that a good estimate for the number of times a person blinks in one day is 9000.
Use your calculator to explain how she got that estimate. [2]

4 Rebecca says 'the square root of 46 is bigger than one integer but smaller than another'.

a What are these two integers? [1]

b Write down the squares of these two numbers. [2]

c Use your answer to estimate the square root of 46 to the nearest integer. [1]

5 a Patrick estimates the value of this calculation to be $1\frac{1}{3}$, find his mistakes and give a sensible estimate for the calculation. [3]

$$\frac{5.8^2 - 1.5 \times 1.8^2}{7.59 + 6.89} \approx \frac{6^2 - 2 \times 2^2}{8 + 7} \approx \frac{36 - 15}{15} = 1\frac{1}{3}$$

b Seth estimates the value of this calculation to be 0.0775

$$\frac{28.9 + \sqrt{0.827}}{(23.4 - 17.6)^2} \approx 0.0775$$

Use your calculator to find the difference between the exact value and Seth's estimate. Say why the difference is large. [2]

6 A chocolate fountain was eaten at a wedding reception.
Everyone ate some and none was left over!
The chocolate in the fountain had a mass of 3.56 kg and each person ate, on average, 88 g of chocolate.
Estimate the number of people at the reception. [2]

7 Here is a sketch of Mr Digweed's garden.

a Find the area of the garden. [4]

6.8 m, 9.6 m, 8.5 m

He has a load of soil delivered and spreads it evenly over the whole garden.

b The depth of the soil is 12.5 cm. What volume of soil was delivered? [3]

c The new soil weighs 0.3 g/cm³.

i Calculate its weight in kg/m³. [2]

ii Calculate the weight in tonnes, of soil delivered. [2]

8 Naomi writes these expressions. Put brackets into each expression to make them correct.

    a $4.6 + 4.1 + 1.2 \times 2.6 = 18.38$ [1]   **b** $14.9 - 6.8 \div 3.7 - 1.2 = 12.18$ [1]

    c $3.4 \times 2.4 - 0.8 + 5.9 - 2.8 = 8.54$ [2]   **d** $2.6 + 5.52 + 2.04 \div 1.8 - 0.72 = 6.08$ [2]

    e $12.3 - 5.2 \times 1.6 + 3.4 \times 2 = 14.76$ [2]   **f** $5.9 + 2.2 \div 3.6 - 2.4 \times 0.3 = 1.53$ [2]

9

tonne	kilogram	gram	kilometre	metre	centimetre	millimetre	litre	centilitre	millilitre

Estimate the following values using an appropriate unit from the list in each case.

    a Capacity of a paddling pool.   [2]   **b** Mass of a full suitcase.   [2]

    c Mass of an eyeliner pencil.   [2]   **d** Length of a rugby pitch.   [2]

    e Weight of a plane.   [2]   **f** Amount in a dose of cough mixture.   [2]

    g Thickness of a DVD case.   [2]   **h** Distance run in a marathon.   [2]

    i Length of a hairbrush.   [2]   **j** Amount of tea in one cup.   [2]

10 Here is a calculator display.   `25.3`

    a If the unit is £ what is the written answer?   [1]

    b If the unit is metres write the answer in metres and centimetres.   [1]

    c If the unit is kilograms write the answer in kilograms and grams.   [1]

    d If the unit is centilitres write the answer in centilitres and millilitres.   [1]

    e If the unit is hours write the answer in hours and minutes.   [2]

11 **a** Bradley cycles $165\,$km in $2\frac{1}{2}$ hrs. What is his average speed?   [2]

    b Lewis' car travels $9500\,$m in $3.8\,$min. What is its average speed in km/h?   [2]

    c A train travels for $2\frac{3}{4}$ hrs at an average speed of $175\,$km/h. How far does it go?   [2]

    d A plane travels for $7\frac{1}{3}$ hrs at an average speed of $465\,$mph. How far has it flown?   [2]

    e A snail crawls travels $14\,$m at $47\,$m/h. How long does his journey take?   [2]

12 **a** An cricket ball has a mass of 158 grams and volume of $195\,$cm³?
Would it float in water? Explain your answer.   [3]

    b A gold bracelet has a mass of $44\,$g and a density of $19.3\,$g/cm³. What is its volume?   [2]

    c A concrete girder measuring $25\,$m by $15\,$m by $6\,$m has a density of $2.4\,$g/cm³.
What is the mass of the bar in tonnes?   [3]

13 Write down the upper and lower limits of these measurements.

    a The distance from the moon to the earth varies from $221\,460$ to $251\,970$ miles to the nearest ten miles.   [1]

    b Farakh's leg is $90\,$cm long to the nearest $10\,$cm.   [1]

    c A box of chocolates weighs $455\,$g to the nearest $5\,$g.   [1]

    d In July 2013 Mo Farah beat Steve Cram's 28 year old British $1500\,$m record with a time of $3\,$min $28.8\,$s to the nearest $0.1\,$s.   [1]

    e The number of tea bags in a box is 240 to the nearest 4 bags.   [1]

    f A lorry has a mass of 28 tonnes to the nearest tonne.   [1]

    g A walking stick has a length of $586\,$mm to the nearest mm.   [1]

14 A rectangular tray measures $45\,$cm by $24\,$cm by $0.3\,$cm correct to the nearest cm.
Its mass is $76\,$g correct to the nearest gram.

    a Calculate the biggest and smallest possible values of its volume.   [2]

    b Calculate the biggest and smallest possible values of its density.   [3]

10 Equations and inequalities

Introduction

Maths is not all about whether something is equal to something else. Sometimes it can be useful to know when a quantity needs to be less than or more than a particular value. An example could be a recipe for a soft drink, where the proportion of cane sugar needs to be within certain limits to conform to standards and to customers' appetites. Some companies are very secretive about the formula for their particular branded soft drink. All you are allowed to know is that the proportions of ingredients lie within a certain range. Inequalities are statements that allow us to work mathematically with this kind of restriction.

What's the point?

In the food and drink industries it is important to be able to work within restrictions on particular ingredients, which are often imposed by health legislation. In the pharmaceutical industry, it can be critical as too much of one particular chemical can have harmful effects.

Objectives

By the end of this chapter you will have learned how to …

- Set up and solve simple linear equations.
- Solve quadratic equations algebraically by factorising.
- Derive and solve two linear simultaneous equations in two variables.
- Find approximate solutions to two linear simultaneous equations using a graph.
- Solve linear inequalities in one variable and represent the solution on a number line.

Check in

1 Copy and complete these equations.

   a $4 \times 3 = 12$ $12 \div \square = 4$ $12 \div \square = 3$

   b $7 \times 5 = 35$ $35 \div \square = 7$ $35 \div \square = 5$

2 Simplify these expressions.

   a $4x \div 4$ **b** $3m \div 3$ **c** $6n \div 2$ **d** $8p \div 4$

3 Work out the value of each expression when $y = 3$.

   a $2y$ **b** $4y - 1$ **c** y^2 **d** $4y \div 2$

4 Work out the value of these expressions when $x = 3$ and $y = 4$.

   a $x + y$ **b** $x - y$ **c** $x \times y$ **d** $\dfrac{3x}{y}$

Chapter investigation

This L shape is drawn on a 10×10 grid numbered from 1 to 100. It has 5 numbers inside it. We can call it L_{35} because the largest number inside it is 35.

What is the sum total of the numbers inside L_{35}?

Find a connection between the L-number and its total.

1	2	3	4	5	6	7	8	9	10
11	12	13	14	15	16	17	18	19	20
21	22	23	24	25	26	27	28	29	30
31	32	33	34	35	36	37	38	39	40
41	42	43	44	45	46	47	48	49	50
51	52	53	54	55	56	57	58	59	60
61	62	63	64	65	66	67	68	69	70
71	72	73	74	75	76	77	78	79	80
81	82	83	84	85	86	87	88	89	90
91	92	93	94	95	96	97	98	99	100

10.1 Solving linear equations 1

- A **linear equation** is an equation of degree one.

$$2x + 1 = 83 \qquad 3(x - 2) = 23 \qquad \frac{x}{2} + 3 = 31$$

A linear equation can be solved by algebraic methods.

- You *must* do the same to the expressions on *both* sides of the = sign.

EXAMPLE

Solve these equations.

a $x + 13 = 32$ **b** $y - 23 = 14$ **c** $12x = 84$ **d** $\frac{p}{4} = 17$

a $x + 13 - 13 = 32 - 13$ **b** $y - 23 + 23 = 14 + 23$ **c** $\frac{12x}{12} = \frac{84}{12}$ **d** $\frac{p}{4} \times 4 = 17 \times 4$

$x = 19$ $y = 37$ $x = 7$ $p = 68$

EXAMPLE

Solve these equations.

a $3x + 12 = 57$

b $\frac{y}{5} - 3 = 8$

a $3x + 12 - 12 = 57 - 12$ **b** $\frac{y}{5} - 3 + 3 = 8 + 3$

$3x = 45$ $\frac{y}{5} = 11$

$\frac{3x}{3} = \frac{45}{3}$ $\frac{y}{5} \times 5 = 11 \times 5$

$x = 15$ $y = 55$

Expressions on each side of the equals sign must be kept balanced at all times.

For part a

x	x	x	12
	57		

x	x	x
	45	

x
15

EXAMPLE

Solve these equations.

a $2a - 5 = 25$ **b** $4(b - 5) = -24$ **c** $4c + 5 = 2$

a $2a - 5 = 25$ $+5$ **b** Expand the bracket   **c** $4c + 5 = 2$ -5

$2a = 30$ $\div 2$ $4b - 20 = -24$ $+20$ $4c = -3$ $\div 4$

$a = 15$ $4b = -4$ $\div 4$ $c = -\frac{3}{4}$

                       $b = -1$

For part **b** you could first divide by 4.
$4(b - 5) = -24$
$b - 5 = -6$
$b = -1$

Linear equations can also be solved approximately using graphs.

EXAMPLE

Use this graph of $y = 3x - 2$ to solve these equations.

a $3x - 2 = 7$

b $3x - 2 = 12$

c $3x - 2 = -2$

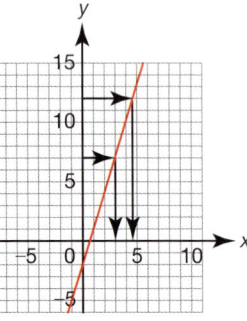

Read off where the graph takes the given y value.

a $x = 3$ $y = 7$

b $x \approx 4.5$ $y = 12$

c $x = 0$ $y = -2$

Exercise 10.1S

1 Solve these equations.

    a $x + 17 = 31$ **b** $x - 31 = 22$

    c $6x = 78$ **d** $\dfrac{x}{5} = 14$

2 Solve $3x + 5 = 32$.

3 Solve $4(x - 2) = 48$.

4 **a** Solve $2x + 7 = 23$.

    b This is the graph of $y = 2x + 7$.

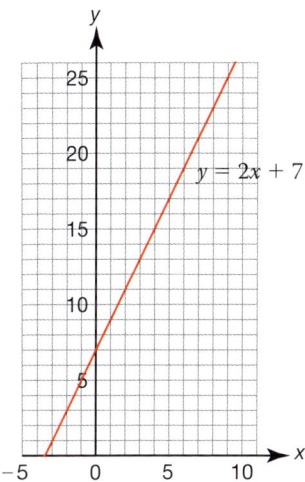

        Use the graph to solve $2x + 7 = 23$.

    c Compare your solution to part **a**.

5 Solve these equations.

    a $4x + 7 = 81$ **b** $5x + 13 = 53$

    c $8x - 3 = 37$ **d** $7x - 15 = 20$

    e $20 = 2x + 6$ **f** $25 = 4x - 7$

6 Solve these equations.

    a $4y + 7 = 3$ **b** $5y + 13 = 8$

    c $3y - 3 = -3$ **d** $7y - 15 = -8$

    e $2 = 2y + 6$ **f** $-3 = 4y - 7$

7 Solve these equations.

    a $4(m + 7) = 44$ **b** $5(m + 6) = 45$

    c $8(p - 3) = 48$ **d** $7(p - 15) = 63$

    e $20 = 2(f + 6)$ **f** $36 = 4(h - 7)$

8 Solve these equations.

    a $4(b + 7) = 20$ **b** $5(a + 6) = 15$

    c $8(x - 3) = -8$ **d** $7(y - 15) = 14$

    e $-10 = 2(w + 6)$ **f** $-12 = 4(x - 7)$

9 Solve these equations.

    a $\dfrac{x}{2} + 3 = 13$ **b** $\dfrac{m}{3} + 6 = 21$

    c $\dfrac{p}{4} - 3 = 12$ **d** $\dfrac{2(m + 4)}{5} = 8$

10 Use this graph of $y = 3x - 2$ to solve these equations.

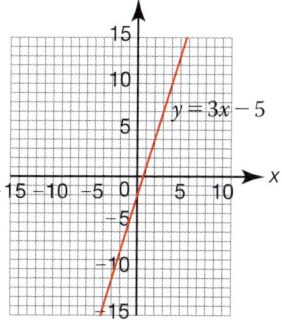

    a $3x - 2 = 5$

    b $3x - 2 = -4$

    c $3x - 2 = 4$

    d $11 = 3x - 2$

11 This is the graph of $y = 2x + 3$.

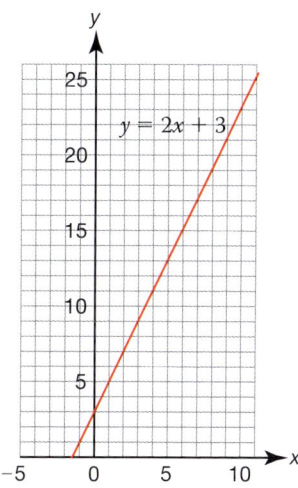

Use the graph to solve these equations.

    a $2x + 3 = 13$ **b** $2x + 3 = 12$

    c $2x + 3 = 19$ **d** $2x + 3 = 3$

*12 **a** Draw the graph of $y = \dfrac{x}{3} - 1$, using axes $-10 \leqslant x \leqslant 10$

    b Using the graph, solve these equations.

        i $\dfrac{x}{3} - 1 = 2$ **ii** $\dfrac{x}{3} - 1 = -4$

        iii $-1 = \dfrac{x}{3} - 1$ ***iv** $\dfrac{x}{3} - 3 = 0$

*13 Solve $5x + 2 = 3x + 12$

14 The perimeter of the rectangle is 30 cm.

    Find the length of the smallest side.

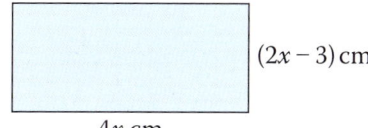

        $(2x - 3)$ cm

        $4x$ cm

1154, 1395, 1925   SEARCH

10.1 Solving linear equations 1

RECAP

- You can use the balance method to solve one-step and two-step equations.
- Do the same to each side to keep the equation balanced until you find the value of the unknown letter.

$$3x - 5 = 16$$
$$3x - 5 + 5 = 16 + 5$$
$$3x = 21$$
$$\frac{3x}{3} = \frac{21}{3}$$
$$x = 7$$

HOW TO

To solve linear equations in real-life problems
1. RTQ and write an equation for the problem in the question.
2. Use the balance method to solve the equation.
3. ATQ and check the answer by substituting the solution into the equation.

EXAMPLE

I think of a number.

I divide the number by 6 and subtract 10.

The answer is -6.

What is my number?

① Write an equation for the problem in the question.

Let my number be x. $\frac{x}{6} - 10 = -6$

② $\frac{x}{6} - 10 + 10 = -6 + 10$ Add 10 to both sides.

$$\frac{x}{6} = 4$$

$$\frac{x}{6} \times 6 = 4 \times 6 \quad \text{Multiply both sides by 6.}$$

$$x = 24$$

③ My number is 24.

Check $24 \div 6 - 10 = 4 - 10 = -6$ ✔

EXAMPLE

The perimeter of this rectangle is 24 cm.

Find x, the length of the rectangle.

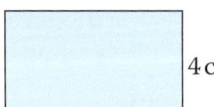

4 cm

x cm

① The perimeter is the distance around the edge of the rectangle.

$$4 + x + 4 + x = 24$$
$$2x + 8 = 24$$

② $2x + 8 - 8 = 24 - 8$ Subtract 8 from both sides.

$$2x = 16$$

$2x \div 2 = 16 \div 2$ Divide both sides by 2.

$$x = 8$$

③ The length of the rectangle is 8 cm.

Check   4 cm + 8 cm + 4 cm + 8 cm = 24 cm ✔

Exercise 10.1A

1 Maisie has 5 boxes of pencils.

Each box contains n pencils.

a Write an expression for the number of pencils Maisie has.

b Maisie counts the pencils. She has 40 in total.
Use your expression from part **a** to write an equation.

c Solve your equation to find n, the number of pencils in a box.

2 Write each 'think of a number' problem as an equation.

Solve the equation to find the number.

a I think of a number and subtract 12. The answer is 11.

b I think of a number and divide by 5. The answer is 8.

3 One of these equations has a different solution to the other two. Which is it?

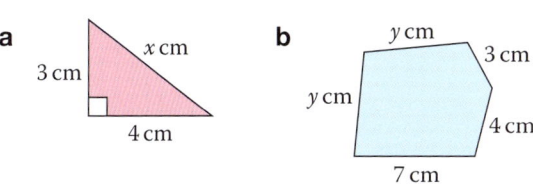

$\frac{m}{2} + 12 = 16$ $6p + 12 = 16$ $\frac{n}{4} + 11 = 13$

4 In each wall, add two bricks to find the number on the brick above.

Write and solve equations to find the unknown letter in each wall.

a

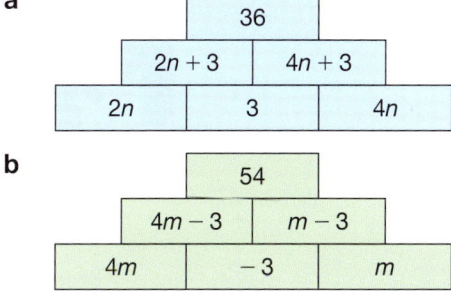

| 36 |
| $2n + 3$ | $4n + 3$ |
| $2n$ | 3 | $4n$ |

b

| 54 |
| $4m - 3$ | $m - 3$ |
| $4m$ | -3 | m |

5 Find the missing side length for each shape.

a

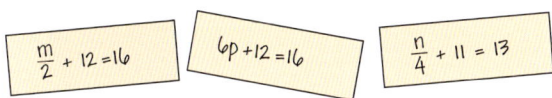

3 cm, x cm, 4 cm

Perimeter = 12 cm

b

y cm, 3 cm, y cm, 4 cm, 7 cm

Perimeter = 26 cm

5 c

8 cm, 8 cm, 8 cm, z cm, 8 cm, z cm

Perimeter = 48 cm

What do you notice about shape **c**?

6 Match each equation in box A with its solution in box B.

There are four 'spare' solutions. Make up an equation with each of these spare solutions.

Box A	Box B	
$2x + 6 = 9$	$x = 1$	$x = -1$
$3x + 1 = 7$	$x = 2$	$x = -2$
$3x + 15 = 6$	$x = 3$	$x = \frac{1}{2}$
$4x - 3 = -5$	$x = 4$	$x = \frac{3}{2}$
$4x + 11 = 3$		
$4x + 7 = 19$	$x = 5$	$x = -\frac{1}{2}$
$5x - 12 = 13$	$x = 6$	$x = -3$
$7x + 8 = 15$		

7 Sarah, Josh and Millie bring sandwiches to a picnic. Sarah brings y sandwiches. Josh brings twice as many sandwiches as Sarah. Millie brings 4 less than Josh.

Work out the number of sandwiches each person brings to the picnic.

8 Four people go out for a meal. The meal costs £x each. All the drinks cost £15.

The total bill comes to £65.

Find the cost of one meal.

9 The perimeter of this shape is 30 mm. Find the length of each side.

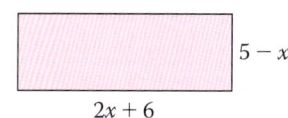

$5 - x$

$2x + 6$

10 This shape is a quadrilateral. Work out the value of x.

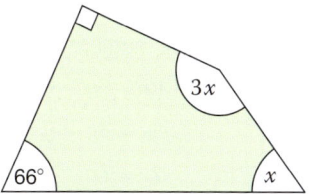

$3x$

$66°$

x

10.2 Solving linear equations 2

Linear equations can have unknowns on one side or unknowns on both sides.

$$3x - 2 = 10$$
$$3x - 2 = 5x + 4$$

- If an equation has unknowns on both sides then you can
  - either use algebra to get an exact solution
  - or use graphs to find an approximate solution.

EXAMPLE

Solve these equations.

a $4x + 3 = 2x + 12$ **b** $10 - 5x = 6 + 3x$

Rearrange to get all the unknowns on one side of the equation.

a $\quad 4x + 3 = 2x + 12 \quad -2x$
$\quad\quad 2x + 3 = 12 \quad\quad\quad -3$
$\quad\quad\quad 2x = 9 \quad\quad\quad\quad \div 2$
$\quad\quad\quad\; x = 4\frac{1}{2}$

b $\quad 10 - 5x = 6 + 3x \quad +5x$
$\quad\quad\quad 10 = 6 + 8x \quad\quad -6$
$\quad\quad\quad\; 4 = 8x \quad\quad\quad\quad \div 8$
$\quad\quad\quad \frac{4}{8} = x$
$\quad\quad\quad\; x = \frac{1}{2}$

In part **b** you could subtract $3x$ from both sides to get $10 - 8x = 6$ but it is best to avoid negatives.

Sometimes you need to construct an equation before you can solve it and interpret the solution.

EXAMPLE

Three angles around a point are: $2x°$, $(3x - 20)°$ and $110°$.

Find the value of the largest angle.

$2x + (3x - 20) + 110 = 360$ Angles around a point add up to $360°$.
$\quad\quad\quad\quad 5x + 90 = 360 \quad -90$
$\quad\quad\quad\quad\quad\quad 5x = 270 \quad \div 5$
$\quad\quad\quad\quad\quad\quad\; x = 54$

The angles are: $2 \times 54 = 108°$,
$3 \times 54 - 20 = 142°$ and $110°$.

The largest angle is $142°$.

$110°$ $2x°$ $(3x - 20)°$

p.48

EXAMPLE

Angela thinks of a number. If she doubles the number and adds 16, she gets the same answer as multiplying the number by 5 and subtracting 5.

Find Angela's number.

Let the number = n
$2n + 16 = 5n - 5 \quad -2n$
$\quad\quad 16 = 3n - 5 \quad +5$
$\quad\quad 21 = 3n \quad\quad\; \div 3$
$\quad\quad\; 7 = n$

Angela's number is 7.

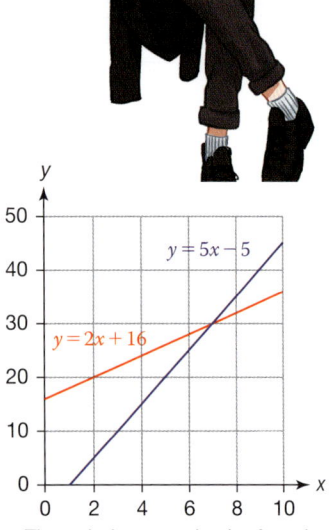

▲ The solution can also be found by finding where the lines $y = 2x + 16$ and $y = 5x - 5$ cross.

Algebra Equations and inequalities

Exercise 10.2S

1 Solve $3x = x + 6$

2 Solve $4x + 2 = 2x + 12$

3 Solve these equations.

    a $5x + 6 = 2x + 27$

    b $3y + 2 = y + 12$

    c $4p + 4 = p + 10$

    d $5f + 6 = 2f + 3$

    e $2x + 14 = 4x + 4$

    f $8 + x = 15 + 2x$

    g $2b + 4 = 6b + 12$

    h $3p = 3 + 9p$

4 Solve these equations.

    a $5x - 6 = 3x - 2$

    b $3y - 2 = y - 12$

    c $4p - 10 = 2p + 4$

    d $5f + 6 = 2f - 3$

    e $2f + 6 = 6f - 22$

    f $p - 2 = 3p - 3$

5 Solve these equations.

    a $11 - 3x = 2x - 4$

    b $15 - 4x = x + 10$

    c $6 - 2x = 8 - x$

    d $12 - 2x = 18 - 5x$

    e $10 - 4p = 2p - 8$

    f $20 - 4h = 10 - 2h$

6 Use the graphs of $y = 5x - 2$ and $y = 6 - x$ to find an *approximate* solution for the equation $5x - 2 = 6 - x$.

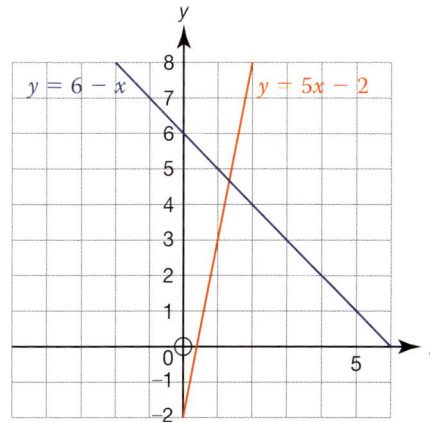

7 Say why the equation $3 - 2x = 10 - 2x$ does not have a solution.

8 **a** Use the sum of the angles in a quadrilateral to construct an equation.

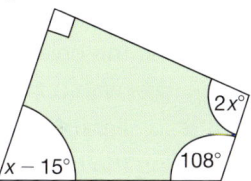

    b Solve the equation to find the smallest angle.

9 The rectangle and triangle have the same perimeter.

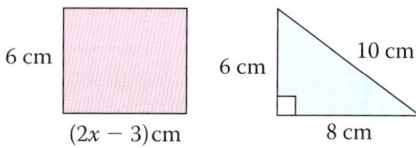

    a Set up an equation to help find the value of x.

    b Solve the equation.

    c Which shape has the largest area?

10 The rectangle and triangle have the same area.

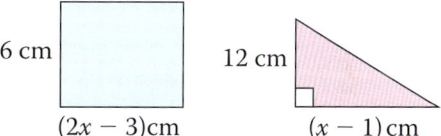

    a Construct an equation to help find the value of x.

    b Solve the equation.

    c Which shape has the shortest base?

*11 Solve $4 - 3(b + 2) = 14 + 5b$

12 Ben thinks of a number.

If he doubles the number and adds 4, he gets the same answer as multiplying the number by 10 and subtracting 8.

Find Ben's number.

*13 **a** Draw the graph of $y = 4 - x$.

    b Draw the graph of $y = 2x + 3$.

    c Use the graphs to find an approximate solution for the equation $4 - x = 2x + 3$.

Q 1182, 1928   SEARCH

10.2 Solving linear equations 2

- To solve an equation, do the same operation to both sides until you find the unknown.
- To solve an equation with brackets
  - expand the brackets first
  - then solve using the balance method.
- To solve an equation with unknowns on both sides
  - subtract the smaller unknown term from both sides
  - then solve using the balance method.
- To solve an equation with fractions
  - arrange the equation so the fraction is on its own on one side of the equation
  - multiply both sides by the denominator (bottom number)
  - then solve using the balance method.

$$2x + 5 = 6x - 3$$
$$2x + 5 - 2x = 6x - 3 - 2x$$
$$5 = 4x - 3$$
$$5 + 3 = 4x - 3 + 3$$
$$8 = 4x$$
$$\frac{8}{4} = \frac{4x}{4}$$
$$2 = x, \text{ or } x = 2$$

HOW TO

To solve a problem involving a more complex linear equation
1. Read the question and form an equation.
2. Simplify the equation by
   - expanding brackets
   - subtracting the smaller unknown term from both sides
   - multiplying both sides by the denominator of the fraction.
3. Solve the equation and give your answer in the context of the question.

EXAMPLE

Lucy thinks of a number.
She adds 8 and then divides by 3.
Her answer is the number she first thought of.
Find Lucy's number.

1. Let the number = n
$$\frac{n + 8}{3} = n$$
$$n + 8 = 3n \qquad$$ ② Multiply both sides by 3.
$$8 = 2n \qquad\qquad$$ Subtract n from both sides.
$$4 = n \qquad\qquad$$ Divide both sides by 2.
3. Lucy's number is 4.

EXAMPLE

This *square* pattern is made from rectangular tiles.

Each tile has length $x + 3$ and width x.

The pattern is 10 tiles high and 4 tiles across.

Find the dimensions of the square.

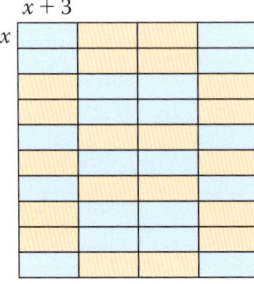

$x + 3$
x

1. The pattern is a square so length = width.
   Length = $4(x + 3) = 4x + 12$
   Width = $10x$
   $$4x + 12 = 10x$$
2. $12 = 6x \qquad -4x$
3. $x = 2 \qquad\quad \div 2$

The square is 20 units long by 20 units wide.

Exercise 10.2A

1 Charlie thinks of a number.

He multiples the number by 2 and subtracts 6. His answer is the number he thought of.

Find Charlie's number.

2 Karena thinks of a number.

She divides the number by 5 and adds 7. Her answer is 10.

Find Karena's number.

3 The diagram shows a square with sides $2y + 5$ cm.

The perimeter of the square is 28 cm. Find the value of y.

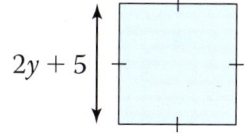

4 The area of the rectangle is 8 cm².

Find the value of z.

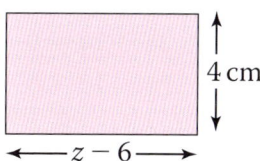

5 **a** Choose one expression from each set of cards.

Set 1

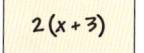 $2(x + 3)$ | $4(2x - 1)$ | $3(4x + 1)$

Set 2

$3x - 2$ | $4x + 1$ | $8x - 3$

Write them as an equation: expression from set 1 = expression from set 2.

b Solve your equation to find the value of x.

c Repeat for different pairs of expressions.

6 The triangle and the square have equal perimeter.

Find the value of x.

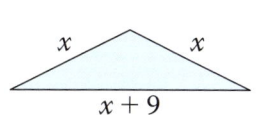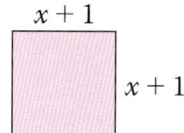

7 Square A and square B have equal area.

Find the value of x.

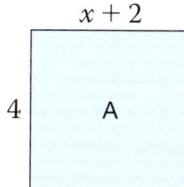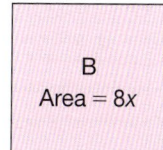

8 A blouse has m buttons. A shirt has $m + 2$ buttons.

Three shirts have the same number of buttons in total as four blouses.

How many buttons are there on a shirt?

9 **a** I think of a number, multiply by 2 and add 7. I get the same answer when I multiply the number by 4 and subtract 13. Find my number.

b I think of a number, multiply by 5 and subtract 8. I get the same answer when I double the number and add 10. Find my number.

c I think of a number, multiply by 3 and add 4. I get the same answer when I multiply by 5 and add 12. Find my number.

10 The perimeter of this equilateral triangle is $2x + 6$.

The length of one side of the triangle is 8 cm. Find the value of x.

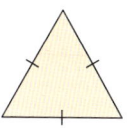

11 The perimeter of this square is $4 + x$.

The length of one side is 10 cm.

Find the value of x.

10.3 Quadratic equations

- A **quadratic equation** is an equation of degree two.

$$x^2 = 25$$
$$2x^2 = 12 - 8x$$
$$3x^2 + 8x + 12 = 0$$

Always start by writing the quadratic equation so that it equals zero: $ax^2 + bx + c = 0$, where a, b and c are constants.

Quadratic equations can be solved by **factorising**.

EXAMPLE

Solve these equations. Give the solutions to one decimal place where appropriate.

a $x^2 = 30$ **b** $x^2 + 3x = 0$

a $x = \pm\sqrt{30}$ The inverse of 'square' is 'square root'.
$x = 5.5$ (1 dp) or
$x = -5.5$ (1 dp)

b $x(x + 3) = 0$ $x^2 + 3x$ factorised.
$x = 0$ or $x + 3 = 0$
$x = 0$ or $x = -3$

> If the product of two numbers is zero then at least one of the numbers must be zero.

EXAMPLE

Solve these equations.

a $x^2 + 7x + 10 = 0$ **b** $x^2 + 3x - 10 = 0$
c $x^2 - 3x = 10$ **d** $x^2 - 7x + 10 = 0$

a $x^2 + 7x + 10 = 0$
$(x + 5)(x + 2) = 0$
$x + 5 = 0$ or $x + 2 = 0$
$x = -5$ or $x = -2$

b $x^2 + 3x - 10 = 0$
$(x + 5)(x - 2) = 0$
$x + 5 = 0$ or $x - 2 = 0$
$x = -5$ or $x = 2$

c $x^2 - 3x - 10 = 0$
$(x - 5)(x + 2) = 0$
$x - 5 = 0$ or $x + 2 = 0$
$x = 5$ or $x = -2$

d $x^2 - 7x + 10 = 0$
$(x - 5)(x - 2) = 0$
$x - 5 = 0$ or $x - 2 = 0$
$x = 5$ or $x = 2$

> The equation in part **c** needed rewriting first.

Approximate solutions to quadratic equations can also be found using graphs.

EXAMPLE

This is the graph of $y = x^2 + 5x$.

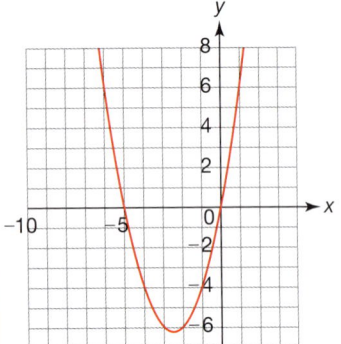

Use the graph to find approximate solutions for this equation.
$x^2 + 5x = 2$

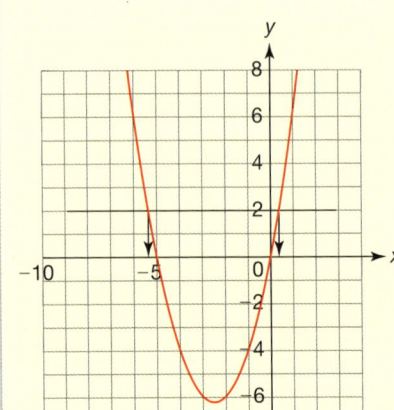

Find the points where the graph of $y = x^2 + 5x$ has the value of $y = 2$.

$x = 0.4$ and
$x = -5.4$

Exercise 10.3S

1 Solve $x^2 = 20$, giving your solutions to one decimal place.

2 Solve $x^2 + 2x = 0$

3 Solve these quadratic equations.

   a $x^2 - 4 = 0$

   b $x^2 + 4x = 0$

   c $x^2 - 4x = 0$

4 Solve these quadratic equations.

   a $x^2 + 5x + 6 = 0$

   b $x^2 + 7x + 12 = 0$

   c $x^2 + 8x + 15 = 0$

   d $x^2 + 8x + 16 = 0$

   e $0 = x^2 + 18x + 17$

   f $0 = x^2 + 15x + 26$

5 Solve these quadratic equations.

   a $x^2 + x - 6 = 0$

   b $x^2 + x - 12 = 0$

   c $x^2 + 2x - 15 = 0$

   d $x^2 + 6x - 16 = 0$

   e $0 = x^2 + 3x - 18$

   f $0 = x^2 + 9x - 22$

6 Solve these quadratic equations.

   a $x^2 - x - 6 = 0$

   b $x^2 - x - 12 = 0$

   c $x^2 - 2x - 15 = 0$

   d $0 = x^2 - 6x - 16$

   e $0 = x^2 - 13x - 30$

   f $0 = x^2 - 3x - 28$

7 Solve these quadratic equations.

   a $x^2 + 9x + 20 = 0$

   b $x^2 + 13x + 12 = 0$

   c $x^2 + 8x - 20 = 0$

   d $x^2 + 6x - 27 = 0$

   e $0 = x^2 - 10x + 24$

   f $0 = x^2 - 12x + 35$

   g $2x^2 + 12x + 10 = 0$

   h $3x^2 - 18x + 27 = 0$

8 Solve these quadratic equations.

   a $x^2 - x = 20$ **b** $x^2 + 11x = 12$

   c $8 = x^2 + 2x$ **d** $x^2 + 4x = 21$

   e $x^2 + 45 = 14x$ **f** $12x = x^2 + 32x$

9 This is the graph of $y = x^2 + 3x$.

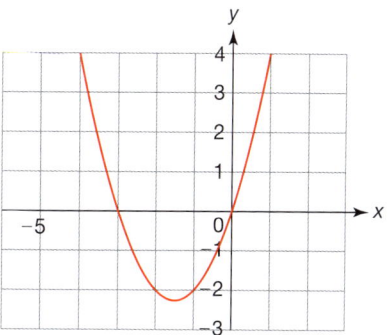

Use the graph to find approximate solutions, to one decimal place when appropriate, for these equations.

   a $x^2 + 3x = 0$ **b** $x^2 + 3x = 2$

   c $x^2 + 3x = 4$ **d** $x^2 + 3x = -2$

10 This is the graph of $y = x^2 - 4x$.

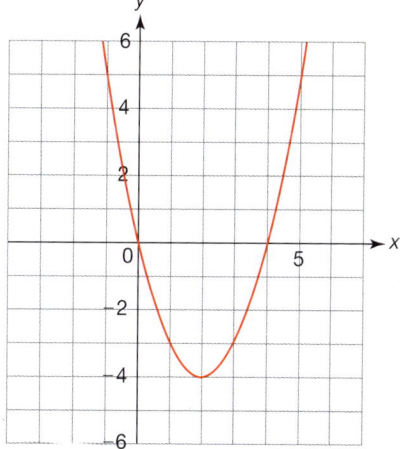

Use the graph to explain why the equation $x^2 - 4x = -5$ does not have a solution.

*11 Explain why $x^2 + 20 = 0$ does not have a solution.

12 The area of the rectangle is $18\,\text{cm}^2$.

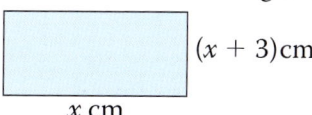

$(x + 3)\,\text{cm}$

$x\,\text{cm}$

Find the length of the smallest side.

Q 1169, 1181   SEARCH

10.3 Quadratic equations

p.144

RECAP

- **Quadratic** equations is of the form $ax^2 + bx + c$ for some numbers $a \neq 0$, b, and c.
- Many quadratic equations can be solved by:
  - rearranging so that one side equals zero
  - **factorising**.

$$x^2 - 7x = 18$$
$$x^2 - 7x - 18 = 0$$
$$(x + 2)(x - 9) = 0$$
$$x + 2 = 0$$
$$x = -2$$
$$\text{or } x - 9 = 0$$
$$x = 9$$

HOW TO

To solve a problem that involves a quadratic equation
1. Use the information in the question to form a quadratic equation.
2. Rearrange the quadratic so that it equals zero.
3. Solve the quadratic by factorising.
4. Check that your answers make sense.
   You may need to reject one solution, depending on the context.

EXAMPLE

Two numbers have a product of 105 and a difference of 8.
If the larger number is x

a show that $x^2 - 8x - 105 = 0$ **b** solve this equation to find the two numbers.

a ① The two numbers are x and $x - 8$.

So $x(x - 8) = 105$

② $x^2 - 8x - 105 = 0$ (as required)

b $x^2 - 8x - 105 = 0$

$(x + 7)(x - 15) = 0$

$x + 7 = 0$ or $x - 15 = 0$

$x = -7$ and $x - 8 = -15$

or $x = 15$ and $x - 8 = 7$.

The two numbers are -7 and -15 or 7 and 15.

③ Factorise the quadratic.
Find two numbers that add to -8 and multiply to give -105.
The two numbers are 7 and -15.

④ Two answers for x lead to two answers for $x - 8$.

EXAMPLE

If the area of the trapezium is $144\,\text{cm}^2$, show that $x^2 + 10x = 144$ and find the value of x.

① Area of trapezium $= \dfrac{(a + b)}{2}h$

$A = \dfrac{1}{2}(x + 10) \times 2x = x(x + 10)$

$144 = x(x + 10)$

$x^2 + 10x = 144$ (as required)

② $x^2 + 10x - 144 = 0$

$(x - 8)(x + 18) = 0$

Either $x - 8 = 0$ so $x = 8$

Or $x + 18 = 0$ so $x = -18$

③ Factorise the quadratic.
Find two numbers that add to 10 and multiply to give -144.
The two numbers are -8 and 18.

④ The trapezium cannot have sides with negative lengths.

The only solution is $x = 8\,\text{cm}$.

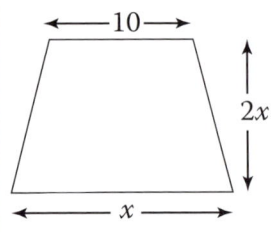

Exercise 10.3A

1 Vicky is trying to solve $(x + 4)(x - 2) = 7$. Here is her attempt.

> $(x + 4)(x - 2) = 7$
>
> Either $x + 4 = 7$ or $x - 2 = 7$
>
> $x = 3$ or $x = 9$

   a What is wrong with her method?

   b Solve this equation correctly to show that the two solutions should be $x = 3$ and $x = -5$.

2 Simplify and then solve this equation.

$$x^2 + 4x - 5 = 2x(x - 1)$$

3   **a** Two numbers which differ by 4 have a product of 60.
Find the numbers.

   b Two numbers which differ by 8 have a product of -12.
Find the numbers.

4   **a** Solve the equation $x^2 + 3x - 40 = 0$.

   b Two numbers have a product of 80. The larger number is 6 more than twice the smaller number.
Find two possibilities for the pair of numbers.

5   **a** Solve the equation $x^2 + 5x - 50 = 0$.
A rectangle has sides t cm and $t + 5$ cm.
The area of the rectangle is 50 cm².

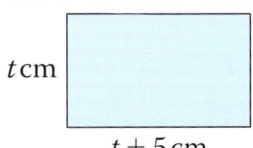

t cm

$t + 5$ cm

   b Show that t satisfies the equation $t^2 + 5t - 50 = 0$

   c Explain why there is only one possible value for t.

6 A triangle has angles $x°$, $6x°$ and $(x^2 + 10)°$.
Find the value of x.

7 The area of the trapezium is 36 cm².
Find the value of x.

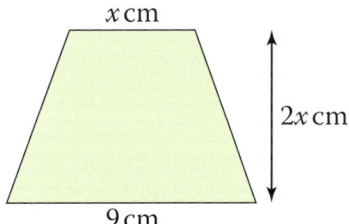

x cm

$2x$ cm

9 cm

8 The sides of a right-angled triangle are x cm, $x + 1$ cm and 5 cm.

p.388

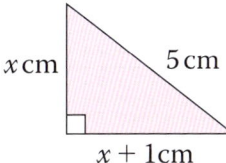

x cm

5 cm

$x + 1$ cm

   a Show that x satisfies the equation $x^2 + x - 12 = 0$.

   b Find the perimeter of the right-angled triangle.

9 If this square and rectangle have equal areas, find the side length of the square.

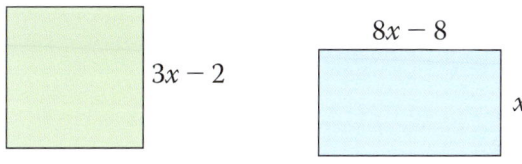

$3x - 2$

$8x - 8$

x

10 Two consecutive integers multiply together to give 4290.
Find the two integers.

10.4 Simultaneous equations

- **Simultaneous** equations are a set of equations that have the same solution.

Simultaneous equations can be solved algebraically using the methods of **elimination** or **substitution**.

EXAMPLE

Solve these simultaneous equations $x + 2y = 17$ and $3x + 2y = 19$

a using elimination   **b** using substitution.

For both methods, it is useful to label the equations.

a
$$x + 2y = 17 \quad ①$$
'Eliminate' y. $\quad 3x + 2y = 19 \quad ②$
$$② - ① \quad 2x = 2 \quad ÷ 2$$
$$x = 1$$
$x = 1$ in ① $\quad 1 + 2y = 17 \quad$ Solve
$$2y = 16$$
$$y = 8$$

Check the solution in ②.
$(3 \times 1) + (2 \times 8) = 19$ ✓

Solution: $x = 1$ and $y = 8$

b
$$x + 2y = 17 \quad ①$$
$$3x + 2y = 19 \quad ②$$
From ① $\quad x = 17 - 2y \quad ③$
③ in ② $\quad 3(17 - 2y) + 2y = 19 \quad$ Expand
$$51 - 6y + 2y = 19 \quad \text{Like terms}$$
$$51 - 4y = 19 \quad + 4y$$
$$51 = 19 + 4y \quad - 19$$
$51 - 19 = 32 \qquad 32 = 4y \quad ÷ 4$
$$y = 8$$
Sub in ③ $\quad x = 17 - (2 \times 8) = 1$

EXAMPLE

Solve these simultaneous equations.

a $\quad 2x - y = 9 \quad ①$
$\quad\quad 5x - y = 21 \quad ②$

b $\quad 3x + 2y = 20 \quad ①$
$\quad\quad 2x - y = 18 \quad ②$

Always check the solution works for both equations.

a Eliminate y.
$② - ① \quad 3x = 12 \quad ÷ 3$
$$x = 4$$
$2 \times 4 - y = 9 \qquad x = 4$ in ①
$$8 - y = 9$$
$$8 = 9 + y$$
$$y = -1$$

b Eliminate y.
$2 \times ② \quad 4x - 2y = 36 \quad ③$
$① + ③ \quad 7x = 56 \quad ÷ 7$
$$x = 8$$
$x = 8$ in ① $\quad 24 + 2y = 20 \quad - 24$
$$2y = -4 \quad ÷ 2$$
$$y = -2$$

Approximate solutions to simultaneous equations can also be found by drawing graphs.

EXAMPLE

Use the graphs to find an approximate solution for these simultaneous equations.

$y = 2x - 7$

$2y + x = 3$

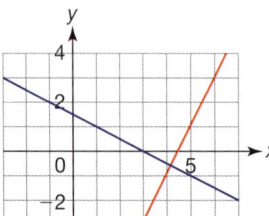

The solution is where the graphs intersect.

$x = 3.4$

$y = -0.2$

Algebra   Equations and inequalities

Exercise 10.4S

1 Solve these simultaneous equations.

$3x + y = 24$

$2x + y = 17$

2 Solve these simultaneous equations.

$3x + 3y = 24$

$3x + y = 20$

3 a Find three pairs of solutions for this equation.
$2x + y = 11$

b Find three pairs of solutions for this equation.
$6x + y = 27$

c Solve these simultaneous equations.

$2x + y = 11$

$6x + y = 27$

4 Solve these simultaneous equations.

a $4x + 4y = 16$ **b** $3x + 2y = 19$

   $x + 4y = 13$ $x + 2y = 9$

c $5p + 3q = 31$ **d** $4a + 3b = 17$

   $5p + q = 17$ $4a + 5b = 15$

5 Solve these simultaneous equations.

a $4m + 4n = 24$ **b** $3x + 2y = 16$

   $m + 2n = 8$ $2x + y = 9$

c $5x + 3y = 17$ **d** $4e + 3f = 13$

   $x + 6y = -2$ $8e + 5f = 23$

e $2m + 3n = 14$ **f** $2y + 3x = 5$

   $m = 14 - 5n$ $y = 7 - 3x$

6 Solve these simultaneous equations.

a $4x - 4y = 20$ **b** $x - 2y = 11$

   $x - 4y = 2$ $3x - 2y = 25$

c $5p - 3q = 27$ **d** $a - 2b = 5$

   $5p - q = 29$ $4a - 5b = 23$

7 Solve these simultaneous equations.

a $4x + 2y = 26$ **b** $x + 3y = 13$

   $x - 2y = 4$ $3x - 3y = 15$

c $5p + 3q = 7$ **d** $3a + 2b = 9$

   $2p - q = 5$ $6a - 5b = 45$

8 Use this graph to find approximate solutions for these simultaneous equations.

$x + 3y = 4$

$y = 2x + 2$

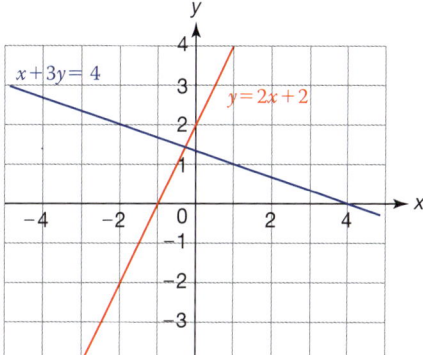

*9 By drawing graphs, find approximate solutions for these simultaneous equations.

a $x + 3y = 4$ **b** $x + 2y = 6$

   $y = 2x - 2$ $y = 3x + 1$

c $y = 3x - 1$ **d** $y = 3x - 7$

   $y = x - 2$ $x = y + 2$

> You can use graphing software to help you or to check your answers.

*10 Solve these simultaneous equations.

$4x + y = 3$

$x^2 + y = 15$

11 The sum of the ages of Bob's grandparents is 135 years.

The difference between their ages is 11.

What are the possible ages of Bob's grandparents?

10.4 Simultaneous equations

RECAP

- You can solve a pair of simultaneous equations exactly by
  - eliminating one of the two variables
  - by rearranging substituting for one variable in the second equation.
- You can solve a pair of simultaneous equations approximately by plotting a graph. The solution is where the two lines intersect.

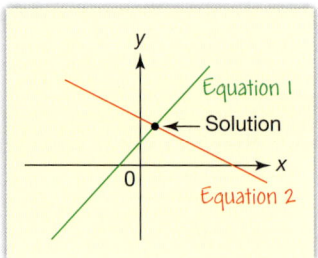

HOW TO

To solve problems that involve a pair of simultaneous equations

① Use the information in the question to form a pair of simultaneous equations.

② Solve the simultaneous equations using elimination or by drawing a graph.

③ Give your answers and check that they make sense.

EXAMPLE

In a sweet shop, Tyler spends £3.20 on three cans of soft drink and four bars of chocolate. The next day, he buys a can of soft drink and four bars of chocolate for £2. How much does each item cost?

① Need to find the price of both items. Give all the variables letters.

Let the number of cans $= c$
and the number of bars of chocolate $= b$.

Write the information as equations using algebra.

$$3c + 4b = 320 \quad (1)$$
$$c + 4b = 200 \quad (2)$$

② You can eliminate the b-terms by subtracting (1) – (2).

$$2c = 120$$
$$c = 60$$

Substitute $c = 60$ into equation (2) to find b.

$$c + 4b = 200$$
$$60 + 4b = 200$$
$$4b = 140$$
$$b = 35$$

③ Check your answer in (1). $3c + 4b = 3 \times 60 + 4 \times 35 = 320$ ✓

Answer the question. I can of drink costs 60p
and I chocolate bar costs 35p

> Don't forget to number your equations. This helps to avoid confusion.

Exercise 10.4A

1 Which pairs of equations have the solution $x = 2$ and $y = 7$?

A
$$x + y = 9$$
$$x - y = -5$$

B
$$2x + y = 11$$
$$3x - y = -1$$

C
$$2x + 2y = 15$$
$$4x - y = 1$$

D
$$x + 2y = 16$$
$$x - 2y = 8$$

E
$$5x - y = 3$$
$$y - x = 5$$

2 Solve these simultaneous equations.

a $2x + y = 8$
  $5x + 3y = 12$

b $3x + 2y = 19$
  $4x - y = 29$

c $8a - 3b = 30$
  $3a + b = 7$

d $2v + 3w = 12$
  $5v + 6w = 27$

e $9p + 5q = 15$
  $3p - 2q = -6$

f $3x - 2y = 11$
  $2x - y = 8$

3 Make as many pairs of simultaneous equations as you can using these three cards. Solve your pairs.

A $2x + y = 12$

B $y - x = 15$

C $3x - 4y = 7$

4 Solve these problems by using simultaneous equations.

a How much does a lemon cost?

> 3 lemons
> 4 oranges
> £1.27

> 4 oranges
> 5 lemons
> £1.61

b The perimeter of this triangle is 30 cm.

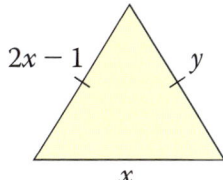

$2x - 1$ y

x

How long is the base?

5 For each question, set up a pair of simultaneous equations and solve them to find the required information.

a Two numbers have a sum of 23 and a difference of 5. What numbers are they?

b Two numbers have a difference of 6. Twice the larger plus the smaller number also equals 6. What numbers are they?

6 Use simultaneous equations to find the value of each symbol in the puzzle.

| 🌙 | ☀ | 🌙 | ☀ | 92 |
| ☀ | 🌙 | ☀ | ☀ | 104 |

7 Tickets for a theatre production cost £3.50 per child and £5.25 per adult. 94 tickets were sold for a total of £365.75 How many children attended the production?

8 Use a graphical method to solve these problems.

a Twice one number plus three times another is 4.
  Their difference is 2.
  What are the numbers?

b The sum of the ages of James and Isla is 4.
  The difference between twice Isla's age and treble James' age is 3.
  How old are they?

9 a By plotting graphs if necessary, explain why the simultaneous equations $y = 2x - 1$ and $y = 2x + 4$ have no solution.

b Is it possible to have a pair of simultaneous equations with more than one solution?

1175, 1176, 1319   SEARCH

10.5 Inequalities

p.4

- **Inequalities** are statements involving the symbols
  - $<$ less than $\leqslant$ less than or equal to
  - $>$ greater than $\geqslant$ greater than or equal to

Linear inequalities can be solved algebraically similar to solving linear equations.

Solutions can be represented using a number line such as $x \geqslant 3$.

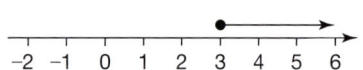

EXAMPLE

List all the integer values of n such that satisfy these linear inequalities.

a $-3 < n \leqslant 5$ **b** $5 < 2n < 11$

a $-2, -1, 0, 1, 2, 3, 4, 5$

     You include 5 because the inequality includes the $\leqslant$ sign.

b $2.5 < n < 5.5$ Divide both sides of
     n can be 3, 4, 5   the inequality by 2.

EXAMPLE

For each of these linear inequalities

   i   solve it   **ii**   represent the solution on a number line.

   a $3x < 15$ **b** $4x + 3 \geqslant 21$

   c $6 < 3(x - 2) < 18$ **d** $\dfrac{5x - 2}{3} \leqslant 6$

> Use a solid dot if the end point is included in the interval and an open dot if it isn't.

a   **i** $3x < 15$ $\div 3$

       $x < 5$

   ii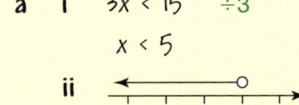

b   **i** $4x + 3 \geqslant 21$ -3

          $4x \geqslant 18$ $\div 4$

           $x \geqslant 4.5$

   ii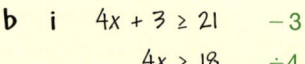

c   **i** $6 < 3(x - 2) < 12$

       $6 < 3x - 6 < 12$ $+6$

       $12 < 3x < 18$ $\div 3$

       $4 < x < 6$

   ii

d   **i** $\dfrac{5x - 2}{3} \leqslant 6$ $\times 3$

       $5x - 2 \leqslant 18$ $+2$

          $5x \leqslant 20$ $\div 5$

           $x \leqslant 4$

   ii

You must be careful when solving an inequality if the coefficient of the unknown is negative.

$-2x > 6$ $-2x > 6$ $-2x > 6$

$0 > 6 + 2x$ $x > 6 \div -2$ $x < 6 \div -2$

$-6 > 2x$ $x > -3$ ✗ $x < -3$ ✓

$-3 > x$ ✓

- When you divide an inequality by a negative number you must reverse the inequality sign.

Algebra   Equations and inequalities

Exercise 10.5S

1 List all the integer values of n such that $-2 < n < 5$.

2 Write an inequality to represent the solution shown on this number line.

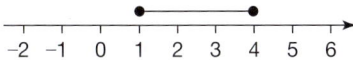

3 List all the integer values of n that satisfy these inequalities.

   a $1 < n \le 6$ **b** $4 < 2n \le 9$

   c $3 < 2n + 1 < 11$

4 Solve these linear inequalities.

   a $2x > 12$ **b** $2x + 1 < 17$

   c $3x - 2 < 19$ **d** $5x - 2 \le 28$

   e $24 \le 7x + 3$ **f** $29 \le 8x - 3$

   g $5x + 13 \le 68$ **h** $-5 < 10 + 3x$

5 Solve these linear inequalities.

   a $10 > 2x > 2$

   b $1 < 2x + 1 < 17$

   c $7 \le 3x - 2 < 19$

   d $3 < 5x - 2 \le 28$

   e $-2 \le 5x + 3 \le 23$

   f $-5 < 8x + 3 < 19$

   g $-20 < 3x - 5 < 19$

   h $-5 < 10 + 3x < -2$

6 Solve these linear inequalities.

   a $2(x - 3) > 18$ **b** $3(x + 1) < 27$

   c $3(x - 2) < 36$ **d** $5(x - 2) \le 20$

   e $27 \le 9(x - 2)$ **f** $30 \le 4(x + 3)$

   g $-15 \le 5(y - 2)$ **h** $-5 < 5(1 + 3x)$

7 Solve these linear inequalities.

   a $\dfrac{x}{2} + 3 > 7$ **b** $\dfrac{x}{2} - 5 < 4\dfrac{1}{2}$

   c $\dfrac{3x - 6}{4} > 3$ **d** $\dfrac{5x - 2}{3} \le 6$

   e $3 < \dfrac{9x - 12}{5}$ **f** $8 > \dfrac{4(x + 3)}{3}$

8 Solve these linear inequalities.

   a $-2x < 6$ **b** $-3x \ge 6$

   c $-7x > 49$ **d** $16 \le -4x$

8   **e** $-5x > -15$ **f** $-3x \le -51$

      g $81 \le -9x$ **h** $-81x > -27$

9 Solve these linear inequalities.

   a $10 - 4x < 2$ **b** $24 - 3x \ge 12$

   c $20 - 7x > 27$ **d** $16 - 4x > -4$

   e $27 - 6x > 39$ **f** $123 \le 36 - 3x$

   g $54 - 16x \ge 46$ **h** $100 < 75 - 10x$

10 Using your solutions to question **4**, find the smallest or largest integer that satisfies these inequalities.

   a $2x > 12$ **b** $2x + 1 < 17$

   c $3x - 2 < 19$ **d** $5x - 2 \le 28$

   e $24 \le 7x + 3$ **f** $29 \le 8x - 3$

   g $5x + 13 \le 68$ **h** $-5 < 10 + 3x$

11 Show each of these inequalities on a number line.

   a $2 < x < 10$ **b** $1 \le x < 4$

   c $7 \le x \le 9$ **d** $-3 < x \le 4$

12 Show the solution for each of these inequalities on a number line.

   a $10 \le 3x - 2 < 22$

   b $13 < 5x - 2 \le 18$

   c $-7 \le 5x + 3 \le 3$

   d $-3 < 4x + 3 < 2$

13 Using your solutions to question **6**, find the smallest or largest integer that satisfies these inequalities.

   a $2(x - 3) > 18$ **b** $3(x + 1) < 27$

   c $3(x - 2) < 36$ **d** $5(x - 2) \le 20$

   e $27 \le 9(x - 2)$ **f** $30 \le 4(x + 3)$

   g $-15 \le 5(y - 2)$ **h** $-5 < 5(1 + 3x)$

14 Solve these linear inequalities.

   a $2(8 - 3x) < 4$ **b** $5(17 - 3x) \ge 25$

   c $3(14 - 7x) > -21$ **d** $24 \le 8(7 - 4x)$

   e $7(24 - 8x) \ge 0$ **f** $75 \le 5(15 - 5x)$

   g $8(4 - 7x) \le 4$ **h** $-12 < 3(29 - 7x)$

Q 1161, 1162, 1930   SEARCH

10.5 Inequalities

RECAP

- An **inequality** is a mathematical statement including one of these symbols.

$<$	$>$	$\leqslant$	$\geqslant$
less than	more than	less than or equal to	more than or equal to

- You can solve an inequality by rearranging and using inverse operations, in a similar way to solving an equation.

- The solution to an inequality can be a range of values, which you can show on a number line.

- If you multiply or divide an inequality by a negative number you need to reverse the inequality sign to keep it true.

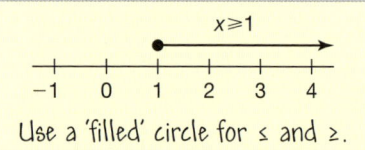

$$3x + 2 > 5x - 1$$
$$2 > 2x - 1$$
$$3 > 2x$$
$$1.5 > x$$

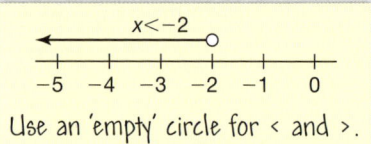

$x \geqslant 1$

Use a 'filled' circle for $\leqslant$ and $\geqslant$.

$x < -2$

Use an 'empty' circle for $<$ and $>$.

$4 < 6$ but $-4 > -6$
$5 > -2$ but $-5 < 2$

> Algebra isn't just about things being equal. Inequalities allow us to work with quantities that are unequal.

HOW TO

To solve an inequality

① Use the information in the question to form an inequality. The inequality could be one-sided or two-sided.

② Use the balance method to solve the inequality. Remember that multiplying or dividing by a negative number changes the direction of the inequality sign.

③ Give the range of values for the answer.

EXAMPLE

The area of the rectangle is less than $24\,\text{cm}^2$. Explain why x must satisfy $4 < x < 12$.

$(12 - x)\,\text{cm}$

$3\,\text{cm}$

① Area < 24

  Area of a rectangle = length $\times$ width

② $3(12 - x) < 24$ Expand brackets.
  $36 - 3x < 24$ -36
    $-3x < -12$ $\div -3$ and reverse the inequality sign.
      $x > 4$

③ The sides of the rectangle must be positive lengths.
    $12 - x > 0$
      $12 > x$
  So $4 < x < 12$

Exercise 10.5A

1 A number x can take the possible integer values $-1, 0, 1, 2, 3$. Which of these inequalities could be true?

a $x > -2$ **b** $-1 \leqslant x \leqslant 3$

c $-2 < x < 4$ **d** $-1 \leqslant x < 3$

e $-2 < x \leqslant 3$

2 The area of this square is less than the area of this triangle.

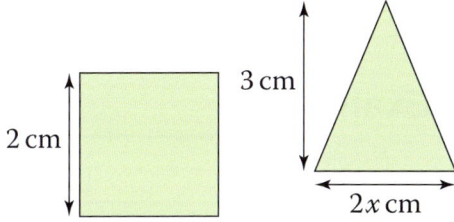

Find the range of possible values that x can take and show your solution on a number line.

3 **a** The area of this rectangle exceeds its perimeter. Write an inequality and solve it to find the range of values of x.

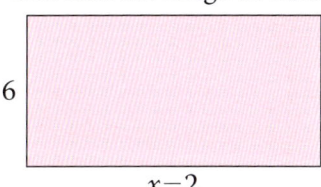

b Given that x is an integer, find the smallest possible value that x can take.

4 Solve the inequality $-2 < 3x - 1 \leqslant 5$ and show the solution on a number line.

*5 Is it possible to write a single linear inequality to represent this solution?

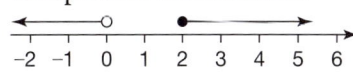

Give your reasons.

6 **a** The angle $(2x + 30)°$ is obtuse.

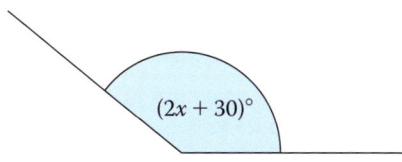

Find the range of values that x can take.

6 **b** The angle $(5y - 45)°$ is reflex.

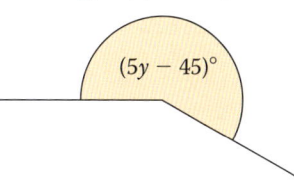

Find the range of values that y can take.

*7 Solve $x^2 > 9$.

8 Find all the positive pairs of integers that satisfy this inequality.

$2x + 5y \leqslant 16$

9 Find the range of values of x that satisfy *both* inequalities.

a $3x + 6 < 18$ and $-2x < 2$

b $10 > 5 - x$ and $3(x - 9) < 27$

10 Explain why it is not possible to find a value of y such that $3y \leqslant 18$ and $2y + 3 > 15$.

11 This triangle is isosceles.

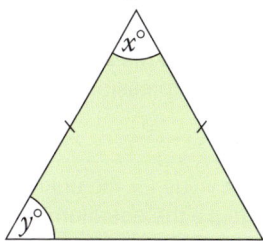

Angle x is less than $30°$.

a Say why $180 - 2y < 30$.

b Complete this statement for the range of possible values that y can take.
$y > \square$

c Give a reason why y must be less than $90°$.

12 Here is a parallelogram.

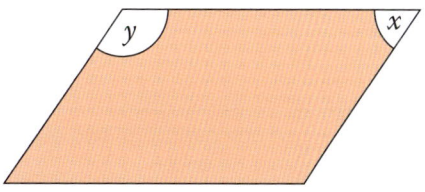

Angle y is greater than $120°$.
Show that x must be less than $60°$.

Summary

Checkout
You should now be able to...

Test it
Questions

	Questions
✔ Derive and solve simple linear equations.	1 – 3
✔ Solve quadratic equations algebraically by factorising.	4, 5
✔ Derive and solve two linear simultaneous equations in two variables.	6, 7
✔ Find approximate solutions to two linear simultaneous equations using a graph.	8
✔ Solve linear inequalities in one variable and represent the solution on a number line.	9, 10

Language	Meaning	Example
Balance method	A method for solving an equation by performing the same operation on each side.	$x - 6 = 7$ $x - 6 + 6 = 7 + 6$ add 6 to both sides $x = 13$
Quadratic	A quadratic expression contains a square term such as x^2 as the highest power.	$6x^2 - 7x - 3$
Factorising	Writing an expression as two or more different expressions multiplied together.	$3x^2 + 6x = 3(x^2 + 2x)$ $= 3x(x + 2)$ $x^2 - 3x - 10 = (x + 2)(x - 5)$
Solve **Solution**	Find a value for the unknown variable that will make the equation true.	$4x - 3 = 9$ is true when $x = 3$. $x = 3$ is the solution to the equation.
Simultaneous equations	Two or more equations that are true at the same time for the same values of the variables.	$y = 3x - 2$ and $2x + y = 8$ are both true when $x = 2$ and $y = 4$.
Inequality	A comparison of two quantities that may not be equal.	$5 < 9$ 5 is less than 9. $x \geqslant 6$ The value of x is greater than or equal to 6.

Review

1 Solve these equations.

    a $a - 13 = 35$ **b** $9b = 54$

    c $5c + 8 = 43$ **d** $2d - 8 = 25$

    e $\frac{e}{5} = 20$ **f** $7f + 19 = 5$

2 Solve these equations with unknowns on both sides.

    a $2x + 5 = x + 11$

    b $5x - 7 = 9x - 15$

    c $20 - 3x = 34 - x$

3 Vicky is x years old. Her mum is 24 years older and the sum of their ages is 42.

    a Write an equation in x for the sum of their ages.

    b Solve your equation to find the ages of Vicky and her mum.

4 Solve these quadratic equations by factorising.

    a $x^2 + 8x + 15 = 0$

    b $x^2 - 6x + 5 = 0$

    c $x^2 + x - 6 = 0$

    d $x^2 - 64 = 0$

    e $x^2 - 12x = 0$

5 Solve this quadratic equation.

    $x^2 - 8x = 20$

6 Solve these pairs of simultaneous equations.

    a $2x + 3y = 21$ **b** $5a - b = 7$

       $2x - y = 1$ $10a + b = 8$

    c $x + 4y = 21$ **d** $5v + 3w = 1$

       $2x - y = 15$ $4v - 9w = 35$

7 The combined mass of two boxes of chocolates and one bottle of water is 3.75 kg. The mass of three boxes of chocolates and two bottles of water is 6 kg.

    a Write simultaneous equations to describe this situation.

    b Find the mass of a box of chocolates and the mass of a bottle of water.

8 Plot a graph to solve these simultaneous equations.

    $x + y = 4$

    $y = 2x + 1$

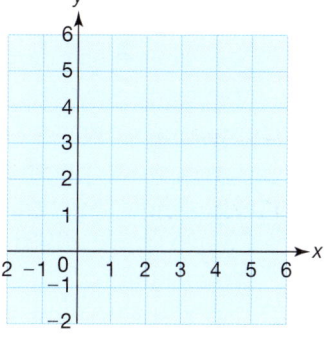

9 Show these inequalities on number lines.

    a $x > 4$ **b** $x \leqslant 6$ **c** $1 < x \leqslant 6$

10 Solve these inequalities. Show each solution on a number line.

    a $3x + 7 > 22$ **b** $5x - 2 \leqslant 13$

    c $2x + 9 \geqslant 12 - x$

    d $8 - 2x < 10$

What next?

Score			
	0 – 4		Your knowledge of this topic is still developing.
			To improve look at MyMaths: 1154, 1161, 1162, 1169, 1175, 1176, 1181, 1182, 1319, 1395, 1925, 1928, 1930
	5 – 8		You are gaining a secure knowledge of this topic.
			To improve your fluency look at InvisiPens: 10Sa – o
	9 – 10		You have mastered these skills. Well done you are ready to progress!
			To develop your problem solving skills look at InvisiPens: 10Aa – g

Assessment 10

1 A square of side 7 cm has the same area as a rectangle with sides 5 cm and $2s$ cm. Set up and solve an equation in s. [3]

2 Avel's dad is 4 times as old as Avel.
The sum of their ages is 55. How old is Avel? [2]

3 Jacques and Gilles went up the hill to fetch a pail of water. The mass of the water was 22 kg more than the pail. In total the mass was 27.5 kg. How heavy was the pail? [2]

4 Oliver says he is three times as old as his brother Albert. Albert says Oliver is 6 years older than he is. They are both right. How old are Oliver and Albert? [4]

5 Alice bought a cake and cut it into 3 pieces. Tweedledum's piece was 35 g heavier than Tweedledee's. Tweedledee's piece was 22 g lighter than Alice's. The total mass was 454 g. Calculate the mass of Alice's piece of cake. [4]

6 Brendan, Arsene and José go on holiday. Brendan takes €x, Arsene takes half as much as José and José takes €150 more than Brendan. They took €2000 spending money in total.

   a Write down an equation involving x. [1]

   b Solve the equation to find the value of x. [3]

   c How much money did each person take? [2]

7 $PQRS$ and $PYZS$ are rectangles.
The area of $YQRZ$ is 45 cm^2.
Grace works out that the value of a is 5 cm.
Show how she worked out the value of a.
When Grace solved her equation, she got
two solutions.
Say why she could ignore the other solution. [8]

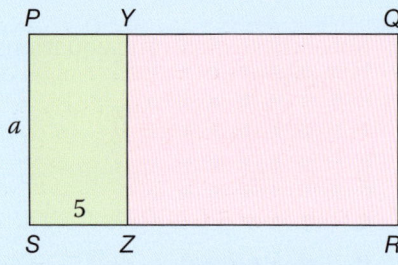

8 Mario attempted to kayak 30 km on Lake Garda to raise money for charity. Sophia said she would give him €5 for each complete kilometre he covered, as long as Mario gave her €2 for each complete kilometre he failed to kayak. They gave €115 to charity.

   a Mario kayaked k kilometres in the event. Write down an equation involving k and solve it to find the number of kilometres he kayaked. [4]

   b Find the minimum number of kilometres that Mario needed to complete to be sure of making a contribution to charity. [3]

9 Patrick factorised this expression incorrectly.
$w^2 - w - 72 = (w - 9)(w - 8)$

   a Show how to factorise the expression correctly. [2]

   b Solve the equation $w^2 - w - 72 = 0$ [1]

10 Maria says that $89^2 - 11^2 = 7800$ and $6.89^2 - 3.11^2 = 37.8$
Without using your calculator, show that both of her answers are correct. [6]

11 A square has a side length s cm. Another square has side length 2 cm shorter.
The total area of the two squares is 202 cm^2. Find s. [6]

12 Two consecutive odd numbers have a product of 63.
There are two 'sets' of answers. What are they? [4]

13 The formula $S = \dfrac{n(n + 1)}{2}$ represents the sum of the numbers $1 + 2 + 3 + \cdots\cdots + n$.
The total of the numbers 1 to n is 5050. What is n? [4]

14 Eoin hits a cricket ball straight upwards. The formula $h = 20t - 5t^2$ represents
its height, h, above the ground, t seconds after he throws it.
Find the two times when the height of the ball is 15 m above the ground.
Say why there are two possible answers. [5]

15 The formula $2d = n(n - 3)$ represents the number of diagonals, d, in a polygon
with n sides. A dodecagon has 54 diagonals. How many sides does it have? [4]

16 Batman and Robin are b and r years old. Their ages sum to 46 and Robin is
10 years younger than Batman. How old are they? [4]

17 On a coach trip, an adult ticket is £a and a child ticket £c.
Tickets for 2 adults and 2 children are £22. Tickets for 2 adults and 5 children are £35.50.
How much is   **a**   an adult's ticket   [3]   **b**   a child's ticket? [1]

18 A magic goose lays brown eggs and once a year a gold egg! 5 brown eggs and 1 gold
egg have a mass of 150 g, while 8 brown and 1 gold egg have a mass of 210 g.
Find the weight of a gold egg. [4]

19 Liz and Graham had dinner. Graham's meal was £2.50 more expensive than Liz's.
The final bill was £40.50. How much did each meal cost? [4]

20 Five packets of 'doggibix' and three packets of 'cattibix' cost £5.49.
Three packets of 'doggibix' and one packet of 'cattibix' cost £2.79.
Find the cost of one packet of   **a**   Doggibix   [4]   **b**   Cattibix. [1]

21 A DVD costs £D and a CD £c. 3 DVDs and 2 CDs cost £45.83 in total.
1 DVD and 3 CDs cost £26.92. How much does each cost? [5]

22 Farmer Scott can buy 2 sheep and 6 cows, or 10 sheep and 2 cows, for £3500.
What is the price of   **a**   a sheep   [4]   **b**   a cow? [1]

23 Titus Lines is going fishing and needs bait. He can buy 5 maggots and 6 worms for
38p or 6 maggots and 12 worms for 60p. How much are maggots and worms? [5]

24 Shaun has incorrectly represented the inequality $-1 < x \leqslant 3$ on a number line as shown.

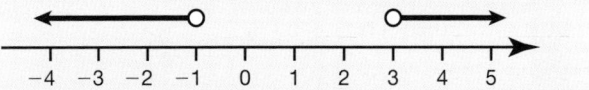

Draw the expression correctly on a number line. [3]

25 Craig incorrectly says that all of the integer solutions to $14 \geqslant 4c \geqslant -3$ are $c = 4, 3, 2$
and 1. List all of the integer solutions. [3]

26 A triangle has angles x, y and z, $x > 35$, $y > 61$. Write an inequality for z. [2]

27 **a**   If $a > 1$ write down the inequality for $\dfrac{1}{a}$. [1]

    b   Is it always true that $b^2 > b$? Give reasons for your answer. [2]

Life skills 2: Starting the business

Abigail, Mike, Juliet and Raheem, have completed their business plan, and decide they want to locate their restaurant in an area near the railway station in Newton-Maxwell. To choose the ideal location, they need to think about how close they are to the High Street and competitor restaurants. They also start planning other key items: designing promotional material, what tables they need to buy and how many, and contacting potential suppliers.

Task 1 – Location

The friends make a list of conditions that the location must meet. Raheem draws a scale map on 5 mm square paper.

a What is the scale of the map?

b Draw an accurate copy of the map and shade the areas that meet all three conditions.

They decide on a location that is on a bearing of 315° from T and 014° from A.

c On your copy of the map, label their desired location with R.

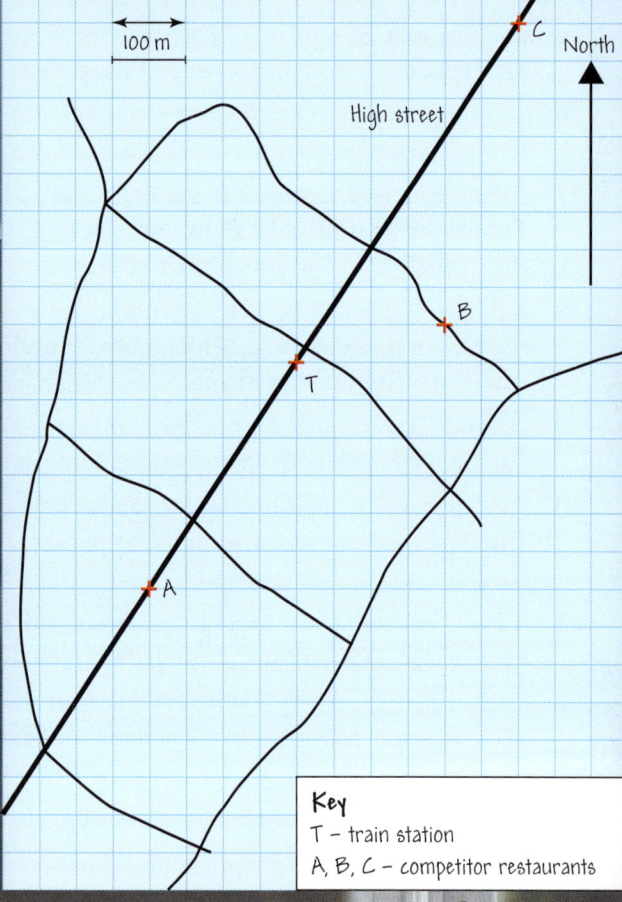

Conditions for location

- No more than 500 m from the railway station
- At least 200 m from each competitor restaurant
- No more than 150 m from the high street

Key
T – train station
A, B, C – competitor restaurants

Task 2 – The restaurant logo

The diagram shows the dimensions used for a logo to appear on the restaurant's business cards.

p.230

a Find the area of the logo.

For use on A5 flyers, the logo is enlarged by a scale factor of 2.5

b Find the area of the enlarged logo.

The logo is the only part of the business card and the flyer that uses coloured ink.

c A colour ink cartridge lasts long enough to print 4000 business cards. How many A5 flyers would you expect to print using one colour ink cartridge?

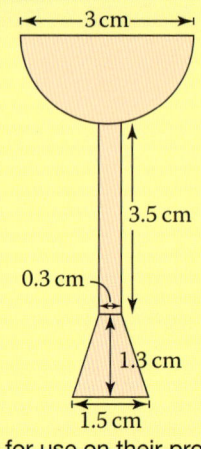

▲ Juliet's logo for use on their promotional material. It consists of a semicircle, a rectangle and a trapezium, as shown in the second figure.

FIRE

Fire regulations

- Total number of staff and non-staff must not exceed 32.
- There must be at least one member of staff for every 8 non-staff.

Table	Shape of top	Dimensions
Style A	Circular	Diameter = 1.4 m
Style B	Regular octagon	Side length = 58 cm
		Length between opposite sides = 1.4 m

▲ Possible styles of table for restaurant.

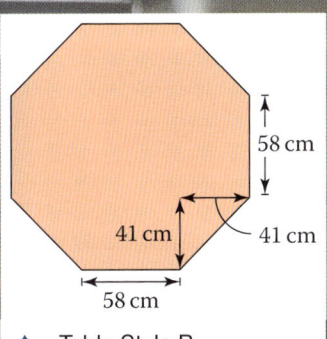

▲ Table Style B.

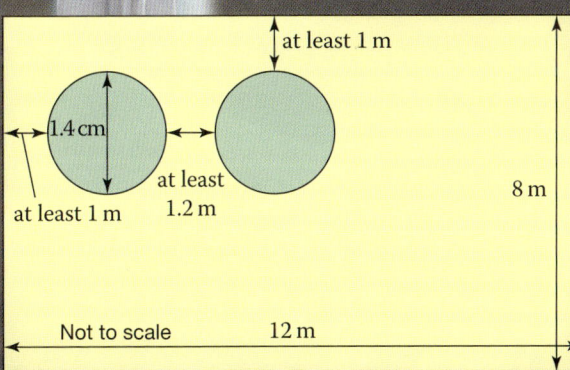

Caller	Total number of calls made	Number of unanswered calls
Abigail	30	10
Mike	22	8
Juliet	40	18
Raheem	18	8

▲ Number of calls made in the first week.

Task 3 – Safety regulations

The restaurant must adhere to fire safety regulations.

Find the maximum number of customers allowed in the restaurant at any time.

Task 4 – Choosing tables

They narrow down their choice of table based on the maximum number of customers allowed at one time.

a Show that the regular octagon in the diagram measures 1.4 m across.

b Without doing any calculations, which of the table tops (style A or style B) has the larger area?

c Calculate the area of each table top.

Style A is cheaper, so they decide to buy that one.

d Is this a reasonable decision? Give reasons for your answer.

Task 5 – Number of tables

The dining space is rectangular and measures 12 m by 8 m. Abigail draws a sketch to indicate the gap required from each wall, and between each table.

How many tables of Style A can fit in the dining space?

Task 6 – Calling potential suppliers

Each member of the team makes telephone calls to potential suppliers. They logged how many calls they made in the first week, and how many went unanswered.

a Which person had the highest proportion of unanswered calls?

b Estimate how many unanswered calls you would expect out of the next 50 calls made by the team.

c What assumptions have you made in answering part b?

11 Circles and constructions

Introduction

The invention of the wheel was most definitely a landmark event in human technological development, giving people the ability to travel at speed. However it was the use of gears and cogs on a massive scale during the Industrial Revolution that really accelerated advancement, not just in technology but also in social and economic development.

What's the point?

Without mankind's understanding of circles, and how their properties can be exploited in marvellous ways, we would still be living in largely agricultural communities in a pre-industrial state, with no computers, mobile phones, cars, …

Objectives

By the end of this chapter you will have learned how to …

- Identify and apply circle definitions, properties and formulae.
- Construct triangles.
- Use the standard ruler and compass constructions.
- Solve loci problems.

Check in

1   Using a protractor, measure these angles.

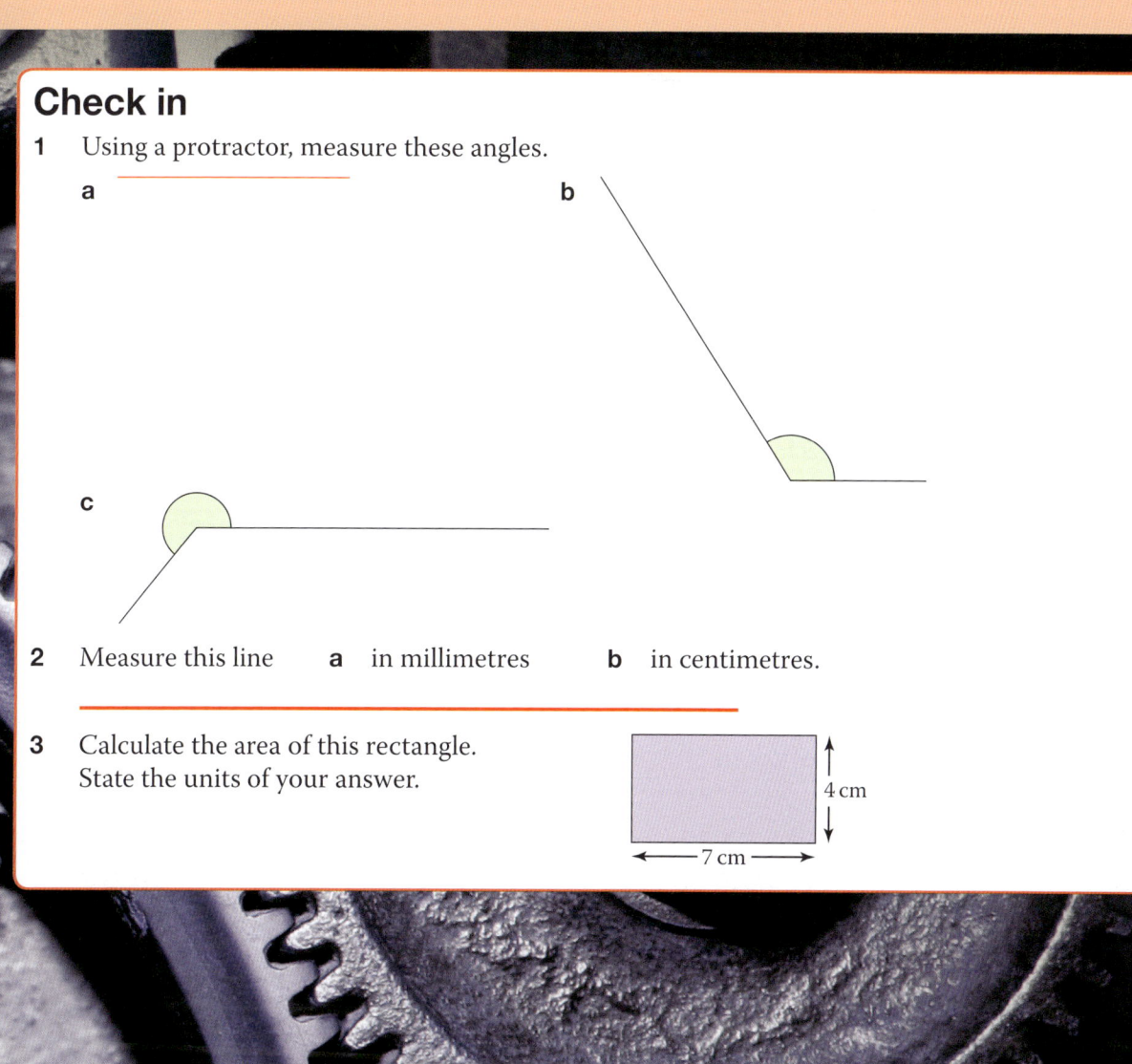

a

b

c

2   Measure this line   **a**   in millimetres   **b**   in centimetres.

3   Calculate the area of this rectangle.
    State the units of your answer.

4 cm

7 cm

Chapter investigation

Sketch a circle, and draw a straight line through it. How many pieces have you divided the circle into?

Draw a second straight line thought the circle. What is the maximum number of pieces you can divide the circle into?

Continue drawing straight lines through the circle (the lines can cut at any angle, and the circle can be as big as you want it to be). Is there a relationship between the number of cuts and the maximum number of pieces? Investigate.

11.1 Circles 1

Perimeter is the total distance around the sides of a 2D shape.

For a circle, the perimeter has a special name: **circumference**.

Diameter of a **circle** = 2 × **radius**

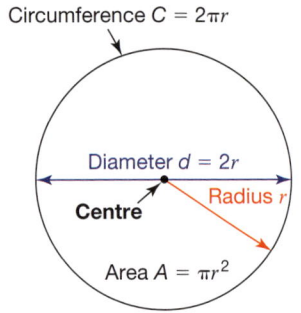

Circumference $C = 2\pi r$

Diameter $d = 2r$

Radius r

Centre

Area $A = \pi r^2$

p.116

- Circumference of a circle, $C = \pi d = 2\pi r$ where $\pi = 3.14159\ldots$

  Rearranging gives $d = \dfrac{C}{\pi}$ or $r = \dfrac{C}{2\pi}$

- Area inside a circle, $A = \pi r^2 = \pi \times r \times r$

  Rearranging gives $r = \sqrt{\dfrac{A}{\pi}}$

EXAMPLE

This shape is made from a semi-circle and a rectangle.

a Find the perimeter of the shape.

b Find the area of the shape.

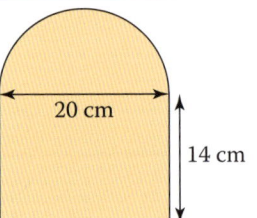

20 cm

14 cm

a Arc length of semi-circle $= \dfrac{1}{2} \times \pi d$

   $= \dfrac{1}{2} \times \pi \times 20$

   $= 10 \times \pi = 31.415\ldots\,cm$

A semi-circle is half of a circle.

You can use $\pi \approx 3$ to check this value.

$10 \times 3 = 30$

Total perimeter $= 31.415\ldots + 14 + 20 + 14$

   $= 79.415\ldots$

   $= 79.4\,cm$ (3 sf)

An *exact* answer is $(10\pi + 48)\,cm$

p.144

b Area of rectangle $= 20 \times 14 = 280\,cm^2$

Area of semi-circle $= \dfrac{1}{2} \times \pi r^2$

   $= \dfrac{1}{2} \times \pi \times 10 \times 10$

   $= 50 \times \pi = 157.079\ldots\,cm^2$

Area of rectangle = length × width

Radius = diameter ÷ 2

p.316

Total area $= 157.079\ldots + 280 = 437.079\ldots$

   $= 437\,cm^2$ (3 sf)

An *exact* answer is $(50\pi + 280)\,cm^2$

EXAMPLE

The area of a circle is $200\,m^2$. Calculate the radius of the circle.

Put 200 into the area formula

$200 = \pi r^2$

$r^2 = \dfrac{200}{\pi}$ Rearrange to find r

   $= 63.6619\ldots$

$r = \sqrt{63.6619\ldots} = 7.9788\ldots$

The radius of the circle is $8.00\,m$ (3 sf)

Carry on the working on your calculator – just round the final answer.

Exercise 11.1S

1 Find the values of these expressions to three significant figures where appropriate.

    a 7^2 **b** 12.5^2 **c** $\sqrt{81}$ **d** $\sqrt{436}$

    e 5π **f** $4^2 \times \pi$ **g** $\sqrt{10\pi}$ **h** $\sqrt{8\pi}$

2 This circle is drawn on a centimetre grid.

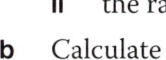

    a Write down

      i the diameter

      ii the radius.

    b Calculate

      i the circumference

      ii the area.

    c **i** Use string to check whether your answer to part **bi** is reasonable.

      ii Use the grid to check whether your answer to part **bii** is reasonable.

3 For each circle

    i calculate the circumference

    ii use $\pi \approx 3$ to check your answer.

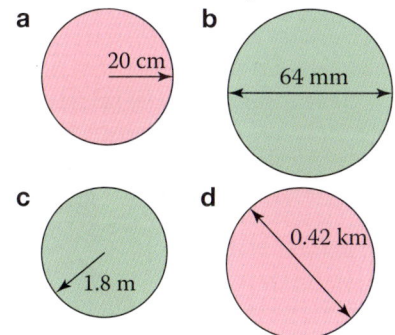

4 Find the area of each circle above. Use approximations to check your answers.

5 The circumference of a circle is 88 cm. Find the diameter of the circle.

6 The area of a circle is $1.72\,\text{m}^2$. Find the radius of the circle.

7 Copy and complete the table for 4 circles.

	Radius	Diameter	Circumference	Area
a	6 cm			
b		4.6 m		
c			98 mm	
d				15.2 m²

8 A pond is in the shape of a semi-circle.

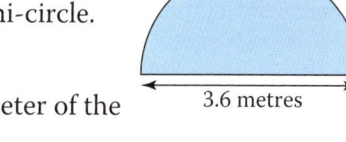

    3.6 metres

    Find

    a the perimeter of the pond

    b the area of the pond.

9 For each shape, calculate

    i the perimeter **ii** the area.

    a 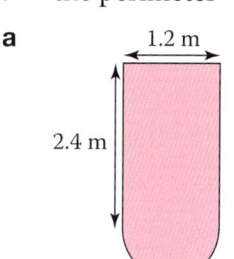 **b**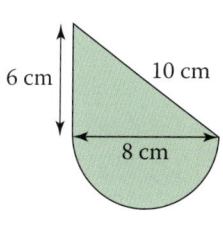

10 A circular hole is cut in a square card. Find the area of card that is left.

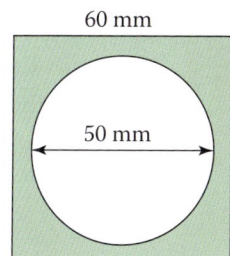

*11 This pendant is made from an isosceles trapezium and a semi-circle. The circular hole has diameter 4 mm.

Find the area of the pendant.

12 How many times larger is circle A than circle B when comparing their

    a circumferences

    b areas?

    A   B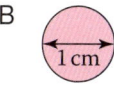

    3 cm   1 cm

 1083, 1088   SEARCH

11.1 Circles 1

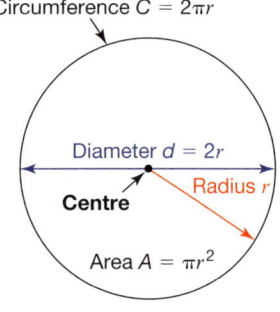

Circumference $C = 2\pi r$

Diameter $d = 2r$

Centre

Radius r

Area $A = \pi r^2$

RECAP

- Diameter of a circle $= 2 \times$ radius
- Circumference of a circle, $C = \pi d$ or $2\pi r$

  Rearranging gives $\qquad d = \dfrac{C}{\pi}$ or $r = \dfrac{C}{2\pi}$
- Area inside a circle, $A = \pi r^2 = \pi \times r \times r$

  Rearranging gives $r = \sqrt{\dfrac{A}{\pi}}$

$1\,m = 100\,cm$

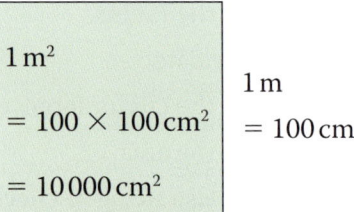

$1\,m^2$

$= 100 \times 100\,cm^2$

$= 10\,000\,cm^2$

$1\,m$

$= 100\,cm$

HOW TO

To calculate perimeters or areas involving circles

① Decide which formulae and units to use.

② Calculate the perimeters or areas needed.

③ Answer the question, giving units where necessary.

You may need to convert area units: $1\,m^2 = (100\,cm)^2 = 10\,000\,cm^2$.

EXAMPLE

A play area consists of a triangle and a semi-circle.
Jake wants to put bark chippings on this area.
He has enough bark to cover $1\frac{1}{2}\,m^2$.
Does Jake need more bark?
Explain your answer.

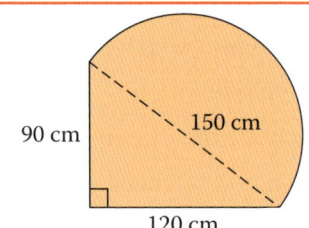

90 cm

150 cm

120 cm

p.144

① Use the area formulae for a triangle and semi-circle.
  Work in metres.

$\frac{1}{2} \times$ base $\times$ *perpendicular* height

② Area of triangle $= \frac{1}{2} \times 1.2 \times 0.9 = 0.54\,m^2$

  Radius of semi-circle $= 1.5 \div 2 = 0.75\,m$

  Area of semi-circle $= \frac{1}{2} \times \pi r^2 = 0.5 \times \pi \times 0.75 \times 0.75$

  $\qquad\qquad\qquad = 0.8835729...\,m^2$

Or you could work in cm and then convert to m using $10\,000\,cm^2 = 1\,m^2$.

  Total area $= 0.8835729... + 0.54 = 1.4235729...\,m^2$

③ $1.5\,m^2$ is more than the area of the play area, so Jake does not need more bark.

EXAMPLE

Show that the perimeter of this shape is equal to the circumference of a circle with diameter 40 cm.

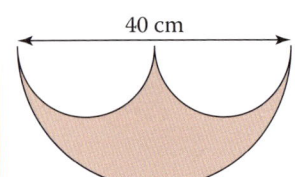

40 cm

① A semi-circle is half of a circle.

  Arc length of a semi-circle $= \frac{1}{2} \times \pi d$

② Find the perimeters needed.

  Total perimeter $= 20\pi + 10\pi + 10\pi = 40\pi$

  Circumference of a circle with diameter 40 cm $= \pi d = 40\pi$.

③ The total perimeter of the shape is equal to this.

Exercise 11.1A

1 Yasmin wants to make this figure of eight with ribbon.

Is 1 metre of ribbon long enough?

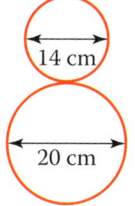
14 cm
20 cm

2 The diagram shows a simple archery target. Tom says that the gold area is greater than the green area.

Is Tom correct? Show your working.

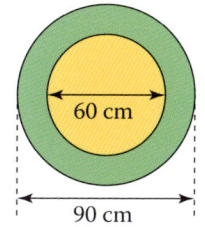
60 cm
90 cm

3 The diagram shows the dimensions of a flowerbed in Yusuf's garden.

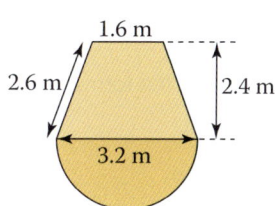
1.6 m
2.6 m
2.4 m
3.2 m

a Yusuf says that 12 m of edging is enough to go around the flowerbed. Is Yusuf correct? Show your working.

b A bag of fertiliser will fertilise 100 m². How many times can Yusuf fertilise his flowerbed with this bag?

4 This shape is formed by 3 semi-circles.

a Show that the perimeter of the shape is 30π cm.

b Find the area of the shape in terms of π.

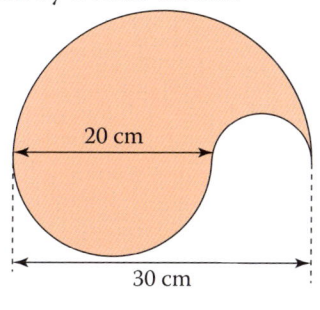
20 cm
30 cm

5 How much longer is the circumference of the circle than the perimeter of the regular hexagon?

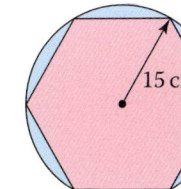
15 cm

6 The perimeter of an aircraft window must be less than $1\frac{1}{2}$ metres. The area must be less than 0.1 m².

Does this window satisfy these requirements? Show all your working.

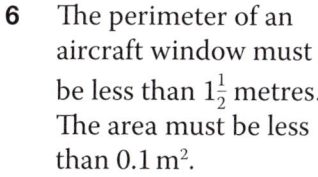
24 cm
32 cm

7 The diameter of the wheels on Maisy's bike is 60 cm. Maisy says the wheels go round over 500 times for every kilometre she travels.
Is Maisy correct? Show your working.

*8 Show that the area of a circle diameter 70 cm is more than three-quarters of the area of a square with sides 70 cm.

*9 A sports club wants a running track to have 2 straight sides and 2 semi-circular ends with diameter 49 metres. The total length of the running track must be 400 metres.

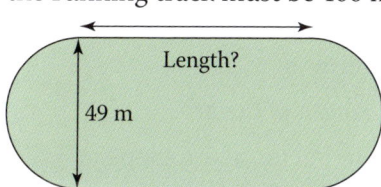
Length?
49 m

Find the length of the straight sides of the track.

*10 A rectangular sheet of metal is 1.2 metres long and 80 centimetres wide. A badge-making machine cuts circles of metal from this sheet.

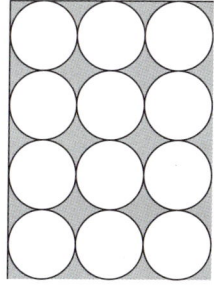

a How many circles of diameter 4 cm can be cut from the sheet?

b Sally says that more than 20% of the metal is wasted.
Is Sally correct?
You must show your working.

1083, 1088   SEARCH

11.2 Circles 2

- **Arc** length $s = \dfrac{\theta}{360°} \times 2\pi r$ or $\dfrac{\theta}{360°} \times \pi d$

- Area of **sector** $A = \dfrac{\theta}{360°} \times \pi r^2$

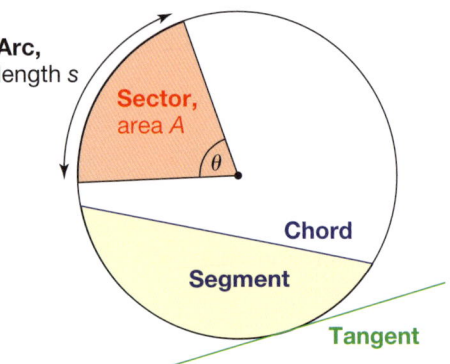

The perimeter of a sector consists of an arc and two radii.

The perimeter of a segment consists of an arc and a chord.

▶ An **arc** is a part of the circumference.
A sector is an area between two radii and an arc.
A **chord** is a straight line between two points on the circumference.
A **segment** is an area between a chord and an arc.
A **tangent** is a straight line that touches the circle at a single point.

EXAMPLE

Find the area of this sector.

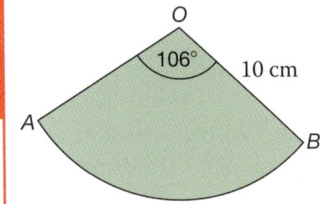

Substitute the values in the formula.

$$\text{Area of sector} = \dfrac{106°}{360°} \times \pi \times 10^2$$

$$= 92.502\ldots \qquad \text{Using a calculator.}$$

$$= 92.5\,\text{cm}^2 \ (3\,\text{sf})$$

EXAMPLE

Find, in terms of π,

a the length of the arc

b the perimeter of the shape.

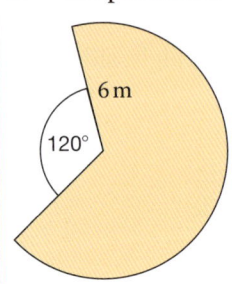

p.92

a Take care to use the correct angle.

Angle in the sector = 360° − 120°   Angles at a point.

$$= 240°$$

$$\text{Arc length} = \dfrac{240°}{360°} \times 2\pi \times \overset{2}{\cancel{6}}$$

Simplify by cancelling.

$$= 8\pi$$

b The perimeter includes the arc and two radii.

Perimeter = 8π + 6 + 6

$$= (8\pi + 12)\,\text{m}$$

EXAMPLE

A circle has a radius of 20 cm. A sector of this circle has an area of 160 cm².

Calculate the angle of the sector.

$$160 = \dfrac{\theta}{360°} \times \pi \times 20^2$$ Substitute the values in the formula.

$$160 = \dfrac{\theta}{360°} \times \pi \times \overset{1}{\cancel{400}}$$ Divide both sides by 10.

p.202

$$\dfrac{16 \times 9}{\pi} = \theta$$

Then cancel by 40, or by 10, then by 4.

Multiply by 9 and divide by π.

$$\theta = 45.836\ldots = 45.8° \ (3\,\text{sf})$$

Remember to do the same to both sides.

Exercise 11.2S

1 **a** Simplify these fractions.

 i $\dfrac{150°}{360°}$ **ii** $\dfrac{200°}{360°}$ **iii** $\dfrac{144°}{360°}$

 b Simplify, leaving each in terms of π.

 i $\dfrac{60°}{360°} \times 2\pi \times 24$ **ii** $\dfrac{120°}{360°} \times 2\pi \times 15$

 iii $\dfrac{80°}{360°} \times \pi \times 6^2$ **iv** $\dfrac{135°}{360°} \times \pi \times 12^2$

2 This quarter-circle has a radius of 8 cm.
Calculate

 a the length of arc AB

 b the total perimeter

 c the area of sector OAB.

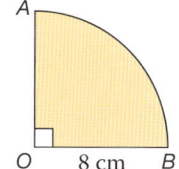

3 Calculate the length of each arc.

 a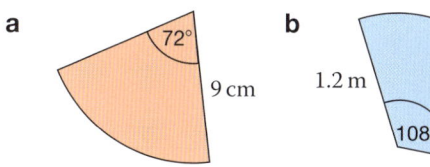

 b

 c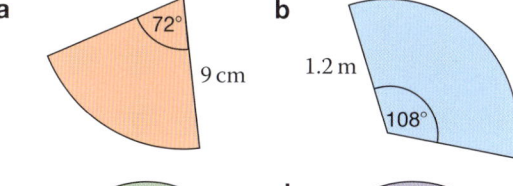

 d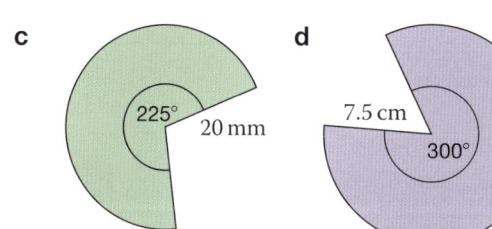

4 Find the area of each sector in question **3**.

5 Copy and complete the table for four sectors.

	angle	diameter	arc length	area of sector
a	40°	18 mm		
b	135°	1.6 m		
c	252°	3.5 cm		
d	312°	0.46 km		

6 For each sector, find in terms of π,

 i the perimeter **ii** the area.

 a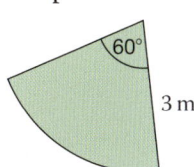

 b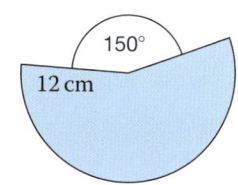

7 The radius of a circle is 30 cm.
An arc of this circle is 75 cm long.

 Calculate the angle θ.

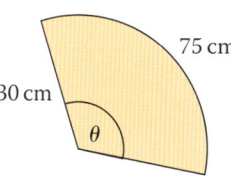

8 The area of a sector is 8 m².
The radius of the circle is 2 m.
Calculate the angle of the sector.

*9 A square card has sides of length 24 cm. A quarter-circle of radius 24 cm is cut from the square.

 Find the area that is left.

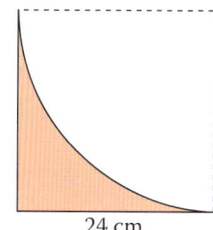

*10 A circle has diameter 20 cm.
The arc length of a sector of this circle is 10 cm.
Calculate the area of the sector.

11 The diagram shows the shape of a flowerbed.

 a Find the length of edging needed to go around the perimeter of the flowerbed.

 b Find the area of the flowerbed.

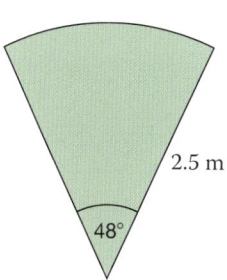

Did you know…

A rainbow is formed by light from the sun reflecting off water droplets in the atmosphere. A full rainbow would form a circle but usually you can only see an arc.

11.2 Circles 2

- **Arc** length $\quad s = \dfrac{\theta}{360°} \times 2\pi r \quad$ or $\quad \dfrac{\theta}{360°} \times \pi d$
- Area of **sector** $A = \dfrac{\theta}{360°} \times \pi r^2$

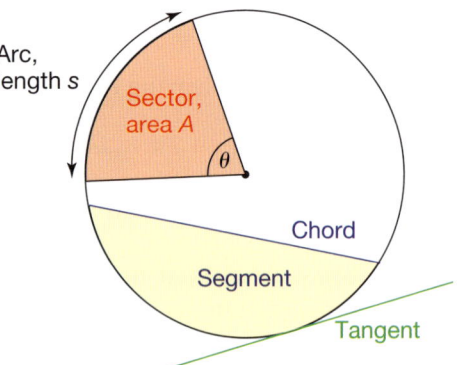

Arc, length s

Sector, area A

θ

Chord

Segment

Tangent

HOW TO

To find areas

① Draw a diagram, label useful points and given dimensions.

② Decide which formula to use.

③ Work out the answer to the question, giving units where necessary.

EXAMPLE

The diagram shows the shape of a window.
ABC is an equilateral triangle.
AC is an arc of a circle, centre *B*.
BC is an arc of a circle, centre *A*.

p.52

p.92

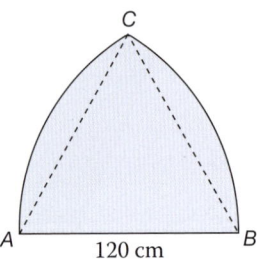

120 cm

Amy says the perimeter of the window is less than 4 metres. Is Amy correct?

You must show your working.

① Each angle in an equilateral triangle is 60°.
The diagram shows sector *ABC*.

② Use arc length $s = \dfrac{\theta}{360°} \times 2\pi r$

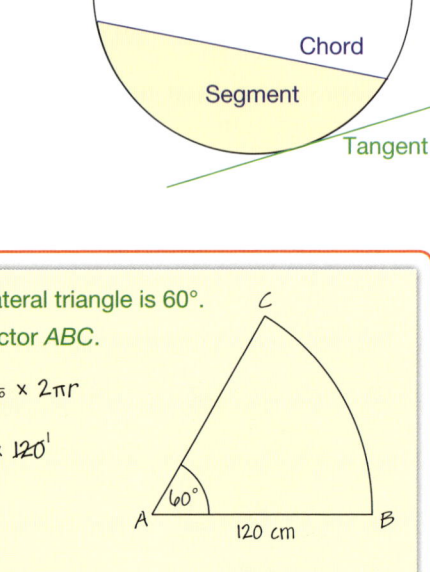

$60°$

120 cm

Arc $BC = \dfrac{\cancel{60}^{20}}{\cancel{360}_{\cancel{3}}} \times 2\pi \times \cancel{120}^{1}$

$\qquad = 40\pi \text{ cm}$

Arc AC is also 40π cm

Perimeter $= 40\pi + 40\pi + 120$

$\qquad\qquad = 80\pi + 120$

$\qquad\qquad = 371.327... \text{ cm}$

4 metres $= 400$ cm

The perimeter is less than 400 cm.

③ Amy is correct.

> You can use your calculator earlier if you prefer.

EXAMPLE

Show that the shaded area is equal to 24π cm²

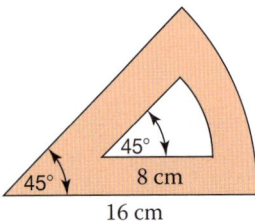

$45°$
$45°$
8 cm
16 cm

② Use area of sector, $A = \dfrac{\theta}{360°} \times \pi r^2$

Area of large sector $= \dfrac{45°}{360°} \times \pi \times 16^2 \qquad$ Cancel the fraction.

$\qquad\qquad = \dfrac{1}{8} \times \pi \times \cancel{16}^{2} \times 16 \qquad$ Simplify again.

$\qquad\qquad = 32\pi$

Area of small sector $= \dfrac{1}{8} \times \pi \times 8 \times 8 \qquad$ The fraction is the same.

$\qquad\qquad = 8\pi$

③ Shaded area $= 32\pi - 8\pi = 24\pi$ cm²

Exercise 11.2A

1 The diagram shows the fabric that is used to make a skirt.
The angle of the sector is 240° and the radius is 60 cm.
Is $2\frac{1}{2}$ metres of ribbon enough to go around the hem of the skirt? Show your working.

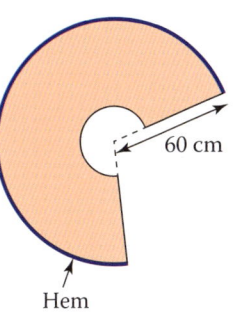
60 cm
Hem

2 The diagram shows a plan for a lawn.

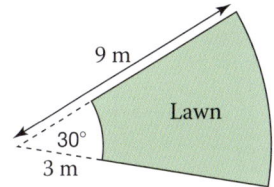
9 m
Lawn
30°
3 m

a Find the area of the lawn.

b Is 20 metres of edging enough to go around the lawn?
Show your working.

3 The diagram shows the landing sector for a javelin. The distance between the arcs is 10 m.

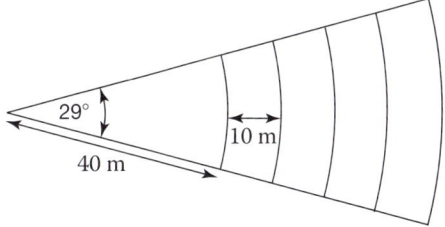
29°
10 m
40 m

The groundsman says that the total length of the lines is over 300 metres.
Is the groundsman correct? Show your working.

4 A continuous belt is needed to go around four identical oil drums as shown. The diameter of each drum is 24 inches. Work out the length of the belt. Give your answer in terms of π.

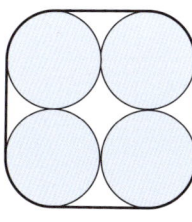

5 **a** How many times does the small cog rotate when the large cog rotates once?

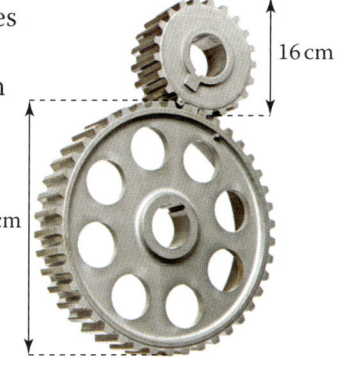
16 cm

b Through what angle does the large cog turn when the small cog rotates once?

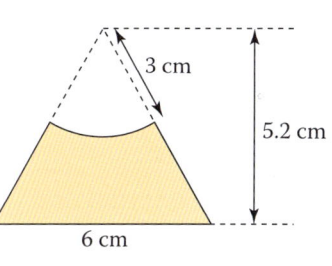
48 cm

*6 A sector of radius 3 cm is cut from one corner of an *equilateral* triangle.
Find the area of the remaining shape.

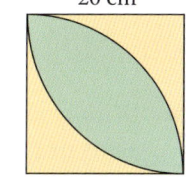
3 cm
5.2 cm
6 cm

*7 Two quarter-circles are used to draw the leaf on this square tile.
Show that

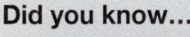
20 cm

a the perimeter of the leaf is 20π cm

b the area of the leaf is $200(\pi - 2)$ cm².

Did you know...

The diameter of the London Eye is 135 m. It has 32 capsules spaced equally along its circumference and takes 30 minutes to complete 1 revolution.

Can you work out the length of the arc between each capsule and the next?

How far does a capsule travel in ten minutes?

1118   SEARCH

11.3 Constructions

● The **perpendicular bisector** of a **line segment**, *AB*, joins all the points that are **equidistant** (the same distance) from *A* and *B*.

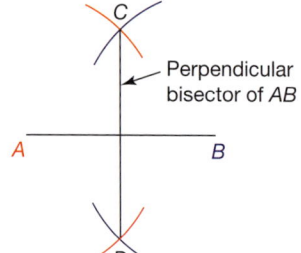

● With centre *A* and radius greater than half the length *AB*, draw arcs on each side of *AB*.

● With centre *B* and the same radius, draw two more arcs cutting the first pair of arcs at *C* and *D*.

● Join *CD* – this is the perpendicular bisector of *AB*.

You can also construct a perpendicular to a line *at* a point, *P*, on the line or *from* a point, *P*, not on the line.

● The perpendicular is the shortest distance from a point to a line.

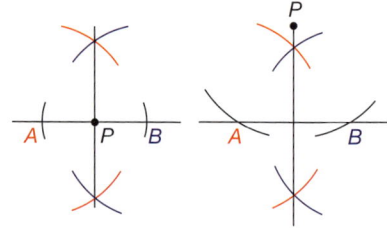

● With centre *P*, draw arcs to cross the line at *A* and *B*.

● Then draw the perpendicular bisector of *AB*, as described above. Sometimes you may need to extend the line first.

▲ The perpendicular to a line at the point P.
▲ The perpendicular to a line *from* the point *P*.

● An **angle bisector** is a line that cuts an angle in half. It joins all the points that are equidistant from two lines.

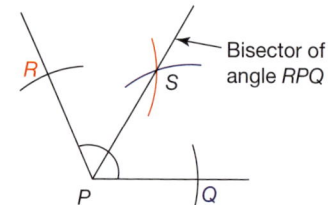

● With centre on the vertex of the angle, *P*, draw arcs to cut the arms of the angle at *Q* and *R*.

● With centres *Q* and *R*, draw arcs to cross inside the angle at *S*.

● Join *P* to *S*, this is the bisector of angle *P*.

---

p.52

EXAMPLE

Using a *pencil, ruler and compasses only*

a construct an angle of 60°

b bisect the angle to give an angle of 30°

a Draw a line *AB*.
With centre *A* and radius equal to *AB*, draw an arc above *AB*.
With centre *B* and the *same* radius, draw another arc to intersect the first arc at *C*.
Join *AC* to give a 60° angle.

> You can use a protractor to check angles, but *not* to draw them.

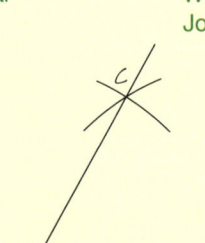

b With centre *A*, draw arcs at *X* and *Y*.
With centre *X*, then *Y*, draw arcs to meet at *Z*.
Join *AZ* to give 30°.

Exercise 11.3S

Use a *pencil, ruler and compasses only* for each construction.

1 **a** **i** Draw a horizontal line *AB*, 8 cm long.

   ii Construct the perpendicular bisector of *AB*

   iii Use a ruler and protractor to check that your perpendicular bisector cuts *AB* in half at 90°.

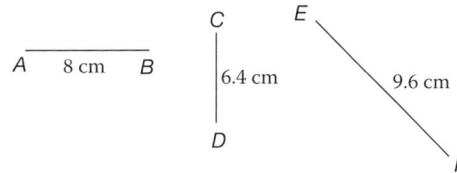

   Repeat part **a** using

   b a vertical line *CD*, 6.4 cm long

   c a diagonal line *EF*, 9.6 cm long.

2 **a** Draw a line *XY*, 10 cm long. Mark point *P* on the line so that *XP* = 3 cm.

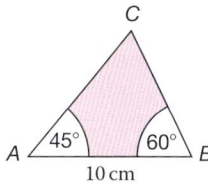

   b Construct the line at *P* which is perpendicular to *XY*.

   c Use a protractor to check that the angle formed is 90°.

3 **a** Draw any line *MN*.

   b Construct a line at *M* which is perpendicular to *MN*.

   c Construct a line at *N* which is perpendicular to *MN*.

   d Give a mathematical description of the two lines you have constructed.

4 **a** Draw any vertical line *QR*. Mark a point *P* at one side as shown.

   b Construct the shortest line from *P* to *QR*.

   c Use a protractor to check that the angle formed is 90°.

5 **a** **i** Draw any acute angle and construct its bisector.

   ii Use a protractor to check your answer.

   b Repeat part **a** for any obtuse angle.

6 **a** Draw any line *XY* and construct its perpendicular bisector.

   b Bisect one of the 90° angles.

   c Use a protractor to check that you have constructed an angle of 45°.

7 **a** Construct an angle, *CAB*, of 60° as described in the worked example.

   b **i** Extend *BA* to point *D* as shown.

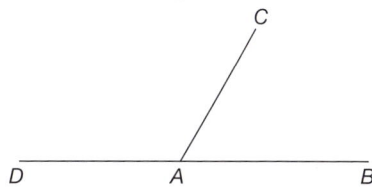

   ii What is the size of angle *DAC*?

   iii Bisect the angles and use a protractor to check your answers.

*8 Construct each of the following angles. Use a separate diagram for each part.

   a 15°   **b** 135°   *c 75°

9 Construct this triangle.

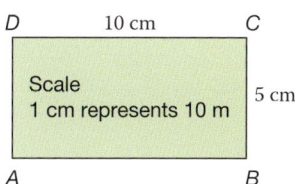

*10 The sketch shows the dimensions needed to draw a rectangle to represent a football pitch. p.140

   Draw an accurate rectangle with the dimensions shown.

11.3 Constructions

RECAP

- A pencil, pair of compasses and a ruler can be used to construct
  - the perpendicular bisector of a line segment
  - a perpendicular to a line from or at a given point
  - an angle bisector
  - an angle of 60°
- The perpendicular is the shortest distance from a point to a line.

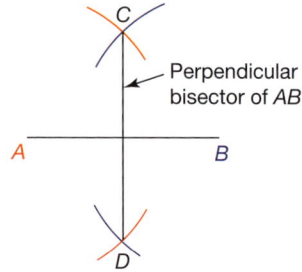

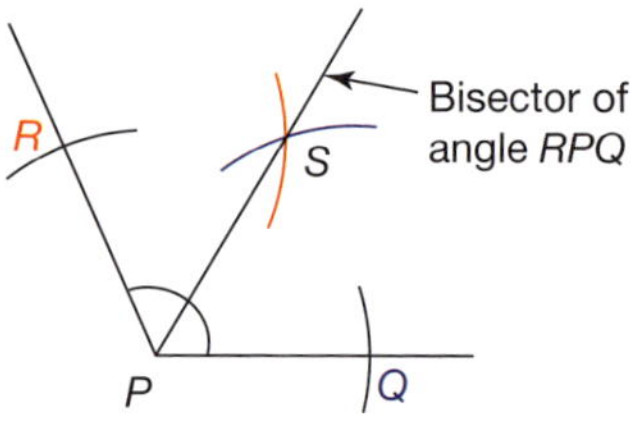

Perpendicular bisector of *AB*

HOW TO

To solve a constructions problem

① Draw a diagram or copy a given diagram.

② Decide what you need to construct.

③ Use a pencil, pair of compasses and ruler to carry out the construction.

④ Answer any questions asked.

Bisector of angle *RPQ*

EXAMPLE

p.140

The scale diagram shows a pipeline *PQ* and a new house *H*.

The builder needs to lay a pipe to connect *H* to *PQ*.

Using pencil, compasses and ruler only, find the length of the shortest possible pipe.

House, *H*

Q

pipeline

P

Scale
1 cm represents 50 m

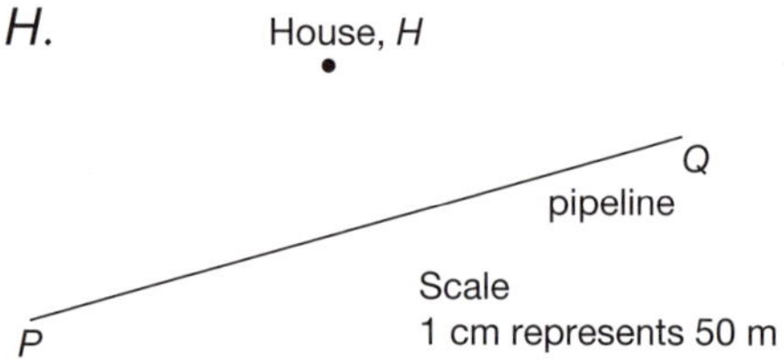

② The shortest line between *H* and *PQ* is the perpendicular from *H* to *PQ*.

③ Construct the perpendicular from the house, *H*, to the pipeline *PQ*.
The shortest possible pipe is represented by *HR*.

④ Measure the distance *HR*, then use the scale.

HR = 1.2 cm

Actual length of pipe = 1.2 × 50

= 60 metres

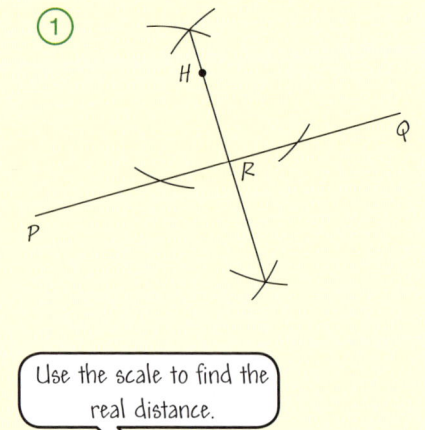

Use the scale to find the real distance.

Exercise 11.3A

> Use a *pencil*, *ruler and compasses only* for each construction.

1 The diagram shows the position of a new factory, *F*, and the nearest main road, *MN*.

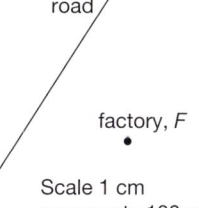

road *N*

factory, *F*
•

M Scale 1 cm represents 100 m

   a Copy the diagram.

   b Find the shortest possible length for a road from *F* to *MN*.

2 The sketch shows two plots of land.

   a Draw a scale diagram of these plots. Use 1 cm to represent 5 m.

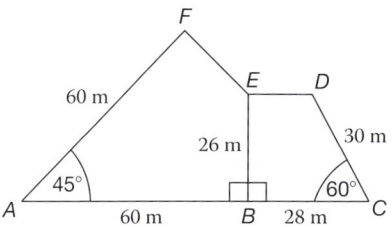

F

60 m

E *D*

26 m

30 m

45°

60°

A 60 m *B* 28 m *C*

   b Find the actual length of *DE* and *EF*.

3 In the construction of an angle of 60°, the radii of both arcs are equal to *AB*.
How does this ensure that the angle drawn is 60°?

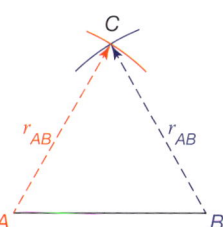

C

r_{AB} r_{AB}

A *B*

4 Jamie says that the construction of an angle bisector works because of congruent triangles.
Is this true? Give your reasons.

5 Sue says that, when you construct a perpendicular bisector, the arcs drawn from one end of the line do not need to have the same radius as those drawn from the other end.
Is Sue correct? Give your reasons.

6 For each part use *x* and *y* axes from −6 to 6.

   a **i** Join the points *A* (2, 6) and *B* (−4, −2).

     ii Construct the perpendicular bisector of *AB*.

   a **iii** Write down the coordinates of *M*, the mid-point of *AB*.

   b **i** Draw the line $y = x$.

     ii Construct the perpendicular from the point (3, −5) to the line $y = x$.

     iii Write down the coordinates of the point where the perpendicular meets $y = x$.

*7 **a** Draw a *large* triangle *PQR*.

   b Construct the angle bisector of

     i angle *P* **ii** angle *Q* **iii** angle *R*.

   c Label the point where all the angle bisectors meet as *O*.

   d With centre *O*, draw a circle that just touches all the sides of triangle *PQR*. This is the *in-circle* of triangle *PQR*.

*8 **a** Draw a *large* acute-angled triangle *ABC*.

   b Construct the perpendicular bisector of

     i *AB* **ii** *BC* **iii** *CA*.

   c Label the point where all the perpendicular bisectors meet as *O*.

   d With centre *O*, draw a circle that passes through *A*, *B* and *C*. This is called the *circumcircle* of triangle *ABC*.

   e Draw the circumcircle of a large obtuse-angled triangle.

   f **i** Draw the circumcircle of a large right-angled triangle.

     ii What is special about the circle on the longest side of the triangle?

> **Did you know...**
>
>
>
> In constructions, Euclid used a straight edge with no markings, rather than a ruler. He could not bisect a line by just measuring so he had to construct the midpoint.

1089, 1090   SEARCH

11.4 Loci

There are two different ways to think of a **locus**.

- A set of points that follow one or more rules.
- The path followed by a moving point.

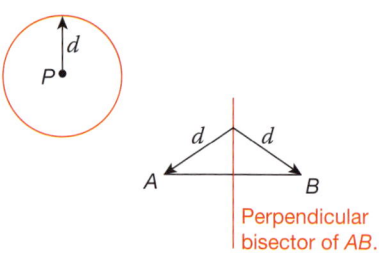

Loci is the plural of locus.

- The locus of points that are equidistant from a point, *P*, is a circle with centre *P*.

- The locus of points that are equidistant from two given points, *A* and *B*, is the **perpendicular bisector** of *AB*.

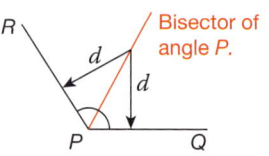

Perpendicular bisector of *AB*.

- The locus of points that are equidistant from line segments *PQ* and *PR* is the **angle bisector** of angle *P*.

Bisector of angle *P*.

A locus may be one or more points, a line or a region.

Sometimes the rules for a locus give overlapping regions.

EXAMPLE

In rectangle *ABCD* sketch the locus of points that are less than 2 cm from *A* and nearer to *B* than to *C*.

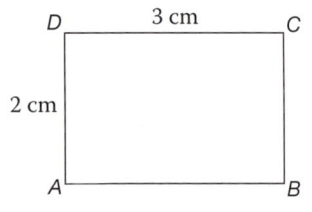

Sketch a circle, centre *A*, radius 2 cm. Use a dotted line to show that it is not included in the locus. Shade *inside* the circle to show that these are the points that are *less than* 2 cm from *A*.

Draw a line to show the points that are equidistant from *B* and *C* (the perpendicular bisector of *BC*). Then shade the region of points nearer to *B*.

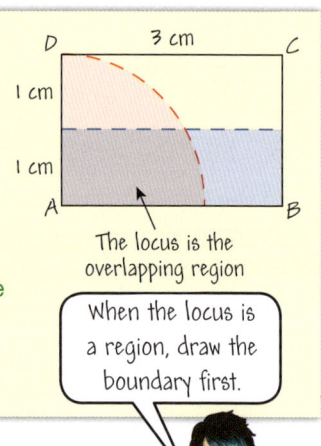

The locus is the overlapping region

When the locus is a region, draw the boundary first.

EXAMPLE

A point *P* moves so that its distance from *AB* is always equal to its distance from *CD*.

Using a *pencil, ruler and compasses only,* construct the locus of *P*.

Construct the bisector of each angle.

The locus is in two parts.

Label the answer clearly.

Exercise 11.4S

> Unless told otherwise, use a pencil, ruler and compasses only for each construction.

1 Sketch each of the following loci.

a The locus of points that are 2 cm from a fixed point *P*.

b The locus of points that are less than 3 cm from a fixed point *Q*.

c The locus of points that are more than 4 cm from a fixed point *R*.

2 Draw sketches to show the locus of a point that moves so that

a its distance from line *PQ* is equal to its distance from a parallel line *RS*

b its distance from line segment *XY* is always equal to a fixed length *d*

c it is outside square *ABCD* and a distance *d* from its sides.

3 **a** Draw a line, *AB*, that is 7.8 cm long.

b Construct the locus of points that are equidistant from *A* and *B*.

4 **a** Use a protractor to draw angle *PQR* = 76°.

b Construct the locus of points that are equidistant from *PQ* and *QR*.

5 **a** Draw a line, *AB*, of length 6.4 cm.

b Draw the locus of points that are 3 cm from *AB*.

6 Using a line *XY*, 7 cm long.

a Draw the locus of points that are 4 cm from *X*.

b Draw the locus of points that are 5 cm from *Y*.

c Shade the area containing the points that are less than 4 cm from *X* and less than 5 cm from *Y*.

7 **a** Draw a rectangle *ABCD* with the dimensions shown.

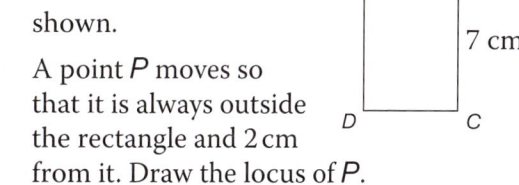

b A point *P* moves so that it is always outside the rectangle and 2 cm from it. Draw the locus of *P*.

8 Draw an accurate diagram to show the locus of points that are more than 3.5 cm, but less than 4.5 cm from a fixed point *P*.

9 **a** Draw a square with sides of length 6 cm. Label the vertices *A*, *B*, *C* and *D*.

b On your diagram show the locus of points inside *ABCD* that are less than 3 cm from *A* and more than 4 cm from *BC*.

10 **a** Draw a right-angled triangle *ABC* with the dimensions shown.

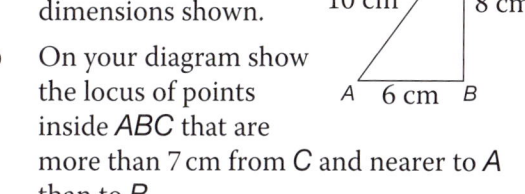

b On your diagram show the locus of points inside *ABC* that are more than 7 cm from *C* and nearer to *A* than to *B*.

11 **a** Construct an equilateral triangle, *ABC*, with sides of length 10 cm.

b Draw the locus of points that are equidistant from *AB* and *BC*.

c Draw the locus of points that are equidistant from *AB* and *AC*.

d Mark the point that is equidistant from *AB*, *BC* and *AC*. Label this point *X*.

*12 Draw sketches to show

a the locus of a point *P* on the circumference of a coin as it rolls along the ruler

b the locus of a vertex *Q* of the square card as it is rotated against the ruler.

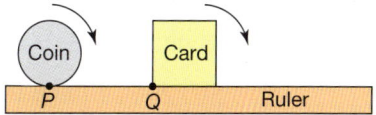

11.4 Loci

- The locus of points that are equidistant from a point, *P*, is a **circle** with centre *P*.
- The locus of points that are equidistant from two given points, *A* and *B*, is the **perpendicular bisector** of *AB*.
- The locus of points that are equidistant from line segments *PQ* and *PR* is the **angle bisector** of angle *P*.

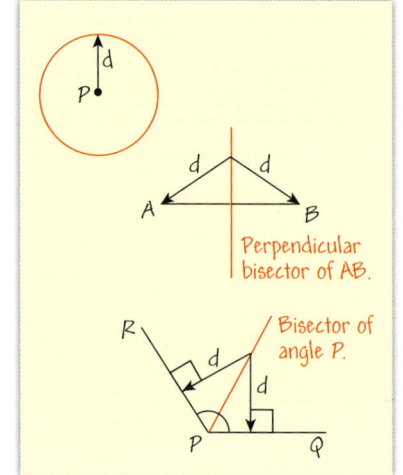

Perpendicular bisector of AB.

Bisector of angle P.

HOW TO

To solve a loci construction problem

① Use a sketch or a scale diagram.
   If a scale is not given, choose one that is easy to use.

② Use a pencil, pair of compasses and ruler to carry out constructions. (Leave all construction lines on your diagram.) Label clearly.

③ Answer any questions asked.

EXAMPLE

A buoy, *B*, lies 90 metres from a lighthouse, *L*, on a bearing of 120°.

a Draw a scale diagram to show the position of the lighthouse and the buoy. Use a scale of 1 cm to represent 20 metres.

b A ship sails between the lighthouse and buoy, on a course that is equidistant from them. Using a *pencil, ruler and compasses only*, show the course of the ship.

c The buoy can be seen from the ship when it is within 50 metres. The captain says he will be able to see the buoy for over 40 metres of the journey. Is the captain correct? Show how you decide.

p.140

p.50

a ① The bearing 120° is measured clockwise from North.

Distance *LB* on the scale diagram = 90 ÷ 20 = 4.5 cm

The diagram shows the position of *L* and *B*.

Leave all construction lines on the diagram.

b ② The course that is equidistant from *L* and *B* is the perpendicular bisector of *LB*. Construct this on the diagram.

c ② The points within 50 metres of *B* lie in a circle.

50 metres on the scale diagram
= 50 ÷ 20 = 2.5 cm

The shaded circle shows the points within 50 metres of *B*. It can be seen from the ship between *P* and *Q*.

③ On the diagram *PQ* = 2.2 cm

The actual distance *PQ* = 2.2 × 20 = 44 m

This is more than 40 m, so the captain is correct.

N

120°

L

Scale 1 cm represents 20 m

Q

P

B

Course equidistant from L and B

Exercise 11.4A

1 Two straight roads, *AB* and *CD*, intersect at *X*.

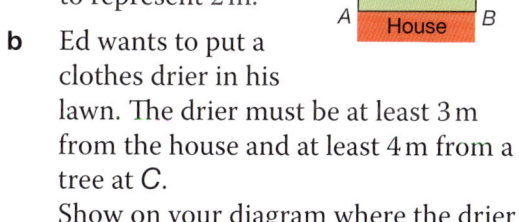

Harry rides his mountain bike from *X* on a course that is equidistant from *XD* and *XB*.

Draw an accurate diagram to show Harry's route.

2 Two radio beacons, *A* and *B*, are 80 km apart. The bearing of *B* from *A* is 240°

  a Using a scale of 1 cm to represent 10 km, draw a diagram to show *A* and *B*.

  b An aircraft flies so that it is equidistant from *A* and *B*. Draw the aircraft's route.

  c Beacon *A* can transmit signals up to 50 km.
  Show clearly where on its route the aircraft can receive signals from *A*. How far is this?

3 The sketch shows Ed's lawn.

  a Draw a scale diagram of Ed's lawn. Use 1 cm to represent 2 m.

  b Ed wants to put a clothes drier in his lawn. The drier must be at least 3 m from the house and at least 4 m from a tree at *C*.
  Show on your diagram where the drier could be placed.

4 A lifeboat *L* is 10 km from another lifeboat *K* on a bearing of 045°. They both receive a distress call from a ship. The ship is within 7 km of *K* and within 5 km of *L*.

  a Draw a scale drawing to show the positions of *K* and *L*.

  b On your diagram shade the area in which the ship could be.

5 Meera's lawnmower cable can reach 12 m from the power point at *P*.

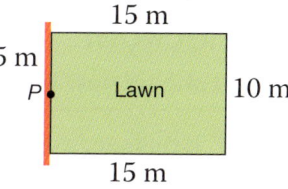

  a Use a scale diagram to show which parts of the lawn Meera can cut.

  b What is the minimum length of cable Meera needs to reach all of the lawn?

6 An island has 2 mobile phone masts, *A* and *B*. Mobile phone mast *A* has a range of 30 km. Mobile phone mast *B* has a range of 40 km.

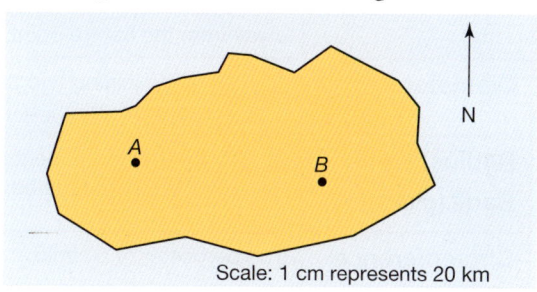

Scale: 1 cm represents 20 km

  a On a copy of the map, show the area covered by the mobile phone masts.

  b Mobile phone mast *B* is to be replaced. Suggest ways in which the coverage of the island could be improved by the new mobile phone mast.

*7 Pat wants to tether her goat on this grass using a 4 metre long chain.

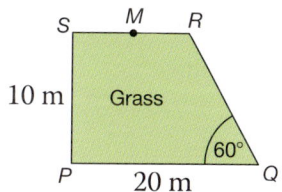

  a Draw a scale drawing to show the grass the goat can reach if it is tethered to

   i *P*   **ii** *Q*

   iii *R*   **iv** *S*

   v *M*, the mid-point of *RS*.

  b **i** Which point allows the goat to reach the greatest area of grass?

   ii Show that your answer to part **bi** is correct by calculating the areas reached from each point.

Summary

Checkout
You should now be able to...

Test it
Questions

	Test it Questions
✔ Identify and apply circle definitions, properties and formulae.	1 – 5
✔ Construct triangles.	6
✔ Use the standard ruler and compass constructions.	7 – 9
✔ Solve loci problems.	10

Language	Meaning	Example
Circle	The 2D shape formed by all the points that are the same distance away from the centre point.	circumference, diameter, centre, radius
Diameter	A straight line joining two points on a circle that pass through its centre.	
Radius **Radii (plural)**	A line drawn from the centre of the circle to the circumference.	
Circumference	The perimeter of a circle.	
Arc	Any part of the circumference.	arc, sector, θ, chord, segment, tangent
Chord	A straight line joining two points on the circumference of a circle.	
Tangent	A line which touches the circle at one point only.	
Segment	The shape enclosed by an arc and a chord.	
Sector	The shape enclosed by two radii and an arc.	
Construct	Draw something accurately using compasses and a ruler.	Equilateral triangle
Construction lines	Lines drawn during a construction that are not part of the final object.	
Bisect	Cut into two parts of the same shape and size.	30°, angle bisector, 30°, A, M, B
Angle bisector	A line which bisects an angle.	
Perpendicular bisector	A line which bisects another line at right angles.	
Locus **Loci (plural)**	A set of points which satisfy a set of rules. The path followed by a moving point.	The set of points less than 1 cm from P is the interior of a circle. ·P

Geometry Circles and constructions

Review

1 What term is used to describe

    a a line segment that joins the centre to a point on the circumference of a circle

    b a line segment that joins two points on the circumference of a circle?

2 Calculate the area of these circles.

    a 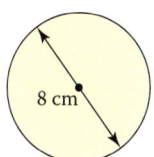 8 cm   **b** 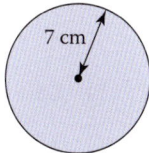 7 cm

3 Calculate the circumference of the circles in question **2**.

4 Calculate

    a the area

    b the perimeter of this semi–circle.

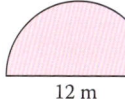 12 m

5 Calculate

    a the arc length

    b the area of the sector.

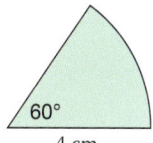 60° 4 cm

6 Use a ruler and a pair of compasses or a protractor to construct these triangles.

    a 5 cm, 4 cm, 7 cm   (SSS)

    b 5 cm, 70°, 6 cm   (SAS)

    c 6 cm, 3 cm, 90°   (RHS)

    d 35°, 7 cm, 25°   (ASA)

7 **a** Draw a line exactly 7 cm long.

    b Use a ruler and pair of compasses to accurately construct the perpendicular bisector of this line.

8 Trace this line and the point marked x. Use a ruler and a pair of compasses to accurately construct a perpendicular from the point x to the line

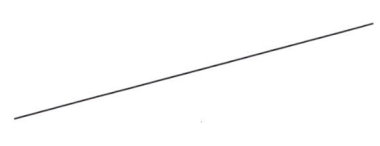

    • x

9 **a** Use a protractor to draw an angle of 70°.

    b Use a ruler and a pair of compasses to accurately construct an angle bisector of this angle.

10 A dog is attached to a lamp post by a rope which is 2 m long and can rotate freely around the post.

Use a scale of 3 cm : 1 m to draw to scale the loci of the furthest points the dog can reach.

What next?

Score			
	0 – 4		Your knowledge of this topic is still developing. To improve look at MyMaths: 1083, 1088, 1089, 1090, 1118, 1147
	5 – 8		You are gaining a secure knowledge of this topic. To improve your fluency look at InvisiPens: 11Sa – h
	9 – 10		You have mastered these skills. Well done you are ready to progress! To develop your problem solving skills look at InvisiPens: 11Aa – e

Assessment 11

1 **a** Draw a circle with radius 5 cm.   [1]

  b Draw a circle with diameter 90 mm.   [1]

  c Draw a chord of length 7 cm across each circle.   [2]

  d Kay says she can draw a chord 11 cm long across each circle.
       Is she right? Give reasons for your answer.   [1]

2 **a** Paul calculates the radii of three different circles from their circumferences.
       Find the missing radius and match the correct radius to the correct circumference.
       Show your working.

  Radii   **i** x cm   **ii** 0.320 cm   **iii** 50.0 cm

  Circumferences **A** 61.58 cm   **B** 314.16 cm   **C** 2.01 cm   [3]

  b He then calculates the radii of three different circles from their areas. Find the missing
       radius and match the correct radius to the correct area. Show your working.

  Radii   **i** 5.08 m   **ii** 7.50 m   **iii** y m

  Areas   **A** 176.7 m²   **B** 81.1 m²   **C** 0.407 m²   [3]

3 Bengt works out the shaded areas in these shapes to 3 sf.

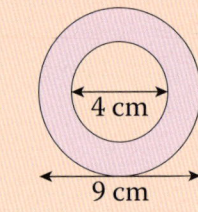

  A   **B**   **C**

  10 cm   6 cm   4 cm
                                                                 9 cm

  These are his answers.

  a 29.3 cm²   [3] **b** 51.1 cm²   [4] **c** 21.5 cm²   [2]

  One of his answers is incorrect.
  Match each correct answer to the correct shape and identify the incorrect answer.
  Show your working.

4 A reel of electrical cable contains 50 m of cable. The diameter of the reel is 25 cm.
   An electrician completely unwinds the reel.
   Calculate the number of times the reel rotates (ignore the thickness of the cable).   [4]

5 The British Kite Mark on a toy consists of
   two semicircles over an equilateral triangle
   of side 4 cm.

   3.46 cm

  a Find the perimeter.   [3]

  b Find the area.   [4]

6 A sector has both radius and arc length equal to 10 cm.
   Jacob says that the angle at the centre is 57.3° to 1 dp.
   Show how he worked out the value of this angle.   [3]

7 A football club has this logo.

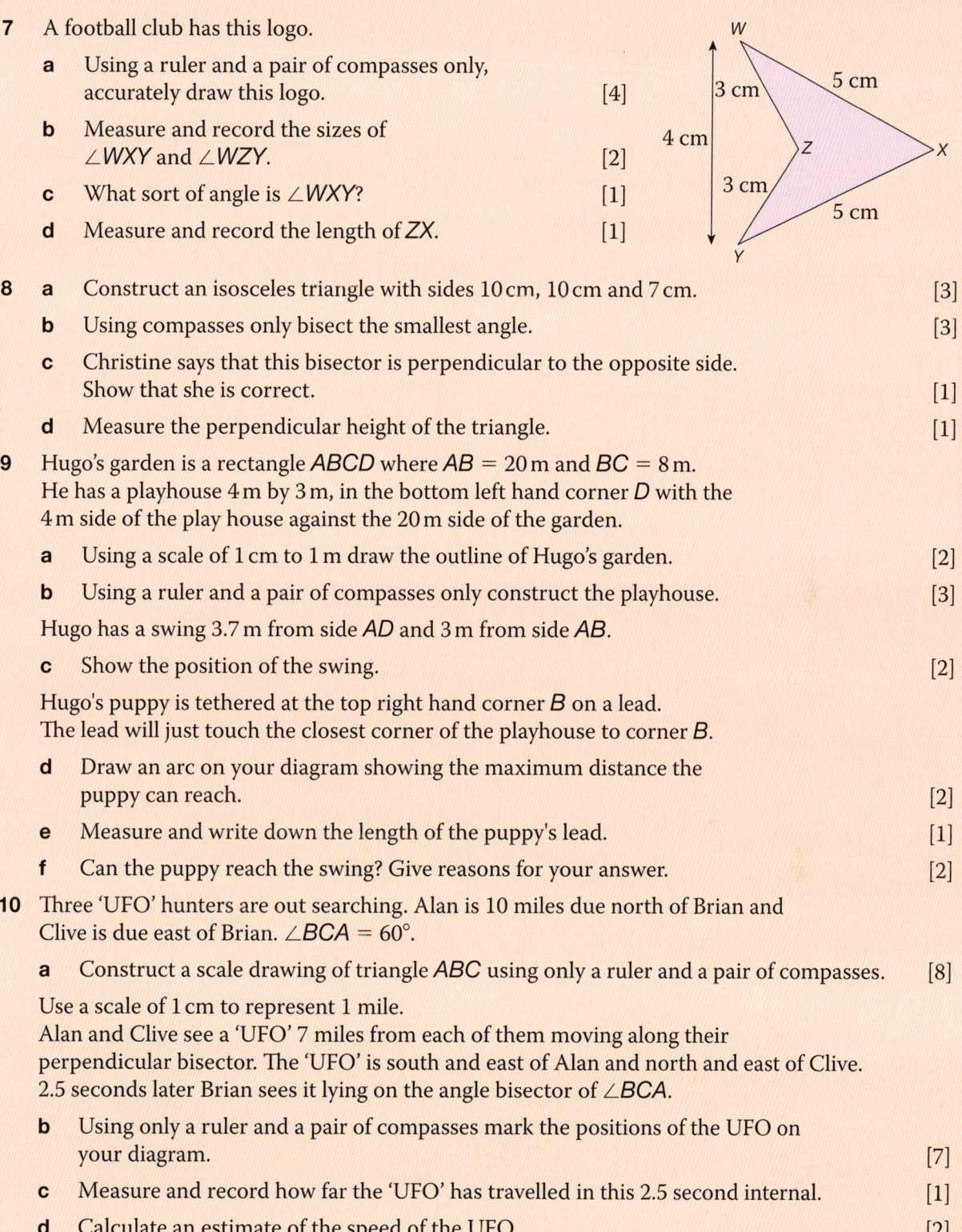

  a Using a ruler and a pair of compasses only, accurately draw this logo. [4]

  b Measure and record the sizes of ∠*WXY* and ∠*WZY*. [2]

  c What sort of angle is ∠*WXY*? [1]

  d Measure and record the length of *ZX*. [1]

8 **a** Construct an isosceles triangle with sides 10 cm, 10 cm and 7 cm. [3]

  b Using compasses only bisect the smallest angle. [3]

  c Christine says that this bisector is perpendicular to the opposite side. Show that she is correct. [1]

  d Measure the perpendicular height of the triangle. [1]

9 Hugo's garden is a rectangle *ABCD* where *AB* = 20 m and *BC* = 8 m. He has a playhouse 4 m by 3 m, in the bottom left hand corner *D* with the 4 m side of the play house against the 20 m side of the garden.

  a Using a scale of 1 cm to 1 m draw the outline of Hugo's garden. [2]

  b Using a ruler and a pair of compasses only construct the playhouse. [3]

Hugo has a swing 3.7 m from side *AD* and 3 m from side *AB*.

  c Show the position of the swing. [2]

Hugo's puppy is tethered at the top right hand corner *B* on a lead. The lead will just touch the closest corner of the playhouse to corner *B*.

  d Draw an arc on your diagram showing the maximum distance the puppy can reach. [2]

  e Measure and write down the length of the puppy's lead. [1]

  f Can the puppy reach the swing? Give reasons for your answer. [2]

10 Three 'UFO' hunters are out searching. Alan is 10 miles due north of Brian and Clive is due east of Brian. ∠*BCA* = 60°.

  a Construct a scale drawing of triangle *ABC* using only a ruler and a pair of compasses. [8]

Use a scale of 1 cm to represent 1 mile.
Alan and Clive see a 'UFO' 7 miles from each of them moving along their perpendicular bisector. The 'UFO' is south and east of Alan and north and east of Clive. 2.5 seconds later Brian sees it lying on the angle bisector of ∠*BCA*.

  b Using only a ruler and a pair of compasses mark the positions of the UFO on your diagram. [7]

  c Measure and record how far the 'UFO' has travelled in this 2.5 second internal. [1]

  d Calculate an estimate of the speed of the UFO. [2]

12 Ratio and proportion

Introduction

Colour theory is based on the idea that all possible colours can be created from three 'primary colours'. Video screens use the additive colours red, green and blue. Artists often use red, yellow and blue as the basis for mixing paints. Whilst the printing industry uses the subtractive colours cyan, magenta and yellow. The ratio in which these primary colours are mixed determines the colour of the result.

What's the point?

An understanding of ratio is essential for artists in mixing colours on a palette to achieve a desired result. Furthermore, and understanding of proportion is essential for artists in ensuring that the elements of their artwork are in the 'right proportions'.

Objectives

By the end of this chapter you will have learned how to …

- Use fractions and percentages to describe a proportion.
- Write a ratio in its simplest form and divide a quantity in a given ratio.
- Use scale factors, scale diagrams and maps.
- Solve problems involving percentage change.

Check in

1 Convert these fractions to decimals.

   a $\frac{3}{4}$ **b** $\frac{2}{5}$ **c** $1\frac{1}{4}$

2 Convert these decimals to fractions in their simplest form.

   a 0.6   **b** 0.25   **c** 0.2

3 The table shows the favourite types of sandwiches in the school canteen.
What proportion of the class surveyed chose ham sandwiches?
Write your answer in its simplest form.

Type	Frequency
Cheese	12
Salad	8
Ham	10
Total	30

4 10 litres of white paint cost £12.
Work out the cost of 20 litres of paint.

Chapter investigation

The ratio of height to head circumference for the average person is said to be around 3:1.
Investigate.

13 Factors, powers and roots

Introduction

Cryptography is the study of codes, with the aim of communicating in a secure way without messages being deciphered. Modern cryptography is often based on prime numbers. In particular finding very large numbers that can be written as the product of two not-quite-as-large prime factors. Finding what these two prime numbers are is a challenge, even with modern computers; but you need to find them in order to crack the code.

What's the point?

In the modern day, illegal computer-based syndicates use increasingly sophisticated techniques to access sensitive digitally-held information, including bank accounts. Prime number encryption increases the security of stored data, making it harder for the hackers.

Objectives

By the end of this chapter you will have learned how to ...

- Use mathematical language to describe factors, multiples and primes.
- Use Venn diagrams or factor trees to systematically list the prime factors of a number.
- Use prime factor decomposition to calculate the HCF and LCM of two or more numbers.
- Write the HCF and LCM using product notation.
- Calculate positive integer powers and their roots.
- Recognise powers of 2, 3, 4 and 5.

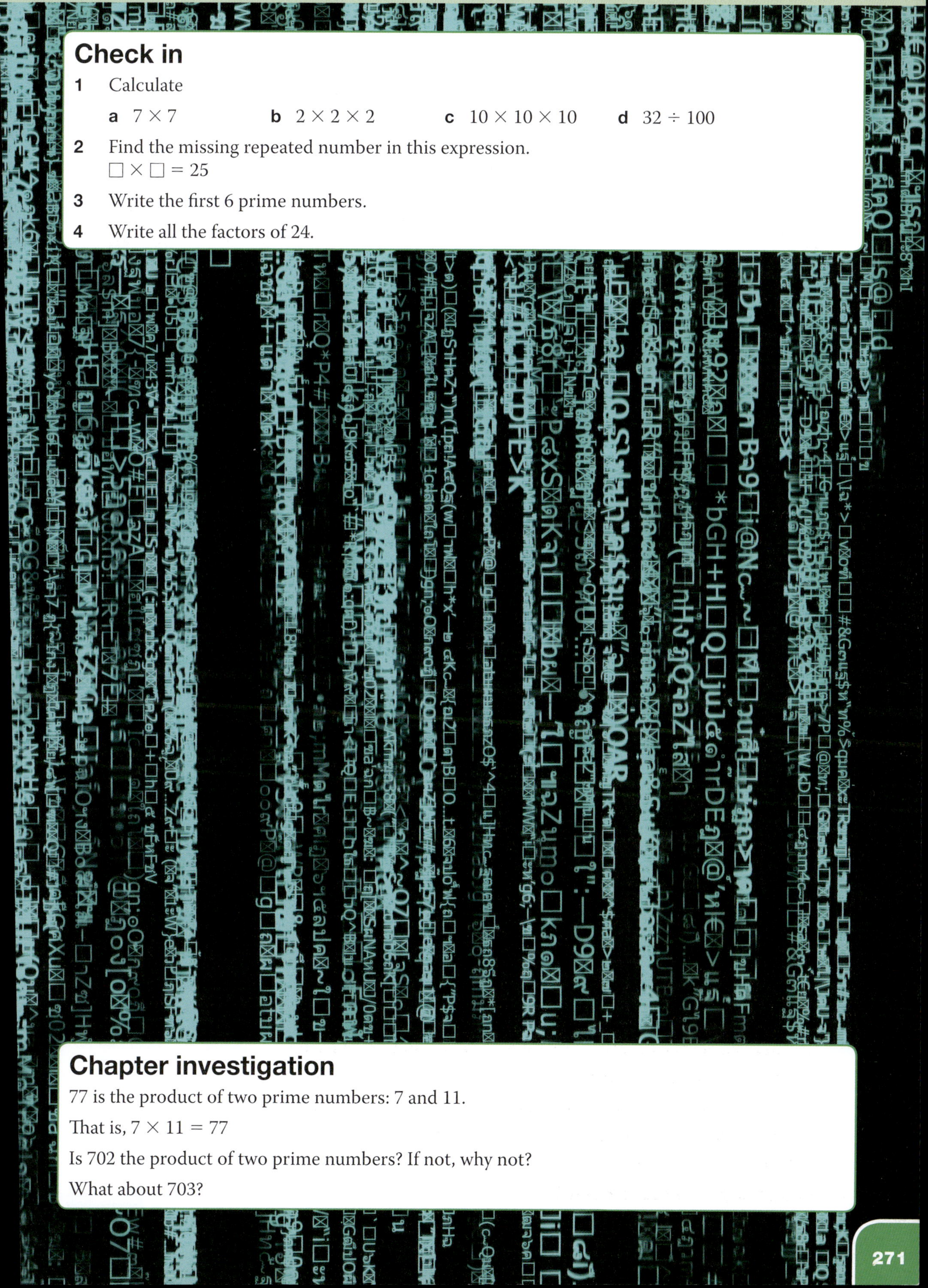

Check in

1 Calculate

    a 7×7 **b** $2 \times 2 \times 2$ **c** $10 \times 10 \times 10$ **d** $32 \div 100$

2 Find the missing repeated number in this expression.

    $\square \times \square = 25$

3 Write the first 6 prime numbers.

4 Write all the factors of 24.

Chapter investigation

77 is the product of two prime numbers: 7 and 11.

That is, $7 \times 11 = 77$

Is 702 the product of two prime numbers? If not, why not?

What about 703?

13.1 Factors and multiples

P.38

● The factors of a number are those numbers that divide into it exactly, leaving no remainder.

You can write a number as the **product** of pairs of factors.

Product means multiply together.

24 = 1 × 24 = 2 × 12 = 3 × 8 = 4 × 6
The factors of 24 are 1, 2, 3, 4, 6, 8, 12, 24.

You can use simple divisibility tests to find the factors of a number.

Factor	Test
2	The number ends in a 0, 2, 4, 6 or 8.
3	The sum of the digits is divisible by 3.
4	The last two digits of the number are divisible by 4.
5	The number ends in 0 or 5.

Factor	Test
6	The number is divisible by 2 *and* by 3.
7	There is no simple check for divisibility by 7.
8	Half of the number is divisible by 4.
9	The sum of the digits is divisible by 9.
10	The number ends in 0.

● A **prime number** is a number with only two factors. These are 1 and the number itself.

The first ten prime numbers are 2, 3, 5, 7, 11, 13, 17, 19, 23 and 29.

One is not a prime number because it only has one factor, 1.

● The **multiples** of a number are those numbers that divide by it exactly, leaving no remainder.

You can think of multiples as being the numbers in the 'times tables'.

The first five multiples of 18 are 18, 36, 54, 72, 90, ...

● The **highest common factor** (HCF) of two numbers is the largest number that is a factor of both numbers.

● The **lowest common multiple** (LCM) of two numbers is the smallest number that they both divide into.

Be systematic. Listing the multiples and factors makes it easy to spot the HCF and LCM.

EXAMPLE

Find the HCF and LCM of 10 and 15.

The factors of 10 are   1   2   ⑤   10
The factors of 15 are   1   3   ⑤   15

1 and 5 are **common factors** of 10 and 15.
5 is the highest common factor of 10 and 15.

The first six multiples of 10 are   10   20   ㉚   40   50   60
The first six multiples of 15 are   15   ㉚   45   60   75   90

30 and 60 are **common multiples** of 10 and 15.
30 is the lowest common multiple of 10 and 15.

Exercise 13.1S

1 Write all the factor pairs for each number.

 a 8 **b** 16 **c** 23 **d** 34

 e 39 **f** 44 **g** 42 **h** 48

2 Write the first three multiples of each number.

 a 7 **b** 9 **c** 12 **d** 15

 e 30 **f** 32 **g** 45 **h** 50

3 Use the divisibility tests to answer each of these questions.
In each case explain your answer.

 a Is 5 a factor of 135?

 b Is 5 a factor of 210?

 c Is 2 a factor of 321?

 d Is 3 a factor of 231?

 e Is 9 a factor of 91?

 f Is 4 a factor of 434?

 g Is 10 a factor of 710?

 h Is 9 a factor of 321?

 i Is 3 a factor of 451?

 j Is 6 a factor of 98?

 k Is 4 a factor of 3428?

 l Is 8 a factor of 312?

 m Is 6 a factor of 528?

4 Write all the factor pairs of each of these numbers.

 a 45 **b** 100 **c** 120 **d** 132

 e 160 **f** 180 **g** 324 **h** 224

 i 264 **j** 312 **k** 325 **l** 432

5 Write the first three multiples of each of these numbers.

 a 17 **b** 29 **c** 42 **d** 25

 e 47 **f** 35 **g** 90 **h** 120

 i 95 **j** 208 **k** 144 **l** 111

6 Find the common factors of these numbers.

 a 10 and 20 **b** 12 and 15

 c 20 and 25 **d** 8 and 20

 e 21 and 28 **f** 24 and 30

 g 12 and 28 **h** 9 and 36

7 Find the first two common multiples of these numbers.

 a 6 and 10 **b** 9 and 12

 c 6 and 8 **d** 10 and 15

 e 14 and 21 **f** 20 and 30

8 Find the highest common factor of these numbers.

 a 6 and 4 **b** 25 and 40

 c 18 and 30 **d** 24 and 56

 e 30 and 75 **f** 36 and 54

 g 50 and 125 **h** 24, 36 and 72

 i 30, 75 and 105 **j** 48, 112 and 114

9 Find the lowest common multiple of these numbers.

 a 6 and 4 **b** 5 and 8 **c** 12 and 18

 d 15 and 25 **e** 18 and 27 **f** 30 and 75

10 a Copy the 1 to 100 number square.

 b Follow these instructions.

  ● 1 is not prime, cross it out.

  ● 2 is the lowest prime number. Circle 2 and cross out all the other multiples of 2.

  ● 3 is the next number not crossed out. It is the next prime number. Circle 3 and cross out all the other multiples of 3.

  ● 5 is the next number not crossed out. It is the next prime number. Circle 5 and cross out all the other multiples of 5.

 c Carry on until only prime numbers are left.

 d Make a list of all the prime numbers from 1 to 100.

1	2	3	4	5	6	7	8	9	10
11	12	13	14	15	16	17	18	19	20
21	22	23	24	25	26	27	28	29	30
31	32	33	34	35	36	37	38	39	40
41	42	43	44	45	46	47	48	49	50
51	52	53	54	55	56	57	58	59	60
61	62	63	64	65	66	67	68	69	70
71	72	73	74	75	76	77	78	79	80
81	82	83	84	85	86	87	88	89	90
91	92	93	94	95	96	97	98	99	100

13.1 Factors and multiples

RECAP

- The **highest common factor** (HCF) of two numbers is the largest number that is a factor of both numbers.
- The **lowest common multiple** (LCM) of two numbers is the smallest number that they both divide into.
- You can use divisibility tests to find the factors of a number.
- A prime number has two factors, 1 and itself.

Primes 2, 3, 5, 7, 11, 13, 17, 19, 23, 29, ...

	Factors	Multiples
12	1, 2, 3, <u>4</u>, 6, 12	12, 24, 36, <u>48</u>, ...
16	1, 2, <u>4</u>, 8, 16	16, 32, <u>48</u>, 64, ...
	HCF = 4	LCM = 48

HOW TO

To answer a question involving factors and multiples

① RTQ – decide how to use your knowledge of factors and multiples.

② Be systematic – use divisibility tests and listing to help you to find multiples and factors.

③ ATQ – make sure that you explain your answers fully.

EXAMPLE

n is an even positive number. $3n + 1$ is a multiple of 5. Find a possible value for n.

① Even numbers are multiples of 2.

② Using a table helps to organise the answer.

Find $3n + 1$ for each multiple of 2. Decide if the answer is a multiple of 5.

n	$3n + 1$	Multiple of 5?
2	$3 \times 2 + 1 = 6 + 1 = 7$	No
4	$3 \times 4 + 1 = 12 + 1 = 13$	No
6	$3 \times 6 + 1 = 18 + 1 = 19$	No
8	$3 \times 8 + 1 = 24 + 1 = 25$	Yes, $5 \times 5 = 25$

③ 8 is a possible value for n.

EXAMPLE

A pilot can see two beacons.

One beacon flashes every 15 seconds and the other flashes every 25 seconds.

If the beacons flash together, how long will it be before they next flash together?

① The question is about multiples of numbers.

② Write a list of multiples for 15 and 25. Find the LCM.

Beacon 1:   0   15   30   45   60   (75)   90   ...

Beacon 2:   0   25   50   (75)   100   ...

③ The beacons will flash together after 75 seconds.

EXAMPLE

A large rectangular play area measures 36 metres by 42 metres.

The council wants to lay the play area with square concrete tiles.

Tiles are available with lengths of whole numbers of metres.

If the tiles are to be as large as possible, find the size of the tiles.

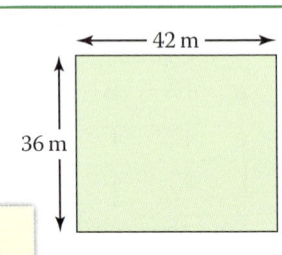

① The question is about factors of numbers.

② Write a list of factors for 36 and 42. Find the highest common factor.

Factors of 36: 1   2   3   4   ⑥   9   12   18   36

Factors of 42: 1   2   3   ⑥   7   14   21   42

③ The tiles are 6 metres by 6 metres.

Exercise 13.1A

1 a Write a multiple of 20 that is bigger than 200.

b Write a multiple of 15 that is between 100 and 140.

c Write a multiple of 6 that is bigger than 70 but less than 100.

2 Jo's age this year is a multiple of 6.
Next year it will be a multiple of 5.
How old might Jo be?

3 n is a whole number.
$2n + 3$ is a prime number.
Find two possible values of n.

4 n is an odd number.
$4n - 5$ is a multiple of 3.
Find two possible values of n.

5 a Amos says 'all odd numbers are prime numbers'. Give two examples that show he is wrong.

b Aya says 'all prime numbers are odd numbers'. Give an example to show she is wrong.

c Arik says 'take one off any multiple of 6 and you always get a prime number'.

　i Give three examples where this is true.

　ii Give one example where it is false.

6 The warning light on a TV mast flashes every eight seconds.

A nearby TV mast flashes every 14 seconds.

If they flash together at a particular moment, how long will it be before they flash together again?

7 Jamie is tiling his bathroom wall.
The wall measures 100 cm by 160 cm.
He wants to tile the wall with square tiles.
Tiles are available in whole numbers of centimetres.
If the tiles are to be as large as possible, what size tile should Jamie buy?

8 Marisa is making flower arrangements for the tables in a restaurant.
Marisa has 36 roses and 60 daisies.
The flower arrangements must each contain the same flowers. She must make as many flower arrangements as possible.

8 She must use all of the flowers.
How many flower arrangements can Marisa make?

9 Simon is having a barbecue.
He buys burgers and burger buns.
The burgers come in packets of 10.
The burger buns come in packets of 24.
Simon needs to buy the same number of burgers as burger buns.
He wants to spend as little money as possible.
How many packets of burgers and burger buns does Simon need to buy?

10 Mai and Luca are playing a game.
They choose a number from 1 to 99 and find the sum of all of its factors.
Whoever has the highest total wins the game.

a Mai chooses 32 and Luca chooses 55. Who wins the game?

b Mai chooses 80.
Suggest a number that Luca could choose to win the game.

11 a Copy and complete this table.

Numbers	Product	HCF	LCM
6 and 4	6 × 4 = 24	2	12
8 and 10	8 × 10 =		
12 and 18			
6 and 9			
15 and 20			
15 and 25			

b Write anything you notice about the numbers in your table.

c Write a quick way to find the LCM if you know the HCF.

*12 Add up all the factors of a number apart from the number itself. If your answer is less than the number it is **deficient**, equal to the number it is **perfect** and more than the number it is **abundant**. For example, 6 is perfect because $1 + 2 + 3 = 6$.
Are these numbers deficient, perfect or abundant?

a 5　　**b** 12　　**c** 27

d 28　　**e** 100　　**f** 496

1032, 1034, 1044　SEARCH

● A **prime factor** is a factor of a number which is also prime.

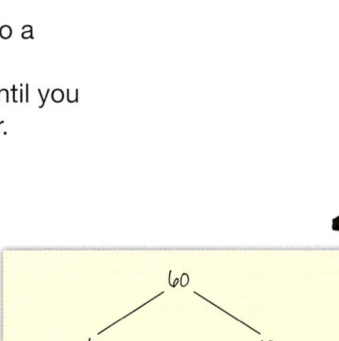

Remember, one is not a prime number.

The factors of 28 are {1, 2, 4, 7, 14, 28}.

The prime factors of 28 are {2, 7}.

● Every whole number can be written as the *unique* product of its **prime factors**.

There are two common methods to find the prime factors.

Division by prime numbers

Divide the number by the smallest **prime number**. Repeat dividing by larger prime numbers until you reach a prime number.

Factor trees

Split the number into a **factor** pair. Continue splitting until you reach a prime factor.

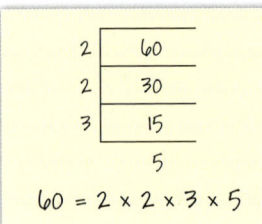

$60 = 2 \times 2 \times 3 \times 5$

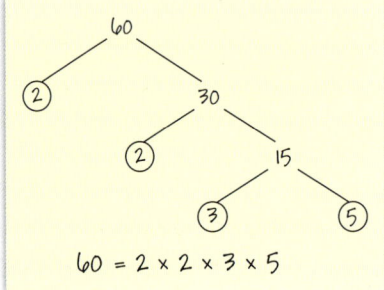

$60 = 2 \times 2 \times 3 \times 5$

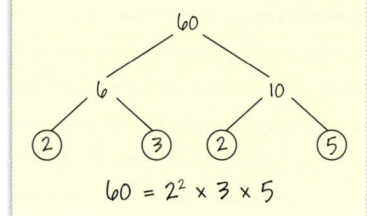

$60 = 2^2 \times 3 \times 5$

You can find the HCF and LCM of two numbers by writing their **prime factors** in a Venn diagram.

It doesn't matter how you do the factorising, you always get the same answer!

● The HCF is the product of the numbers in the **intersection**.

  ● The intersection is where the circles in the Venn diagram overlap.

● The LCM is the product of all the numbers in the diagram.

EXAMPLE

Find the HCF and LCM of 60 and 280.

$60 = 2^2 \times 3 \times 5$ $280 = 2^3 \times 5 \times 7$ 60 and 280 have been written as a product of powers.

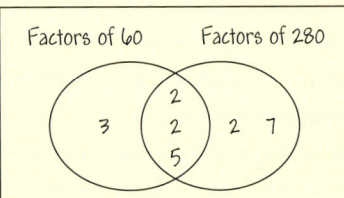

The HCF is the product of the numbers in the intersection.

$HCF = 2^2 \times 5 = 20$

An alternative is to write:

$60 = 2^2 \times 3 \times 5$

$280 = 2^3 \times 5 \times 7$

Identify the highest power of each factor.

$LCM = 2^3 \times 3 \times 5 \times 7 = 840$

The LCM is the product of all the numbers in the diagram.

$LCM = 2^3 \times 3 \times 5 \times 7 = 840$

p.410

Exercise 13.2S

1 Write all the factor pairs of each of these numbers.

    a 18 **b** 14 **c** 30 **d** 64

2 Find the highest common factor of these numbers.

    a 8 and 10   **b** 12 and 18

    c 14 and 16   **d** 28 and 35

    e 30 and 54   **f** 56 and 64

    g 60 and 72   **h** 27 and 45

3 Find the least common multiple of these numbers.

    a 14 and 16   **b** 8 and 10

    c 12 and 16   **d** 25 and 35

    e 21 and 28   **f** 26 and 39

    g 40 and 50   **h** 21 and 35

4 Copy and complete these calculations to show the different ways that 24 can be written as a product of its factors.

    a $24 = \square \times 2$ **b** $24 = 3 \times \square$

    c $24 = 2 \times 3 \times \square$ **d** $24 = 4 \times \square$

5 Each of these numbers has just two prime factors, which are not repeated. Write each number as the product of its prime factors.

    a 77   **b** 51   **c** 65

    d 91   **e** 119   **f** 221

6 Copy and complete the factor tree to find the prime factor decomposition of 18.

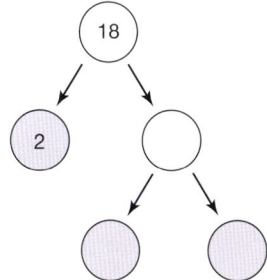

7 Work out the value of each of these expressions.

    a 2^3

    b 3×5^2

    c $2^3 \times 5$

    d $3^2 \times 7$

    e $2^2 \times 3^2 \times 5$

    f $3^2 \times 7^2$

8 Express these numbers as products of their prime factors.

    a 18   **b** 24   **c** 40

    d 39   **e** 48   **f** 82

    g 100   **h** 144   **i** 180

    j 315   **k** 444   **l** 1350

9 For each number, find its prime factors and write it as the product of powers of its prime factors.

    a 36   **b** 120   **c** 34

    d 25   **e** 75   **f** 90

    g 27   **h** 64   **i** 72

10 Write the prime factor decomposition for each of these numbers.

    a 1052 **b** 2560 **c** 630

    d 825   **e** 715   **f** 1001

    g 219   **h** 289   **i** 2840

    j 2695 **k** 1729 **l** 3366

    m 9724 **n** 2852 **o** 10 179

11 Find the highest common factor (HCF) of each pair of numbers, by drawing a Venn diagram or otherwise.

    a 35 and 20   **b** 48 and 16

    c 21 and 24   **d** 25 and 80

    e 28 and 42   **f** 45 and 60

12 Find the least common multiple (LCM) of each pair of numbers, by drawing a Venn diagram or otherwise.

    a 24 and 16   **b** 32 and 100

    c 22 and 33   **d** 104 and 32

    e 56 and 35   **f** 105 and 144

13 Find the LCM and HCF of each pair of numbers.

    a 180 and 420   **b** 77 and 735

    c 240 and 336   **d** 1024 and 18

    e 762 and 826   **f** 1024 and 1296

14 Find the LCM and HCF of these numbers.

    a 15, 20 and 35   **b** 12, 18 and 28

Q 1032, 1044   SEARCH

13.2 Prime factor decomposition

RECAP

- A number can be written as the product of its prime factors. This is called prime factor decomposition.
- You can use a factor tree to find the prime factors of a number.
- The **highest common factor** (HCF) of two numbers is the largest number that is a factor of them both.
- The **least common multiple** (LCM) of two numbers is the smallest number that they both divide into.
- You can find the HCF and LCM of two numbers by writing their **prime factors** in a Venn diagram.
- The HCF is the product of the numbers in the intersection.
- The LCM is the product of all the numbers in the diagram.

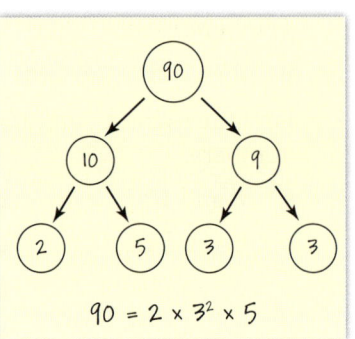

$90 = 2 \times 3^2 \times 5$

HOW TO

To solve problems involving factors and multiples

1. RTQ – decide how to use your knowledge of factors and multiples.
2. Be systematic – use listing, factor trees and/or Venn diagrams to help you to find multiples and factors.
3. ATQ – make sure that you explain your answers.

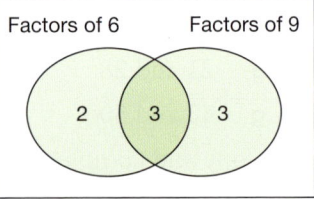

HCF = 3, LCM = 18

EXAMPLE

Remi has four numbers given as prime factorisations.

$2^2 \times 3$ $\qquad$ $2 \times 3^2 \times 5$

$5^2 \times 13$ $\qquad$ $2^2 \times 7 \times 11$

Which of the numbers are

a multiples of 6

b multiples of 10

c multiples of 8?

Give reasons for your answers.

1. Can you make 6, 10 or 8 from the factors in each decomposition?

a
2. Multiples of 6 have 2 <u>and</u> 3 as prime factors.
3. $2^2 \times 3$ and $2 \times 3^2 \times 5$ are multiples of 6.

b
2. Multiples of 10 have 2 <u>and</u> 5 as prime factors.
3. $2 \times 3^2 \times 5$ is a multiple of 10.

c
2. Multiples of 8 have 2^3 as a prime factor.
3. There are no multiples of 8.

EXAMPLE

The highest common factor of two numbers is 6.

The lowest common multiple of the two numbers is 540.

One of the numbers is 54.

Find the other number.

1. Use a Venn diagram and prime factors.
2. Find the prime factor decompositions of 6, 54 and 540.

$6 = 2 \times 3$

Write 2 and 3 in the intersection.

$54 = 2 \times 3 \times 3 \times 3$

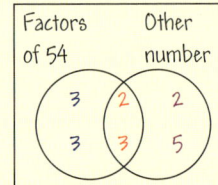

Add 3 and 3 to the diagram.
The product of the numbers in the Venn diagram is 540.

$540 = 2 \times 3 \times 3 \times 3 \times 2 \times 5$

Add 2 and 5 to the diagram.

3. Find the other number.
   The other number is $= 2^2 \times 3 \times 5 = 60$.

Exercise 13.2A

1 Jack has been asked to find the prime factors of two numbers.

    i Mark Jack's work for him.

    ii Correct any of Jack's mistakes.

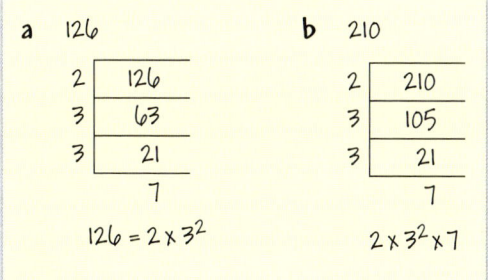

a 126

2	126
3	63
3	21
	7

$126 = 2 \times 3^2$

b 210

2	210
3	105
3	21
	7

$2 \times 3^2 \times 7$

2 Allison has four numbers given as prime factorisations.

$$2^2 \times 7 \qquad 2 \times 3^2 \qquad 2^3 \times 5 \qquad 2 \times 3 \times 5^2$$

List every number which is

    a a multiple of 6

    b a multiple of 10

    c a multiple of 50

    d a multiple of 8.

3 The number 18 can be written as $2 \times 3 \times 3$. You can say that 18 has three prime factors.

    a Find three numbers with exactly three prime factors.

    b Find five numbers with exactly four prime factors.

    c Find four numbers between 100 and 300 with exactly five prime factors.

    d Find a two-digit number with exactly six prime factors.

4 A cuboid is made from 210 small cubes. The prime factor decomposition of 210 is $2 \times 3 \times 5 \times 7$.

One way of combining the prime factors is $(2 \times 3) \times 5 \times 7 = 6 \times 5 \times 7$. The dimensions of the cuboid could be $6 \times 5 \times 7$.

List the dimensions of all the cuboids that could be made with 210 cubes.

5 A metal cuboid has a volume of $1815 \, cm^3$. Each side of the cuboid is a whole number of centimetres and longer than 1 cm.

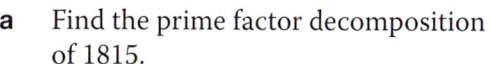

    a Find the prime factor decomposition of 1815.

    b Find all the possible dimensions of the cuboid.

6 Two needles move around a dial. One needle moves around in 24 seconds, and the other in 30 seconds. If they start together at the top of the dial, how many seconds does it take before they are next together at the top?

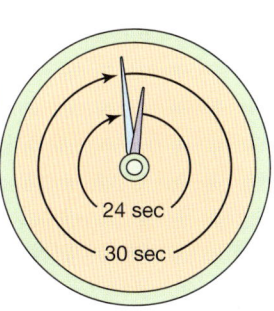

24 sec

30 sec

7 A wall measures 234 cm by 432 cm. What is the largest size of square tile that can be used to cover the wall, without needing to cut any of the tiles?

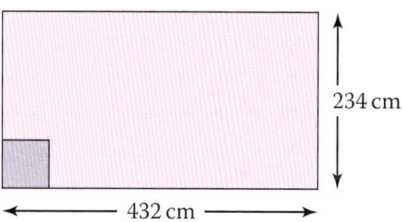

234 cm

432 cm

8 Usha says that if you double two numbers, you double their lowest common multiple.

    a Show that this is true for 6 and 9.

    b Is this true for all pairs of numbers? Give your reasons.

    c Is there a similar result for the HCF? Show your working.

9 Two numbers have HCF = 15 and LCM = 90.
One of the numbers is 30.
What is the other number?

10 $108 = 2^2 \times 3^3$
How many factors does 108 have?

p.312

13.3 Powers and roots

- A **square number** is the result of multiplying a whole number by itself.

$$1^2 = 1 \times 1 = 1 \qquad 2^2 = 2 \times 2 = 4 \qquad 3^2 = 3 \times 3 = 9$$

- The **square root** of a given number is a number that when multiplied by itself equals the given number.

The square roots of 225 are 15 and -15 because
$15 \times 15 = 225$ and $-15 \times -15 = 225$.
You write $\sqrt{225} = 15$.

- A **cube number** is the result of multiplying a whole number by itself and then multiplying by that number again.

$$1^3 = 1 \times 1 \times 1 = 1 \qquad 2^3 = 2 \times 2 \times 2 = 8 \qquad 3^3 = 3 \times 3 \times 3 = 27$$

- The **cube root** of a given number is a number that when multiplied by itself and then multiplied by itself again equals the given number.

$\sqrt[3]{4913} = 17$ because $17^3 = 17 \times 17 \times 17 = 4913$.

- A positive number has a positive cube root and a negative number has a negative cube root.

$\sqrt[3]{-125} = -5$ because $(-5)^3 = -5 \times -5 \times -5 = -125$

- The **index**, or **power**, tells you how many times the number must be multiplied by itself.

$$7^4 = 7 \times 7 \times 7 \times 7 = 2401$$
$$6^3 = 6 \times 6 \times 6 = 216$$

Work out how to do this on your calculator.

See if your calculator has these keys.

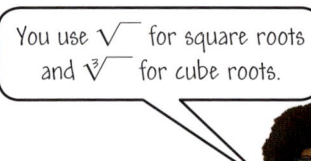

You use $\sqrt{}$ for square roots and $\sqrt[3]{}$ for cube roots.

Take care when finding powers of negative numbers on your calculator. Remember to put brackets around the negative number.

$13^5 \qquad 371293$

EXAMPLE

Use your calculator to find the missing powers.

a $\quad 18^{\square} = 5832$ b $\quad 1.7^{\square} = 8.3521$ c $\quad (-1.2)^{\square} = -1.728$

For each question test different powers.

a $\quad 18^2 = 324$ ✗ b $\quad 1.7^2 = 2.89$ ✗ c Square numbers are always positive.

Try again. Try again. $(-1.2)^3 = -1.728$ ✓

$18^3 = 5832$ ✓ $1.7^3 = 4.913$ ✗ Correct.

Correct. $1.7^4 = 8.3521$ ✓

Correct.

Did you know...

French mathematician and philosopher Rene Descartes was the first person to use index form to write powers of numbers.

Exercise 13.3S

1 **a** Write out the first 20 square numbers.

b Write out the first 10 cube numbers.

2 **a** Find a square number that is the sum of two cube numbers.

b Find a cube number that is the sum of two square numbers.

3 Use your calculator to work out each of these squares and cubes.

a 6^2 **b** 11^2 **c** 14^2 **d** 23^2

e 31^2 **f** 47^2 **g** 4^3 **h** 6^3

i 8^3 **j** 13^3 **k** 18^3 **l** 21^3

4 Use your calculator to work out each of these numbers.

a $3^2 + 2^2$ **b** $5^2 - 3^2$

c $6^2 - 2^3$ **d** $4^3 + 4^2$

e $10^2 - 8^2$ **f** $13^2 + 4^3$

g $14^2 - 5^3$ **h** $6^3 - 13^2$

i $12^2 + 13^2 + 14^2$ **j** $6^3 + 7^3 + 8^3$

5 Use your calculator to work out each of these powers. Give your answer to 2 decimal places where appropriate.

a 2.5^2 **b** 49^2 **c** 3.2^3

d 4.8^2 **e** 7.3^2 **f** 4.9^3

g 1.2^2 **h** 0.5^2 **i** 9.9^2

j 9.9^3 **k** $(5\,\text{cm})^2$ **l** $(4\,\text{m})^3$

6 Find these square roots without using a calculator.

a $\sqrt{25}$ **b** $\sqrt{9}$ **c** $\sqrt{16}$

d $\sqrt{1}$ **e** $\sqrt{4}$ **f** $\sqrt{64}$

7 Use your calculator to find these square roots. Give your answers to 3 sf where appropriate.

a $\sqrt{40}$ **b** $\sqrt{61}$ **c** $\sqrt{180}$

d $\sqrt{249}$ **e** $\sqrt{5.4}$ **f** $\sqrt{7.6}$

8 Find these cube roots without using a calculator.

a $\sqrt[3]{8}$ **b** $\sqrt[3]{125}$ **c** $\sqrt[3]{-1}$

d $\sqrt[3]{64}$ **e** $\sqrt[3]{1000}$ **f** $\sqrt[3]{-27}$

9 Use your calculator to workout each of these cube roots. Give your answers to 3 sf where appropriate.

a $\sqrt[3]{40}$ **b** $\sqrt[3]{512}$ **c** $\sqrt[3]{3375}$

d $\sqrt[3]{-100}$ **e** $\sqrt[3]{3.6}$ **f** $\sqrt[3]{-26.7}$

10 Copy each equation and fill in the missing powers.

a $3^\square = 9$ **b** $5^\square = 25$ **c** $4^\square = 64$

d $2^\square = 8$ **e** $3^\square = 27$ **f** $10^\square = 1000$

11 Find the value of these powers.

a 3^4 **b** 1^5 **c** 2^7

d 3^6 **e** 10^6 **f** $(-5)^4$

12 Use your calculator to find these powers.

a 12^3 **b** 6^6 **c** 21^3

d 16^5 **e** 13^3 **f** $(-4)^4$

13 Use your calculator to work out these powers. In each case copy the question and fill in the missing numbers.

a $24^\square = 576$ **b** $1.5^\square = 3.375$

c $(-2.25)^\square = 5.0625$

d $(-0.5)^\square = -0.125$

e $5^\square = 3125$ **f** $(-3)^\square = 729$

14 Use a calculator to find these powers. Give your answers to 2 decimal places where appropriate.

a 15.5^3 **b** $(-2.7)^6$ **c** 2.1^{10}

d 21.6^4 **e** 13.5^3 **f** $(-16.8)^5$

15 Find the value of these powers.

a $\left(\frac{1}{10}\right)^2$ **b** $\left(\frac{1}{10}\right)^3$ **c** $\left(\frac{1}{10}\right)^4$

d $\left(\frac{1}{2}\right)^2$ **e** $\left(\frac{1}{4}\right)^2$ **f** $\left(\frac{1}{3}\right)^3$

16 Use a mental, written or calculator method to work out each of these.

a $2^4 + 3^2$ **b** $10^3 \div 5^2$

c $8^6 - 13^3$ **d** $10^6 \div 5^3$

e $\left(\frac{1}{4}\right)^4 + \left(\frac{1}{8}\right)^2$ **f** $\sqrt{\frac{4}{9}} + \sqrt[3]{\frac{1}{8}}$

13.3 Powers and roots

- The index, or power, tells you how many times the number must be multiplied by itself.
- A square number is the result of multiplying an integer by itself.
- A positive number has a negative and positive square root.
- A cube number is the result of multiplying an integer by itself and then multiplying by that integer again.
- A positive number has a positive cube root and a negative number has a negative cube root.

Squares – Square roots
$6^2 = 6 \times 6 = 36$
$(-6)^2 = -6 \times -6 = 36$
$\sqrt{36} = 6$

Cubes – Cube roots
$6^3 = 6 \times 6 \times 6 = 216$
$\sqrt[3]{216} = 6$
$(-6)^3 = -6 \times -6 \times -6 = -216$
$\sqrt[3]{-216} = -6$

HOW TO

To solve problems involving powers and roots

① RTQ. You may need to use your knowledge from other areas of maths.

② Apply your knowledge of squares and cubes.

③ ATQ and include any units.

EXAMPLE

Elsa has square tiles each with area 81 cm².
She uses six tiles to make this shape.
Find the perimeter of the shape.

① The area of a square = length²

   length² = 81

② Find the square root of 81.

   length = $\sqrt{81}$ = 9 cm

③ The perimeter is the distance around the shape.

   Perimeter = 9 + 27 + 9 + 18 + 9 + 27 + 9 + 18

   = 126 cm

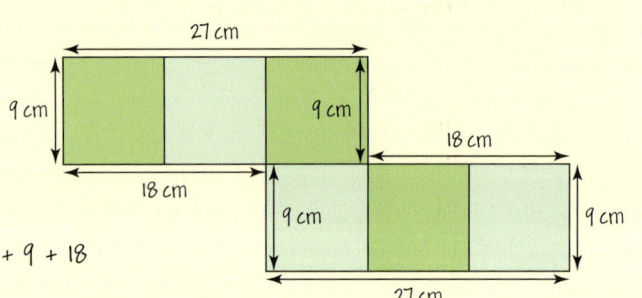

EXAMPLE

Find three whole numbers n, such that $\sqrt{5 + 4n}$ is a whole number.

① If $\sqrt{5 + 4n}$ is a whole number, then $5 + 4n$ must be a square number.

   5 + 4n = square number

② Find a square number that is 5 more than a multiple of 4.

   Square numbers   1, 4, 9, 16, 25, 36, 49, 64, 81, 100 ...

   Multiples of 4   4, 8, 12, 16, 20, 24, 28, 32, 36, 40, 44, 48 ...

   9 = 5 + 4   = 5 + 4 × 1

   25 = 5 + 20 = 5 + 4 × 5

   49 = 5 + 44 = 5 + 4 × 11

③ Three possible values for n are 1, 5 and 11.

This approach is more efficient than trying different values of n until you find a correct solution.

Exercise 13.3A

1 Some numbers can be written as the sum of two square numbers.
For example $1^2 + 2^2 = 1 + 4 = 5$.
Find all the numbers less than 50 that can be written this way.

2 Harry has mixed up his answers to these questions.

  a Without using a calculator, match each of these questions to the correct answer.

Questions		Answers	
1	$\sqrt{169}$	A	9
2	$\sqrt[3]{343}$	B	8
3	$\sqrt{121}$	C	7
4	$\sqrt{81}$	D	4
5	$\sqrt{64}$	E	11
6	$\sqrt[3]{1000}$	F	13
7	$\sqrt{196}$	G	10
8	$\sqrt[3]{64}$	H	14

  b Check your answers using your calculator.

How many of the questions and answers did you match correctly?

3 Lewis has square tiles with area 144 cm².
He uses nine tiles to make a larger square.
Find the perimeter of the larger square.

4 Jack has a set of cubes each with volume 216 cm³.
Jack wants to build a tower by stacking the cubes on top of each other.
He wants the tower to be at least 1 metre tall.
Find the minimum number of cubes that Jack needs to build the tower.

5 Do not use a calculator for this question.
$\sqrt{95}$ lies between 9 and 10.
Find two consecutive numbers that these square roots lie between.

  a $\sqrt{150}$ **b** $\sqrt{300}$ **c** $\sqrt{80}$

6 $\sqrt{7} = 2.645751$
Calculate $(2.645751)^2$.
Why is the answer not 7?

7 Two consecutive numbers are multiplied together.
The answer is 3192.
What are the two numbers?

8 The cube of a particular integer lies between 3000 and 4000.
Without using a calculator, find the integer.
Show your working.

9 The product of three consecutive numbers is 2184.
What are the numbers?

10 In a school the teachers never reveal their ages but always give clues.
Mr Earle gave three clues.

> ● My age is a multiple of 9
> ● My age is a cube number
> ● My age is less than 40

  a What is Mr Earle's age?

  b Show how you can work out Mr Earle's age from two clues.

11 Find a pair of prime numbers a and b such that $\sqrt{a^2 - b}$ is an integer.

12 Find a pair of integers p and q such that $\sqrt{p^3 - 3q}$ is an integer.

13 a Give two **counter examples** for each of the following to show they are not always true.

    i James says that when you square a number the answer always gets bigger.

    ii Lauren says that when you find the square root of a number the answer always gets smaller.

  b Describe what happens to the size of a number when you find the square or the square root of it.

14 Do not use a calculator.

  a Find the prime factor decomposition of 1444.

  b Use your answer to find the square root of 1444.

Q 1053, 1924   SEARCH

Summary

Checkout

You should now be able to...

Test it

✔ Use mathematical language to describe factors, multiples and primes.	**1 – 4**
✔ Use Venn diagrams or factor trees to systematically list the prime factors of a number.	**5 – 7**
✔ Use prime factor decomposition to calculate the HCF and LCM of two or more numbers.	**8, 9**
✔ Write the HCF and LCM using product notation.	**10**
✔ Calculate positive integer powers and their roots.	**11, 12**
✔ Recognise powers of 2, 3, 4 and 5.	**13**

Language	Meaning	Example
Multiple	The original number multiplied by an integer (a whole number).	$2 \times 6 = 12$ $3 \times 6 = 18$ 12 and 18 are both multiples of 6.
Factor	A number that divides exactly into another number.	$15 = 3 \times 5$ 3 and 5 are both factors of 15.
Prime number	A number that has only two factors, itself and one.	$13 = 1 \times 13$
Prime factor	A factor that is a prime number.	13 is a prime factor of 52.
Prime factor decomposition	Writing a number as a product of its prime factors.	$52 = 2^2 \times 13$
Common factor	A factor that is shared by two or more numbers.	$30 = 2 \times 3 \times 5$ $70 = 2 \times 5 \times 7$ 2 and 5 are common factors of 30 and 70
Highest common factor (HCF)	The largest factor that is shared by two or more numbers.	Factors of 30 Factors of 70
Least common multiple (LCM)	The smallest multiple that is shared by two or more numbers.	2 3 5 7 HCF $= 2 \times 5 = 10$ LCM $= 3 \times 2 \times 5 \times 7 = 210$
Square number	To square a number multiply it by itself. Square numbers have integer square roots.	$1^2 = 1$ $2^2 = 4$ $3^2 = 9$ $4^2 = 16$
Square root	A number that when multiplied by itself is equal to the number underneath the square root symbol.	$4 \times 4 = 16$ $\sqrt{16} = 4$
Cube number	To cube a number multiply it by itself and then by itself again. Cube numbers have cube roots that are integers.	$1^3 = 1$ $2^3 = 8$ $3^3 = 27$ $4^3 = 64$
Cube root	A number that when multiplied by itself and then by itself again is equal to the number underneath the cube root symbol.	$4 \times 4 \times 4 = 64$ $\sqrt[3]{16} = 4$

Review

1 List all the factors of these numbers.

    a 12   **b** 19   **c** 48

2 List the first 5 multiples of these numbers.

    a 5   **b** 13   **c** 18

3 State whether or not each of these is a prime and explain your answers.

    a 25   **b** 29   **c** 57

4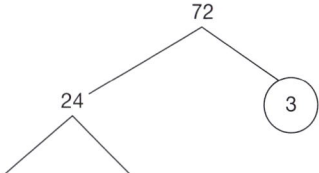

List the numbers in the box that are

    a prime   **b** factors of 36

    c multiples of 6.

5 Copy and complete this factor tree to show the prime factors of 72.

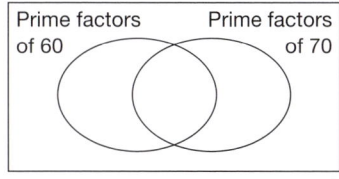

6 Find the prime factor decomposition of these numbers.

    a 18   **b** 28

7 **a** Copy and complete this Venn diagram to show the prime factors of 60 and 70.

    Prime factors of 60   Prime factors of 70

7 **b** Use the diagram to work out the

    i highest common factor

    ii lowest common multiple

    of 60 and 70.

8 Find the highest common factor of these pairs of numbers.

    a 8 and 12   **b** 27 and 54

    c 7 and 9   **d** 11 and 13

9 Find the lowest common multiples of these pairs of numbers.

    a 3 and 4   **b** 6 and 8

    c 11 and 9   **d** 11 and 13

10 **a** Write these numbers as products of their prime factors, using product notation.

    i 54   **ii** 200

    b Find the LCM and HCF of 54 and 200. Write your answers using product notation.

11 Calculate the value of these roots.

    a $\sqrt{16}$ **b** $\sqrt{81}$

    c $\sqrt[3]{8}$ **d** $\sqrt[3]{1000}$

12 Calculate the value of the following expressions.

    a 3^2 **b** 7^2

    c 5^3 **d** 2^5

13 What is the value of x in these equations?

    a $4^x = 64$ **b** $5^x = 125$

What next?

Score			
	0 – 5		Your knowledge of this topic is still developing. To improve look at MyMaths: 1032, 1034, 1044, 1053, 1924
	6 – 10		You are gaining a secure knowledge of this topic. To improve your fluency look at InvisiPens: 13Sa – f
	11 – 13		You have mastered these skills. Well done you are ready to progress! To develop your problem solving skills look at InvisiPens: 13Aa – h

Assessment 13

1   Leo says that all of the pairs of factors of 42 are 2×21 and 3×14.
Is he correct? If not, write down all of the pairs of factors of 42.   [2]

2   **a**   3454 is not divisible by 6. Why is this statement true?   [2]

   b   Write a single digit on the end of 10 to make a number that is divisible by

   i 2   **ii** 3   **iii** 4   **iv** 5   **v** 6   **vi** 7   **vii** 8   **viii** 9   **ix** 10.   [9]

3   **a**   Is 4 a factor of 110? Say how you know.   [2]

   b   I am thinking of a number that is greater than 4 and less than 20.
My number is a factor of 110.
Write down all the possible integer answers.   [3]

4   Gareth says that the HCF of 54 and 99 is 3.
Is he correct? If not, show how to work out the HCF of these two numbers.   [2]

5   Hannah says that the LCM of 36 and 60 is 180.
Show how she could have worked this out.   [3]

6   At a party it was discovered that John, Paul, George and Ringo had birthdays
on the 6th, 15th, 21st and 30th of the month. Mick joined the group and
it was discovered that his birthday was a factor of everyone else's.
What days of the month could Mick have been born on?   [2]

7   Trains from Birmingham go to Glasgow, Liverpool and Leeds. Trains to Glasgow
leave once an hour, to Liverpool every 45 minutes and to Leeds every 25 minutes.
At 6 am three trains leave for all these destinations.
When will trains to all these destinations next leave at the same time?   [5]

8   A 'perfect' number is one where all its factors, except for the number itself, add up to that
number. The factors of 6 are 1, 2, 3 and 6 and $1 + 2 + 3 = 6$ so 6 is the first perfect number.
The next perfect number is 28. Show that 28 is a 'perfect' number.
Show your working.   [3]

9   Henry knows an alternative method to find the HCF of two numbers.

   1.   Write down the numbers side by side.

   2.   Cross out the largest and write underneath it the 'difference' between the two.

   3.   Repeat step 2 until the two numbers left are the same.

   4.   The remaining number is the HCF of the two original numbers.

   Try Henry's method with the following numbers

   a   16 and 28   [3]   **b**   30 and 66   [3]   **c**   252 and 588.   [3]

10   Isa and Josh both write 19 800 as a product of its prime factors using index notation.

   Isa writes $19\,800 = 2^3 \times 3^2 \times 5^2 \times 11$ Josh writes $19\,800 = 2 \times 3^3 \times 5 \times 11$
   Who is correct? Show your working.   [3]

11   Amanda says that

   a 4^5 is greater than 5^4 **b** 2^{10} is greater than 10^2 **c** 2^4 is greater than 4^2.

   Decide for each pair of values if she is correct or not. Give reasons for your answers.   [6]

12 The number grid below has 4 columns. The *prime numbers* in the grid are *shaded*.

Column A	Column B	Column C	Column D
1	2	3	4
5	6	7	8
9	10	11	12
13	14	15	16
17	18	19	20
21	22	23	24
25	26	27	28
29	30	31	32
33	34	35	36
37	38	39	40

a 33 is not shaded. Explain why 33 is **not** a prime number. [2]

b Hasan says that 100 will eventually be in Column A.
Is he right? Give reasons for your answer. [1]

c There is only one prime number in column B.
Will there ever be another? Explain how you know. [2]

d What will be the next two prime numbers in Column C? [2]

e There are no prime numbers in column D.
Will there ever be a prime number in column D. Give reasons for your answer. [1]

13 Tope says that all square numbers over 1 can be written as the sum of two prime numbers.

a Show this is true for the number 25. [2]

b Show there are two ways of doing this for the number 16. [2]

c Show there are four ways of doing this for the number 36. [4]

14 a Kerry says that to work out $3^3 \times 3^2$ you multiply the indices. Glen says you
add the indices. Who is correct? Use the correct rule to work out $3^3 \times 3^2$. [2]

b Giorgia says that to work out $14^8 \div 14^2$ you subtract the indices. Jonas says you
divide the indices. Who is correct? Use the correct rule to work out $14^8 \div 14^2$. [2]

15 a Sam says numbers have two square roots. George says some numbers
have no square roots. Who is right? Give reasons for your answer. [2]

b Amelia joins in the conversation and says that all numbers have 2 cube roots.
Is she right? Give reasons for your answer. [2]

16 Ms Connell took her class to the monkey house at the zoo, but a naughty
child opened the cage door and let the monkeys out with the children.
Ms Connell counted about 70 heads and tails but there were twice as many
tails as heads. She also knew that the number of children was a prime number.
How many children were in Ms Connell's class? [4]

17 Arthen thinks that all the prime numbers can be found by substituting $p = 0, 1, 2$, etc. into the
formula $P = 41 - p + p^2$.

a Check Arthen's formula for $p = 0, 3$, and 6. [4]

b Write down one value which shows that Arthen is not correct. Give your reasons. [2]

14 Graphs 1

Introduction

Kinematics is the topic within maths that deals with motion. By writing equations and drawing graphs that describe the relationship between distance, speed and acceleration, it is possible to calculate things such as the speed of a vehicle at a particular time during its journey.

What's the point?

Having the mathematical tools to describe how objects move means that we can better understand our world, which is in continual motion.

Objectives

By the end of this chapter you will have learned how to …

- Work with coordinates in all four quadrants.
- Plot straight-line graphs including diagonal, vertical and horizontal lines.
- Identify gradients and intercepts of straight lines graphically and algebraically.
- Use the form $y = mx + c$ to identify parallel lines.
- Use one point and the gradient of the line to find its equation.
- Use two points to find the equation of a line.
- Interpret the gradient of a straight line graph as a rate of change.
- Plot and interpret graphs involving distance, speed and acceleration.

Check in

1   Work out the value of each expression when $x = 2$.

a $x + 6$ b $3x - 1$ c $2x + 4$ d $\frac{x}{4} + 3$

2   Substitute $x = -1$ into these expressions and evaluate them.

a $2x + 1$ b $3x - 2$ c $4x + 5$ d $2x - 5$

3   a Draw a coordinate grid with x and y-axes from -5 to 5, using squared paper.

b Plot these coordinates on your grid.

i $(0, 4)$ ii $(2, -5)$ iii $(-4, 1)$ iv $(-2, -1)$

Chapter investigation

The points $(2, 4)$ and $(5, 13)$ both lie on the same straight line.

Find the equation of this line.

Does the point $(21, 18)$ lie on this same line? How can you tell without drawing the line?

14.1 Drawing straight-line graphs

A **coordinate grid** can be used to describe the position of points.

The grid uses a horizontal **x-axis** and a vertical **y-axis**.

The point where these **axes** cross is called the **origin**.

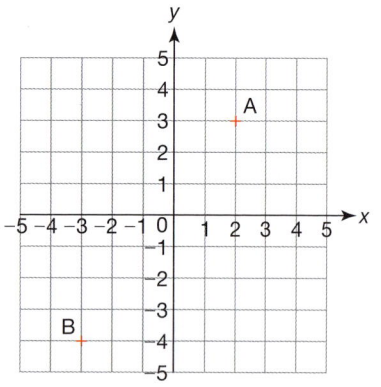

- A point is described by its horizontal and vertical distance from the origin, (x, y).

Coordinates can also represent inputs and outputs for a function.
A set of these points is used to plot the graph for a function.

▲ A has coordinates (2, 3).
   B has coordinates (−3, −4).

p.116

EXAMPLE

a Complete each table of values for the given function.

$y = 2x + 1$

x	0	1	2	3	4
y				7	

b On the same set of axes plot the graphs of
$y = 2x + 1$, $y = 5 − x$ and $y = 6$.

$y = 5 − x$

x	3	−1	0	2	7
y					

c Add the graph of $x = 4$ to your diagram.

$y = 6$

x	−2	0	2	4	6
y					

a Work out the value of y for each value of x.

$y = 2x + 1$

$0 \rightarrow 2 \times 0 + 1 = 1$ $1 \rightarrow 2 \times 1 + 1 = 3$

$2 \rightarrow 2 \times 2 + 1 = 5$ $4 \rightarrow 2 \times 4 + 1 = 9$

x	0	1	2	3	4
y	1	3	5	7	9

b $y = 5 − x$

x	3	−1	0	2	7
y	2	6	5	3	-2

$y = 6$

x	−2	0	2	4	6
y	6	6	6	6	6

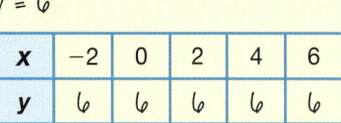

Beware! The x-values might not be in order. Take care with negative numbers too.

Draw axes that allow you to plot all your points.
Draw a line through each set of points.

c Just as $y = 6$ is a horizontal line joining all the points with y-coordinate equal to 6,
$x = 4$ is a vertical line joining all the points with x-coordinate equal to 4.

Exercise 14.1S

1 Write the coordinates of the points A to H.

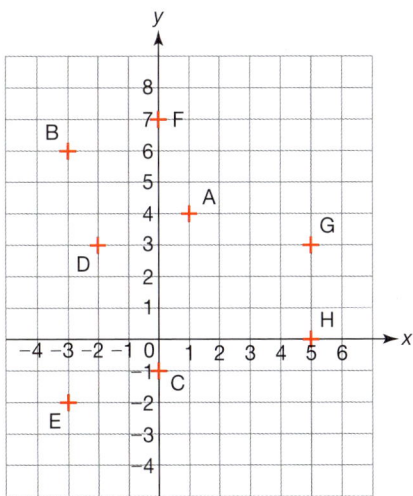

2 Look at the coordinate grid in question **1**.

   a What is the name given to the line through C and F?

   b Abbi says, 'The horizontal line through the point H is called the x-axis'. Do you agree? Say why.

3 Copy and complete the tables of values for each of these functions.

   a $y = x + 3$

x	1	2	3	4	5
y					

   b $y = 3$

x	1	2	3	4	5
y					

   c $y = 2x + 3$

x	3	1	0	2	4
y					

   d $y = 3x - 2$

x	−1	0	1	2	3
y					

   e $y = 8 - x$

x	2	0	5	4	1
y					

4 For each part of question **3**

   i construct a coordinate grid that can be used to plot all the points found

   ii plot the graph of the function.

5 For each equation

   i create a table of values

   ii construct a coordinate grid that can be used to plot all the points found

   iii plot the graph of the function.

   a $y = 2 - x$ **b** $y = 9 - x$

   c $y = 8 - 2x$ **d** $y = 4 - \frac{1}{2}x$

6 Plot the graphs of these functions on the same coordinate grid.
Use axes from −10 to 10 in both directions.

   a $y = 2x$ **b** $y = x + 6$

   c $y = 2x + 6$ **d** $y = \frac{1}{2}x$

   Label each of your lines clearly.

*7 Plot the graph of the function $y = -2x - 5$.

8 Write down the equations of these horizontal and vertical lines.

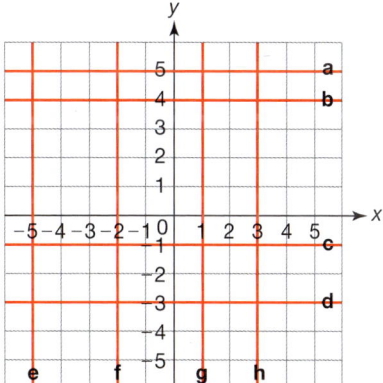

9 On a coordinate grid plot these lines.

   a $y = 3$ **b** $y = -2$

   c $x = 1$ **d** $x = -3$

   e $y = 0$ **f** $x = 0$

10 All of these points except one lie on the same straight line.

   $(6, -2) \ (0, 10) \ (3, 4) \ (4, 3) \ (-2, 14)$

   Which point does not lie on the line?

Q 1093, 1394, 1395, 1396   SEARCH

14.1 Drawing straight-line graphs

RECAP

- A coordinate grid can be used to describe the position of points.
- Coordinates can also represent inputs and outputs for a function. A set of these points will help plot the graph for a function.

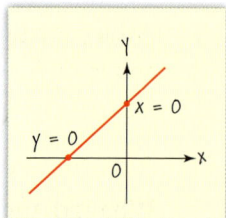

The equation of a straight line may *not* be written in the form $y = $ something.

In this case find the points where the line crosses the two axes and join these points.

EXAMPLE p.116

Plot the graph of **a** $3x + y = 6$ **b** $x - 2y = 4$ **c** $2x + 5y = 10$

a $x = 0 \Rightarrow y = 6$ $\quad\quad y = 0 \Rightarrow 3x = 6$
$\quad\quad\quad\quad\quad\quad\quad\quad\quad\quad \Rightarrow x = 2$
The points (0, 6) and (2, 0) lie on the line.
Check (1, 3) $3 \times 1 + 3 = 6$ ✓ $\quad\quad 3x + y = 6$

b $x = 0 \Rightarrow -2y = 4$ $\quad\quad y = 0 \Rightarrow x = 4$
$\quad\quad\quad \Rightarrow y = -2$
The points (0, –2) and (4, 0) lie on the line.
Check (2, –1) $2 - 2 \times -1 = 2 - -2 = 4$ ✓ $\quad\quad x - 2y = 4$

c $x = 0 \Rightarrow 5y = 10$ $\quad\quad y = 0 \Rightarrow 2x = 10$
$\quad\quad\quad \Rightarrow y = 2$ $\quad\quad\quad\quad\quad\quad \Rightarrow x = 5$
The points (0, 2) and (5, 0) lie on the line.
Check (2.5, 1) $2 \times 2.5 + 5 \times 1 = 5 + 5 = 10$ ✓ $\quad\quad 2x + 5y = 10$

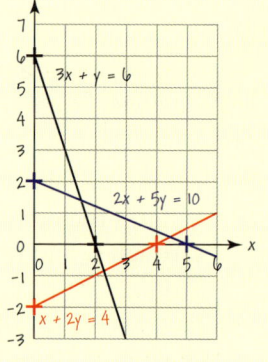

HOW TO

To solve a real-life problem using a straight line graph
① Identify the variables and write down the equation connecting them.
② Calculate points and draw a graph.
③ Use your graph to answer the question.

EXAMPLE

You are given two options for your weekly pocket money.
   A   £3.00 + 40p per chore   **B**   £5.00 + 30p per chore
Which option should you chose if
   a   you do 15 chores per week   **b**   you want to earn £13 per week?
   c   Is there a number of chores where it makes no difference which option you chose?

Let x = number of chores ① Choose variables.   ② Draw the graphs.
   y = pocket money (pence)

A $\quad y = 300 + 40x$ $\quad\quad$ Set up equations.
B $\quad y = 500 + 30x$

a A £9.00 B £9.50 $\quad$ ③ Read up and then across
   Choose option B $\quad\quad\quad$ Remember to ATQ.
b A 25 chores
   B 27 chores $\quad\quad$ ③ Read across and then down.
   Choose option A
c Yes. If you do 20 chores you get £11.00 with
   both options. $\quad\quad$ This is where the lines cross.

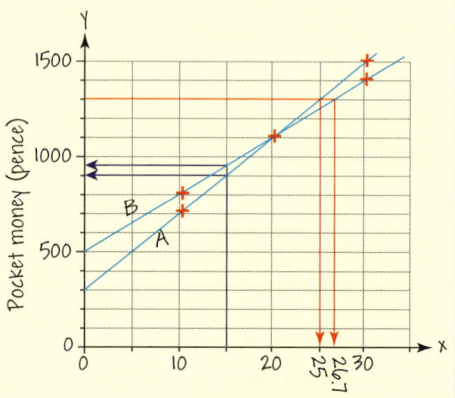

Algebra Graphs 1

Exercise 14.1A

1 a Draw a square coordinate grid and plot and label these three points.

A (1, 4)   B (5, 4)   C (−1, 2)

b A fourth point D forms a parallelogram. What are the coordinates of D?

*c How many other solutions can you find?

2 Plot the graphs of these functions.

a	$x + y = 8$	**b**	$x - y = 2$
c	$x - y = 4$	**d**	$x + y = 3$
e	$x + y = -5$	**f**	$x - y = -4$

For each of your graphs write the coordinates of three more points on the line. Show how each point fits the function.

3 Plot the graphs of these functions.

a	$2x + y = 1$	**b**	$3x - y = 4$
c	$5x - y = 2$	**d**	$2x + y = -2$

For each of your graphs write the coordinates of three more points on the line. Show how each point fits the function.

4 Plot the graphs of these functions.

a	$5x + 2y = 10$	**b**	$3x + 4y = 12$
c	$7x + 2y = 14$	**d**	$3x - 2y = 6$
e	$5x - 4y = 20$	**f**	$3x + 2y = -6$
g	$3x - 4y = -12$	**h**	$2x + 4y = -5$
i	$6x - 2y = -9$	**j**	$2x - 5y = 8$

5 Find the coordinates of the point where each of these pairs of lines cross.

a $y = 3x + 3$ and $x + y = 3$

b $y = 3x + 1$ and $x + y = -3$

c $y = 2x - 6$ and $2x + y = 2$

d $y = 3x + 5$ and $x + 2y = 3$

> Use graphing software to verify your solutions to questions **4** and **5**.

6 The point (a, b), where a and b are integers, satisfies these conditions.

Right of $x = 2$

Above $y = x - 2$ and $y = 1$

Below $x + y = 6$

Find the point.

7 The graphs of six functions are plotted.

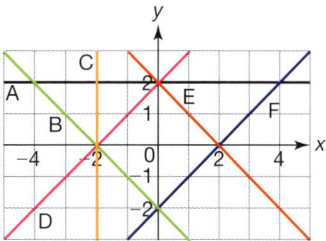

Match each function to a graph.

a	$y = 2$	**b**	$x + y = 2$
c	$x = -2$	**d**	$y = x + 2$
e	$x + y = -2$	**f**	$x - y = 2$

*8 The graph shows two straight lines. Find the equation of the red line.

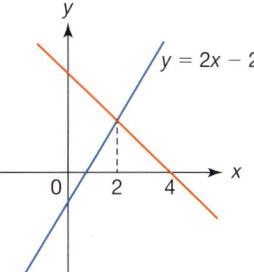

9 Two companies have the following charges for organising a party.

	Venue hire	Cost per guest
Bill's Parties	£100	£5
Ben's Events	£40	£8

a It there are 30 guests, then Bill's Parties charges £100 + 30 × £5 = £250. How much would Ben's Events charge?

b Draw a graph to show the cost of a party against the number of guests for the two companies. Allow up to 50 guests.

c Use your graph to find the cost of a party for 25 people for the two companies.

d If you had £160 to spend, how many guests could you have using the two companies?

e If you were planning a party for 25 guests, which company would you use? Give your reasons.

Q 1093, 1394, 1395, 1396 SEARCH

14.2 Equation of a straight line

The **gradient** of a straight line is its steepness.

It is found by choosing and comparing two points on the line. For every unit right, the gradient is the number of units up.

The **y-intercept** of a line is the value of y where the line crosses the y-axis.

● The equation $y = mx + c$ describes a straight line with gradient m and y-intercept of c.

The gradient of the line is 2. The y-intercept of the line is 4. When written using algebra, the function is $y = 2x + 4$.

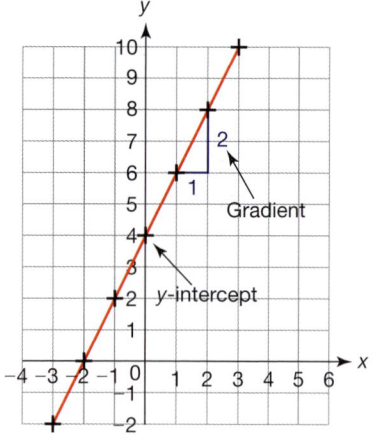

▲ $y = 2x + 4$
gradient = 2, y-intercept = 4

EXAMPLE

For each graph

i state the y-intercept   **ii** find the gradient   **iii** give the equation of the line.

a 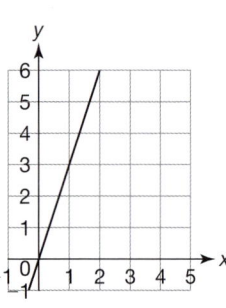   **b** 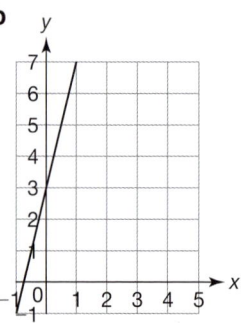   **c** 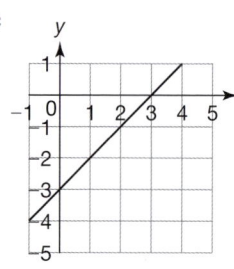   **d**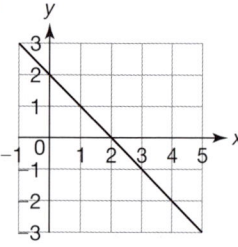

---

a 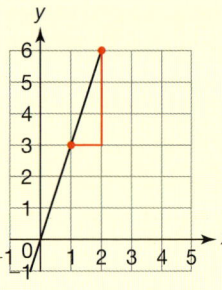   **b** 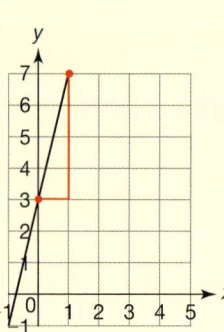   **c** 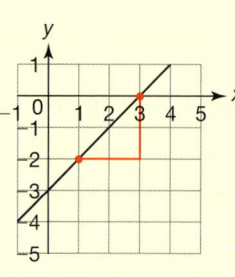   **d**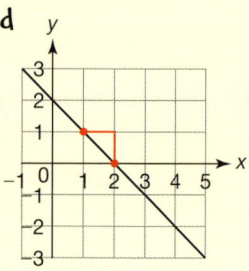

i   Find where the line crosses the y-axis.

  y-intercept = 0   y-intercept = 3   y-intercept = -3   y-intercept = 2

ii   Choose a pair of points to compare in each case

  Gradient = 3   Gradient = 4   Gradient = 1   Gradient = -1

                                                              The graph slopes down
                                                              so the gradient is negative.

iii   Write the equation as $y = mx + c$, that is, $y = $ 'gradient' $\times x + $ 'y-intercept'.

  $y = 3x$ $y = 4x + 3$ $y = x - 3$ $y = -x + 2$

Exercise 14.2S

1 Write the *y*-intercepts of these lines.

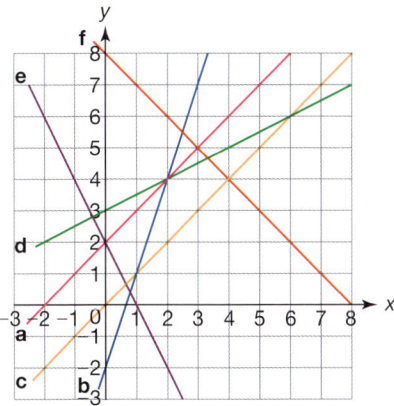

2 For each line in question **1** find its gradient.

3 For each line in question **1** give its equation.

4 Find the equations of these lines.

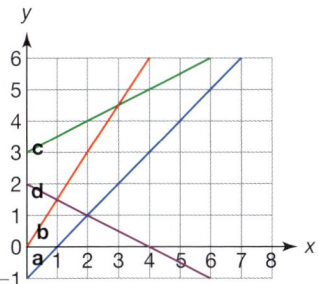

5 Find the equations of these lines.

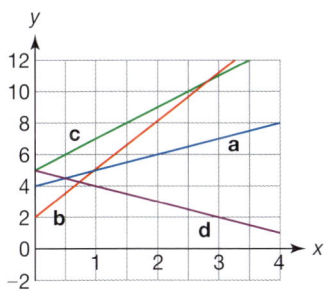

6 For each of these functions write the

   i   *y*-intercept   **ii**   gradient.

   a $y = 9x + 7$ **b** $y = \frac{1}{2}x - 3$

   c $y = x + 23$ **d** $y = 7x - 2$

   e $y = 1 + 2x$ **f** $y = 10 - x$

7 Write the equations of these lines.

   a   Gradient $= 3$ *y*-intercept $= 1$

   b   Gradient $= 1$ *y*-intercept $= 7$

   c   Gradient $= -2$ *y*-intercept $= 0$

   d   Gradient $= \frac{1}{2}$ *y*-intercept $= -2$

   e   Gradient $= -\frac{1}{3}$ *y*-intercept $= -3$

   f   Gradient $= 0$ *y*-intercept $= 4$

8 For each function

   i   Create a table of values and draw a graph.

   From the graph find the

   ii   *y*-intercept   **iii**   gradient.

   a $y = 1 + 2x$ **b** $y = 3x + 2$

   c $y = 4 - x$ **d** $y = -5 + 2x$

9 How do your answers to question **8** parts **ii** and **iii** compare to the equation of the function?

10 Draw the graphs of these functions *without* working out a table of values.

   a $y = 2 + x$ **b** $y = 1 + 2x$

   c $y = 3x$ **d** $y = 4 - x$

11 Imagine the graphs of these functions.

   A $y = 3x + 1$ **B** $y = 2x - 1$

   C $y = 3x - 7$ **D** $y = 7x + 3$

   E $y = 3x + 7$ **F** $y = 2x - 3$

   Which lines are parallel? How do you know?

12 Match each graph with its equation.

   a $y = 2$ **b** $y = 2x - 2$

   c $y = -2x$ **d** $y = 2x + 2$

   e $y = 2x$ **f** $y = -2x + 2$

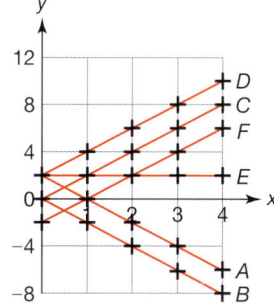

13 **a**   Write down four functions whose graphs show parallel lines.

   b   Write four functions whose graphs cross the *y*-axis at the point (0, 6).

   *****c**   Write two functions whose lines cross at right-angles.

1153, 1312, 1314   SEARCH

14.2 Equation of a straight line

RECAP

- The equation of a straight line is $y = mx + c$.
  - c is the y-intercept, where the line crosses the y-axis.
  - m is the gradient, the steepness of the line.
- The gradient tells you how many units up, or down if it is negative, the line rises for each unit right.
- Parallel lines have the same gradient.

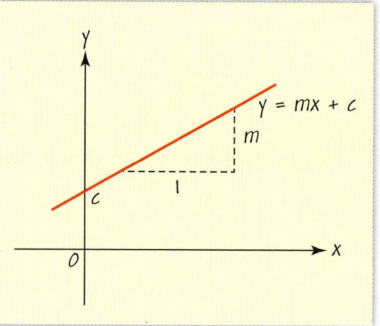

HOW TO

To find the equation of a line from given information
1. Identify the gradient, m.
2. Identify the y-intercept, c.
3. Use $y = mx + c$ where m = gradient and c = y-intercept.

EXAMPLE

Use the information given to find the equation of each line.

a The line is parallel to $y = 7x - 3$ and passes through the point (0, 2).

b The line has gradient -2 and passes through the point (3, 5).

a $m = 7$ ① $y = 7x - 3$ has gradient 7.

   $c = 2$ ② (0, 2) is where the line crosses the y-axis.

   $y = 7x + 2$ ③ $y = mx + c$

b $m = -2$ ① Given, so $y = -2x + c$.

   ② To find c, substitute $x = 3$ and $y = 5$ into $y = -2x + c$.

   $5 = -2 \times 3 + c = -6 + c \Rightarrow c = 5 + 6 = 11$

   $y = -2x + 11$ ③ $y = mx + c$

You can find the gradient of a line given any two points on the line.

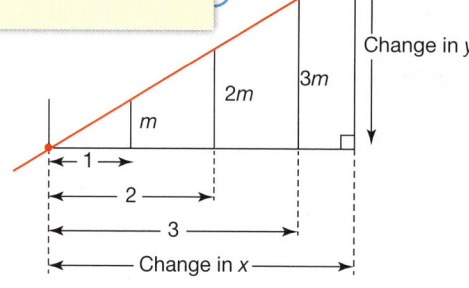

p.392
- Gradient $= \dfrac{\text{Change in } y}{\text{Change in } x}$

EXAMPLE

Find the equation of the line that passes through the points (2, 1) and (6, -1).

$\dfrac{\text{change in } y}{\text{change in } x} = \dfrac{-1 - 1}{6 - 2} = -\dfrac{2}{4} = -\dfrac{1}{2}$ ① Use the gradient formula.

$m = -\dfrac{1}{2}$ Negative, since line slopes down.

$1 = -\dfrac{1}{2} \times 2 + c = -1 + c$ ② Substitute $x = 2$ and $y = 1$

$c = 1 + 1 = 2$ into $y = -\dfrac{1}{2}x + c$.

$y = -\dfrac{1}{2}x + 2$ ③ $y = mx + c$ You can also read the y-intercept off your graph.

It often helps to draw a graph.

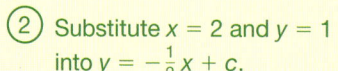

Algebra Graphs 1

Exercise 14.2A

1 Use the information given to find the equation of the line.

  a Gradient is 6. Passes through $(0, -9)$.

  b Gradient is -10. Passes through $(0, 18)$.

  c Gradient is $\frac{1}{4}$. Passes through $(0, 1)$.

  d Parallel to $y = 2x + 6$. y-intercept is -7.

  e Parallel to $y = 10 - 2x$. Passes through $(0, 1)$.

  f The same y-intercept as $y = 7x - 7$. Parallel to $y = 5x + 1$.

  g Parallel to $2y = 6x + 3$. Passes through $(0, 8)$.

2 Work out the equations of these lines.

  a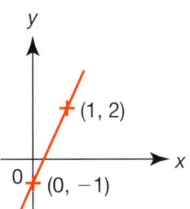

  b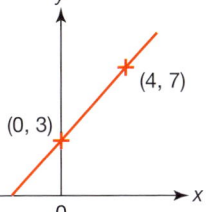

  c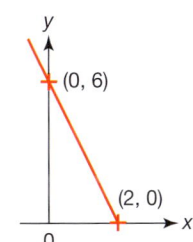

  d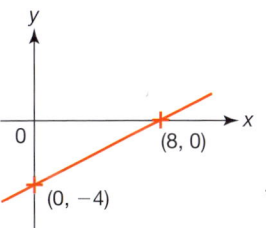

3 Find the equations of the lines that pass through the given pairs of points.

  a $(1, 2)$ and $(3, 4)$ **b** $(2, 8)$ and $(5, 17)$

  c $(2, 0)$ and $(4, 1)$ **d** $(1, 4)$ and $(2, 2)$

  e $(-1, 5)$ and $(4, -5)$

  f $(6, -3)$ and $(12, -5)$

  Use graph plotting software for question **4**.

4 **a** Plot the graph of $y = x + a$.

  b Use the constant controller to vary the value of a. What do you notice?

  c Plot the graph of $y = ax$

  d Use the constant controller to vary the value of a. What do you notice?

5 The graph has had the labels removed from the axes.

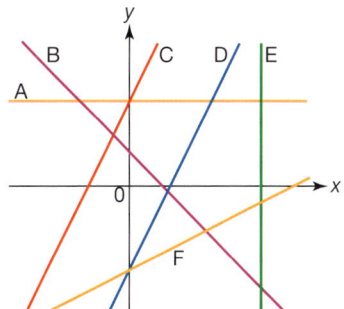

Match each function to its graph.

  a $x = 8$ **b** $y = 2x - 5$

  c $y = \frac{1}{2}x - 5$ **d** $y = 2 - x$

  e $y = 5$ **f** $y = 2x + 5$

6 *ABCD* is a parallelogram. The equation of the line through *A* and *B* is $y = 3x - 7$. *D* is the point $(7, 5)$. Find the equation of the line through *C* and *D*.

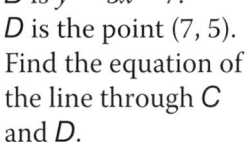

7 A line passes through the points $(-2, 16)$ and $(2, 8)$. The point $(p, -4)$ also lies on this line. Find the value of p.

*8 A line passes through the points $(3, 1)$ and $(6, 3)$. The point $(36, q)$ also lies on this line.

  a Find the value of q.

  b Write the equation of the line.

9 **a** Find the equation of this line in the form $y = mx + c$.

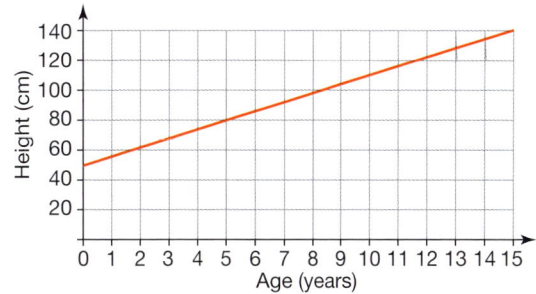

  b How would you interpret the line's

    i gradient   **ii** y-intercept?

14.3 Distance–time graphs

A distance–time graph shows information about a journey.

Time is always displayed on the horizontal axis and distance on the vertical axis.

EXAMPLE

Jason went on a cycle ride.
The distance–time graph shows information about his journey.

a How far did Jason travel in the first 30 mins?

b At what time did Jason reach 30 km from the start?

c How far did Jason travel in total?

d What happened between 08:30 and 08:45?

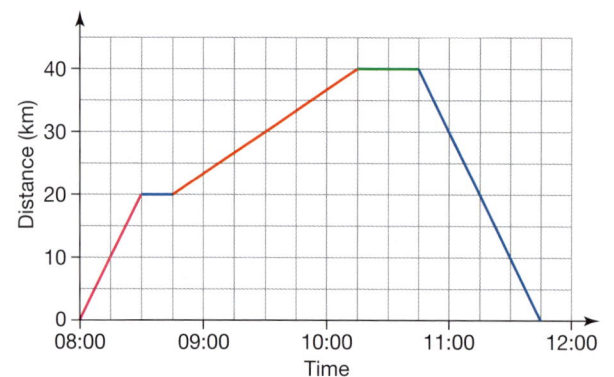

a 20 km   Read up from 08:30 and then across.   **b** 09:30   Read across from 30 and then down.

c 2 × 40 = 80 km   The graph shows both an outward 40 km *and* a homeward 40 km journey.

d Jason stopped   The graph is horizontal so the distance does not change.

The gradient of a distance–time graph tells you about the **speed**.

p.192

- Speed = $\dfrac{\text{Distance travelled}}{\text{Time taken}}$
- Average speed = $\dfrac{\text{Total distance}}{\text{Total time}}$

EXAMPLE

a For Jason's journey calculate the speed, in kilometres per hour, during these times.

  i 08:00 to 08:30   **ii** 08:30 to 08:45   **iii** 08:45 to 10:15

  iv 10:15 to 10:45   **v** 10:45 to 11:45

b Show this information on a speed–time graph.

p.376

c What is Jason's average speed between 08:00 and 10:15?

> In reality Jason's speed wouldn't suddenly jump from 0 to 40 km/hr.
> He would **accelerate** and there would be a smooth curve.

a **i** Speed = $\dfrac{20\,\text{km}}{30\,\text{mins}} = \dfrac{20\,\text{km}}{\frac{1}{2}\,\text{hr}} = 40\,\text{km/h}$ **b**

  ii Speed = 0   Distance travelled = 0

  iii Speed = $\dfrac{20\,\text{km}}{1\frac{1}{2}\,\text{hr}} = \dfrac{40\,\text{km}}{3\,\text{hr}} = 13.3\,\text{km/h}$

  iv Speed = 0

  v Speed = $\dfrac{40\,\text{km}}{1\,\text{hr}} = 40\,\text{km/h}$

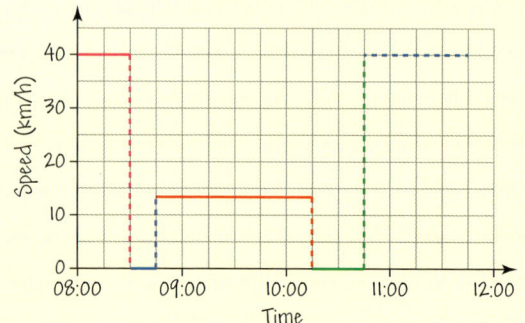

  c Average speed = $\dfrac{40\,\text{km}}{2\frac{1}{4}\,\text{hr}} = 40 \div 2.25 = 17.8\,\text{km/h}$

Exercise 14.3S

1 The distance–time graph shows information about Lisa's coach journey.

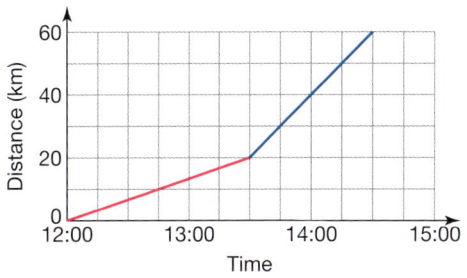

a How far does she travel between
 i 12:00 and 13:30
 ii 13:30 and 14:30?
b How long does it take to travel
 i 10 km from the start
 ii 50 km from the start?

2 Tamera is riding a bike. Information about her journey is shown in the graph.

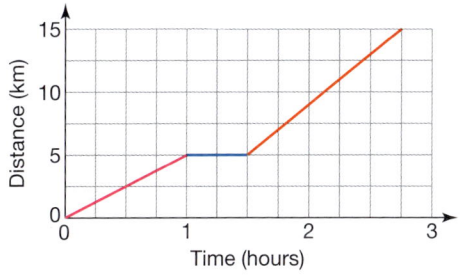

a Tamera's journey starts at 8 a.m. At what time does her journey finish?
b What happens after Tamera has cycled for one hour?
c At what time does Tamera start cycling for the second time?

3 Mark sets off from home in his car at 2 pm. He stops to get petrol, and then continues on his journey. Mark returns home later in the afternoon. The distance–time graph shows more information about this journey.

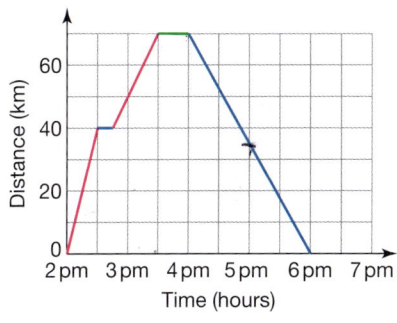

3 a How far was the petrol station from Mark's home?
b For how long did Mark wait at the petrol station?
c After stopping for petrol, how long did Mark take to drive to his destination?
d At what time did Mark set off on his journey home?
e What was Mark's speed during the first section of his journey?
f Between what times was Mark travelling at his greatest speed? How do you know?
g How far did Mark travel
 i between 4 pm and 5 pm
 ii between 3 pm and 5 pm
 iii in total?

4 Work out the overall average speed for each of these journeys.

a
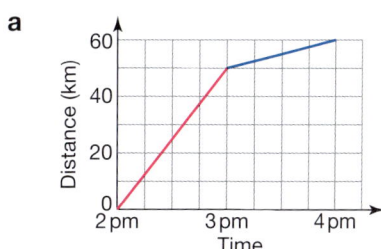

b
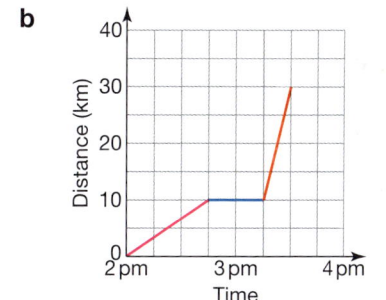

*c
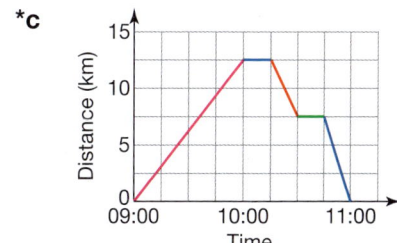

5 Draw speed–time graphs for the three journeys shown in question **4**.

6 Bertha flies an aeroplane in a straight line at 200 km/h for 3 hours. Draw a distance–time graph to show this.

14.3 Distance–time graphs

RECAP

- Information about a journey can be shown on a distance–time or speed–time graph.
- The gradient of a line on a distance–time graph is the speed.

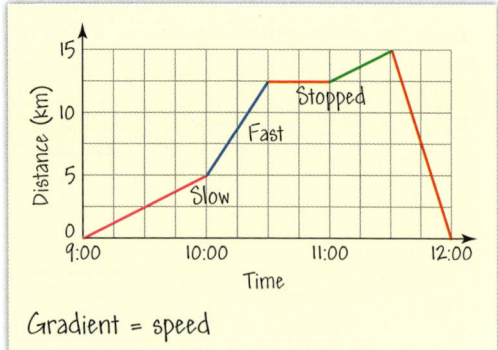

Gradient = speed

HOW TO

p.192

To solve kinematics problems
1. RTQ – decide what to do.
2. Use the formula speed = $\dfrac{\text{distance}}{\text{time}}$
3. ATQ – remember to include any units.

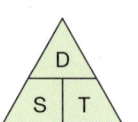

EXAMPLE

a Linda goes on a bicycle ride. Draw a distance–time graph for her journey.

b What was Linda's average speed on her return journey?

c If Linda stopped for a 30 min rest on her return journey, how would this affect your answer to part **b**?

- Linda started her journey at 11:00.
- She travelled 15 km in the first 30 mins.
- She travelled at a constant speed of 20 km/hr for the next 2 hours.
- She then stopped for 1 hour.
- She returned home at 17:00.

a First point is (11:00, 0 km) ①

   Second point is (11:30, 15 km)

   $20 \text{ km/hr} = \dfrac{\text{distance}}{2 \text{ hr}}$ ②

   distance = 2 × 20 = 40 km

   Third point is (13:30, 55 km)
   Fourth point is (14:30, 55 km)
   Last point is (17:00, 0 km)

b Average speed = $\dfrac{55 \text{ km}}{2\frac{1}{2} \text{ hr}}$ = 22 km/hr ②

c The answer does not change. The distance travelled and time taken are both the same.

③ Axes need to include 11:00 to 17:00 and 0 to 55 km.

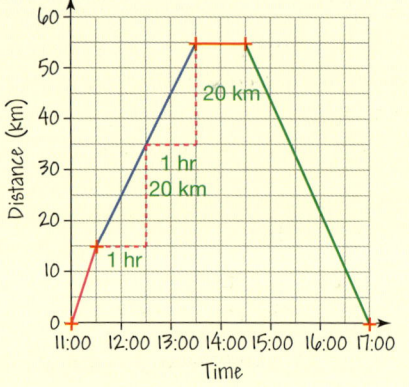

The gradient of a speed–time graph is the **acceleration**. It tells you how the speed is changing.

- Acceleration = $\dfrac{\text{Change in speed}}{\text{Time taken}}$

EXAMPLE

The graph shows the speed–time graph for a train leaving a station.

a The gradient of a speed–time graph is the acceleration. What is the acceleration in m/s²?

b If the train accelerates, at the same rate, for one minute what will be its final speed?

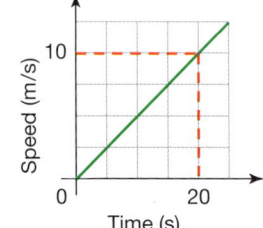

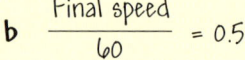

a Acceleration = $\dfrac{10 \text{ m/s}}{20 \text{ s}}$ = 0.5 m/s²

b $\dfrac{\text{Final speed}}{60}$ = 0.5

   Final speed = 0.5 × 60 = 30 m/s

Algebra Graphs 1

Exercise 14.3A

1 James walks 15 km in 3 hours. He rests for 30 minutes and then walks home at a constant speed of 6 km/h. Draw a distance–time graph to show his journey.

2 Jodie visited her gran. Use the information given to draw a distance–time graph for her journey.

- Jodie sets out at 14:00.
- She travels 400 m in the first 5 mins before realising she has forgotten a present.
- She goes straight back home, travelling at the same speed.
- Jodie quickly picks up the present and takes 25 mins to travel the 2 km to her gran's house.
- She spends 30 mins at her grans.
- She goes straight back home at a speed of 0.833 m/s.

3 A train leaves a station at 11 a.m. and travels at a constant speed of 80 km/h. A second train leaves the same station 30 minutes later. It travels at 140 km/h.

a Construct a distance-time graph to show this.

b At what time are the two trains the same distance from the starting station?

4 The graph shows information about a race between Asif and Sian.

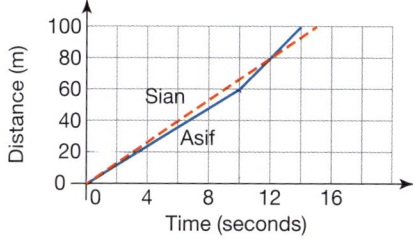

a Who was quickest at the start of the race?

b What happened 12 seconds into the race?

c Who won the race?

4 d i At what speed was Asif running for the first 10 seconds of the race?

ii If he had continued at this speed for the whole race who would have won and by how many seconds?

5 Explain why each of these distance-time graphs are impossible.

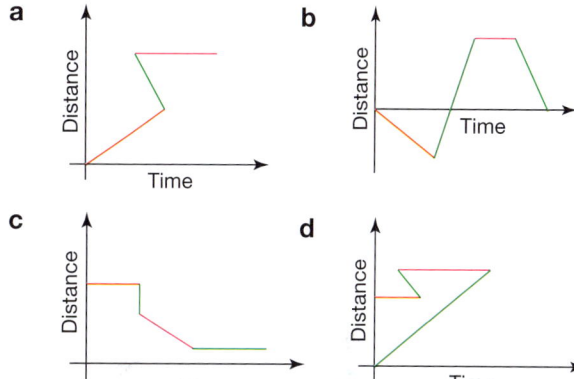

6 Match each distance–time graph to its matching speed–time graph.

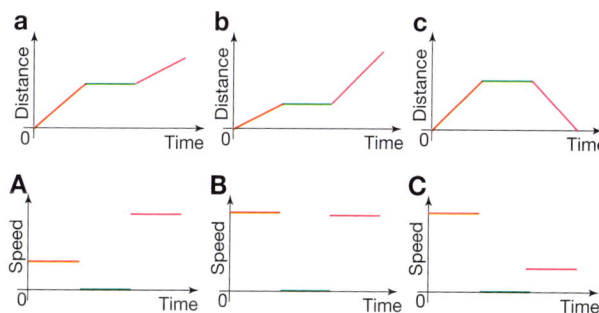

7 A rocket accelerates in two stages as shown in the speed–time graph.

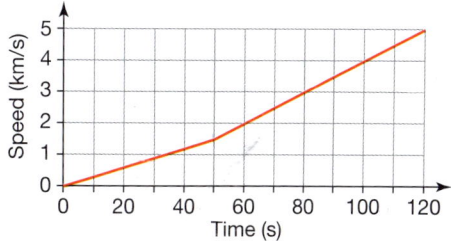

a Calculate the acceleration, in km/s², for

i stage 1   **ii** stage 2.

b Calculate the average acceleration for the whole journey.

Summary

Checkout
You should now be able to...

Test it
Questions

✔ Work with coordinates in all four quadrants.	**1**
✔ Plot straight-line graphs including diagonal, vertical and horizontal lines.	**2 – 4**
✔ Identify gradients and intercepts of straight lines graphically and algebraically.	**5, 6**
✔ Use the form $y = mx + c$ to identify parallel lines.	**7**
✔ Use one point and the gradient of the line to finds its equation.	**8**
✔ Use two points to find the equation of a line.	**9**
✔ Interpret the gradient of a straight line graph as a rate of change.	**10**
✔ Plot and interpret graphs involving distance, speed and acceleration.	**10**

Language   Meaning   Example

Language	Meaning	Example
Coordinate grid	An **origin** and a set of **x** and **y** **axes** that allow you to specify a point.	$(1, -3)$ is a point one unit right and three units down from the origin $(0, 0)$.
Gradient	A measure of the slope of a line on a graph found by dividing the change in y by the change in x.	Gradient, $m = \frac{2}{1} = 2$
y-intercept	The point at which a straight line graph crosses the y-axis	
y = mx + c	The standard way to write the equation of a straight line. m = gradient, c = y-intercept	y-intercept, $c = -2$ $y = 2x - 2$
Distance–time graph	Shows the relationship between distance moved by an object and time.	Speed $= \frac{12}{6} = 2$ m/s
Speed	How for an object travels in a unit of time. It is given by the gradient of a line on a distance–time graph.	
Acceleration	How much an object's speed changes in a unit of time.	0 to 10 m/s in 2 s is an acceleration of $10 \div 2 = 5$ m/s².

Review

1 Write down the coordinates of points A, B, C, D and E.

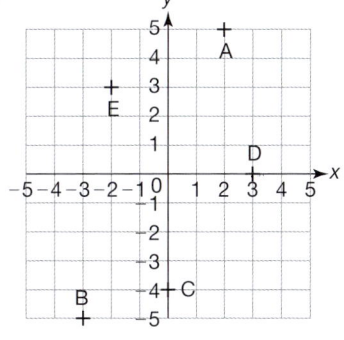

2 Copy and complete this table of values for each equation.

   a $y = 3x$ b $y = 2x + 7$

x	−2	−1	0	1	2
y					

3 Draw appropriate axes and plot the lines for each of the equations in question **2**.

4 Copy the grid from question **1** without the points. Draw these graphs on the same grid.

   a $y = 3$ b $x = -2$

5 a What is the gradient of the line shown?

    b What is the equation of the line?

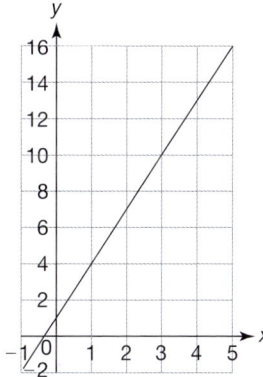

6 Complete a table of values and plot the graphs of these equations for values of x between −1 and 4. For each line state the gradient and y-intercept.

   a $y = 3x - 1$ b $y = 20 - 5x$

7 Write down the equation of any line that is parallel to the line with equation $y = 3x + 4$.

8 What is the equation of the line that has gradient 7 and crosses through the y-axis at the point $(0, -4)$?

9 What is the equation of the line that passes through the points $(0, 2)$ and $(3, 14)$?

10 The distance–time graph shows a bus journey from the depot to the station and back again.

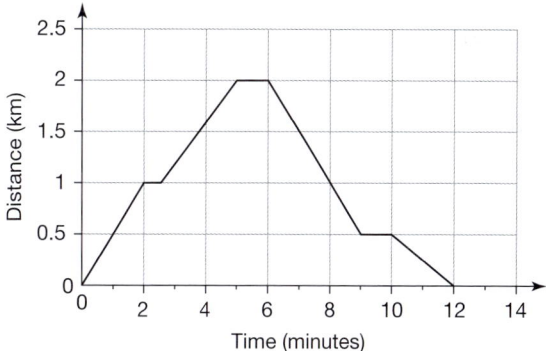

   a What is the total distance travelled by the bus?

   b What does the gradient of the curve represent in this context?

   c What is the average speed of the bus over the whole 12 minute journey? Give your answer in km/h.

What next?

Score			
	0 – 4		Your knowledge of this topic is still developing. To improve look at MyMaths: 1093, 1153, 1312, 1314, 1322, 1323, 1394, 1395, 1396
	5 – 8		You are gaining a secure knowledge of this topic. To improve your fluency look at InvisiPens: 14Sa – k
	9 – 10		You have mastered these skills. Well done you are ready to progress! To develop your problem solving skills look at InvisiPens: 14Aa – f

Assessment 14

1 **a** Draw a grid with x-axis from -2 to 3 and y-axis from -2 to 3.

Plot these sets of points on a copy of this grid.

$(-1, 2)$ $(2, 2)$ $(2, 1)$ $(1, 1)$

$(1, -1)$ $(0, -1)$ $(0, 1)$ $(-1, 1)$ [4]

b Join the set of points in order.
What letter have you drawn?   [1]

2 **a** Complete this table for the
four graphs.   [8]

b Draw a grid with x-axis from -1 to 4.
Draw all four graphs on the
same axes. Label each graph.   [8]

c Write down the coordinates of the
points of intersection of these graphs.

x	−1	0	1	2	3	4
A $y = 2x + 1$						
B $y = 7 - x$						
C $y = \frac{1}{2}x + 1$						
D $2x + y = 8$						

    i A and B   **ii** A and C   **iii** B and C   **iv** B and D   [4]

3 Draw a grid with the x- and y-axes from 0 to 6.
Draw these graphs on the same grid.

    a $x = 5$ **b** $x = 2$ **c** $y = 1$ [3]

d Plot the points $(0, 6)$ and $(6, 0)$ on the grid. Join the points with a straight line.
What is the equation of the line?   [2]

e Find the area of the triangle enclosed by the four lines?   [2]

4 Carli says that $(-2, -1)$, $(0, 2)$ and $(3, 9)$ are all points on the line $y = 2x + 3$.
Is Carli correct? Give reasons for your answer.   [3]

5 **a** Does the point $(-4, 5)$ lie on the line $3y - 4x = 30$?   [1]

b Does the point $(7, 6)$ lie on the line $y = 1 - x$ or $y = x - 1$?   [1]

6 Work out the gradient and y-intercept for each of the following straight lines.

    a $y = 2x + 7$ [2]   **b** $y = 9 + 4x$ [2]   **c** $y = 6x - 11$ [2]

    d $y = 12 - 4x$ [2]   **e** $4x - 7y = 14$ [3]   **f** $15x + 14y = 35$ [3]

7 The gradient and y-intercept of four lines are
shown in the table.
Find the equations of each line in the
form $y = mx + c$.   [4]

	gradient	y-intercept
a	1	1
b	−1	1
c	2	6
d	−4	13

8 **a** The sketch shows Demelza's journey on Monday morning.

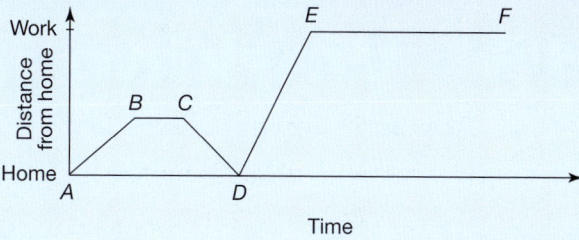

Match each stage in Demelza's journey, A to B, B to C, C to D, D to E and E to F, with the correct description.

    i   Demelza drives from her home to work.

    ii Demelza leaves her car in the carpark at her work.

    iii Demelza returns home to pick up her daughter's maths homework.

    iv Demelza stops to answer a phone call.

    v Demelza starts her journey from home to work.   [4]

  b On which section of her journey is she travelling fastest? Give reasons for your answer.   [2]

9 Jess goes for a walk.

- The park is 600 m away and the walk takes 10 minutes.

- Jess stays in the park for an hour.

- Jess walks to a shop which is a further 100 m away. The walk to the shop takes 5 minutes.

- Jess spends 10 minutes in the shop.

- Jess walks home. The walk home takes 15 minutes.

- Jess walks at a constant speed for each stage of her journey.

Draw a distance–time graph of Jess's journey.   [10]

10 Calculate the gradients of the straight lines which pass through the following pairs of points.

  a $(1, 5)$ and $(5, 9)$ [2]   **b** $(6, 7)$ and $(8, -9)$ [2]

  c $(-3, 5)$ and $(-4, 6)$ [2]   **d** $(-11, 0)$ and $(5, -8)$ [2]

11 Find the equation of each line in the form $y = mx + c$.

  a Gradient $= -5$, passing through $(0, -61)$ [1]

  b Gradient $= 3$, passing through $(1, 2\frac{1}{2})$ [2]

  c Gradient $= \frac{1}{3}$, passing through $(3, -1)$ [2]

  d Gradient $= \frac{1}{4}$, passing through $(0, -\frac{3}{4})$ [1]

12 Find the equation of each line in the form $y = mx + c$.

  a Line parallel to $y - 7x - 9 = 0$ that intercepts the y-axis at $(0, 5)$.   [2]

  b Line parallel to $3x + y - 77 = 0$ that intercepts the y-axis at $(0, -7.25)$.   [2]

  c Line parallel to $2x + 6y = 15$ that intercepts the y-axis at $(0, 2\frac{2}{3})$.   [2]

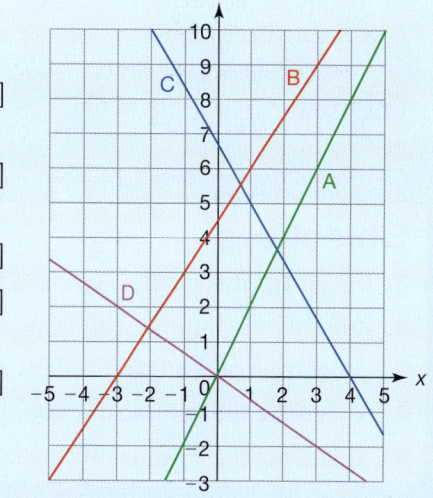

13 **a** State the gradient of each line (A–D) on the grid.   [4]

  b Match each equation to one of the lines A, B, C and D drawn on the graph.   [3]

    i $5x + 3y = 20$ **ii** $2y - 3x = 9$

    iii $2x - y = 0$ **iv** $2x + 3y = 0$

15 Working in 3D

Introduction

In the Cave of Crystal Giants in Mexico, vast crystals of selenite grow to lengths of up to 11 metres with masses up to 55 tons – the largest crystals to have yet been discovered. The cave was only discovered because there was a mine nearby – it is highly likely that there are even more awesome geological structures lying somewhere undiscovered beneath our feet!

What's the point?

We live in a three-dimensional world, and 3D geometry allows us to describe the wonderful things that we can see in the natural world, as well as helping us to devise increasingly sophisticated structures in the manmade world.

Objectives

By the end of this chapter you will have learned how to …

- Identify the numbers of faces, edges and vertices of 3D shapes.
- Construct and interpret plans and elevations of 3D shapes.
- Calculate the volume of cuboids, cylinders and other prisms.
- Calculate the volume and surface area of spheres, pyramids, cones and composite solids.

Check in

1 Calculate the area of each shape.

a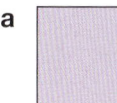
5 cm

8 cm

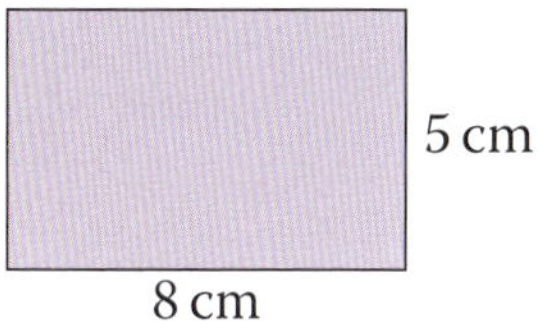

b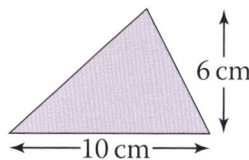
6 cm

10 cm

2 Give the mathematical name for these distances in a circle.

a 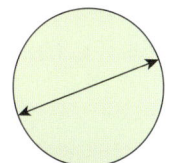　　b 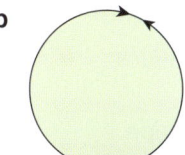　　c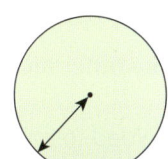

Chapter investigation

A cylindrical drinks container is to have a capacity of 1 litre.

Give a possible set of dimensions for the container.

Find another possible set of dimensions – which of your two sets has the lower surface area?

The manufacturer wants to minimise the amount of packaging used. Investigate different dimensions, to find the lowest surface area. Could the manufacturer sell this to consumers? Give reasons for your answer.

15.1 3D shapes

A **solid** is a **three-dimensional** shape. The diagrams show some examples. You may find it easier to draw them on square or isometric paper.

A **prism** has the same **cross-section** throughout its length.

Cubes, **cuboids** and **cylinders** are all prisms.

Other prisms are named according to the shape of their cross-sections.

The base of a **pyramid** is a polygon. The other faces are triangles.

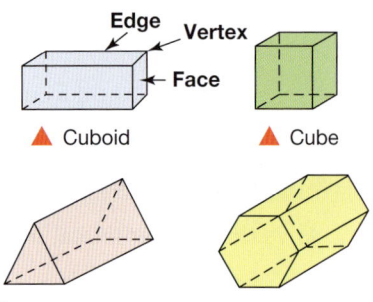
▲ Cuboid   ▲ Cube

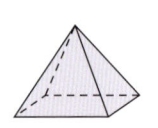 ▲ Triangular prism   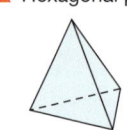 ▲ Hexagonal prism

● A **net** is a 2D shape that can be folded to make a 3D shape.

Imagine cutting along some of the edges of the cuboid and opening it out to give this net.

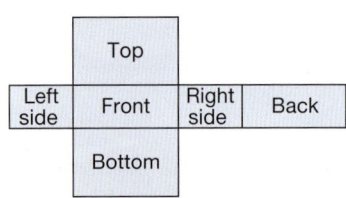

	Top		
Left side	Front	Right side	Back
	Bottom		

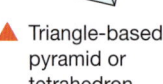 ▲ Square-based pyramid   ▲ Triangle-based pyramid or tetrahedron

● 3D shapes can be represented by **plans** and **elevations**.

 ▲ Cylinder   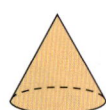 ▲ Cone    ▲ Sphere

EXAMPLE

Nine cubes are joined to make this 3D shape.

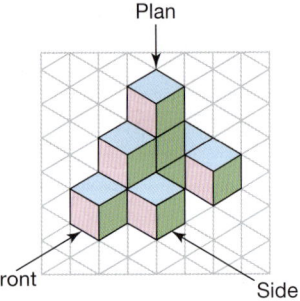

Plan

Front   Side

On squared paper, draw

a   a **plan** of the shape

b   a **front elevation**

c   a **side elevation**.

a   The plan is the view from above.
You see the blue faces.

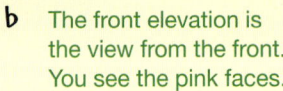
Plan

b   The front elevation is the view from the front.
You see the pink faces.

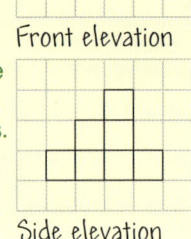
Front elevation

c   The side elevation is the the view from the side.
You see the green faces.

Side elevation

Imagine standing at the front and looking down.
You don't see any of the front or side faces.

Exercise 15.1S

1 a Copy and complete the table.

3D solid	Number of		
	Edges E	Vertices V	Faces F
Cube			
Triangular prism			
Hexagonal prism			
Triangle-based pyramid			
Square-based pyramid			

    b Copy and complete the following
$V + F = E$ .....

2 Sketch and name a 3D shape that has

    a two circular ends and a curved surface between them

    b a circular base and one curved surface

    c no plane (flat) surfaces.

3 Sketch and name the 3D shape that can be made from each net.

    a   **b**

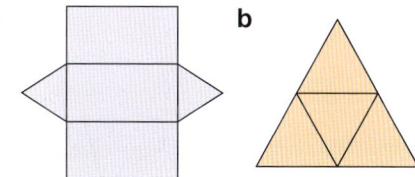

4 Which of these are possible nets for a cube?

    a   **b**   **c**

    d   **e**

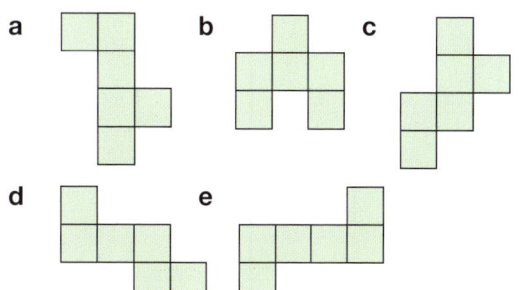

5 Draw an accurate net of this cuboid.

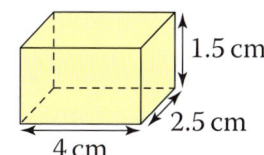

1.5 cm
2.5 cm
4 cm

6 The table gives the plans and elevations of some solids. Sketch and name each solid.

	Plan	Front elevation	Side elevation
a			
b			
c			

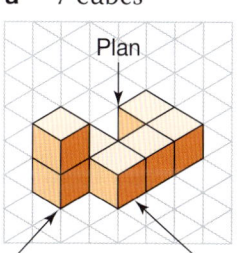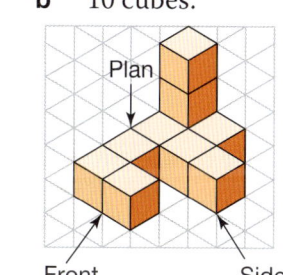

7 Sketch the plan, front and side elevations of

    a a sphere   **b** a cuboid   **c** a cone

8 On squared paper draw the plan, front and side elevation of each of these shapes made from

    a 7 cubes   **b** 10 cubes.

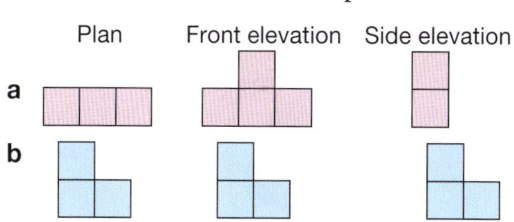

Plan   Plan

Front   Side   Front   Side

9 The plan, front elevation and side elevation are given for these solids made from cubes, draw a 3D sketch of each shape.

    Plan   Front elevation   Side elevation

    a

    b

*10 Use card to make accurate models of some 3D shapes.

15.1 3D shapes

- A net is a 2D shape that can be folded to make a 3D shape.
- 3D shapes can be represented by plans and elevations.

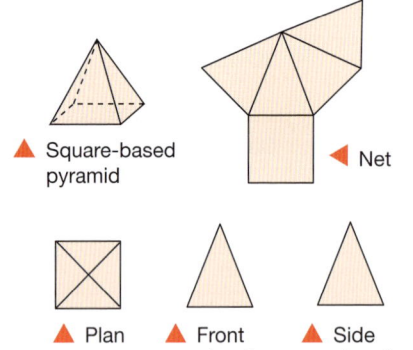

▲ Square-based pyramid ◀ Net

HOW TO

When asked to create a 2D representation of a 3D shapes
1. Think about which type of diagram you need to draw.
2. When asked for a scaled diagram, choose a scale that is easy to use and work out the dimensions you need.
3. Use the diagram to answer the questions.

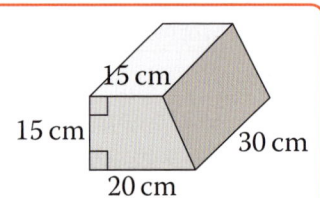

▲ Plan ▲ Front elevation ▲ Side elevation

EXAMPLE

A manufacturer wants to make a tool box in the shape of a prism. The cross section is a trapezium. The diagram gives the dimensions.

a Sketch a plan, front and side elevation of the box.

p.140

b Draw a scale diagram of the net. Use 1 cm to represent 10 cm.

c The tool box is to be made from a rectangular sheet of material. Find the minimum dimensions of the sheet needed for your net.

15 cm, 15 cm, 20 cm, 30 cm

a ① Imagine looking at the prism from above, then from the front, then from the side.

You can see just two faces from above.

Plan, Front elevation, Side elevation

b ② Using 1 cm to represent 10 cm

Long parallel edge of trapezium = 2 cm
Short parallel edge of trapezium = 15 ÷ 10
= 1.5 cm

Length of prism = 3 cm.
Draw the front, then measure to find the length of the slant edge.

Using squared paper or graph paper makes it easier to draw right angles and edges.

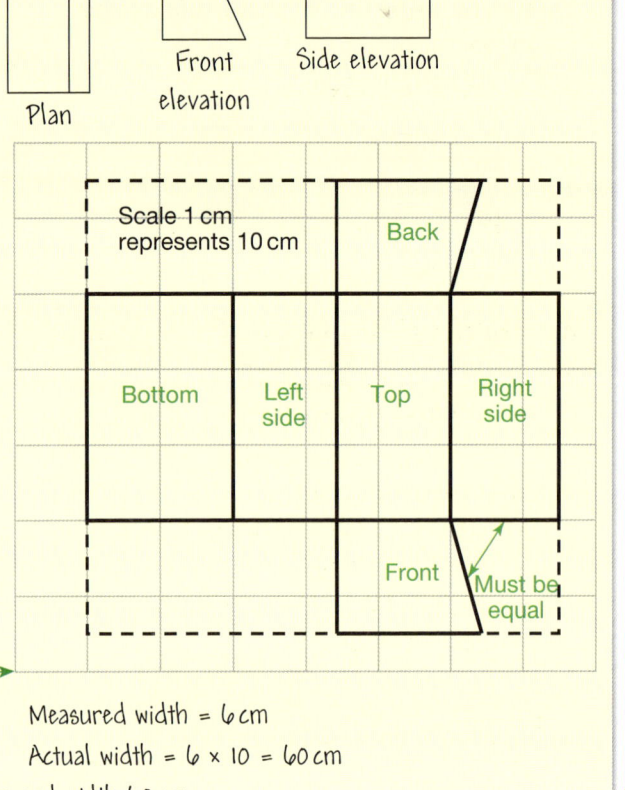

Scale 1 cm represents 10 cm

Back

Bottom | Left side | Top | Right side

Front

Must be equal

c ③ Measure the smallest rectangle that will go round the net.

Measured length = 6.6 cm   Measured width = 6 cm
Actual length = 6.6 × 10 = 66 cm   Actual width = 6 × 10 = 60 cm
The minimum dimensions are length 66 cm and width 60 cm.

Task 4 – Maintenance

They need to hire a plumber to fix some pipes in the restaurant.

There are two local plumbers who have been recommended by friends.

a On the same axes, draw graphs showing the cost (excluding parts) of hiring each plumber against the time taken.

b For what range of times is Bill cheaper than Alan?

Both quote that the job will take about 6.5 hours.

c Who should they get to do the job? Explain your reasons.

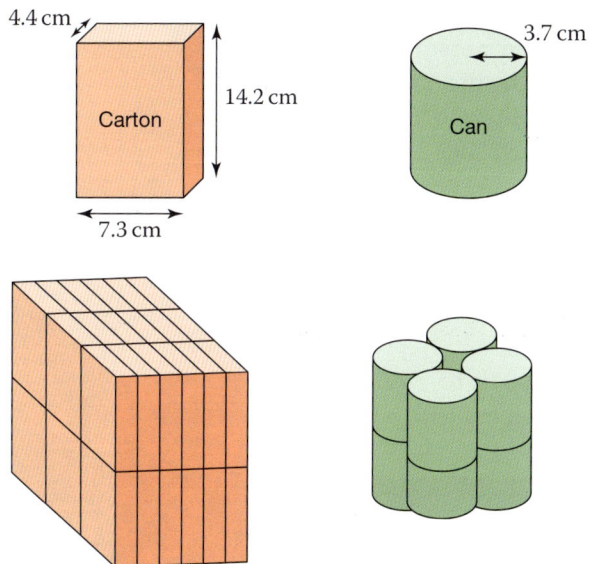

▲ Top: dimensions of one carton and one can.
Bottom: stacked arrangement of cartons in a box and an example of how cans are able to stack.

Task 5 – Stock keeping

The friends need to decide between buying cartons or cans of tomatoes.

The cartons are stacked in 2 layers in a box. Each layer is 6 cartons deep and 3 cartons across.

a Find the volume of a box.
Add an extra 5% to the volume for packaging.

The boxes are stored in a cuboid cupboard with dimensions

70 cm deep, 115 cm wide and 180 cm high.

b Calculate the volume of the cupboard in cm^3 and m^3.

c What percentage of the space in the cupboard do 30 cardboard boxes take up?

d Find the surface area of one carton.

The cans are cylindrical with a radius of 3.7 cm.

e Find the area of the circular base of the can.

f One can has the same volume as a carton. Find the height of a can.

g The cans can be stacked as shown in the diagram. Work out how many cans could fit in the same space taken up by one cardboard box.

h If the cans and cartons cost the same which would you buy? Give your reasons.

Annual amount spent eating out (£)	Number of people
0–400	36
Over 400–800	59
Over 800–1200	61
Over 1200–1600	24
Over 1600–2000	20

▲ Annual spend on eating out in Newton-Maxwell.

Task 6 – Repeat business

The table shows some results from the larger survey.

a Calculate an estimate of the mean annual amount spent eating out.

b Draw a histogram showing the data.

c Does you answer for the mean agree with what your histogram shows? Give your reason.

Juliet assumes that a given person will spend the same amount at each of the four local restaurants.

d If Juliet is right, how much should the friends expect the average person to spend at their restaurant in a year?

16 Handling data 2

Introduction

In the UK, literacy is almost taken for granted: most adults can read and write. In many other countries this is not true. Statistics show that there is a correlation, or link, between the wealth per person of a country and the adult literacy rate. In 2013, in the UK the average person 'earned' $36 000 and 99% of adults were literate. By comparison, in Sierra Leone the average person earned $2000 and only 35% of adults were literate.

The branch of statistics that deals with the relationship between variables is called correlation.

What's the point?

Understanding the relationship between quantities helps us to make sensible decisions on a global scale. Literacy problems will not be solved unless poverty is also tackled.

Objectives

By the end of this chapter you will have learned how to …

- Interpret and construct tables, graphs and charts for discrete, continuous and grouped data.
- Use the median, mean, modal class and range to interpret and compare distributions.
- Use correlation to describe scatter graphs but know that it does not imply causation.
- Draw estimated lines of best fit and make predictions but understand their limitations.
- Interpret and construct line graphs for time series data.

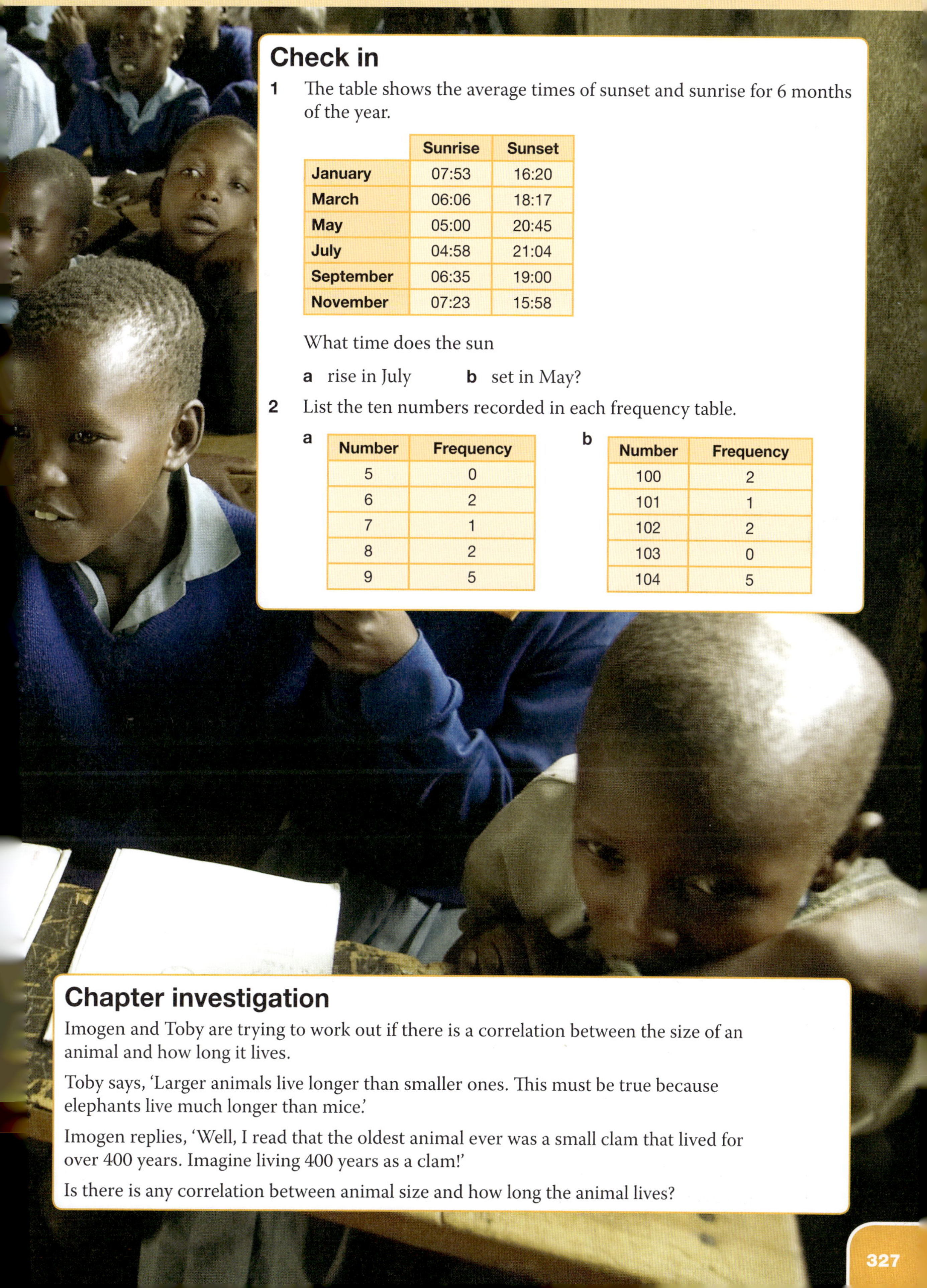

Check in

1 The table shows the average times of sunset and sunrise for 6 months of the year.

	Sunrise	Sunset
January	07:53	16:20
March	06:06	18:17
May	05:00	20:45
July	04:58	21:04
September	06:35	19:00
November	07:23	15:58

What time does the sun

a rise in July **b** set in May?

2 List the ten numbers recorded in each frequency table.

a

Number	Frequency
5	0
6	2
7	1
8	2
9	5

b

Number	Frequency
100	2
101	1
102	2
103	0
104	5

Chapter investigation

Imogen and Toby are trying to work out if there is a correlation between the size of an animal and how long it lives.

Toby says, 'Larger animals live longer than smaller ones. This must be true because elephants live much longer than mice.'

Imogen replies, 'Well, I read that the oldest animal ever was a small clam that lived for over 400 years. Imagine living 400 years as a clam!'

Is there is any correlation between animal size and how long the animal lives?

16.1 Frequency diagrams

Some surveys produce data with many different values.

You can **group** data into **class intervals** to avoid too many categories.

The exam marks of class 10A are shown:

35	47	63	25	31	8	19	55	47	14
24	36	56	61	15	43	22	50	66	10
36	45	18	20	53	31	40	60	44	47

Complete a **grouped frequency table**.

Mark	Tally	Frequency
1–20	IIII II	7
21–40	IIII IIII	9
41–60	IIII IIII I	11
61–80	III	3

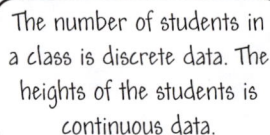 IIII = 5

Check that the frequencies add to 30.

- **Discrete** data can only take exact values (usually collected by counting).

- **Continuous** data can take any value in a given range (usually collected by measuring).
  Continuous data cannot be measured exactly.

The number of students in a class is discrete data. The heights of the students is continuous data.

You must be careful if the data is **grouped**.

You can use a **bar chart** to display grouped discrete data.

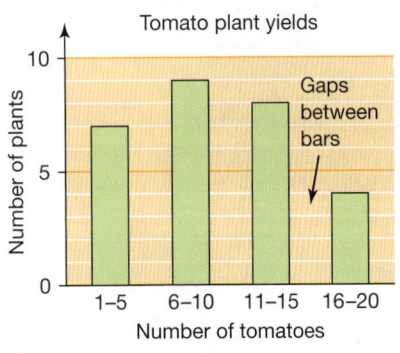

Tomato plant yields

Gaps between bars

You can use **a histogram** to display grouped continuous data.

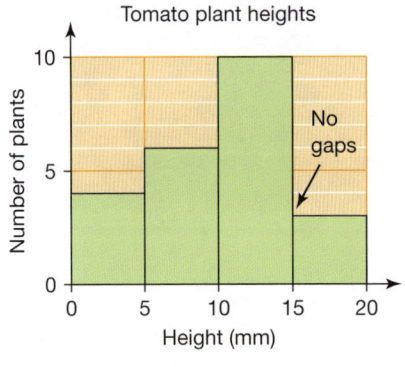

Tomato plant heights

No gaps

The times taken, in seconds, to run 100 m are shown in the table.

Time (seconds)	Number of people
$0 < t \leqslant 10$	0
$10 < t \leqslant 20$	4
$20 < t \leqslant 30$	6
$30 < t \leqslant 40$	3

Draw a frequency diagram to illustrate this information.

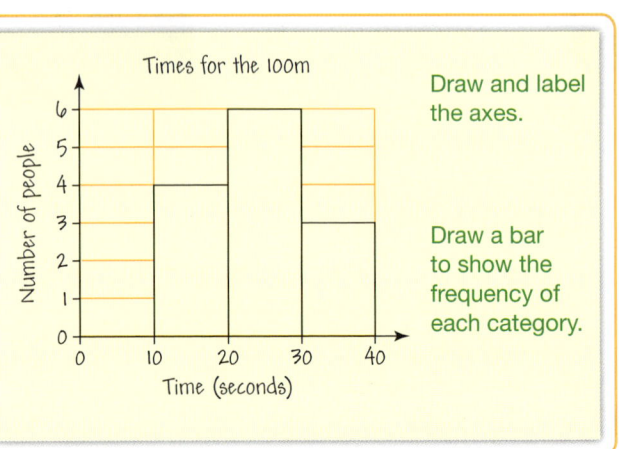

Times for the 100m

Draw and label the axes.

Draw a bar to show the frequency of each category.

Exercise 16.1S

1 **a** Copy and complete the frequency table using these test marks.

8	14	21	4	15	22	25	24	15	11
10	17	24	20	13	16	12	9	3	14
20	10	16	15	7	23	23	14	15	16
8	2	9	19	12	10	10	20	13	13
15	17	11	14	19	20	23	23	24	5

Test mark	Tally	Frequency
1 – 5		
6 – 10		
11 – 15		
16 – 20		
21 – 25		

b Calculate the number of people who took the test.

2 **a** Copy and complete the frequency table using these masses of people, given to the nearest kilogram.

48	63	73	55	59	61	70	63	58	67
46	45	57	58	63	71	60	47	49	51
53	61	68	65	70	60	52	59	50	49
48	47	63	61	58	71	53	51	60	70

Mass (kg)	Tally	Number of people
$45 \leqslant w < 50$		
$50 \leqslant w < 55$		
$55 \leqslant w < 60$		
$60 \leqslant w < 65$		
$65 \leqslant w < 70$		
$70 \leqslant w < 75$		

b Calculate the number of people in the sample.

c Draw a histogram to show the data.

d Copy and complete the frequency table.

Mass (kg)	Tally	Number of people
$45 \leqslant w < 55$		
$55 \leqslant w < 65$		
$65 \leqslant w < 75$		

e Draw a grouped bar chart to show the data.

Compare your two graphs in question **2**. The second graph is easier and faster to draw, but contains less detail.

3 **a** Copy and complete the frequency table using these heights of people.

153	134	155	142	140	163	150	135
170	156	171	161	141	153	144	163
140	160	172	157	136	160	134	154
176	154	173	179	160	152	170	148
151	165	138	143	147	144	156	139

Height (cm)	Tally	Number of people
$130 < h \leqslant 140$		
$140 < h \leqslant 150$		
$150 < h \leqslant 160$		
$160 < h \leqslant 170$		
$170 < h \leqslant 180$		

b Draw a histogram to show the data.

4 The depth, in centimetres, of a reservoir is measured daily throughout April.

Draw a histogram to show the depths.

Depth (cm)	Number of days
$0 < d \leqslant 5$	1
$5 < d \leqslant 10$	5
$10 < d \leqslant 15$	14
$15 < d \leqslant 20$	8
$20 < d \leqslant 25$	2

5 The exam marks of 36 students are shown in the frequency table.

Draw a bar chart to show the exam marks.

Exam mark (%)	Number of students
1 to 20	4
21 to 40	8
41 to 60	13
61 to 80	6
81 to 100	5

1193, 1196 | SEARCH

16.1 Frequency diagrams

- You can group data into class intervals when there are many values.
- Discrete data can only take exact values (usually collected by counting), for example the number of students in a class.
- Continuous data can take any value (collected by measuring), for example the heights of the students in a class.
- You can use a bar chart to display grouped discrete data.
- You can use a grouped bar chart to display continuous data.

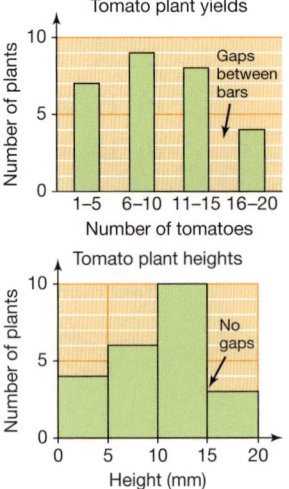

HOW TO

To interpret a frequency diagram

① Read information from the diagram.
   Make sure that you read the scale carefully.

② Use the information from the diagram to answer the question.

EXAMPLE

The table shows some information about the times taken by a bus to travel from Cookham to Marlow.

Time (minutes)	Frequency
$6 \leqslant x < 8$	
$8 \leqslant x < 10$	8
$10 \leqslant x < 12$	
$12 \leqslant x < 14$	7
$14 \leqslant x < 16$	2

Bus travel times

a Copy and complete the histogram and the table.
Include a scale on the vertical axis of the histogram.

b The bus company claims that

> Half of our journeys from Cookham to Marlow take less than 10 minutes!

Is the bus company correct?

① The $8 \leqslant x < 10$ category has frequency 8, so the scale on the axis goes up in twos.

② Complete the table and the graph.

a

Time (minutes)	Frequency
$6 \leqslant x < 8$	5
$8 \leqslant x < 10$	8
$10 \leqslant x < 12$	13
$12 \leqslant x < 14$	7
$14 \leqslant x < 16$	2
Total	35

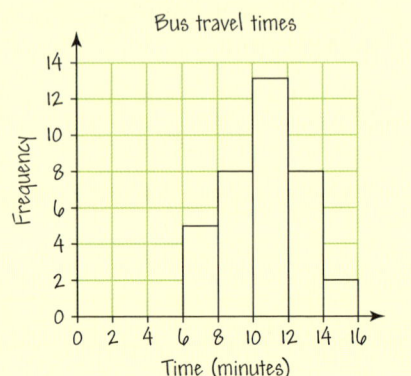

b Add up the frequencies to find the total number of buses.

Less than 10 minutes $= \frac{13}{35} < \frac{1}{2}$

The bus company is wrong.

Exercise 16.1A

1 The chart shows the best distances, in metres, that athletes threw a javelin in a competition.

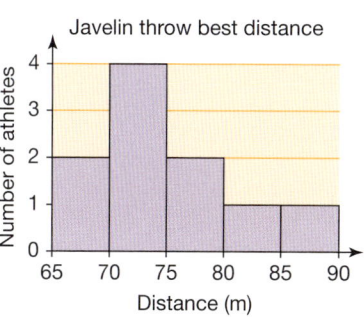

Javelin throw best distance

a State the number of athletes who threw the javelin

  i between 65 and 70 metres

  ii between 80 and 85 metres.

b In which class interval was the winner?

c Calculate the total number of athletes who threw a javelin.

2 A garden centre gave a discount to its staff. It kept a record of how much each staff member spent during one month. The values are recorded on the histogram.

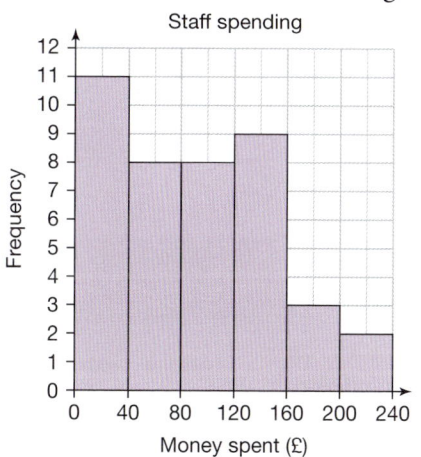

Staff spending

a Is it possible for the actual range of money spent to be

  i £159   **ii** £160   **iii** £161?

  Explain how you worked out your answers.

b How many people work at the garden centre in total?

c A person at the garden centre is chosen at random.
  What is the probability that they spent more than £120 in one month?

3 Ali surveyed 40 students to find out how much time they spent doing homework in the past two weeks.
Find the value of x and complete the frequency table and the histogram.

Time spent doing homework, t, hours	Frequency
$0 \leqslant t < 5$	
$5 \leqslant t < 10$	x
$10 \leqslant t < 15$	
$15 \leqslant t < 20$	13
$20 \leqslant t < 25$	$2x$

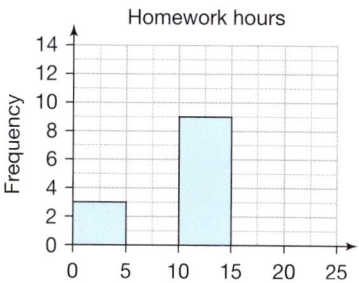

Homework hours

4 David carries out a survey to find the distance travelled to work by the staff at a school.
His results are shown on the histogram.

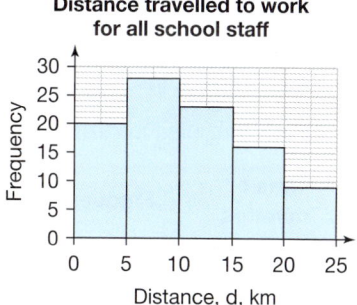

Distance travelled to work for all school staff

David kept his data for the teachers, but lost the data for the non-teachers he surveyed.

Distance, d (km)	Teachers	Non-teachers
$0 \leqslant d < 5$	13	
$5 \leqslant d < 10$	17	
$10 \leqslant d < 15$	14	
$15 \leqslant d < 20$	11	
$20 \leqslant d < 25$	5	

Complete the frequency table to show the data for the non-teachers at the school.

Q 1193, 1196   SEARCH

16.2 Averages and spread 2

● You can put large amounts of continuous data into a **grouped frequency table**.

A grouped frequency table does not tell you the actual data values so you can only find estimates of the averages.

● You use **estimates** of averages to summarise the data.

● For grouped data in a frequency table, you can calculate

  ● the **estimated mean**

  ● the **modal class**

  ● the **class interval** in which the median lies.

> The modal class is the class with the greatest frequency.

EXAMPLE

The table shows the time taken, to the nearest minute, by a group of students to solve a crossword puzzle.

Time, t, minutes	Frequency
$5 < t \leqslant 10$	2
$10 < t \leqslant 15$	14
$15 < t \leqslant 20$	13
$20 < t \leqslant 25$	6
$25 < t \leqslant 30$	1

For these data, work out

a the **modal class**

b the class containing the **median**

c an estimate for the **mean**.

a Modal class is $10 < t \leqslant 15$

b Class containing median is $15 < t \leqslant 20$

c

Time t minutes	Frequency	Midpoint	Midpoint × frequency
$5 < t \leqslant 10$	2	7.5	7.5 × 2 = 15
$10 < t \leqslant 15$	14	12.5	12.5 × 14 = 175
$15 < t \leqslant 20$	13	17.5	17.5 × 13 = 227.5
$20 < t \leqslant 25$	6	22.5	22.5 × 6 = 135
$25 < t \leqslant 30$	1	27.5	27.5 × 1 = 27.5
Total	**36**		580

Add two extra columns in the table.

Use the midpoint of each interval.

Find the totals of the Frequency and Midpoint × frequency columns.

Total number of students

Total time

Estimated mean = $\dfrac{\text{Estimated total time}}{\text{Total number of students}}$

Estimated mean = $\dfrac{580}{36}$ = 16.1 (1 dp)

Exercise 16.2S

1 The masses, to the nearest kilogram, of 25 men are shown.

69	82	75	66	72
73	79	70	74	68
84	63	69	88	81
73	86	71	74	67
80	86	68	71	75

Note:
You can use a scientific calculator to work out the mean of grouped data. You should find out how to do this on your calculator.

a Copy and complete the frequency table.

Mass (kg)	Tally	Number of men
60 to 64		
65 to 69		
70 to 74		
75 to 79		
80 to 84		
85 to 89		

b State the modal class.

c Find the class interval in which the median lies.

2 The speeds of 10 cars in a 20 mph zone are shown in the frequency table.

Speed (mph)	Mid-value	Number of cars	Mid-value × Number of cars
11 to 15		1	
16 to 20		6	
21 to 25		2	
26 to 30		1	

a Calculate the number of cars that are breaking the speed limit.

b Copy the frequency table and calculate the mid-values for each class interval.

20

c Complete the last column of your table and find an estimate of the mean speed.

3 The grouped frequency tables give information about the time taken to solve four different crosswords.

For each table, copy the table, add extra working columns and find

i the modal class

ii the class containing the median

3 iii an estimate of the mean.

a

Time, t, minutes	Frequency
$5 < t \leqslant 10$	2
$10 < t \leqslant 15$	14
$15 < t \leqslant 20$	13
$20 < t \leqslant 25$	6
$25 < t \leqslant 30$	1

b

Time, t, minutes	Frequency
$0 < t \leqslant 10$	3
$10 < t \leqslant 20$	6
$20 < t \leqslant 30$	4
$30 < t \leqslant 40$	5
$40 < t \leqslant 50$	2

c

Time, t, minutes	Frequency
$5 < t \leqslant 10$	8
$10 < t \leqslant 15$	5
$15 < t \leqslant 20$	7
$20 < t \leqslant 25$	4
$25 < t \leqslant 30$	0
$30 < t \leqslant 35$	1

d

Time, t, minutes	Frequency
$5 < t \leqslant 15$	3
$15 < t \leqslant 25$	9
$25 < t \leqslant 35$	7
$35 < t \leqslant 45$	8
$45 < t \leqslant 55$	2
$55 < t \leqslant 65$	1

4 The heights of 50 Year 10 students were measured. The results are shown in the table.

Height, h, cm	Number of students
$150 \leqslant h < 155$	3
$155 \leqslant h < 160$	5
$160 \leqslant h < 165$	15
$165 \leqslant h < 170$	25
$170 \leqslant h < 175$	2

a What is the modal group?

b Estimate the mean height.

c Which class interval contains the median?

Q 1201, 1202   SEARCH

16.2 Averages and spread 2

RECAP

- For large amounts of data presented as grouped data in a frequency table, you cannot calculate the exact mean, mode or median. Instead, you can calculate:
  - the estimated mean
  - the modal class
  - the class interval in which the median lies.

> The mean, mode and median are measures of average. The range is a measure of spread.

HOW TO

Compare grouped data
1. Compare a measure of average such as the modal class, median class or estimated mean.
2. Compare the ranges of the data sets.

EXAMPLE

The tables show the ages of people attending concerts to see the bands Badness and Cloudplay.

Badness	
Ages, a, years	**Frequency**
$20 \leqslant a < 30$	1600
$30 \leqslant a < 40$	4300
$40 \leqslant a < 50$	2100
$50 \leqslant a < 60$	1000
$60 \leqslant a < 70$	300

Cloudplay	
Ages, a, years	**Frequency**
$15 \leqslant a < 20$	2800
$20 \leqslant a < 30$	4600
$30 \leqslant a < 40$	3300

> You could also compare the median class or the estimated mean.

Compare the ages of the people attending the two concerts.

1. Compare a measure of average.

The modal class of the the ages is greater at Badness concerts then Cloudplay concerts.

The highest frequency for Badness is in the class interval 30–40 years old, whereas the highest frequency for Cloudplay is in the interval 20–30 years old.

2. Compare the range in ages.

There is more variation in the ages of the people at Badness concerts than Coldplay concerts.

The maximum range for Cloudplay is 39 – 15 = 24 years. The minimum range for Badness is 60 – 29 = 32

Exercise 16.2A

1 Jayne kept a daily record of the number of miles she travelled in her car during two months.

	December	January
Miles travelled, *m*	Frequency	Frequency
$0 < m \leqslant 20$	3	0
$20 < m \leqslant 40$	8	5
$40 < m \leqslant 60$	10	12
$60 < m \leqslant 80$	6	8
$80 < m \leqslant 100$	4	6

  a Estimate the mean number of miles for each month.

  b Find the modal class and median class for each month.

  c Compare the number of miles Jayne travelled in December and January.

2 David carried out a survey to find the time taken by 120 teachers and 120 office workers to travel home from work.

Teachers

Time taken, *t*, minutes	Frequency
$0 < t \leqslant 10$	12
$10 < t \leqslant 20$	33
$20 < t \leqslant 30$	48
$30 < t \leqslant 40$	20
$40 < t \leqslant 50$	7

Office workers

Time taken, *t*, minutes	Frequency
$10 < t \leqslant 20$	2
$20 < t \leqslant 30$	21
$30 < t \leqslant 40$	51
$40 < t \leqslant 50$	28
$50 < t \leqslant 60$	18

  a Estimate the mean number of minutes for the teachers and office workers.

  b Find the modal class and median class for the teachers and office workers.

  c Make comparisons between the time taken by the teachers and office workers to travel home from work.

3 The masses of some apples are shown in the table.

Mass, *m* (g)	Frequency
$30 \leqslant w < 40$	6
$40 \leqslant w < 50$	28
$50 \leqslant w < 60$	21
$60 \leqslant w < 70$	25

Granny Smith apples have a mean mass of 45 g and a range of 39 g.

Is the data in the table about Granny Smith apples?

4 A company produces three million packets of crisps each day. It states on each packet that the bag contains 25 g of crisps. To test this, a sample of 1000 bags are weighed.

The table shows the results.

Mass, *m* (g)	Frequency
$23.5 \leqslant w < 24.5$	20
$24.5 \leqslant w < 25.5$	733
$25.5 \leqslant w < 26.5$	194
$26.5 \leqslant w < 27.5$	53

Is the company justified in stating that each bag contains 25 g of crisps?

Show your working and justify your answer.

5 Two machines are each designed to produce paper 0.3 mm thick.

The tables show the actual output of a sample from each machine.

	Machine A	Machine B
Thickness, *t* (mm)	Frequency	Frequency
$0.27 \leqslant t < 0.28$	2	1
$0.28 \leqslant t < 0.29$	7	50
$0.29 \leqslant t < 0.30$	32	42
$0.30 \leqslant t < 0.31$	50	5
$0.31 \leqslant t < 0.32$	9	2

Compare the output of the two machines using suitable calculations.

Which machine is producing paper closer to the required thickness?

Scatter graphs and correlation

You can use a **scatter graph** to compare two sets of data,
for example, height and mass.

● The data is collected in pairs and plotted as coordinates.

● If the points lie roughly in a straight line, there is a **linear
relationship** or **correlation** between the two **variables**.

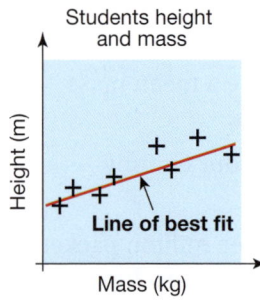

Students height
and mass

Height (m)

Mass (kg)

Line of best fit

▲ Positive correlation: as
height increases, mass
also increases.

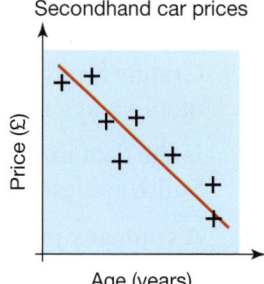

Secondhand car prices

Price (£)

Age (years)

▲ Negative correlation: as
the age of a car increases,
the price decreases.

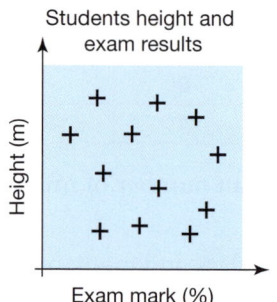

Students height and
exam results

Height (m)

Exam mark (%)

▲ No correlation: there is
no relationship between
height and exam mark.

You cannot draw
a line of best
fit for no
correlation.

If the points lie close to the line of best fit, the correlation is strong.

The exam results (%) for Paper 1 and Paper 2 for 10 students are shown.

Paper 1	56	72	50	24	44	80	68	48	60	36
Paper 2	44	64	40	20	36	64	56	36	50	24

a Draw a scatter graph and line of best fit.

b Describe the relationship between the Paper 1 results and
Paper 2 results.

a Plot the exam marks as
coordinates. The line of best
fit should be close to all the
points, with approximately the
same number of crosses on
either side of the line.

b There is a positive correlation.
Students who did well on
Paper 1 did well on Paper 2.
Students who did not do well
on Paper 1 did not do well
on Paper 2 either.

Paper 1 and 2 exam results

Paper 2 (%)

Paper 1 (%)

Points that are an
exception to the
general pattern
of the data are
called **outliers**.

Performing well on
Paper 1 does not cause
you to perform well
on Paper 2. The tests
could be on RE and
sport.

● If two variables are correlated it does not always mean that one **causes**
the other.

The events could both be a result of a common cause, or there could be no
connection at all between the events.

Exercise 16.3S

1 Match each scatter graph to its type of correlation.

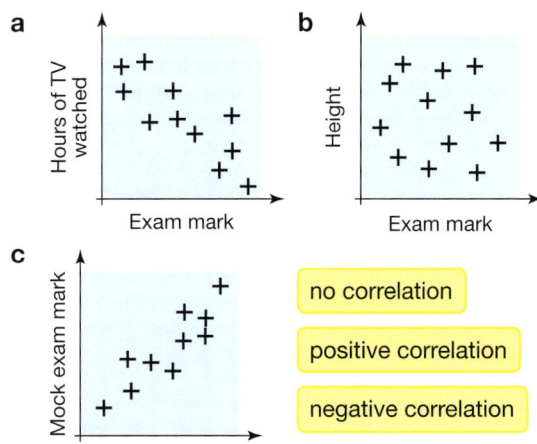

a

b

c

no correlation

positive correlation

negative correlation

2 Describe the points A, B, C, D and E on each scatter graph.

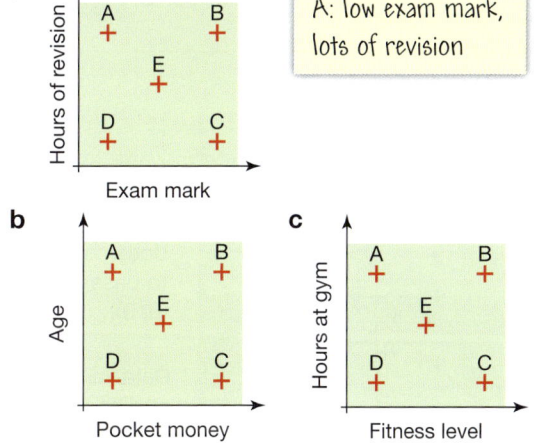

a

A: low exam mark, lots of revision

b

c

3 Match a type of correlation to each scatter graph.

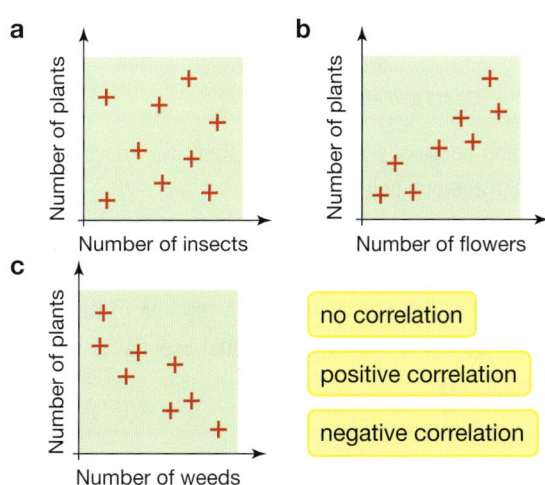

a

b

c

no correlation

positive correlation

negative correlation

4 The table shows the amount of water used to water plants and the daily maximum temperature.

Water (litres)	25	26	31	24	45	40	5	13	18	28
Maximum temperature (°C)	24	21	25	19	30	28	15	18	20	27

a Copy and complete the scatter graph for this information.

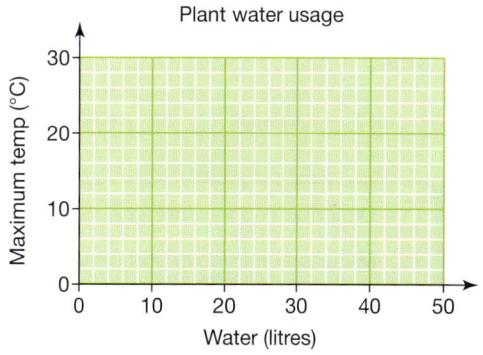

Plant water usage

b State the type of correlation shown in the scatter graph.

c Copy and complete these sentences:

   i As the temperature increases, the amount of water used _____.

   ii As the temperature decreases, the amount of water used _____.

5 The times taken, in minutes, to run a mile and the shoe sizes of ten athletes are shown in the table.

Shoe size	10	$7\frac{1}{2}$	5	9	6	$8\frac{1}{2}$	$7\frac{1}{2}$	$6\frac{1}{2}$	8	7
Time (mins)	9	8	8	7	5	13	15	12	5	6

a Draw a scatter graph to show this information.

   Use 2 cm to represent 1 shoe size on the horizontal axis.

   Use 2 cm to represent 5 minutes on the vertical axis.

b State the type of correlation shown in the scatter graph.

c Describe, in words, any relationship that the graph shows.

 1213, 1250   SEARCH

16.3 Scatter graphs and correlation

- You can compare two sets of data on a scatter graph.
- The data is collected in pairs and plotted as coordinates.
- If the points lie roughly in a straight line, there is a correlation between the two variables.

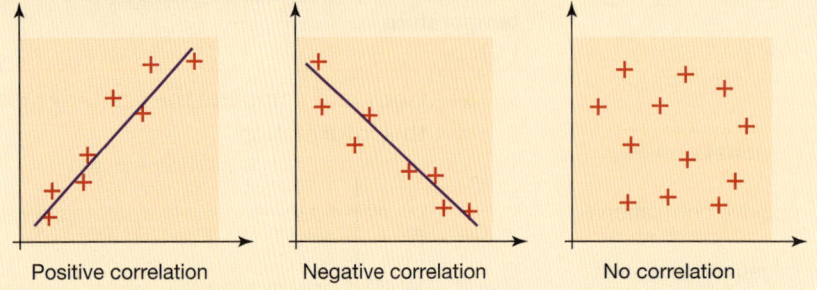

Positive correlation   Negative correlation   No correlation

You can use the line of best fit to make predictions for a value that falls in the range of the data - this is called **interpolation**. Interpolation is reliable, especially if the correlation is strong.

You can use the line of best fit to make predictions for a value that falls outside the range of the data - this is called **extrapolation**. Extrapolation is not always reliable as you cannot be sure that pattern holds for values outside of the data collected.

HOW TO

(1) Describe and interpret the relationship shown by the diagram.

(2) Make predictions based on the correlation shown.

EXAMPLE

The scatter graph shows the number of goals scored by 21 football teams in a season plotted against the number of points gained.

a Describe the relationship between the goals scored and the number of points.

b Describe the goals and points for team A.

c If a team scored 45 goals, how many points would you expect it to have?

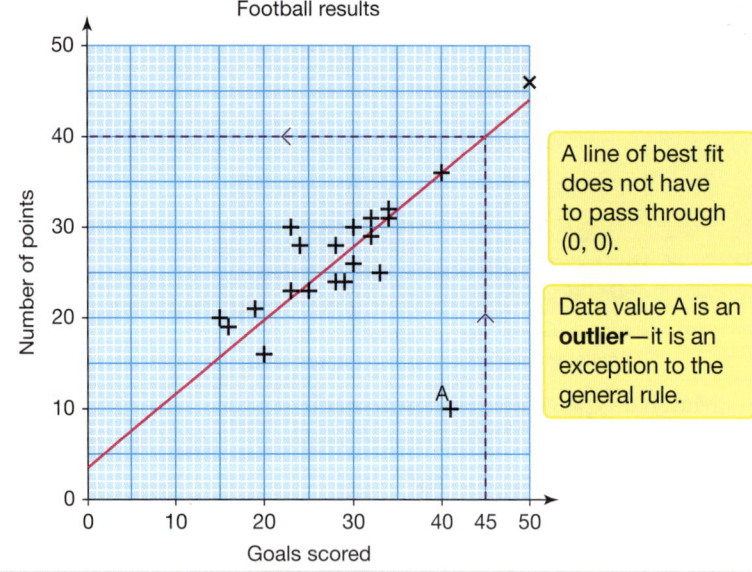

A line of best fit does not have to pass through (0, 0).

Data value A is an **outlier**—it is an exception to the general rule.

(1) Describe and interpret the relationship shown by the diagram.

a The graph shows a positive correlation: the more goals scored, the more points gained.

b Team A has scored lots of goals but has gained very few points.

(2) Make predictions based on the correlation shown.

c Reading from the graph, and using the line of best fit as a guide, you could expect a team that scored 45 goals to gain 40 points.

Exercise 16.3A

1 The scatter graph shows the exam results (%) for Paper 1 and Paper 2 for 11 students.

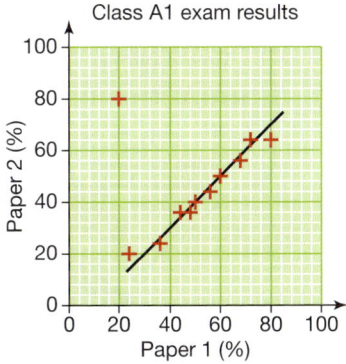

Class A1 exam results

a Describe the relationship between Paper 1 and Paper 2 results.

b Identify an outlier and describe how this student performed in the papers.

c If a student scored 40 in Paper 1, what score would you expect them to gain in Paper 2?

d Jenna extends the line of best fit on the graph and tries to predict the score that a student who scored 100 on Paper 2 will score on Paper 1.

Will her estimate make sense?

2 The graph shows the marks in two papers achieved by ten students.

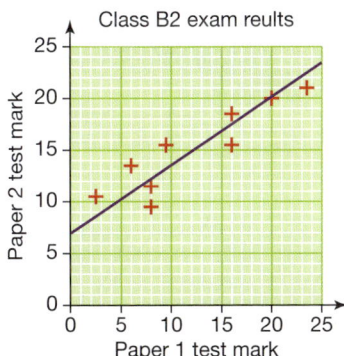

Class B2 exam reults

Use the line of best fit to estimate

a the Paper 2 mark for a student who scored 13 in Paper 1

b the Paper 1 mark for a student who scored 23 in Paper 2.

c Describe the outlier in the scatter graph. How well did this student perform on the two papers?

3 The table shows the age and diameter, in centimetres, of trees in a forest.

Age (years)	10	27	6	22	15	25	11	16	21	19
Diameter (cm)	20	78	9	65	38	74	25	44	59	50

a Draw a scatter diagram to show the information.

b State the type of correlation between the age and diameter of the trees.

> Use 2 cm to represent 10 centimetres on the horizontal axis, numbered 0 to 80.
>
> Use 2 cm to represent 5 years on the vertical axis, numbered 0 to 30.

c Draw a line of best fit.

d If the diameter of a tree is 55 cm, estimate the age of the tree.

e Use your graph to estimate the diameter of a tree that is one year old.

f Explain why the estimate in part **d** is more reliable than the estimate in part **e**.

4 The table gives the marks earned in two exams by 10 students.

Maths %	70	76	61	70	89	65	59	58	73	82
Statistics %	78	82	74	75	93	70	66	62	77	89

a Draw a scatter graph for the data.

> Choose a sensible scale for the graph. Try to make your graph as large as possible.

b Describe the correlation and the relationship shown.

c Draw a line of best fit.

d Predict the Statistics mark for a student who scored 62% in Maths.

e Could you use your graph to predict the Maths mark for a student who scored 32% in Statistics? Give your reasons.

16.4 Time series

You can use a **line graph** to show how data changes as time passes.

The data can be discrete or continuous.

The temperature of a liquid is measured every minute.

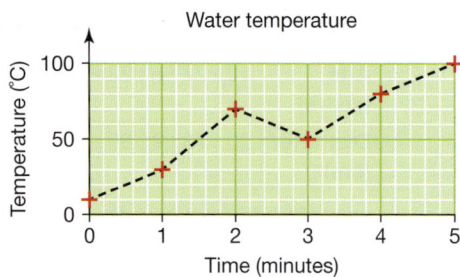

This is an example of a **time series graph**.

- A time series graph shows
  - how the data changes over time, or the **trend**
  - each individual value of the data.

Time is always the **horizontal** axis. Time could be seconds, minutes, hours, days, weeks, months or years.

EXAMPLE

The table shows the average monthly rainfall, in centimetres, in Sheffield over the last 30 years.

Month	Jan	Feb	Mar	Apr	May	Jun	Jul	Aug	Sep	Oct	Nov	Dec
Rainfall (cm)	8.7	6.3	6.8	6.3	5.6	6.7	5.1	6.4	6.4	7.4	7.8	9.2

Draw a time series graph to show this information.

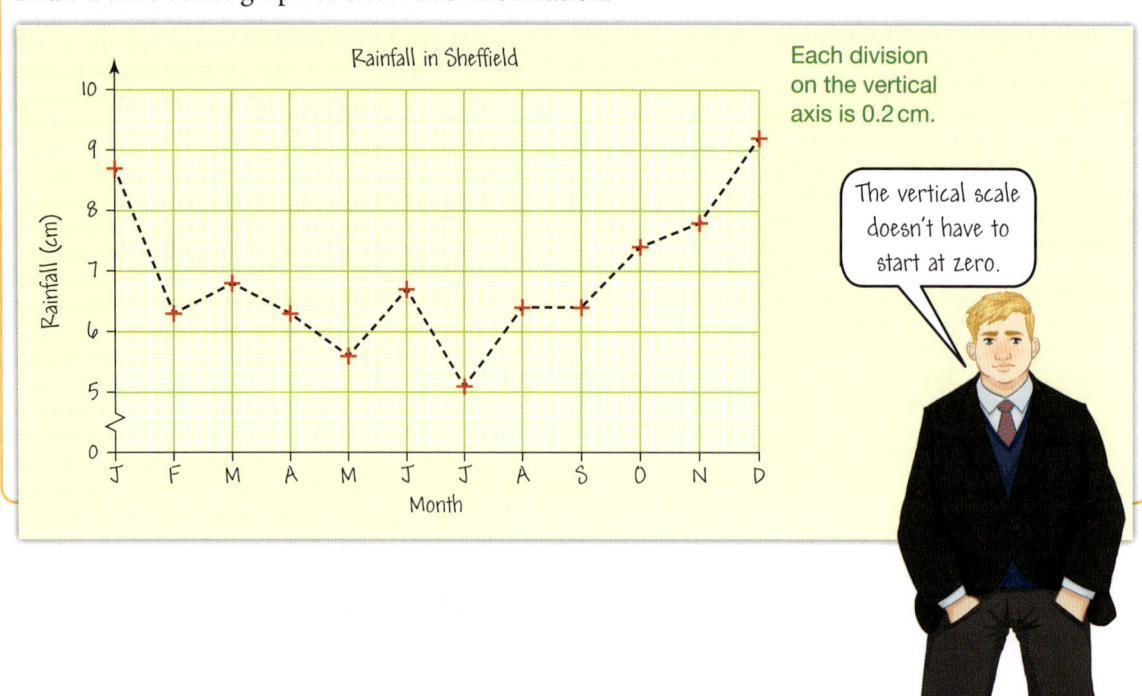

Each division on the vertical axis is 0.2 cm.

The vertical scale doesn't have to start at zero.

Exercise 16.4S

1 The number of photographs taken each day during a 7-day holiday is given.

Sunday	Monday	Tuesday	Wednesday	Thursday	Friday	Saturday
8	12	11	16	19	2	13

Copy and complete the line graph to show this information.

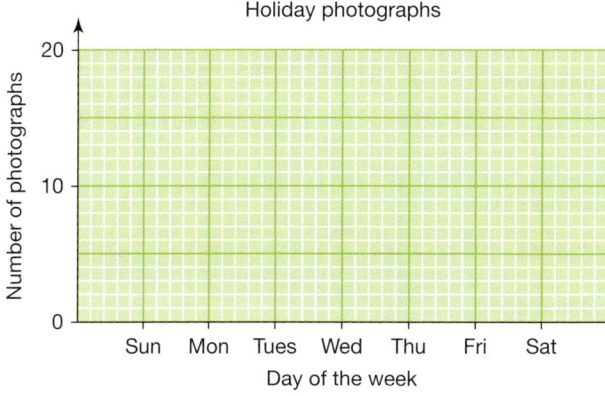

2 The numbers of DVDs rented from a shop during a week are shown.

Sunday	Monday	Tuesday	Wednesday	Thursday	Friday	Saturday
18	9	7	11	15	35	36

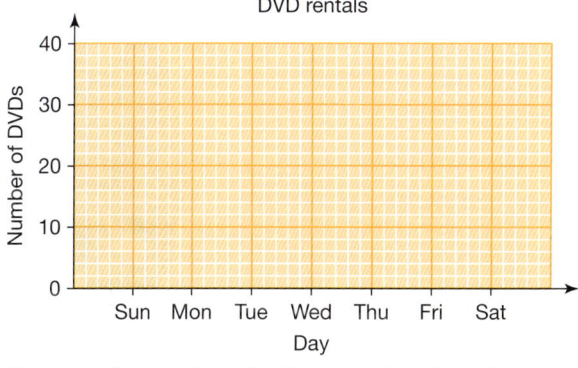

Copy and complete the line graph, choosing a suitable vertical scale.

3 Every year on his birthday, Peter's mass in kilograms is measured.

Age	2	3	4	5	6	7	8	9	10	11	12	13
Mass (kg)	14	16	18	20	22	25	28	31	34	38.5	40	45

Draw a line graph to show the masses.

4 The hours of sunshine each month are shown in the table.

Jan	Feb	Mar	Apr	May	Jun	Jul	Aug	Sep	Oct	Nov	Dec
43	57	105	131	185	176	194	183	131	87	53	35

Draw a line graph to show the hours of sunshine.

5 The daily viewing figures, in millions, for a reality TV show are shown.

Day	Sat	Sun	Mon	Tue	Wed	Thu	Fri
Viewers (in millions)	3.2	3.8	4.3	4.5	3.1	5.2	7.1

Draw a line graph to show the viewing figures.

1198 SEARCH

16.4 Time series

RECAP

- You can use a line graph to show how data changes over time.
- This is sometimes called a time series graph. It shows:
  - how the data changes over time, or the trend
  - each individual value of the data.
- This graph shows the how the temperature changes over time.

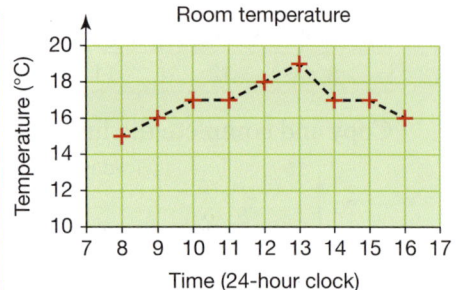

Room temperature

HOW TO

1. Draw axes on graph paper and label with suitable scales.
2. Plot the data from the table as coordinate pairs. Join the plotted points with straight lines.
3. Discuss any short term trends, seasonal variation and any longer term trends.

EXAMPLE

Jenny's quarterly gas bills over a period of two years are shown in the table.

	Jan–March	April–June	July–Sept	Oct–Dec
2003	£65	£38	£24	£60
2004	£68	£42	£30	£68

Plot the data on a graph and comment on any pattern in the data.

The graph shows seasonal variation.

1. Draw axes on graph paper with time on the horizontal axis.
   Plot time on the horizontal axis, J–M means Jan–March.
2. Plot the coordinates as crosses on the grid.
   Join them up with straight lines.

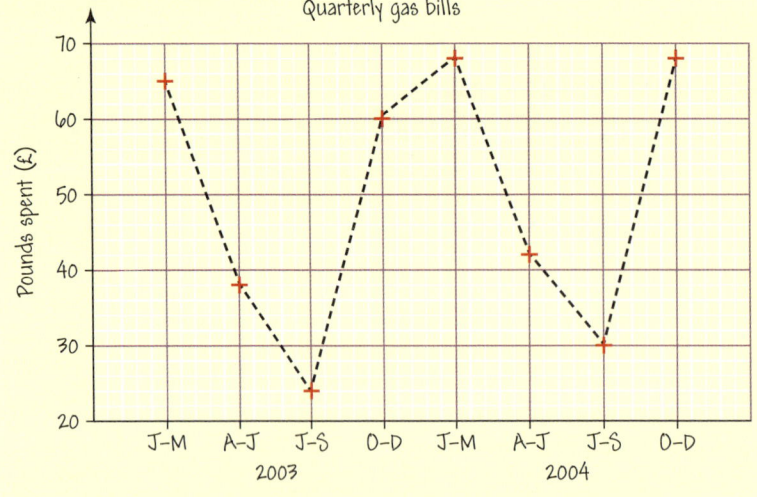

Quarterly gas bills

3. Gas bills are highest in the winter months and lowest in the summer months. This annual pattern appears to repeat itself.

   There is a slight trend for the bills to rise from year to year.

Exercise 16.4A

For each of questions **1–4**

    a   Plot the data on a graph

    b   Comment on any patterns in the data.

1   The table shows Ken's monthly mobile phone bills.

Jan	Feb	Mar	April	May	June	July	Aug	Sept	Oct	Nov	Dec
£16	£12	£15	£18	£16	£18	£12	£10	£12	£15	£16	£20

2   The table shows Mary's quarterly electricity bills over a two-year period.

	Jan–March	April–June	July–Sept	Oct–Dec
2004	£45	£20	£15	£48
2005	£54	£24	£18	£50

3   The table shows monthly ice-cream sales at Angelo's shop during one year.

Jan	Feb	Mar	April	May	June	July	Aug	Sept	Oct	Nov	Dec
£16	£12	£15	£18	£38	£48	£52	£58	£18	£15	£16	£40

4   A town council carried out a survey over a number of years to find the percentage of local teenagers who used the town's library. The table shows the results.

year	1998	1999	2000	2001	2002	2003	2004	2005
%	14	18	24	28	25	20	18	22

5   This news report was written about sales representatives of a small firm.

> On average, sales representatives at the firm travel 77 km per day. The range of distances travelled is 116 km.

These two graphs were drawn to summarise the distances travelled daily by representatives at two different firms.

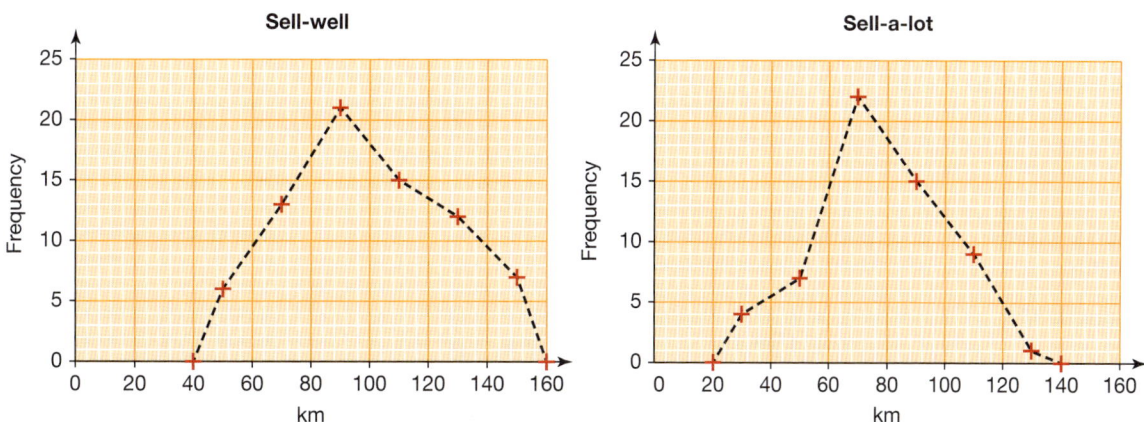

Use the graphs to identify which firm was being reported on. Give a reason for your choice.

Summary

Checkout

You should now be able to...

✓ Interpret and construct tables, graphs and charts for discrete, continuous and grouped data.	1
✓ Use the median, mean, modal class and range to interpret and compare distributions.	2
✓ Use correlation to describe scatter graphs but know that it does not imply causation.	3
✓ Draw estimated lines of best fit and make predictions but understand their limitations.	4
✓ Interpret and construct line graphs for time series data.	5

Language · Meaning · Example

Language	Meaning
Modal class	The most commonly occurring class in a set of grouped data
Estimated mean	An estimate for the mean of the data using an approximation for the total of the values. This approximation is found by multiplying the midpoint of each group by the frequency.

Time, t, minutes	Frequency	Midpoint	Mid-point · Frequency
$5 < t \leqslant 10$	2	7.5	15
$10 < t \leqslant 15$	3	12.5	37.5
$15 < t \leqslant 20$	6	17.5	105
$20 < t \leqslant 25$	1	22.5	22.5

Modal class $= 15 < t \leqslant 20$

$$\text{Estimated mean} = \frac{15 + 37.5 + 105 + 22.5}{2 + 3 + 6 + 1}$$

$$= \frac{180}{12} = 15 \text{ minutes}$$

Language	Meaning
Scatter graph	A graph that shows how two sets of numerical data are related.
Line of best fit	The single line that best represents the general direction of a set of points
Correlation (Linear relationship)	If the points lie roughly on a straight line there is a correlation between the two variables. It is a measure of how strongly they appear to be related.

Positive correlation   Negative correlation   No correlation

Language	Meaning
Time series graph	A graph that shows how a measurement changes with time.
Line graph	A graph where points are joined with straight lines.
Trend	The direction in which data appears to head as it changes over time.

Water temperature

Review

1 Chloe measures the length of 16 worms (in cm).

8.5, 10.7, 12.4, 12.8, 9.9, 11.5, 10, 8.9, 9.3, 12.2, 11.8, 12.5, 14.4, 10.2, 13.1, 12.2

Copy and complete the frequency table for this data.

Length, L (cm)	Frequency
$8 \leqslant L < 10$	
$10 \leqslant L < 12$	
$12 \leqslant L < 14$	
$14 \leqslant L < 16$	

2 For the data in question **1**

a what is the modal class?

b In which class does the median lie?

c Estimate the mean length.

3

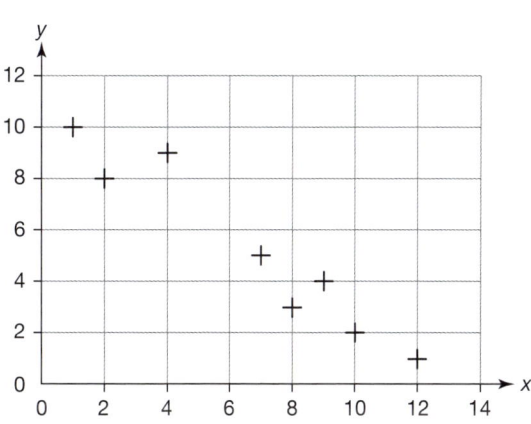

a Describe the type of correlation shown in this scatter diagram.

b Evie says 'this diagram proves that a lower value of x causes a higher value of y'. Is Evie correct?

4 The table shows the results of a group of students in two different maths tests.

Name	Test A	Test B
Aaron	12	15
Brooke	13	16
Colton	8	8
Daisy	5	8
Ellis	5	9
Frankie	8	10
Glen	10	14
Harriet	7	11
Isaac	8	14
Jasmine	6	9

a Plot this data on a scatter diagram.

b What type of correlation does the diagram show?

c Draw a line of best fit on the graph.

d Estimate the result in Test B of a student who scored 9 on Test A.

e Kevin has only sat Test B and scored 4.

    i Estimate his result in Test A.

    ii What could be wrong with this estimate and why?

5 Draw a line graph for this time series data.

Time	Temperature (°C)
06:00	15
09:00	17
12:00	20
15:00	22
18:00	24
21:00	21

What next?

	Score		
Score	0 – 2		Your knowledge of this topic is still developing. To improve look at MyMaths: 1193, 1196, 1198, 1201, 1202, 1213, 1250
	3 – 4		You are gaining a secure knowledge of this topic. To improve your fluency look at InvisiPens: 16Sa – i
	5		You have mastered these skills. Well done you are ready to progress! To develop your problem solving skills look at InvisiPens: 16Aa – d

Assessment 16

1 The table shows the age distribution of the members of a club.

Age (Y) (yrs)	10 ⩽ Y < 15	15 ⩽ Y < 20	20 ⩽ Y < 25	25 ⩽ Y < 30	30 ⩽ Y < 35	35 ⩽ Y < 40
People	16	19	27	34	29	25

    a Draw a grouped bar chart to illustrate this data. [4]

    b How many members of the club are between 10 and 20 years old? [1]

2 The grouped frequency table shows the speeds
of 250 vehicles.
No vehicles were travelling at less than 10 mph.
The national speed limit is 70 mph.
A police officer wants to work how many vehicles
were breaking the speed limit.
Give two ways to improve the table. [2]

Speed (mph)	No. of Vehicles
0 < V ⩽ 20	6
20 < V ⩽ 40	42
40 < V ⩽ 60	104
60 < V ⩽ 80	87
80 < V ⩽ 100	11

3 The grouped bar chart shows the ages of all the people
staying in a hotel.
It does not include a scale on the vertical axis.

    a Write down the modal class. [1]

    b Say how you can work out the modal class
without a scale. [1]

There are 24 guests were under 20 years old.

    c How many people are in the hotel? [3]

    d Which class contains the median? [1]

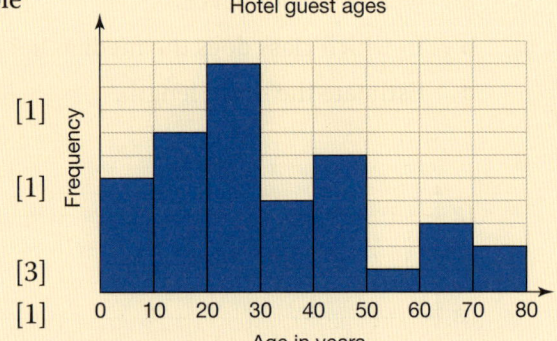

4 Boxes of tea are supposed to contain 400 tea bags. The table shows the difference between actual
number of tea bags and 400 for 30 boxes.

Difference (d)	0 ⩽ d < 5	5 ⩽ d < 10	10 ⩽ d < 15	15 ⩽ d < 20
Frequency	13	11	5	1

    a Draw a grouped bar chart to show this information. [4]

    b Estimate the mean value of the difference, *d*. [5]

    c The exact value for the mean number of teabags in the 30 boxes is 400.
Say how this could be possible even though your answer to part **b** is not zero. [1]

5 The table shows the extension of a spring when various masses are added to it.

Mass (g)	50	100	150	200	250	300	350	400	450
Extension (cm)	3.1	6.0	9.2	12.5	15.3	18.8	21.7	25.0	28.0

    a Plot a scatter diagram of the data. [4]

    b Draw the line of best fit and comment on the type of correlation. [3]

    c Use your line of best fit to estimate

        i the extension when a mass of 325 g is added to the spring [1]

        ii the mass added when an extension of 14 cm is produced. [1]

    d To estimate the extension when a mass of 600 g is added to the spring.
Should you use the line of best fit? Give reasons for your answer. [2]

6 The table shows the results of a group of key Stage 2 children's English and Science tests.

Science	25	31	33	37	48	54	56	61	75	80	89	90	93	96	98
English	86	47	58	12	76	55	82	60	59	52	66	80	15	72	68

a Plot a scatter diagram for the data. [4]

b Draw the line of best fit and comment on the correlation. [3]

c Mr Patel thinks that the results for the English test for two students are mixed up. Which two results do not fit the pattern of the data? [1]

d Use your line to estimate

i an Science Mark corresponding to an English mark of 52 [1]

ii an English Mark corresponding to a Science mark of 52. [1]

7 The number of patients in the waiting room of a Health Centre was recorded at hourly intervals and the results recorded on this time series graph.

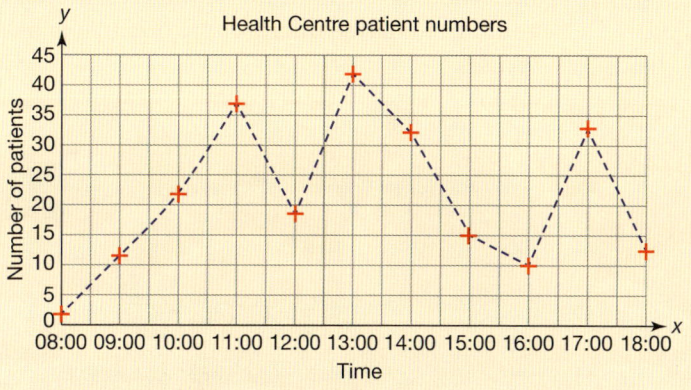

a How many patients were in the waiting room at 10:00? [1]

b How many patients were in the waiting room at 12:00? [1]

c How many patients were in the waiting room at 14:30? [1]

d Why is your answer to **c** is only an approximation? [1]

e What is the maximum number of patients recorded in the waiting room? [1]

f Give a possible reason for the three peaks in the time series. [1]

8 Scream4IceCream issued their sales figures for the past two years.
The figures in the table are in thousands of pounds to the nearest £1000.

	Jan	Feb	Mar	Apr	May	Jun	Jul	Aug	Sep	Oct	Nov	Dec
Year 1	100	95	125	150	176	203	251	266	204	131	101	88
Year 2	70	70	105	135	165	176	215	189	217	145	133	110

a Draw a time series diagram for both sets of data on the same axes. [6]

b Compare the summer and winter sales for these periods. [4]

17 Calculations 2

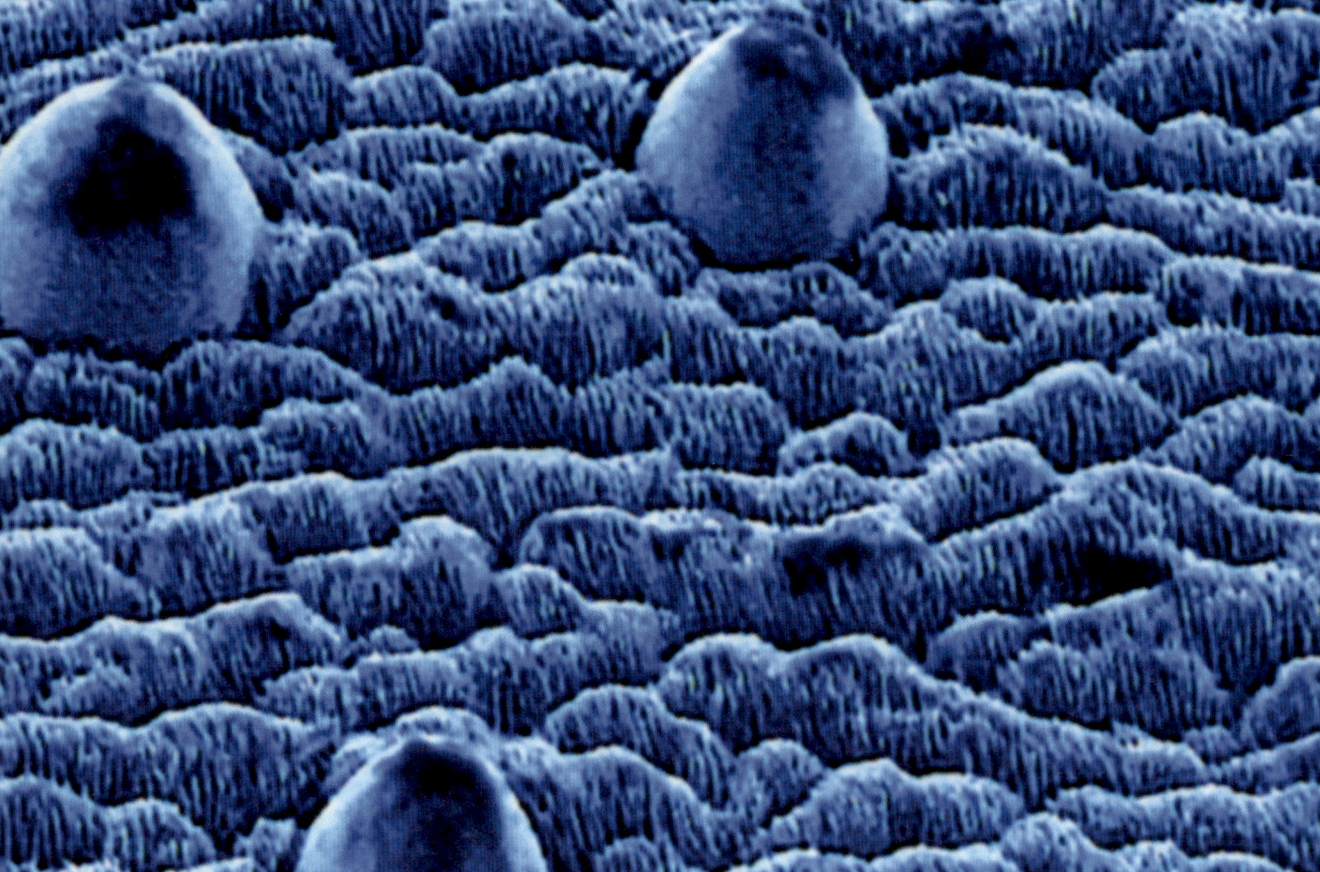

Introduction

An electron microscope is much more powerful than a normal microscope, and is used to look at very small objects. Biologists use electron microscopes to look at micro-organisms such as viruses. Materials scientists might use one to analyse crystalline structures. Nowadays there are electron microscopes that can detect objects that are smaller than a nanometre (0.000000001 m), so they can 'see' molecules and even individual atoms.

What's the point?

Mathematics needs to be able to describe very small quantities, such as lengths measured in nanometres. Without this ability, researchers could not analyse microscopic organisms like viruses.

Objectives

By the end of this chapter you will have learned how to …

- Calculate with roots and with integer indices.
- Calculate exactly with fractions and multiples of π.
- Calculate with and interpret numbers written in standard form.

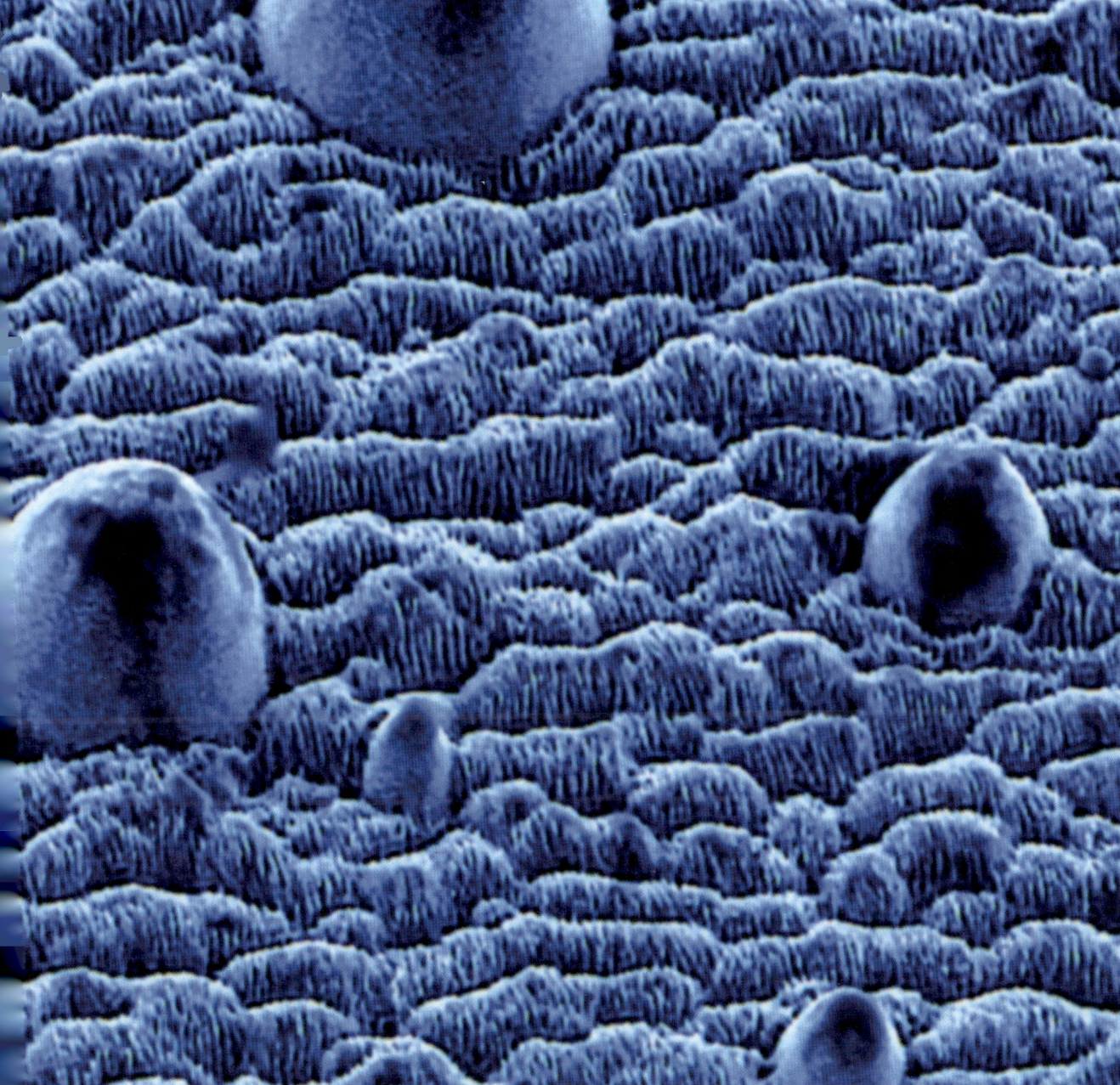

Chapter investigation

A picometre is a unit of length that is used to describe atomic distances.
It is 0.000000000001 m in length.

See if you can find out about the smallest metric unit of length that has a name, and if it has any possible real-world uses. What about the largest unit of length?

17.1 Calculating with roots and indices

The **square root** of a given number is a number that when multiplied by itself equals the given number.

p.280

The **cube root** of a given number is a number that when multiplied by itself and then multiplied by itself again equals the given number.

$$\sqrt{9} = 3$$
because $3^2 = 3 \times 3 = 9$
$$\sqrt[3]{8} = 2$$
because $2^3 = 2 \times 2 \times 2 = 8$

Powers of the same **base** number can be multiplied and divided.

● When multiplying, you add the indices.

$$5^3 \times 5^4 = (5 \times 5 \times 5) \times (5 \times 5 \times 5 \times 5)$$
$$= 5^{3+4} = 5^7$$

● When dividing, you subtract the indices.

$$3^5 \div 3^2 = \frac{3 \times 3 \times 3 \times \cancel{3} \times \cancel{3}}{\cancel{3} \times \cancel{3}} = 3 \times 3 \times 3 = 3^{5-2} = 3^3$$

● To raise to a power, multiply the indices.

$$(9^2)^3 = (9 \times 9) \times (9 \times 9) \times (9 \times 9)$$
$$= 9^{2 \times 3} = 9^6$$

EXAMPLE

i Write the answers to these calculations in index form.

ii Use your calculator to work out each answer.

a $5^5 \times 5^3$

b $4^7 \div 4^3$

c $(2^3)^4$

a i $5^5 \times 5^3 = 5^{5+3} = 5^8$

ii
```
5^8

        390625
```

b i $4^7 \div 4^3 = 4^{7-3} = 4^4$ ii 256

c i $(2^3)^4 = 2^{3 \times 4} = 2^{12}$ ii 4096

Make sure you know how to use your calculator. Does it have these keys?

$3^2 \times 3 = (3 \times 3) \times 3 = 3^3$.
You can write this as $3^2 \times 3^1 = 3^{2+1} = 3^3$, so $3 = 3^1$.

● Any number to the power 1 is just the number itself: $x^1 = x$ for any value of x.

Using the rule for multiplication, $7^3 \times 7^0 = 7^3$.
Since multiplying by 7^0 leaves the 7^3 unchanged, 7^0 must be equal to 1.

● Any number (except 0) to the power 0 is 1: $x^0 = 1$ for any value of x, if $x \neq 0$.

EXAMPLE

Find the value of

a 19^0

b 8^1

c $(16^3 - 81 \times 17)^0$

d $(4.8)^1$

a $19^0 = 1$.

b $8^1 = 8$.

c There is no need to find the number in the brackets.
$(16^3 - 81 \times 17)^0 = 1$.

d $(4.8)^1 = 4.8$

Exercise 17.1S

1 Find these square roots.

   a $\sqrt{25}$ **b** $\sqrt{9}$ **c** $\sqrt{16}$

   d $\sqrt{1}$ **e** $\sqrt{4}$ **f** $\sqrt{36}$

2 Calculate these using a calculator. Give your answers to 2 dp where appropriate.

   a $\sqrt{40}$ **b** $\sqrt{61}$ **c** $\sqrt{180}$

   d $\sqrt{249}$ **e** $\sqrt{676}$ **f** $\sqrt{1234}$

3 Calculate these cube roots using your calculator. Give your answers to 2 dp where appropriate.

   a $\sqrt[3]{27}$ **b** $\sqrt[3]{512}$ **c** $\sqrt[3]{3375}$

   d $\sqrt[3]{100}$ **e** $\sqrt[3]{24389}$ **f** $\sqrt[3]{16129}$

4 Find the value of these expressions.

   a 5^2 **b** 2^3 **c** 3^3

   d 8^2 **e** 12^2 **f** 13^2

5 Find the value of these expressions.

   a 3^4 **b** 1^5 **c** 2^7

   d 3^6 **e** 10^6 **f** 4^4

6 Use your calculator to work these out.

   a 12^3 **b** 6^6 **c** 21^3

   d 16^5 **e** 13^3 **f** 14^6

7 Evaluate these expressions.

   a 5^1 **b** 6^1

   c 6^0 **d** 7^0

   e $(4 + 88^2)^0$ **f** $(4^2 + 5^2)^1$

   g $(92.5)^0$ **h** 0^1

8 Write the answers to these multiplications in index form.

   a $7 \times 7 \times 7$ **b** 3×3^2

   c 5×5^2 **d** $6 \times 6 \times 6^2$

   e $5^3 \div 5$ **f** $8^4 \times 8$

   g $9^3 \times 9^2 \times 9$ **h** $8^7 \times 8$

9 Simplify these expressions, giving your answers in index form.

   a $6^2 \times 6^3$ **b** $4^5 \times 4^4$

   c $2^6 \times 2^7$ **d** $11^5 \times 11^2$

   e $1^{17} \times 1^{13}$ **f** $7^8 \times 7^4$

   g $3^6 \times 3^6$ **h** $9^9 \times 9^1$

10 Simplify these expressions, giving your answers in index form where appropriate.

   a $7^8 \div 7^6$ **b** $8^6 \div 8^2$

   c $3^3 \div 3^2$ **d** $9^{11} \div 9^8$

   e $4^7 \div 4^1$ **f** $2^9 \div 2^9$

   g $12^8 \div 12^6$ **h** $6^{13} \div 6^{13}$

11 Use the relationship $(x^m)^n = x^{mn}$ to simplify these expressions, giving your answers in index form.

   a $(2^3)^2$ **b** $(4^2)^5$ **c** $(7^2)^2$

   d $(5^5)^3$ **e** $(3^4)^4$ **f** $(6^2)^2$

   g $(5^7)^3$ **h** $(10^4)^4$ **i** $(8^7)^4$

12 Simplify these expressions, giving your answers in index form.

   a $3^4 \times 3^2 \div (3^3 \times 3^2)$

   b $(5^6 \div 5^2) \times 5^4 \times 5^2$

   c $(4^5 \div 4^2) \div (4^6 \div 4^5)$

   d $(7^9 \div 7^2) \div (7^2 \times 7^3)$

   e $(8^7 \div 8^4) \times 8^5 \times 8^3$

   f $9^3 \times (9^5 \div 9^2) \times 9^4$

13 Simplify these expressions, giving your answers in index form.

   a $\dfrac{4^2 \times 4^2}{4^2}$ **b** $\dfrac{6^3 \times 6^4}{6^5}$

   c $\dfrac{9^8}{9^2 \times 9^4}$ **d** $\dfrac{8^6 \div 8^3}{8^2}$

   *e $\dfrac{5^9 \times 5^4}{5^3 \times 5^7}$ ***f** $\dfrac{6^3 \times 6^4}{6^5 \div 6^3}$

   *g $\dfrac{8^9 \div 8^2}{8^7 \div 8^2}$ ***h** $\dfrac{10^6 \div 10^2}{10^2 \times 10^2}$

*14 Simplify these expressions, giving your answers in index form.

   a $\left(\dfrac{4^4}{4^2}\right)^2$ **b** $\dfrac{(3^3)^2 \times 3^5}{3^7}$

   c $\dfrac{6^3 \times (6^2)^4}{(6^3)^2}$ **d** $\left(\dfrac{5^8 \div 5^2}{5^2}\right)^3$

   e $\left(\dfrac{2^4 \div 2^2}{2^9 \div 2^8}\right)^2$ **f** $\left(\dfrac{7^{15} \div 7^2}{7^8 \times 7^2}\right)^4$

   g $\dfrac{(3^3)^5}{3^4 \times 3^2} \div 3$ **h** $\left(\dfrac{9^7 \div 9}{9^2 \times 9^3}\right)^3 \times 9$

17.1 Calculating with roots and indices

RECAP

- Add indices when multiplying powers of the same number.
- Subtract indices when dividing powers of the same number.
- Multiply indices when finding the power of a power.
- Any number (except 0) to the power of 0 is 1.

$3^2 \times 3^4 = (3 \times 3) \times (3 \times 3 \times 3 \times 3)$
$= 3^{2+4} = 3^6$
$5^5 \div 5^3 = \dfrac{5 \times 5 \times \cancel{5} \times \cancel{5} \times \cancel{5}}{\cancel{5} \times \cancel{5} \times \cancel{5}} = 5^{5-3} = 5^2$
$(7^2)^3 = (7 \times 7) \times (7 \times 7) \times (7 \times 7)$
$= 7^{2 \times 3} = 7^6$
$(4.86)^0 = 1$

HOW TO

To solve problems involving calculating with powers

① RTQ carefully; decide how to use your knowledge of powers and roots.

② Apply the index laws and look out for chances to use known facts about squares and cubes.

③ ATQ; make sure that you leave your answer in index form if the question asks for it.

EXAMPLE

For the calculation $2^3 \times 5^4$, Jack wrote

> $2 \times 5 = 10$, and $3 + 4 = 7$.
> So, the answer is 10^7.

Jack is incorrect, say why. Find the correct answer.

① They need to have the same base.

② You should work out each term then multiply them together.

③ $2^3 = 8$ and $5^4 = 625$ $2^3 \times 5^4 = 8 \times 625 = 5000$

EXAMPLE

$2^{12} = 4^a$. Find the value of a.

① Rewrite 2^{12} as a power of 4. This will give you the value of a.

② Use the index law for the power of a power and the fact that $4 = 2^2$.
$2^{12} = (2^2)^6 = 4^6$

③ $a = 6$

EXAMPLE

Explain why the square of a prime number has exactly three factors.

① A prime number, p, has two factors, 1 and p (itself).

② List the factors of the square of a prime number.
$p^2 = 1 \times p^2$ and $p^2 = p \times p$

③ A prime number squared has three factors.
One, the prime number, p, and the number itself, p^2.

> The base is the number you are multiplying.

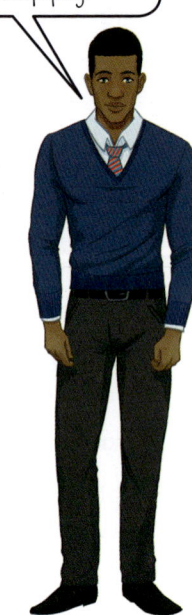

Exercise 17.1A

1 For the calculation $3^2 + 3^2 + 3^2$, Kira wrote

$$3^2 + 3^2 + 3^2 = (3 + 3 + 3)^2 = 9^2$$

Explain why Kira is incorrect.
Find the correct answer.

2

$6^0 = 1$	$6^1 = 6$	$6^2 = 36$
$6^3 = 216$	$6^4 = 1296$	$6^5 = 7776$
	$6^6 = 46\,656$	

Use the table to

a explain why $36 \times 216 = 7776$

b work out $46\,656 \div 36$.

3 Find the value of the letter in each equation.

a $3^4 \times 9 = 3^a$ **b** $5^b = 5^8 \div 25$

c $9^5 = 3^c$ **d** $4^{10} = 16^d$

e $(2^e)^3 = 2^9$ ***f** $3^5 \times 3^f = 9^2 \times 3^4$

4 Say why the cube of a prime number has exactly four factors.

5 Which of these are correct?
Change one symbol or number to correct any mistakes.

a $2^3 + 2^5 = 2^8$ **b** $3^2 \times 3^3 = 3^5$

c $(5^2)^3 = 5^6$ **d** $(7^3)^4 = 7^7$

e $4^5 \div 4^3 = 4^8$ **f** $6^4 \div 6^3 = 6$

6 Arrange these numbers in order of size, starting with the smallest first.

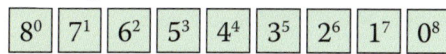

$$8^0 \quad 7^1 \quad 6^2 \quad 5^3 \quad 4^4 \quad 3^5 \quad 2^6 \quad 1^7 \quad 0^8$$

7 Myia runs a secret maths club. To get in to the club, you have to give her two numbers. Myia squares the numbers, adds these together, then square roots the answer. If the answer is a whole number you can join the club.

a Which of these pairs of numbers would get you into her club?

    i 3 and 4   **ii** 4 and 5

    iii 6 and 12   **iv** 8 and 15

b Can you find two more that work?

8 The Richter scale measures how powerful an earthquake is. Each level is ten times more powerful than the previous value. So level 4 is 10 times as powerful as level 3.

a How many times more powerful is a level 4 earthquake compared to a level 1?

b How many times more powerful is level 9 (Complete devastation) compared to level 5 (Buildings shake)?

c What level is an earthquake 10^5 times more powerful than level 3?

***9** Here are twelve number cards.

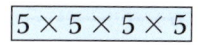

$5 \times 5 \times 5 \times 5$	10^2	4^3	8^2
0.2	2^{10}	6^1	125
2^9	$\frac{1}{4^3}$	$3^2 + 3^2$	1000

a Find a pair of cards where

    i they are both the same value

    ii one is twice the other

    iii one is ten times as big as the other

    iv they are both odd numbers

    v they multiply together to make 1

    vi they are reciprocals

    vii they multiply together to make 5^7

    viii they multiply together to make 5^2.

b Which card

    i is three times the cube root of 216

    ii has the largest value

    iii has the smallest value?

***10 a** If $7^8 = 5\,764\,801$, what is $\sqrt[8]{5\,764\,801}$?

    b If $(-3)^9 = -19\,683$, what is $\sqrt[9]{-19\,683}$?

11 Find the missing indices in these expressions.

a $2^2 \times 2^{\square} \times 2^5 = 2^{11} \div 2^2$

b $(3^4 \times 3^{\square})^3 = 3^{18}$

1033, 1924   **SEARCH**

17.2 Exact calculations

Some numbers do not look nice when written as decimals.

$\frac{1}{7} = 0.1428571428\ldots$ $\pi = 3.141592654\ldots$

In practice you would round your answer and it would no longer be **exact**.

$\frac{1}{7}$ is a recurring decimal, you can write it as $0.\dot{1}4285\dot{7}$ using 'dot notation' but it still doesn't look nice.

> ● If a calculation involves fractions or π, it is better to work with the numbers in fraction form and with multiples of π in order to give an exact answer.

If a question says exact then avoid using your calculator.

EXAMPLE

Calculate

a $\frac{1}{2} + \frac{2}{3}$

b $\frac{3}{4} \times \frac{2}{9}$

c $\frac{4}{5} \div \frac{3}{10}$

a $\frac{1}{2} + \frac{2}{3} = \frac{1 \times 3 + 2 \times 2}{2 \times 3} = \frac{3+4}{6} = \frac{7}{6} = 1\frac{1}{6}$

b $\frac{\cancel{3}^1}{\cancel{4}_2} \times \frac{\cancel{2}^1}{\cancel{9}_3} = \frac{1 \times 1}{2 \times 3} = \frac{1}{6}$

c $\frac{4}{5} \div \frac{3}{10} = \frac{4}{\cancel{5}_1} \times \frac{\cancel{10}^2}{3} = \frac{4 \times 2}{1 \times 3} = \frac{8}{3} = 2\frac{2}{3}$

p.100

EXAMPLE

Which of these calculations gives a different answer to the others?

a $1\frac{3}{4} - 1\frac{1}{2}$ **b** $\frac{1}{2} \times 1\frac{1}{2}$ **c** $3\frac{1}{8} - 2\frac{7}{8}$ **d** $2\frac{1}{2} \div 10$

You need to work out all four parts to see which is different.

a $1\frac{3}{4} - 1\frac{1}{2}$ **b** $\frac{1}{2} \times 1\frac{1}{2}$ **c** $3\frac{1}{8} - 2\frac{7}{8}$ **d** $2\frac{1}{2} \div 10$

$= \frac{3}{4} - \frac{1}{2} = \frac{1}{4}$ $= \frac{1}{2} \times \frac{3}{2} = \frac{3}{4}$ $= 1\frac{1}{8} - \frac{7}{8} = \frac{2}{8} = \frac{1}{4}$ $= \frac{\cancel{5}^1}{2} \times \frac{1}{\cancel{10}^2} = \frac{1}{4}$

Part **b** is different to the others.

p.312

EXAMPLE

Find an exact value for the total volume of these three shapes.

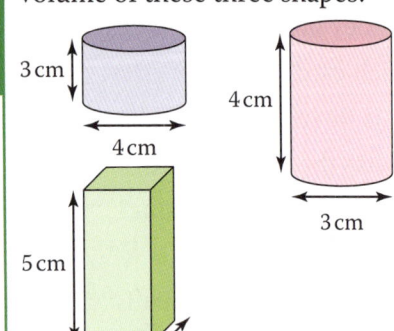

Use volume formulae and leave π in the answers.

Blue volume $= \pi \times 2^2 \times 3 = 12\pi$ Volume of a cylinder $= \pi r^2 h$.

Red volume $= \pi \times (1\frac{1}{2})^2 \times 4$

$= \pi \times \frac{3}{2} \times \frac{3}{2} \times 4$

$= 9\pi$

Green volume $= 3 \times 3 \times 5 = 45$ Volume $= l \times w \times h$.

Total volume $= 45 + 21\pi\,\text{cm}^2$ Add the 3 parts.

Exercise 17.2S

Give exact answers to all questions in this exercise.

1 Evaluate these calculations exactly.

a $\dfrac{1}{3} + \dfrac{1}{5}$ b $\dfrac{2}{5} + \dfrac{3}{7}$

c $\dfrac{3}{8} - \dfrac{2}{7}$ d $\dfrac{8}{15} + \dfrac{4}{9}$

e $6\dfrac{1}{2} - 1\dfrac{5}{8}$ f $\dfrac{2}{5} + \dfrac{1}{3} + \dfrac{1}{4}$

g $\dfrac{3}{8} + \dfrac{1}{2} - \dfrac{2}{5}$ h $7\dfrac{2}{9} + 2\dfrac{1}{4}$

2 Evaluate these calculations exactly.

a $\dfrac{2}{3} \times \dfrac{3}{4}$ b $\dfrac{5}{9} \div \dfrac{1}{3}$

c $2\dfrac{1}{2} \times \dfrac{5}{8}$ d $\dfrac{8}{9} \div 1\dfrac{2}{3}$

e $3\dfrac{1}{2} \div 2\dfrac{1}{4}$ f $5\dfrac{1}{5} \times 2\dfrac{3}{4}$

g $8\dfrac{2}{5} \div 3\dfrac{1}{7}$ h $3\dfrac{1}{2} \times 7\dfrac{5}{9}$

3 The decimal equivalent of a fraction will terminate if the only prime factors of the denominator are 2 or 5. Otherwise the decimal equivalent will be recurring.

Without using a calculator, say whether your answers to questions **1** and **2** will be terminating or recurring.

4 Evaluate these calculations exactly.

a $\dfrac{2}{5} \times \left(\dfrac{3}{4} + \dfrac{2}{3} \right)$

b $\left(\dfrac{2}{3} + \dfrac{4}{5} \right) \div \left(\dfrac{2}{5} + \dfrac{3}{7} \right)$

c $\dfrac{5}{6} \div \dfrac{3^2 + 4^2}{10}$

d $\left(\dfrac{3}{7} \right)^2 \times \left(\dfrac{4}{5} - \dfrac{1}{7} \right)$

e $\left(\dfrac{1}{2} + \dfrac{5}{9} \right)^2 + \left(\dfrac{2}{3} \right)^3$

f $\left(\dfrac{5}{6} + \dfrac{1}{2} \right) - \left(\dfrac{4}{7} \times 3 \right)$

5 Simplify these expressions.

a $\pi + \pi$ b $2\pi + \pi$

c $4 + \pi - 2 + \pi$ d $\pi(4^2 - 6)$

e $2\pi(4^2 - 7)$ f $4\pi(2 + \sqrt{4})$

g $\dfrac{42\pi}{4}$ h $3 \times (7 - \sqrt{9})\pi$

6 Simplify these expressions.

a $\pi(6^2 - 4)$ b $\sqrt{49} + \pi(7 - 5)$

c $4(7 + \sqrt{2})$ d $\pi(8^2 - \sqrt{4})$

e $2\pi(3 - 2\pi) + (2\pi)^2$

f $\pi^2 - \pi(\pi - 4)$

g $\dfrac{2}{3} + \dfrac{1}{2}\pi - \dfrac{1}{6}\pi$ h $\dfrac{1}{3}\left(\dfrac{3}{4} - \dfrac{\pi}{2} \right)$

7 A dolls house is made on a scale of one inch to one foot. This means everything is $\dfrac{1}{12}$ normal size. What size are these items?

a A table 3 feet wide by $4\dfrac{1}{2}$ feet long.

b A pot 10 inches tall.

c A bowl $7\dfrac{1}{2}$ inches wide.

8 Geoff is laying a rectangular lawn with semi circular flowerbeds at each end.

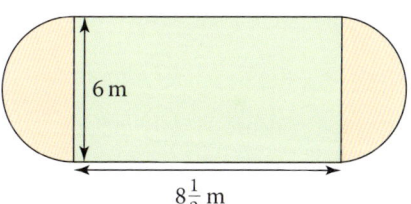

6 m

$8\dfrac{1}{2}$ m

The rectangle is 6 m wide by $8\dfrac{1}{2}$ m long.

a What is the perimeter of the rectangle?

b What is the perimeter of the entire shape?

c What is the area of the lawn?

d What is the area of a flowerbed?

e What is the area of the entire shape?

9 Julia says that all these four calculations give the same answer. Is she correct? Show your working.

a $\dfrac{3}{4} + \dfrac{1}{2}$ b $\dfrac{1}{2} \times 2\dfrac{1}{2}$

c $2\dfrac{3}{4} - 1\dfrac{1}{2}$ d $3 \div 2\dfrac{2}{5}$

p.230

17.2 Exact calculations

p.100

▲ 14 March is celebrated as 'pi day' each year. (3rd month, 14th day, so 3.14). Some people say 22 July should be pi day, because the fraction $\frac{22}{7}$ is a very good approximation to π.

RECAP

- To find an exact answer do not use a calculator. Instead work with fractions and multiples of π, as appropriate, throughout the calculation.
- If an expression contains π, treat it like a variable and collect like terms, etc.
- In the answer any fractions should be given in their simplest form.

HOW TO

To find an exact answer to a question

① RTQ; decide what to do and which formulae to use.

② If they occur, work with fractions and treat π as a variable.

③ ATQ; simplify expressions and collect terms with π together.

EXAMPLE

A designer is making a model of a chair at one-quarter of actual size.
What is the size on the model for actual sizes of

a the height 34 inches

b the width $16\frac{1}{2}$ inches

c the leg thickness $1\frac{3}{4}$ inches

d the seat thickness $\frac{7}{8}$ inch?

① To reduce the scale divide by 4 or multiply by $\frac{1}{4}$.

a $34 \div 4 = \frac{34}{4} = \frac{17}{2}$ ② Write as a fraction.   **b** $16\frac{1}{2} \div 4 = 16 \div 4 + \frac{1}{2} \times \frac{1}{4}$ ②

$= 8\frac{1}{2}$ inches   ③ Convert to mixed number. $= 4 + \frac{1}{8}$

Width is $4\frac{1}{8}$ inches   ③ Whole + fraction

c $1\frac{3}{4} \div 4 = \frac{7}{4} \times \frac{1}{4}$ ② Improper fraction $\times \frac{1}{4}$ **d** $\frac{7}{8} \div 4 = \frac{7}{8} \times \frac{1}{4}$

$= \frac{7}{16}$ inches   ③ Include units. $= \frac{7}{32}$ inches   ③

EXAMPLE

A bicycle wheel rotates 3000 times.

a Write an exact expression for how far the bicycle travels.

b The same bicycle travels $6\frac{1}{2}$ km. Write an exact expression for how many times the wheel rotates.

50 cm

① Multiply the circumference of the wheel by the number of rotations.

a $C = 50 \times \pi \text{ cm} = \frac{1}{2}\pi \text{ m}$ ② 50 cm $= \frac{1}{2}$ m use exact fractions.

Distance $= \frac{1}{2}\pi \times 3000$ The wheel rotates 3000 times.

$= 15000\pi \text{ m}$ ③ Simplify.

b Divide the distance by the circumference.

② Distance $= 6\frac{1}{2}$ km $= 6500$ m   Change to metres.

Number of rotations $= \dfrac{6500}{\frac{1}{2}\pi}$

$= \dfrac{13000}{\pi}$ ③ Simplify.

Leaving π in your answers actually makes working easier!

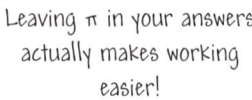

Exercise 17.2A

Give exact answers to all the questions in this exercise.

1 a Joshua's bicycle wheel has a diameter of 0.4 m. On his journey to work it rotates 5500 times. Write an exact expression for how far he travels.

b Kim's wheel has a radius of 35 cm and it rotates 4000 times. Which expression is correct for how far he travels?

    A $140\,000\,\pi$ m   **B** $2800\,\pi$ m

    C $280\,000\,\pi$ m   **D** $1400\,\pi$ m

c Lex's wheel has a diameter of 600 mm and she travels 9 km. Write an exact expression for how many times her wheel rotates.

d Whose wheel rotates the most: Joshua, Kim or Lex?

e Who has the longest journey?

2 Simplify each expression and arrange them in order of size from smallest to largest.

a
$$4\tfrac{2}{3} + 1\tfrac{1}{2}, \quad 4\tfrac{7}{8} \div \tfrac{3}{4}, \quad 2\tfrac{1}{4} \times 2\tfrac{5}{6}, \quad 7\tfrac{1}{3} - \tfrac{7}{8}$$

b
$$8 \times (\sqrt{49} - 12 \div 3)\pi, \quad \tfrac{400\pi}{12},$$
$$(4\pi)^2 + 2\pi(15 - 8\pi), \quad (50 - 2^2 \times 4)\pi$$

3 Paper used to be produced in *foolscap folio* size, which is $13\tfrac{1}{2}$ inches tall by $8\tfrac{1}{2}$ inches wide. Other sizes were calculated from this.

foolscap   same height, double width

foolscap quanto   same width, half height

foolscap octavo   half width, half height

a Work out the height and width of

    i foolscap **ii** quarto **iii** octavo.

b 16 octavo sheets are used for a display. They are lined up in 2 rows of 8 sheets.

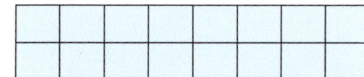

    i How long is the display?

    ii How high is the display?

    iii What is the total area?

c Colin says that 8 octavo sheets make one foolscap sheet. Is he correct? Give your reasons.

---

4 A square tile has area $4\,\text{cm}^2$. A quadrant has area $\pi\,\text{cm}^2$.

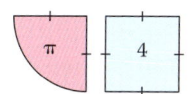

a Choose an expression to describe the total area of each of these shapes.

    i   **ii**

    iii   **iv**

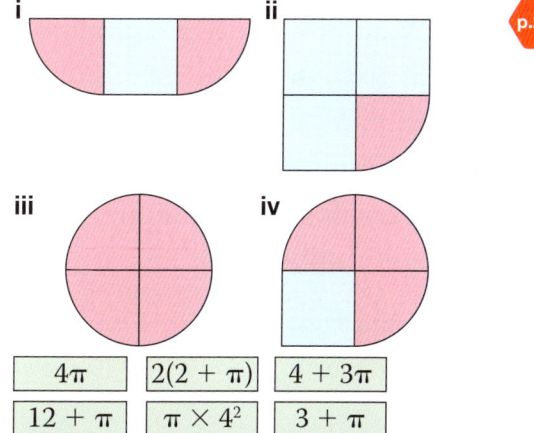

4π	$2(2 + \pi)$	$4 + 3\pi$
$12 + \pi$	$\pi \times 4^2$	$3 + \pi$

p.230

*b Explain why the quadrant and square fit together exactly.

5 A tin can has a base but no lid. It is made from metal which is $\tfrac{1}{32}$ of an inch thick. Work out

p.312

a the area of the base

b the circumference

c the curved surface area

*d the volume of metal used

*e the volume of the can.

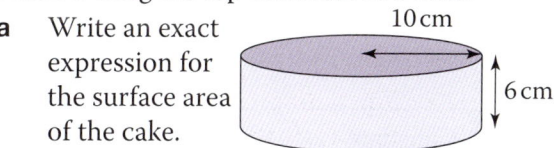

3 inches

$4\tfrac{1}{4}$ inches

6 Anita is icing the top and sides of a cake.

a Write an exact expression for the surface area of the cake.

10 cm

6 cm

b She has a circular sheet of icing with diameter 300 mm. Does she have enough icing to cover the cake? Give your reasons.

*c She adds another layer to the cake. James says that the new SA is $430\pi\,\text{cm}^2$. Is he correct? Give your reasons.

5 cm

6 cm

*d She adds a third layer to the cake, of height 6 cm and radius 15 cm. She buys four sheets of icing to cover the whole cake. How much icing will she have left?

1017, 1040, 1047   SEARCH

17.3 Standard form

You can use **standard form** to represent large and small numbers.

- In standard form, a number is written as $A \times 10^n$, where $1 \leq A < 10$ and n is an integer.

EXAMPLE

Write these numbers in standard form.

a 235 b 0.23×10^6

c 0.45 d 0.000 000 416

a $235 = 2.35 \times 10^2$

b $0.23 \times 10^6 = 2.3 \times 10^5$

c 4.5×10^{-1}

d 4.16×10^{-7}

EXAMPLE

Write these numbers in order, starting with the smallest.

6.35×10^4, 5.44×10^4, 6.95×10^3, 7.075×10^2, 9.9×10^{-1}

The correct order is

9.9×10^{-1}, 7.075×10^2, 6.95×10^3, 5.44×10^4, 6.35×10^4

First order the powers of 10.
$10^{-1} < 10^2 < 10^3 < 10^4$

Then compare numbers with the same powers of 10.
$5.44 < 6.35$

- To multiply or divide numbers given in standard form, multiply or divide the numbers and then add or subtract the powers.

EXAMPLE

Calculate and give your answers in standard form.

a $(3.6 \times 10^5) \div (1.2 \times 10^3)$ b $(5.4 \times 10^4) \times (2 \times 10^{-3})$

a $(3.6 \times 10^5) \div (1.2 \times 10^3) = (3.6 \div 1.2) \times (10^5 \div 10^3)$
$= 3 \times 10^{5-3} = 3 \times 10^2$

b $(5.4 \times 10^4) \times (2 \times 10^{-3}) = (5.4 \times 2) \times (10^4 \times 10^{-3})$
$= 10.8 \times 10^{4-3} = 1.08 \times 10^2$

- To add or subtract numbers given in standard form, write the numbers as ordinary numbers and then add or subtract as usual. Give your answer in standard form.

EXAMPLE

Calculate
$3.2 \times 10^5 + 7.1 \times 10^4$.
Give your answer in standard form.

$3.2 \times 10^5 + 7.1 \times 10^4 = 320\,000 + 71\,000$
$= 391000$
$= 3.91 \times 10^5$

- You need to know how to enter standard form calculations into your calculator and how to interpret the display.

EXAMPLE

Use your calculator to work out $(6.43 \times 10^6) \div (4.21 \times 10^{-2})$.

$6.43 \times 10^6 \div 4.21 \times 10^{-2}$
152731591.4

1.53×10^8 (3 sf)

Exercise 17.3S

1 Write these numbers as powers of 10.

   a   100   **b**   10

   c   1000   **d**   1

   e   10 000   **f**   1 000 000

   g   100 000   **h**   100 000 000

2 Write these numbers as powers of 10.

   a   0.01   **b**   0.1

   c   0.001   **d**   0.00001

   e   0.0001   **f**   0.000 000 1

   g   0.000 001   **h**   1.0

3 Convert these powers to ordinary numbers.

   a 10^3 **b** 10^6

   c 10^5 **d** 10^9

   e 10^4 **f** 10^1

   g 10^2 **h** 10^7

4 Convert these powers to ordinary numbers.

   a 10^0 **b** 10^{-2}

   c 10^{-5} **d** 10^{-3}

   e 10^{-7} **f** 10^{-1}

   g 10^{-4} **h** 10^{-6}

5 Work out these calculations and give your answers in index form.

   a $10^2 \times 10^3$ **b** $10^4 \times 10^5$

   c $10^5 \times 10^3$ **d** $10^6 \div 10^3$

   e $10^8 \div 10^4$ **f** $10^6 \div 10^2$

6 Work out these calculations and give your answers in index form.

   a $10^4 \div 10^6$ **b** $10^3 \div 10^7$

   c $10^2 \div 10^{10}$ **d** $10 \div 10^9$

   e $1 \div 10^8$ **f** $10^7 \div 10$

7 Write these numbers in standard form.

   a   200   **b**   800

   c   9000   **d**   650

   e   6500   **f**   952

   g   23.58   **h**   255.85

   i   0.3   **j**   0.0047

   k   0.000 078   **l**   0.4485

8 These numbers are in standard form. Write each of them as an 'ordinary' number.

   a 5×10^2 **b** 3×10^3

   c 1×10^5 **d** 2.5×10^2

   e 4.9×10^3 **f** 3.8×10^6

   g 7.5×10^{11} **h** 8.1×10^{18}

9 Although they are written as multiples of powers of 10, these numbers are not in standard form. Rewrite each of them correctly in standard form.

   a 60×10^1 **b** 45×10^3

   c 0.65×10^1 **d** 0.05×10^8

   e 28×10^{-2} **f** 0.4×10^{-1}

   g 13.5×10^{-4} **h** 12×10^{-8}

10 Evaluate these calculations. Give your answers in standard form. Do not use a calculator.

   a $(2 \times 10^2) \times (2 \times 10^3)$

   b $(3 \times 10^4) \times (3 \times 10^3)$

   c $(5 \times 10^3) \times (5 \times 10^4)$

   d $(8 \times 10^7) \times (3 \times 10^5)$

   e $(2.5 \times 10^{-3}) \times (2 \times 10^2)$

   f $(4.6 \times 10^{-6}) \times (2 \times 10^{-2})$

11 Evaluate these calculations, showing your working. Do not use a calculator. Give your answers in standard form.

   a $(4 \times 10^4) \div (2 \times 10^2)$

   b $(8.4 \times 10^9) \div (4.2 \times 10^5)$

   c $(2 \times 10^6) \div (4 \times 10^4)$

   d $(3 \times 10^5) \div (4 \times 10^2)$

12 Use a calculator to find the value of these in standard form.

   a $(6.4 \times 10^{-4}) + (7.1 \times 10^{-3})$

   b $(9.9 \times 10^5) - (2.7 \times 10^4)$

   c $(4.8 \times 10^{-6}) + (3.9 \times 10^{-5})$

   d $(3.3 \times 10^2) - (7.5 \times 10^1)$

   e $(9.8 \times 10^5) - (6.4 \times 10^5)$

   f $(3.5 \times 10^{-2}) + (9.7 \times 10^{-3})$

1049, 1050, 1051   SEARCH

17.3 Standard form

- You can write a number in standard form as $A \times 10^n$, where n is a positive or negative integer and $1 \leqslant A < 10$.

- To multiply or divide numbers written in standard form, either use a calculator or multiply or divide the number part and add or subtract the powers of 10.

- To add or subtract numbers written in standard form, either use a calculator or write the numbers as 'ordinary' numbers.

$5\underset{\frown}{3\,7\,0} = 5.37 \times 10^3$

$0.\underset{\frown}{0\,3\,8} = 3.8 \times 10^{-2}$

$4 \times 10^5 \times 3 \times 10^2$

$= (4 \times 3) \times 10^{5+2}$

$= 12 \times 10^7$

$= 1.2 \times 10^8$

HOW TO

To solve problems involving standard form

① RTQ and decide which calculation you need to carry out.

② Calculate using your knowledge of standard form.

③ Give your answer in standard form and check that your answer is sensible.

$4 \times 10^5 + 3 \times 10^2$

$= 400\,000 + 300$

$= 400\,300$

$= 4.003 \times 10^5$

EXAMPLE

The mass of a carbon atom is 2×10^{-23} g.

How many atoms are there in one gram of carbon?

① Divide 1 g by the mass of one carbon atom.

② Write both numbers in standard form.

$1 \div (2 \times 10^{-23}) = 1 \times 10^0 \div (2 \times 10^{-23})$

Divide the numbers and subtract the powers of 10.

$= (1 \div 2) \times (10^0 \div 10^{-23})$

$= 0.5 \times 10^{0 - {}^{-23}}$

$= 0.5 \times 10^{23}$

③ Give your answer in standard form.

There are 5×10^{22} atoms in one gram of carbon.
You would expect there to be a very large number of atoms, so the answer is sensible.

Take care with negative powers!

EXAMPLE

A butterfly has a mass of 4.5×10^{-4} kg. A ladybird has a mass of 2.1×10^{-5} kg.

An ant has a mass of 3.4×10^{-5} kg. An ant can carry 20 times its own mass.

Could an ant carry a butterfly and a ladybird?

① Compare the combined mass of a butterfly and a ladybird to 20 times the mass of an ant.

Mass of butterfly and ladybird

$= 4.5 \times 10^{-4} + 2.1 \times 10^{-5}$

$= 0.00045 + 0.000021$

$= 0.000471$

$= 4.71 \times 10^{-4}$ kg

Mass an ant can carry

$= 20 \times 3.4 \times 10^{-5}$

$= (2 \times 3.4) \times (10 \times 10^{-5})$

$= 6.8 \times 10^{-4}$ kg

② Writing the answers in standard form makes it easy to compare the masses.

$6.8 \times 10^{-4} > 4.71 \times 10^{-4}$

③ An ant could carry a butterfly and a ladybird.

Number Calculations 2

Exercise 17.3A

1 Write these measurements using standard form.

   a Ten metres in kilometres.

   b Two milligrams in grams

   c Five millimetres in kilometres

   d 11 millilitres in litres

2 Orla thinks she has answered her maths homework correctly in standard form. Check her answers and correct any mistakes.

   a $3.2 \times 10^5 \times 2.5 \times 10^6 = 8 \times 10^{11}$

   b $5.2 \times 10^8 \times 6.3 \times 10^4 = 32.8 \times 10^{12}$

   c $4 \times 10^5 + 3 \times 10^4 = 7 \times 10^9$

   d $8 \times 10^3 \div 5 \times 10^6 = 1.6 \times 10^{-3}$

3 The American value for a billion is 1×10^9. The British value for a billion is 1×10^{12}. How many times smaller is an American billion than a British billion?

4 The mass of the Sun is 2×10^{30} kg. The mass of the Earth is 6×10^{24} kg. Joe says that the Sun is six million times heavier than the Earth. Is he correct? Give your reasons.

5 The mass of one atom of the element Mercury is 3.3×10^{-22} g. The mass of the planet Mercury is 3.3×10^{23} kg. How many Mercury atoms are there in 3.3×10^{23} kg?

6 The width of a plant cell is 60 micrometres. A micrometre is 1×10^{-6} m (one millionth of a metre). The diagram of a plant cell in a science textbook has width 3 cm. How many times bigger is the diagram than the real plant cell?

7 Light travels about 3×10^8 metres per second.

   a Find the time it takes for light to travel 1 metre.

   b Find the distance light travels in 1 year. Give your answers in standard form.

   c The star Vega is about 2.37×10^{17} m away from Earth. How long will it take the light from Vega to reach Earth? Give your answer in standard form to 3 sf.

8 A mass of 14 grams of nitrogen contains about 6.02×10^{23} nitrogen atoms.

   Find the mass of one nitrogen atom, giving your answer in kg in standard form, correct to 2 sf.

9 A bumblebee has a mass of 5.2×10^{-5} kg. An adult man has a mass of 70 kg. A bumblebee can carry 75% of its mass. How many bumblebees would it take to lift a man? Give your answer in standard form to 3 sf.

10 As the moon orbits Earth the distance between them varies between 4.06×10^5 km and 3.63×10^5 km. Find the difference between these two distances.

11 The masses of the eight planets in our solar system are listed in the table.

Mercury	3.30×10^{23}
Venus	4.87×10^{24}
Earth	5.97×10^{24}
Mars	6.42×10^{23}
Jupiter	1.90×10^{27}
Saturn	5.68×10^{26}
Uranus	8.68×10^{25}
Neptune	1.02×10^{26}

The masses are given in kg.

Carrie calculated the total mass of the planets. Her answer is 7.68612×10^{26} kg.

   a Which planet did Carrie forget to include in her total?

   b Use your calculator to work out the correct total mass.

   c Carrie says that Saurn is about three times the mass of Jupiter. Is she correct? Give your reasons.

   d Which planet is about 3000 times the mass of Mars?

   e Which two planets have the most mass?

   f How many 'Earth masses' is Mercury?

1049, 1050, 1051   SEARCH

Summary

Checkout

You should now be able to...

Test it

Questions

✔ Calculate with roots and with integer indices.	**1, 2**
✔ Calculate exactly with fractions and multiples of π.	**3 – 6**
✔ Calculate with and interpret numbers written in standard form.	**7 – 9**

Language	Meaning	Example
Index **Base** **Power**	In index notation, the index or power shows how many times the base has to be multiplied by itself. The plural of index is **indices**.	Index $2^3 = 2 \times 2 \times 2 = 8$ Base
Index laws	The rules for how to multiply, divide or raise to a power numbers written as powers of the *same* base.	$5^3 \times 5^2 = 5^{3+2} = 5^5$ $5^3 \div 5^2 = 5^{3-2} = 5^1 = 5 \qquad 5^0 = 1$ $(5^3)^2 = 5^{3 \times 2} = 5^6$
Square root	A number that when multiplied by itself is equal to the number underneath the square root symbol.	$4 \times 4 = 16$ is the square of 4 $\sqrt{16} = 4$ is the square root of 16
Cube root	A number that when multiplied by itself and then by itself again is equal to the number underneath the cube root symbol.	$4 \times 4 \times 4 = 64$ is the cube of 4 $\sqrt[3]{64} = 4$ is the cube root of 64
Terminating	A decimal with a definite number of digits.	$\frac{1}{8} = 0.125$
Recurring	A decimal with a repeating pattern that goes on forever.	$\frac{1}{3} = 0.333... = 0.\dot{3}$ $\frac{9}{11} = 0.818181... = 0.\dot{8}\dot{1}$
Pi, π	The number obtained by dividing the circumference of any circle by its diameter.	Circumference = π × diameter π = 3.141 592 653 589 793 (15 dp)
Exact calculation	A calculation that does not involve decimals that have been rounded or other approximations. Exact answers are given in terms of integers, fractions and π.	$2 \times \pi \times \left(\frac{1}{3}\right)^2 + 2\pi \times \frac{1}{3}$ $= 2\pi \times \frac{1}{9} + \frac{2}{3}\pi - \left(\frac{2}{9} + \frac{2}{3}\right)\pi$ $= \frac{8}{9}\pi$
Standard form	A number written as a decimal between 1 and 10 multiplied by a power of 10.	$498\,000 = 4.98 \times 10^5$ $0.0056 = 5.6 \times 10^{-3}$

Review

1 Simplify these expressions.

    a $5^3 \times 5^7$ **b** $3^{12} \div 3^8$

    c $(7^4)^3$ **d** $2^4 \times 2^3 \div 2^7$

    e $\dfrac{3^2 \times 3^7}{3^3 \times 3^3}$ **f** $\left(\dfrac{4^8 \div 4^3}{4^3}\right)^2$

2 Write down the numerical value of these expressions.

    a $2^3 \times 2^4$ **b** $4^3 \div 4$

    c 5^1 **d** $(2^3)^2$

    e 8^0 **f** $3^{12} \div 3^9$

    g $\sqrt{36}$ **h** $\sqrt[3]{8}$

3 Find the exact answers to these calculations.

    a $\dfrac{3}{10} \times 2$ **b** $\dfrac{2}{7} \times \dfrac{3}{4}$

    c $\dfrac{2}{5} + \dfrac{1}{5}$ **d** $\dfrac{7}{8} - \dfrac{1}{3}$

    e $\dfrac{9}{8} + \dfrac{1}{6}$ **f** $3\dfrac{1}{4} - 1\dfrac{1}{5}$

    g $\dfrac{11}{3} \times \dfrac{2}{3}$ **h** $3\dfrac{1}{7} \times \dfrac{1}{4}$

    i $\dfrac{6}{7} \div 3$ **j** $\dfrac{1}{9} \div \dfrac{2}{3}$

4 Calculate the areas of these shapes. All measurements are in metres.

    a 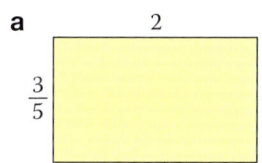   **b**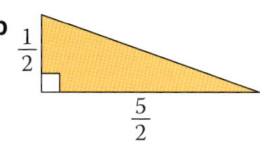

5 Simplify these expressions.

    a $5 + 2\pi - 3 + \pi$ **b** $\pi(3^2 - \sqrt{16})$

    c $\dfrac{1}{2} \times \dfrac{\pi}{3} + \dfrac{\pi}{4}$ **d** $\dfrac{4}{3}\pi\left(\dfrac{3}{2}\right)^3$

6 **a** Calculate the exact areas of the circle and the semi-circle.

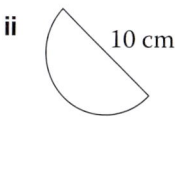

    b Calculate the exact circumference of the circle and the perimeter of the semi-circle.

7 Write these approximations in standard form.

    a The population of China in 2014: 1 370 000 000

    b The closest Earth comes to Mars: 54 600 000 miles

    c The amount of vitamin C in an orange: 0.0697 g

    d The size of a grain of sand: 0.0000625 m

8 Write these as ordinary numbers.

    a 3.5×10^5 **b** 8.21×10^8

    c 2.7×10^{-3} **d** 2.07×10^{-7}

9 Calculate these and leave your answer in standard form.

    a $(2 \times 10^7) \times (3 \times 10^3)$

    b $(3.5 \times 10^6) \times (2 \times 10^{-2})$

    c $(8 \times 10^9) \div (2 \times 10^4)$

    d $(1.2 \times 10^{-3}) \div (2 \times 10^{-2})$

    e $2.4 \times 10^5 + 2.4 \times 10^3$

    f $3.6 \times 10^3 - 1.8 \times 10^2$

What next?

Score			
	0 – 4		Your knowledge of this topic is still developing. To improve look at MyMaths: 1017, 1033, 1040, 1047, 1049, 1050, 1051, 1924
	5 – 8		You are gaining a secure knowledge of this topic. To improve your fluency look at InvisiPens: 17Sa – c
	9		You have mastered these skills. Well done you are ready to progress! To develop your problem solving skills look at InvisiPens: 17Aa – c

Assessment 17

1   Peter says that   **a** $0^2 = 1$ **b** $49^{\frac{1}{2}} = 7$ **c** $(-3)^2 = -9$ [1]

Soraya says that   **a** $0^2 = 0$ [1]   **b** $49^{\frac{1}{2}} = \frac{1}{49}$ [1]   **c** $(-3)^2 = 9$

Without using your calculator, for each value who is correct?
Give your reasons.

2   Work out the following using the appropriate keys on a calculator.

   a 2.1^3 [1]   **b** 7.2^4 [1]   **c** 1.99^{10} [1]

   d $\sqrt{841}$ [1]   **e** $\sqrt{77}$ [1]   **f** $\sqrt[3]{654.321}$ [1]

3   **a** Paige says that to work out $15^7 \times 15^5$ you multiply the indices. Is she correct?
Work out the expression and leave your answer as a single power.   [1]

   b Hayley says that to work out $(3^4)^5$ you multiply the indices. Is she correct?
Work out the expression and leave your answer as a single power.   [1]

   c Matt says that $(3^4)^0 = 3^4$. Eliza says that $(3^4)^0 = 1$. Lois says that $(3^4)^0 = 3$.
Who is correct? Give reasons for your answer.   [2]

   d Mike says that $7^7 \times 7^2 \div 7^6 = 7^8$. Show that Mike is incorrect.   [1]

4   A dice is a cube of side 2.3 cm. Calculate the volume of the dice.   [2]

5   **a** Joe is batting in a cricket tournament. He scores 6 sixes in each of 6 games
every day for 6 days. How many sixes does Joe score?   [2]

   b Joss scores x sixes in each of x games every day for x days, getting 125 sixes overall.
Find x.   [2]

6   Solve each of these equations to find the unknown.

   a $4^2 \times 64 = 4^a$ [2]   **b** $8^b = 4^6$ [2]   **c** $(3^2)^4 = 9^c$ [2]

7   Use four *different* digits to make this calculation correct. $\frac{\Box}{\Box} + \frac{\Box}{\Box} = \frac{1}{2}$
Write down four different answers.   [4]

8   240 pupils sat an exam and only 48 answered the last question correctly.

   a What fraction of the pupils got the question right?   [2]

   b How many pupils got the question wrong?   [3]

9   The Chiefs played the Saints at rugby. There were 55 points scored in the match.

The Chiefs scored $\frac{8}{11}$ of those points.

   a How many points did the Chiefs score?   [2]

   b How many points did the Saints score?   [1]

10 A vegetable box has potatoes, carrots and leeks in it.

$\frac{3}{5}$ of the box is potatoes and $\frac{1}{3}$ is carrots. There are 4 leeks.

   a What fraction of the box is made up of potatoes and carrots together?   [2]

   b What fraction of the box is leeks?   [1]

   c How many potatoes are there in the box?   [4]

11 Romeo and Juliet are saving up to get married.

Romeo saves $\frac{7}{25}$ of his salary and Juliet saves $\frac{7}{20}$ of hers.

Romeo's annual salary is £15 500 and Juliet's is £13 400.

    a Who saves most annually and by how much? [4]

    b How much do they save in total during the year? [2]

12 An athlete has a mass of $13\frac{3}{4}$ stone when he starts training.

After six months, his mass is $9\frac{5}{8}$ stone.

What fraction of his original mass is his new mass? [3]

13 a The areas of some of our oceans and seas in mi^2 are shown.
Convert them to ordinary numbers and write them in increasing order of size.

Malay Sea	$3.14 \times 10^6 \, \text{mi}^2$	Indian Ocean	$2.84 \times 10^7 \, \text{mi}^2$
Bering Sea	$8.76 \times 10^5 \, \text{mi}^2$	Caribbean Sea	$1.06 \times 10^6 \, \text{mi}^2$
English Channel	$2.9 \times 10^4 \, \text{mi}^2$	Baltic Sea	$1.46 \times 10^5 \, \text{mi}^2$

[4]

    b Which is bigger: 1×10^9 or 999 999 999 and by how much? [2]

    c Find the value of n in each of the following equations.

        i $4.7 \times 10^n = 47\,000$ [2]   **ii** $6.81 \times 10^n = 681$ [1]

        iii $3.467 \times 10^n = 3\,467\,000$ [1]   **iv** $27.5 \times 10^n = 0.0275$ [2]

    d Rewrite each of these sentences using ordinary numbers.

        i The energy released by the wingbeat of a bee is 8×10^{-4} Joules/second. [1]

        ii The distance from Mexico City to Moscow is 1.0763×10^4 km. [1]

    e Rewrite this sentence using standard form.

    The average length of a bedbug is $\frac{4}{1\,000\,000}$ km thick. [1]

    f What is this length in mm? [1]

14 a The Wright Brothers Flyer I, the world's first aircraft, had a mass of 3.4×10^2 kg.
The Saturn V Rocket had a mass of 2.96×10^6 kg.
How many times heavier is a Saturn V than Flyer I? [2]

    b Flyer I attained a speed of $3.04 \times 10^0 \, \text{ms}^{-1}$ on its first flight and the supersonic airliner, Concorde, attained a speed of $2.179 \times 10^3 \, \text{kmh}^{-1}$.
How much faster was the Concorde than Flyer I? [2]

15 There are approximately 4.336×10^9 stars in our galaxy and about 5.776×10^3 stars visible to the naked eye.

    a What fraction of the galaxy can we see? [2]

    b Write your fraction in the form $\frac{1}{x}$. [1]

16 A triathlon has 3 stages, the largest triathlon has a 3.8×10^2 km swim, 1.8×10^2 km cycle ride and 0.42195×10^2 km run.

How far is the race in full? Give your answer to 3 sf

    a as an ordinary number [3]

    b in standard form. [1]

18 Graphs 2

Introduction

When you look at a ball in flight, you already know certain things about its path, or trajectory. It will travel in a smooth curve. It will reach a maximum point. Its downward path will tend to be a mirror image of the upwards path. In football, a goalkeeper knows this as well, so a striker might put spin on the ball to make its flight less predictable.

The path of the ball can be modelled mathematically by a type of equation called a quadratic equation, which is described in this chapter.

What's the point?

An appreciation of quadratic equations and their graphs enables us to understand how an object moves under gravity, and tells us where it's likely to land!

Objectives

By the end of this chapter you will have learned how to …

- Draw graphs to identify and interpret roots, intercepts and turning points of quadratic functions.
- Solve a quadratic equation by finding approximate solutions using a graph.
- Recognise, sketch and interpret graphs of linear, quadratic cubic and reciprocal functions.
- Plot and interpret real-life graphs.

Check in

1 Substitute the following values into the expression $x^3 - 2x$.

    a 2   **b** 2.1   **c** -3.2

2 Plot these points on a coordinate grid.

    a $(3, 6)$ **b** $(-4, 2)$ **c** $(5, -3)$ **d** $(-3, -1)$

3 Copy and complete this table of values for the equation $y = 2x + 5$

x	−2	−1	0	1	2
y					

Chapter investigation

It takes 20 people 18 days to build an extension to a sports centre.

How long would it take one person? State any assumptions that you have made.

Find out how long the job would take for different numbers of people.
Can you draw a graph to show this information?

18.1 Properties of quadratic functions

Graphs of linear functions, $y = mx + c$, are always straight lines.

p.210

Graphs of **quadratic functions**, $y = ax^2 + bx + c$, give ∪- or ∩-shaped curves that are symmetrical about a vertical line.

- A point where the function changes from rising to falling, ∩, or from falling to rising, ∪, is called a **turning point**.

- The **x-intercepts** of the graph $y = f(x)$ are the **roots** of the function $f(x) = 0$.

  - An input that gives zero as the output.

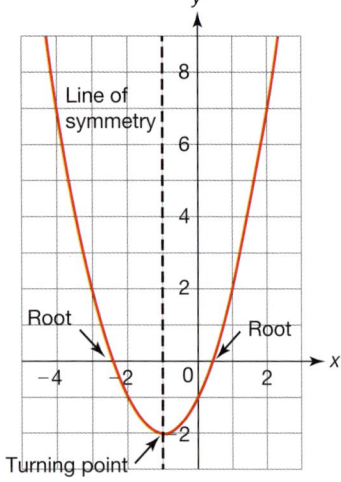

▲ The quadratic function $y = x^2 + 2x - 1$

p.290

EXAMPLE

a Complete the table of values for the function $y = x^2 - 4x$.

x	−1	0	1	2	3	4	5	6
y	5		−3				5	

To draw a graph you must plot the points accurately.

b Draw the graph of $y = x^2 - 4x$.

c Using the graph, or otherwise, find

    i the y-intercept   **ii** the coordinates of the turning point

    iii the roots   **iv** the equation of the line of symmetry.

---

a Work out the value of y for each value of x.

$0^2 - 4 \times 0 = 0$, $2^2 - 4 \times 2 = 4 - 8 = -4$,

$3^2 - 4 \times 3 = 9 - 12 = -3$, ...

x	−1	0	1	2	3	4	5	6
y	5	0	−3	−4	−3	0	5	12

b Plot the eight points on a set of axes.

c **i** 0   The graph crosses the y-axis at $y = 0$.

    ii $(2, -4)$ The turning point is a minimum.

    iii 0 and 4   The graph crosses the x-axis at $x = 0$ and $x = 4$.

    iv $x = 2$ The vertical line of symmetry goes through the turning point.

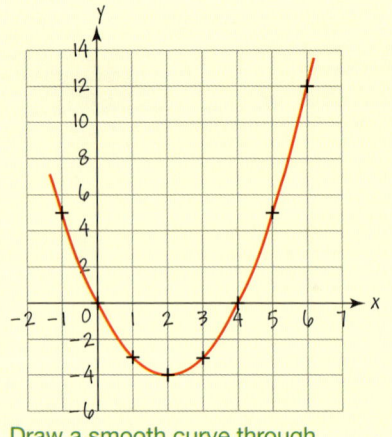

Draw a smooth curve through your points.

Exercise 18.1S

1 Copy and complete the tables of values for each of these functions.

a $y = x^2 + 2$

x	−3	−2	−1	0	1	2	3
y			3				

b $y = x^2 - 2$

x	−2	−1	0	1	2	3	4
y						7	

c $y = x^2 + x$

x	−3	−2	−1	0	1	2	3
y			0				

d $y = x^2 + 3x$

x	−4	−3	−2	−1	0	1	2
y							10

e $y = 2 - x^2$

x	−3	−2	−1	0	1	2	3
y	−7						

f $y = x^2 + 2x + 1$

x	−4	−3	−2	−1	0	1	2
y		4					

g $y = 2x^2$

x	−3	−2	−1	0	1	2	3
y							18

h $y = 2x^2 + x$

x	−2	−1.5	−1	−0.5	1	0.5	1
y		3					

2 For each part of question **1**

i Construct a coordinate grid that can be used to plot all the points found.

ii Draw the graph of the function.

3 For each part of question **1**

i write the value of the y-intercept

ii write the coordinates of the turning point

iii write the roots of the function.

4 a One of the graphs in question **1** has a highest point which graph is it?

b How is the function for this graph different from the functions of the graphs with lowest points?

5 i Draw the graphs of these functions on separate coordinate grids. Choose your axes carefully.

Write an estimate for

ii the coordinates of the turning point

iii the roots of the function.

    a $y = x^2 + x - 3$ **b** $y = x^2 - 3x + 1$

6 i Draw the graphs of these functions on separate coordinate grids.

Label each graph with

ii the value of the y-intercept

iii the coordinates of the turning point

iv the roots of the function.

    a $y = 4 - x^2$ **b** $y = 2 + 2x - x^2$

*7 Draw the graph of $y = 5 + 4x - 2x^2$.

8 Use algebra to find the roots of these functions.

    a $y = x^2 + 7x$ **b** $y = x^2 - 7x$

    c $y = x^2 + 5x + 6$ **d** $y = x^2 + 7x + 12$

    e $y = x^2 + 4x + 4$ **f** $y = 6 + 5x - x^2$

9 a Draw graphs of these four functions on the same coordinate grid.

    A $y = 4 + 3x - x^2$

    B $y = x^2 + 2x - 4$

    C $y = 2x^2 - 4$

    D $y = x^2 - 3x - 4$

b i Which graphs have the same y-intercept?

ii Which graphs have the same roots?

Q 1168, 1169   SEARCH

18.1 Properties of quadratic functions

RECAP

- The graph of a quadratic function, $y = ax^2 + bx + c$, is a ∪- or ∩-shaped curve with a vertical line of symmetry.
- The highest or lowest point on the curve is called a turning point.
- The turning point lies on the line of symmetry.
- The x-intercepts of the graph are the roots of the function.

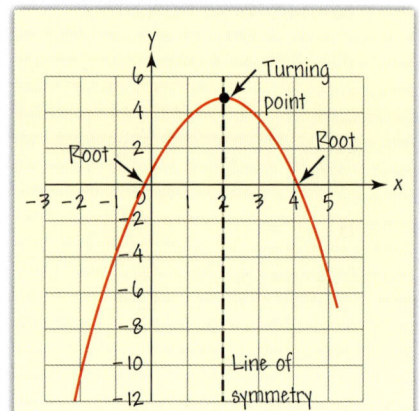

HOW TO

To solve problems involving quadratic functions, $f(x) = ax^2 + bx + c$
1. **Draw** or **sketch** the function $y = f(x)$.
2. To find the roots of the function either find the x-intercepts of the graph or solve $f(x) = 0$ by factorising.
3. To find the y-intercept either find where the graph crosses the y-axis or use $f(0) = c$.
4. Work out the coordinates of the turning point.
5. Interpret these values in the context of the problem.

EXAMPLE

a   Show that $x = 1$ is one of the roots of $y = x^2 - 7x + 6$.
b   Find the second root of the function.
c   Find the coordinates of all the intercepts of the graph of $y = x^2 - 7x + 6$.
d   Find the coordinates of the turning point.

p.210

a $1^2 - 7 \times 1 + 6 = 0$ ①

b $x^2 - 7x + 6 = (x - 1)(x - 6)$ ② Since $x = 1$ is a root $(x - 1)$ is a factor.

   $= 0 \implies (x - 1) = 0$ or $(x - 6) = 0$

   $x = 1$ or $x = 6$

c   The x-intercepts are $(1, 0)$ and $(6, 0)$.

   $x = 0 \implies y = 6$. The y-intercept is $(0, 6)$.   ③

d $(6 + 1) \div 2 = 3.5 \implies x = 3.5$ ④ Use symmetry.

   $y = 3.5 \times 3.5 - 7 \times 3.5 + 6 = -6.25$

   The turning point is $(3.5, -6.25)$

EXAMPLE

Tayla kicks a ball in the air. Its height is given by the equation $y = 25x - 5x^2$ where y is the height in metres and x is the time in seconds.

p.290

a   Draw a graph to show the ball's path from 0 to 6 seconds.

b   When does the ball hit the ground?

c   At what time is the ball highest?

a

x	0	1	2	3	4	5	6
y	0	20	30	30	20	0	−30

① Use a table of values.

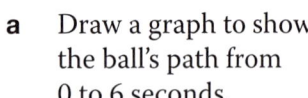

b   ② Find the roots.

   After 5 seconds.   ⑤

c   ④ Find the turning point.

   After 2.5 seconds.   ⑤

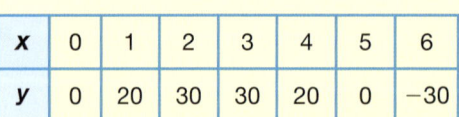

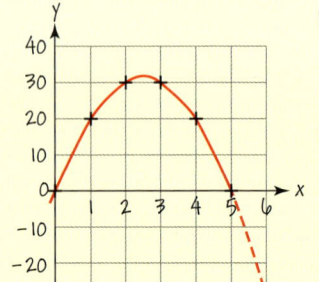

Exercise 18.1A

1　**a**　Is $x = 3$ a root of the function $f(x) = x^2 - 2x - 3$?

　　b　Find the second root of the function.

　　c　Find the coordinates of all the intercepts of the graph of $y = x^2 - 2x - 3$

　　d　Find the coordinates of the turning point.

2　For the function $f(x) = x^2 + 12x + 27$

　　a　find the roots

　　b　find the coordinates of the points where the graph $y = f(x)$ crosses the axes

　　c　find the coordinates of the turning point

　　d　sketch the graph of $y = f(x)$.

3　**a**　Is $x = 0$ a root of $f(x) = 6x - x^2$?

　　b　Find the second root of the function.

　　c　At what point does the function take its highest value?

4　Sketch graphs of these functions showing clearly all their main features.

　　a　$y = 10 + 3x - x^2$

　　b　$y = x^2 - 6x + 9$

　　*c　$y = 2x^2 - 3x - 5$

5　On a copy of this Venn diagram place each of these functions in the correct position.

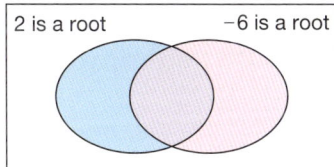

　　A　$y = x^2 + 2x - 8$　**B**　$y = x^2 + x - 42$
　　C　$y = x^2 - 7x + 10$　**D**　$y = x^2 + 12x + 20$
　　E　$y = x^2 + 7x + 6$　**F**　$y = x^2 + 4x - 12$

Use graphing software for questions **6** to **9**.

6　The height above ground of a javelin is modelled by the function $h = 100 + 48t - t^2$ where h = height in centimetres and
　　　　t = length of throw in metres.

　　a　Plot the graph of the function.

　　b　Find the maximum height of the javelin.

　　c　State the length of the throw.

7　Shot balls used to be made by dropping molten lead from the top of a tower into a pool of water. After lead is dropped, its height (in metres) above ground is given by the formula $h = 44.1 - 4.9t^2$, where t is time in seconds.

　　a　Draw the graph of the function.

　　b　What height is the lead dropped from?

　　c　After how many seconds does the lead land in the pool of water?

8　Some quadratic equations have no solutions. A graph can be used to work out when this is the case. Write the number of solutions to each of these equations.

A quadratic equation can have 0, 1 or 2 solutions.

　　a　$x^2 + 8x + 2 = 0$

　　b　$3x^2 - 9x + 8 = 0$

　　c　$2x^2 + 28x + 98 = 0$

　　d　$x^2 + x - 3 = 0$

9　**a**　Plot the graph of $y = x^2 + a$. Use the constant controller to vary the value of a. What do you notice?

　　b　Repeat for $y = (x + a)^2$.

10　The roots of the equation $x^2 + 6x + 8 = 0$ are where the graph of the function $y = x^2 + 6x + 8$ crosses the x-axis, $y = 0$.

The roots of the equation $x^2 + 6x + 8 = 3$ are where the graph of $y = x^2 + 6x + 8$ crosses the line $y = 3$.

Use a graph to solve these equations.

　　a　$x^2 + 6x + 8 = 0$

　　b　$x^2 + 6x + 8 = 3$

　　c　$x^2 + 6x + 8 = 15$

　　d　$x^2 + 6x + 8 = -1$

　　e　$x^2 + 6x + 8 = 1$

p.410

18.2 Sketching functions

p.290

You should be able to use your knowledge of linear and quadratic functions to **sketch** and interpret their graphs.

There are two other types of function you should recognise.

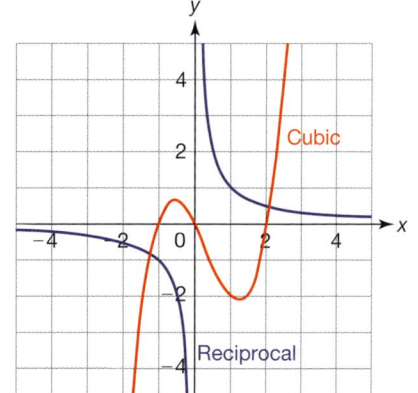

- A **cubic** function is of the form $f(x) = ax^3 + bx^2 + cx + d$.
  - A graph of $y = f(x)$ has an S-shape and the function can have 1, 2 or 3 roots.
- A **reciprocal** function is of the form $f(x) = \frac{c}{x}$.
  - A graph of $y = f(x)$ has two parts for positive and negative x values.

EXAMPLE

a Complete the table of values for the function $y = x^3 - 2x^2$.

x	−2	−1	0	1	2	3	4
y		−3					32

b Plot the graph of $y = x^3 - 2x^2$. **c** Find the roots of the function.

a $(-2)^3 - 2 \times (-2)^2 = -8 - 2 \times 4 = -8 - 8 = -16$,
$0^3 - 2 \times 0^2 = 0$, $1^3 - 2 \times 1^2 = -1$, ...

x	−2	−1	0	1	2	3	4
y	−16	−3	0	−1	0	9	32

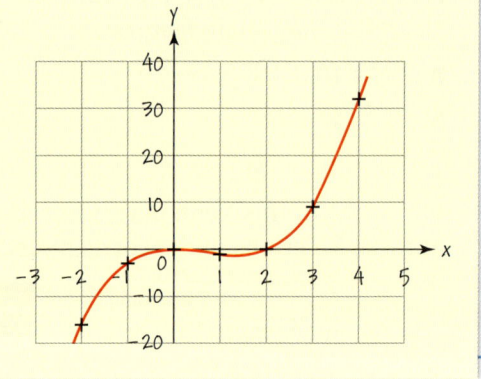

b Plot the seven points on a set of axes.
Draw a smooth curve through the points.

c Find where the curve touches the x-axis.

$x = 0$ and $x = 2$

EXAMPLE

a Complete a table of values for $y = \frac{2}{x}$ using the x-values −4, −3, −2, −1, −0.5, 0.5, 1, 2, 3, 4.

b Use your values to plot the graph of the function $y = \frac{2}{x}$.

a

x	−4	−3	−2	−1	−0.5
y	$-\frac{1}{2}$	$-\frac{2}{3}$	−1	−2	−4

x	0.5	1	2	3	4
y	4	2	1	$\frac{2}{3}$	$\frac{1}{2}$

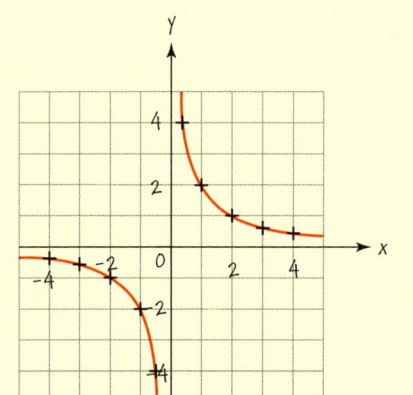

There is no output when x is zero.

b A reciprocal graph has two sections.
The curves do not touch the axes.

Exercise 18.2S

1 Copy and complete the tables of values for each of these functions.

a $y = x^3 + 1$

x	−3	−2	−1	0	1	2	3
y							

b $y = -x^3$

x	−3	−2	−1	0	1	2	3
y							

c $y = \dfrac{1}{x}$

x	−4	−3	−2	−1	1	2	3	4
y								

d $y = \dfrac{5}{x}$

x	−4	−3	−2	−1	1	2	3	4
y								

e $y = x^3 - x$

x	−1.5	−1	−0.5	−0	0.5	1	1.5	2
y								

f $y = -\dfrac{2}{x}$

x	−4	−3	−2	−1	1	2	3	4
y								

2 For each part of question **1**

i Construct a coordinate grid that can be used to plot all the points found.

ii Draw the graph of the function.

3 Sketch the graphs of these three functions on the same axes.

i Label each line.

ii Label the y-intercept of each line.

a $y = 2x$ **b** $y = 2x - 3$ **c** $y = 2x + 6$

4 Sketch the graphs of these functions on the same axes.

a $y = x^2$ **b** $y = x^2 - 2$ **c** $y = -x^2$

5 Use the information given to sketch each quadratic function.

a Roots $x = -3$ and $x = 1$
y-intercept $= 6$

b Roots $x = -2$ and $x = 8$
Highest value at the turning point $(3, 5)$

c y-intercept $= 1$
Lowest value at the turning point $(4, -3)$

6 Sketch these quadratic functions.

a $y = (x - 1)(x + 1)$

b $y = (x - 2)(x + 3)$

c $y = (x - 1)(1 - x)$

d $y = 2(x + 2)(x - 3)$

*7 Sketch this cubic function.

$y = (x - 1)(x - 2)(x - 3)$

8 Sketch the graphs of these three functions on the same axes.

a $y = \dfrac{2}{x}$ **b** $y = \dfrac{1}{x}$ **c** $y = \dfrac{5}{x}$

Label each curve.

*9 Draw the graph of the function $y = x^3 - 3x + 1$. Use your graph to estimate the roots of the function.

> Use graphing software for questions **10** to **11**.

10 a Plot the graph of $y = ax + a$.

b Use the constant controller to vary the value of **a**. What do you notice? Say why this happens.

11 a Plot the graph of $y = x^3 + a$. Use the constant controller to vary the value of a. What do you notice?

b Repeat part **a** for each of the following functions.

i $y = (x + a)^3$ **ii** $y = \dfrac{a}{x}$

iii $y = \dfrac{1}{x} + a$ **iv** ax^3

18.2 Sketching functions

RECAP

- Linear

  $y = mx + c$

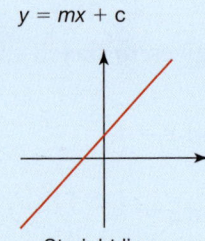

Straight line

- Quadratic

  $y = ax^2 + bx + c$

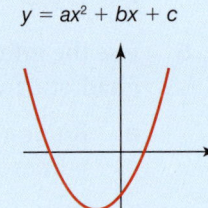

U-shape

- Cubic

  $y = ax^3 + bx^2 + cx + d$

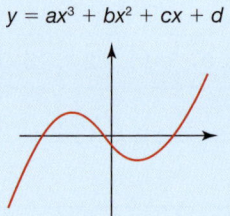

S-shape

- Reciprocal

  $y = \dfrac{c}{x}$

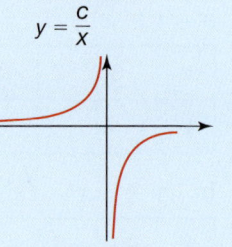

Two parts

HOW TO

To solve problems involving quadratic graphs and other types of function

1. Identify the roots and *y*-intercept of the function.
2. Sketch a graph.
3. Identify whether the roots or turning point help solve the problem.
4. Interpret the solution.

EXAMPLE

A company manufactures and sells packs of batteries. If the selling price of their batteries is too low they make no profit. If the selling price is too high they do not sell enough to make a profit. The company works out that that profit is a function of price.

$P = (2 - s)(s - 5)$ where P = profit (p) and s = selling price (£)

a   At what selling price will the company start losing money?

b   What selling price will maximise their profit? How much profit will they make in this case?

(1) The roots of the function are $s = 2$ and $s = 5$.

$x = 0 \Rightarrow y = 2 \times -5 = -10$

The P-intercept = -10

(2)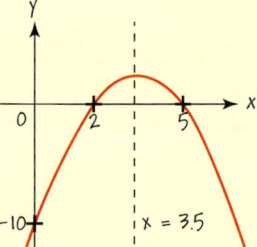

a   If the company charges more than £5 they will make a loss. (4)

b   (3) The highest point is needed. Since a parabola is symmetrical this is found halfway between the roots: $(5 + 2) \div 2 = 3.5$

(4) A selling price of £3.50 will maximise their profits. By substitution, $P = (2 - 3.5) \times (3.5 - 5) = 2.25$. Their profit is 2.25 p.

EXAMPLE

Oona is asked to plot the graph of $y = (x + 3)(14 - x)$.
The diagram shows her correct solution.
Oona is then asked to write the function in the form $y = ax^2 + bx + c$.
She multiplies out the brackets in the function and writes

$y = (x + 3)(14 - x) \Rightarrow y = x^2 + 11x + 42$

How does the graph show you that Oona's solution is *wrong*?

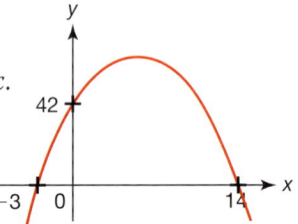

When Oona multiples out the brackets she has written an x^2 term that is positive.

When a parabola has a highest point it has an x^2 term that is negative.

The correct answer is $y = -x^2 + 11x + 42$. So $a = -1$, $b = 11$ and $c = 42$.

Exercise 18.2A

1 This graph has had the axes removed.

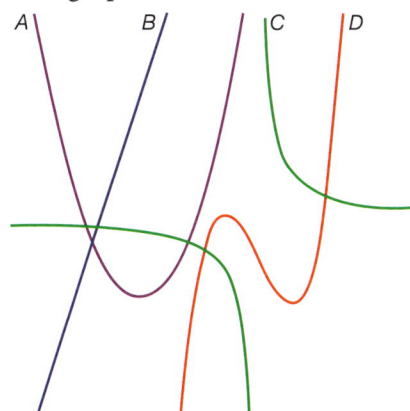

The graph shows these four functions.
Match each function to its graph.

a $y = 3x + 12$ **b** $y = x^3 - 2x$

c $y = \dfrac{1}{x}$ **d** $y = x^2 + 6x + 7$

2 Two companies establish these profit functions.

a $P = (1 - s)(s - 5)$

b $P = (50 - s)(s - 120)$

where P = profit (£) and s = selling price (£).

For each company

i sketch the graph of the profit function

ii state at what prices the profit would be zero

iii find the price that maximizes their profit

iv find their maximum profit.

3 A company uses a profit function of $P = (15 - s)(s - 60)$ where P = profit (£) and s = selling price (£).

a What is the maximum profit they expect to make?

b What other information is provided by the profit function?

4 An arrow is fired. Its height above ground is modelled by the function $h = (39 - t)(t + 1)$ where h = height in centimetres and t = length of shot in metres.

a Find the maximum height of the arrow.

b State the length of the shot.

*5 The jet of water in a fountain is modelled by the function $y = 10x(0.5 - x)$ where x = distance from source (m) and y = height (m)

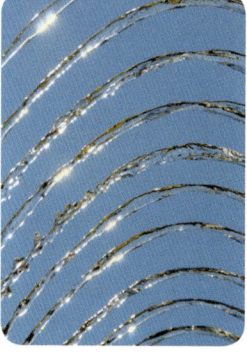

a At what distance from the source does the jet enter the water again?

b What is the maximum height reached by the jet?

c Write the function in the form $y = ax^2 + bx + c$

6 Match each of the functions **a** to **d** to its graph and its factorised form.

a $y = 7x - x^2$ **b** $y = x^2 - 8x + 16$

c $y = -x^2 + 3x + 4$ **d** $y = x^2 - 4$

A 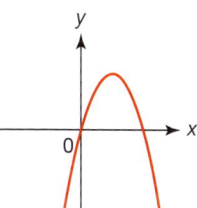   **B**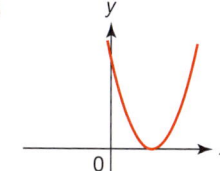

C 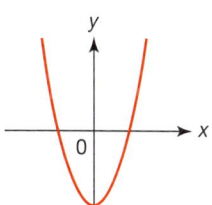   **D**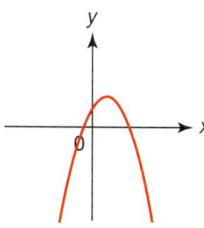

1 $y = (4 - x)(1 + x)$

2 $(x - 4)(x - 4)$

3 $x(7 - x)$

4 $y = (x + 2)(x - 2)$

*7 A cubic function is written as $f(x) = x(x + 2)(x + 1)$.
Sketch the graph of $y = f(x)$.

*8 Another cubic function is written as $f(x) = (x + 1)(x + 4)(x - 2)$.
Sketch the graph of $y = f(x)$.

1071, 1172, 1180, 1316   SEARCH

18.3 Real-life graphs

Many situations can be represented using line graphs.

EXAMPLE

The graph shows the average head circumference of a baby born at 'x' weeks of pregnancy.

a What is the average head circumference of a baby born at
  i   42 weeks
  ii   40 weeks
  iii 36 weeks.

b During which week of pregnancy would you expect the head circumference to be 33 cm?

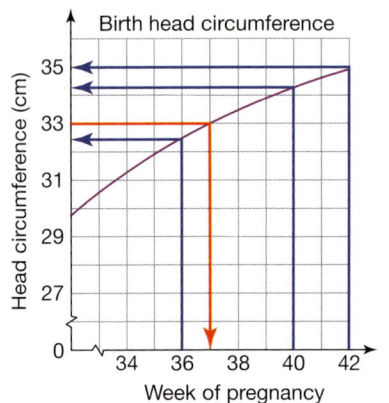

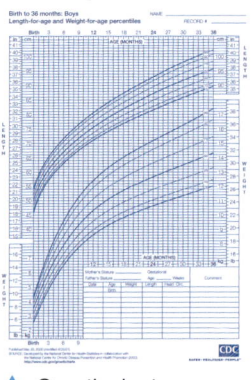

▲ Growth chart

a Start the reading from the horizontal axis.
  Use the scale to estimate the values.

  i   35.0 cm   ii   34.3 cm   iii   32.5 cm

b Start the reading from the vertical axis.

  37 weeks

● A straight line shows a quantity is changing at a **steady rate**.

● The steepness of a line shows how fast a quantity is changing.

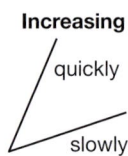

| Increasing | Not changing | Decreasing |
| quickly, slowly | — | slowly, quickly |

EXAMPLE

The graph shows the depth of water in Ali's bath.

a How long did it take to fill the bath?

b When did Ali get in the bath?

c How long was Ali in the bath?

d What happened between 18:35 and 18:40?

e At what rate did the depth decrease when the bath was emptied? Give your answer in cm/min.

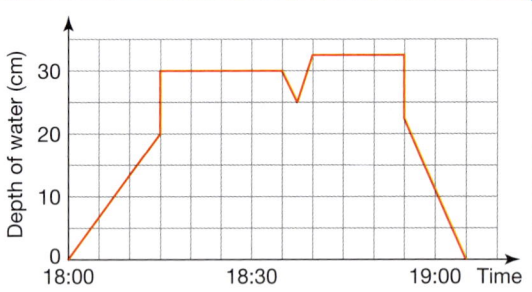

a 15 mins   The depth increased between 18:00 and 18:15.

b 18:15   The depth suddenly increased, by 10 cm.

c 40 mins   Between 18:15 and 18:55.

d The water level went down, 5 cm, and then increased by 7.5 cm.

  Water could have been let out of the bath and more hot water added.

p.448

e Rate = $\dfrac{22.5\ cm}{10\ min}$ = 2.25 cm/min

Algebra   Graphs 2

Exercise 18.3S

1 A call-centre worker earns a wage of £900 per month plus 12% of all sales they achieve. The graph shows this information.

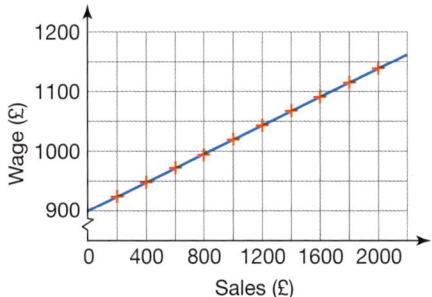

a How much do you earn with sales of

    i £500   **ii** £1400   **iii** £450.

b Miranda wants to earn at least £1000. What value of sales should she aim for?

2 The speed-time graph shows information about a train.

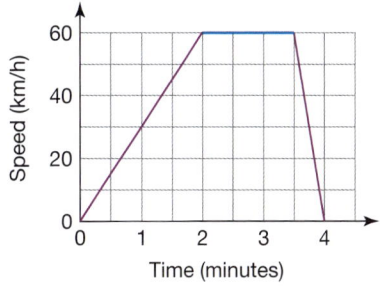

a What is the speed of the train at

    i 1 min   **ii** $2\frac{1}{2}$ mins?

b What happens between $3\frac{1}{2}$ and 4 mins?

3 Bret goes on a cycle ride. The graph shows his distance from home in kilometres.

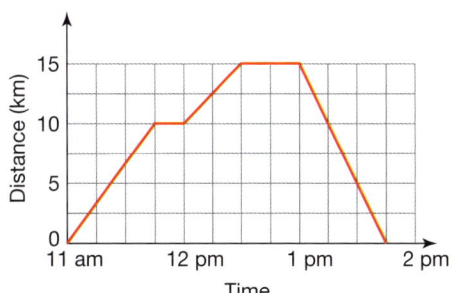

a When did Bret return home?

b How long was Bret stationary?

c **i** Between which times was Bret moving fastest?

    ii What was his speed in km/hr?

4 The graph shows information about a race between Matt and Ben.

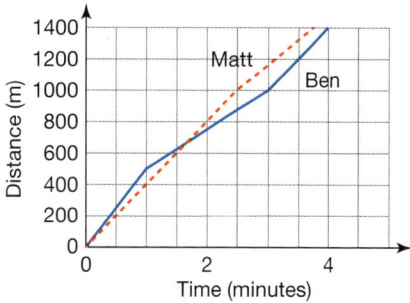

a How far was the race?

b Who was quickest at the start of the race?

c At what time were Matt and Ben level?

d Who finished the race first?

5 The graph shows information about the damping in a mountain bike's suspension.

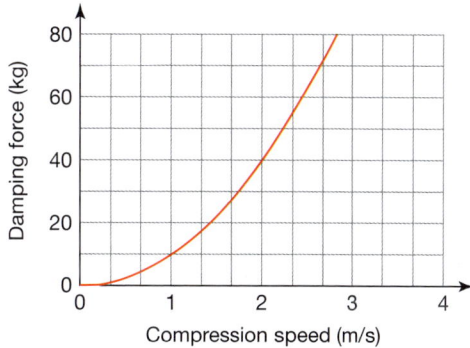

a What is the damping force when the compression speed is 2 m/s?

b Estimate the compression speed when the damping force is 70 kg.

c What type of function is this graph? Choose one from these options.

> Linear   Quadratic   Cubic   Reciprocal

*6 **a** For the distance–time graph in question **3** draw a speed–time graph.

  **b Do you think the graph is realistic? If not say how you would change it and how these changes would alter the distance–time graph.

18.3 Real-life graphs

- The gradient of a straight line shows how fast a quantity is changing.
  - A straight line implies a constant rate of change.
  - A horizontal line means there is no change.

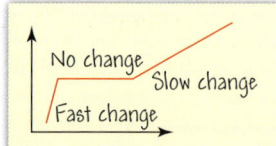

EXAMPLE

Water is poured into this container at a constant rate. Sketch the depth of water versus time.

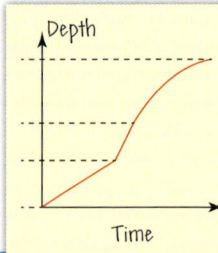

Container widens – rate of filling slows.

Container narrow – fills quickly.

Container wide – fills slowly.

HOW TO

To identify the type of function that can be used to model some data

1. Plot the points on a graph and draw a smooth curve through them.
2. Compare the shape of the data with the common types of function. You may need to stretch or translate the curve.
3. ATQ

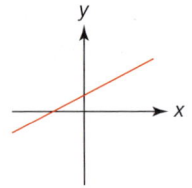

Linear

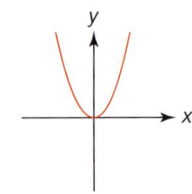

Quadratic

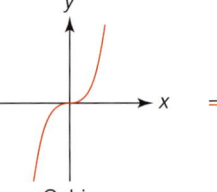

Cubic

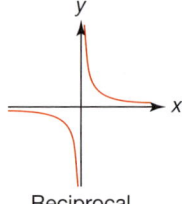

Reciprocal

EXAMPLE

Beth is investigating the features of an engine.
The table shows the power generated by different amounts of torque.

Torque (N/m)	1	2	3	4	5	6	7
Power (W)	22	42	53	63	62	55	40

Torque is a turning force. You can still complete the question without knowing that detail though.

a Plot the data on a graph.
b What type of function could be used to *model* the data?
c Estimate the power when a torque of 5.5 N/m is applied.

a

1. Choose axes that allow you to plot all the data and give you a roughly square-shaped graph.

b 2. The function used to model the data doesn't have to pass through the points but it does have to have the right shape.

$y = -x^2$ has the right type of shape, ∩.
A quadratic function.

c 3. Read up from 5.5 N/m and then across.

58 W

Algebra Graphs 2

Exercise 18.3A

1 The diagrams show four different types of beaker.

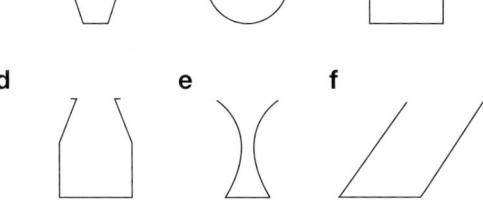

Liquid is poured into a beaker at a constant rate. A graph is plotted to show 'time' on the horizontal axis and 'depth of liquid' on the vertical axis.

i Sketch the graph for each beaker.

ii Match each beaker to the type of function that best describes the graph.

Cubic	Quadratic
No standard function	Linear

*2 Naomi is planning a conference. As part of the catering deal she orders 300 sandwiches.

a The average number of sandwiches per person depends on how many people come to the conference. Copy and complete this table.

People	20	40	60	80	100	120
Sandwiches					3	

b Use this data to plot a line graph.

c What type of function is the graph?

d If the average is less than 3.5 Naomi will order more sandwiches.
Up to how many people can come to the conference before she has to order more sandwiches?

3 Steve is a scientist developing energy-efficient LCD displays. Power is required to update his display. When this happens, the 'pulse size' (in volts) is connected to the temperature. Steve has this set of data

°C	−10	−5	0	10	23	40
Volts	20.5	17.7	16	13.6	12.5	12.1

3 **a** Plot the data on a graph.

b What type of function best describes the data?

c Steve has worked out a function that connects temperature and pulse size. He uses his formula to work out that if temperature $= -20°C$, then pulse size $= 30\,V$. He also knows that the pulse size cannot be less than $11.8\,V$. Do these facts confirm your reasoning in **b**? Give your reasons.

4 Look back at lesson **18.3S**.

a Write a description of Bret's cycle ride in question **3**.

b Write a race commentary for question **4**.

c Write a Short story to describe Ali's bath from the second example.

Use graph plotting software for question **5**.

*5 During the 1960s British mathematicians developed radar that could track the position of artillery fire. From this it could be worked out where the fire had been launched.
Artillery fire follows a parabolic path.

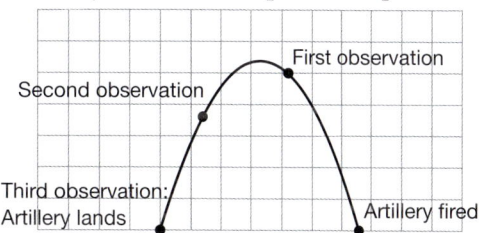

a Plot the points (6, 12), (2, 8) and (0, 0). Use these as three observations. The artillery lands at (0, 0).

b Starting with the values $a = 1$, $b = 3$ and $c = 7$, plot the graph of $y = -a(x + b)^2 + c$.

c Experiment with the constant controller to vary the values of a, b and c. Can you find a quadratic curve that passes through all three points?

d Write the coordinates of the point where the artillery is fired from.

e Work with a partner. Choose three points, including (0, 0). Challenge them to find a quadratic curve that passes through those points.

Q 1184, 1322   SEARCH

Summary

Checkout

You should now be able to...

✓	Draw graphs to identify and interpret roots, intercepts and turning points of quadratic functions.	1, 2
✓	Solve a quadratic equation by finding approximate solutions using a graph.	3
✓	Recognise, sketch and interpret graphs of linear, quadratic, cubic and reciprocal functions.	4, 5
✓	Plot and interpret real-life graphs.	6, 7

Language	Meaning	Example
Quadratic function	A function of the form $ax^2 + bx + c$. They have a characteristic $\cup$- or $\cap$-shape.	
Cubic function	A function of the form $ax^3 + bx^2 + cx + d$. They have a characteristic S-shape.	
Reciprocal function	A function of the form $\frac{c}{x}$. They have two parts for negative and positive x.	
Turning point	A point on a curve where the curve changes from rising to falling, $\cap$ or falling to rising, $\cup$	
Root	The points where curve crosses the x-axis.	
y-intercept	The point where a curve crosses the y-axis.	
Solve solution	Find a value for the unknown variable that will make the equation true.	$4x - 3 = 9$ is true when $x = 3$. $x = 3$ is the solution to the equation.

Review

1 Here is the graph of $y = x^2 - 4x - 5$.

Write down

a the y-intercept of the graph

b the x-intercepts of the graph

c the roots of the equation $x^2 + 4x - 5 = 0$

d the coordinates of the turning point of the curve.

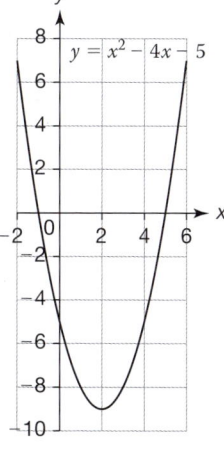

2 a Copy and complete this table of values for $y = x^2$.

x	−4	−3	−2	−1	0	1	2	3	4
y									

b Draw the graph of $y = x^2$ for values of x between −4 and 4.

c State the coordinates of the turning point on the curve.

3 a Use a table of values, with x from −4 to 4, to draw the graph of $y = x^2 - 3x - 6$.

b By adding lines, use your graph to find approximate solutions to

    i $x^2 - 3x - 6 = 0$

    ii $x^2 - 3x - 6 = -4$

4 Sketch the graph of $y = \frac{1}{x}$ for $x \neq 0$.

5 The graph shows these four functions.

$$y = x^3 - 4x \qquad y = \frac{2}{x} \qquad y = 2x + 3 \qquad y = x^2 + 3x$$

Match each graph to its mathematical name and equation.

5

Linear   Reciprocal   Cubic   Quadratic

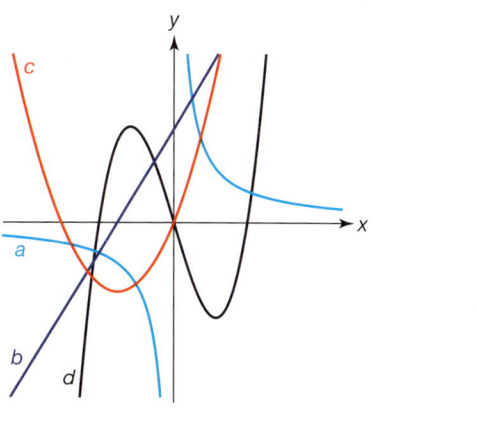

6

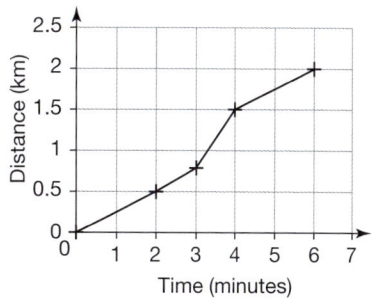

a Find the speed during the first two mins.

b What was the average speed, in km/h, during the 6 minute journey?

7 a Plot a speed–time graph for the information given in this table.

Time (s)	0	4	7	10
Speed (m/s)	5	3	3	0

b What is happening to the speed during the first 4 seconds?

c What is happening between 4 s and 7 s?

d What is the rate of change of speed during the final 3 seconds?

What next?

Score			
	0 – 3		Your knowledge of this topic is still developing. To improve look at MyMaths: 1168, 1169, 1071, 1172, 1180, 1184, 1316, 1322
	4 – 6		You are gaining a secure knowledge of this topic. To improve your fluency look at InvisiPens: 18Sa – d
	7		You have mastered these skills. Well done you are ready to progress! To develop your problem solving skills look at InvisiPens: 18Aa – c

Assessment 18

1 **a** Draw the graph of $y = x^2 - x - 6$
     and $y = x$ on a copy of this grid.   [6]

    b Estimate the coordinates of the
       points where the graphs intersect.   [1]

    c Estimate the coordinates of the turning
       point on the quadratic curve.   [1]

    d Use your graph to estimate the solutions of
       the equation $x^2 - x - 6 = 0$ [2]

    e By factorising, solve the equation $x^2 - x - 6 = 0$.
       Compare your answer with your estimate in part **d**.   [2]

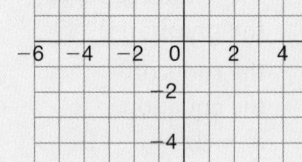

2 A metal spring stretches when a mass of m grams hung on the end.

    The distance stretched, d cm, is given by the formula $d = 8 + \dfrac{m}{15}$.

    a Complete the table of values for m and d.
       Write your answers to 1 dp where appropriate.   [2]

m	10	20	30	40	50	60	70	80	90	100
d										

    b Plot the graph of d against m for values of m from 0 to 100.   [2]

    c Use your graph to

      i   find the stretch, d, when a mass of 75 g is hung on the spring   [1]

      ii find the mass hung on the spring when $d = 9$ cm.   [1]

    d Complete this sentence.
       If the mass of the weight hung on the spring increases by 1 g,
       the distance the spring stretches increases by ☐ cm.   [1]

    e What is the length of the spring when no weights are hung on it?   [1]

3 The average safe stopping distance for cars consists of two parts:

    a thinking distance $\dfrac{v}{3}$ and a braking distance $\dfrac{v^2}{40}$.

    The formula for the safe stopping distance, d metres, is given by the equation

    $d = \dfrac{v^2}{40} + \dfrac{v}{3}$, where v is the speed of the car in km/h.

    a Complete the table of values for d and v. Write your distances to the nearest metre.   [3]

v	0	10	20	30	40	50	60	70	80	90	100
d											

    b Draw the graph of $d = \dfrac{v^2}{40} + \dfrac{v}{3}$.   [3]

    c Use your graph to find the safe stopping distance when a car travels at

      i   25 km/h   [1] **ii** 42 km/h   [1] **iii** 77 km/h.   [1]

4 A crystal glass making firm makes y thousand pounds profit from the sale of x thousand glasses.

The formula they use to estimate their profits is $y = 5x - 2 - x^2$

a Complete the table of values. [3]

x	0	0.5	1	1.5	2	2.5	3	3.5	4	4.5	5
y											

b Use appropriate axes and scales, plot the graph of this function. [3]

c Use your graph to estimate

 i the profit when 900 glasses are made [1]

 ii the profit when 3750 glasses are made [1]

 iii the number of glasses made when the profit is £3750 [1]

 iv the maximum profit the firm can make [1]

 v how many glasses they must sell to make the maximum profit [1]

 vi the smallest number of glasses the firm must make to avoid making a loss [1]

 vii the greatest number of glasses the firm must make to avoid making a loss [1]

 viii the range of values of x for which the profit is more than £2500. [1]

5 Match the graphs **A – E** to one of the five statements **i–v**. [4]

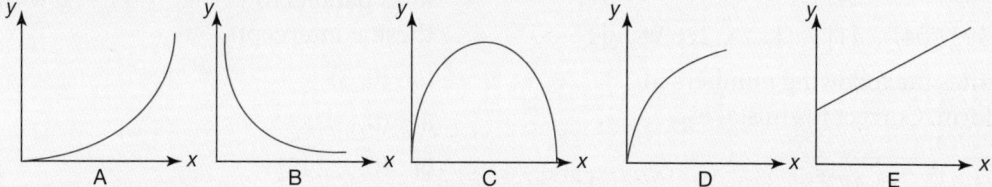

 i The cost of gas, y, is a set amount added to a price per unit multiplied by the number of units used, x.

 ii The height, y, of a cricket ball a time, x, after being thrown.

 iii The time taken to drive a set distance, y, compared to the average speed, x.

 iv The height of water in a conical flask, y, being filled at a steady rate, at time x.

 v The height of water in a hemispherical flask, y, being filled at a steady rate, at time x.

6 **a** Taking values of x from -4 to 4, draw the graph of $y = x^3 - 16x$. [5]

 b On the same grid, draw the graph of $y = 12 - 3x$ [2]

 c Use your graph to solve the equation $12 - 3x = 0$ [1]

 d Use your graph to solve the equation $x^3 - 16x = 0$ [3]

 e Find and record the coordinates of the points where the two graphs intersect. [3]

 f Complete the sentence: The solutions to $x^3 - \square x - \square = 0$ are the points of intersection of the graphs of $y = x^3 - 16x$ and $y = 12 - 3x$. [2]

7 The product of x and y is 36.

 a Write down a formula for y in the form $y = \ldots$ [1]

 b Draw a suitable graph the equation in part a for x values between -12 and 12. [6]

 c Use your graph to find a pair of numbers that multiply to give 36 and add together to give 13. Why is there only one solution but two points of intersection? [4]

Revision 3

1 a An orchard owner packs apples in boxes. Each box contains the same number of apples. She delivers 240 apples to one shop and 144 to another. What is the largest number of apples in any box? [4]

b Kevin says that the LCM of the numbers 180 and 210 is 36 and the HCF is 1260.
Correct his mistakes. [7]

c Show that the least positive whole number that 180 must be multiplied by to make the result a perfect square is 5. [1]

2 Find the value of these expressions.

a 11^2 [1] **b** 2.6^2 [1]

c $\sqrt{441}$ [1] **d** $\sqrt{2.25}$ [1]

e 1^7 [1] **f** $(-3)^3$ [1]

g 3.2^3 [1] **h** $\sqrt[3]{46.656}$ [1]

i $\sqrt[3]{(-592.704)}$ [1] **j** $\sqrt[3]{226.981}$ [1]

3 Shivani writes the following numbers in standard form. Correct his mistakes.

a	8000	8×10^4	[1]
b	75000	75×10^5	[1]

c Each side of a dice is 1.6 cm. Write the volume

 i in index form [1]

 ii as an ordinary number. [1]

d Hal says that $2^3 \times 2^{-3} = 2^{-9}$. Is he correct? Give your reasons. [2]

4 Zak is filling his watering can. The sketch graph shows the height of water in the can at a particular time.

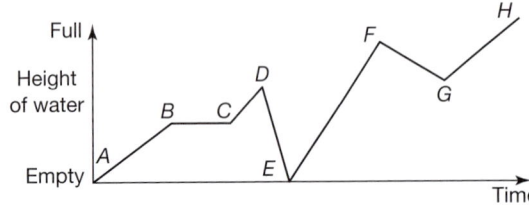

Give reasons for all your answers.

a At what point does Zak's dog knock over the can and spill all the water? [2]

b In which section does Zak stop filling the can and water some plants? [2]

c In which section of the graph is the can filling fastest? [2]

d In which section does Zak stop filling to answer his mobile? [2]

5 a Show that the point $(7, -2)$ lies on the line $3y + 4x = 22$. [2]

b Match each equation of a line to its gradient and y-intercept. [8]

 i $y = -9 - 5x$ **ii** $y = 2 - 6x$

 iii $3x - 2y = 6$ **iv** $4x + 7y = -21$

 A $m = -5, c = -9$ **B** $m = 1\frac{1}{2}, c = -3$

 C $m = -6, c = 2$ **D** $m = -\frac{4}{7}, c = -3$

c Write down the equations of the straight lines parallel to $y + 4x + 11 = 0$ with these y-intercepts.

 i $(0, 3)$ [1]

 ii $(0, -7)$ [1]

 iii $(0, -245)$ [1]

6 Sarah says that these nets all form cubes. Is she correct? Give reasons for your answers. [3]

7 A ball has an 8.6 cm diameter. Calculate its

 a surface area [2] **b** volume. [2]

8 A trophy consists of a cuboid, 15 cm by 10 cm by 6 cm. On top of it is a pyramid with base 15 cm by 10 cm. The pyramid is 20 cm high. Find the trophy's total volume. [5]

9 A sphere has a volume of 4000 m³. Using $\pi = 3.142$ Denise calculates its radius to be 9.85 m to 3 sf. Show how she worked this out. [5]

10 The cross section of a prism is a right-angled triangle with hypotenuse 53 cm and the other two sides 28 cm and 45 cm. The thickness of the block is 15 cm. Find

 a its volume [4]

 b its surface area. [6]

11 A garage checked 104 vehicles for signs of tyre wear.

Depth of tyre tread (mm)	Frequency
$1.0 \leqslant t < 1.6$	19
$1.6 \leqslant t < 2.1$	13
$2.1 \leqslant t < 2.6$	27
$2.6 \leqslant t < 3.1$	25
$3.1 \leqslant t < 3.6$	17
$3.6 \leqslant t < 4.0$	3

   a Write down

      i the modal class [1]

      ii the median interval. [1]

   b The legal limit for tread on tyres is 1.6 mm. How many tyres have less than the legal limit? [1]

   c Calculate an estimate for the mean. [6]

12 The data shows the distances thrown compared with the lengths of the arms of some discus throwers.

Arm length (cm)	58	61	63	67	69	70	72	75	76	77
Distance thrown (m)	52	55	57	58	62	64	67	70	72	74

   a Draw a scatter diagram of the data with a line of best fit. [4]

   b Comment on the type of correlation. [2]

   c Use your line to estimate

      i the distance thrown corresponding to an arm length of 65 cm. [1]

      ii the arm length corresponding to a distance of 65 m. [1]

13 The table shows UK unemployment rates for the period Oct 2012 to Sept 2014.

Month	Oct 2012	Nov 2012	Dec 2012	Jan 2013	Feb 2013	Mar 2013
Rate (%)	7.8	7.7	7.8	7.8	7.9	7.8
Month	Apr 2013	May 2013	Jun 2013	Jul 2013	Aug 2013	Sep 2013
Rate (%)	7.8	7.8	7.8	7.7	7.7	7.6

Month	Oct 2013	Nov 2013	Dec 2013	Jan 2014	Feb 2014	Mar 2014
Rate (%)	7.6	7.4	7.1	7.2	7.2	6.9
Month	Apr 2014	May 2014	Jun 2014	Jul 2014	Aug 2014	Sep 2014
Rate (%)	6.8	6.6	6.5	6.4	6.2	6.0

13 **a** On the same graph draw a time series graph for Oct 2013 – Sep 2014 and Oct 2012 – Sep 2013. [6]

   b Compare unemployment rates in these periods. [1]

14 **a** A gym has 1.2×10^2 members, each of whom use, on average, 4.7×10^3 units of electricity per year. How many units does the gym use in a year? [3]

   b The mass of a nitrogen atom is 2.326×10^{-23} g. One litre of air contains 4.3602×10^{-1} g of nitrogen. How many nitrogen atoms are there in one litre of air? [2]

   c Fiona says that 3×10^{-2} is bigger than -3×10^2. Is she correct? Why? [2]

   d Work out n for each of the following equations.

      i $2.6 \times 10^n = 2600$ [1]

      ii $3.76 \times 10^n = 3760000$ [1]

      iii $20.4 \times 10^n = 20400$ [1]

15 A bridge goes over a river. y is the height (m), of the arch of the bridge above the river level and x is the distance (m), from the north bank of the river.

$$y = \frac{3x}{2} - \frac{x^2}{40}$$

   a Draw the graph of $y = \frac{3x}{2} - \frac{x^2}{40}$ for values of x in tens from 0 to 90. [6]

   b The top of the mast of a yacht is 19 m above water level. How far from the bank does the yacht have to be to be able to pass under the bridge? [2]

   c What width of river is available if the top of the funnel of a boat is 23 m above water level? [2]

19 Pythagoras and trigonometry

Introduction

The highest mountain in the world is Mount Everest, located in the Himalayas. Its peak is now measured to be 8848 metres above sea level.

The mountain was first climbed in 1953, by Edmund Hilary and Sherpa Tenzing, almost 100 years after its height was first measured as part of the Great Trigonometrical Survey of India in 1856. The original surveyors, who included George Everest, obtained a height of 8840 metres by measuring the distance and angle of elevation between Mount Everest and a fixed location.

What's the point?

Once a right-angled triangle is seen in a particular problem then a mathematician only needs two pieces of information to be able to calculate all the other lengths and angles.

Objectives

By the end of this chapter you will have learned how to …

- Use the formulae for Pythagoras' theorem: $a^2 + b^2 = c^2$,
- Use the trigonometric ratios and apply them to find angles and lengths in right-angled triangles:

$$\sin\theta = \frac{\text{Opp}}{\text{Hyp}} \qquad \cos\theta = \frac{\text{Adj}}{\text{Hyp}} \qquad \tan\theta = \frac{\text{Opp}}{\text{Adj}}$$

- Know the exact values of $\sin\theta$ and $\cos\theta$ for $\theta = 0°, 30°, 45°, 60°$ and $90°$.
- Know the exact value of $\tan\theta$ for $\theta = 0°, 30°, 45°$ and $60°$.
- Write column vectors and draw vector diagrams.
- Add, subtract and find multiples of vectors.

Check in

1 Substitute these numbers into the expression $3x^2$.

    a 2   **b** 5   **c** 3.1   **d** -2

2 Calculate each of these quantities.

    a $3^2 + 4^2$ **b** $7^2 - 5^2$ **c** $\sqrt{70}$ (to 1 dp)

Chapter investigation

An engineering company is building a ski lift.
The height of the lift is exactly 45 m.
An engineer suggests using a cable of length 200 m.

The maximum angle of incline is 12°.
Does the engineer's lift meet the criteria?
Design a lift that meets the criteria using the smallest possible length of cable.

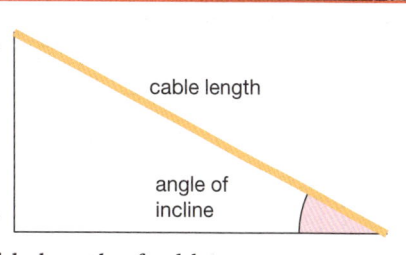

19.1 Pythagoras' theorem

The longest side of a **right-angled triangle** is called the **hypotenuse**.

In a right-angled triangle the square of the length of the hypotenuse is equal to the sum of squares of the lengths of the other two sides.

The hypotenuse is always opposite the right angle and labelled c.

Hypotenuse

- In symbols, Pythagoras' theorem is
  $c^2 = a^2 + b^2$ where c is the length of the hypotenuse.

EXAMPLE

Calculate the unknown length in each triangle.

a
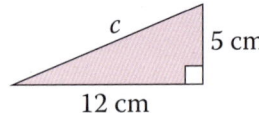
c
5 cm
12 cm

b
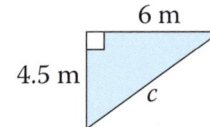
6 m
4.5 m
c

Label the sides.

a $a = 5, b = 12, c = ?$

Substitute into $c^2 = a^2 + b^2$.

$c^2 = 5^2 + 12^2$

$c^2 = 25 + 144$

$c^2 = 169$

$c = \sqrt{169} = 13\,cm$ | $\sqrt{\ }$ means square root.

b $a = 4.5, b = 6, c = ?$

$c^2 = 4.5^2 + 6^2$

$c^2 = 20.25 + 36$

$c^2 = 56.25$

$c = \sqrt{56.25} = 7.5\,m$

p.280

EXAMPLE

Work out the unknown sides in these triangles.

a
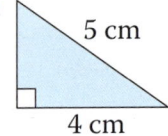
5 cm
4 cm

b
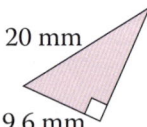
20 mm
9.6 mm

Label the sides.

a $a = ?, b = 4, c = 5$

Use Pythagoras to find the unknown side.

$a^2 = c^2 - b^2$

$a^2 = 5^2 - 4^2$ This is sometimes called the 3, 4, 5 triangle.

$a^2 = 25 - 16$

$a^2 = 9$

$a = \sqrt{9}$

$a = 3\,cm$

b $a = 9.6, b = ?, c = 20$

$b^2 = c^2 - a^2$

$b^2 = 20^2 - 9.6^2$ Round the answer when it is not exact. One decimal place is sensible here.

$b^2 = 400 - 92.16$

$b^2 = 307.84$

$b = \sqrt{307.84}$

$b = 17.5\,cm$ (1 dp)

p.8

Exercise 19.1S

1 Write down the first 15 square numbers.

2 Find the length represented by each letter.

   a $6^2 + 8^2 = p^2$ **b** $15^2 + q^2 = 17^2$

   c $7^2 + 7^2 = r^2$ **d** $1^2 + s^2 = 3^2$

3 Calculate the area of these squares. State the units of your answers.

a 9 cm   **b** 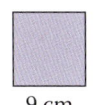 12 mm   **c** 1.5 m

d  35 cm   **e**  4.8 m   **f**  0.6 km

4 Find the length of a side of each square. State the units of your answers.

a Area = 1 cm²   **b** Area = 64 m²   **c** Area = 16 mm²

d Area = 256 cm²   **e** Area = 5.29 m²   **f** Area = 0.25 km²

5 In each part find the unknown area.

a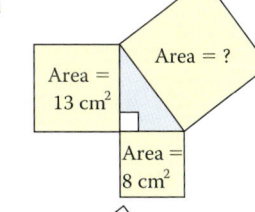
Area = ?   Area = 13 cm²   Area = 8 cm²

b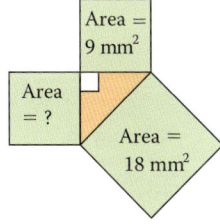
Area = 9 mm²   Area = ?   Area = 18 mm²

c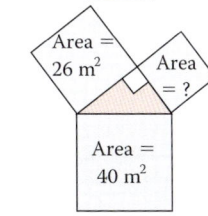
Area = 26 m²   Area = ?   Area = 40 m²

d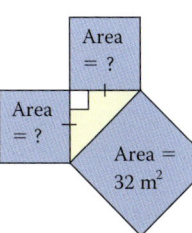
Area = ?   Area = ?   Area = 32 m²

6 Calculate the length of the hypotenuse in these right-angled triangles. State the units.

a 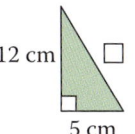 12 cm, 5 cm   **b** 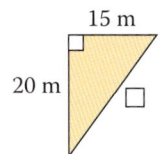 15 m, 20 m

6 **c** 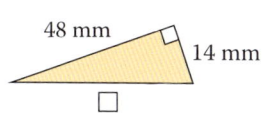 48 mm, 14 mm   **d** 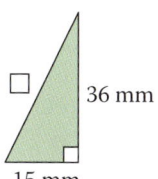 36 mm, 15 mm

e 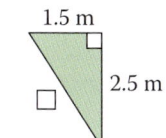 1.5 m, 2.5 m   **f** 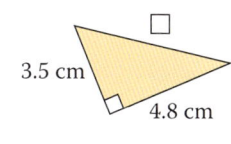 3.5 cm, 4.8 cm

7 Calculate the lengths marked by letters. State the units of your answers.

a 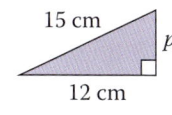 15 cm, 12 cm, p **b** 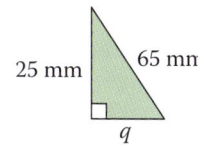 25 mm, 65 mm, q

c 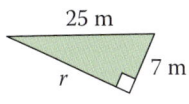 25 m, 7 m, r **d** 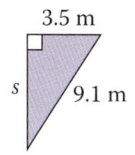 3.5 m, 9.1 m, s

e 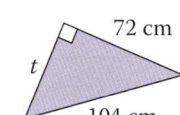 72 cm, 104 cm, t **f** 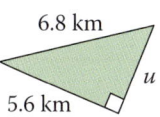 6.8 km, 5.6 km, u

*8 Find the height of this isosceles triangle.

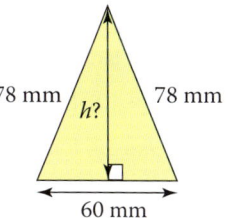 78 mm, 78 mm, h?, 60 mm

9 Find the missing lengths in these triangles.

a 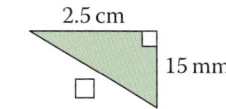 2.5 cm, 15 mm   **b** 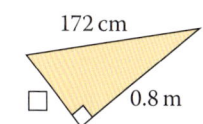 172 cm, 0.8 m

10 Work out the length of the support for the shelf.

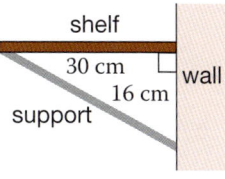 shelf, 30 cm, wall, 16 cm, support

19.1 Pythagoras' theorem

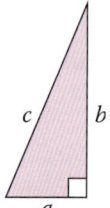

RECAP

- In a right-angled triangle the square of the length of the hypotenuse is equal to the sum of squares of the lengths of the other two sides.
  - To find the length of the hypotenuse use $c^2 = a^2 + b^2$
  - To find the length of a shorter side use $a^2 = c^2 - b^2$ or $b^2 = c^2 - a^2$

HOW TO

To solve a problem involving the sides of a right-angled triangle

1. Sketch a diagram (if needed) decide which side is the hypotenuse.
2. Use Pythagoras to find the side you need.
3. Round your answer to an appropriate degree of accuracy and include any units.

EXAMPLE

A rectangle measures 6.8 cm by 3.6 cm. Find the length of its diagonal.

1. The diagonal is the hypotenuse of both right-angled triangles.
2. Use the formula $c^2 = a^2 + b^2$.

$c^2 = 3.6^2 + 6.8^2$

$c^2 = 59.2$

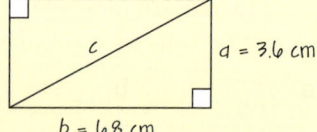

3. Round to the same accuracy as the given measurements.

$c = \sqrt{59.2}$

$c = 7.7\,cm$ (1 dp)

$\sqrt{\ }$ means square root.

p.8

EXAMPLE

A ladder of length 6 metres leans against a wall. Its base is 1.5 metres from the wall. How far up the wall does the ladder reach?

1. The ladder is the hypotenuse of a right-angled triangle.

$a = 1.5, b = ?, c = 6$

2. Use the formula $b^2 = c^2 - a^2$

$b^2 = 6^2 - 1.5^2 = 33.75$

3. Round sensibly.

$b = \sqrt{33.75} = 5.8\,m$ (1 dp)

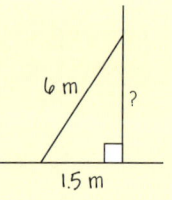

p.290

EXAMPLE

Calculate the distance between the points (2, 5) and (6, 3).

1. The line between (2, 5) and (6, 3) is the hypotenuse of a right-angled triangle.
2. Use $c^2 = a^2 + b^2$

$c^2 = 2^2 + 4^2$

$c^2 = 20$

$c = \sqrt{20}$

3. Round sensibly.

$c = 4.5\,units$ (1 dp)

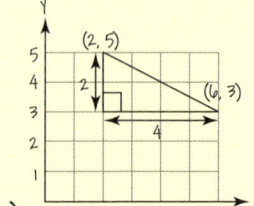

Use 1 dp unless the given measurements suggest a different level of accuracy.

Exercise 19.1A

1 Find the length of the diagonals in each rectangle. Give each answer to two significant figures.

	Length of rectangle	**Width of rectangle**
a	21 mm	28 mm
b	5.8 cm	4.3 cm
c	24.6 m	15.7 m
d	156 km	89 km

2 A rectangle is 8.5 cm long. The length of its diagonals is 9.6 cm.

Find the width of the rectangle.

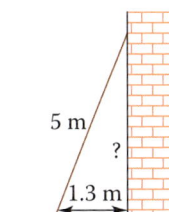

9.6 cm

8.5 cm

3 **a** A square has sides of length 10 cm. Find the length of its diagonals.

   *b A square has diagonals of length 10 cm. Find the length of its sides.

4 A 5 metre ladder leans against a wall with its base 1.3 metres from the wall.

How far up the wall does the ladder reach?

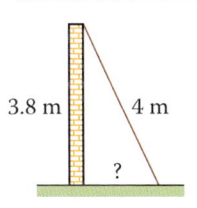

5 m
?
1.3 m

5 The top of a 4 metre ladder leans against the top of a wall. The wall is 3.8 metres high.

How far from the wall is the bottom of the ladder?

3.8 m 4 m
?

6 Calculate the distance between these points.

   a (1, 3) and (4, 7)   **b** (2, 1) and (6, 4)

   c (3, 2) and (4, 5)   **d** (0, 5) and (3, 2)

   e (3, 7) and (8, 0)   **f** (−2, 7) and (2, 4)

   g (2, 0) and (−3, 2)   **h** (−5, 4) and (−1, −2)

7 The diagram shows a path across a rectangular field.

How much further is it from A to C along the sides of the field than along the path?

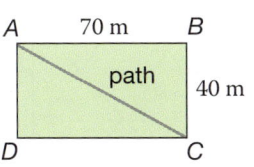

A   70 m   B
path
40 m
D   C

8 Find the areas of these triangles.

a

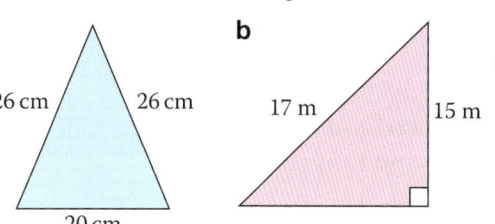

26 cm   26 cm

20 cm

b

17 m   15 m

p.144

9 **a** Show that a triangle with sides 6 cm, 8 cm and 10 cm is right-angled.

   b Explain why a triangle with sides 8 cm, 10 cm and 12 cm cannot be right-angled.

10 Jake draws this triangle.

   a Explain why angle ABC cannot be 90°.

   *b Is angle ABC acute or obtuse? Explain your answer.

C
9 cm   4 cm
A   8 cm   B

*11 **a** Find the length AD.

   b Find the length PQ.

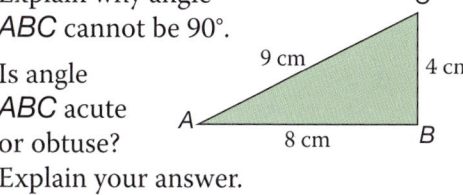

D
1.8 cm
C
1.8 cm
B   5.4 cm   A

S
8.2 cm
17.5 cm   R
6.4 cm
P   Q

12 The area of triangle ABD is 17.5 m². What is the length of side BC?

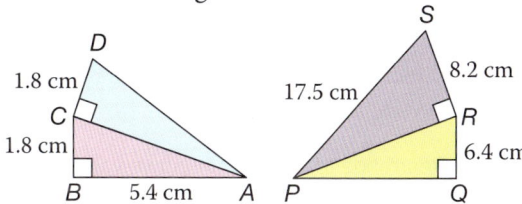

B
A   7 m   D   12 m   C

*13 When three whole numbers a, b, and c satisfy $a^2 + b^2 = c^2$ then a, b and c are called a **Pythagorean triple**.

   a How many Pythagorean triples can you find with $c < 100$?

   b Lola says that you can find other Pythagorean triples by multiplying those you have found by 2 or 3 or 4…. Is this true? Give your reasons.

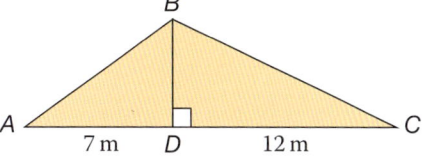

🔍 1053, 1112   SEARCH

19.2 Trigonometry 1

All right-angled triangles that include the same acute angle θ are similar. Their angles are equal and corresponding pairs of sides are in the same ratio.

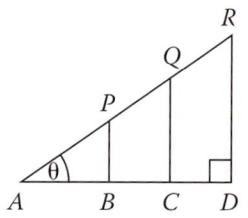

For example,
$$\frac{PB}{AB} = \frac{QC}{AC} = \frac{RD}{AD}$$

 p.56

These **trigonometric** ratios are called the **sine** (sin), **cosine** (cos) and **tangent** (tan) of θ.

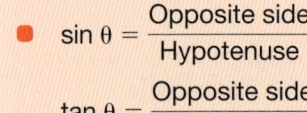
$$\sin θ = \frac{\text{Opposite side}}{\text{Hypotenuse}} \qquad \cos θ = \frac{\text{Adjacent side}}{\text{Hypotenuse}}$$
$$\tan θ = \frac{\text{Opposite side}}{\text{Adjacent side}}$$

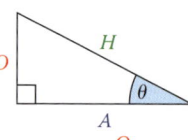
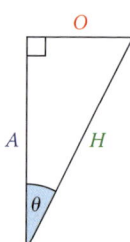

$$\sin θ = \frac{O}{H}$$
$$\cos θ = \frac{A}{H}$$
$$\tan θ = \frac{O}{A}$$

When you know an acute angle and a side in a right-angled triangle, you can find the other sides.

EXAMPLE

Find the sides marked by letters.

a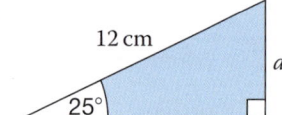

b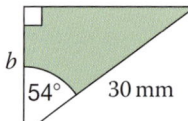

c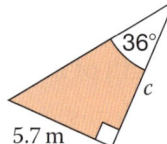

Identify which values you know and require.

a θ = 25°, O = a, H = 12   **b** θ = 54°, A = b, H = 30   **c** θ = 36°, O = 5.7, A = c

Choose the trigonometric ratio and substitute values.

$$\sin 25° = \frac{a}{12} \qquad\qquad \cos 54° = \frac{b}{30} \qquad\qquad \tan 36° = \frac{5.7}{c}$$

Check that your answers seem reasonable.

Rearrange to find the unknown value.

$$12 × \sin 25° = a \qquad\qquad 30 × \cos 54° = b \qquad\qquad c = \frac{5.7}{\tan 36}$$

$$a = 5.1 \text{ cm} \quad (1 \text{ dp}) \qquad b = 17.6 \text{ mm} \quad (1 \text{ dp}) \qquad c = 7.9 \text{ m} \quad (1 \text{ dp})$$

You could use a different trigonometric ratio or Pythagoras theorem to find the third side.

When you know a side and an angle in an isosceles triangle, you can find the other sides.

EXAMPLE

Calculate *AB*.

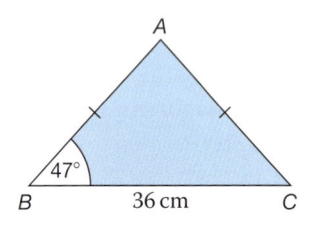

Draw *AD* perpendicular to *BC* to give a right-angled triangle.

In triangle *ABD*

θ = 47°, A = 18, H = AB

$$\cos 47° = \frac{18}{AB}$$

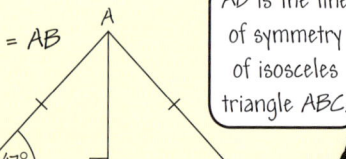

AD is the line of symmetry of isosceles triangle *ABC*.

$$AB = \frac{18}{\cos 47°}$$

AB = 26.4 cm (1dp)

Exercise 19.2S

1 Copy these triangles.
Label the sides O, A and H.

a 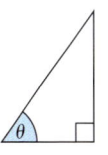 **b** 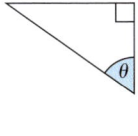 **c**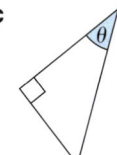

2 Find the value of each letter.

a $\sin 25° = \dfrac{a}{6}$ **b** $\cos 54° = \dfrac{b}{14}$

c $\tan 43° = \dfrac{c}{3.2}$ **d** $\cos 75° = \dfrac{24}{d}$

e $\sin 16° = \dfrac{4.9}{e}$ **f** $\tan 68° = \dfrac{230}{f}$

3 Find the sides marked by letters.

a 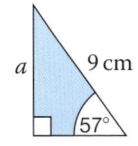 **b** 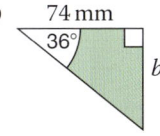 **c**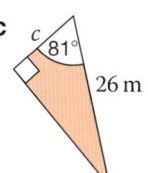

4 Find the sides marked by letters.

a 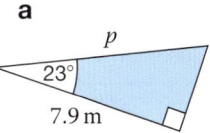 **b** 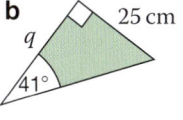 **c**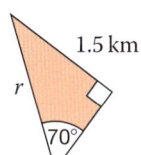

5 Find the sides marked by letters.

a 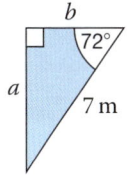 **b** 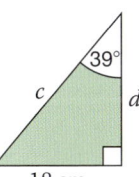 **c**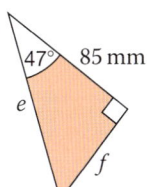

d 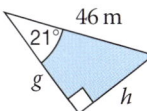 **e**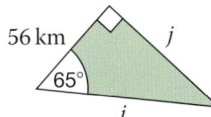

6 Find the sides

a AD

b BC.

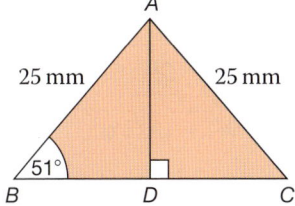

7 Find the other sides of triangle PQR when angle P is 90° and

a $QR = 40$ cm and angle $Q = 17°$

b $PQ = 5$ m and angle $Q = 49°$

c $QR = 142$ km and angle $R = 31°$

d $PQ = 82$ mm and angle $R = 64°$

*8 Find

a AB

b BD

c CD

d BC

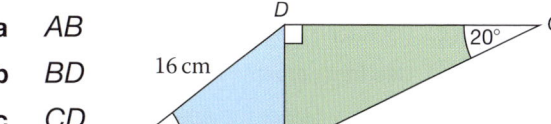

*9 **a** $AB = BC = 1$ unit
Show that

  i $\sin 45° = \dfrac{1}{\sqrt{2}}$

  ii $\cos 45° = \dfrac{1}{\sqrt{2}}$

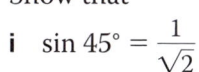

b $PQ = PR = 2$ units
Show that

  i $\cos 60° = \dfrac{1}{2}$

  ii $\sin 60° = \dfrac{\sqrt{3}}{2}$

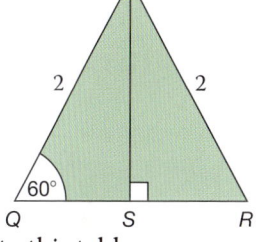

c Copy and complete this table.

Angle	sin	cos	tan
0°			
30°			
45°			
60°			
90°			∞

You need to learn these *exact* ratios.

10 A ladder is 5 metres long. Safety advice says that the angle between the ladder and the ground must be 75°. For this angle, find

a how far up the wall the ladder reaches

b how far the bottom of the ladder is from the wall.

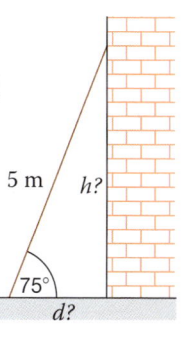

19.2 Trigonometry 1

- In a *right-angled* triangle

$$\sin \theta = \frac{\text{Opposite side}}{\text{Hypotenuse}} \qquad \cos \theta = \frac{\text{Adjacent side}}{\text{Hypotenuse}} \qquad \tan \theta = \frac{\text{Opposite side}}{\text{Adjacent side}}$$

$$\sin \theta = \frac{O}{H}$$
$$\cos \theta = \frac{A}{H}$$
$$\tan \theta = \frac{O}{A}$$

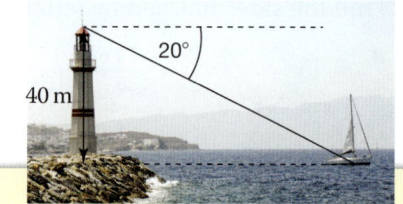

HOW TO

To solve a trigonometry problem

1. Draw a sketch (if useful). Include all the information you know.
2. Look for a right-angled triangle. Decide which sides are O, A and H.
3. Use trigonometric ratio(s) to work out the answers.
4. Check whether the answer(s) seem sensible.

Angle	sin	cos	tan
0°	0	1	0
30°	$\frac{1}{2}$	$\frac{\sqrt{3}}{2}$	$\frac{1}{\sqrt{3}}$
45°	$\frac{1}{\sqrt{2}}$	$\frac{1}{\sqrt{2}}$	1
60°	$\frac{\sqrt{3}}{2}$	$\frac{1}{2}$	$\sqrt{3}$
90°	1	0	∞

EXAMPLE

Jan looks at a boat from the top of a 40 metre lighthouse. She says the boat is over 100 metres from the lighthouse. Is Jan correct?

State any assumptions you make.

Assume the sea is calm and Jan is 40 m above it.

1. Draw a sketch to show more information.

3. $\tan 20° = \dfrac{40}{d}$ Using $\tan \theta = \dfrac{O}{A}$

$d = \dfrac{40}{\tan 20°}$ Rearrange to find d.

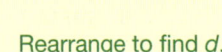

Use a letter like d for the distance you want to find.

$d = 109.89... \text{ m}$

2. Alternate angle from parallel horizontal lines.

This is over 100 m, so Jan is correct.

The answer assumes the base of the tower is level with the boat.

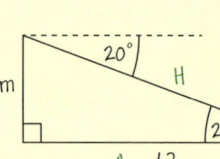

p.48

p.144

EXAMPLE

Find the exact area of rectangle $PQRS$.

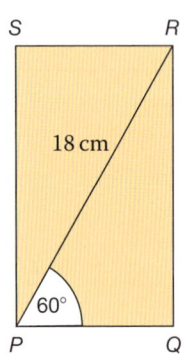

3. Area = length × width = $PQ \times QR$
   Find PQ and QR.

$$\cos 60° = \frac{A}{H} \qquad\qquad \sin 60° = \frac{O}{H}$$

$$\frac{1}{2} = \frac{PQ}{18} \qquad\qquad \frac{\sqrt{3}}{2} = \frac{QR}{18}$$

$$18 \times \frac{1}{2} = PQ \qquad\qquad 18 \times \frac{\sqrt{3}}{2} = QR$$

$$PQ = 9 \text{ cm} \qquad\qquad QR = 9\sqrt{3} \text{ cm}$$

Area of $PQRS = 9 \times 9\sqrt{3}$

$$= 81\sqrt{3} \text{ cm}^2$$

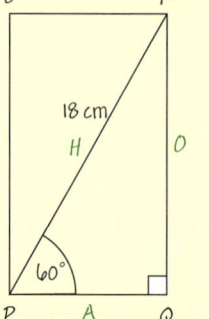

The *exact values* of cos 60° and sin 60° are in the table above.

Exercise 19.2A

1 A boat is observed from the top of a 50 metre tower. How far is the boat from the cliff?

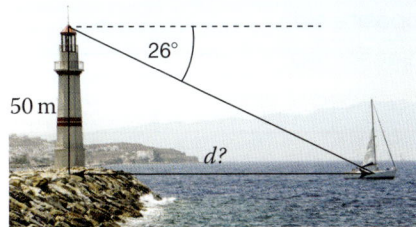

2 Carl looks at Blackpool Tower from the beach. He estimates these measurements.

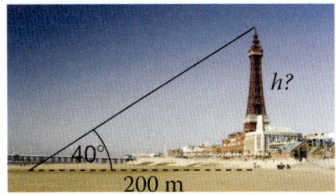

a Use Carl's estimates to find the height of Blackpool Tower.

b Why might the answer to **a** be inaccurate?

3 Jack is 1.8 m tall.

a Find the length of Jack's shadow when θ is

i 20°

ii 40°

iii 60° **iv** 80°

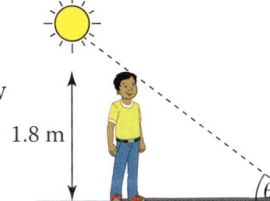

b Describe what happens to the shadow as the sun rises in the sky.

4 A see-saw is 2.8 m long. When one end touches the ground it makes an angle of 35° with it.

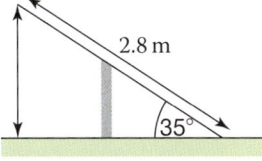

Tina says the other end is less than $1\frac{1}{2}$ metres above the ground. Is she correct? Give your reasons.

5 *PQRS* is a rectangle. Find the *exact* area of *PQRS*.

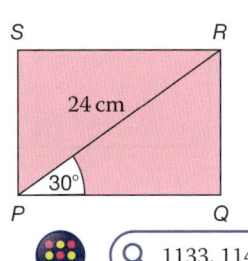

6 This flowerbed is in the shape of a parallelogram. Find the *exact* area of the flowerbed.

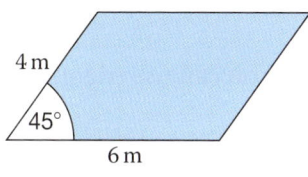

7 After take-off, a plane climbs at an angle of 32° to the horizontal. Find the height of the plane above the ground when it has travelled 1 kilometre.

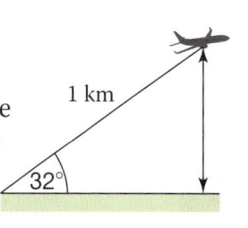

8 Beth is 1.6 m tall. She looks at the top of a statue that is 8 metres from her. Beth says the statue is over 10 metres tall. Is she correct? Show how you decided.

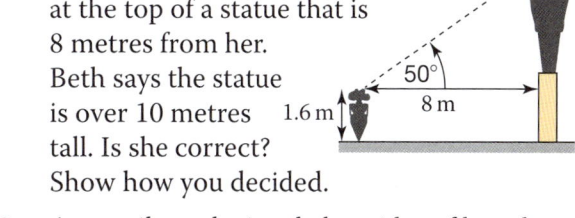

*9 An equilateral triangle has sides of length 18 mm. Find exact values for

a the height of the triangle

b the area of the triangle.

*10 The diagram shows a flag on top of a building. How tall is the flag?

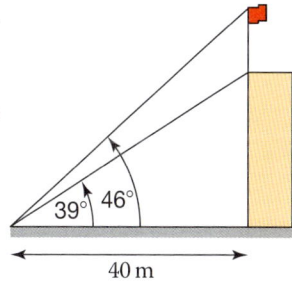

*11 The angle between the legs of a stepladder must be 74°. A metal bar is needed to keep the legs the correct distance apart. Is a 2 metre bar long enough?

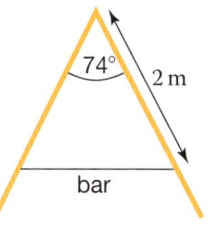

19.3 Trigonometry 2

You can use trigonometric ratios to find angles as well as sides.

- $\sin \theta = \dfrac{\text{Opposite side}}{\text{Hypotenuse}}$ $\cos \theta = \dfrac{\text{Adjacent side}}{\text{Hypotenuse}}$ $\tan \theta = \dfrac{\text{Opposite side}}{\text{Adjacent side}}$

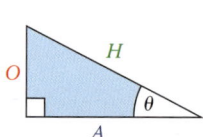

$$\sin \theta = \frac{O}{H}$$
$$\cos \theta = \frac{A}{H}$$
$$\tan \theta = \frac{O}{A}$$

EXAMPLE

Find the angles marked by letters.

a 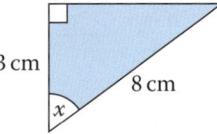 3 cm, 8 cm, x

b 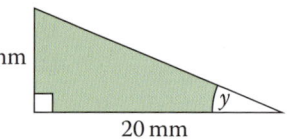 6 mm, 20 mm, y

c 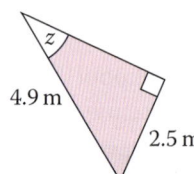 z, 4.9 m, 2.5 m

> You can subtract from 180° to find the third angle if you need it.

Identify which sides you know.

a $A = 3$, $H = 8$ **b** $O = 6$, $A = 20$ **c** $O = 2.5$, $H = 4.9$

Choose the ratio, substitute values, then write as a decimal.

$\cos x = \dfrac{3}{8} = 0.375$ $\tan y = \dfrac{6}{20} = 0.3$ $\sin z = \dfrac{2.5}{4.9} = 0.5102...$

Use the inverse trigonometric function.

$x = \cos^{-1} 0.375$ $y = \tan^{-1} 0.3$ $z = \sin^{-1} 0.5102...$

$x = 68.0°$ (1 dp) $y = 16.7°$ (1 dp) $z = 30.7°$ (1 dp)

> Carry on the working on your calculator.

In a right-angled triangle, given two sides or one side and one angle, you can find all the other sides and angles using trigonometry and Pythagoras' theorem.

EXAMPLE

Find the unknown sides and angles in triangle PQR.

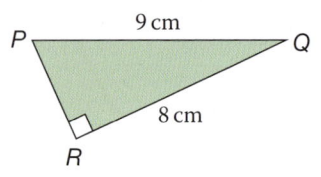
P, 9 cm, Q, 8 cm, R

Use Pythagoras' theorem for the third side and trigonometry for an angle.

$PR^2 = 9^2 - 8^2$ $\sin P = \dfrac{8}{9} = 0.8888...$

$\quad\;\; = 81 - 64$ $P = \sin^{-1} 0.8888...$

$\quad\;\; = 17$ $P = 62.73...$

$PR = \sqrt{17} = 4.123...$ $P = 62.7°$ (3 sf)

$PR = 4.1$ cm (1 dp)

> It is better to use the sides you were given rather than the one you have found.

Subtract from 180° to find the third angle.

$Q = 180° - 90° - 62.7°$ Angle sum of a triangle.

$\quad = 27.3°$ (1 dp)

> If you have time, use the other trigonometric ratios to check your answers.

> Check that the smallest angle is opposite the shortest side and the largest angle opposite the longest side.

p.52

Exercise 19.3S

1 Find the size of each angle.

    a $\cos A = 0.5$ **b** $\sin B = 0.6$

    c $\tan C = 0.43$ **d** $\cos D = 0.37$

    e $\sin E = 0.892$ **f** $\tan F = 1.645$

2 Calculate the size of each angle.

    a $\tan P = \dfrac{2}{5}$ **b** $\sin Q = \dfrac{37}{50}$

    c $\cos R = \dfrac{21}{25}$ **d** $\sin S = \dfrac{120}{150}$

    e $\tan T = \dfrac{7.8}{5.4}$ **f** $\cos U = \dfrac{19.8}{43.5}$

3 Find the angles marked by letters.

    a 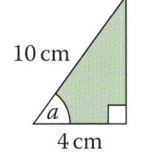   **b**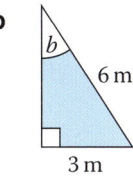

    c 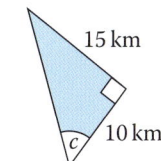   **d**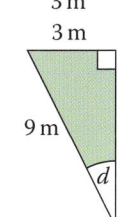

    e 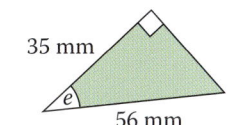   **f**

4 Calculate all the unknown sides and angles.

    a 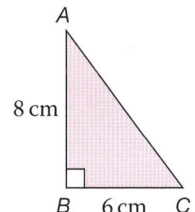   **b**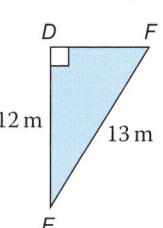

    c 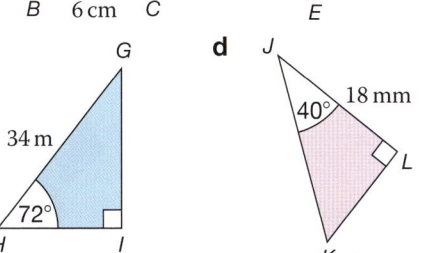   **d**

    e 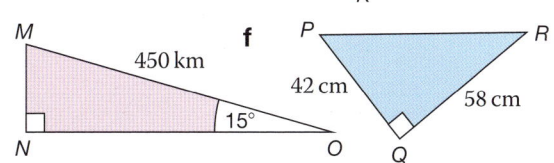   **f**

5 In each part, find the other side and angles of triangle *RST*.

    a $RS = 7\,\text{cm}, RT = 10\,\text{cm}$ and angle $R = 90°$

    b $RS = 36\,\text{m}, RT = 54\,\text{m}$ and angle $S = 90°$

    c $RS = 3.5\,\text{km}, TS = 1.67\,\text{km}$ and angle $T = 90°$

6 *PQRS* is a rectangle. Calculate

    a QS

    b angle *PQS*

    c angle *PSQ*

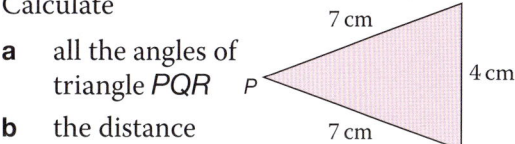

7 The diagonals of kite *KLMN* intersect at *X*. $KX = 15\,\text{cm}$, $XM = 27\,\text{cm}$ and $LN = 24\,\text{cm}$.

    Calculate the sides and angles of kite *KLMN*.

8 Calculate

    a all the angles of triangle *PQR*

    b the distance from *P* to *QR*.

*9 A rhombus has sides of length 10 cm. The length of the longest diagonal is 17 cm. Find

    a the angles of the rhombus

    b the length of the shortest diagonal.

10 The 20% on this road sign means the hill goes down by 20 metres for every 100 metres in the horizontal direction.

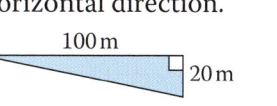

    a Find the angle between the road and the horizontal.

    b How far along the road do you travel as it falls by 20 metres?

19.3 Trigonometry 2

In a *right-angled* triangle

$$\sin \theta = \frac{\text{Opposite side}}{\text{Hypotenuse}} \qquad \cos \theta = \frac{\text{Adjacent side}}{\text{Hypotenuse}} \qquad \tan \theta = \frac{\text{Opposite side}}{\text{Adjacent side}}$$

$$\sin \theta = \frac{O}{H}$$
$$\cos \theta = \frac{A}{H}$$
$$\tan \theta = \frac{O}{A}$$

HOW TO

To solve problems involving trigonometry

① Draw a sketch (if needed). Include all the information you know.

② Look for a right-angled triangle. Decide which sides are O, A and H.

③ Choose to use Pythagoras or a trigonometric ratio.

④ Work out the value required. Check whether it seems reasonable.

Angle	sin	cos	tan
30°	$\frac{1}{2}$	$\frac{\sqrt{3}}{2}$	$\frac{1}{\sqrt{3}}$
45°	$\frac{1}{\sqrt{2}}$	$\frac{1}{\sqrt{2}}$	1
60°	$\frac{\sqrt{3}}{2}$	$\frac{1}{2}$	$\sqrt{3}$

EXAMPLE

On leaving a port, P, a boat sails 20 kilometres due south to a point Q.
The boat then sails 50 kilometres due west to point R.
Find the distance and 3-figure bearing for the shortest route back from R to P.

Bearings are measured clockwise from north.

② *PQR* is a right-angled triangle.
The hypotenuse, RP, is the shortest route back.
The bearing, θ, equals angle RPQ
(Alternate angles from parallel north lines.)

①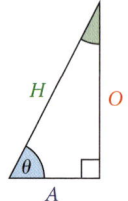

③ Use Pythagoras to find RP.

④ $RP^2 = PQ^2 + QR^2$

$= 20^2 + 50^2$

$= 400 + 2500$

$= 2900$

$RP = \sqrt{2900} = 53.85...$

The distance on the shortest route back is 53.9 km (1 dp)

This seems reasonable as it is the longest side of the triangle.

③ Use trigonometry to find the bearing.

④ $\tan \theta = \frac{50}{20} = 2.5$

$\theta = \tan^{-1} 2.5$

$= 68.198...°$

The bearing of the shortest route back is 068° (nearest degree)

The diagram suggests this is reasonable.

Use the given values 20 and 50. 53.9 is not as accurate and may be incorrect if an error was made.

p.48

Exercise 19.3A

Check that each answer seems reasonable.

1 On leaving an airport, *A*, a plane flies 100 km due north to a point *B*. The plane then flies 75 km due east to point *C*.

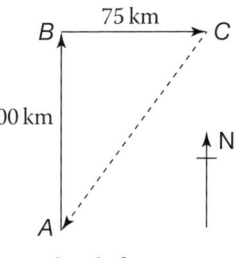

Find the distance and bearing for the shortest route back from *C* to *A*.

2 Find a single journey that is equivalent to

a 90 km due south followed by 60 km due east.

b 50 km due west followed by 120 km due north

3 Simton is 34 km west of Rinley. Taxham is 57 km north of Rinley.

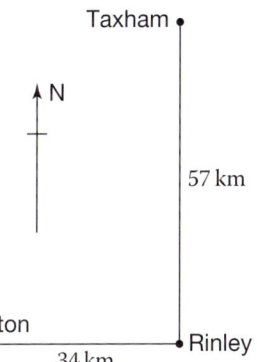

a Find the distance and three-figure bearing of Simton from Taxham.

b What is the bearing of Taxham from Simton?

4 Safety advice says that the angle between a ladder and the ground must be 75°.

a Show that this ladder is not safe.

b Explain how the ladder should be moved to make it safe.

5 The diagram shows the end view of a shed. Find

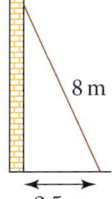

a *h*

b the angle the roof makes with the vertical.

6 A ramp is needed to help wheelchairs go up three steps. Each step is 12 cm high and 80 cm wide.

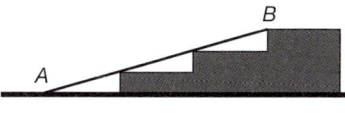

a Find the length of the ramp *AB*.

b Safety advice says the angle between the ramp and horizontal must be less than 4°. Is this ramp safe? Explain your answer.

7 a i Draw the line *y* = 2*x* on squared or graph paper.

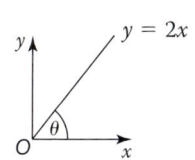

p.294

ii Use your graph to find tan θ.

b Repeat part **a** for the line *y* = 3*x*.

c Repeat part **a** for the line $y = \frac{1}{2}x$.

d What do you notice?

*8 The diagonals of a rectangle are twice as long as the shorter sides. Calculate

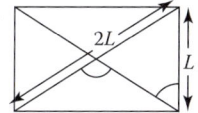

a the angle between a diagonal and a short side

*b the obtuse angle between the diagonals.

*9 The diagram shows the end view of a house. Find

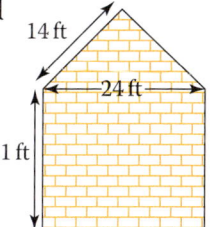

a the angle the roof makes with the horizontal

b the height of the house

c the area of this end of the house.

p.144

*10 Using a diagram, show that sin 40° = cos 50°.

*11 a Given $\tan \theta = \frac{3}{4}$, show that

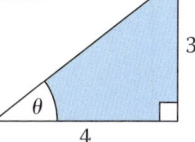

i $\sin \theta = \frac{3}{5}$

ii $\cos \theta = \frac{4}{5}$

b Given that $\sin \theta = \frac{12}{13}$, find as fractions

i cos θ   **ii** tan θ.

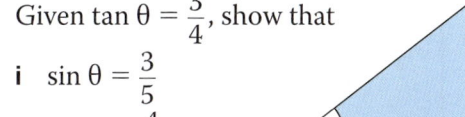

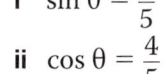 1131, 1133, 1145   SEARCH

19.4 Vctors

19.4 Vectors

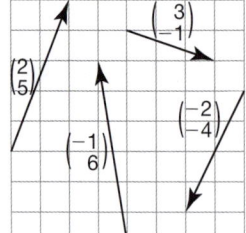

p.148

- A **scalar** has size, but no direction.
- A **vector** has size and direction.

Scalars
Distance, mass, temperature

Vectors
Translations, velocities, forces

Vectors can be represented in several ways:

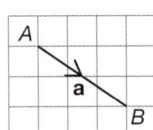

- on a diagram;
- using letters

 a, a or $\overrightarrow{AB}$, the vector from point A to point B;

- as a **column vector** $\begin{pmatrix} 3 \\ -2 \end{pmatrix}$, 3 right and 2 down.

You can add and subtract vectors or multiply them by a scalar.

$\mathbf{a} + \mathbf{b} = \mathbf{b} + \mathbf{a}$ is vector **a** followed by vector **b** (or **b** followed by **a**)

$\mathbf{a} - \mathbf{b} \neq \mathbf{b} - \mathbf{a}$ is vector **a** followed by vector $-\mathbf{b}$ (or $-\mathbf{b}$ followed by **a**)

$3\mathbf{a} = \mathbf{a} + \mathbf{a} + \mathbf{a}$ is a vector parallel to **a**, but 3 times as long

$-\mathbf{b}$ \qquad is a vector equal in size to **b** but in the opposite direction

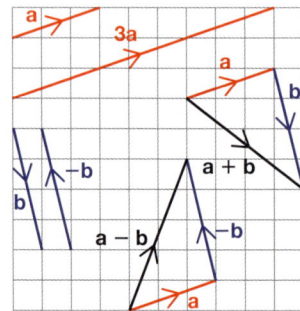

EXAMPLE

p.290

A is the point (1, 4) and B is the point (5, 2).

Write as column vectors

a $\overrightarrow{AB}$

b $\overrightarrow{BA}$

The vector from A to B goes 4 right and 2 down.

a $\overrightarrow{AB} = \begin{pmatrix} 4 \\ -2 \end{pmatrix}$

b $\overrightarrow{BA}$ goes in the opposite direction.

 $\overrightarrow{BA} = \begin{pmatrix} -4 \\ 2 \end{pmatrix}$

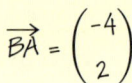

You can work this out from a diagram or the coordinates of the points.

EXAMPLE

$\mathbf{p} = \begin{pmatrix} 2 \\ -1 \end{pmatrix}$, $\mathbf{q} = \begin{pmatrix} 0 \\ 3 \end{pmatrix}$ and $\mathbf{r} = \begin{pmatrix} -3 \\ 4 \end{pmatrix}$

a Calculate **i** $\mathbf{p} + \mathbf{q} + \mathbf{r}$ **ii** $2\mathbf{p} - \mathbf{r}$.

b Use diagrams to check your answers.

The first number in a column vector is called the *x*-**component** and the second is the *y*-**component**.

Work out the *x* component, then the *y* component.

a **i** $\underline{p} + \underline{q} + \underline{r} = \begin{pmatrix} 2 \\ -1 \end{pmatrix} + \begin{pmatrix} 0 \\ 3 \end{pmatrix} + \begin{pmatrix} -3 \\ 4 \end{pmatrix} = \begin{pmatrix} -1 \\ 6 \end{pmatrix}$ **b**

 $2 + 0 - 3 = -1$
 $-1 + 3 + 4 = 6$

ii $2\underline{p} - \underline{r} = 2\begin{pmatrix} 2 \\ -1 \end{pmatrix} - \begin{pmatrix} -3 \\ 4 \end{pmatrix} = \begin{pmatrix} 7 \\ -6 \end{pmatrix}$

 $2 \times 2 + 3 = 7$
 $2 \times -1 - 4 = -6$

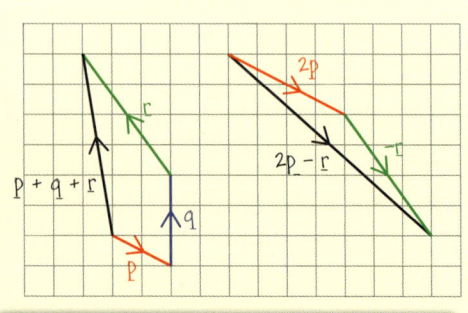

Underline letters to show they are vectors.

Geometry and measures   Pythagoras and trigonometry

Exercise 19.4S

1 Write each vector as a column vector.

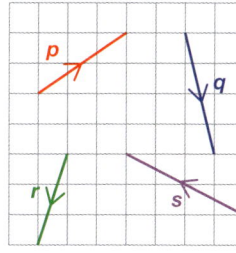

2 Write as column vectors

a $\overrightarrow{AB}$

b $\overrightarrow{BC}$

c $\overrightarrow{CD}$

d $\overrightarrow{DC}$

e $\overrightarrow{AC}$

f $\overrightarrow{EA}$

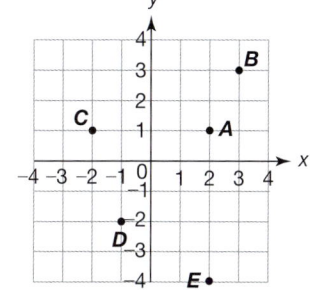

3 P is the point $(5, 3)$, Q is the point $(8, 4)$, R is the point $(5, 8)$ and S is the point $(0, 4)$.

a Calculate these column vectors

i $\overrightarrow{PQ}$ ii $\overrightarrow{QR}$ iii $\overrightarrow{PR}$

iv $\overrightarrow{RS}$ v $\overrightarrow{PS}$ vi $\overrightarrow{SP}$

b Use a diagram to check your answers.

4 Use the vectors from question **1**.
Draw diagrams to show

a $-\mathbf{p}$ b $2\mathbf{q}$ c $-3\mathbf{r}$

d $\mathbf{p} + \mathbf{q}$ e $\mathbf{q} + \mathbf{s}$ f $\mathbf{r} + \mathbf{s}$

g $\mathbf{p} - \mathbf{q}$ h $\mathbf{q} - \mathbf{r}$ i $\mathbf{r} - \mathbf{s}$

5 $\mathbf{p} = \begin{pmatrix} 1 \\ 3 \end{pmatrix}$, $\mathbf{q} = \begin{pmatrix} 2 \\ 0 \end{pmatrix}$ and $\mathbf{r} = \begin{pmatrix} -2 \\ -1 \end{pmatrix}$

a Calculate

i $2\mathbf{p}$ ii $3\mathbf{r}$ iii $\mathbf{p} + \mathbf{q}$

iv $\mathbf{p} + \mathbf{r}$ v $\mathbf{p} - \mathbf{q}$ vi $\mathbf{p} - \mathbf{r}$

vii $\mathbf{q} + \mathbf{r}$ viii $\mathbf{q} - \mathbf{r}$ ix $\mathbf{p} + \mathbf{q} + \mathbf{r}$

b Use diagrams to check your answers.

6 K is the point $(-3, 1)$, L is the point $(-2, -3)$, M is the point $(1, -3)$ and N is the point $(2, 4)$.
Write as column vectors

a $\overrightarrow{KL}$ b $\overrightarrow{KM}$ c $\overrightarrow{KN}$

d $\overrightarrow{ML}$ e $\overrightarrow{LN}$ f $\overrightarrow{NM}$

7 Write down a vector that is

a in the same direction as $\begin{pmatrix} -2 \\ 1 \end{pmatrix}$ but 4 times as long

b half as long as $\begin{pmatrix} -6 \\ 8 \end{pmatrix}$ and in the opposite direction.

8 $\mathbf{a} = \begin{pmatrix} -2 \\ 4 \end{pmatrix}$, $\mathbf{b} = \begin{pmatrix} 3 \\ -1 \end{pmatrix}$ and $\mathbf{c} = \begin{pmatrix} 0 \\ -2 \end{pmatrix}$

a Calculate

i $\mathbf{a} + \mathbf{b} + \mathbf{c}$

ii $2\mathbf{a} + \mathbf{b}$

iii $\mathbf{a} - \mathbf{b} - \mathbf{c}$

iv $\mathbf{b} - 3\mathbf{c}$

v $2\mathbf{a} + 3\mathbf{b}$

vi $3\mathbf{b} + \dfrac{1}{2}\mathbf{a} - \dfrac{1}{2}\mathbf{c}$

b Use diagrams to check your answers.

9 On isometric paper, draw diagrams to show

a $4\mathbf{p}$

b $-2\mathbf{r}$

c $\mathbf{p} + \mathbf{q}$

d $\mathbf{p} - \mathbf{r}$

e $\mathbf{p} + \mathbf{q} + \mathbf{r}$

f $3\mathbf{p} + \mathbf{r}$

g $3\mathbf{r} - \mathbf{q}$

h $\mathbf{r} - \mathbf{q} - \mathbf{p}$

i $2\mathbf{p} - 3\mathbf{r}$

j $-2\mathbf{r} + \mathbf{p} - 3\mathbf{q}$

Q 1134, 1135   SEARCH

19.4 Vurrs Vectors

RECAP

Addition   **a** + **b** is **a** followed by **b** (or **b** followed by **a**)

Subtraction   **a** − **b** is **a** followed by − **b** (or − **b** followed by **a**)

Multiplication gives parallel vectors

                    3**a** is parallel to **a**, but 3 times as long

                    − **b** is equal in size to **b** but in the opposite direction

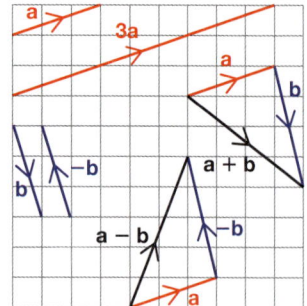

HOW TO

To solve problems involving vectors

① Decide whether to use a diagram or algebra.

② Find the vectors or components required.

EXAMPLE

Write each of the other vectors in terms of **a** and/or **b**.

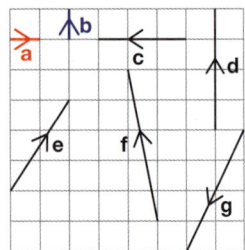

① Use the diagram to compare the other vectors with **a** and **b**.

② $\underline{c} = -3\underline{a}$ **c** is in the opposite direction to **a** and 3 times as long.

   $\underline{d} = 4\underline{b}$ **d** is in the same direction as **b** and 4 times as long.

① Draw your own sketch if it helps.

② $\underline{e} = 2\underline{a} + 3\underline{b}$

   $\underline{f} = -\underline{a} + 5\underline{b}$

   $\underline{g} = -2\underline{a} - 4\underline{b}$

Remember to underline your vectors.

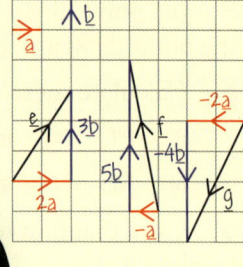

EXAMPLE

Vector $\begin{pmatrix} x \\ -2 \end{pmatrix}$ is parallel to vector $\begin{pmatrix} 12 \\ -8 \end{pmatrix}$. Find x.

You can use a diagram to check that $\begin{pmatrix} 3 \\ -2 \end{pmatrix}$ is parallel to $\begin{pmatrix} 12 \\ -8 \end{pmatrix}$

① When vectors are parallel, one is a scalar multiple of the other. Compare the components, then use algebra.

② $-2 \times 4 = -8$ Comparing y components gives the multiple 4.

   $x \times 4 = 12$ Use the same multiple for the x component.

   $x = 12 \div 4$

   $x = 3$

p.202

Exercise 19.4A

1 Write each of the other vectors in terms of **a** and/or **b**.

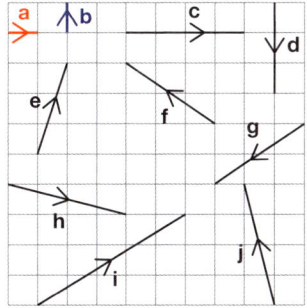

2 Write each of the other vectors in terms of **p** and/or **q**.

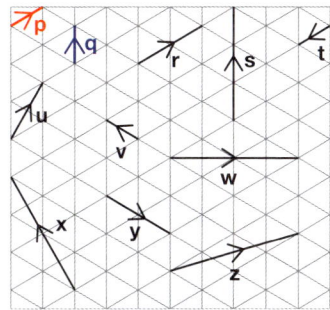

3 Find a, b and c given that

a $\begin{pmatrix} a \\ 3 \end{pmatrix}$ is parallel to $\begin{pmatrix} 20 \\ 15 \end{pmatrix}$

b $\begin{pmatrix} -4 \\ b \end{pmatrix}$ is parallel to $\begin{pmatrix} -1 \\ 3 \end{pmatrix}$

c $\begin{pmatrix} c \\ 2 \end{pmatrix}$ is parallel to $\begin{pmatrix} 12 \\ -6 \end{pmatrix}$

4 Show that $\begin{pmatrix} 3 \\ 2 \end{pmatrix}$, $\begin{pmatrix} 12 \\ 8 \end{pmatrix}$ and $\begin{pmatrix} -6 \\ -4 \end{pmatrix}$ are all parallel.

5 a Starting from the castle, a journey is represented by vector $\begin{pmatrix} 2 \\ -3 \end{pmatrix}$. Where does the journey end?

b Write down vectors that represent journeys from the harbour to

   i the cave   **ii** the cliff

   iii the castle, then the headland.

5 c i Write down a series of vectors for a boat trip around the island. The trip must start and end at the harbour.

   ii Add all of your vectors and comment on the answer.

6 Find the image of point $P(2, 3)$ after the translation given by each vector.

a $\begin{pmatrix} 4 \\ 1 \end{pmatrix}$ **b** $\begin{pmatrix} 5 \\ -2 \end{pmatrix}$ **c** $\begin{pmatrix} -6 \\ 7 \end{pmatrix}$ **d** $\begin{pmatrix} -3 \\ -5 \end{pmatrix}$

7 In each case find the vector $\begin{pmatrix} x \\ y \end{pmatrix}$

a $\begin{pmatrix} x \\ y \end{pmatrix} + \begin{pmatrix} 1 \\ 3 \end{pmatrix} = \begin{pmatrix} 6 \\ 2 \end{pmatrix}$ **b** $\begin{pmatrix} x \\ y \end{pmatrix} - \begin{pmatrix} 2 \\ 5 \end{pmatrix} = \begin{pmatrix} 8 \\ 4 \end{pmatrix}$

8 Find an expression for

a the length of vector $\begin{pmatrix} x \\ y \end{pmatrix}$

b the angle it makes with the horizontal.

*9 A plane travels due south 100 km. It then travels 40 km south east.

Find the distance and bearing of the plane's final position.

(diagram labels: 100 km/h, 40 km/h)

*10 Using diagrams, show that

a $\mathbf{p} + \mathbf{q} = \mathbf{q} + \mathbf{p}$

b $3(\mathbf{p} + \mathbf{q}) = 3\mathbf{p} + 3\mathbf{q}$

c $-(\mathbf{p} + \mathbf{q}) = -\mathbf{p} - \mathbf{q}$

d $(\mathbf{p} + \mathbf{q}) + \mathbf{r} = \mathbf{p} + (\mathbf{q} + \mathbf{r})$

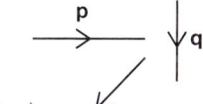

11 The vector $\overrightarrow{AC} = \overrightarrow{AB} + \overrightarrow{BC}$

Given that $\overrightarrow{AB} = \begin{pmatrix} 4 \\ 1 \end{pmatrix}$, $\overrightarrow{BC} = \begin{pmatrix} 2x \\ 3y \end{pmatrix}$ and $\overrightarrow{AC} = \begin{pmatrix} 8 \\ -2 \end{pmatrix}$, find x and y.

12 A translation, T_1, transforms the point $P(2, 3)$ to the point $P_1(-1, 4)$. A translation T_2 takes the point P to the point $P_2(0, -2)$. Find the translation that moves point P_1 to point P_2.

Assessment 19

1 **a** Hannah says that the hypotenuse of this triangle is 181 cm.

Say what mistake Hannah has made and work out the correct hypotenuse. [3]

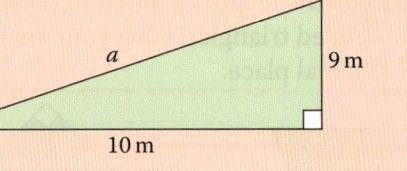

b Paul says that the missing side in this triangle is 15.45 m to 2 dp.

Say what mistake Paul has made and work out the correct value of the missing side. [4]

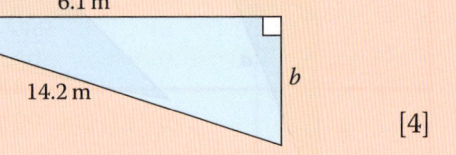

2 **a** Angelina says that 2, 3, 13 is a Pythagorean triple. Say why she is wrong. [2]

b Darshna says that 20, 99 and 100 is a Pythagorean triple. Say which number is wrong and find the correct third value in the triple. [2]

3 Find the distance between these points $(-3, -5)$ and $(6, -1)$. [3]

4 There are three rectangles with area 28 m². The sides of each rectangle are a whole number of metres.

a Find the side lengths of the three possible rectangles. [3]

b Calculate the length of the diagonal of each of these rectangles. Give your answer to 4 sf. [6]

5 Five triangles have sides with these lengths.

a 9, 12, 15   [3]   **b** 9, 14, 17   [3]   **c** 1.6, 3.0, 3.4   [3]

d 11, 19, 22   [3]   **e** 3.6, 7.7, 8.5   [3]

Which of these are right-angled triangles? Give a reason for each answer.

6 A boat leaves the harbour, *H*, and sails 2.5 km on a bearing of 062° until it reaches a buoy, *B*.

The boat then sails for 3.6 km on a bearing of 152° until it reaches the lighthouse, *L*. It then returns to the Harbour.

a Show that the angle *x* is a right angle. [4]

b Calculate the distance from the lighthouse to the harbour. [3]

c Calculate the total distance travelled. [1]

d The angle *y* is 55°. Find the bearing from *L* to *H*. [2]

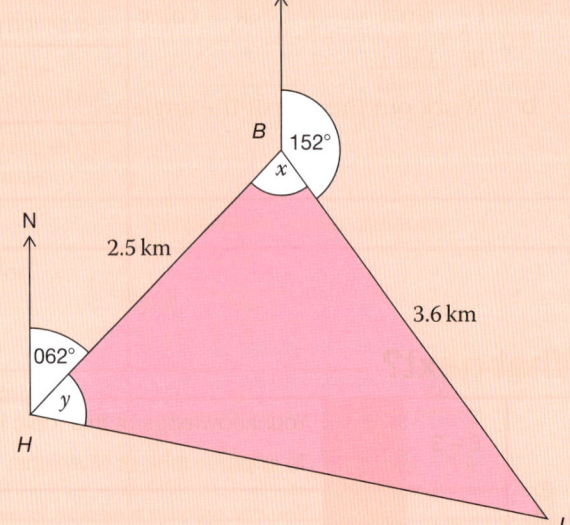

7 a Erica is given the three triangles shown. All lengths are in cm and angles in degrees correct to 1 decimal place.

She says incorrectly that to find the

    i angle in triangle **A**, you use tan

    ii angle in triangle **B**, you use sin

    iii angle in triangle **C**, you use cos.

Write correctly which trigonometric function you need to use to find each angle, and find the angles. [3]

b Use trigonometry to find the missing side in each triangle. [3]

8 a The base of a ladder is on horizontal ground and is leaning against a vertical wall.
The base is 1.75 m from the wall and the ladder makes an angle of 12.5° with the wall.
How long is the ladder? [3]

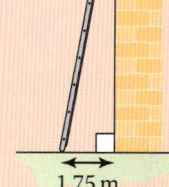

b How far up the wall does the ladder reach? [2]

9 a Draw these vectors on squared paper:

    i $\begin{pmatrix} 3 \\ 4 \end{pmatrix}$ [2]   **ii** $\begin{pmatrix} -5 \\ 12 \end{pmatrix}$ [2]   **iii** $\begin{pmatrix} 3 \\ -1 \end{pmatrix}$ [2]   **iv** $\begin{pmatrix} -2 \\ -3 \end{pmatrix}$ [2]

b If $\mathbf{x} = \begin{pmatrix} 4 \\ 6 \end{pmatrix}$ and $\mathbf{y} = \begin{pmatrix} -2 \\ 7 \end{pmatrix}$. Draw these vectors on squared paper:

    i $\mathbf{x} + \mathbf{y}$ [2]   **ii** $\mathbf{y} - \mathbf{x}$ [2]   **iii** $\mathbf{x} - \mathbf{y}$ [2]

    iv $2\mathbf{x}$ [2]   **v** $-2\mathbf{y}$ [2]   **vi** $2\mathbf{x} - 3\mathbf{y}$ [2]

c Use Pythagoras' theorem to find the lengths of each of the vectors in part **b**.
Leave your answers in exact form. [12]

d Compare your drawings to parts **ii** and **iii**.
Write down two things you notice. [2]

10 On this grid a vector one square to the right is given by **x** and one square vertically upwards by **y**.

The six vectors have been written in terms of the vectors **x** and **y**.

    a $3\mathbf{x}$ **b** $2\mathbf{x} + 7\mathbf{y}$

    c $6\mathbf{y}$ **d** $\mathbf{x} - 3\mathbf{y}$

    e $3\mathbf{x} + 2\mathbf{y}$ **f** $3\mathbf{x} + 7\mathbf{y}$

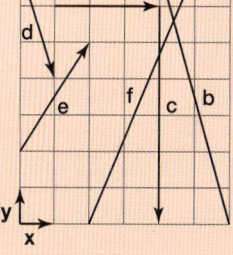

Which vectors are written down correctly?
For the vectors written incorrectly, write down the correct vector in terms of the vectors **x** and **y**. [10]

11 If $\mathbf{x} = \begin{pmatrix} 4 \\ 1 \end{pmatrix}$, $\mathbf{y} = \begin{pmatrix} 7 \\ -2 \end{pmatrix}$ and $\mathbf{z} = \begin{pmatrix} -8 \\ 5 \end{pmatrix}$, write down these vectors:

    a $5\mathbf{y}$ [1]   **b** $-2\mathbf{z}$ [1]   **c** $\mathbf{y} + \mathbf{z}$ [1]

    d $\mathbf{y} - \mathbf{x}$ [2]   **e** $\mathbf{x} + \mathbf{y} + \mathbf{z}$ [2]   **f** $4\mathbf{z} - 2\mathbf{x}$ [2]

    g $4\mathbf{y} + 3\mathbf{x} - 2\mathbf{z}$ [3]

20 The probability of combined events

Introduction

There is an old British myth that says if it rains on St Swithin's Day (15th July) it will rain for 40 days afterwards. Weather statistics show that it is untrue. Unfortunately long-scale weather forecasting is fairly unreliable, particularly in the UK, so people look for tell-tale signs to help their predictions. In probability language, if a particular event occurs (rain on St Swithin's Day), does this increase the probability of another event occurring (a wet summer)?

What's the point?

Quite often, the occurrence of one event will significantly increase the probability of another event occurring. For example, lung cancer appears unpredictably, but its occurrence is greatly increased if a particular person is a smoker. Understanding the probabilities attached to linked events helps us to evaluate everyday risks.

Objectives

By the end of this chapter you will have learned how to ...

- Use Venn diagrams to record outcomes and calculate probabilities of events.
- Construct possibility spaces and use these to calculate probabilities.
- Use tree diagrams to show the frequencies or probabilities of two events.
- Use tree diagrams to calculate the probability of independent and dependent events.

Check in

1 Cancel these fractions to their simplest form.

a $\frac{6}{9}$ **b** $\frac{5}{10}$ **c** $\frac{15}{20}$ **d** $\frac{2}{8}$ **e** $\frac{20}{20}$

2 Calculate

a $\frac{3}{4} + \frac{1}{4}$ **b** $\frac{7}{10} + \frac{3}{10}$ **c** $1 - \frac{7}{10}$ **d** $1 - \frac{3}{5}$

3 Calculate

a $1 - 0.1$ **b** $1 - 0.6$ **c** $1 - 0.15$

4 Choose a number from the rectangle that is

a prime **b** square

c triangular **d** a multiple of 4

e a factor of 10.

```
        6
   5           2
        4
  10        8
1
   7   9      3
```

Chapter investigation

Two six-sided dice, labelled 1 to 6 are thrown. The score is given by adding the two numbers that appear uppermost.

Which result is most likely to occur, or are all possible results equally likely?

What if, instead of adding the two numbers, you multiply them together?

20.1 Sets

- A **set** is a collection of numbers or objects.
- The objects in the set are called the **members** or **elements** of the set.

If the set X is 'the factors of 6', then you can write $X = \{1, 2, 3 \text{ and } 6\}$.

$3 \in X$ means that 3 is an element (member) of the set X.

If the set Y is 'the even numbers', then you can write $Y = \{2, 4, 6, 8, ...\}$.

- The universal set, which has the symbol ξ, is the set containing all the elements.
- The empty set, $\varnothing$, is the set with no elements.

You can use a Venn diagram to show the relationship between sets.

- The **intersection** of two sets, A∩B, consists of the elements common to both sets.

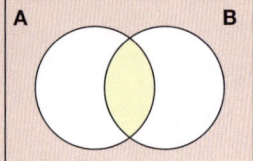

- The **union** of two sets, A∪B, consists of the elements which appear in at least one of the sets.

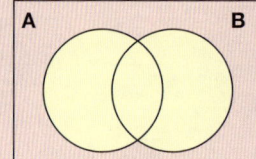

- The **complement** of a set, A′, consists of the elements which are not in A.

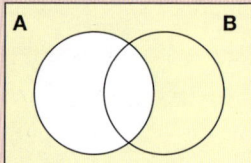

EXAMPLE

$\xi = \{1, 2, ... 11, 12\}$

$A = \{\text{factors of } 12\}$

$B = \{2, 3, 5, 6, 11\}$.

Find **a** A ∩ B

b A ∪ B

c (A ∪ B)′

a $A = \{1, 2, 3, 4, 6, 12\}$
$A \cap B = \{2, 3, 6\}$

b $A \cup B = \{1, 2, 3, 4, 5, 6, 11, 12\}$

c $(A \cup B)' = \{7, 8, 9, 10\}$

- You list the elements of each set or show the number of elements in each region

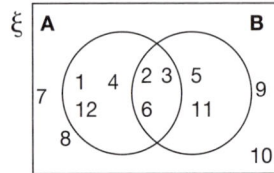

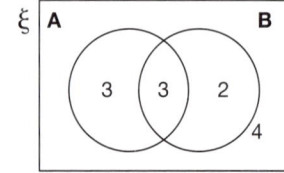

In the example,
$P(A) = \dfrac{6}{12} = \dfrac{1}{2}$

- You can use Venn diagrams to work out probabilities.
- $P(A) = \dfrac{\text{number of elements in set A}}{\text{total number of elements in } \xi}$

Probability   The probability of combined events

Exercise 20.1S

1 List the elements of these sets.

   a P = the first ten square numbers

   b R = countries in North America

   c S = the first ten prime numbers

   d T = factors of 36

2 Using the sets in question **1**, give the sets

   a P ∩ T   **b** S ∩ T

   c P ∩ S   **d** P ∪ S

3 Give a precise description of each set.

   a {1, 2, 5, 10}

   b {2, 4, 6, 8, 10, 12, ...}

   c {a, e, i, o,u}

   d {HH, HT, TH, TT}

   e {1p, 2p, 5p, 10p, 20p, 50p, £1, £2}

   f {3, 6, 9, 12, 15, 18, 21, 24, 27, 30}

4 List the elements of these sets.

   a A = the first ten positive integers

   b B = single digit odd numbers

   c C = single digit prime numbers

   d D = single digit square numbers

5 Using the sets in question **4**, give the sets

   a B ∩ C   **b** B ∩ D

   c B ∪ D   **d** C ∪ D

6 Say why B, C and D must be subsets of A for the sets in question **4**.

7 A = {even numbers}

   B = {odd numbers}

   C = {multiples of 5}

   D = {prime numbers}

   E = {multiples of 3}

   F = {square numbers}

   G = {factors of 36}

   For these pairs of sets, state if they have any elements in common, if they do, then list them.

7 **a** A, C   **b** A, D   **c** F, G

   d E, F   **e** A, B   **f** C, G

   g D, F   **h** D, E   **i** D, E and F

8 The Venn diagram shows information about the sport that 25 students are playing in PE this term.

The results are shown on the Venn diagram.

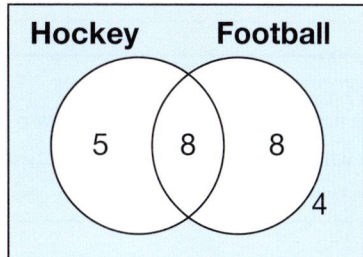

   a How many students are

      i in the intersection of hockey and tennis

     ii playing hockey

    iii not playing tennis

     iv in the union of hockey and tennis.

   b Describe the shaded region in words.

   c What fraction of students are playing either hockey or tennis, but not both?

9 An insurance company surveys 50 customers. The customers are sorted into

   C = {car insurance}

   H = {home insurance}

   The results are shown on the Venn diagram.

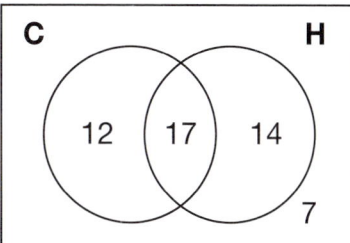

   a A customer is chosen at random. Find

      i P(C)   **ii** P(H′)

    iii P(C ∩ H)   **iv** P(C ∪ H)

   b How many customers had home insurance but not pet insurance?

20.1 Sets

RECAP

- The **intersection** of two sets, A ∩ B, consists of the elements common to both sets.
- The **union** of two sets, A ∪ B, consists of the elements which appear in at least one of the sets.
- The **compliment** of a set consists of all the elements that are not in the set.
- The **universal set**, which has the symbol ξ, is the set containing all the elements.
- The **empty set**, {}, is the set with no elements.
- **Venn diagrams** can be used to represent the relationships between sets.

HOW TO

To solve a problem involving several sets
1. If one is not given, draw a Venn diagram.
2. Decide which elements belong to each set.
3. Use the Venn diagram to calculate probabilities.

Descriptions often use properties of numbers like primes, multiples etc.

EXAMPLE

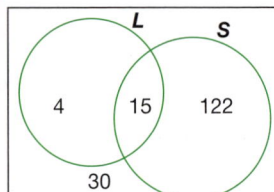

This Venn Diagram shows the *numbers* of pupils in a primary school classified by whether they are left handed (L) and whether they can swim (S).

a How many pupils are in the school?

b How many pupils can swim?

c What do you know about a pupil in L ∩ S?

d How many pupils are in L ∩ S?

a 171 = 4 + 15 + 122 + 30 ②   **b** 137 = 15 + 122 Add up just the numbers falling within S. ②

c They are left handed and can swim. **d** 15   It is just the overlapping area in both L and S. ②

EXAMPLE

A teacher is organising a school trip to Rome for the students in year 11.

There are 120 students in the school year.

72 students study history. 90 students are visiting Rome.

12 students don't study history and aren't visiting Rome.

Find the probability that a student is studying history and visiting Rome.

① Draw a Venn diagram to show the number of students in each region.
Let the number of students studying history and visiting Rome be x.

$72 - x + x + 90 - x + 12 = 120$ The total number of students is 120.

$\qquad 174 - x = 120$

$\qquad\qquad x = 174 - 120$

$\qquad\qquad\quad = 54$

P (History and Rome) $= \dfrac{54}{120}$ ③

$\qquad\qquad\qquad\quad = 0.45$

① ②

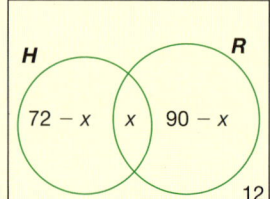

Exercise 20.1A

1 P = {factors of 42} and Q = {factors of 63}

   a Find the set P ∩ Q

   b What is the largest element in P ∩ Q?

2 The Venn diagram shows pupils in a primary school class. Girls are in set Q and pupils with dark hair are in set P.

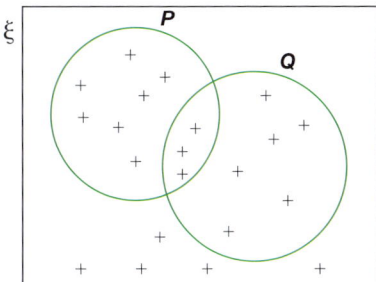

   a How many pupils are in the class?

   b How many girls have dark hair?

   c How many boys have dark hair?

   d Describe in words a pupil in P′ ∩ Q′

3 ξ = {all triangles}, E = {equilateral triangles}

   I = {isosceles triangles},

   R = {right-angled triangles}

   a Sketch a member of I ∩ R

   b Explain why E ∩ R = {}

4 Elsie sorts a group of objects into the sets A and B.

   She draws a Venn diagram to show her results.

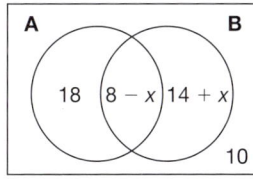

   a Show that Elsie must have started with 50 objects.

   b Use the value $x = 4$ to find

     i P(A)   **ii** P(B′)

     iii P(A ∪ B)   **iv** P(A ∩ B)

   c If A and B are mutually exclusive, find the value of x.

   d If $x = -14$, what can you say about the sets A and B?

5 David is organising a family reunion.

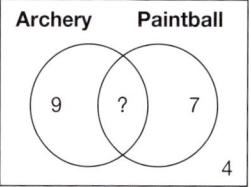

   His relatives can take part in two activities, archery or paintball.

   He shows the results on a Venn diagram.

   Archery costs £22 and paintball costs £18.

   David collects £524 to pay for the activities.

   How many relatives sign up for both activities?

6 At Newtown School there are 27 students in class 11B.

   17 students play tennis, 11 play basketball and 2 play neither game.

   How many students play both tennis as basketball?

7 There are 80 people in a tennis club.

   24 people have green eyes, 32 people have red hair and 32 people have neither red hair nor green eyes.

   a How many people have green eyes and red hair?

   b If a member is picked at random what is the probability that they have

     i green eyes

     ii red hair but *not* green eyes

     iii *not* red hair?

8 There are 64 people in a football club.

   32 people have blue eyes, 12 people have blue eyes and grey hair and half the people have either blue eyes or grey hair.

   If a member is picked at random what is the probability that they have

   a blue eyes

   b blue eyes and *not* grey hair

   c *not* grey hair?

1262, 1921, 1922   SEARCH

20.2 Possibility spaces

● The list or table of all of the possible outcomes of a trial is called a **possibility space** or **sample space**.

	1	2	3	4	5	6
1	1,1	1,2	1,3	1,4	1,5	1,6
2	2,1	2,2	2,3	2,4	2,5	2,6
3	3,1	3,2	3,2	3,4	3,5	3,6
4	4,1	4,2	4,3	4,4	4,5	4,6
5	5,1	5,2	5,3	5,4	5,5	5,6
6	6,1	6,2	6,3	6,4	6,5	6,6

When two ordinary dice are thrown, there are 36 possible outcomes.

If the dice are fair then the 36 outcomes are equally likely.

● You can use a sample space to calculate probabilities.

EXAMPLE

Jasper throws two dice and adds the results.

Lena throws two dice and multiplies the results.

a Draw a possibility space for Jasper's and Lena's experiments.

b Find the probability that Jasper scores 8.

c Find the probability that Lena scores 6.

d Find the probability that Lena scores 5 or less.

a Jasper

+	1	2	3	4	5	6
1	2	3	4	5	6	7
2	3	4	5	6	7	8
3	4	5	6	7	8	9
4	5	6	7	8	9	10
5	6	7	8	9	10	11
6	7	8	9	10	11	12

Lena

×	1	2	3	4	5	6
1	1	2	3	4	5	6
2	2	4	6	8	10	12
3	3	6	9	12	15	18
4	4	8	12	16	20	24
5	5	10	15	20	25	30
6	6	12	18	24	30	36

b The same sum appears on each diagonal.

$$P(8) = \frac{5}{36}$$

c $P(6) = \frac{4}{36} = \frac{1}{9}$

d $P(5 \text{ or less}) = \frac{10}{36} = \frac{5}{18}$

● You can use the possibility space to write the set of all possible outcomes.

EXAMPLE

A fair coin is tossed and a fair die is thrown.

● If a head is seen then the score on the dice is doubled.

● If a tail is seen then the score is just the number on the dice.

a Show the possibility space in a grid.

b Write the set of all possible outcomes.

c Find P(6).

a

	1	2	3	4	5	6
T	1	2	3	4	5	6
H	2	4	6	8	10	12

b The set of all outcomes is {1, 2, 3, 4, 5, 6, 8, 10, 12}

c $P(6) = \frac{2}{12} = \frac{1}{6}$

Probability The probability of combined events

Exercise 20.2S

1 Using Jasper's table shown opposite find the probability that the **sum of the scores** seen on two fair dice is

  a exactly 10 **b** at least 10

  c a square number **d** less than 5.

  e Write the set of all possible outcomes.

2 Using Lena's table shown opposite find the probability that the **product of the scores** seen on two fair dice is

  a exactly 10 **b** at least 10

  c a square number **d** less than 5.

  e Write the set of all possible outcomes.

3 Two fair dice are thrown and the difference between the scores showing on the two dice is recorded.

  a Make a table to show the possibility space.

  b Write the set of all possible outcomes.

  c Find the probability that the difference is

  i 0 **ii** 3

  iii 6 **iv** a prime number.

4 A fair coin is tossed three times and the outcome recorded (for example HHT).

  a Write the set of the 8 possible outcomes.

  b In how many of these are exactly two heads seen?

  c In how many do you see three of the same?

5 Two fair spinners are used. On one the possible scores are 1, 2 and 4, on the other the scores are 1, 3 and 5. The sum of the scores on the two spinners is recorded.

  a Make a table to show the possibility space.

  b Write the set of all possible outcomes.

  c Find the probability that the score is

  i 2 **ii** 3 **iii** even

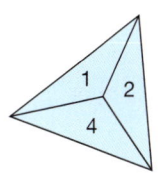 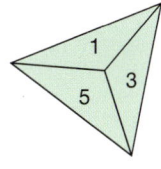

6 The two spinners in question **5** are used again but the score recorded is the product of the two scores.

  a Make a table to show the possibility space.

  b Write the set of all possible outcomes.

  c Find the probability that the score is

  i 2 **ii** 3 **iii** even.

7 For the spinners used in questions **5** and **6**, what is the probability to get an even number when you

  a add **b** multiply

  the scores on the two spinners?

8 A pair of fair dice are thrown and the sum and product of the scores are recorded in two lists. The dice are thrown 100 times.

  a Estimate the number of times a sum of exactly 10 will be seen.

  b Estimate the number of times a product of exactly 10 will be seen.

  c Would you expect to see

  i 6 **ii** 3

  in the list of sums more often, less often or about the same number of times as in the list of products?

9 Two fair dice are thrown together. One is an ordinary dice with the numbers 1 to 6, and the other has faces labelled 1, 2, 2, 3, 3, 3.

The possibility space for the sum of scores is shown.

	1	2	3	4	5	6
1	2	3	4	5	6	7
2	3	4	5	6	7	8
2	3	4	5	6	7	8
3	4	5	6	7	8	9
3	4	5	6	7	8	9
3	4	5	6	7	8	9

  a Find the probability that the score is

  i 6 **ii** 7 **iii** 9 **iv** 3

  b What other scores have the same probability as 6?

1199, 1263 SEARCH

20.2 Possibility spaces

RECAP

- When dealing with equally likely outcomes for a single or a combined experiment, a list or table showing the outcomes is helpful.

> Once you're drawn a table, the individual cells give you the outcomes.

HOW TO

① If you are constructing a list, work systematically so you can generate them all in sequence.
For tossing three coins, THH, HTH, TTT, HHT, HHH, THT are 6 of the 8 possibilities – but what are the other 2?

② When combining two simple experiments, a table allows you to enter the 'score' in the cell while the row and header column still tells you what each outcome was.

EXAMPLE

Tara has 6 cards with the numbered from 1 to 6.
She takes two cards without replacement.
A = {product is even} B = {sum is even}
Show that P(A) = 2P(B)

> Could you work out what the probabilities of even sum or product is without finding all the sums and products?

① Draw a possibility space.

You cannot take the same card twice so there are 30 outcomes.

② Use the sample space to calculate probabilities.

$P(A) = \dfrac{24}{30} = \dfrac{4}{5}$ There are 24 cells with an even product.

	1	2	3	4	5	6
1		2	3	4	5	6
2	2		6	8	10	12
3	3	6		12	15	18
4	6	8	12		20	24
5	5	10	15	20		30
6	6	12	18	24	30	

$P(B) = \dfrac{12}{30} = \dfrac{2}{5}$ There are 12 cells with an even sum.

$P(A) = 2P(B)$

	1	2	3	4	5	6
1		3	4	5	6	7
2	3		5	6	7	8
3	4	5		7	8	9
4	5	6	7		9	10
5	6	7	8	9		11
6	7	8	9	10	11	

EXAMPLE

A fair coin is tossed and a fair die is thrown. If a head is seen then the score on the dice is squared. If a tail is seen then the score is just the number on the dice.

Show the possibility space in a grid. Explain why 1 and 4 are the most likely scores to be recorded.

① Draw a possibility space.

	1	2	3	4	5	6
T	1	2	3	4	5	6
H	1	4	9	16	25	36

② 1 and 4 are the only square numbers that fall within the range 1-6.

Exercise 20.2A

1 Two cards are taken from a set of cards showing the numbers 1 to 6. Find the probability that the **difference** in the value of the two cards is

   a   3

   b   a factor of 6

   c   at least 1.

2 A fair coin is tossed until a head is seen. The number of times it has been tossed is recorded as the score on that trial.

   What is the probability that the score recorded is at least 3?

> Hint: consider the possible outcomes if a fair coin is tossed twice.

3 A fair dice in thrown and the spinner shown is spun. If the spinner lands on yellow you miss a turn (move 0 squares) otherwise you move the number of squares given by the product of the scores on the spinner and the dice.

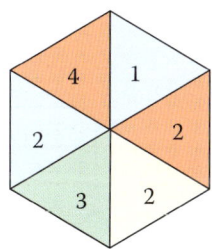

   a   Using labels R1, Y2 etc for the outcomes on the spinner construct a table to show the possible scores.

   b   What is the probability that the score recorded is

      i   4   **ii**   more than 6?

4 Cards numbered 1 to 100 are put in a box and Alessandra is asked to pick one at random. What is the probability that she chooses

   a   a single digit number

   b   a two digit number

   c   a number containing at least one 3?

> Don't write out the whole list, but imagine how many cards satisfy each condition.

5 Two fair spinners are used – one has sections showing the numbers 1, 1, 2, 3, 5 and the other has sections showing 3, 4, 5, 7, 8, 9.

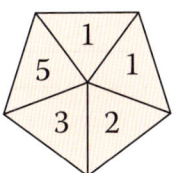 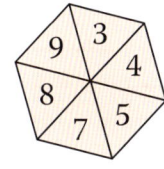

   a   What is the probability that the total score on the two spinners is

      i   6   **ii**   even?

   b   What is the probability that the score on one spinner is at least twice the score on the other spinner?

*6 Jack and Jill each roll a fair dice. Whoever gets the larger score wins the game.

   a   If a draw is allowed, what is the probability that Jill wins the game?

   b   A draw is not allowed and if the two dice show the same they roll again until one wins.
        What is the probability it will be Jill?

*7 A blue and a red dice are both fair and are thrown together. The following events are defined on the scores seen

   A – the dice show the same score

   B – the total score is at least 10

   C – the total score is odd

   D – the high score is a 4

   E – the score on one dice is a proper factor of the score on the other (a proper factor is a factor which is not 1 or the number itself)

   a   Calculate the probabilities of the five events A – E.

   b   find these probabilities

      i $P(A \cap B)$

      ii $P(C \cap D)$

   c   Find at least two pairs of mutually exclusive events (there are 4 pairs).

1199, 1263   SEARCH

20.3 Tree diagrams

● You can use a **frequency tree** to show the outcomes of two events.

EXAMPLE

A factory employs 300 workers. 100 workers are skilled and the rest are unskilled.
25 skilled workers work part time. 120 unskilled workers work full time.

a Draw a frequency tree to show this information.

b How many part-time workers are there altogether?

c If a worker is chosen at random what is the probability they are full-time?

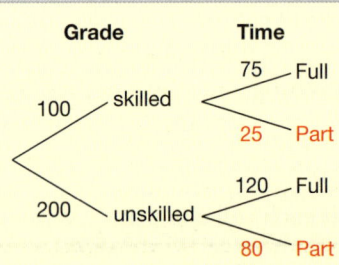

The frequency tree shows the number of workers along each branch.

The probability of each outcome is the relative frequency.

b 25 skilled and 80 unskilled workers are part time. $25 + 80 = 105$

c 75 skilled and 120 unskilled workers are full time. $\dfrac{75 + 120}{300} = \dfrac{13}{20}$

You can use a tree **diagram** to show the probabilities of two events.

● Write the outcomes at the end of each branch

● Write the probability on each branch

● The probabilities on each set of branches should add to 1

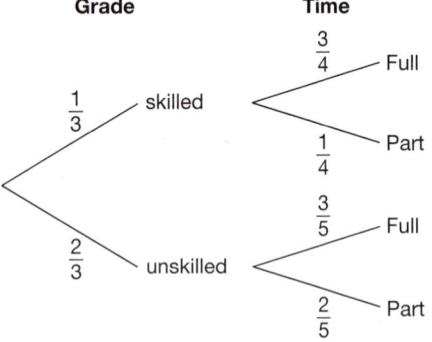

80 out of 200 unskilled workers work part time.
$\dfrac{80}{200} = \dfrac{2}{5}$

When you give a probability as a fraction, try to reduce it to its simplest form.

To find probabilities when an event can happen in different ways

● Multiply the probabilities along the branches.

● Add the probabilities for the different ways of getting the chosen event.

$P(\text{full time}) = P(\text{skilled and full time}) + P(\text{unskilled and full time})$

$= \left(\dfrac{1}{3} \times \dfrac{3}{4} \right) + \left(\dfrac{2}{3} \times \dfrac{3}{5} \right)$

$= \dfrac{3}{12} + \dfrac{6}{15} = \dfrac{13}{20}$

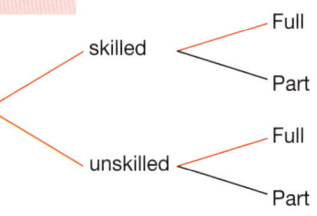

Probability   The probability of combined events

Exercise 20.3S

1 A company wants to classify its employees by whether they are male or female, and by whether or not they are part of the company's pension scheme.

   a Draw a frequency tree that could be used.

       The company has 360 employees, of whom 120 are male. 80% of the male employees and 70% of the female employees are in the pension scheme.

   b Show the numbers of employees on the branches of your diagram.

2 Lydia travels to work by car two days a week and by train on the other three. She is late for work 10% of the time when she travels by car, and late 20% of the time when she travels by train. She works 150 days during the first 8 months of a year.

   a Draw a possibility tree that could be used.

   b Show the numbers of days Lydia travels to work by car and train, and on which days she is late and on time.

   c Estimate how many days she is late for work during this period.

   d What is the probability that Lydia is late for work on a day chosen at random during this period?

3 A university collects some data about student debt among their first year undergraduates. There are 3250 students of whom approximately 2000 live at home. In their sample, the university found that a quarter of those living at home, and three in five of those living away from home, felt that debt was a problem.

   a Draw a frequency tree that could be used.

   b If the whole year is similar to the sample data the university collected, put in the numbers on the branches.

   c If a first year undergraduate is selected at random what is the probability that they are concerned about debt?

4 A red and a blue dice are thrown together. A is the event 'the red dice shows an even number'. B is the event 'the blue dice shows a multiple of 3'.

   a Describe the event $A \cap B$ in words.

   b Calculate $P(A \cap B)$ directly.

   c Calculate $P(B) \times P(A)$

   d Are events A and B independent of one another?

5 The MOT test examines whether cars are roadworthy. 10% of cars fail because there is something wrong with their brakes. 40% of the cars with faulty brakes also have faulty lights, while 20% of cars whose brakes are satisfactory have faulty lights.
In March an MOT test centre deals with approximately 3000 cars.

   a Draw a frequency tree showing the numbers of cars failing with brakes and lights in March.

   b How many cars fail on at least one of brakes and lights?

   c Tommy says that this means the rest of the 3000 cars passed the MOT test. Why is Tommy wrong?

6 Athletes are regularly tested for performance drugs. If an athlete is taking the drug, the test will give a positive result 19 times out of 20. However one in fifty tests on athletes who are not taking the drug is also positive. It is thought that around 20% of athletes in a particular event are taking the drug. The authorities carry out tests on 500 athletes during the season.

   a Draw a frequency tree showing the numbers of athletes expected to be taking or not taking the drug, and the results of the test.

   b Calculate the total number of positive tests expected to be seen.

   c How many of the positive tests came from athletes not taking the drug?

1208, 1334, 1935   SEARCH

20.3 Tree diagrams

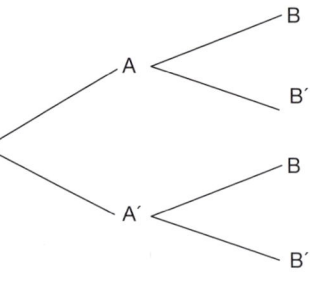

RECAP

- You can use tree **diagrams** to show the probabilities of two events.
  - Write the outcomes at the end of each branch
  - Write the probability on each branch
  - The probabilities on each set of branches should add to 1
- To find probabilities when an event can happen in different ways
  - Multiply the probabilities along the branches.
  - Add the probabilities for the different ways of getting the chosen event

- If the outcome of one event does not affect what happens in another then the events are **independent**.
  If A and B are independent events then
  $P(A \text{ and } B) = P(B) \times P(A)$

If you can shows that
$P(A \text{ and } B) = P(B) \times P(A)$
then A and B are independent.

HOW TO

To prove that two events are independent
1. Draw a tree diagram showing the possible outcomes and probabilities of each outcome
2. Use the tree diagram to find the probability of P(A and B), P(A) and P(B).
3. Test to see if $P(A \text{ and } B) = P(B) \times P(A)$ is satisfied.

EXAMPLE

A bag has 5 red and 3 blue balls in it. A ball is taken from it at random and not replaced, and then a second ball is taken out.
Show that choosing a red ball on the second attempt is dependent on whether or not the first ball was red.

1. Draw a tree diagram.

First	Second

$\frac{5}{8}$ Red — $\frac{4}{7}$ Red
— $\frac{3}{7}$ Blue
$\frac{3}{8}$ Blue — $\frac{5}{7}$ Red
— $\frac{2}{7}$ Blue

2. Find the probabilities of each event.

$P(\text{first ball red}) = \frac{5}{8}$

$P(\text{second ball red}) = P(R, R) + P(B, R)$

$$= \frac{5}{8} \times \frac{4}{7} + \frac{3}{8} \times \frac{5}{7}$$

$$= \frac{20}{56} + \frac{15}{56} = \frac{35}{56}$$

$$= \frac{5}{8}$$

$P(\text{both balls are red}) = \frac{5}{8} \times \frac{4}{7}$

$$= \frac{20}{56} = \frac{5}{14}$$

Don't cancel fractions to lowest form when you multiply the probabilities along the path – you are likely to have to add them!

3. If the events are independent then

$P(\text{first ball red}) \times P(\text{second ball red}) = P(\text{both balls are red})$

$P(\text{first ball red}) \times P(\text{second ball red}) = \frac{5}{8} \times \frac{5}{8} = \frac{25}{64} \neq \frac{5}{14}$

So the events are dependent.

Exercise 20.3A

1 In a league, teams are awarded 3 points for a win, one for a draw and none for a loss. Amelie thinks that

- her team has a probability of 0.6 of winning any match, and a probability of 0.3 for a draw
- the result of any game is independent of other results.

a Find the probability that her team has at least three points after two games.

b How have you used Amelie's assumption that the results of the game are independent in your answer to part **a**?

c Do you think that Amelie's assumption hat the results of the game are independent is reasonable? Give a reason for your answer.

2 A Year 13 pupil is taking their driving test. Records show that people taking the test at that age have a 70% chance of passing on the first attempt and 80% on any further attempts needed.

a Show this information on a tree diagram showing up to three attempts.

b Find the probability that

   i the pupil passes at the second attempt

   ii the pupil has still not passed after three attempts.

3 Denzel is going to the airport to catch a flight. He needs to travel on a bus and then catch a train to the airport.
He catches a bus which has a probability of 0.8 of making a connection with a train which always gets to the airport on time. The next train has a probability of 0.7 of getting him to the airport on time.

a Show this information on a tree diagram.

b Find the probability that Denzel gets to the airport in time for his flight.

4 A bag has 4 red and 4 blue balls in it. A ball is taken from it at random and not replaced and then a second ball is taken out.

X = the second ball is red

Y = the two balls are the same colour

Z = both balls are red

a Ellie says that $Z = (X \cap Y)$.
Is she correct?

b Construct a tree diagram and show that X and Y are independent events.

5 A bag has 10 white and 5 black balls in it. A ball is taken from it at random, the colour noted and the ball is replaced and then a second ball is taken at random.

A = the two balls are different colours

B = at least one ball is black.

Construct a tree diagram and decide if A and B are independent events.

*6 A spinner has a probability of $\frac{1}{6}$ of landing on blue.
Green is three times as likely to occur as blue.
Red, black and yellow are equally likely to occur.

a Calculate the probability that the spinner lands on red.

Sara is playing a game where she spins the spinner and if it lands on red, she takes double the score seen when she throws a fair dice.
Sara needs a 4 to finish.

b Draw a probability tree showing the outcomes of the spinner and the dice.

c What is the probability Sara finishes on her go?

d Is the use of the spinner a help or a hindrance to Sara getting a 4 to finish, or does it not matter? Give a reason.

e How would Sara's experiment change if she needed

   i a 5 to finish?

   ii an 8 to finish?

1208, 1334, 1935   SEARCH

Summary

Checkout

You should now be able to...

	Test it Questions
✔ Use Venn diagrams to record outcomes and calculate probabilities of events.	1
✔ Construct possibility spaces and use these to calculate probabilities.	2
✔ Use tree diagrams to show the frequencies or probabilities of two events.	3
✔ Use tree diagrams to calculate the probability of independent and dependent events.	4 – 5

Language	Meaning	Example
Set	A collection of numbers or objects	{a, e, i, o, u} is the set of all vowels
Member **Element**	A member or element of a set is one of the objects contained in that set.	a, e, i, o and u are all members of the set of all vowels.
Universal set, ξ	Once defined this is the set containing all the elements.	ξ = {positive numbers less than ten}
Empty set, {}	The empty set has no members.	The set {all odd numbers divisible by 2} is empty.
Intersection, ∩	The intersection of 2 or more sets is the single set containing only members that are common to all.	A = {even numbers less than ten} B = {multiples of 3 less than nine}
Union, ∪	The union of 2 or more sets is the single set containing all of the members of the original sets.	A ∩ B = {6} A ∪ B = {2, 3, 4, 6, 8, 9, 10}
Complement	The complement of a set is all members which are in that set but are in the universal set. The complement of A is A′.	ξ = {positive numbers less than ten} A = {even numbers less than ten} A′ = {1, 3, 5, 7, 9}
Venn diagram	Shows the relationship between sets.	See lesson 20.1
Possibility space **Sample space**	A list or table that shows all the possible outcomes of one or two events.	See lesson 20.2
Frequency tree	A tree diagram which shows the outcomes of two events.	See lesson 20.3
Tree diagram	Shows the probabilities of two or more events. To find the probability of an event happening multiply the probabilities along the branches.	See lesson 20.3
Independent	The outcome of one event does not affect what happens to the other. P(A and B) = P(A) × P(B)	There are 2 red sweets and 3 green sweets in a bag. I take one sweet, put it back and then take another. Taking one sweet does not affect taking the other so they are **independent** events.
Dependent	The outcome of one event affects what happens to the other. P(A and B) ≠ P(A) × P(B)	I take one sweet and then another. Taking the first sweet affects the taking the second as there is now one less sweet. These are **dependent** events.

Review

1 a Copy and complete the Venn diagram by adding all the integers up to and including 12.

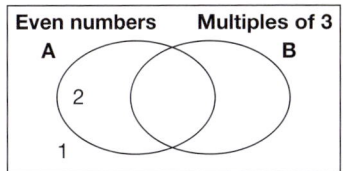

b How many numbers are in the intersection of A and B?

c A number is chosen at random. Find P(A and B).

2 Plates and cups come in red, blue and green One plate and one cup are chosen randomly.

a Draw a table to show all the possible combinations of colours.

There are the same number of each colour of plate and the same number of each colour of cup.

b What is the probability of choosing

i a red plate and a green cup

ii a blue cup

iii a plate and a cup of the same colour?

3 A road has two sets of traffic lights. The probability the first set is red is 0.2 and the probability the second set is red is 0.3

a Copy and complete the tree diagram to show all the possible outcomes.

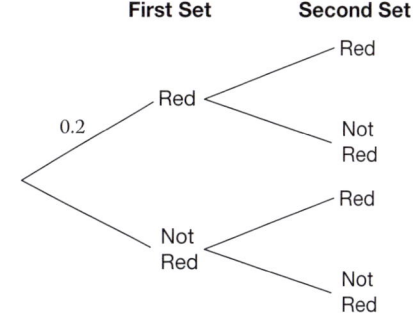

b In a 150 journeys, show the number of times the lights are red or not red.

c Calculate the probability that

i both sets of lights are on red

ii neither light is on red.

4 The probability that Stevie jumps more than 4 m in a competition is 0.6 The probability that Rachel jumps more than 4 m in the competition is 0.1 The probabilities are independent. Calculate the probability that

a Stevie and Rachel both jump more than 4 m

b Only one of them jumps more than 4 m.

5 A bag contains 3 red and 2 black counters. A counter is removed at random and *not* put back in. A second counter is then removed. Calculate the probability that

a both counters are red

b both counters are black

c the counters are different colours.

What next?

Score			
	0 – 2		Your knowledge of this topic is still developing. To improve look at MyMaths: 1199, 1208, 1262, 1263, 1334, 1921, 1922, 1935
	3 – 4		You are gaining a secure knowledge of this topic. To improve your fluency look at InvisiPens: 20Sa – g
	5		You have mastered these skills. Well done you are ready to progress! To develop your problem solving skills look at InvisiPens: 20Aa – d

Assessment 20

1 a List the elements of each set. [4]

    i   N = {the first 10 prime numbers}

    ii   M = {the first 10 multiples of 3}

    iii   P = {the individual letters in the word *parallelepiped*}

    iv   F = {the elements of the title of the book *Fahrenheit 451*}

b List the elements in the sets

    i   N ∩ M   **ii**   N ∪ M   **iii**   N ∩ F   **iv**   P ∩ F

    v   P ∪ F   **vi**   N ∩ M ∩ F   **vii**   M ∩ P   **viii**   N ∪ M ∪ P ∪ F   [8]

2 A residential area contains 34 cats. A survey found that 18 cats ate Kattibix and 13 cats ate Mice Pudding. Two ate both.

    a Draw a Venn diagram to show this information. [4]

    b How many cats did not eat either Kattibix or Mice Pudding? [1]

    c How many cats ate either Kattibix or Mice Pudding, but not both? [2]

3 The table shows the groupings of 80 people in a local Gymnastics club.

	Children under 12		Teenagers 13 to 19		Adults 20 to 30	
	Male	Female	Male	Female	Male	Female
Number of people	4	9	17	24	15	11

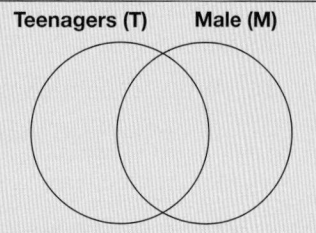

Teenagers (T)   Male (M)

    a Use the information in the table to complete the Venn diagram. [4]

    b Calculate these probabilities.

       i   P(T)   **ii**   P(M′)   **iii**   P(T ∪ M)   **iv**   P(T ∩ M)   [4]

4 Dominos is game that consists of a set of tiles divided into two squares, each with a number of dots on it.
Alyssa has a fair coin and these five domino tiles.
Alyssa throws the coin and chooses a domino tile.
If the coin shows heads, then she adds the dots on the two squares to give a score.
If the coin shows tails, then she multiples the dots on the two squares to give a score.

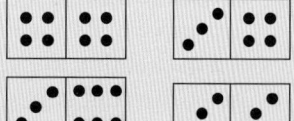

    a Complete the sample space diagram to show the possible scores. [4]

		Domino tile				
Coin	Heads	8				
	Tails	16				

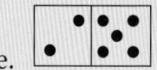

    b What is the probability that Alyssa score is 7? [1]

    c What is the probability that Alyssa's score is even? [1]

    d What is the probability that Alyssa's score is 10 or more? [1]

    e What is the probability that Alyssa's score is a factor of 3? [1]

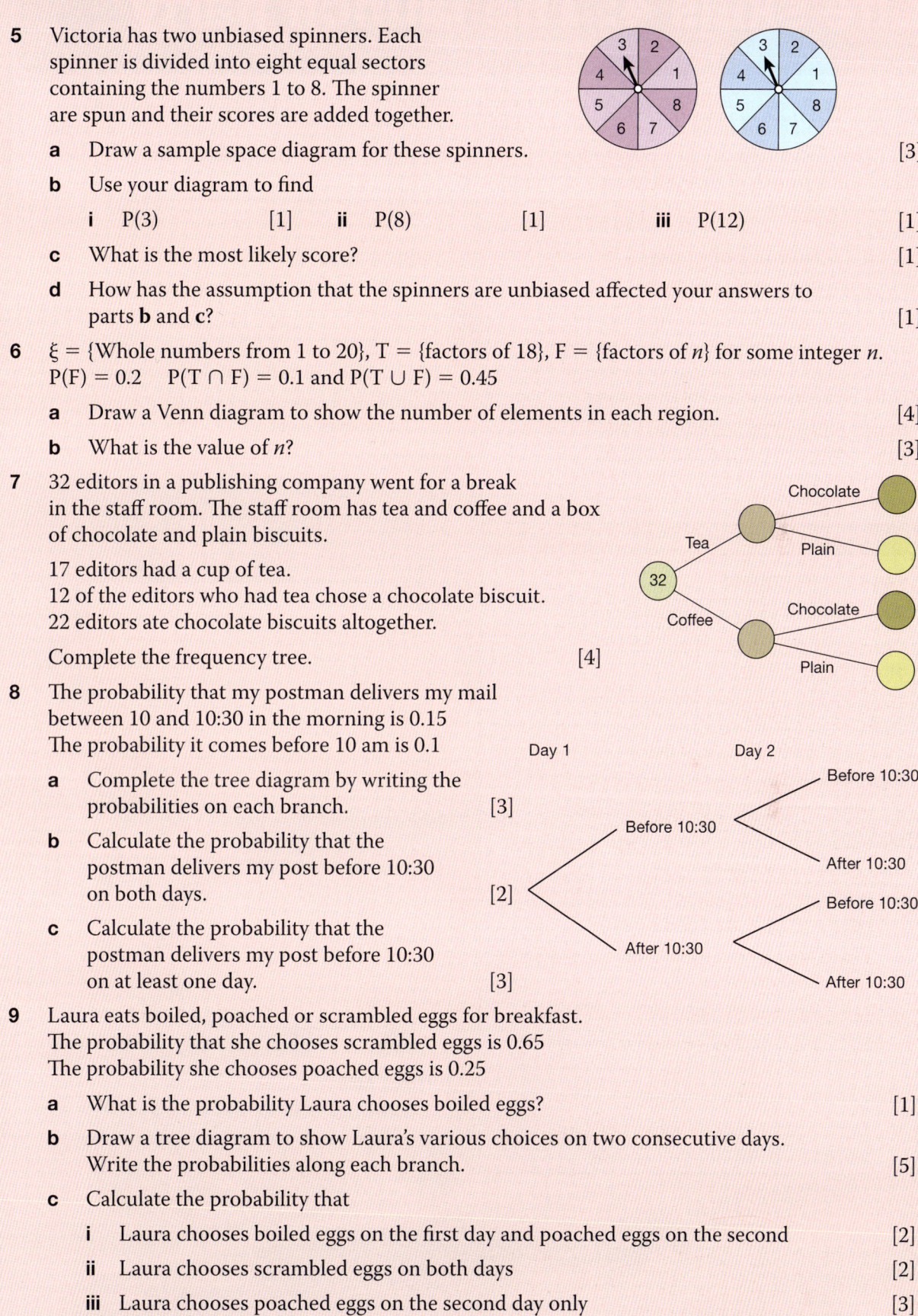

5   Victoria has two unbiased spinners. Each
    spinner is divided into eight equal sectors
    containing the numbers 1 to 8. The spinner
    are spun and their scores are added together.

    a   Draw a sample space diagram for these spinners.   [3]

    b   Use your diagram to find

        i   P(3)   [1]   ii   P(8)   [1]   iii   P(12)   [1]

    c   What is the most likely score?   [1]

    d   How has the assumption that the spinners are unbiased affected your answers to
        parts **b** and **c**?   [1]

6   ξ = {Whole numbers from 1 to 20}, T = {factors of 18}, F = {factors of n} for some integer n.
    P(F) = 0.2   P(T ∩ F) = 0.1 and P(T ∪ F) = 0.45

    a   Draw a Venn diagram to show the number of elements in each region.   [4]

    b   What is the value of n?   [3]

7   32 editors in a publishing company went for a break
    in the staff room. The staff room has tea and coffee and a box
    of chocolate and plain biscuits.

    17 editors had a cup of tea.
    12 of the editors who had tea chose a chocolate biscuit.
    22 editors ate chocolate biscuits altogether.

    Complete the frequency tree.   [4]

8   The probability that my postman delivers my mail
    between 10 and 10:30 in the morning is 0.15
    The probability it comes before 10 am is 0.1

    a   Complete the tree diagram by writing the
        probabilities on each branch.   [3]

    b   Calculate the probability that the
        postman delivers my post before 10:30
        on both days.   [2]

    c   Calculate the probability that the
        postman delivers my post before 10:30
        on at least one day.   [3]

9   Laura eats boiled, poached or scrambled eggs for breakfast.
    The probability that she chooses scrambled eggs is 0.65
    The probability she chooses poached eggs is 0.25

    a   What is the probability Laura chooses boiled eggs?   [1]

    b   Draw a tree diagram to show Laura's various choices on two consecutive days.
        Write the probabilities along each branch.   [5]

    c   Calculate the probability that

        i   Laura chooses boiled eggs on the first day and poached eggs on the second   [2]

        ii   Laura chooses scrambled eggs on both days   [2]

        iii Laura chooses poached eggs on the second day only   [3]

        iv   Laura chooses boiled eggs on at least one day.   [4]

Life skills 4: The launch party

Now that the business is set up, the restaurant is ready to start receiving customers. Abigail, Raheem, Mike and Juliet plan a grand opening. They expect more people to come than they can fit in the restaurant, so they plan to hire a marquee for the car park. They also continue to plan the future growth of their business.

Task 1 – Number of guests

The friends send emails to 128 people about the opening night.

They ask each person contacted to forward the email to 5 other people in exchange for entry into a prize draw for a free meal for two.

Based on email invite only, and the assumptions in the box on the right, how many people would you expect to attend the opening night?

Assumptions

- Half of the 128 people forward the email to 5 others
- $\frac{1}{4}$ of these forward the email to 5 more people
- No one receives the email more than once
- 10% of the people who receive the email go to the opening night

Task 2 – Marquees

They consider various marquees to hire.

One marquee is in the shape of a cuboid, the other is cylindrical.

The marquee company charges a flat fee of £500 for hire of all marquees, plus £1 per cubic metre for marquees larger than the standard of size of 300 m³

a Which marquee is cheaper to hire?

Mike wants to work out how much marquee they get for their money.

b What is the total surface area of each marquee (roof and sides)?

c Using Mike's ratio, which marquee is better value for money?

d Suggest a better calculation for Mike to work out how much marquee he gets for every pound spent.

They chose the cuboid marquee.

e Draw a plan, front elevation and side elevation of the chosen marquee.

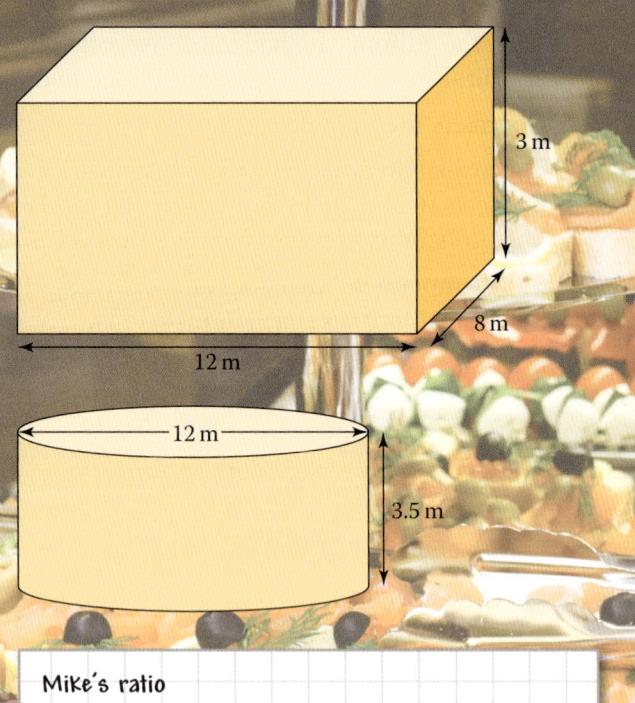

Mike's ratio
Total hire cost ÷ Surface area of marquee

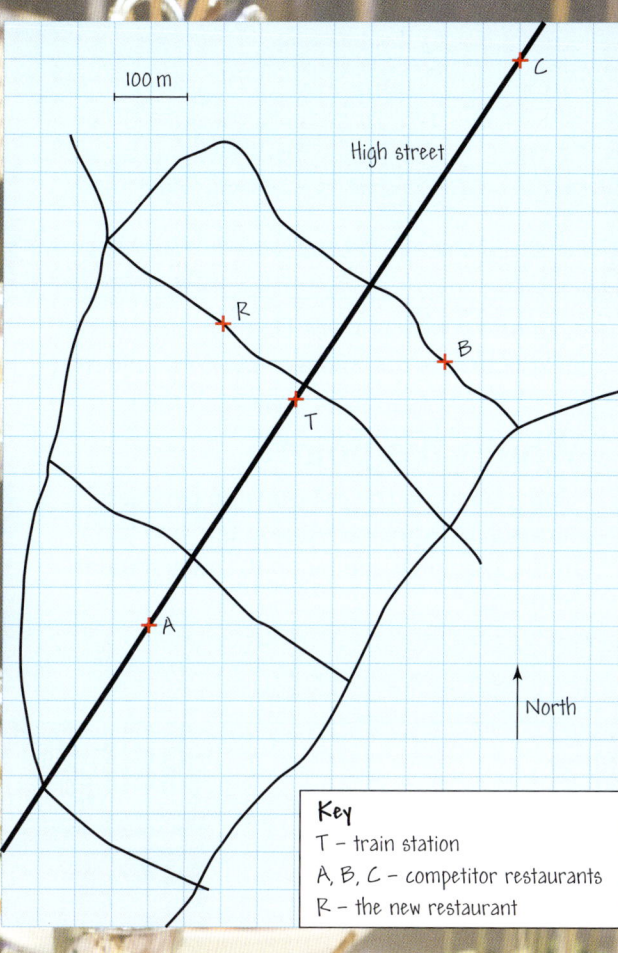

On the map:
- C (top right, on High street)
- High street
- 100 m (scale bar)
- R
- B
- T
- A
- North

Key
T – train station
A, B, C – competitor restaurants
R – the new restaurant

Task 3 – Marketing slogan

The friends make the following claim in their advertising for the opening night.

'Closest restaurant to the train station!'

Their restaurant is shown on the map by R.

a By drawing a suitable triangle, use trigonometry to find the acute angle between the High Street and North.

b Use Pythagoras' theorem to find the distances RT, BT, CT and AT, and so determine if their marketing claim is true.

c If someone walks at 4 km/hour, how long will it take them to walk directly from the station to restaurant C?

Task 4 – Forecasting

Main courses will be available at three different prices.

£15 (expensive)

£13 (medium)

£11 (cheap)

Assume that all customers will have a main course.

Suppose 100 customers visit the restaurant. Use the market research to estimate how many you expect to order expensive, medium and cheap main courses.

Use your estimated numbers and the prices to estimate the mean average amount that a customer spends on a main course.

Task 5 – Future growth

The friends have used the outcome of the opening night, and some additional market research, to make some projections about the growth of the business. They create a table as part of a report to the bank who gave them the business loan.

Based on their projections, copy and complete the following table for their report.

Number of customers in 4th month	
Total number of customers in the first six months	

21 Sequences

Introduction

Musical scales are typically written using eight notes: the C Major scale uses C D E F G A B C. The interval between the first and last C is called an octave.

The pitch of a musical note, measured in Hertz (Hz), corresponds to the number of vibrations per seconds.

The frequencies of the corresponding notes in each octave follow a geometric sequence. If the C in one octave is 130.8 Hz then the C in the next octave is $2 \times 130.8 = 261.6$ Hz, the next C is $2 \times 261.6 = 523.2$ Hz, etc.

What's the point?

Understanding the relationship between terms in a sequence lets you find any term in the sequence and begin to understand its properties.

Objectives

By the end of this chapter you will have learned how to ...

- Find terms of a linear sequence using a term-to-term or position-to-term rule.
- Recognise special types of sequence and find terms using either a term-to-term or position-to-term rule.
- Find terms of a quadratic sequence using a term-to-term or position-to-term rule.

Check in

1   Work out the difference between each of these pairs of numbers.

    a 5 and 8   **b** 3 and 9   **c** 6 and 11   **d** −4 and +2

2   Find the difference between each of these pairs of numbers.

    a 10 and 7   **b** 9 and 5   **c** 12 and 7   **d** 3 and −1

3   Write down the first six multiples of each of these numbers.

    a 4   **b** 3   **c** 5   **d** 6

4   Substitute $n = 10$ into each of these expressions.

    a $4n$ **b** $1 + 3n$ **c** $3n - 4$ **d** n^2

Chapter investigation

Abi, Bo and Cara are making patterns with numbers.

Abi makes a sequence by adding a fixed number onto a starting number.

1, 4, 7, 10, 13, 16, 19, 22, 25, 28, …   Start with 1, add 3 each time

Bo makes a second sequence by adding Abi's sequence onto his starting number, 1.

1, 2, 6, 13, 23, 36, 52, 71, 93, 118, … $1 + 1 = 2, 2 + 4 = 6, 6 + 7 = 13, 13 + 10 = 23$, etc.

Cara takes the first two numbers in Bo's sequence and makes a third term by adding these together, then a fourth term by adding the second and third terms, etc.

1, 2, 3, 5, 8, 13, 21, 34, 55, 89, 144, … $1 + 2 = 3, 2 + 3 = 5, 3 + 5 = 8, 5 + 8 = 13$, etc.

How do sequences like these behave?

21.1 Sequence rules

A **sequence** is an ordered list of numbers.

The individual numbers are called **terms** and you identify them by giving their **position**: first, second, 3rd, 4th, … 10th,…

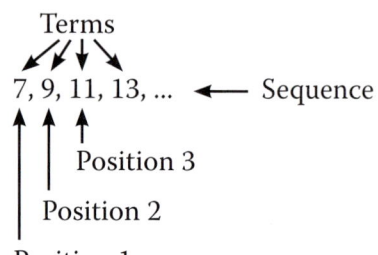

Terms

7, 9, 11, 13, … ← Sequence

Position 3

Position 2

Position 1

- Sequences can be generated and described using **term-to-term** rules.

EXAMPLE

Find the first five terms of these sequences using the given term-to-term rules.

a First term 5   Rule Add 3   **b** First term 20   Rule Subtract 7

c First term 4   Rule Multiply by 3   **d** First term 80   Rule Divide by 2

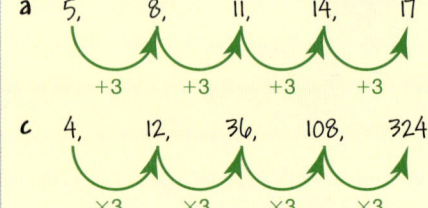

a 5,   8,   11,   14,   17
    +3   +3   +3   +3

c 4,   12,   36,   108,   324
    ×3   ×3   ×3   ×3

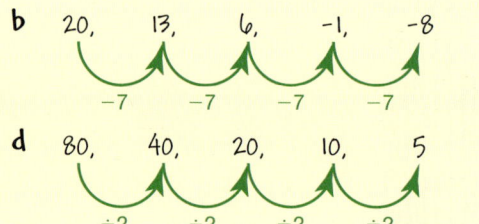

b 20,   13,   6,   -1,   -8
    -7   -7   -7   -7

d 80,   40,   20,   10,   5
    ÷2   ÷2   ÷2   ÷2

EXAMPLE

Find the missing terms in these sequences.

a 2, 6, □, 14, 18   **b** 70, 65, 60, □, 50, □   **c** 1, 2, □, 8, □, 32

a Look at the difference between terms. Rule is add 4.

10

2,   6,   10,   14,   18
   +4   +4   +4   +4

b Look at the difference between terms. Rule is subtract 5.

55, 45

70,   65,   60,   55,   50,   45
   -5   -5   -5   -5   -5

c Look at the ratio of terms. Rule is multiply by 2.

4, 16

1,   2,   4,   8,   16,   32
   ×2   ×2   ×2   ×2   ×2

- Sequences can be generated and described using **position-to-term** rules.

$T(n) = 2n + 3$ is the formula for the term in the nth position in the sequence.

EXAMPLE

The rule for a sequence is $T(n) = 2n + 3$, where $T(n)$ is the nth term of the sequence and n is the position of the term in the sequence.

Find the first three terms of the sequence.

$2n$ means $2 \times n$

p.116

$T(1) = 2 \times 1 + 3 = 5$

$T(2) = 2 \times 2 + 3 = 7$

$T(3) = 2 \times 3 + 3 = 9$

Sequence is 5, 7, 9, …

T(1) is the first term of the sequence.

T(2) is the second term of the sequence.

Algebra Sequences

Exercise 21.1S

1 Find the first four terms of these sequences using the term-to-term rules.

    a First term 2   Rule Add 6

    b First term 3   Rule Add 5

    c First term 25   Rule Add 10

    d First term 25   Rule Add 3

    e First term 2.5   Rule Add 5

    f First term $1\frac{3}{4}$ Rule Add $\frac{1}{2}$

    g First term 8   Rule Add 2.5

    h First term -10 Rule Add 3

2 Find the first four terms of these sequences using the term-to-term rules.

    a First term 20   Rule Subtract 2

    b First term 100   Rule Subtract 5

    c First term 30   Rule Subtract 4

    d First term 12.5   Rule Subtract 3

    e First term 5   Rule Subtract 4

    f First term -1 Rule Subtract 4

3 Find the first four terms of these sequences using the term-to-term rules.

    a First term 2   Rule Multiply by 2

    b First term 3   Rule Multiply by 2

    c First term 5   Rule Multiply by 10

    d First term 10   Rule Multiply by 5

    e First term 0.5   Rule Multiply by 2

    f First term -5 Rule Multiply by 2

4 Find the first four terms of these sequences using the term-to-term rules.

    a First term 64   Rule Divide by 2

    b First term 100   Rule Divide by 2

    c First term 27   Rule Divide by 3

    d First term 64   Rule Divide by 4

    e First term 26   Rule Have the previous term

    f First term 2   Rule Divide by 0.5

5 For each of these sequences

    i write down the rule for going from one term to the next term

    ii find the missing terms.

    a 3, 8, □, 18, 23   **b** 4, 7, □, 13, 16

    c 20, □, 26, □, 32   **d** 11, 15, □, 23, □

    e 30, 24, □, 12, 6   **f** 10, 7, □, 1, □

    g 3, 6, □, 24, 48   **h** 2, 20, □, 2000

    i 1, 4, □, 64   **j** 2, -4, 8, □, 32

    k 200, 100, □, 25, □   **l** 4, 2, 1, □, □

6 Give a reason why each of these sequences is an odd one out

    A 4, 7, 10, 13, ...   **B** 2, 6, 10, 14, ...

    C 2, 5, 8, 11, ...

7 a The rule for a sequence is $T(n) = 3n + 3$, where $T(n)$ is the nth term of the sequence and n is the position of the term in the sequence.

    Find the first five terms of the sequence.

    b The rule for a sequence is $T(n) = 5n - 4$, where $T(n)$ is the nth term of the sequence and n is the position of the term in the sequence.

    Find the first five terms of the sequence.

8 Generate the first five terms of the sequences with these position-to-term rules.

    a $T(n) = 2n$ **b** $T(n) = 2n + 1$

    c $T(n) = 6 - n$ **d** $T(n) = 18 - 3n$

    e $T(n) = n^2$ **f** $T(n) = 2n^2 - 1$

    g $T(n) = n^3$ **h** $T(n) = \frac{1}{2}(n^2 + n)$

    i $T(n) = \dfrac{32}{n}$ **j** $T(n) = 2^n$

9 A sequence is generated by the rule 'add the two previous terms to create the next term'.

    Using this rule, find the next four terms of the sequence 2, 3, 5, □, □, □, □.

Q 1173   SEARCH

21.1 Sequence rules

RECAP

- Know how to generate sequences using a 'term-to-term' and a 'position-to-term' rule.
- Know how to describe a sequence using a 'term-to-term' rule.

Sequence	3, 8, 13, 18, 23,
Term-to-term	add 5
Position-to-term	$T(n) = 5n - 2$

HOW TO

① To generate a sequence apply the term-to-term rule to the previous term.

② To find a particular term, substitute n = the term's position into the nth term formula, $T(n)$.

③ To find the term-to-term rule look at successive terms: first look at their differences then consider their ratios.

EXAMPLE

Draw the next two diagrams in these sequences.

a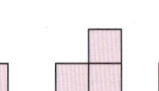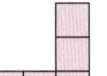

b

③ Look at how one diagram differs from the next.

a Two squares are added, one at the end of each 'leg'.

b One triangle is added.

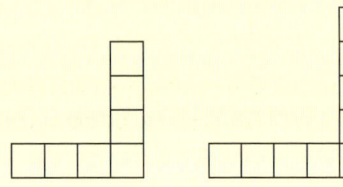

> A good strategy is to 'subtract' the shapes in the previous diagram to see what has been added.

EXAMPLE

133 is in the sequence described by the rule $T(n) = 5n - 2$.

a What term number is 133 in the sequence?

b Find the next term after 133 in the sequence

Each term is generated by the rule 'multiply by 5' and then 'subtract 2'.

a Reverse this process. A more efficient method.

$133 + 2 = 135$	$+ 2$	$5n - 2 = 133$	
$135 \div 5 = 27$	$\div 5$	$5n = 135$	$+ 2$
133 is the 27th term		$n = 21$	$\div 5$

b $T(28) = 5 \times 28 - 2$ ② The next term has $n = 28$.

$= 138$ ① or $133 + 5 = 138$

p.202

Algebra Sequences

Exercise 21.1A

1 Draw the next two diagrams in these sequences.

a

b

c

d

2 a Match each sequence with the correct 'term-to-term' rule.

−2, 1, 4, 7, ..	Add 2
4, 12, 36, 108, …	
3, 5, 7, 9, 11, …	Subtract 2
	Add 3
11, 9, 7, 5, 3, …	Subtract 3

b Complete the missing entries.

3 A sequence of numbers is generated by this rule.

> Start with 5, multiply by 2 and add 3 to generate the next term.

a Find the first five terms.

b The term 1021 is in the sequence. Calculate the term that is immediately *before* 1021.

4 Does the number 1000 appear in these sequences? If it does, at what position does it appear?

a First term 375   Rule add 50

b First term 10 000   Rule subtract 75

c First term 74   Rule add 7

d First term 2345   Rule subtract 25

5 Match each sequence with the correct 'term-to-term' rule and 'position-to-term' rule.

7, 10, 13, 16, …	Subtract 4
4, 1, −2, −5, …	Add 4
6, 10, 14, 18, …	Add 3
3, 10, 17, 24, …	Subtract 3
6, 2, −2, −6, …	Add 7

$T(n) = 4n + 2$
$T(n) = 3n + 4$
$T(n) = 7n - 4$
$T(n) = 7 - 3n$
$T(n) = 10 - 4n$

6 Belle is generating a sequence using the rule $T(n) = 4n - 2$. She thinks that every term will be an even number.
Do you agree with Belle?
Give your reasons.

7 a Is this claim always true?
Give your reasons.

> If you follow this procedure then the 8th number is will be 60 times the first number.
>
> - Pick any whole number
> - Write down 4 times the number
> - Add these two numbers together
> - Add the second and third numbers together
> - Add the third and fourth numbers together.
> - Keep repeating until you get the 8th number.

b Does it work for negative integers, fractions and decimals?

8 A sequence has 31st term 159. The rule for the sequence is add 5. What is the first term in the sequence?

9 Research and produce a presentation about the classic 'Rice and Chessboard' problem.

Q 1173   SEARCH

21.2 Finding the *n*th term

- A **linear** sequence has a constant difference between terms.

This is a linear sequence 5, 8, 11, 14, 17, …

The terms all increase by +3.

Plotting the terms against their position on a graph, the sequence forms a straight line with gradient +3.

You can describe and generate a sequence using an expression for the ***n*th term**.

- For a linear expression the *n*th term takes the form
  $T(n) = an + b$
  where a and b are fixed numbers.

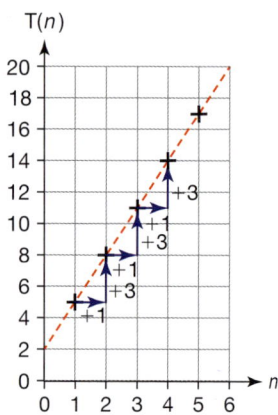

▲ Terms in the sequence $T(n) = 2 + 3n$ plotted against their position, n.

The coefficient of n is the constant difference.

For the sequence 5, 8, 11, 14, 17… $T(n) = 3n + 2$.

EXAMPLE

The *n*th term of a sequence is $5n - 2$.

  a Find the first five terms of the sequence   **b** Find the 50th term.

  a 3, 8, 13, 18, 23

   1st term ($n = 1$): $5 \times 1 - 2 = 3$

   3rd term ($n = 3$): $5 \times 3 - 2 = 13$

   5th term ($n = 5$): $5 \times 5 - 2 = 23$

  b 50th term = $5 \times 50 - 2 = 248$

   2nd term ($n = 2$): $5 \times 2 - 2 = 8$

   4th term ($n = 4$): $5 \times 4 - 2 = 17$

EXAMPLE

Find the *n*th term for these sequences.

  a 5, 9, 13, 17, 21, …

  b 10, 4, −2, −8, −14, …

  a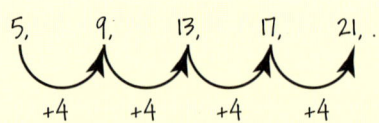

   The difference between each term is + 4.

   nth term $= 4n \pm \square$

   Compare the sequence with the first terms of the sequence $4n$.

   $4n$: 4, 8, 12, 16, 20, …

   $T(n)$: 5, 9, 13, 17, 21,

   $T(n) = 4n + 1$

  b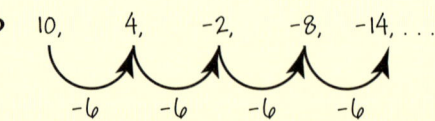

   The difference between each term is − 6.

   nth term $= -6n \pm \square$

   Compare the sequence with the first terms of the sequence $-6n$.

   $-6n$: −6, −12, −18, −24, …   +16

   $T(n)$: 10, 4, −2, −8, …

   $T(n) = -6n + 16$

    $= 16 - 6n$

Exercise 21.2S

1 Find the first five terms of these sequences.

   a $3n$ **b** $4n + 2$ **c** $2n + 5$

   d $5n + 3$ **e** $n + 2$ **f** $4n - 1$

   g $5n - 3$ **h** $2n - 3$ **i** $n - 7$

   j $3n - 10$ **k** $0.5n + 5$ **l** $2.5n - 1$

2 Find the first five terms of these sequences.

   a $-2n$ **b** $15 - 4n$ **c** $10 - n$

   d $25 - 5n$ **e** $12 - 3n$ **f** $15 - 6n$

   g $4 - 0.5n$ **h** $2 - 7n$ **i** $\frac{1}{2} + \frac{1}{6}n$

3 The terms of a sequence can be generated using this rule: $T(n) = 4n + 7$.

   Calculate the

   a   10th term   **b**   15th term

   c   100th term   **c**   1000th term.

4 For each sequence find the

   i   first term   **ii**   fifth term

   iii   fiftieth term   **iv**   100th term.

   a $2n$ **b** $-4n$

   c $3n + 2$ **d** $5n + 7$

   e $12 - n$ **f** $7 - 2n$

   g $0.3n$ **h** $1 + 0.25n$

   i $10 - 0.1n$ **j** $\frac{1}{2} + \frac{1}{2}n$

5 Sort these sequences into a copy of this table.

Linear Sequence	Non-linear sequence

   a $1, 6, 11, 16, 21, \dots$ **b** $1, 3, 6, 10, 15, \dots$

   c $1, 2, 4, 8, 16, \dots$ **d** $3, 6, 9, 12, 15, \dots$

   e $\frac{1}{2}, 1, 1\frac{1}{2}, 2, 2\frac{1}{2}, \dots$ **f** $10, 8, 6, 4, 2, \dots$

   g $3, -3, 3, -3, 3, \dots$ **h** $1, 4, 9, 16, \dots.$

   i $-2, -5, -8, -11, \dots$

6   **a**   Find the constant difference for this sequence.

         $5, 8, 11, 14, 17, \dots$

6   **b**   Copy and complete this table containing the sequence, multiples of the constant difference and their difference.

$T(n)$					
$\square n$					
Difference					

   c   Write the nth term in the form $T(n) = \square n + \square$.

7 Find the nth term of these sequences.

   a $3, 5, 7, 9, 13, \dots$

   b $4, 7, 10, 13, 16, \dots$

   c $5, 8, 11, 14, 17, \dots$

   d $4, 10, 16, 22, 28, \dots$

   e $7, 17, 27, 37, \dots$

   f $2, 10, 18, 26, 34, \dots$

   g $2, 3.5, 5, 6.5, 8, \dots$

   h $1.4, 2, 2.6, 3.2, 3.8, \dots$

   i $2, 2\frac{1}{2}, 3, 3\frac{1}{2}, 4, \dots$

   j $3.1, 3.25, 3.4, 3.55, 3.7, \dots$

8 Find the nth term of these sequences.

   a $11, 9, 7, 5, 3, \dots$

   b $16, 13, 10, 7, 3, \dots$

   c $10, 8, 6, 4, 2, \dots$

   d $5, 2, -1, -4, -7, \dots$

   e $27, 17, 7, -3, -13, \dots$

   f $8, 6.5, 5, 3.5, 2, \dots$

   g $3, 2\frac{1}{2}, 2, 1\frac{1}{2}, 1, \dots$

   h $1, -4, -9, -14, -19, \dots$

   i $5, 3\frac{1}{2}, 2, \frac{1}{2}, -1, \dots$

   j $2, -0.5, -3, -5.5, -8, \dots$

9   **a**   Predict the 10th term for the sequence $2, 6, 10, 14, 18, \dots$

   b   Find the nth term for the sequence $2, 6, 10, 14, 18, \dots$

   c   Use your answer to part **b**, to evaluate the accuracy of your prediction.

10 Is 75 a term in the sequence described by the nth term $5n - 3$?

   Show your working.

21.2 Finding the *n*th term

- Know how to generate and describe linear sequences using the *n*th term.

2, 6, 10, 14, 18, ...

all differences = +4

*n*th term rule = $4n \pm \square$

Compare $4n$

$$\begin{array}{cccccc} & 2, & 6, & 10, & 14, & 18 \\ 4n & 4, & 8, & 12, & 16, & 20 \\ \hline & -2, & -2, & -2, & -2, & -2 \end{array}$$

*n*th term rule = $4n - 2$

HOW TO

To describe a linear sequence using the *n*th term

① Find the *constant* difference between terms.

② This difference is the first part of the *n*th term.

③ Add or subtract a constant to adjust the expression for the *n*th term.

EXAMPLE

Eliza and Esme make two claims about this sequence of patterns. Do you agree with their claims? Give your reasons.

a Eliza says

'*There will be 20 squares in the 8th diagram because there are 10 squares in the 4th diagram.*'

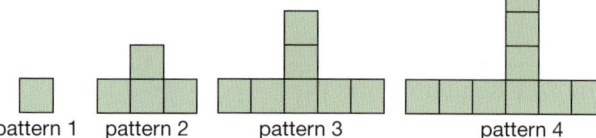

pattern 1   pattern 2   pattern 3   pattern 4

b Esme says

'*There is a pattern with 345 squares.*'

a

Pattern, *n*	1	2	3	4
Number of squares, $T(n)$	1	4	7	10

           $+3$ $+3$ $+3$ ① Calculate the differences.

*n*th term, $T(n) = 3n \pm \square$ ② Difference is a constant = +3

③ Compare the sequence with the first term of the sequence, $3n$.

$$\begin{array}{ccccc} T(n) & 1, & 4, & 7, & 10, & ... \\ 3n & 3, & 6, & 9, & 12, & ... \\ \hline & -2 & -2 & -2 & -2 \end{array}$$

*n*th term, $T(n) = 3n - 2$

> The eighth term in a sequence doesn't usually equal twice the value of fourth term.

$T(8) = 3 \times 8 - 2$

      $= 24 - 2$

      $= 22$

No, I do not agree with Eliza.

There are 22 squares in the 8th diagram.

b If Esme is right then $T(n) = 345$ for some position *n*.

$3n - 2 = 345$ $+2$

   $3n = 347$ $\div 3$

     $n = 115\frac{2}{3}$ Not a whole number.

Esme is wrong.

Exercise 21.2A

1 Kate thinks that the tenth term of the sequence 3, 7, 11, 15, 18, ... will be 36.

Do you agree with Kate?
Give your reasons.

2 Anika says that the nth term of the sequence
13, 17, 21, 24, 29, ... is $n + 4$.

Do you agree with Anika?
Give your reasons.

3 **a** Match each sequence with the correct nth term.

1, 3, 5, 7, ...	$3n - 2$
	$2n - 1$
3, 5, 7, 9, 11, ...	
3, 7, 11, 15, ...	$n + 2$
3, 4, 5, 6, 7, ...	$2n + 1$

b Complete the missing entries.

4 Sam is making patterns using matches.

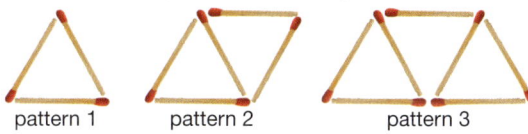

pattern 1   pattern 2   pattern 3

a Complete a copy of this table.

Pattern, n	1	2	3	4
Number of matches, m	3			

b Find a formula for the number of matchsticks, m, in the nth pattern.

c How many matches will Sam need for the 50th pattern?

d Which is the first pattern that will need more than 100 matches to make?

5 How many squares are there in the 15th diagram?

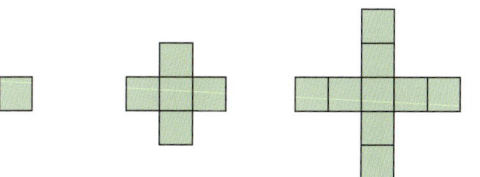

6 Draw a set of repeating patterns to represent the sequence described by the nth term $4n + 3$.

7 Matthew is building a 11.2 m long fence.

40 cm

a How many posts does Matthew need?

b How many planks does Matthew need?

8 Hannah is placing paving slabs around different size ponds.

She notices that the sequence can be described by the nth term $2n + 6$.

a Justify that the nth term is $2n + 6$.

b Explain how the expression for the nth term relates to the structure of Hannah's patterns.

9 Sandra is generating a sequence using the rule $T(n) = 10n - 15$. She thinks that every term will be a multiple of 5. Do you agree with Sandra? Give your reasons.

10 Do these sequences contain the given term? Give your reasons.

a $T(n) = 6n + 4$ Term = 94

b $T(n) = 70 - 3n$ Term = -48

c $T(n) = 7 + 0.7n$ Term = 32.5

d $T(n) = 60 - 12n$ Term = -1766

11 Write down the nth term formula for the sequences shown in these graphs.

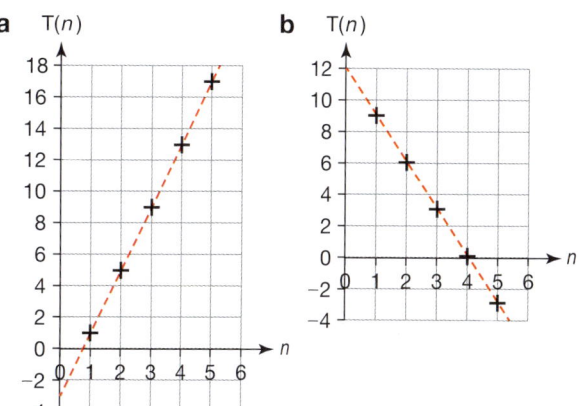

21.3 Special sequences

● Square, cube and triangular numbers are associated with geometric patterns.

Square numbers

1, 4, 9, 16, 25, ...

Cube numbers

1, 8, 27, 64, 125, ...

Triangular numbers

1, 3, 6, 10, 15, ...

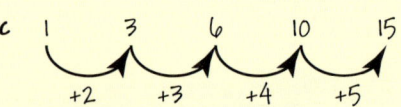

EXAMPLE

Write down the fifth   **a** square number   **b** cube number   **c** triangular number.

a $5 \times 5 = 25$ **b** $5 \times 5 \times 5 = 125$ **c**
  1   3   6   10   15
     +2   +3   +4   +5

● **Arithmetic** (linear) sequences have a constant difference between terms.
● **Geometric** sequences have a constant ratio between terms.

EXAMPLE

Are these sequences arithmetic or geometric?

  a   5, 8, 11, 14, ...   **b**   1, −2, 4, −8, ...

  a $8 - 5 = 11 - 8 = 14 - 11 = 3$ **b** $-2 \div 1 = 4 \div -2 = -8 \div 4 = -2$
     Arithmetic   Constant difference, +3   Geometric   Constant ratio, −2

● In a **Fibonacci**-type sequence each term is a sum of previous terms.

1, 1, 2, 3, 5, 8, ... $2 = 1 + 1$, $3 = 1 + 2$, $5 = 2 + 3$, $8 = 3 + 5$

1, 1, 3, 7, 17, 41, ... $3 = 1 + 2 \times 1$, $7 = 1 + 2 \times 3$, $17 = 3 + 2 \times 14$, $41 = 7 + 2 \times 17$

EXAMPLE

a Is this a Fibonacci-type sequence?
2, 1, 3, 4, 7, 11

b What is next term?

a Check that each term, beyond the second,   **b** $7 + 11 = 18$
is the sum of the previous two terms.
   $3 = 2 + 1$ ✓ $4 = 1 + 3$ ✓
   $7 = 3 + 4$ ✓ $11 = 4 + 7$ ✓

● In a **quadratic** sequence the differences between terms form an arithmetic sequence; the second differences are constant.

EXAMPLE

Is this a quadratic sequence?
2, 6, 12, 20, 30, 42, 56

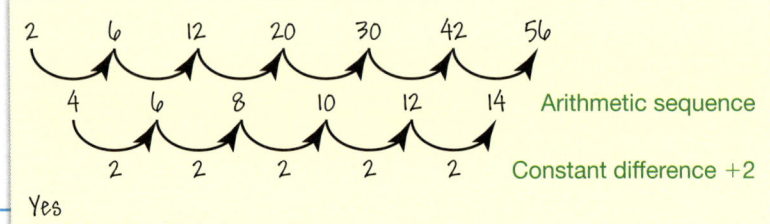

  2   6   12   20   30   42   56
     4   6   8   10   12   14   Arithmetic sequence
        2   2   2   2   2   Constant difference +2

Yes

Algebra Sequences

Exercise 21.3S

1 Write down the sixth term of the

    a triangular number sequence

    b square number sequence

    c cube number sequence.

2 Draw the first five terms of the triangular number sequence.

3 Describe these sequences using one of the words in the coloured panel.

> Arithmetic   Geometric
> Quadratic   Fibonacci-type

    a $2, 5, 8, 11, …$ **b** $7, 11, 15, 19, …$

    c $2, 3, 5, 8, 13, …$ **d** $2, 5, 10, 17, …$

    e $2, 6, 18, 54, …$ **f** $18, 15, 12, 9, …$

    g $1, 4, 5, 9, …$ **h** $1, 2, 4, 8, …$

    i $3, 7, 13, 21, …$ **j** $0.5, 2, 3.5, 5, …$

    k $\frac{1}{4}, \frac{1}{2}, 1, 2, …$ **l** $3, 1, \frac{1}{3}, \frac{1}{9}, …$

4 Find the next three terms of the following sequences using the properties of the sequence.

    a Arithmetic $2, 4, \square, \square, \square$

    b Geometric $2, 4, \square, \square, \square$

    c Fibonacci-type $2, 4, \square, \square, \square$

    d Quadratic $2, 4, \square, \square, \square$

5 Find the missing term in each of these sequences.

    a $5, 10, \square, 20, 25, …$

    b $5, 10, \square, 40, 80, …$

    c $5, 10, \square, 25, 40, …$

    d $5, 10, \square, 26, 37, …$

6 Alice thinks that the sequence $1, 4, 9, 16, 25, …$ is a quadratic sequence.

Bob thinks that the sequence $1, 4, 9, 16, 25, …$ is a square sequence.

Who is correct?

Explain your reasoning.

7 Drew thinks that the square number sequence can be generated using this rule.

    $T(n) = n^2$

Do you agree with Drew? Give your reasons.

8 Emily thinks that the triangular number sequence can be generated using this rule.

    $T(n) = n^3$

Do you agree with Emily? Give your reasons.

9 Tyler thinks that the triangular number sequence can be generated using this rule.

    $T(n) = \frac{1}{2}n(n+1)$

Do you agree with Tyler? Give your reasons.

10 Generate the first four terms of a geometric sequence using the following facts.

	First term	Multiplier
a	3	2
b	10	5
c	3	0.5
d	2	-3
e	$\frac{1}{2}$	$\frac{1}{2}$
f	-3	-2
g	4	$\sqrt{3}$

11 In a Fibonacci-type sequence, the next term is given by the sum of the previous two terms.

If the first term is a, the second b and the third 13, find three possible values for a and the corresponding values for b.

12 In this Fibonacci-type sequence the next term is given by the sum of the previous three terms.

$1, 1, 1, 3, \square, \square, \square, \square$

Find the next four terms in the sequence.

13 **a** Find the total of the first five terms of the geometric sequence $4, 2, 1, \frac{1}{2}, \frac{1}{4}, …$

    b Find the total of the first ten terms of the geometric sequence $4, 2, 1, \frac{1}{2}, \frac{1}{4}, …$

    c Comment on your results.

14 Write these numbers as the sum of not more than three triangular numbers.

    a 30   **b** 31   **c** 32

15 Write these numbers as the sum of not more than four square numbers.

    a 15   **b** 32   **c** 56

🔍 1053, 1054, 1920   SEARCH

21.3 # Special sequences

- Square numbers   1, 4, 9, 16, …   general form 'number × number'
- Cube numbers   1, 8, 27, 64, …   general form 'number × number × number'
- Triangular numbers   1, 3, 6, 10, 15, …   differences 2, 3, 4, 5, …
- Arithmetic sequence   constant difference
- Geometric sequences   constant ratio
- Quadratic sequence   differences form an arithmetic sequence
- Fibonacci-type sequence   each term is a sum of previous terms

HOW TO

① Know the square, triangular and cube number sequences.

② Generate an arithmetic sequence by adding the same constant to terms.

③ To generate a quadratic sequence, first create a linear sequence then add the linear sequence to successive terms in the quadratic sequence.

④ Generate geometric sequences by multiplying terms by the same constant.

⑤ Generate Fibonacci-type sequences by adding the previous terms to create the next term.

EXAMPLE

Create two sequences with the following properties.

a Arithmetic sequence with starting term 5   **b** Geometric sequence with starting term 3

c Fibonacci-type sequence with starting term 4   **d** Quadratic sequence with starting term 4

---

a $5, 8, 11, 14, 17, …$ ② Constant difference, $+ 3$

   $5, 2, -1, -4, -7, …$ Constant difference, $- 3$

b $3, 6, 12, 24, 48, …$ ④ Constant ratio, $× 2$

   $3, 30, 300, 3000, 30000, …$ Constant ratio, $× 10$

c $4, 5, 9, 14, 23, …$ ⑤ Pick a second term, 5, then $9 = 4 + 5, 14 = 5 + 9, 23 = 9 + 14$

   $4, 8, 12, 20, 32, …$ Pick a second term, 8, then $12 = 4 + 8, 20 = 8 + 12, 32 = 12 + 20$

d $4, 7, 12, 19, 28, …$ ③ Pick a linear sequence,   3,   5,   7,   9

                                      then $7 = 4 + 3, 12 = 7 + 5, 19 = 12 + 7, 28 = 19 + 9$

   $4, 10, 18, 28, 40, …$ Pick a linear sequence,   6,   8,   10,   12

                                      then $10 = 4 + 6, 18 = 10 + 8, 28 = 18 + 10, 40 = 28 + 12$

EXAMPLE

Hannah is designing square frames for different sized pictures.

How many squares will she need to make a frame for a picture with dimensions $2 × 15$?

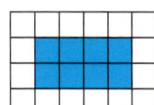

Pattern, n	1, 2, 3, …	Identify the pattern.

Number of squares, $T(n)$ 12, 14, 16, …   ② Constant difference, 2 linear.

$T(n) = 2n + 10$

$2 × 15$ frame has $n = 14$ $T(14) = 2 × 14 + 10 = 28 + 10 = 38$

Hannah needs 38 squares

Algebra Sequences

Exercise 21.3A

1 Hannah would like to create arithmetic sequences using the rule 'add 4'.

Write down four possible sequences.

2 George would like to create geometric sequences using the rule 'multiply by 2'.

Write down four possible sequences.

3 Dan has found a number that is in both the square number and cube number sequences.

  a What is the number?

  b Write down its position in

    i the square number sequence

    ii the cube number sequence.

4 Is this statement true or false?

> The sum of two terms of the triangular number sequence equals one term of the square number sequence.

Show your reasoning.

5 Jenny thinks that the triangular number sequence can be created by starting with the number 1 and then adding on 2, adding on 3 and so on.
Do you agree with Jenny?
Give your reasons.

6 The first term of a sequence is 4.

  a Create five terms of a sequence which is

    i arithmetic   **ii** geometric

    iii Fibonacci-type   **iv** quadratic.

  b Say how you created each sequence.

7 **a** Research and produce a presentation about how Leonardo Fibonacci, also known as Leonardo of Pisa, discovered the Fibonacci sequence 1, 1, 2, 3, 5, …

  b How is the Fibonacci sequence linked to

    i Pascal's Triangle

    ii the golden ratio?

8 Research other special sequences, such as pentagonal, hexagonal and tetrahedral numbers.

9 Carl Gauss is one of the world's most famous mathematicians. One day at school, he was asked to find the sum of $1+2+3+ … + 100$. Within seconds, he gave the correct total of 5050. Explore how he could calculate the sum so quickly.

10 Research and produce a presentation about the 'Handshakes' problem.

11 Hannah and Sam are given these options for money as a gift to celebrate their birthdays.
Option 1 £500
Option 2 £100 for the first month, £200 the next month, £300 the next month until the end of the year.
Option 3 During the month of their birthday 1p on Day 1, 2p on Day 2, 4p on Day 3, 8p on Day 4, …
Hannah's birthday is 20th February.
Sam's birthday is 5th September.

  a Which option should Hannah choose? Explain your reasoning.

  b Which option should Sam choose? Explain your reasoning.

12 Using the cards below, match the sequence with the correct description, rule and visual representation. Complete any missing cards.

$T(n) = n^2 - n + 1$	………………..	Triangular
$T(n) = ……………$	1, 3, 6, 10, …	Square
$T(n) = 4n - 3$	1, 4, 9, 16, …	……………….
$T(n) = \frac{1}{2}n(n+1)$	1, 3, 7, 13, …	Arithmetic

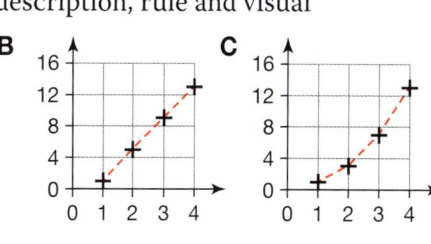

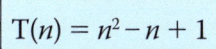

Summary

Checkout

You should now be able to...

You should now be able to...	Test it Questions
✔ Generate terms of a sequence from both a term-to-term and a position-to-term rule.	1 – 3
✔ Write a formula for the *n*th term of a linear sequence.	4, 5
✔ Recognise special sequences and use them to solve problems.	6 – 8

Language   Meaning   Example

Language	Meaning	Example
Sequence	A set of numbers or other objects arranged in order that follow a rule.	Sequence: square numbers 1 4 9 16 ...
Term	One of the separate items in a sequence.	$T(1) = 1$ First term, position 1
Position	A number that counts where a term appears in a sequence.	$T(2) = 4$ Second term, position 2
Term-to-term rule	A rule that links a term in a sequence with the previous term.	Sequence: 3, 5, 7, 9, 11, 13 …. Term-to-term rule, 'add 2'
Position-to-term rule	A rule that links a term in a sequence with its position in the sequence.	Position-to-term rule, $T(n) = 2n + 1$ $T(n)$ is the term in position n.
Linear/ Arithmetic	A sequence that has a constant difference between the terms. If plotted on a graph, a linear relationship gives a straight line.	4, 9, 14, 19, 24, ... +5 +5 +5 +5 Common difference = 5 $T(n) = 5n - 1$ is the nth term
Common difference	The difference between each term and the previous term in a linear sequence	$T(1) = 5 \times 1 - 1 = 4$ is the 1st term $T(2) = 5 \times 2 - 1 = 9$ is the 2nd term
General term *n*th term	A general expression that can be used to find all the terms of the sequence.	$T(10) = 5 \times 10 - 1 = 49$ is the 10th term
Cube numbers	The sequence formed by multiplying the position number by itself three times.	1, 8, 27, 64, 125, ... $1 = 1 \times 1 \times 1 = 1^3$ $8 = 2 \times 2 \times 2 = 2^3$
Triangular numbers	Form a triangle. Each successive term is formed by adding on another layer to the triangle.	1, 3, 6, 10, 15, ...
Geometric sequence	A sequence that has a constant ratio between terms.	1, 3, 9, 27, 81, ... The constant ratio is 3.

Fibonacci-type sequence	Each term is a sum of previous terms.	1, 1, 2, 3, 5, 8, 13, ...
		   $2 = 1 + 1, 3 = 1 + 2, 5 = 2 + 3$, etc.
		 1, 1, 2, 4, 7, 13, 24, ...
		   $4 = 1 + 1 + 2, 7 = 1 + 2 + 4$,
		   $13 = 2 + 4 + 7$, etc.

Quadratic sequence	A sequence in which the differences between terms form an arithmetic sequence.	4, 9, 16, 25, ... +5, +7, +9, ...

Review

1 a What are the next three terms of these sequences?

    i   5, 9, 13, 17, ...

    ii   44, 34, 24, 14, ...

    iii 0.3, 0.9, 1.5, 2.1, ...

  b Write the term–to–term rule for each of the sequences in part **a**.

2 Calculate the 9th term for the sequences with these position–to–term rules.

  a $T(n) = 3n + 2$

  b $T(n) = 8n - 11$

  c $T(n) = 12 - n$

  d $T(n) = 4 + \frac{1}{2}n$

3 The nth term of a sequence is given by $2 + n^2$.

  Calculate

  a the 5th term   b the 12th term

  c the 100th term.

4 Write a rule for the nth term of these sequences.

  a 3, 6, 9, 12...   b 4, 9, 14, 19, ...

  c 8, 14, 20, 26, ... d 20, 18, 16, 14, ...

5 Matchsticks are arranged into squares as shown.

  a Draw the fourth pattern in the sequence.

5 b Copy and complete the table using the pattern.

Number of Squares	Number of Matchsticks
1	4
2	
3	
4	
5	
6	

  c Write down a formula that links the number of squares, s, with the number of matchsticks, m.

6 This sequence is formed by doubling the current term to get the next term.
3, 6, 12, ...

  a Write down the next three terms of the sequence.

  b Name the type of sequence.

7 a Write down the next two terms of these sequences.

    i   1, 8, 27, 64, ...

    ii 1, 3, 6, 10, ...

  b Name the type of sequence.

8 Describe these sequences using one of the words in the coloured panel.

Arithmetic	Quadratic
Fibonnaci-type	Geometric

  a $\frac{3}{4}$, $1\frac{1}{2}$, 3, 6, 12, ...

  b 0, 2, 2, 4, 6, ...

  c 4, 5.5, 7, 8.5, 10, ...

  d 5, 6, 9, 14, 21, ...

What next?

Score			
	0 – 3		Your knowledge of this topic is still developing. To improve look at MyMaths: 1053, 1054, 1165, 1173, 1920
	4 – 7		You are gaining a secure knowledge of this topic. To improve your fluency look at InvisiPens: 21Sa – h
	8		You have mastered these skills. Well done you are ready to progress! To develop your problem solving skills look at InvisiPens: 21Aa – e

Assessment 21

1 Jill is given the following sequences, all with terms missing.
For each sequence find and write down

   i   the next two terms

   ii any missing terms

   iii if the sequence is ascending or descending

   iv the differences between consecutive terms

   v the rule explaining how the sequence works.

   a $\square$,   20,   10,   0, -10 [5]

   b 44,   66,   88,   110,   132, $\square$, 166   [5]

   c -15, -12, $\square$, -6, $\square$,   0   [5]

   d 45,   34,   23, $\square$,   1   [5]

   e 24,   29, $\square$, $\square$,   44   [5]

   f 1.1,   1.5, $\square$,   2.3,   2.7   [5]

   g 12, $\square$,   36,   48, $\square$,   72   [5]

   h 1.12, 1.11, 1.1, $\square$,   1.08   [5]

   i -0.90, -0.95, $\square$, -1.05, $\square$ [5]

   j 999 993, 999 996, $\square$, $\square$, 1 000 005   [5]

   k 3,   2.25,   1.5, $\square$,   0   [5]

   l -10, -17, $\square$, -31 [5]

   m 6, $\square$,   6.25, 6.375, 6.5   [5]

2 Chris has the terms and descriptions of some sequences.
He can't remember which description matches which group of terms.
Match each group of terms to the correct description.   [7]

   a The largest five even numbers less than 5.   **i** -3 -1 1   3   5

   b The first five multiples of 7.   **ii** 5   10 15 20 25 30

   c The first five factors of 210.   **iii** 11 13 17 19 23 29

   d The first five triangular numbers.   **iv** 7   14 21 28 35

   e Odd numbers from -3 to 5.   **v** 4   2   0 -2 -4

   f Prime numbers between 10 and 30.   **vi** 1   3   6   10 15

   g $15 <$ square numbers $\leqslant 81$.   **vii** 1   2   3   5   7

   h Multiples of 5 between 3 and 33.   **viii** 16 25 36 49 64 81

3 Nathan is given this sequence.

     1   11   21   31   41 $\square$ $\square$

He says that the common difference of this sequence is $+11$.

   a Work out the correct common difference.   [1]

   b Complete this sentence. Show your working.
       'The nth term of this sequence is...'   [2]

4 Jack and Dawn are looking at the start of this sequence of patterns.

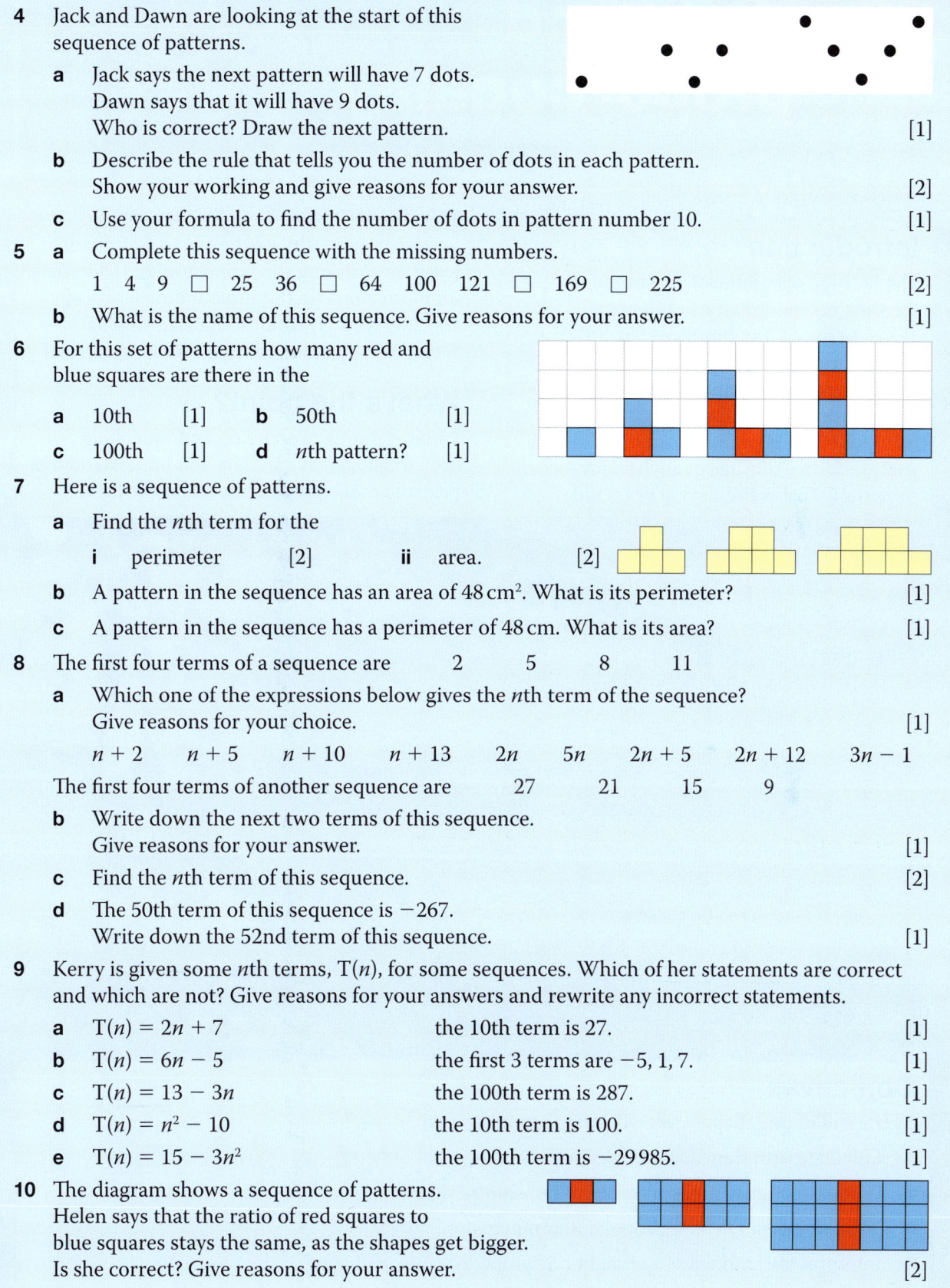

    a Jack says the next pattern will have 7 dots.
Dawn says that it will have 9 dots.
Who is correct? Draw the next pattern. [1]

    b Describe the rule that tells you the number of dots in each pattern.
Show your working and give reasons for your answer. [2]

    c Use your formula to find the number of dots in pattern number 10. [1]

5 **a** Complete this sequence with the missing numbers.

      1   4   9 ☐ 25 36 ☐ 64 100 121 ☐ 169 ☐ 225 [2]

    b What is the name of this sequence. Give reasons for your answer. [1]

6 For this set of patterns how many red and blue squares are there in the

    a 10th   [1]   **b** 50th   [1]

    c 100th   [1]   **d** nth pattern?   [1]

7 Here is a sequence of patterns.

    a Find the nth term for the

       i perimeter   [2]   **ii** area.   [2]

    b A pattern in the sequence has an area of 48 cm². What is its perimeter? [1]

    c A pattern in the sequence has a perimeter of 48 cm. What is its area? [1]

8 The first four terms of a sequence are   2   5   8   11

    a Which one of the expressions below gives the nth term of the sequence?
Give reasons for your choice. [1]

      $n + 2$ $n + 5$ $n + 10$ $n + 13$ $2n$ $5n$ $2n + 5$ $2n + 12$ $3n - 1$

    The first four terms of another sequence are   27   21   15   9

    b Write down the next two terms of this sequence.
Give reasons for your answer. [1]

    c Find the nth term of this sequence. [2]

    d The 50th term of this sequence is -267.
Write down the 52nd term of this sequence. [1]

9 Kerry is given some nth terms, T(n), for some sequences. Which of her statements are correct and which are not? Give reasons for your answers and rewrite any incorrect statements.

    a T(n) = $2n + 7$ the 10th term is 27. [1]

    b T(n) = $6n - 5$ the first 3 terms are $-5, 1, 7$. [1]

    c T(n) = $13 - 3n$ the 100th term is 287. [1]

    d T(n) = $n^2 - 10$ the 10th term is 100. [1]

    e T(n) = $15 - 3n^2$ the 100th term is $-29\,985$. [1]

10 The diagram shows a sequence of patterns.
Helen says that the ratio of red squares to blue squares stays the same, as the shapes get bigger.
Is she correct? Give reasons for your answer. [2]

22 Units and proportionality

Introduction

The half-life of a radioactive isotope is the time taken for half its radioactive atoms to decay. The number of radioactive isotopes remaining after each half-life forms a geometric sequence. Comparing the proportion of remaining radioactive isotopes to the geometric sequence, allows you to estimate the age of an artefact – even something that is millions of years old.

Scientists can estimate the age of a dinosaur fossil by analysing the proportion of radioactive uranium atoms in the surrounding layers of volcanic rock. The oldest dinosaur fossils are thought to be more than 240 million years old.

What's the point?

Understanding proportion and modelling growth and decay can help you understand the past and make predictions about the future.

Objectives

By the end of this chapter you will have learned how to …

- Calculate with standard and compound units.
- Compare lengths, areas and volumes of similar shapes.
- Solve direct and inverse proportion problems.
- Interpret the gradient of a straight line graph as a rate of change.
- Interpret graphs that illustrate direct and inverse proportion.
- Set up, solve and interpret growth and decay problems.

Check in

1   Neil buys 2 pizzas at a cost of £7.00.
    What is the cost of 8 pizzas?

2   John has a mass of 50 kg. Kevin has a mass of 75 kg.
    How many times more massive than John is Kevin?

3   Krishna earns £8.50 per hour.
    How much does he get paid for 8 hours work?

Chapter investigation

Ian and Jeannie both earn £20 000 per year, in different companies.

Jeannie's pay increases by 8% per year.

Ian's pay increases by £2000 per year.

Who earns the most after one year?

At what point, if ever, does the other person's salary overtake?

22.1 Compound units

Compound measures describe one quantity in relation to another.

These are examples of compound measures.

- **Speed** = $\dfrac{\text{Total distance travelled}}{\text{Total time taken}}$ Units such as m/s; km/h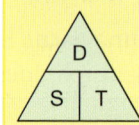

- **Density** = $\dfrac{\text{Mass}}{\text{Volume}}$ Units such as g/cm³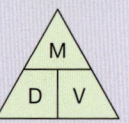

- **Pressure** = $\dfrac{\text{Force}}{\text{Area}}$ Units such as N/m²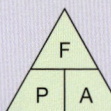

> Use the triangle to work out which calculation to use.
>
> Cover D (for distance)
> You multiply
> S (speed) × T (time)

> Density = $\dfrac{\text{mass}}{\text{volume}}$

EXAMPLE

Kerry jogs at an average speed of 5 km/h for $1\frac{1}{2}$ hours.
What distance does she jog?

> Distance = $5 \times 1\frac{1}{2}$
> = 7.5 km

EXAMPLE

Find the density of a piece of wood with cross-section area 42 cm², length 12 cm and mass 693 g.

> Volume = 42 × 12 = 504 cm³
> Density = 693 ÷ 504 = 1.375 g/cm³
> Mass in grams Volume in cm³
> So Density in g/cm³.

Force is measured in newtons (N).

EXAMPLE

A force of 18 N acts over an area of 5 m².
What is the pressure?

> Pressure = $\dfrac{18}{5}$ = 3.6 N/m²

A rate is also a compound unit. It tells you how many units of one quantity there are compared with one unit of another quantity.

- **Rate of pay** = $\dfrac{\text{Amount of money}}{\text{Time}}$ Units such as £/h

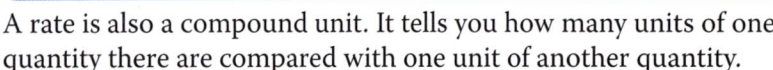

Rate of flow is a compound measure. It is the volume of liquid that passes through a container in a unit of time.

- **Rate of flow** = $\dfrac{\text{Volume}}{\text{Time}}$ Units such as cm³/s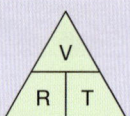

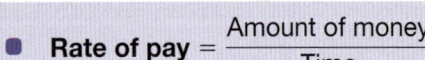

EXAMPLE

Water empties from a tank at a rate of 1.5 litres per second.
It takes 10 minutes to empty the tank.
How much water was in the tank?

> Use the triangle to work out which formula to use.
> Volume = rate × time
> Convert the time to seconds.
> 10 minutes = 10 × 60 s = 600 s
> Amount of water = 1.5 × 600 = 900 litres

Exercise 22.1S

1 The winners' times in some of the races at a sports day are

   a 100 metres in 13 seconds

   b 200 metres in 28 seconds

   c 400 metres in 58.4 seconds

   d 1500 metres in 4 minutes 52 seconds.

Calculate the speed of each winner in m/s, correct to 1 dp.

2 Work out the distance travelled in

   a 2 hours, at 80 km/h

   b 7 hour, at 23 mph

   c 6 seconds, at 9 m/s

   d 1 day, 12 mph.

3 Work out the time it takes to travel

   a 180 kilometres, at 60 km/h

   b 280 miles, at 70 mph

   c 8 kilometres, at 24 km/h

   d 15 miles at 60 mph.

4 A cube with volume 640 cm³ has a mass of 912 g. Find the density of the cube in g/cm³.

5 An emulsion paint has a density of 1.95 kg/litre. Find

   a the mass of 4.85 litres of the paint.

   b the number of litres of the paint that would have a mass of 12 kg.

6 The table shows the densities of different metals.

Metal	Density
Zinc	7130 kg/m³
Cast iron	6800 kg/m³
Gold	19320 kg/m³
Tin	7280 kg/m³
Nickel	8900 kg/m³
Brass	8500 kg/m³

Use the information in the table to find

   a the mass of 0.8 m³ of zinc

   b the mass of 0.5 m³ of cast iron

   c the mass of 3.2 m³ of gold

   d the volume of 910 g of tin

   e the volume of 220 g of nickel

   f the volume of a brass statue that has mass 17 kg.

7 The table shows the pressure, force and area of different materials.

Complete the table. Include the correct units in your answers.

	Pressure	Force	Area
a		12.9 N	10 m²
b		482.5 N	25 cm²
c	2560 N/m²	1200 N	
d	512 N/mm²		14.5 mm²
e	17.8 N/cm²	225 N	
f	24.6 N/m²		2.8 m²

8 If 4 metres of fabric costs £8.40, find the price of the fabric in pounds per metre.

9 Jane is paid £478 a week. Each week she works 40 hours. What is her hourly rate of pay?

10 Find the rate of flow for pipes A and B in litres/s.

   a Pipe A: 20 litres of water in 8 seconds.

   b Pipe B: 48 litres of water in 30 seconds.

11 Water empties from a tank at a rate of 2 litres/s. It takes 10 minutes to empty the tank.

How much water was in the tank?

12 An electric fire uses 18 units of electricity over a period of 7.5 hours.

   a What is the hourly rate of consumption of electricity in units per hour?

   b How many units of electricity are used in 24 hours?

13 A car has fuel efficiency of 8 litres per 100 km.

   a How far can the car travel on 40 litres of petrol?

   b How many litres of petrol would be needed for a journey of 250 miles?

   c What is the car's rate of fuel consumption in km per litre?

14 Rose received 75 US dollars in exchange for £50.

   a Calculate the rate of exchange in US dollars per £.

   b How many US dollars would she get for £125?

   c If Rose received $120 US dollars, how many pounds did she exchange?

Q 1061, 1121, 1246　SEARCH

22.1 Compound units

RECAP

Compound units describe one quantity in relation to another.

- The density of a material is its mass divided by its volume.
- Speed is the distance travelled divided by the time taken.
- Pressure is the force divided by the area.
- A formula triangle is a useful way to remember the relationships between the different parts.
- A rate is also a compound unit.

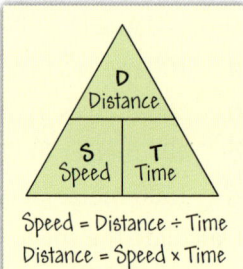

Speed = Distance ÷ Time
Distance = Speed × Time
Time = Distance ÷ Speed

HOW TO

① Draw a formula triangle and write the correct formula.

② Convert units or work out quantities to apply the formula.

③ Work out the answer, making sure the units are correct.

EXAMPLE

A train leaves Norwich at 13:40 and arrives in Cambridge at 15:00. If the distance is 90 km find the average speed of the train.

① Draw a formula triangle and write the formula for speed.

② Work out the time in hours.

③ Work out the answer using the correct units.

$\text{Speed} = \dfrac{\text{distance}}{\text{time}}$

Time = 1 hour 20 min = $1\frac{1}{3}$ h = 1.333... h

$\text{Speed} = \dfrac{90}{1.333} = 67.5$

The average speed of the train is 67.5 km/h.

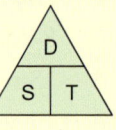

EXAMPLE

Sand was falling from the back of a lorry at a rate of 0.4 kg/s. It took 20 minutes for all the sand to fall from the lorry.

How much sand was the lorry carrying?

① Draw a formula triangle and write the formula for mass.

② Convert the time to seconds.

③ Work out the answer using the correct units.

The rate of flow is in kg/s, which is mass divided by time.

Mass = rate × time

20 minutes = 20 × 60 s = 1200 s

Mass = 0.4 × 1200 = 480

The lorry was carrying 480 kg of sand.

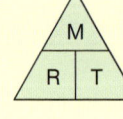

EXAMPLE

A metal cuboid has a length of 7 cm, a width of 5 cm and a height of 4 cm. It has a mass of 1.470 kg. Find its density in g/cm³.

① Draw a formula triangle and write the formula for density.

② Work out the volume of the cuboid and convert mass to grams.

③ Work out the answer using the correct units.

$\text{Density} = \dfrac{\text{mass}}{\text{volume}}$

Volume of cuboid = length × width × height
= 7 × 5 × 4 cm³ = 140 cm³

Mass = 1.470 kg = 1470 g

$\text{Density} = \dfrac{1470}{140} = 10.5 \text{ g/cm}^3$

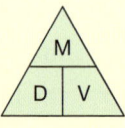

Exercise 22.1A

1. A train leaves Euston at 8:57 a.m. and arrives at Preston at 11:37 a.m.
   If the distance is 238 miles find the average speed of the train.

2. Sand falls from the back of a lorry at a rate of 0.2 kg/s.
   It took 25 minutes for all the sand to fall from the lorry.

   How much sand was the lorry carrying?

3. A metal cuboid has a length of 9 cm, a width of 5 cm and a height of 4 cm.
   It has a mass of 1.53 kg.

   Find its density in g/cm³.

4. A car travels 24 miles in 45 minutes.
   Find the average speed of the car in miles per hour (mph).

5. Copy and complete the table to show speeds, distances and times for five different journeys.

Speed (kmph)	Distance (km)	Time
105		5 hours
48	106	
	84	2 hours 15 minutes
86		2 hours 30 minutes
	65	1 hours 45 minutes

6. A cube of side 2 cm has a mass of 40 grams.

   a. Find the density of the material from which the cube is made, giving your answer in g/cm³.

   > Volume of cube = length³.

   b. A cube of side length 2.6 cm is made from the same material.
      Find the mass of this cube, in grams.

7. A solid block has a length and width of 22.50 mm, and a height of 3.15 mm. It has a mass of 9.50 g.

   a. Find the density of the metal from which the block is made, giving your answer in g/cm³.

   b. How many blocks can be made from 1 kg of the material?

8. In this question, give your answers in kg/m³.

   a. The volume of 31.5 g of silver is 3 cm³.
      Work out the density of silver.

   b. The volume of 18 g of titanium is 4 cm³.
      Work out the density of titanium.

   c. A sheet of aluminium foil has volume 0.4 cm³ and mass 1.08 g. Work out the density of aluminium foil.

9. Grace earns £340 per week for 40 hours work.
   If Grace works overtime, she is paid 1.5 times her standard hourly rate.

   a. How much is Grace paid for 7 hours of overtime work?

   b. Grace earned £531.25 last week. How many hours of overtime did she work?

10. The toll charged for a car travelling on a motorway was £33.60 for a journey of 420 km. Cars with trailers are charged double.

    How much would it cost for a car with a trailer to travel 264 km?

11. A yacht race has three legs of 8 km, 6 km and 10 km.
    The average speed for the winning yacht was 6.2 km/h.
    The second yacht finished 8 minutes after the winner.
    How long did it take the second place yacht to finish the race?

12. a. Julia is wearing high-heeled shoes.
       Each heel has an area of 1 cm². Julia has a mass of 55 kg.

       How much pressure does Julia's heel exert when she has one heel on the ground?

    b. An elephant's foot is 45 cm across is approximately circular.
       An elephant walks with two feet on the ground at a time.
       An elephant has a mass of 5500 kg.

       How much pressure does an elephant's foot exert when the elephant has two feet on the ground?

22.2 Direct proportion

A ratio tells you how many times bigger one number is compared with another number. You can calculate the ratio by dividing one number by the other.

- Two variables are in **direct proportion** if the ratio between then stays the same as the actual values vary.

EXAMPLE

A pipe 2.5 metres long has a mass of 35 kilograms. What would be the mass of 5.5 metres of the same pipe?

Here are two different ways to solve this problem.

2.5 m	mass	35		2.5 m	mass	35
$\Downarrow \div 5$		$\Downarrow \div 5$		$\Downarrow \div 2.5$		$\Downarrow \div 2.5$
0.5 m	mass	7		1 m	mass	14 kg
$\Downarrow \times 11$		$\Downarrow \times 11$		$\Downarrow \times 5.5$		$\Downarrow \times 5.5$
5.5 m	mass	77 kg		5.5 m	mass	77 kg

This is an informal scaling method.

This is called the unitary method.

5.5 metres of pipe have a mass of 77 kg.

- When two quantities are in direct proportion, then their graph is a straight line that passes through the origin.
- The line has the equation $y = kx$.
- The gradient of the line, k, is the rate of change between the two quantities.

> Two points is enough to plot a straight line. The third point checks the line is accurate.

> The graph has £s on the horizontal axis and euros on the vertical axis. You could use either axis for either currency.

Davina buys some euros for a trip to France. The **exchange rate** is £1 = €1.20.

She draws a conversion graph to help her convert prices.

> £1 = €1.20
>
> £0 = €0
> £10 = €12
> £20 = €24

First she works out some simple **conversions** to plot on the graph by writing a formula: number of euros = 1.20 × number of pounds

She wants to include prices up to £40. Using the formula, 1.20 × 20 = 24, and 1.20 × 40 = 48

£20 = €24, and £40 = €48. The euros scale needs to go up to at least €50.

She chooses the scale so her graph fits her paper.

This point represents £10 = €12

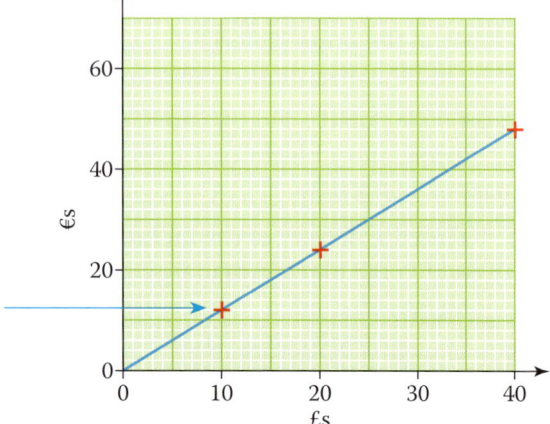

Exercise 22.2S

1 Ribbon costs £2.75 per metre. Find the cost of these lengths of ribbon.

   a 3 m   **b** 4.5 m

   c 6.85 m   **d** 27.55 m

2 A shop sells shelving at £3.45 per metre. Find the cost of these lengths of shelving.

   a 5 m   **b** 3.45 m

   c 2.25 m   **d** 4.85 m

3 A 2 m length of pipe has a mass of 8 kg. What is the mass of a 3 m length of the same pipe?

4 Four buckets of water have a mass of 60 kg. What is the mass of 5 buckets of water?

5 400 g of powder paint costs £2.40.

   a Find the cost of 100 g of the paint.

   b Use your answer to part **a** to find the cost of 300 g of the paint.

6 300 g of sherbet drops cost £1.20.

   a How much do 100 g of sherbet drops cost?

   b How much do 700 g of sherbet drops cost?

7 A pack of 250 tea bags contains 130 g of tea and costs £7.50. Calculate

   a the cost of one tea bag

   b the mass of tea in one bag.

8 A shop sells five different types of luxury tea. Calculate the cost of 100 g of each brand, given that

   a 200 g of brand A costs £3.75

   b 500 g of brand B costs £7.40

   c 300 g of brand C costs £5.20

   d 250 g of brand D costs £5.10

   e 350 g of brand E costs £6.50.

9 Alan and Barry buy sand from a builders' merchant.

Alan buys 35 kg of sand for £4.55.

Barry buys 28 kg of the same sand.

How much does Barry pay?
Show your working.

10 The conversion rate for millimetres to centimetres is 1 cm = 10 mm

   a Work out two simple conversions you could plot for a millimetres to centimetres conversion graph.

   b The graph needs to convert distances up to 10 cm. What is the highest value needed on the mm scale?

11 The conversion rate for pounds (lb) to kilograms (kg) is 1 kg = 2.2 lbs

   a Copy and complete these conversions.

      i 0 kg = ☐ lbs

      ii 10 kg = ☐ lbs

      iii 5 kg = ☐ lbs

   b Use your results from part **a** to draw a conversion graph from pounds to kilograms on a grid from 0 kg to 10 kg.

   c Use your graph to convert

      i 10 lbs to kg   **ii** 5 lbs to kg

      iii 3 kg to lbs   **iv** 2.5 kg to lbs.

12 a This graph is a conversion graph for miles to kilometres and vice versa.

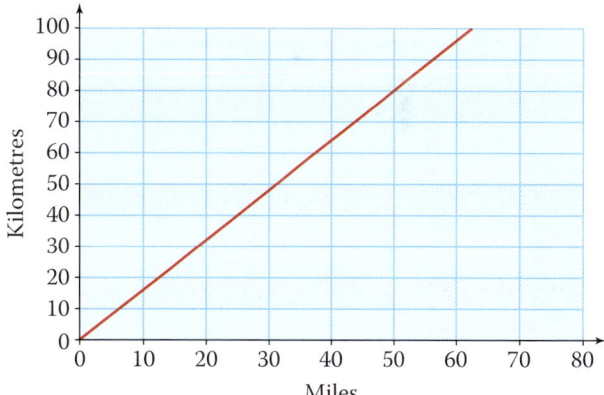

     Use the graph to convert

     i 20 miles into kilometres

     ii 60 kilometres into miles.

   b If Dan ran 30 miles and Charlie ran 50 kilometres, who ran further?

   c By finding the gradient of the line, give a formula to connect the number of miles (x) with the number of kilometres (y).

22.2 Direct proportion

- Numbers or quantities are in **direct proportion** when the ratio of each pair of corresponding values is the same.
- When two quantities are in direct proportion, then their graph is a straight line that passes through the origin.
- The line has the equation $y = kx$.
- The gradient of the line, k, is the rate of change between the two quantities.

HOW TO

① Use the values in the question or the gradient of the graph to find the rate of conversion for the quantities.

② Divide and/or multiply to keep the ratio between the quantities the same.

EXAMPLE

Two bottles of olive oil are on sale in a supermarket.

The 500 ml bottle is priced at £2.99.

The 750 ml bottle is priced at £4.29.

Which is better value?

The line y = kx is a straight-line that passes through the origin.

① Find the cost per ml, k, for each bottle.

500 ml bottle   299 = k × 500, so k = 0.598

750 ml bottle   429 = k × 750, so k = 0.572

② Compare the cost per ml.

The 750 ml bottle costs less per ml, so it is better value.

EXAMPLE

Here is a recipe for blackcurrant squash for 5 people.

Work out the number of grams of blackcurrants needed to make squash for 8 people.

Blackcurrant squash (for 5 people)

400 g of blackcurrants
1200 ml of water
100 g sugar
250 ml blackcurrant juice

The ratio of people to blackcurrants stays the same.

① Set out the problem.

Number of people   Grams of blackcurrants

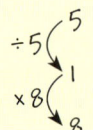

÷5 ⟨ 5
       1
×8   8

400 ⟩ ÷5
80 ⟩ ×8
640

② Divide by 5 to find the value for 1 person.
Then multiply by 8 to find the value for 8 people.

③ Write the answer.

So for 8 people you need 640 g blackcurrants.

Exercise 22.2A

1 A shop sells drawing pins in two different sized packs.

 Pack A contains 120 drawing pins and costs £1.45.

 Pack B contains 200 of the same drawing pins, and costs £2.30.

 Calculate the cost of one drawing pin from each pack, and say which pack is better value.

2 A store sells packs of paper in two sizes.

> **Regular**
>
> 150 sheets
>
> Cost £1.05

> **Super**
>
> 500 sheets
>
> Cost £3.85

 Which of these two packs gives better value for money?

 You must show all of your working.

3 A recipe for cake uses 400 g of sugar for 5 people.

 What mass of sugar is needed for

 a 8 people **b** 12 people

 c 14 people **d** 30 people?

4 A recipe for three bean chilli uses 840 g of beans for 7 people.

 What mass of beans is needed for

 a 6 people **b** 17 people

 c 24 people **d** 100 people?

5 Although builders often mix materials according to their volumes, these materials are usually bought according to their mass. The conversions they must make are based upon mass being directly proportional to the volume.

 1 cubic metre of sand has mass 1.7 tonnes.

 1 cubic metre of cement has mass 1.4 tonnes.

 Mortar is made by mixing sand and cement in the ration 3 : 1.

 a What is the total mass of mortar made using 1 cubic metre of cement?

 b Find and simplify the ratio *by mass* of cement to sand in a standard mortar mix.

6 Vince works for 4 hours. He gets paid £24.

 Vince works for 8 hours. He gets paid £50.

 Is Vince's pay directly proportional to the time he works? Give a reason for your answer.

7 Barry can fit 12 radiators in one day. The number of radiators that Barry can fit is directly proportional to the number of days he works.

 Does this formula calculate the number of radiators that Barry can fit?

 number of radiators = days worked + 12

 Give a reason for your answer.

8 The table shows standard paper sizes.

Paper size	Width (mm)	Height (mm)
A0	841	1189
A1	594	841
A2	420	594
A3	297	420
A4	210	297
A5	148	210

 a Find the ratio of width : height for each paper size. Is the height directly proportional to the width?

 b Plot a graph with width on the *x*-axis and height on the *y*-axis.
 Does the graph show that the height is directly proportional to the width?

22.3 Inverse proportion

- When variables are in **inverse proportion**, one of the variables increases as the other one decreases, and vice-versa.

For example, the number of bricklayers building a wall and the time taken are in inverse proportion.

5 bricklayers $\times$ 4 hours $= 20$ 5 people take 4 hours.

10 bricklayers $\times$ 2 hours $= 20$ 10 people take 2 hours.

When one quantity doubles, the other halves.

1 bricklayer $\times$ 20 hours $= 20$ 1 person takes 20 hours.

When one quantity is divided by 10, the other is multiplied by 10.

The amount of labour needed is often called 'man-hours'.

- If two variables are in inverse proportion, the *product* of their values will stay the same.

- 'y is inversely proportional to x' can be written as 'y is proportional to $\frac{1}{x}$' or $y = \frac{k}{x}$, where k is the constant of proportionality.
- When two quantities are in inverse proportion, then their graph is a reciprocal curve.
- The product of each x- and y-coordinate on the curve is the same.

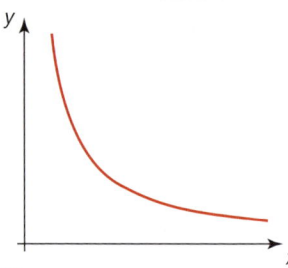

EXAMPLE

The variable x is inversely proportional to y with $y = \frac{20}{x}$.

a Complete the table.

x	1	2	4	5	10	20
y						

b Plot each pair of values for x and y. Join the points with a smooth curve.

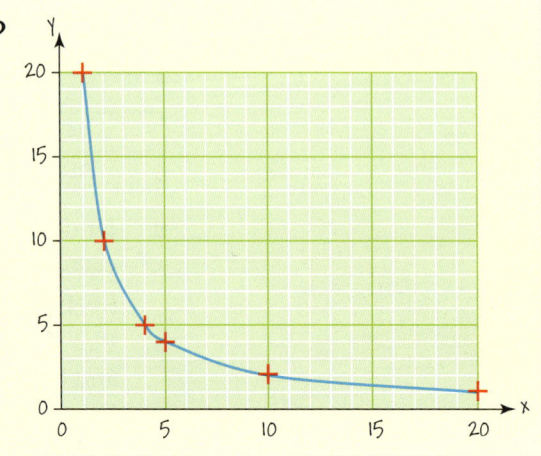

a Complete the table.

x	y
1	20
2	10
4	5
5	4
10	2
20	1

b

The product of each x- and y-coordinate on the curve is 20.

You could use the equation to work out the number of bricklayers needed to build the wall in the previous example.

Exercise 22.3S

1 You are told that y is **directly** proportional to x.
Explain what will happen to the value of y when the value of x is

  a doubled   **b** halved

  c multiplied by 6   **d** divided by 10

  e multiplied by a factor of 0.7.

2 You are told that w is *inversely* proportional to z.
Explain what will happen to the value of w when the value of z is

  a doubled   **b** halved

  c multiplied by 6   **d** divided by 10

  e multiplied by a factor of 0.7.

3 It takes 12 hours for 5 builders to build a wall.

  a How many 'man-hours' does it take to build the wall?

  b How many builders are needed to build the same wall in 6 hours?

  c How long will it take 15 builders to build the same wall?

4 A ship has enough food to supply 600 passengers for 3 weeks.

  a How long would the food last for one passenger?

  b How long would the food last for 300 passengers?

  c How long would the food last for 1200 passengers?

5 It takes 20 hours for 5 people to decorate 100 cakes.

  a How many people are needed to decorate 100 cakes in 10 hours?

  b How many hours will it take 4 people to decorate 100 cakes?

  c How long will it take 5 people to decorate 200 cakes?

> Take care! Hours and people are inversely proportional. People and cakes are directly proportional.

6 Sort each of the formulae into one of the categories in the table.

$$y = 4x \qquad y = \frac{20}{x} \qquad y = \frac{x + 1}{10}$$

$$y = \frac{13.5}{x} \qquad y = 5x + 1 \qquad y = \frac{x}{10}$$

Inverse proportion	Direct proportion	Neither

> Hint: There are two of each type.

7 The variable x is inversely proportional to y with $y = \frac{16}{x}$.

  a Complete the table

x	$\frac{1}{4}$	1	2	4	8	16
y						

  b Plot each pair of values for x and y. Join the points with a smooth curve.

  c Pick any point on your line. Is the product of the x and y-coordinate 16?

8 h is inversely proportional to b, with $h = \frac{60}{b}$.

  a Find b when $h = 5$.

  b Explain why you can use this equation to check your answers to question **3**.

Q 1048   SEARCH

22.3 Inverse proportion

RECAP

- When variables are in **inverse proportion**, one of the variables increases as the other one decreases, and vice-versa.
- 'y is inversely proportional to x' can be written as $y = \frac{k}{x}$, where k is the **constant** of proportionality.

The graph of $y = \frac{k}{x}$ is a reciprocal graph.

EXAMPLE

The time taken for a journey is
- directly proportional to the distance travelled
- inversely proportional to the average speed.

A pilot flies at a moderate speed and completes a journey of 1200 km in 3 hours.

Can the pilot complete a 3600 km journey in less than 5 hours, if he doubles the average speed of the plane?

The distance has tripled. Time and distance are directly proportional.

The time will also triple.   3 hours × 3 = 9 hours

The speed has doubled. Time and speed are inversely proportional.

The time will halve.   9 hours ÷ 2 = 4.5 hours

Yes, the pilot can complete the journey in less than 5 hours.

HOW TO

Show that two quantities are inversely proportional
1. Use the fact that if two variables are in inverse proportion, the product of their values will stay the same.
2. Plot of graph of the variables. The graph should be a reciprocal curve.

EXAMPLE

Stefan carries out an experiment to confirm Ohm's law. This states that for a constant voltage, the current in a circuit is inversely proportional to the resistance. Stefan varies the resistance and records the current. His results are shown in the table.

Stefan thinks that one of the readings is wrong. Which reading could be wrong?

Resistance (ohms)	Current (amps)
5	16
10	8.1
15	4.2
20	4.0
25	3.2
30	2.6

1. If two quantities are inversely proportional, then the product of the values will stay the same.

Resistance (ohms)	Current (amps)	Resistance × current
5	16	80
10	8.1	81
15	4.2	63
20	4.0	80
25	3.2	80
30	2.6	78

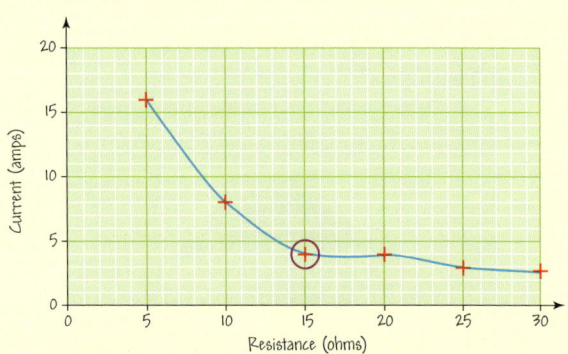

The reading at 15 ohms could be wrong as all the products of the other readings are approximately 80.

2. The graph should be a reciprocal curve – the point (15, 4.2) does not fit the pattern.

Exercise 22.3A

1 Rebecca travels at 40 km/h, the journey takes her 3 hours.

How long will it take Carly to make the same journey if she travels at 60 km/h?

2 Masood assumes that when cooking in a microwave, the cooking time is inversely proportional to the power level.

Level	Power (watts)
Full	600
Roast	400
Simmer	200
Defrost	100
Warm	50

a A pizza takes 8 minutes on 'Roast'. How long will it take on 'Full'?

b Lasagna takes 12 minutes on 'Warm'. How long will it take on 'Simmer'?

3 The safeload of a wooden beam is

● directly proportional to the height

● inversely proportional to the distance between its supports.

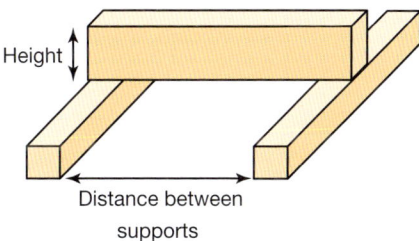

Height

Distance between supports

a The safeload of a particular beam is 500 kg.

i What will the safeload be if the height of the beam is halved?

ii What will the safeload be if the distance between the supports is doubled?

b The safeload of a beam is 800 kg. An engineer plans to double the height of the beam and halve the distance between the supports. Will the modifications be enough to support 3500 kg?

4 z is inversely proportional to y.

y	0.5	1.5					
z			12		6		3.2

Use the numbers on the cards to complete the table. Which card is left over?

2	3	4	4.8	5

7.5	8	10	16	48

5 A rectangle has area 48 cm².

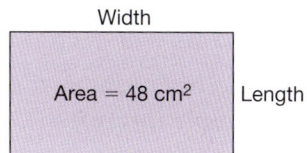

Width

Area = 48 cm²   Length

a Explain why the length of the rectangle is inversely proportional to the width.

b Complete the table to show some of the possible dimensions and perimeter for the rectangle.

Length (cm)	Width (cm)	Perimeter (cm)
1	48	98
2	24	52
3	16	38
4		
6		
8		
12		
16		
24		
48		

Jennifer says

> The perimeter of the shape decreases as the length increases.
> The perimeter must be inversely proportional to the length.

Explain why Jennifer is wrong.

22.4 Growth and decay

An increase of 20% means that the new amount is 120% of the old amount. You can find this by multiplying by 1.2

A decrease of 20% means the new amount is 80% of the old amount. You can find this by multiplying by 0.8

1.2 and 0.8 are called **multipliers**.

- To increase something by r%, multiply by $\dfrac{100 + r}{100} = 1 + \dfrac{r}{100}$
- To decrease something by r%, multiply it by $\dfrac{100 - r}{100} = 1 - \dfrac{r}{100}$

Sometimes percentage increases or decreases are repeated.

When interest is calculated on the interest it is called **compound interest** otherwise it is **simple interest**.

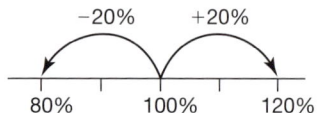

▲ To increase something by 20%, multiply by 1.2
To reduce something by 20%, multiply by 0.8

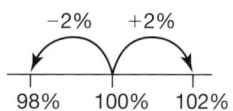

▲ To increase something by 2%, multiply by 1.02
To reduce something by 2%, multiply by 0.98

EXAMPLE

100 red squirrels are delivered to a forest as part of a re-introduction program. This population increases by 45% each year.

This is called **exponential growth**.

a Draw a table to show how the population increases in the next 6 years.

b In which year does the squirrel population first reach

  i 500   **ii** 1000   **iii** 10 000

The population at the end of the year is 100% + 45% = 145% of the population at the start of the year.

a Multiplier = 1.45

Years	Population
0	100
1	100 × 1.45 = 145
2	145 × 1.45 = 210
3	210 × 1.45 = 305
4	305 × 1.45 = 442
5	442 × 1.45 = 641
6	641 × 1.45 = 929

Try using the 'Ans' key on your calculator to repeat calculations.

b i At the end of year 4 population = 442
At the end of year 5 population = 641
Year 5

ii Extend the table
At the end of year 7 population = 1348
Year 7

iii At the end of year 8 population = 1954
At the end of year 9 population = 2833
Year 9

EXAMPLE

The value of a car is £15 000. The value decreases by 12% each year.

a Find a formula for the value of the car after n years.

This is called **exponential decay**.

b Use your formula to find the value of the car after 10 years.

a 100% − 12% = 88%, the value at the end of each year is 88% of its value at the start of that year.

Multiplier = 0.88

Value after n years = £15 000 × 0.88^n Multiply by 0.88 for each year.

b Value after 10 years = £15 000 × 0.88^{10}

You can check this by multiplying by 0.88 ten times.

  = £4178 (nearest £)

Ratio and proportion   Units and proportionality

Exercise 22.4S

1 Find the multiplier that gives the amount after each change. Write your answers as decimals.

    a Increase of 30%   **b** Decrease of 30%

    c Increase of 3%   **d** Decrease of 3%

    e Increase of 25%   **f** Decrease of 25%

    g Increase of 2.5%   **h** Decrease of 2.5%

2 A bacteria population doubles every 20 mins.

    a Copy and extend this table to 180 mins.

Minutes	Population
0	1
20	2
40	4

    b In which hour did the population first pass

       i 300   **ii** 1000   **iii** 10 000?

    c Estimate the number of bacteria after

       i 150 minutes   **ii** 170 minutes.

    d What has happened to the population between 150 minutes and 170 minutes?

3 The half-life of a radioactive substance is the time it takes for the amount to go down by a half. The half-life of caesium-137 is 30 years.

    a Copy and extend this table to 6 half-lives.

No of half-lives	Time (years)	Amount (grams)
0	0	1000
1	30	500
2	60	

    b After how many half-lives does the amount first become less than

       i 100 g   **ii** 10 g   **iii** 1 g?

4 **a** A shop decreases its prices by 20% on each day of a sale. Copy and complete this table for Monday to Saturday.

Day	Price of boots
Monday	£50
Tuesday	

    b By what percentage is the Monday price reduced on Saturday?

5 The number of trout in a lake is 800. The number decreases by 15% each year.

    a Copy and extend this table to 8 years.

Year	Number of trout
0	800

    *b Draw a graph to illustrate your values.

    c **i** Show that the number of trout after n years will be 800×0.85^n

       ii Use this formula to check the values in your table.

6 The rate at which a bacteria colony increases depends on the conditions. The table gives information about two colonies.

Colony	Population now	Increase per hour
A	200	50%
B	400	35%

The population of Colony A after n hours is 200×1.5^n.

    a Find an expression for the population of Colony B after n hours.

    b When does the population of Colony A become bigger than that of Colony B?

7 Sadie invests £2000 in a savings account. The bank adds 4% **compound** interest at the end of each year. Sadie does not add or take any money from the account for 10 years.

    a Copy and extend this table to show how Sadie's investment grows.

End of year	Amount in the account (£)
1	2000 × 1.04 = 2080
2	

    b Work out the percentage interest that Sadie's investment earns in 10 years.

    *c £P is invested with compound interest r% added at the end of each year.

       i Show that the total amount at the end of n years is $A = P\left(1 + \dfrac{r}{100}\right)^n$

       ii Use this formula to check the last amount in your table in part **a**.

22.4 Growth and decay

RECAP

- To increase something by $r\%$, multiply by $1 + \dfrac{r}{100}$
- To reduce something by $r\%$, multiply it by $1 - \dfrac{r}{100}$
- When a principal amount $£P$ is invested with compound interest $r\%$ added at the end of each year, the total amount at the end of n years is

$$A = P\left(1 + \frac{r}{100}\right)^n$$

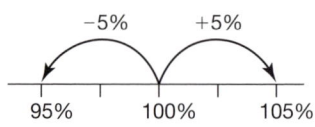

▲ To increase something by 5%, multiply by 1.05
To decrease something by 5%, multiply by 0.95

Money earns **simple** interest when the interest is *taken out* of the account at the end of each time period.

Money earns **compound** interest when the interest is *left in* the account.

HOW TO

To solve a repeated percentage change problem
(1) Find the multiplier.
(2) Work out the value you need using the formula
(3) Answer any related questions.

EXAMPLE

Harry invests £4000 in an account. Compound interest of 3% is added at the end of each year.

a **i** How much is in the account at the end of 5 years?
   ii How much compound interest has been earned?

b How much longer will it take for Harry to have more than £5000 in the account?

c What assumptions have you made?

Banks round money to the nearest pence.

(1) Multiplier $= 1 + \dfrac{3}{100} = 1.03$

a **i** Amount after 5 years $= 4000 \times 1.03^5 = £4637.10$ (2)

   ii Interest $= £4637.10 - £4000 = £637.10$ (3) Take away the original amount.

b Continuing to multiply by 1.03 gives
4776.21, then 4919.50, then 5067.08 (2)
It takes 3 more years.

c This assumes Harry does not add or take out any money from the account. (3)

EXAMPLE

Town planners use the formula 240×0.9^n to estimate the population of a village n years from now.

a What is the population of the village now?

b Explain what the 0.9 tells you about the planners' assumptions.

a Population now $= 240 \times 0.9^0 = 240 \times 1$
240

Any number to the power 0 is 1.

b $0.9 = 90\%$ is the multiplier. (1)
This means the population at the end of each year will be 90% of what it was at the beginning of the year.

The planners assume the population (2)
will decrease by 10% each year.

Ratio and proportion Units and proportionality

Exercise 22.4A

1 For each account in the table

   i find the amount in the account after the given number of years

   ii find the compound interest earned.

Account	Original amount	Compound interest rate	Number of years
a	£1000	5% per year	3
b	£250	4% per year	6
c	£840	2.5% per year	10
d	£45000	3.25% per year	12

2 Lily works out the compound interest earned by £500 invested for 4 years at a rate of 6%. Here is Lily's working.

> Interest in 1 year = 6% of £500
> = 0.06 × £500 = £30
> Interest for 4 years = 4 × £30 = £120

   a Why is Lily's working incorrect?

   b Work out the correct answer.

3 The value of a new car is £16000. Each year the car loses 15% of its value at the start of the year.

   a Work out the value of the car when it is 4 years old.

   b After how many complete years will the car's value drop below £4000?

4 The population of a town is 52000. Assume that the population increases by 1.5% each year.

   a Work out the population 6 years from now.

   b When will the population reach 60000?

   c Why might the answers to parts **a** and **b** be incorrect?

5 A company uses the formula 350×0.96^n to estimate the number of workers needed in their factory n years from now.

   a How many workers are there now?

   b Explain what the 0.96 tells you about the company's assumptions.

6 A road planner uses the formula 2400×1.08^n to estimate the number of vehicles per day that will travel on a new road n months after it opens.

   a Describe two assumptions the planner has made.

   b Give reasons why the planner's assumptions may not be appropriate.

7 Mark has £50000 to invest.
He finds the interest rates for two accounts.

Account A	Account B
4.5% per year interest	Year 1 5.0% interest
	Year 2 4.5% interest
	Year 3 4.0% interest

   a Calculate which account would give more compound interest on £50000 invested for 3 years.

   b Give reasons why Mark may not decide to use this account.

*8 Match a graph to each statement.

   a Sales went down by 40 each month.

   b Sales went up by 40% each month.

   c Sales went down by 40% each month.

   d Sales went up by 40 each month.

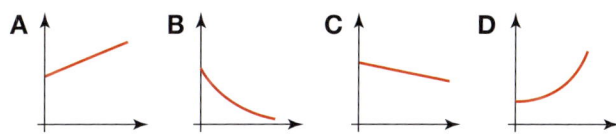

*9 A building society offers two accounts. Karen says that these accounts would give the same interest on an investment. Is Karen correct? Explain your answer.

> **Easy Saver**
> 4% interest added at the end of each year
> **Half-yearly saver**
> 2% interest added at the end of every 6 months

Summary

Checkout
You should now be able to...

Test it
Questions

✔ Calculate with standard and compound units.		**1 – 4**
✔ Compare lengths, areas and volumes of similar shapes.		**5**
✔ Solve direct and inverse proportion problems.		**6, 7**
✔ Interpret the gradient of a straight line graph as a rate of change.		**8**
✔ Interpret graphs that illustrate direct and inverse proportion.		**9**
✔ Set up, solve and interpret growth and decay problems.		**10**

Language	Meaning	Example
Rate	One quantity measured per unit of time or per unit of another quantity.	Total pay for 3 hours work = £16.50 Rate of pay $= \dfrac{16.50}{3}$ $= £5.50$ per hour
Proportion	Proportion is the size of something compared with the size of something else.	If there are 6 eggs in a carton Total number of eggs = 6 × number of cartons The number of cartons is proportional to the number of eggs.
Proportional	Two quantities are proportional if one is always the same multiple of the other.	
Direct proportion	A set of quantities is in direct proportion with another set of quantities if the ratio of each pair of corresponding values is the same.	The number, n, of eggs is in direct proportion to the number, c, of cartons.
Constant of proportionality	The multiplier, k, between the quantities.	
Inverse proportion	Two quantities are in inverse proportion if one increases as the other decreases.	For a rectangle of width w and length l and area $8\,m^2$, w ⬚ Area 8 m^2 　　　l w is inversely proportional to l
Varies	If x varies with y then x changes when y changes.	$y = kx$　direct proportion $y = \dfrac{k}{x}$　inverse proportion

Review

1 Convert

   a 140 mm to m

   b 1000 s to minutes and seconds

   c 3 litres to cm^3

   d 5 m/s to km/h.

2 A force of 10 N acts over an area of $4\,m^2$. What is the pressure?

3 A box of cereal costs £1.80 for 750 g. What is the cost per 100 g?

4 Convert $1.8\,m^2$ into cm^2.

5 A cube of side length 4 cm is enlarged by scale factor 3.

   a What is the volume of

      i the small cube

      ii the enlarged cube?

   b What is the surface area of

      i the small cube

      ii the enlarged cube?

6 Two variables, x and y are directly proportional to each other.
$y = 4x$
What is the value of

   a y when $x = 8$

   b x when $y = 44$?

7 X is inversely proportional to Y.
$Y = \dfrac{2}{x}$
What is the value of

   a Y when $X = 3$ **b** X when $Y = \frac{2}{9}$?

8 The graph shows the price paid for a given number of magazines.

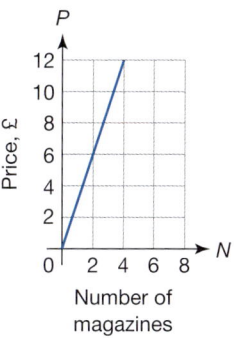

   a Write an equation linking price, P, and the number of magazines, N.

   b What is the gradient of the graph?

   c What does the gradient represent?

9

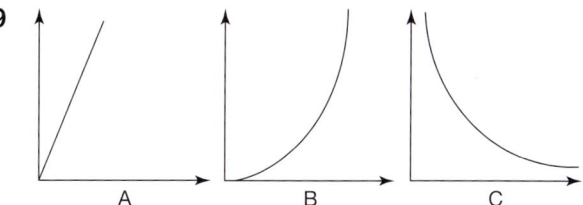

Which of these graphs shows the relationship in

   a question **6**   **b** question **7**?

10 Isabella invests £2000 in a bank account that pays 2% interest per year.

   a How much is in the account after

      i one year   **ii** three years?

   b Write a formula to find the value of the account, £V, after, t years.

What next?

Score	0 – 4		Your knowledge of this topic is still developing.
			To improve look at MyMaths: 1036, 1048, 1059, 1061, 1070, 1121, 1238, 1246
	5 – 8		You are gaining a secure knowledge of this topic.
			To improve your fluency look at InvisiPens: 22Sa – e
	9 – 10		You have mastered these skills. Well done you are ready to progress!
			To develop your problem solving skills look at InvisiPens: 22Aa – f

Assessment 22

1   The distance from Carlisle through Lancaster to Preston is 88 miles.
    Carlisle to Lancaster is 68 miles.

    a   Keanu drives from Lancaster to Preston at an average speed of 40 mph.
    How long does Keanu's journey take? Give your answer in hours and minutes. [2]

    b   Demi meets Keanu in Preston and they drive back to Lancaster together.
    They leave Preston at 1815 and arrive in Lancaster at 1855.
    Calculate their average speed. [2]

    c   Keanu and Demi leave Lancaster at the same time.
    Keanu drives back to Carlisle and Demi travels back to Preston by train.
    They both get to their destinations at the same time.
    Demi's journey takes 1 hour and 20 minutes.
    Calculate the difference in their average speeds. [3]

2   Light travels at 186 000 miles per second. The Sun is 93 000 000 miles from Earth.
    Calculate how long it takes for light from the Sun to reach the Earth.
    Give your answer in minutes and seconds. [3]

3   a   1 inch is equivalent to 2.54 cm. How many inches are there in 1 m? [2]

    b   A pack of 8 batteries cost £6.50. How much do 12 batteries cost? [2]

    c   It takes 15 minutes to cook 8 omelettes. How long would it take
    to cook 28 omelettes? [2]

4   The Olympic sprinter Usain Bolt can run 100 m in 9.8 seconds.
    A wombat can maintain a speed of 40 km/h for 150 m.
    Who would win a 100 m race, Usain Bolt or a wombat?
    Show your workings. [4]

5   a   A statue has a mass of 3850 grams and volume of 529 cm³? What is its density? [2]

    b   A silver bar has a mass of 250 g and a density of 10.5 g/cm³. What is its volume? [2]

    c   A litre of milk has a density of 1.03 g/cm³. What is the mass of a litre of milk? [2]

6   A dog eats three tins of dog food in two days.

    a   How many tins does it eat in 30 days? Give your answer to the nearest tin. [2]

    b   Will 100 tins feed the dog for 67 days? Give your reasons. [2]

    c   A second dog eats twice as much as the first. Will 30 tins be enough to feed
    the two dogs for two weeks? Show your working. [4]

7   A cylindrical container of water emptied in $17\frac{1}{2}$ minutes. The container was full.
    The water flowed out at an average rate of 150 ml/second.

    a   How many litres of water did the container hold? [3]

    b   The base of the cylinder is 23 cm in diameter. How high is it? [3]

8 A cuboid has volume $9600 \, cm^3$. The density of the cuboid is $7.2 \, g/cm^3$.

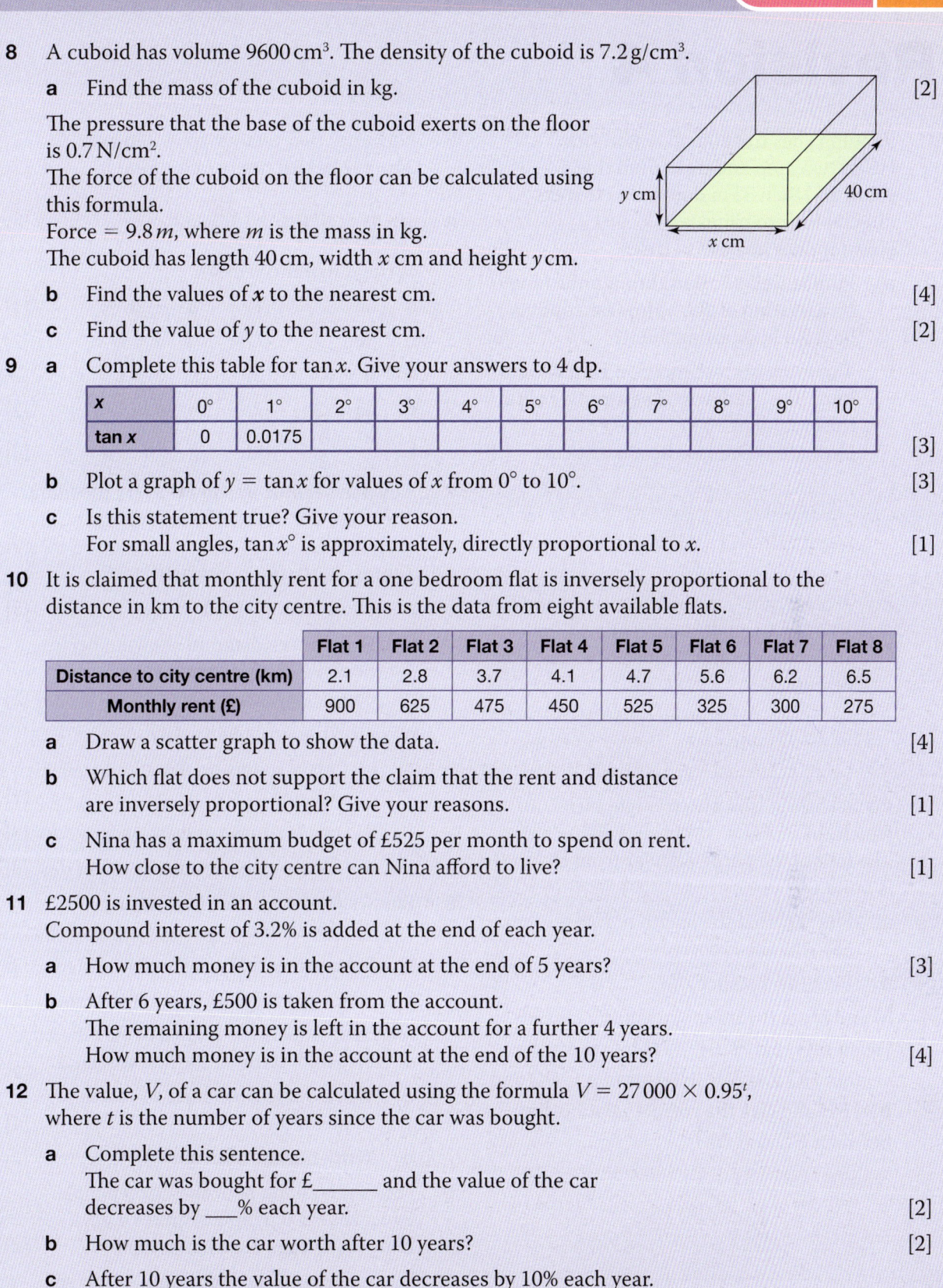

    a Find the mass of the cuboid in kg.   [2]

The pressure that the base of the cuboid exerts on the floor is $0.7 \, N/cm^2$.
The force of the cuboid on the floor can be calculated using this formula.
Force $= 9.8m$, where m is the mass in kg.
The cuboid has length 40 cm, width x cm and height y cm.

    b Find the values of x to the nearest cm.   [4]

    c Find the value of y to the nearest cm.   [2]

9   **a** Complete this table for $\tan x$. Give your answers to 4 dp.

x	0°	1°	2°	3°	4°	5°	6°	7°	8°	9°	10°
$\tan x$	0	0.0175									

                                                                      [3]

    b Plot a graph of $y = \tan x$ for values of x from 0° to 10°.   [3]

    c Is this statement true? Give your reason.
       For small angles, $\tan x°$ is approximately, directly proportional to x.   [1]

10 It is claimed that monthly rent for a one bedroom flat is inversely proportional to the distance in km to the city centre. This is the data from eight available flats.

	Flat 1	Flat 2	Flat 3	Flat 4	Flat 5	Flat 6	Flat 7	Flat 8
Distance to city centre (km)	2.1	2.8	3.7	4.1	4.7	5.6	6.2	6.5
Monthly rent (£)	900	625	475	450	525	325	300	275

    a Draw a scatter graph to show the data.   [4]

    b Which flat does not support the claim that the rent and distance are inversely proportional? Give your reasons.   [1]

    c Nina has a maximum budget of £525 per month to spend on rent.
       How close to the city centre can Nina afford to live?   [1]

11 £2500 is invested in an account.
Compound interest of 3.2% is added at the end of each year.

    a How much money is in the account at the end of 5 years?   [3]

    b After 6 years, £500 is taken from the account.
       The remaining money is left in the account for a further 4 years.
       How much money is in the account at the end of the 10 years?   [4]

12 The value, V, of a car can be calculated using the formula $V = 27\,000 \times 0.95^t$, where t is the number of years since the car was bought.

    a Complete this sentence.
       The car was bought for £_____ and the value of the car decreases by ____% each year.   [2]

    b How much is the car worth after 10 years?   [2]

    c After 10 years the value of the car decreases by 10% each year.
       How much will the car be worth after a further 5 years?   [3]

Revision 4

1 Ranulph goes trekking in Antarctica. He leaves camp and treks 6 km due south and then 3 km due west. He treks directly back to camp. How far does he have to trek? [3]

2 a Pumba walks 7.6 km through the jungle on a bearing of 200°. How far south is he from his starting point? [3]

b The sun casts a shadow on Dan 230 cm long. The shadow makes an angle with the ground of 35°. How tall is Dan, to the nearest cm? [2]

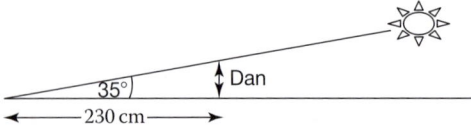

c A plane at P is coming into land and has 10 miles to go. It is currently 1.5 miles above the ground. Calculate its angle of descent, θ. [2]

3 Rachel has an isosceles triangle, ABC, in which $AB = AC = 15$ cm, and $BC = 8$ cm. The foot of the perpendicular from BC to A is point M.

a She says that the length of $BM = 4$ cm. Say why she is correct. [1]

b Find the distance AM. [3]

4 M and N are the midpoints of PQ and PR. Helen says that $\overrightarrow{RQ} = -\overrightarrow{NM}$. Express the vectors $\overrightarrow{RQ}$ and $\overrightarrow{NM}$ in terms of $\mathbf{a} = \overrightarrow{PQ}$ and $\mathbf{b} = \overrightarrow{PR}$ and give the correct relationship between $\overrightarrow{RQ}$ and $\overrightarrow{NM}$. [3]

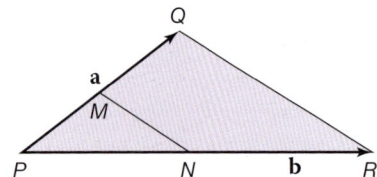

5 Louis has two vectors $a\mathbf{p} + 35\mathbf{q}$ and $24\mathbf{p} + 21\mathbf{q}$ that are parallel. Find a. [2]

6 $\mathbf{a} = \begin{pmatrix} 3 \\ -6 \end{pmatrix}$, $\mathbf{b} = \begin{pmatrix} -3 \\ 9 \end{pmatrix}$ and $\mathbf{c} = \begin{pmatrix} -3 \\ -3 \end{pmatrix}$.

Write down the vectors

a $5\mathbf{a}$ [1]   **b** $-2\mathbf{b}$ [1]

c $3\mathbf{c}$ [1]   **d** $\mathbf{a} - \mathbf{b}$ [2]

e $\mathbf{b} - \mathbf{a}$ [2]   **f** $\mathbf{a} + \mathbf{b} - \mathbf{c}$ [2]

g $4\mathbf{a} + 2\mathbf{c}$ [2]   **h** $\mathbf{c} - 12\mathbf{a}$ [2]

i $\dfrac{\mathbf{a}}{2} + \dfrac{\mathbf{b}}{3}$ [2]   **j** $2\mathbf{a} + \mathbf{b} + \mathbf{c}$. [2]

7 Ally plays tennis or squash every morning. These are the probabilities of choosing each sport.

Tennis   0.63   Squash P

a Show that $P = 0.37$ [2]

b Draw a tree diagram showing her possible choices over two consecutive days. [3]

c Find the probability Ally

   i plays tennis on day 1 and squash on day 2 [2]

   ii plays squash at least once [3]

   iii doesn't play tennis on either day. [3]

8 Ben writes a sequence that begins 2 6 12 20 .... and says that $T(n) = n^2 + n$.

a Is he correct? Show your working. [2]

b Work out the 50th and 100th terms. [2]

A second sequence is formed of the differences of each pair of terms in the first sequence.

c Work out the nth term of this new sequence. [4]

d Work out the 50th and 100th terms. [2]

9 James is cooking pancakes for 9 people. Pancakes for 6 people requires 4 eggs. How many eggs does James need? [2]

10 The supermarkets 'Liddi' and 'Addle' sell teacakes. 'Liddi' sells a pack of 5 for £1.35 and 'Addle' sells a pack of 4 for £1.10. Which supermarket offers the better value? [2]

11

Kilometres	Metres	Centimetres
Millimetres	Tonnes	Kilograms
Grams	Litres	Millilitres

Write down which of the above units you would use to measure the

a length of a trainer [1]

b mass of a suitcase [1]

c volume of a raindrop [1]

d thickness of a sheet of glass [1]

e capacity of a swimming pool [1]

f distance from Earth to Mars [1]

g mass of a mouse [1]

h length of a rugby pitch [1]

i mass of the London Eye. [1]

12 The diagram shows a pattern of black and white triangles.

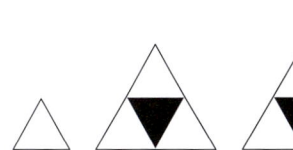

a Copy and complete the table. [5]

Rows (r)	1	2	3	4	5
White triangles	1	3			
Black triangles	0	1			
Total number of triangles	1	4			

b The formula for the number of white triangles is $n = \dfrac{r(r + 1)}{2}$.

Find the number of white triangles in a pattern with

  i 50 rows [2]   ii 100 rows. [1]

c Write down the formula for the total number of triangles in a pattern with r rows. [1]

d Use the formulae from parts **b** and **c** to derive a formula for the number of black triangles in any row. Give your answer in its simplest form. [3]

e Test your formula by working out the number of black triangles in row 2. [1]

12 f Use your formula to find the number of black triangles in a pattern with 60 rows. [1]

13 a Ellie says that there are 13 ml in 1.3 l. Correct her answer. [1]

b Lewis says that there are 76 g in 7.6 kg. Correct his answer. [1]

14 Norwich to Harwich is 72 miles.

a Richard drives this journey in 108 minutes. What is his average speed? Give your answer in mph. [2]

b Jeremy drove this journey at an average speed of 50 mph. How long, to the nearest minute, did his journey take? [2]

c Jeremy drove the return journey at an average speed 4 mph faster than his outward journey. How many minutes did he save on the return journey? [3]

15 a An object has a mass of 1.26 kg and volume 180 cm³. What is its density? [2]

b A cylindrical metal rod with radius 3.5 cm and length 12.5 cm has a density of 11.4 g/cm³. What is the mass of the bar? [3]

c A silver bar has a mass of 30 g and density 10.5 g/cm³. What is its volume? [2]

16 A team of 4 scaffolders can build 6 m of scaffolding in 4 hours.

a How many metres of scaffolding could the team build in 7 hours? [2]

b How long would it take a team of 8 scaffolders to build 6 m of scaffolding? [1]

c How long would it take a team of 12 scaffolders to build 18 m of scaffolding? [3]

d Gina says that the size of the team and the size of the scaffolding are directly proportional to each other. Is Gina correct? Give a reason for your answer. [1]

e Gina says that the size of the team and the time taken to build 10 m of scaffolding are directly proportional to each other. Is Gina correct? Give a reason for your answer. [1]

Formulae

Make sure that you know these formulae.

Circles

Circumference of a circle $= 2\pi r = \pi d$

Area of a circle $= \pi r^2$

Pythagoras' theorem

In any right-angled triangle
$$a^2 + b^2 = c^2$$

Trigonometry formulae

In any right-angled triangle $\quad \sin A = \dfrac{a}{c} \quad \cos A = \dfrac{b}{c} \quad \tan A = \dfrac{a}{b}$

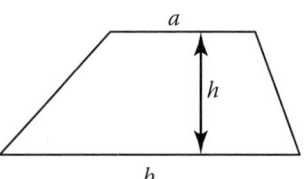

Make sure that you know and can derive these formulae.

Perimeter, area, surface area and volume formulae

Area of a trapezium $= \frac{1}{2}(a+b)h$
Volume of a prism $=$ area of cross section $\times$ length

Compound interest

If P is the principal amount, r is the interest rate over a given period and n is number of times that the interest is compounded then

Total accrued $= P\left(1 + \dfrac{r}{100}\right)^n$

Probability

If P(A) is the probability of outcome A and P(B) the probability of outcome B then P(A or B) = P(A) + P(B) − P(A and B)

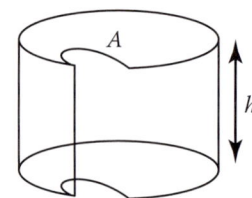

Make sure that you can use these formulae if you are given them.

Perimeter, area, surface area and volume formulae

Curved surface area of a cone $= \pi rl$

Surface area of a sphere $= 4\pi r^2$

Volume of a sphere $= \frac{4}{3}\pi r^3$

Volume of a cone $= \frac{1}{3}\pi r^2 h$

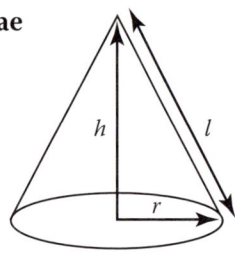
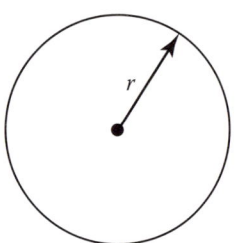

Kinematics formulae

If a is constant acceleration, u is initial velocity, v is final velocity, s is displacement from the position when $t = 0$ and t is time taken then

$$v = u + at \qquad\qquad s = ut + \tfrac{1}{2}at^2 \qquad\qquad v^2 = u^2 + 2as$$

Key phrases and terms

Use our handy grid to understand what you need to do with a maths question.

Circle	Draw a circle around the correct answer from a list.
Comment on	Give a judgement on a result. This could highlight any assumptions or limitations or say whether a result is sensible.
Complete	Fill in any missing information in a table or diagram.
Construct	Draw a shape accurately.
... using ruler and compasses	Compasses must be used to create angles. Do not erase construction lines.
Criticise	Make a negative judgement. This could highlight any assumptions or limitations or say why a result is not sensible.
Describe	Usually describing a graph, one mark per descriptive sentence. When describing transformations give all relevant information.
Diagram not accurately drawn	Calculate any angles or sides, do *not* measure them on the diagram.
Draw	Accurately plot a straight line, bar chart or transformation.
Draw and label	Draw... and mark on values.
Estimate	Simplify a calculation, by rounding, to find an approximate value
Expand	Multiply out brackets.
Factorise fully	Put in brackets with the highest common factor outside the brackets.
Give a reason	Write an explanation of your argument, including the information you use and your reasons or rules used.
Give the exact value	Do not use rounding or approximations in your calculations. You should give your answers as fractions, surds and multiples of π.
Give your answer in terms of π.	Do not use a numerical approximation to π, instead treat it as a letter (algebraic variable).
Give your answer in its simplest form/ simplify	Collect all like terms, any fractions or ratios should be in lowest terms.
Give your answer to an appropriate degree of accuracy	For example, if the numbers in the question are given to 2 decimal places, give your answer to 2 decimal places.
Measure	Use a ruler or protractor to accurately measure lengths or angles.
Rearrange	Change the subject of a formula.
Shade	Use hatching to indicate an area on a diagram.
Show	Present information by drawing the required diagram.
Show that	Obtain a required result showing each stage of your working.
Show working	Usually asked for to support a decision.
Sketch	Represent using a diagram or graph. This should show the general shape and any important features, such as the position of intercepts, using labels as necessary. It does not require an accurate drawing.
Solve	Find an answer using algebra or arithmetic. This often means find the value of x in an equation.

State	Write one sentence answering the question.
Tick a box	Choose the correct answer from a list.
Work out	Find the answer showing your working.
Write down	Write down your answer, written working out is not usually required.
Use an approximation	Estimate using rounding.
Use a line of best fit	Draw a line of best fit and use it.
Use	Use supplied information or the preceding result(s) to answer the question.
You **must** show your working.	Marks will be lost for not writing down how you found the answer to the question.

Chapter 1

Check in 1

1   304
2   **a** 70   **b** −12   **c** 5
3   −8, −5, −3, −1, 2, 4

1.1S

1   **a** 87   **b** 143   **c** 406   **d** 460
    e 2053   **f** 8503   **g** 8530   **h** 34640
    i 30464   **j** 206503
2   **a** Four hundred and fifty-six
    b Thirteen thousand, two hundred
    c One hundred and fifteen thousand and twenty
    d Four hundred and sixty thousand, three hundred and forty
    e Four million, three hundred and twenty-five thousand, four hundred
    f Fifty-five million, six hundred and seventy thousand, three hundred and forty-five
    g Forty-five point eight
    h Three hundred and sixty-seven point zero three
    i Four thousand, five hundred and three point three four
    j Two thousand, seven hundred point zero two
3   **a** 9, 34, 56, 89, 112, 139, 178
    b 1784, 1990, 2372, 2386, 3022, 3233
    c 4005, 4555, 40500, 40545, 44054, 45045
    d 24445, 42024, 204044, 240440, 242404, 245004
4   **a** 5.007, 5.099, 5.103, 5.12, 5.2
    b 0.5, 0.509, 0.525, 0.545, 0.55
    c 7.058, 7.302, 7.35, 7.387, 7.403
    d 0.4, 0.42, 2.4, 4.2, 42
    e 26.9, 26.97, 27.06, 27.1, 27.6
    f 13.19, 13.3, 13.43, 14.03, 14.15
5   **a** =   **b** ·   **c** ·   **d** =
6   **a** <   **b** >   **c** <   **d** <
    e >   **f** <
7   **a** 120   **b** 400   **c** 32   **d** 46
    e 300   **f** 46   **g** 2.3   **h** 65.9
    i 34000   **j** 356   **k** 2.36   **l** 34.5
8   **a** −13, −12, −6, 0, 15, 17   **b** −8, −7, −6, −5, −3, 0
    c −5, −2, 1, 2, 3, 4   **d** −8, −3, −1.5, 2, 3, 9
    e −5, −4.5, −3, −2, 2, 3   **f** −9, −1, 2, 3, 6, 8
    g −4.5, −3, −2.5, −1, 0, 5.5
    h −6, −5.8, −5.7, −5.4, −5.1, −5
9   **a** 16   **b** −7   **c** 4   **d** 37
    e 17   **f** −7   **g** 9   **h** −11
    i −8   **j** 21   **k** −8   **l** −2
    m −18   **n** 4   **o** −5   **p** 2
    q −20   **r** 10   **s** −8   **t** −28

1.1A

1   £3.60
2   **a** 1mm
    b **i** 5.1cm   **ii** 3.6cm   **iii** 12.35cm
    c **i** 5.2cm   **ii** 3.7cm   **iii** 12.45cm
3   **a** 11.06s   **b** 12s   **c** 11.03s
4   **a** 3415   **b** The hit took 3300 points off.
5   **a** Money raised = Donation ÷ 10
    b **i** £405.30   **ii** £102.20   **iii** £9.80   **c** £2641.76
*6 **a** **i** £27.30   **ii** £0.92   **iii** £188.30   **iv** £72
    b 1000   **c** £1

1.2S

1   **a** 50   **b** 90   **c** 480   **d** 790
    e 2640   **f** 6190
2   **a** 300   **b** 500   **c** 900   **d** 2700
    e 5700   **f** 16500

3   **a** 3000   **b** 3000   **c** 5000   **d** 37000
    e 63000   **f** 262000
4   **a** **i** 3000   **ii** 3500   **iii** 3470
    b **i** 81000   **ii** 81400   **iii** 81380
    c **i** 1000   **ii** 1200   **iii** 1240
    d **i** 0   **ii** 300   **iii** 280
    e **i** 14000   **ii** 14000   **iii** 14000
    f **i** 10000   **ii** 10000   **iii** 10000
5   **a** 5   **b** 4   **c** 12   **d** 25
    e 16   **f** 436   **g** 4   **h** 9
    i 19   **j** 69   **k** 110   **l** 7
6   **a** 0.3   **b** 2.9   **c** 3.8   **d** 12.5
7   **a** **i** 0.33   **ii** 0.3
    b **i** 2.87   **ii** 3
    c **i** 3.80   **ii** 4
    d **i** 14.46   **ii** 10
8   **a** **i** 3490   **ii** 3500   **iii** 3000
    b **i** 3390   **ii** 3400   **iii** 3000
    c **i** 14850m   **ii** 14900m   **iii** 15000m
    d **i** £57790   **ii** £57800   **iii** £58000
    e **i** 92640kg   **ii** 92600kg   **iii** 93000kg
    f **i** £86190   **ii** £86200   **iii** £86000
    g **i** 3440   **ii** 3400   **iii** 3000
    h **i** 74900   **ii** 74900   **iii** 75000
9   **a** 4   **b** 29   **c** 469   **d** 369
    e 20   **f** 27   **g** 101   **h** 0
10 **a** **i** 3.447   **ii** 3.45   **iii** 3.4
    b **i** 8.948   **ii** 8.95   **iii** 8.9
    c **i** 0.128   **ii** 0.13   **iii** 0.1
    d **i** 28.387   **ii** 28.39   **iii** 28.4
    e **i** 17.999   **ii** 18.00   **iii** 18.0
    f **i** 10.000   **ii** 10.00   **iii** 10.0
    g **i** 0.004   **ii** 0.00   **iii** 0.0
    h **i** 2785.556 **ii** 2785.56   **iii** 2785.6
11 **a** **i** 8.37   **ii** 8.4   **iii** 8
    b **i** 18.8   **ii** 19   **iii** 20
    c **i** 35.8   **ii** 36   **iii** 40
    d **i** 279   **ii** 280   **iii** 300
    e **i** 1.39   **ii** 1.4   **iii** 1
    f **i** 3890   **ii** 3900   **iii** 4000
    g **i** 0.00837   **ii** 0.0084   **iii** 0.008
    h **i** 2400   **ii** 2400   **iii** 2000
    i **i** 8.99   **ii** 9.0   **iii** 9
    j **i** 14.0   **ii** 14   **iii** 10
    k **i** 1400   **ii** 1400   **iii** 1000
    l **i** 140000   **ii** 140000   **iii** 100000
12 **a** 17.0   **b** 2300   **c** 7.0   **d** 10.61
    e 14624.000 **f** 0.050   **g** 100   **h** 10
    i 95000   **j** 9999.6

1.2A

1   **b**, because he rounded after the calculation.
2   **a** CON 26000   LIB 3000
        LAB 26000   UKIP 6000
    b CON and LAB would have the same votes.
    c 100
3   **a** £263
    b She would give £263 − £262 = £1 less.
4   $\sqrt{200} \approx 14.14 \, (2\,\text{dp}) \approx 14.1 \, (3\,\text{sf})$
    $\frac{1}{6} \approx 0.17 \, (2\,\text{dp}) \approx 0.167 \, (3\,\text{sf})$
5   **a** 17.80m²   **b** 18m²
    c **b** It is easier to estimate with this value.
    d **b** It is better to have too much than too little.
*6 **a** 27000 (3 sf)   **b** 100
*7 You would pay more.

1.3S

1 a 270   b 80   c 330   d 40
   e 3800   f 70   g 430   h 390
2 a 8   b −2   c 9   d 0
   e −13   f −14   g −14   h 20
   i 12   j 3   k −5   l −6
   m 4   n −25   o −1   p 1
   q −9   r −23   s −23   t −23
3 a 355   b 560   c 950   d 808
   e 567   f 889   g 14.3   h 20.7
4 a 15   b 15   c 131   d 124
   e 161   f 202   g 48.6   h 48.3
   i 28.6   j 45.2   k 6.2   l 7.4
   m 18.4   n 25.8   o 131   p 141
   q 202   r 248
5 a 8.12   b 16.52   c 23.31   d 104.19
   e 13.08   f 13.13   g 18.84   h 18.88
   i 26.27   j 56.92   k 3.52   l 36.72
   m 27.98   n 18.88
6 a 36.8   b 78.3   c 27.1   d 42.1
   e 6.2   f 7.1   g 8.5   h 27.9
7 a 10.72   b 19.72   c 18.11   d 141.99
   e 8.05   f 4.73   g 38.74   h 18.68
   i 34.06   j 44.71   k 3.51   l 24.59
   m 17.98   n 28.89
8 a 22.08   b 47.34   c 83.17
*9

```
   6 7 . ¹3 6 ¹2
 − 2 ²1̶ . ⁶5̶ 5̶ 4
   4 5 . 9 0 8
```

1.3A

1 23
2 a

```
    2 1 . 7
    9 1 . 5
  + 9   4 . 9
  2 0 8 . 1
        ₂
```

   b

```
   ²3̶ ¹⁵6̶ . ¹⁴5̶ ¹8
 − 2 7 . 7 9
    8 . 7 9
```

3 119.4, 57.97, 26.09, 35.37, 19.15, 3.45
4 a $a + b + c$ b $a + c$ c $b + 2c$
5 a 11 688   b 18 340   c 292
6 a Bryony 12 653, Callum 7350, Edward 7000, Asha 6523, Dora 964, Total 34 490
   b 11 694
*7 a 342 and 1026   b 742 and 1484
   c 342 and 432, 522 and 1026
8 0.2 kg

1.4S

1 a −15   b −18   c −21   d −56
   e −36   f −12
2 a −25   b −32   c −72   d −20
   e +30   f +49   g +16   h −20
3 a −2   b −5   c +5   d +4
   e −22   f −1   g +40   h +4
4 a 98   b 152   c 273   d 323
5 See question **4**.
6 a 24.91   b 4.284   c 105.84   d 130.8985
   e 42.9442   f 369.5328
7 a 3.87   b 0.775   c 0.916   d 7.53
   e 18.13   f 3.45   g 4.15   h 7.74
   i 4.08   j 2.35
8 a 5.26   b 28.88   c 1384.29   d 175.56
   e 28.65   f 111.51
9 See questions **6–8**.
10 a 28.81   b 28.81   c 4.3   d 0.067
   e 10

11 a 26   b 37   c 52   d 10
   e 33   f 5   g 180
12 a 28   b 72   c 5   d 16
   e 2
13 a 1   b 2   c 2   d 14
   e 40   f 7
14 a 14   b 10   c 2   d 12
   e 91   f 112   g 70   h 37
   i 1   j 3
15 a 170   b 0.58   c 16

1.4A

1 a 136   b 72
2 a £24   b £312
   c Other months have 30 or 31 days (more than 4 weeks).
3 a 24 bags   b £113
   c It costs 50 p less.
4 a 40 graph packs, 48 lined packs   b £1052
5 a 6 packs of small, 13 packs of other sizes   b £747.85
6 a

```
        6 4 7
      ×     4 9
      5 8 2 3
  + 2 5 8 8 0
    3 1 7 0 3
```

   b

```
              5 9 1
    4 9 ) 2 8 9 5 9
        − 2 4 5
            4 4 5
          − 4 4 1
              4 9
            − 4 9
                0
```

7 £31.40

Review 1

1 a 6700   b 85.2   c 240   d 5
2 a 45   b 6.21   c 0.079   d 0.006
3 a 905 < 961   b 14.7 < 14.9
   c 0.7 > 0.09   d 0.214 < 0.22
4 a 53 099, 53 909, 503 099, 503 909, 530 909
   b 4.09, 4.289, 4.29, 4.3, 4.32
   c −14, −8, −4, 0, 9
5 a 850   b 25   c 0.8   d 62.94
6 a 400   b 5100   c 45.7   d 0.08
   e 0.090   f 1
7 a −1   b −5   c 6   d 2
   e 0   f −3
8 a 1358   b 38.4   c 914.07   d 7.401
9 a 513   b 268   c 219.2   d 3.67
10 a −21   b 32   c −5   d 13
11 a 240   b 400   c 40   d 8
12 a 2961   b 14 976   c 28   d 26
13 a 17.92   b 11.637   c 2.45   d 5.65
14 a 33   b 17   c 55   d 45
   e 18   f 14
15 a 36   b 2

Assessment 1

1 a −33 < 8; −33, 8, 19, 44, 303, 576
   b −576 < −19; −576, −19, 8, 33, 44, 303
2 Yes, 0.42, 3, 4.236, 51.6, 4200, 216 000.
3 a −2, −2, −4   b −16, −25, −39
4 a 24   b 7, 12, 5; 6, 8, 10; 11, 4, 9
5 a No, lower limit = 271.75 cm < 271 cm.
   b No, Yao Defen upper limit = 233.345 cm > 233.341 cm.
6 a Yes   b No, Dave 40, Jane 50 (1 sf).
   c Dave 45, Jane 53
7 a Abena 13 000 (2 sf), Edward 8100 (2 sf)
   b 15 000 km (14 522.324)
   c The estimate would be 0 because 12 756, 8134 = 10 000 (1 sf).
8 a C   b B   c B   d A

9 A, C, F, E, H, B, D, G
10 2.125 km
11 a 25 sweets **b** 25 000 (nearest 1000)
12 a $(3 + 4) \times 5 + 2 = 37$ **b** $60 \div (5 + 7) + 5 = 10$

Chapter 2

Check in 2

1 a 9 **b** 4 **c** 16
2 15
3 a 7 **b** -1 **c** 2 **d** -3
  e -6 **f** -8 **g** -2 **h** 4
4 a i 1, 2, 3, 6, 9, 18 **ii** 1, 2, 3, 4, 6, 12
   iii 1, 2, 3, 4, 6, 8, 12, 24
  b 1, 2, 3, 6
  c 6

2.1S

1 $n + 2$ A number add 2
   $n - 2$ A number subtract 2
   $2n$ A number multiplied by 2
   $\frac{n}{2}$ A number divided by 2
   $2 - n$ 2 take away a number
2 a $x + 4$ **b** $y + 6$ **c** $p - 5$
  d $d - 7$ **e** $7 - x$ **f** $7 - x$
3 a $4x$ **b** $6y$ **c** $5p$
  d $7d$ **e** $2f$ **f** $3y$
4 a $\frac{x}{4}$ **b** $\frac{y}{6}$ **c** $\frac{p}{5}$ **d** $\frac{x}{2}$
  e $\frac{x}{2}$ **f** $\frac{3z}{4}$
5 $4x + 2$ x multiplied by 4 add 2
   $2x + 4$ x multiplied by 2 add 4
   $2(x + 4)$ x add 4 all multiplied by 2
   $4(x + 2)$ x add 2 all multiplied by 4
6 a No, $m^2 = m \times m$ and $2m = 2 \times m$.
  b No, b is a variable and can take many values.
  c No, $7y$ means '7 times y', $7 \times 3 = 21$.
  d No, $p^2 = p \times p$, $3^2 = 9$.
7 a 14 **b** 17 **c** 15 **d** 13
  e 8 **f** 2 **g** 2 **h** -2
8 a 24 **b** 72 **c** 48 **d** 96
  e 120 **f** 240 **g** 6 **h** 4
  i 2 **j** 10
9 a i 12 **ii** 4
  b i -6 **ii** 6
  c i 6 **ii** -6
  d i 27 **ii** -5
  e i $\frac{1}{3}$ **ii** -5
  f i 3 **ii** $-\frac{1}{5}$
10 a 16 **b** 25 **c** 100 **d** 144
  e 0 **f** 4 **g** 25 **h** 100
11 $2x^2$ x squared multiplied by 2
   $(2x)^2$ x multiplied by 2 squared
   $(x + 2)^2$ x plus 2 squared
   $x^2 + 2^2$ x squared plus 2 squared
12 $P = a + b + a + b = 2a + 2b$

2.1A

1 n oranges at 5p each, $5n$; $4m$, $m \times 4$;
   $3 \times n$, 3 chews at n pence each; $n + n$, $2n$; $7 \times m$, $7m$;
   m toys at £3 each, $3m$; $6n$, 6 stamps at n pence each
2 Carla, £27 < £30
3 a No, it costs £20 per day. **b** £230 **c** 11 days
4 CHOCOLATE
5 $2x - 3$
6 $4x + 8$
7 $6c + 10d$

8 a $50f + 30g$ **b** $80j + 40k$ **c** $50x + 60y + 30z$
  d $60p + 80q + 40r$
9 Paul, $2 \times 6^2 = 2 \times 36 = 72$
10 a Cerys, yes. Audrey, no, she has 3 times less than Billie.
  b £40

2.2S

1 a $4d$ **b** $3q$ **c** $8a$ **d** $12y$
  e $9t$ **f** $10e$ **g** $2 + 2x$ **h** $4y + 9$
2 No, $k + k + k = 3k$.
3 a $4d$ **b** $3q$ **c** $7a$ **d** $4y$
  e t **f** 0 **g** $3k + 4$ **h** $6y + 6$
  i $7 - 4p$ **j** $8b + 4d - 12 + 4bd$
4 Students' answers, for example, $6m + 6m$, $4m + 8m$, etc.
5 No, $7d + 7e$, you can't combine 'd' and 'e' terms.
6 a $2a + 2b$ **b** $6t + 3s$
  c $d + 6e$ **d** $6y + 8w$
  e $5p + 4f$ **f** $e + 2f + 3g + 4h$
  g $7k + 4 + 2d$ **h** $10y + 16$
  i $5p + 4 + 2q$ **j** $8b + 5 - 2a + 6ab$
7 a $7a - 2b$ **b** $6t - 3s$
  c $4d - 5e$ **d** $2y + 5w$
  e $2f - 3p$ **f** $2e + 2f - 3g$
  g $4 + 2d - 3k$ **h** $6b - 5a + 11$
8 a $8ab$ **b** $3st$
  c $5d - 3e - de$ **d** $10wy - 3y$
  e $5fp - 9p + 6f$ **f** $2e + 2ef + 3fg + 4gh$
9 No, $a^2 + a^2 + a^2 = 3a^2$.
10 a $2d^2$ **b** $4b^2$
  c $9a^3$ **d** $3y^2 + 4y$
  e $3t^3$ **f** $3k^3 + 4$
11 a $2ad^2$ **b** $13ab^2$
  c $9xy^3$ **d** $3xy^2 + 4xy$
  e $5s^3t - 2st^3$ **f** $8y^2 - 5y - 2x + 11x^2$
12 Students' answers, for example, $2a^2 + b^2 + b^2 + b^2$, etc.
13 a $\frac{2}{3}x$ **b** $\frac{1}{2}x$ **c** $\frac{1}{3}a + \frac{1}{2}b$ **d** $\frac{1}{4}x + \frac{1}{4}y$
14 a $3a$ **b** $3a$ **c** $3ab$
  d $3ab$ **e** $6p$ **f** $4a^2$
  g $\frac{3}{a}$ **h** $\frac{a}{3}$ **i** $\frac{5}{p}$
  j $\frac{d}{2}$ **k** $11q$ **l** $-14r$ **m** $20t - s$
*15 a $\frac{4b}{a}$ **b** $2b$ **c** $1 + 3b$

2.2A

1 a No, both have $3x + 2y$. **b** $6x + 4y$
2 $3x + 5y - x + 2y = 3y + 7x + 4y - 5x = 7y + 2x$,
   $2x - 4y + 3x + 2y = 5x - 2y = 2x - 4y + 2y + 3x$,
   $2x + 4y - x = 3x + 6y - 2x - 2y = 4x + 4y - 3x$,
   $2y + 3x - x + 3y = 2x + 5y = 7y - 3x + 5x - 2y$
3 Abdul
4 a $7a = 3a + 5a - a$
  b $3x + 2y = 2x + 3y + 2x - 4y - (x - 3y)$
  c $20a + 4b = 3a \times 6 + 2a + 10b \div 2 - b$
5 a i $8(p + 2)$ **ii** $32p$ **b** $3x$ by $2y$
  c Students' answers, for example, $2(4a + b)$.
6 $17xy$ mm^2
7 a $2x + 6y$ **b** Rectangle, $10y + 8x > 10y + 7y$.
  c $4y + 3x$

2.3S

1 a 3^4 **b** 2^6 **c** $4^2 \times 5^3$ **d** $3^3 \times 2^3$
  e p^3 **f** p^3q^2 **g** 2^3r^4 **h** $3^2s^3t^2$
2 a 125 **b** 8 **c** 1
  d 36 **e** -1 **f** $\frac{1}{4}$
3 a 3^6 **b** 2^9 **c** 5^7
  d 6^7 **e** 7^9 **f** 11^{10}

4 a 5^3 **b** 3^4 **c** 8^5
  d 2^7 **e** 6^{-4} **f** 2^2
5 a 2^6 **b** 5^{24} **c** 3^{21}
  d 8^{16} **e** 7^{-8} **f** $\left(\frac{1}{4}\right)^6$
6 See questions 2–5.
7 No, $a^5 \times a^2 = a^7$.
8 a a^6 **b** y^{10} **c** b^8
  d p^{14} **e** h^{12} **f** s^3t^9
9 No, $4y^5 \times 2y^2 = 8y^7$.
10 a $3x^7$ **b** $5y^7$ **c** $12b^8$
  d $10p^{11}$ **e** $30h^{11}$ **f** $12s^3t^4$
11 No, $12p^{12} \div 3p^4 = 4p^8$.
12 a $2y^4$ **b** $2a^6$ **c** $5k^4$
  d $3p^6$ **e** $5x^5$ **f** $\frac{x^8}{2y^4}$
13 a a^6 **b** y^{12} **c** k^{15}
  d p^{56} **e** a^{21} **f** a^{21}
14 a $4a^6$ **b** $729y^{12}$ **c** $25k^6$
  d $216p^{21}$ **e** $128a^{21}$ **f** $256a^{16}$
15 a y^2 **b** x^{-2} **c** a^{-6}
  d h^{-6} **e** p^4 **f** p^{-1}
16 a b^{-6} **b** k^{-12} **c** q^{27}
17 a $12y$ **b** $6a^{-7}$ **c** $9b^{-5}$ **d** $\frac{1}{2}k$
18 a $8a^{12}$ **b** m^7 **c** $4y^6$ **d** $9y^4$

2.3A

1 $5n = 5 \times n$, $2n^3 = 2 \times n^3$, $n^2 = \frac{n^4}{n^2}$, $2n^2 = 2 \times n \times n$
2 $8x^3y^6\,\text{cm}^3$
3 a $32x^2y^3$ **b** $19a^2b^4$
4 B and C
5 A and C
6 $24p^2q$
7 $4ab^3$
8 a $p^5q^4 = (pq)^3 \times p^2q$ **b** $4x^3y^3 = xy \times (xy + xy)^2$
9 $30w^9$
10 a x^2 **b i** $\frac{1}{x^2}$ **ii** x^{-2}
  c $\frac{1}{x^2} = x^{-2}$
11 a 2   **b** 3   **c** 3   **d** 7
  e 5   **f** 5   **g** 4   **h** 3
*12 x^{40}
*13 a False, $(xy^2)^3 = x^3y^6$ **b** False, $(2xy^2)^3 = 8x^3y^6$
  c True   **d** False, $(x^2 + x \times x)^2 = 4x^4$

2.4S

1 a $4y + 8$ **b** $6b + 42$ **c** $7y + 21$
  d $12d + 60$ **e** $3t + 24$ **f** $\frac{w}{2} + 5$
2 No, $5(x + 4) = 5x + 20$.
3 a $-4x - 20$ **b** $-6b - 18$ **c** $-7t - 14$
  d $-3d - 24$ **e** $-10t - 80$ **f** $-8w - 72$
4 a $-3x + 15$ **b** $-2b + 16$ **c** $-7t + 56$
  d $-7d + 70$ **e** $-81 + 9t$ **f** $-48 + 8w$
5 a $y^2 + 2y$ **b** $b^2 + 7b$ **c** $y^2 + 3y$
  d $d^2 + 5d$ **e** $t^2 + 8t$ **f** $w^2 + 9w$
6 No, $x(x + 4) = x^2 + 4x$.
7 a $y^3 - 2y$ **b** $b^2 - 6b$
  c $3y^2 + 9y$ **d** $2d^2 - 10d$
  e $7t^2 - 56t$ **f** $63w - 9w^2$
  g $a^2b + 5ab^2$ **h** $\frac{s^2t}{2} + \frac{3st^2}{4}$
8 a $9x + 42$ **b** $13y + 9$ **c** $13t + 6$
  d $13p - 43$ **e** $8 - b$ **f** $9m - 34$
9 a 2   **b** 3   **c** 10
  d m **e** $2s$ **f** $4z$
10 $5(4x + 7)$
11 $6(x + 2)$
12 a $4(p + 2)$ **b** $5(y + 2)$ **c** $3(d + 7)$
  d $9(k + 8)$ **e** $6(b + 4)$ **f** $6(w + 9)$

13 Students' answers, for example, $7p + 2$, $4a + 3b$, etc.
14 No, $12p + 20pq = 4p(3 + 5q)$.
15 a $p(1 + 8t)$ **b** $y(1 + 6x)$ **c** $d(1 + 7b)$
  d $t(1 - 12s)$ **e** $6b(1 + 4c)$ **f** $6w(1 + 9y)$
  g $8b(2a - 5)$ **h** $15(q - 3p)$
16 a $p(p + 8)$ **b** $y(1 + 6y)$ **c** $w(1 + 4w^2)$
  d $ab(1 - 4b)$ **e** $6b(b + 4b)$ **f** $12xyz(x + 3y)$
  g $fc(mu + r)$ **h** $me(icky + ous)$
17 $4(x + 3) = 4x + 12$, $4x^2 - 3x = x(4x - 3)$, $3(x - 4) = 3x - 12$,
  $4x + 3x^2 = x(4 + 3x)$

2.4A

1 $3n + 15$
2 a n, $n + 4$, $3(n + 4)$, $4n$, $4n - 2$, $8n - 4$
  b $5n - 16$
3 a Kate
  b Debbie, incorrect coefficient of x^2, incomplete factorisation.
    Bryn, too many powers of x, incomplete factorisation.
4 a 4   **b** 6, 10   **c** 3   **d** 4, 4
  e 4, 5   **f** 6, 4   **g** 3, 1   **h** 3, 4
5 a $3(2x - 1)$ **b** $6x - 3$ **c** $6x - 3 = 15$, so $6x - 18 = 0$
6 a Students' answers, for example, $8(3x + 2)$.
  b Students' answers, for example, $(4x + 6) + (20x + 10)$.
7 a $4(x - 1)$ **b** $20(b + 2)$
8 $4(3x + 2) - 2(2x - 1) = 12x + 8 - 4x + 2 = 8x + 10$
  $= 2(4x + 5)$
9 Students' answers, for example, $6\,\text{cm} \times (2ab^2 + 3a^2b)\,\text{cm}$.
10 Students' answers, for example, $5\,\text{cm} \times x\,\text{cm} \times (4y + 10xy)\,\text{cm}$.
11 a Students' answers, for example, $2\,\text{cm} \times x^2\,\text{cm} \times 8y - 27\,\text{cm}$.
  b $16x^2y - 54x^2 > 0$, $y > \frac{54x^2}{16x^2}$, so $y > 3.5\,\text{cm}$
*12 $\frac{1}{2} \times 2y(y + y + 2) = 2y^2 + 2y = 2y(y + 1)$

Review 2

1 a $13y$ **b** $7xy$ **c** $3x$ **d** y^3
  e $\frac{1}{2}x$ **f** $5xy$
2 a £1.35   **b** $15y$
3 a 35   **b** 15   **c** -4 **d** 49
4 a -28 **b** -12 **c** 32   **d** 18
  e 1   **f** 34
5 a 6   **b** -4 **c** -2 **d** 2
  e -2 **f** 6
6 a $3x - 2 = 7$ **b** $S = \frac{D}{T}$
  c $5x - y$
7 a $5f, 6g, -2h$ **b** $5p, -6q, q^2$
8 a $5ab$ **b** $4c$ **c** d^3 **d** $10fg$
  e $8e - 15$ **f** $a^3 + a^2 + 3$
9 a $2r$ **b** $5a + 6b$ **c** $5d - 7e$ **d** $3x^2 + x$
10 a c^6 **b** d^5 **c** r^{12}
  d t^6 **e** $6u^8$ **f** $3v$
11 a $2a + 2$ **b** $32b - 16c$
  c $-15d + 20$ **d** $h^2 + 2h$
12 a $7(2b + 1)$ **b** $4(2 - c)$ **c** $3x(x + 2)$ **d** $4b(a - 3)$

Assessment 2

1 They are both right, $c + c + c + c$ means add c to itself four
  times which is the same as $4c$.
2 a $V - 4$ years old   **b** $\frac{1}{2}V$ or $\frac{V}{2}$ years old
  c $\frac{1}{2}V + 5$ years old   **d** $V + \frac{1}{2}V = \frac{3V}{2}$ years old
3 a The DVD costs £8.
  b The Blu-ray is £7 more expensive than the DVD.
  c The book is half the cost of the DVD.
  d The DVD and the Blu-ray together cost £23.
  e The total cost of all three items is £27.
4 a $2l + 2w$
  b $2(l + 5) + 2(w - 5)$
  c Yes, $2(l + 5) + 2(w - 5) = 2l + 10 + 2w - 10 = 2l + 2w$.

5 **a** 157 cm²   **b** 3.58 m   **c** 3.27 in

6 Fiona is correct, $9x$ and $4x$ are like terms and 2 is not.

7 **a** $C = 62S + 93L$
   b $C = 4 \times 62 + 2 \times 93 = 248 + 186 = 434$ pence = £4.34
   c £5.89
   d £20.77
   e 558p or £5.58

8 $10z^3$

9 **a** $2y^2 + 80y$
   b No, $V = 20y^2 = 200 \text{ cm}^3$, $y^2 = 10$ so $y = \sqrt{10}$.

10 **a** $5(W - 2)$ **b** $4(2W + 1)$

11 No, $(10 - w) + (3w - 5) = w$, $w = -5$, w cannot be negative as it is a length.

12 $2p(16p - 11)$

13 **a** No, $DC = a + 13$.
   b $2(2a + 13)$
   c Area of a rectangle = height × width.
      Area of $AEFD = a(a + 4)$, area of $EBCF = 9a$.

Chapter 3

Angle rules are abbreviated in the following way.

VO	Vertically opposite angles are equal.
CA	Corresponding angles are equal.
AA	Alternate angles are equal.
IA	Interior angles sum to 180°.
ASL	Angles on a straight line sum to 180°.
AP	Angles at a point sum to 360°.
AST	The angle sum of a triangle is 180°.
ASQ	The angle sum of a quadrilateral is 360°.

Check in 3

1 **a** 33 mm   **b** 3.3 cm
2 **a** ≈ 40°   **b** ≈ 130–140°
3 **a** 180°   **b** 270°   **c** 90°   **d** 180°

3.1S

1 **a** 280° reflex   **b** 160° obtuse
   c 98° obtuse   **d** 75° acute
2 **a** **i** 131°   **ii** 25°   **iii** 38°
   b $p = 131°$
3 **a** $u = 108°$ (VO), $v = 72°$ (ASL), $w = 72°$ (VO)
   b $x = 34°$ (VO), $y = 146°$ (ASL), $z = 146°$ (VO)
4 **a** 50° (AA)   **b** 114° (CA)   **c** 60° (AA)   **d** 81° (CA)
5 $p = 36°$ (ASL), $q = 54°$ (VO), $r = 126°$ (ASL)
6 $s = 78°$ (AA), $t = 102°$ (ASL), $u = 78°$ (CA), $v = 102°$ (VO)
7 **a** **i** y **ii** z **iii** x
   b $x = 290°$ (AP), $y = 70°$ (CA), $z = 110°$ (ASL)
8 **a** $a = 57°$ (CA), $b = 57°$ (AA), $c = 123°$ (ASL), $d = 123°$ (CA)
   b $p = 73°$ (CA), $q = 73°$ (VO), $r = 107°$ (ASL), $s = 98°$ (VO), $t = 98°$ (CA), $u = 82°$ (ASL)

3.1A

1 **a**

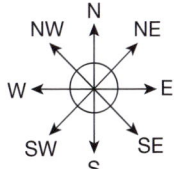

   b N 000° (360°), NE 045°, E 090°, SE 135°, S 180°, SW 225°, W 270°, NW 315°
2 **a** SW   **b** 135° clockwise (or 225° anti-clockwise)
3 **a** **i** 90°   **ii** 120°   **iii** 60°   **iv** 135°
   b **i** 60°   ***ii** 5°

4 Sam 027°, Kate 117°, Neil 222°, Sonja 310°
5 **a** 266°   **b** 288°
6 **a** 234°   **b** 116°
7 347°
8 015°
*9 158° clockwise
*10 **a** L 031°, P 282°, Y 240°
    b from L 221°, from P 102°, from Y 060°
11 **a** $x = a$ (AA), $y = b$ (CA)
   b $a + b + c = x + y + c = 180°$ (ASL)

3.2S

1 **a** Yes   **b** No
2 **a** $a = 98°$, $b = 39°$, $c = 22°$
   b a
3 **a** $p = 80°$, $q = 55°$, $r = 90°$, $s = 214°$
   b **i** r **ii** s
4 **a** $a = 61°$, $b = 119°$ **b** $c = 66°$, $d = 24°$
   c $e = 143°$, $f = 81°$, $g = 62°$ **d** $f = g = 70°$, $h = i = 30°$, $j = 80°$
5 **a** $a = 65°$, $b = 50°$ **b** $c = d = 70°$
   c $e = 45°$ **d** $f = 128°$
6 **a** **i** Parallelogram   **ii** Kite
      iii Trapezium   **iv** Rhombus
   b **i** $p = 56°$, $q = 124°$ **ii** $r = 102°$, $s = 66°$
      iii $t = 82°$, $u = 90°$ **iv** $v = 117°$, $w = 63°$
7 $x = 35°$, $y = 102°$, $z = 55°$
8 **a** **i** 50°   **ii** 50°   **iii** 125°   **iv** 125°
   *b Yes, $\triangle BAC$ and $\triangle BCD$ are isosceles, so
      $\angle BAD = \angle BAC + \angle DAC = \angle BCA + \angle DCA = \angle BCD$.

3.2A

1 **a** 73°   **b** 70°   **c** 116°   **d** 120°, 120°
2

3 **a** $a = e = 66°$, $b = 66°$, $c = d = 48°$
   b $p = 297°$, $q = 63°$, $r = 117°$, $s = t = 76°$
   c $x = z = 60°$, $y = 120°$
   d $j = k = l = m = 61°$, $n = 148°$
4 **a** If more than one angle is obtuse then the total of the angles $> 180°$.
   b Acute
5 Yes, $4x = 360°$ so $x = 90°$.
6 **a** Parallelogram
   b Isosceles triangle
   c Rhombus
   d Kite
   e Right-angled triangle
   f Trapezium
7 **a** **i** Two different isosceles triangles.
      ii 2 pairs of identical right-angled triangles.
   b **i** 1st diagonal gives 2 identical obtuse-angled isosceles triangles. 2nd diagonal gives 2 identical acute-angled isosceles triangles. Both diagonals gives 4 identical right-angled triangles
      ii 1st or 2nd diagonal gives 2 identical right-angled triangles. Both diagonals gives 2 pairs of identical, isosceles triangles – one pair acute-angled and the other obtuse-angled.
      iii 1st or 2nd diagonal gives 2 identical right-angled isosceles triangles. Both diagonals gives 4 identical right-angled isosceles triangles.

8 a Square
  b i Rhombus **ii** Rectangle
    iii Rectangle **iv** Parallelogram
    v Parallelogram
*9 Students' answers, for example, $\angle CBD = 30°$,
  $\angle BCD + \angle ADC = 180°$ (IA), $\angle ADC = 50°$, $\angle ADB = 30°$,
  $\angle ABD = 30°$ (AST), $\angle ADB = \angle ABD$, so $\triangle BAD$ is isosceles.

3.3S

1 a F **b** H **c** E **d** G
2 a Yes, (RHS).
  b No, the equal sides are not corresponding.
  c No, the equal angles are not between the equal sides.
  d Yes, (ASA).
3 B and C, have angles 90°, 37° and 53° (AST), corresponding side
  8 cm hypotenuse (ASA).
4 C and D (AAAA, sf = 2)
5 a $\angle QPR = \angle TSR$ (CA), $\angle PQR = \angle STR$ (CA),
    $\angle QRP = \angle TRS$, $\triangle STR$ is similar to $\triangle PQR$ (AAA).
  b i 12 cm **ii** 4 cm
6 a $\angle ABC = \angle CDE$ (AA), $\angle BAC = \angle CED$ (AA),
    $\angle ACB = \angle ECD$ (VO), $\triangle ABC$ is similar to $\triangle EDC$ (AAA).
  b i 15 cm **ii** 20 cm
7 a $\angle KLX = \angle NMX$ (AA), $\angle LKX = \angle MNX$ (AA),
    $\angle KXL = \angle NXM$ (VO), $\triangle NMX$ is similar to $\triangle KLX$ (AAA).
  b i 27 cm **ii** 36 cm

3.3A

1 33.6 inches
2 a 9.6 cm **b** $6\frac{2}{3}$ cm
3 $AB = AD$, $BC = CD$ and $BM = MD$. $\triangle ABM$ and $\triangle ADM$ (SSS).
  $\triangle BCM$ and $\triangle DCM$ (SSS). $\triangle ABC$ and $\triangle ADC$ (SSS).
4 a $PQ = PR$, $QS = RS$, $PS = PS$,
    $\triangle PQS$ is congruent to $\triangle PRS$ (SSS).
  b i $\angle PSR$
    ii $\angle PSQ = \angle PSR$, $\angle PSQ + \angle PSR = 180°$, $\angle PSQ = 90°$.
5 a $\angle ADB = \angle CBD$ (AA), $\angle ABD = \angle CDB$ (AA), $DB = BD$,
    $\triangle ABD$ is congruent to $\triangle CDE$ (ASA).
  b Congruent triangles means corresponding sides are equal,
    $AB = CD$, $AD = CB$.
6 $AB = AD$, $BC = CD$, $AC = AC$, $\triangle ABC$ is congruent to $\triangle ADC$
  (SSS), so $\angle B = \angle D$
7 Mini and Medium (sf $\frac{5}{3}$), Small and Extra Large (sf 2)
8 Similar (AAA), $\triangle ABC$ and $\triangle PQR$ are isosceles so
    $\angle CAB = \angle CBA = \angle QPR = \angle QRP = 40°$.
    $PR = 6$ cm, $\triangle PQR$ is not equilateral so $PQ = QR \neq 6$ cm.
9 $\angle ACB = 37°$, $\angle DEC = 53°$, $ED = 12$ cm, $AC = 8$ cm
10 A1 594 mm × 841 mm, A2 420 mm × 594 mm,
  A3 297 mm × 420 mm, A4 210 mm × 297 mm,
  A5 148 mm × 210 mm, A6 105 mm × 148 mm.
  Yes, width : height = $\sqrt{2} : 1$. The height of the current size
  becomes the width of the next.

3.4S

1 $(8 - 2) \times 180° = 6 \times 180° = 1080°$
2 $x = 141°$
3 3, 120°, 60°; 4, 90°, 90°; 5, 72°, 108°; 6, 60°, 120°; 7, $51\frac{3}{7}°$, $128\frac{4}{7}°$;
  8, 45°, 135°; 9, 40°, 140°; 10, 36°, 144°
4 144°
5 106°
6 a $2340° \div 15 = 156°$ **b** $180° - 24° = 156°$
7 a $2880° \div 18 = 160°$ **b** $180° - 20° = 160°$
8 a, c Check student's diagram, 7 lines of symmetry.
  b 7
9 a i 8 **ii** 10
  b i 16 **ii** 20

*10 a $x = 40°$, $y = 70°$ **b** $70° \times 2 \times 9 = 1260°$

3.4A

1 30
2 20
3 120
4 a 10, 36°, 144°; 18, 20°, 160°; 24, 15°, 165°; 40, 9°, 171°;
    72, 5°, 175°; 90, 4°, 176°.
  b i The size of exterior angles decreases.
    ii The size of interior angles increases.
5 A Sum of interior angles $= 120° + 108° + 135° = 363° > 360°$.
6 $x = 135°$, $y = 112.5°$
*7 $x = 36°$
*8 $y = 55°$
*9 a $z = 360° - 120° - 90° = 150°$
  b Square or equilateral triangle.
10 12

Review 3

1 $a = 70°$ (ASL), $b = 120°$ (AP), $c = 65°$ (VO), $d = 115°$ (ASL)
2 $a = 55°$ (CA), $b = 100°$ (AA)
3 35°
4 DF
5
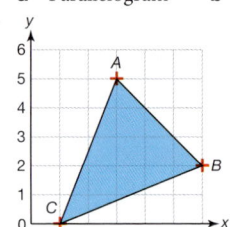
$2 \times 180° = 360°$
6 $a = 50°$ $b = 130°$
7 a Parallelogram **b** Trapezium **c** Kite
8

(graph with triangle plotted, points A at (3, 5), B at (6, 2), C at (1, 0), y-axis labelled 0–6, x-axis labelled 1–6)

Isosceles
9 Yes, $\angle ZXY = \angle WUV$, $XY = UV$, $\angle XYZ = \angle UVW$, (ASA).
10 10 cm
11 540°

Assessment 3

1 a $\frac{2}{3}$ **b** $\frac{3}{4}$ **c** 3
2 a 150° **b** 75°
3 60° clockwise.
4 a 130° **b** 255° **c** 215°
5 Rafa was south of Sunita.
6 No, $64° + 59° + 56° = 179°$ (ASL).
7 a No, $a = 34°$ (AST) **b** Yes, $b = 75°$ (AST)
  c Yes, $c = 46°$ (AST) **d** No, $d = 74°$ (AST)
  e Yes, $e = 121°$ (VO); No, $f = 59°$ (ASL).
8 a False, 2 obtuse angles $> 180°$.
  b True.

(orange triangle diagram)

  c False, One right-angle $= 90°$, obtuse angle $> 90°$,
    right-angle + obtuse-angle $> 180°$.
  d True.

(green right triangle diagram)

9 $\hat{X} = 48°$, $\hat{Y} = 65°$, $\hat{Z} = 67°$
10 15 m
11 a 93° **b** 170° **c** $w = 73°$ **d** 174°
12 a No, UV is not parallel to WY and VW is not parallel to UY.
  b i 60° **ii** 120°
  c 90° **d** No, equilateral triangle.

Chapter 4

Check in 4

1 **a** 6, 17, 19, 26, 29, 30, 37, 42
   b 106, 115, 118, 121, 130, 135
   c 144, 145, 154, 155, 156, 165, 166
2 **a** 121 **b** 144 **c** 252 **d** 413
   e 68 **f** 23 **g** 49 **h** 82
   i 189 **j** 266
3 **a** 90° **b** 130°
4 **a** 120 **b** 45 **c** 60 **d** 72
   e 6 **f** 20 **g** 30 **h** 18
   i 10

4.1S

1 **a** 10, 7, 4, 2 **b** 23
2 **a** 10, 8, 3, 9 **b** 30 **c** Tomato
3 **a** 21, 11, 3, 4, 1 **b** 40 **c** 73
4 **a** 17, 11, 4, 8 **b** 17 **c** 40
5 **a** 10, 12, 9, 9 **b** 9
6 **a** 78 **b** 20 **c** 43 **d** 55
   e 98

4.1A

1 Brown eyed boys 5, blue eyed girls 3
2 8
3 No, there are 17 girls and 19 boys.
4 **a** 28 **b** 67
5 Agree, out of the 60 red cars, 55 were speeding.
6 Advantage, grouping the data makes it easier to spot trends.
   Disadvantage, the raw data is not recorded.

4.2S

1 Library 3 computers, School $4\frac{1}{2}$ computers,
   Work $8\frac{1}{2}$ computers
2 Tea 4 cups, Coffee 3 cups, Hot chocolate $2\frac{1}{2}$ cups, Soup $1\frac{1}{2}$ cups,
   Other 1 cup
3 **a** Italian **b** 7 **c** 5 **d** 23
4
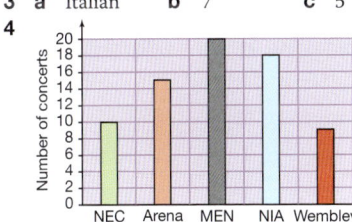
5
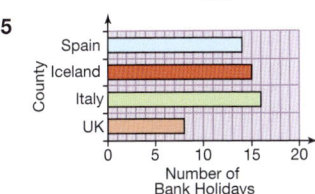
6
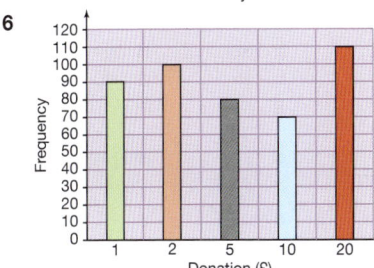
7 **a** 3 **b** Sunday and Wednesday **c** 14

4.2A

1 70
2 **a** 1990
   b Population in village A has increased steadily. Population in village B increased until the 1980s then decreased.
3 Check students' bar charts. Check that total frequencies add up to 28, silver medals has frequency of 7, bronze frequency is greater than gold.
4 **a** Time consuming to draw lots of symbols for large data sets.
   b The vertical scale does not start at 0.
5 World Single Trip = £20 < World Annual = £30.

4.3S

1 Tuna $\frac{1}{8}$, Cheese and tomato $\frac{2}{8}$, Chicken $\frac{3}{8}$, Corned beef $\frac{2}{8}$.
2 **a** 12 **b** 30° **c** Boys 210°, Girls 150°
   d Pie chart with angles given in part **c**.
3 **a** 6°
   b Sunny 90°, Cloudy 108°, Rainy 84°, Snowy 18°, Windy 60°
   c Pie chart with angles given in part **b**.
4 **a** 360 minutes
   b Bat the Rat 30°, Hook a Duck 25°, Smash a Plate 35°, Roll a Coin 80°, Tombola 70°, Break 1 60°, Break 2 60°
5 **a** **i** $\frac{1}{2}$ **ii** $\frac{1}{4}$ **iii** $\frac{1}{4}$
   b **i** 50 **ii** 25 **iii** 25
6 **a** 30°
   b **i** 5 **ii** 3 **iii** 4
7 **a** 20°
   b **i** 7 **ii** 10 **iii** 1

4.3A

1

Region	Number of tourists	Angle
London	225000	225°
Southern	60000	60°
South East	35000	35°
South West	40000	40°
	360000	

2 Students' answers, for example, Too many categories.
3 Students' answers, for example, Frequencies are similar, so difficult to compare on a pie chart.
4 **a** No, the angle for German is the same as the angle for Spanish.
   b No, Lydia doesn't know how many students in the schools are represented in the pie charts.
5 **a** Monday 60°, Tuesday 100°, Wednesday 60°, Thursday 40°, Friday 80°, Saturday 20°.
   b Monday 80°, Tuesday 40°, Wednesday 0°, Thursday 40°, Friday 80°, Saturday 120°.
   c Pie chart, the angles represent the proportion.
   d Bar chart, the heights of the bars show the frequencies.

4.4S

1 **a** 8 **b** 9 **c** 2 **d** 6
   e 2 **f** 24 **g** 18 **h** 104
   i 15 **j** 5
2 **a** **i** 3, 3, 4, 5, 7, 8, 16 **ii** 9, 9, 10, 11, 12
       iii 30, 34, 35, 37, 38 **iv** 95, 97, 97, 98, 99, 101, 103
       v 0, 0, 1, 1, 2, 2, 2, 3, 3
   b **i** 5 **ii** 10 **iii** 35 **iv** 98 **v** 2
   c **i**, no it has not affected the median.
3 **a** Mode = 1, Range = 10 **b** Mode = 8, Range = 3
   c Mode = 11, Range = 4 **d** Mode = 25, Range = 15
   e Mode = 8, Range = 3 **f** Mode = 5, Range = 3
   a, d the outliers affected the range but not the mode.
4 **a** 3, 3, 3, 3, 4, 4, 5, 5
   b Mean = 3.75, Mode = 3, Median = 3.5, Range = 2

5 a 1, 1, 1, 1, 2, 2, 2, 2, 2, 3, 3, 4, 4, 4, 4, 4, 5, 5, 5, 5, 5, 5, 5

b 8

c Mean = 3.28, Mode = 5, Median = 4, Range = 4

6 a i 7   **ii** 6   **iii** 5.82   **iv** 6

   b i 75   **ii** 63   **iii** 60.1   **iv** 63

   c i 8   **ii** 96   **iii** 95.6   **iv** 96

   d i 71   **ii** 22, 37   **iii** 40.4   **iv** 37

   e i 26   **ii** 88, 89   **iii** 84.2   **iv** 87

   f i 72   **ii** 27   **iii** 46.9   **iv** 34

   g i 8   **ii** 105   **iii** 105.2   **iv** 105

7 a 1, 6, 8, 2, 8, 5, 6, 9, 3, 5, 7, 4, 4, 5, 5

   b i 8   **ii** 5   **iii** 5.2   **iv** 5

   c Range stays the same.

4.4A

1 a 2, 4

   b Number 47's data is more spread out than Number 45's.

2 a 1, 3, 4, 1, 1   **b** 1.8, 2, 2

   c On average, houses in Ullswater Drive have more cars than those in Ambleside Close.

3 23 or 42

4 −3.1, −2.6, 3.5, 4.1, 4.1

5 a 3, 4, 0, 3, 3, 3, 3, 0

   b Mean = 5.85, Mode = 4, Median = 6, Range = 7

   c Mean = 6.05, Mode = 4, Median = 6, Range = 8

6 41 years old

7 14 raisins

8 78.6%

Review 4

1

Number	Frequency
1	2
2	4
3	8
4	5

2 a 30   **b** 13   **c** 2   **d** 27

3 a i 16   **ii** 9

   b i $2\frac{1}{2}$ footballs   **ii** $3\frac{3}{4}$ footballs

4

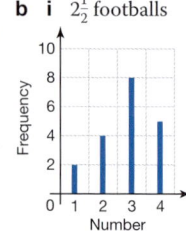

5 a 40   **b** 10

6 Check students' pie charts, Snack 36°, Magazine 108°, Savings 216°.

7 a 6   **b** 6.5   **c** 7   **d** 6

8 a 44   **b** The sample is biased and too small.

9 a 35   **b** 34   **c** 11   **d** 12

   e The 2nd Zoo has a higher median number of birds per aviary and its range is smaller.

Assessment 4

1 a

Milk	Frequency
0	6
1	14
2	13
3	7
4	7
5	2
6	1

b

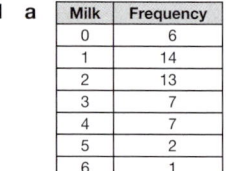

2 a 35   **b** 16   **c** 3

   d Some students access information in more than one way.

3

	Glasses	No glasses
Left-handed	1	4
Right-handed	14	16

4 a i 103 miles   **ii** 99.5 miles   **iii** 88 miles

   b You cannot tell because you do not know how many journeys he makes in one day.

   c 107 miles

   d Mean increases, because the total has increased but not the number of journeys. Median increases, because 98 replaces 90 as one of the two middle numbers. Mode, no mode as each number occurs once. Range, stays the same because the highest and lowest numbers are unaffected.

5 a

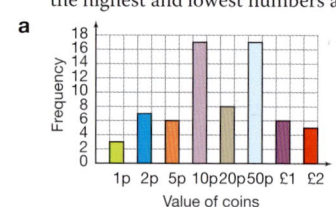

b

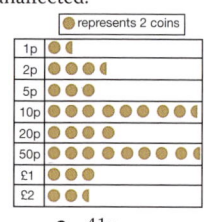

   c 10p and 50p   **d** 20p   **e** 41p

6

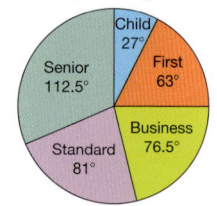

7 a Walk   **b** Car   **c** 25%

   d 150 people   **e** None

8 a 9   **b** 2   **c** 8

   d Zero   **e** 1.2

Chapter 5

Check in 5

1 a 1.3   **b** 0.3

2 a 3.56   **b** 8.04   **c** 0.06

3 0.7, 0.75, 0.8, 0.875

5.1S

1 a $\frac{5}{6}$ **b** $\frac{2}{5}$ **c** $\frac{4}{6}$ **d** $\frac{7}{10}$

2 a A = $\frac{1}{10}$, B = $\frac{4}{10}$, C = $\frac{5}{10}$, D = $\frac{9}{10}$, E = $\frac{11}{10}$

   b F = $\frac{2}{4}$, G = $\frac{3}{4}$, H = $\frac{5}{4}$

   c I = $4\frac{1}{2}$, J = $5\frac{1}{4}$, K = $6\frac{1}{10}$

3 a i $\frac{8}{10}$ **ii** $\frac{4}{5}$

   b i $\frac{14}{16}$ **ii** $\frac{7}{8}$

   c i $\frac{12}{20}$ **ii** $\frac{3}{5}$

   d i $\frac{11}{15}$ **ii** $\frac{11}{15}$

4 a $\frac{1}{3}$ **b** $\frac{3}{4}$ **c** $\frac{3}{5}$ **d** $\frac{4}{9}$

   e $\frac{5}{8}$ **f** $\frac{1}{3}$ **g** $\frac{4}{9}$ **h** $\frac{23}{93}$

5 a $\frac{3}{2}$ **b** $\frac{11}{3}$ **c** $\frac{35}{8}$ **d** $\frac{20}{9}$

   e $\frac{41}{7}$ **f** $\frac{39}{5}$ **g** $\frac{96}{11}$ **h** $\frac{88}{7}$

6 a $1\frac{1}{4}$ **b** $1\frac{3}{5}$ **c** $1\frac{4}{7}$ **d** $2\frac{1}{4}$

   e $2\frac{1}{5}$ **f** $2\frac{6}{7}$ **g** $4\frac{3}{5}$ **h** $3\frac{1}{9}$

7 a 8   **b** 27   **c** 56   **d** 56

   e 2   **f** 90   **g** 85   **h** 7

   i 8

8 a $\frac{1}{10}$ **b** $\frac{3}{5}$ **c** $\frac{3}{4}$ **d** $\frac{5}{4}$

   e $\frac{3}{2}$ **f** $\frac{6}{5}$

9 a 0.3   **b** 0.8   **c** 0.25   **d** 1.1

   e 1.4   **f** 1.6

10 a $\frac{42}{10}$ = 4.2   **b** $\frac{438}{100}$ = 4.38   **c** $\frac{68}{10}$ = 6.8

11 a, b

```
      ii|    i|        iii|
                 ↓     ↓            ↓
  └─┬─┬─┬─┬─┬─┬─┬─┬─┬─┬─┬─┬─┬─┬─┘
  0        0.5        1        1.5
```

12 a $2.07, 2.09, \frac{53}{25}, 2.13, \frac{11}{5}$ **b** $\frac{3}{10}, 0.309, 0.325, 0.345, \frac{7}{20}$
c $\frac{529}{500}, \frac{33}{25}, 1.387, 1.4, 11.35$ **d** $\frac{661}{125}, 5.29, \frac{53}{10}, 5.306, 5.308$

5.1A

1 a $\frac{91}{300}$ **b** $\frac{1}{5}$

2 a $\frac{3}{8}, \frac{7}{16}, \frac{15}{32}, \frac{1}{2}, \frac{17}{32}, \frac{9}{16}$
  b $\frac{13}{32}$, each spanner is $\frac{1}{32}$ bigger than the last.

3 a No, $0.25 < \frac{3}{10}$
  b Light cloud, rain, sunny spells, wind, snow.
  c Students' answers, for example, No, there is a low chance of sun.

4 a Fractions, Geometry and Statistics.
  b ii $\frac{3}{4} = 0.75$ **v** $2\frac{4}{5} = 2.8$

5 a 3, 1, 2 **b** 2, 1, 3

6 $1\frac{1}{2} = 1.5 = \frac{30}{20} = \frac{9}{6}$, $2\frac{1}{5} = \frac{22}{10} = 2.20 = \frac{11}{5}$, $\frac{30}{20} = 0.625 = \frac{5}{8} = \frac{10}{16}$

7 $\frac{1}{3} < \frac{2}{5}$ because $\frac{5}{15} < \frac{6}{15}$

8 Neither, $\frac{69}{40} = \frac{138}{80} = 1\frac{58}{80}$

*9 $\frac{9}{37}$

10 $\frac{1}{2}a + 0.4b - 0.1a + \frac{1}{4}b = 0.4a + 0.65b$,
  $1.8a + 0.4b - 1\frac{2}{5}a + \frac{1}{5}b = 0.4a + 0.6b$,
  $\frac{1}{10}a + \frac{7}{4}a + \frac{2}{5}a + 1.1b = 0.5a + 0.65b$

5.2S

1 a 20 sheep **b** 5 apples **c** 5 shops **d** 12 marks
2 a $2\frac{1}{2}$ **b** 2 **c** $2\frac{2}{3}$ **d** $1\frac{6}{7}$
  e 2 **f** $1\frac{1}{3}$
3 a 4 **b** $3\frac{3}{4}$ **c** $\frac{4}{5}$ **d** $4\frac{2}{3}$
  e $2\frac{1}{4}$ **f** $22\frac{2}{5}$ **g** $13\frac{1}{3}$ **h** $8\frac{5}{9}$
4 a $\frac{4}{10}$ **b** 40% **c** Check 7 squares shaded.
5 a £30 **b** 20 **c** 136p **d** 4
  e 37p **f** £7 **g** 6kg **h** £15.50
  i 2 **j** 4.2m **k** £4 **l** £34
  m £6 **n** 42km **o** $106 **p** £75
  q £6.40 **r** 1.4m
6 a €12 **b** £28 **c** $37\frac{1}{2}$m **d** $36\frac{4}{7}$km
  e £375 **f** $58\frac{1}{3}$mm **g** 1375m **h** $18\frac{6}{13}$g
7 a 264kg **b** $4500 **c** 4.44kg **d** 952 cups
  e 21.67 tonnes **f** 96° **g** 139.35°
  h 0.87 hours **i** £260.67 **j** 377.14cm
8 a £1.98 **b** 4380km **c** £2.40 **d** €60
  e 4.2m **f** 5.2cm **g** 36m **h** £2.10
  i £2.06 **j** 13.92km **k** £3.04 **l** €108.80
  m 11.05m **n** 33.58cm **o** 125.8m **p** £1.528
  q £21.25 **r** 113.75p
9 This is 5%, divide by 10 and double the result.

5.2A

1 a $\frac{4}{5}$kg **b** $\frac{7}{2}$kg
2 £82.90
3 20% of £640 = £128, 25% of £520 = £130, 30% of £450 = £135
4 a Jack 2.4, Jill 6, Dame Dob 9, Jill's mother 4.8
  b 1.8 pints
5 No, $\frac{1}{3} + \frac{3}{8} + \frac{1}{10} + \frac{1}{4} = \frac{8}{24} + \frac{9}{24} + \frac{2.4}{24} + \frac{6}{24} = \frac{25.4}{24} \neq 1$ day
6 $49090.60
*7 5.225 g
*8 a £199.17 **b** 12.92%
*9 24°
10 $\frac{1}{2}$ of 18 = 9, $\frac{1}{3}$ of 18 = 6, $\frac{1}{9}$ of 18 = 2, which leaves 1 horse to return to the neighbour.

5.3S

1 a $\frac{3}{6}, \frac{2}{6}$ **b** $\frac{3}{15}, \frac{5}{15}$ **c** $\frac{5}{10}, \frac{2}{10}$ **d** $\frac{9}{30}, \frac{10}{30}$
  e $\frac{7}{21}, \frac{9}{21}$ **f** $\frac{10}{12}, \frac{9}{12}$
2 a $\frac{2}{3}$ **b** $\frac{3}{5}$ **c** 1 **d** $\frac{3}{2}$
  e 1 **f** $\frac{1}{4}$ **g** $1\frac{1}{3}$ **h** $\frac{3}{4}$
  i $\frac{4}{7}$
3 a $1\frac{1}{4}$ **b** $1\frac{4}{5}$ **c** $1\frac{5}{8}$ **d** $4\frac{1}{4}$
  e $3\frac{1}{7}$ **f** $9\frac{1}{11}$
4 a $\frac{7}{4}$ **b** $\frac{23}{16}$ **c** $\frac{14}{9}$ **d** $\frac{18}{7}$
  e $\frac{26}{5}$ **f** $\frac{19}{4}$
5 a $2\frac{1}{3}$ **b** $3\frac{4}{7}$ **c** $-1\frac{1}{6}$
6 a $\frac{5}{6}$ **b** $\frac{17}{20}$ **c** $\frac{4}{15}$ **d** $\frac{18}{35}$
  e $\frac{23}{24}$ **f** $\frac{38}{45}$ **g** $\frac{59}{99}$ **h** $\frac{94}{105}$
  i $\frac{11}{28}$
7 a $1\frac{1}{2}$ **b** 2 **c** $3\frac{1}{3}$ **d** $2\frac{1}{7}$
  e $2\frac{1}{2}$ **f** $4\frac{1}{3}$
8 a 2 **b** 4 **c** $3\frac{1}{3}$ **d** $\frac{3}{5}$
  e 6 **f** $9\frac{5}{8}$
9 a 8 **b** 10 **c** 14 **d** 20
  e 48 **f** 220
10 a $1\frac{7}{15}$ **b** $2\frac{2}{5}$ **c** $2\frac{7}{12}$ **d** $1\frac{31}{35}$
  e $2\frac{1}{15}$ **f** $1\frac{7}{8}$ **g** $1\frac{7}{12}$ **h** $\frac{43}{63}$
  i $11\frac{3}{4}$
11 a 6 **b** $17\frac{1}{2}$ **c** $2\frac{2}{5}$ **d** 14
  e 48 **f** $6\frac{3}{7}$ **g** 2 **h** $2\frac{1}{7}$
  i $2\frac{14}{15}$
12 a $\frac{3}{10}$ **b** $\frac{9}{20}$ **c** $\frac{15}{28}$ **d** $\frac{12}{35}$
  e $\frac{2}{3}$ **f** $\frac{7}{24}$ **g** $\frac{2}{3}$ **h** $\frac{9}{4}$
  i $\frac{9}{49}$ **j** $\frac{1}{2}$ **k** $\frac{7}{3}$ **l** $1\frac{37}{40}$
13 a 10 **b** $\frac{5}{6}$ **c** $1\frac{1}{15}$ **d** $\frac{6}{7}$
  e $\frac{27}{28}$ **f** $1\frac{4}{5}$ **g** $\frac{1}{4}$ **h** $\frac{4}{35}$
  i $\frac{4}{55}$ **j** $2\frac{5}{8}$ **k** $1\frac{1}{6}$ **l** $1\frac{2}{25}$
  m 2 **n** $3\frac{3}{8}$ **o** $1\frac{13}{15}$
14 a 1 **b** $\frac{1}{5}$

5.3A

1 a $5\frac{3}{8}$ **b** $\frac{5}{8}$
2 a He has added the numerators and the denominators.
  b $\frac{3}{12} + \frac{4}{12} = \frac{7}{12}$
3 4
4 a $11\frac{7}{8}$ hours **b** £178.13
5 a $3\frac{1}{4}$ cm **b** $\frac{21}{32}$ cm²
6 a $\frac{1}{2} + 2\frac{1}{2}$ **b** $\frac{1}{8} + \frac{3}{8}$
  c $\frac{3}{4} \times \frac{4}{3}$ **d** $\frac{3}{5} \times \frac{5}{2}$
  e $\frac{5}{2} = 2\frac{1}{2}$ **f** $\frac{3}{4}$ and $\frac{3}{8}$
  g $\frac{1}{8}$ and $\frac{3}{8}$ **h** $2\frac{1}{2} \div \frac{1}{8} = 20$
  i $2\frac{1}{2} - \frac{1}{8} = 2\frac{3}{8}$ **j** $\frac{3}{4}$ and $\frac{4}{3}$
7 $\frac{3}{35}$
8 $\times$
*9 a $\frac{1}{16}$ **b** 8

5.4S

1 a 0.67 **b** 0.78 **c** 0.99 **d** 0.7
  e 0.39 **f** 0.88 **g** 1.5 **h** 1.25
  i 0.999 **j** 1.1 **k** 0.75 **l** 0.376

2 a 32%   b 22%   c 85%   d 3%
  e 54%   f 63%   g 38%   h 37.5%
  i 33.3% j 125%   k 0.15% l 99.5%
3 a $\frac{4}{5}$ b $\frac{7}{25}$ c $\frac{13}{40}$ d $\frac{1}{20}$
  e $\frac{3}{25}$ f $\frac{3}{8}$
4 a 0.3   b 0.28   c 0.58   d 0.6
  e 2.14   f 3.5
5 a $\frac{1}{4}$ b $\frac{2}{5}$ c $\frac{13}{20}$ d $\frac{3}{20}$
  e $\frac{29}{20}$ f $\frac{11}{20}$
6 a $\frac{3}{10}$ b $\frac{3}{5}$ c $\frac{16}{25}$ d $\frac{9}{20}$
  e $\frac{11}{8}$ f $\frac{27}{25}$ g $\frac{259}{80}$ h $\frac{245}{80}$
  i $\frac{17}{4}$
7 a 0.3   b 0.44   c 1.04   d 0.62
  e 0.45   f 0.52   g 0.28   h 3.35
  i 3.56
8 a 0.44   b 0.67   c 1.35   d 0.73
  e 1.14   f 1.4   g 2.17   h 0.85
  i 3.14
9 a $\frac{3}{5}$ b $\frac{9}{10}$ c $\frac{7}{20}$ d $\frac{7}{20}$
  e $\frac{1}{100}$ f $\frac{181}{50}$ g $\frac{61}{400}$ h $\frac{17}{800}$
  i $\frac{29}{20000}$
10 a 54%   b 40%   c 85%   d 52%
  e 66.6̇% f 24%   g 120%   h 44%
  i 12.5%
11 a 0.37 b 0.07   c 1.89   d 0.45
  e 2   f 0.025
12 a 72%   b 20%   c 200%   d 0.03%
  e 300%   f 1%
13 a 68.6̇% b 64%   c 89.5%   d 191.7%
  e 26.3̇% f 112.5%
14 a 42%, 0.423, $\frac{43}{100}$, $\frac{5}{11}$ b $\frac{13}{25}$, 60%, $\frac{2}{3}$, 0.67
15 a $\frac{7}{2}$, 0.375, 33%, $\frac{13}{40}$ b $1\frac{4}{9}$, $\frac{7}{5}$, 1.33, 45%
*16 a $0.\dot{6}$ b $0.\dot{2}\dot{7}$ c $0.\dot{2}$ d $0.\dot{1}\dot{6}$
  e $0.28514\dot{}$ f $0.42851\dot{}$

5.4A

1 No, $1 - \frac{46}{800} = 94.75\%$
2 a 10D   b 10B
3 a A 10% B 20% C 5%
  b E
4 Online £140 off, Shop sale £70 off.
5 $\frac{4}{29}$
6 a Yes, $6.90 + 3.45 = £10.35$, $2 \times 6.90 \times 0.75 = £10.35$
  b H W Jones, $5.60 + 2.34 = £7.94$, $(5.60 + 4.68) \times 0.75 = £7.71$
*7 a 91.4 g   b 70 g fat, 19.3 g saturates, 6.7 g salt
  c 66.5 g
*8 82.5%

Review 5

1 a $\frac{7}{5}$ b $\frac{25}{7}$
2 a $2\frac{1}{4}$ b $1\frac{5}{6}$
3 a $\frac{3}{10}$ b $\frac{1}{4}$ c $\frac{22}{25}$ d $\frac{1}{20}$
4 a 0.5   b 0.7   c 0.02   d 0.625
5 a $\frac{2}{7}$ b $\frac{8}{3}$
6 a 9   b 12   c 8   d 36
7 a 16   b 48   c 12   d 99
8 a $\frac{7}{11}$ b $\frac{7}{10}$ c $1\frac{1}{24}$ d $1\frac{3}{20}$
  e $2\frac{11}{12}$ f $4\frac{1}{2}$
9 a $2\frac{1}{2}$ b $\frac{2}{35}$ c $\frac{3}{10}$ d $1\frac{7}{18}$
10 a $\frac{2}{25}$ b $\frac{2}{3}$ c 20   d $2\frac{4}{5}$
  e $1\frac{1}{4}$ f $12\frac{1}{4}$
11 $\frac{3}{5}$, 0.6, 60%; $\frac{1}{100}$, 0.01, 1%; $\frac{13}{20}$, 0.65, 65%; $\frac{6}{5}$, 1.2, 120%

Assessment 5

1 0, $\frac{1}{6}$, $\frac{2}{5}$, 0.5, $\frac{2}{3}$, $\frac{8}{10}$, 1
2 a No, $\frac{4}{5} = \frac{12}{15}$.
  b Yes, in a fraction the numerator is divided by the denominator, any number goes into itself exactly once.
3 a $\frac{1}{4}$
  b Ben has treated the 6s as factors. Only factors can be cancelled, 16 is not 1×6 and 64 is not 4×6.
  c Students' answers, for example, $\frac{15}{65} \neq \frac{1}{6}$, $\frac{15}{65} = \frac{3}{13}$.
4 a 5580, 558.0, 55.80, 5.580, 0.5580, 0.05580
  b i Yes   ii No, 9.55   iii No, 2.205   iv No, 6.995
5 a 40%   b 15%   c $\frac{9}{20}$
  d Bananas 45%, Apples 40%, Pears 15%
6 First pair, one quarter = 25% > 20%.
7 a £120   b £37.80   c £31.50   d £14.30
  e £0.36   f £380
8 60
9 a Redyonder b Air
  c Chat-Chat d Incorrect, 8 is 32% of 25 and 40% of 20.
10 a 36.4% (3 sf)   b 14%
11 a 0.12   b 4th   c 3rd
12 $1\frac{1}{2}$ hectares
13 8.5 yards
14 a 145.8 m²   b 142 m²
15 $\frac{10}{33}$
16 Yes, % increase = 60%.
17 a $0.\dot{3}$ b $0.\dot{5}$ c $0.\dot{8}57142\dot{}$ d $0.\dot{6}\dot{3}$
18 33.3%, $33\frac{1}{3}\%$, 0.334, 0.34, $\frac{5}{14}$, $\frac{3}{8}$

Lifeskills 1

1 a
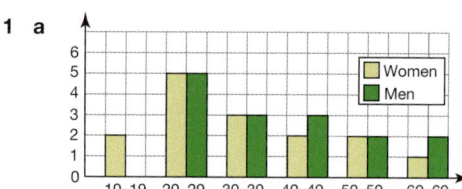
Overall, the men interviewed were older than the women.
  b Yes, Women £28.60, Men £30.
2 a £29.30 (nearest penny)   b £222 222
  c $P = R - G - S - C$ d £52 222
  e $S = R - G - G - P$
3 38, 94, 36, 32
4 a $\frac{7}{40}$ b
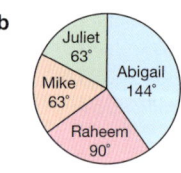
  c Abigail £20000, Raheem £21500, Mike £8750, Juliet £8750
5 a 0.06
  b i £13365.05   ii £12885.48
  c i £11223.29   ii £11641.01

Chapter 6

Check in 6

1 a 5   b 9   c 4   d 10
2 a $3x + 3$ b $2x - 2$ c $8x + 12$ d $12x - 6$
3 a $3(x + 3)$ b $2(3x + 1)$ c $3(y - 3)$ d $4(y + 3)$

6.1S

1 $A = 9$
2 a 4   b 14   c 0   d -2
3 No, $B = 5 \times w = 5 \times 4 = 20$.
4 a 10   b 25   c 35   d 0
  e -10 f -25

5 **a** 3 **b** 5 **c** 4 **d** 10
  e 0 **f** $\frac{1}{2}$
6 **a** 23 **b** 30 **c** 28 **d** 81
  e 162 **f** 3
7 **a** 12 **b** 16 **c** 20 **d** -4
  e 20 **f** -6
8 **a** 14 **b** 8 **c** -1 **d** 2
  e 2
9 77°F
10 **a** $A = p + 6$ **b** $A = m - 5$
  c $A = 2k$ **d** $A = 6 + 5t$
  e $A = 3d + 7$ **f** $A = 3(t + 7)$
  g $A = y^2$
11 **a** $C = 3d$ **b** 30 cm
***12** 10 s (nearest second)
13 **a** 28 cm **b** $2(4 + 10) = 2 \times 14 = 28$ cm
  c Students' answers, $a + b = 20$, for example $a = 5$ and $b = 15$.

6.1A

1 **a** $S = 4\pi = 5026.54824\ldots = 5027$ mm² (4 sf)
  b 37 914 864 km² (nearest km²)
2 **a** 200 **b** 5 **c** 210 **d** 24.696
  e 1.47
3 **a** $s = \frac{d}{t}$ **b** 50 mph **c** 120 km/h **d** 10 km
  e 8 min
4 **a** 1 **b** 200 **c** 6000 **d** 12
5 $t = 2$ and $a = 3$, multiply by $0 = 0$, $2 \times 3^2 = 2 \times 9$, $s = 6$.
6 **a** **i** £13911.29 **ii** £16127.00 **iii** £25125.34
  b £21000

6.2S

1 **a** -2 **b** $+3$ **c** $\times 7$ **d** $\div 6$
  e $\div(-2)$ **f** -6 **g** $+1$ **h** $\times 5$
2 **a** $a = H - 2$ **b** $a = H + 2$ **c** $a = \frac{H}{2}$ **d** $a = 2H$
3 **a** $x = y - 3$ **b** $x = y - c$ **c** $x = y + 4$ **d** $x = y + c$
  e $x = \frac{y}{5}$ **f** $x = \frac{y}{c}$ **g** $x = 6y$ **h** $x = cy$
4 **a** $-1, \div 2$ **b** $\times 4, -3$ **c** $\div 4, +1$ **d** $-2, \times 6$
  e $+3, \div 5$ **f** $\div \frac{1}{2}, -1$ **g** $\times \frac{1}{3}, +2$ **h** $+3, \times 4$
5 **a** $p = \frac{A - 5}{2}$ **b** $p = \frac{A - d}{2}$ **c** $p = \frac{A + 6}{3}$ **d** $p = \frac{A + d}{3}$
  e $p = \frac{H}{4d}$ **f** $p = \frac{H}{ad}$ **g** $p = 4(H - t)$
  h $p = 4H - t$ **i** $p = b(W + t)$ **j** $p = Mb + t$
6 **a** $b = H - a^2$ **b** $b = A + a^2$
  c $b = \frac{Q - a^2}{2}$ **d** $b = \frac{F + a^2}{2}$
  e $b = M - p^2$ **f** $b = \frac{L}{p^2}$
  g $b = \frac{T}{a^2}$ **h** $b = a^2 W$
7 **a** **i** $\times(-1), +2$ **ii** $x = 2 - y$
  b **i** Reciprocal, $\times 3$ **ii** $x = \frac{3}{y}$
  c **i** $\times(-m), +c$ **ii** $x = \frac{c - y}{m}$
  d **i** Reciprocal, $\times k$ **ii** $x = \frac{k}{y}$
8 No, $x = 10 - y$.
9 **a** $x = 5 - y$ **b** $x = 20 - y$
  c $x = m - y$ **d** $x = 2b - y$
  e $x = s^2 - y$ **f** $x = \sqrt{p} - y$
10 $T = \frac{D}{S}$
11 **a** $m = \frac{5}{A}$ **b** $m = \frac{20}{y}$ **c** $m = \frac{x}{A}$ **d** $m = \frac{2b}{H}$
  e $m = \frac{t}{2L}$ **f** $m = \frac{\sqrt{p}}{y}$
12 $y = \frac{2}{A - 3}$
13 **a** $x = y^2$ **b** $x = (y - 2)^2$
  c $x = (3y)^2$ **d** $x = \left(\frac{y}{2b}\right)^2$
  e $x = \pm\sqrt{A}$ **f** $x = \pm\sqrt{P - 2}$
  g $x = \pm\sqrt{3(H - 2)}$ **h** $x = \pm\sqrt{5T - 2}$
***14** $x = \frac{s - q}{p - r}$
15 $L = g\left(\frac{T}{2\pi}\right)^2 = \frac{gT^2}{4\pi^2}$

6.2A

1 **a** 200 **b** 150 **c** 570
2 **a** $t = \frac{mv - mu}{F}$ **b** $a = \frac{F}{m}$
  c $F = AP$ **d** $A = \frac{F}{P}$
  e $v = u + at$ ***f** $m = \frac{Ft}{v - u}$
3 **a** Mass = Volume × Density **b** Volume = $\frac{\text{Mass}}{\text{Density}}$
4 $y = \frac{c - ax}{b}$
5 **a** $h = \frac{V}{\pi r^2}$ **b** $r = \sqrt{\frac{V}{\pi h}}, r > 0$
6 $\pi = \frac{S}{4r^2}$
7 $b = \sqrt{c^2 - a^2}, b > 0$
8 Freddie $\div a, +b$; Asha expanded brackets, $+ab, \div a$. Asha's fraction can be simplified to give Freddie's mixed number.
9 **a**, **c**, **d** Check students' spreadsheets. **b** $C = \frac{5(F - 32)}{9}$

6.3S

1 $x + 3 < 10$ Inequality, $2x + 3x \equiv 5x$ Identity, $2x + 1 = 6$ Equation, $v = u + at$ Formula
2 Students' answers, for example,
  a $3x + 1 = 8$ **b** $F = ma$ **c** $y > 2$ **d** $x + x = 2x$
3 **a** F $4a + 5b$ cannot be simplified **b** F $\equiv 6p$
  c T **d** F $10p - 5$ cannot be simplified
  e T **f** F $3d + 5e$ **g** T
4 **a** $9a$ **b** $5p$ **c** $6a$ **d** $9p$
  e $7a^2$ **f** $4a + 5b$ **g** $8d + e$
  h Students' answers, for example, $4a + 2b - 2a$.
5 **a** F $4a + 8$ **b** T **c** T **d** F $y^2 + 3y$
  e T
6 **a** $\equiv 5 \times a + 5 \times 2 \equiv 5(a + 2)$
  b $\equiv 3 \times x + 3 \times 4 \equiv 3(x + 4)$
  c $\equiv 5 \times y + 5 \times 3 \equiv 5(y - 3)$
  d $\equiv y \times y + 3 \times y \equiv y(y + 3)$
  e $(x + 2)(x - 2) \equiv x^2 + 2x - 2x - 4 \equiv$
7 **a** $\equiv 4a + 8 + 2a + 2 \equiv 6a + 10$
  b $\equiv 3x + 6 + 4x - 4 \equiv 7x + 2$
  c $\equiv 5y - 10 + 3y - 9 \equiv 8y - 19$
  d $\equiv y^2 + 3y + 2y + 6 \equiv y^2 + 5y + 6$
  e $\equiv x^2 - 4x + x^2 + 2x \equiv 2x^2 - 2x$
8 **a** $a = 7, b = 9$ **b** $a = 7, b = 6$
  c $5a - 3b = -19$. Possible solution $a = 1, b = 8$.
  d $a = 1, b = 2$ **e** $a = 3, b = 10$
9 **a** Formula **b** Identity **c** Equation **d** Formula
  e Equation **f** Identity **g** Identity **h** Formula
  i Formula **j** Equation **k** Identity **l** Equation
10 $2(a + b) \equiv 2 \times a + 2 \times b \equiv 2a + 2b$

6.3A

1 **a** **i** $5(a + 2) \equiv 5a + 10$ **ii** $5a + 10 = 80$
  b **i** $7(b + 5) \equiv 7b + 35$ **ii** $7b + 35 = 105$
  c **i** $12(c + 10) \equiv 12c + 120$ **ii** $12c + 120 = 240$
  d **i** $2(d + 1) \equiv 2d + 2$ **ii** $2d + 2 = 24$
  e **i** $4(e + 3) \equiv 4e + 12$ **ii** $4e + 12 = 44$
  f **i** $3(f + 1) \equiv 3f + 3$ **ii** $3f + 3 = 24$
2 **a** $2(x + 7) \equiv 2x + 14$ **b** $2(y + z) \equiv 2y + 2z$
  c $r(p + q) \equiv rp + rq$ **d** $v(s + t + u) \equiv vs + vt + vu$
3 **a** $(2a)^2 = 4a^2$ **b** $3b \times 4b = 12b^2$
4 **a** 12 **b** 9 **c** 6, 72 **d** 4, 4
  e 11, 3 **f** 5, 7
5 **a** 6, 2 **b** 4, 8 **c** 2, 8 **d** 4, 4
  e 2, 1 **f** 8, 3, 67
6 **a** Sometimes true **b** Always true
  c Sometimes true **d** Sometimes true
  e Never true **f** Never true
  g Sometimes true **h** Sometimes true
***7** **a** **i** $x < 9$ **ii** $63 - 3x \equiv 36 + 3(9 - x)$
  b **i** $x < 11$ **ii** $132 - 4x \equiv 88 + 4(11 - x)$

1 a 8, 12 **b** 8, 15 **c** 10, 16
2 a $x^2 + 7x + 10$ **b** $x^2 + 7x + 12$
 c $y^2 + 9y + 14$ **d** $y^2 + 10y + 24$
 e $a^2 + 5a + 6$ **f** $a^2 + 13a + 42$
3 No, $2x \times 5 = 10x$, $3 \times 5 = 15$, $2x^2 + 13x + 15$.
4 a 4, 12 **b** 2, 15 **c** 6, 16 **d** 4, 12
 e 2, 15 **f** 6, 16
5 a $x^2 + 3x - 12$ **b** $x^2 + x - 12$
 c $y^2 - 5y - 14$ **d** $y^2 - 2y - 24$
 e $a^2 - a - 6$ **f** $a^2 - a - 42$
6 a 8, 12 **b** 8, 15 **c** 10, 16
7 a $x^2 - 7x + 10$ **b** $x^2 - 7x + 12$
 c $y^2 - 9y + 14$ **d** $y^2 - 10y + 24$
 e $a^2 - 5a + 6$ **f** $a^2 - 13a + 42$
8 a $x^2 + 6x + 9$ **b** $y^2 + 2y + 4$
 c $y^2 - 8y + 16$ **d** $x^2 - 2x + 1$
9 $x(x + 12)$
10 a $x(x + 5)$ **b** $x(x + 7)$
 c $2x(x + 6)$ **d** $6x(2x + 1)$
11 $(x + 10)(x + 2)$
12 a $(x + 3)(x + 2)$ **b** $(x + 5)(x + 2)$
 c $(y + 4)(y + 3)$ **d** $(y + 5)(y + 3)$
 e $(a + 8)(a + 3)$ **f** $(a + 12)(a + 1)$
 g $(x + 13)(x + 7)$ **h** Cannot be factorised.
13 a $(x + 3)(x - 2)$ **b** $(x + 5)(x - 2)$
 c $(y + 4)(y - 3)$ **d** $(y + 5)(y - 3)$
 e $(x - 12)(x + 2)$ **f** Cannot be factorised.
14 a $(x - 4)(x - 2)$ **b** $(x - 6)(x - 2)$
 c $(y - 6)(x - 3)$ **d** $(y - 9)(y - 3)$
 e $(p - 6)(p - 4)$ **f** $(p - 13)(p - 1)$
 g $(x - 23)(p - 1)$ **h** Cannot be factorised.
15 a $(x + 12)(x + 4)$ **b** $(y - 8)(y + 6)$
 c $(b + 8)(b - 6)$ **d** $(a - 24)(a - 2)$
 e $(x + 30)(x - 2)$ **f** $(x - 15)(x - 4)$
16 a $(x + 3)^2$ **b** $(y - 3)^2$
 c $(y - 2)^2$ **d** $(x + 4)^2$
17 a $(x + 3)(x - 3)$ **b** $(y + 5)(y - 5)$
 c $(b + 10)(b - 10)$ **d** $(h + 9)(h - 9)$
 e $(y + 8)(y - 8)$ **f** $(a + 25)(a - 25)$
 g Cannot be factorised **h** Cannot be factorised
*18 a $2x^2 + 7x + 3$ **b** $3x^2 - 5x - 2$
 c $12x^2 + 7x + 1$ **d** $6x^2 - 5x - 6$

6.4A

1 a

x^2	$3x$
$4x$	12

b

x^2	$2x$
$7x$	14

c

$2x^2$	$3x$
$8x$	12

d

$12x^2$	$8x$
$15x$	10

2 a $(a + 1)(a + 8) \equiv a^2 + 9a + 8$
 b $(b + 7)(b + 8) \equiv b^2 + 15b + 56$
 c $(5c + 1)(c + 3) \equiv 5c^2 + 16c + 3$
 d $(4d + 6)(5d + 2) \equiv 20d^2 + 38d + 12$
3 a 3400 **b** 480 **c** 240 **d** 1000
 e 3596 **f** 1551 **g** 2491 **h** 2484
4 a $(a + b)^2 = a^2 + 2ab + b^2$
 b i 160 000 **ii** 10.24 **iii** 3600
5 a Jason needs two numbers that multiply to 6 and add to 5.
 b $(x + 2)(x + 3)$
6 a 3, 18 **b** 2, 2 **c** 4, 2, 14
7 $(x + 8)(x - 8) = 17$, $x^2 - 64 = 17$, $x^2 - 81 = 0$, $(x - 9)(x + 9) = 0$
8

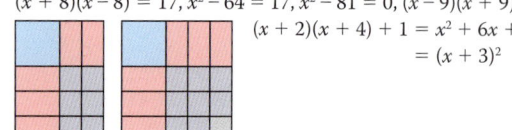

$(x + 2)(x + 4) + 1 = x^2 + 6x + 9$
$= (x + 3)^2$

9 a $3x, 3x + 3, 9x^2 + 9x$ ***b** $x - 1, 2, 2x - 2, 2x$
*10 a $x + 3$ **b** y **c** $y + 2$ **d** $\frac{1}{x + 7}$

1 a $3\,\text{g/cm}^3$ **b** $12.5\,\text{g/cm}^3$
2 a 16 **b** 22.5
3 a i $2x$ **ii** $x - y$
 b £15
4 a $A = b - 3$ **b** $A = \frac{d}{2}$
 c $A = \frac{F + c}{5}$ **d** $A = 2J - h$
 e $A = \pm\sqrt{L + 2K}$ **f** $A = \frac{2}{b}$
5 a $2y + 3 = 7$ **b** $3b \times 4b = 12b^2$
 c $4z + 2$ **d** $F = ma$
 e $y = 3x + 4$
6 $2(3x + 1) - 4 \equiv 6x + 2 - 4 \equiv 6x - 2$
7 a $x^2 + 8x + 15$ **b** $x^2 - 8x + 12$
 c $x^2 - 3x - 28$ **d** $6x^2 + 13x - 5$
8 a $x(x + 5)$ **b** $3x(4x - 1)$
9 a $(x + 4)(x + 1)$ **b** $(x - 6)(x - 1)$
 c $(x - 4)(x + 2)$ **d** $(x + 5)(x - 2)$
10 a $(x + 6)(x - 6)$ **b** $(2x + 5)(2x - 5)$

1 a $m = \frac{t}{5}$ or $t = 5m$
 b i 3 miles **ii** 0.5 miles **iii** $12\frac{1}{2}$ s
2 a $J = 0.45R + 0.55S$ **b** 1.8 kg **c** 3.3 kg
 d Raspberries 4.05 kg, Sugar 4.95 kg **e** 5 kg
3 a i 13 hrs 30 min **ii** 7 hrs 30 min
 b i 12 **ii** 8
 c 1 hr 30 min, no, this is too little sleep.
4 a Yes **b** No, ±11.
5 a i No, $p^2 - 11p + 28$. **ii** No, $26v^2 - 38v - 47$.
 b i Yes **ii** No, $(v + 10)(v - 10)$.
6 a $W = 7D$ **b** $C = 50 + 250D$
7 a i $P = 15m$ **ii** $P = 12m$ **iii** $P = 14n$
 b i £1.72 **ii** £5.47
 c £8.30
8 a F $5 \times 4 = 20$
 b F 5 has two factors, 1 and 5.
 c T Let the two numbers be $2x$ and $2y$. $2x + 2y = 2(x + y)$ so the sum is even.
 d F $0^2 = 0$
 e F $2 \times 7 = 14$
 f T Let the numbers be $2x$, $2x + 2$ and $2x + 4$. $2x + 2x + 2 + 2x + 4 = 6x + 6 = 6(x + 1)$, is divisible by 6.
 g T p^2 has 3 factors, 1, p and p^2.
9 a $y = 4x + 1$ **b** $y = \frac{x}{3} - 2$
10 a 58.8 m (1 dp) **b** 44.7 m **c** 101 km/h

1 5.625 kg
2 Yes, £17.50 − £15 = £2.50
3 a COOL (9) = 9, COOL (28) = 1
 b 1, 4, 7, 9
 c 1, 9
 d 13, 14
 e 20
4 a 95 **b** 355
 c $T_x = 5x + 30$ **d** 15
 e No, it will not fit on the grid because 20 is in the left hand column.
 f $x = 78$, $T_{78} = 77 + 78 + 79 + 88 + 98 = 420$
5 a $y = 50°$ (ASL), $z = 80°$ (Isosceles AST)
 b $y = 32°$, $x = 148°$
 c Equilateral triangle, $y = 60°$ (AST), $x = 60°$ (Isosceles AST).

6 a 55   **b** Mean = 4.85 (3 sf), Median = 5
  c 8
7 a $10y + 8$ **b** $84y + 40$ **c** $3x + 2 = 35, x = 11$
8 a i Correct, *HBD* is an isosceles right-angled triangle.
    ii Incorrect, *HBDF* is a square.
    iii Incorrect, *HBCG* is an isosceles trapezium.
  b i *OPF, OBQ, ODQ*
    ii *HOF, HOB, BOD, FOD, BHF, BDF, HBD, HFD, ACO,*
    CEO, EGO, GAO
    iii Any triangle with letter vertices except *HPG, FPG, BQC,*
    DQC.
    iv *ABC, HBD, GBE, FEG, FHD, FAC, AOC, GOE*
  c i 135°   **ii** 90°   **iii** 67.5°
    iv 45°   **v** 45°   **vi** 135°
9 a £1.50   **b** £2.50   **c** £3.50
10 a 108   **b** 18%   **c** 261
11 a $\frac{1}{5}$ **b** $\frac{4}{5}$ **c** 16 ml acid and 4 ml water.
  d **A** Acid 64 ml, Water 16 ml; **B** Acid 16 ml, Water 84 ml
  e i 4 : 1   **ii** 4 : 21
12 a 2.27 cm (2 dp)   **b** $1 = \sqrt{\frac{A}{4\pi}}, A = 1^2 \times 4\pi = 12.6$ (3 sf)
  c $A = 4\pi r^2$ **d** 2980 cm² (3 sf)
13

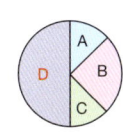

Chapter 7

CW Clockwise
ACW Anti-clockwise

Check in 7

1 a $9\frac{3}{5}$ **b** $22\frac{1}{2}$ **c** $10\frac{1}{2}$
2 a 7.8   **b** 15.5   **c** 9
3 a 45 mm   **b** 4.5 cm

7.1S

1 a 1.9 cm   **b** 4.2 cm   **c** 2.2 cm   **d** 42°
  e 127°   **f** 205°
2 $w = 3.7$ cm, $x = 3$ cm, $y = 5.1$ cm, $z = 1.7$ cm, $a = 105°$,
  $b = 106°, c = 60°, d = 89°$
3 a 112 cm   **b** 7 m   **c** 115 m   **d** 4.2 km
4 C
5 a Distance = 205 km, Bearing = 062°
  b Distance = 125 km, Bearing = 258°
  c Distance = 90 km, Bearing = 113°
  d Distance = 90 km, Bearing = 293°
  e Distance = 60 km, Bearing = 210°
6 a Length = 225 m, Width = 125 m
  b i ≈20 cm   **ii** ≈5 km
7 Check students' measurements, Living room 11.2 cm by
  8.4 cm, Kitchen 8.2 cm by 5.4 cm, Bathroom 5.8 by 3.8 cm,
  Bedroom 6 cm by 4.2 cm.

7.1A

1 a 97 miles (nearest mile), 052° (nearest degree).
  b Use a larger scale.
2 Yes
3 Using a scale of 1 : 50, Bed 3.8 cm × 1.8 cm, Desk 2.4 cm ×
  1.6 cm, Wardrobe 2 cm × 1.16 cm, Drawers 1.74 cm × 0.86 cm,
  Bookshelves 2.4 cm × 0.4 cm.
4 6.4 m
*5 No, total distance ≈ 140 km ≈ 87 miles < 250 miles.

7.2S

1 a 108 cm²   **b** 10 m²   **c** 35 cm²   **d** 800 mm²
2 a 36.9 m²   **b** 2.7 ft²   **c** 864 in²
3 a Triangle, $1\frac{1}{4}$ km²   **b** Trapezium, $48\frac{3}{4}$ m²
  c Parallelogram, $8\frac{1}{4}$ ft²   **d** Trapezium, 4.14 m²
4 Rectangle 15 mm, Parallelogram 40 cm, Triangle 2.5 m
5 a 112.5 cm²   **b** 5.1 m²
6 5 cm
*7 a 9 cm²   **b** 8.025 m²
8 71.4 m²

7.2A

1 a Yes, 2897 m² > 2500 m².
  b The accuracy of the measurements.
2 a **A** = 17300 m², **B** = 15170 m²
  b The shape of the car park and its bays.
3 a Yes, Purple = 990 cm², $6 \times$ Yellow = $6 \times 150 = 900$ cm²
  b 50 cm²
4 a   **b, c** 19 cm²

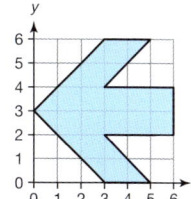

*5 No, Cost = £236 > £200
*6 Yes, $600 - (12 \times 22) = 336 > \frac{1}{2}$ of 600.

7.3S

1

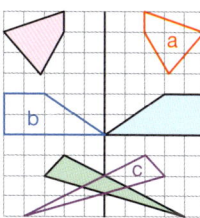

2

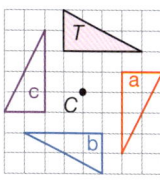

3

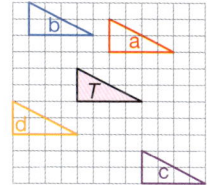

4

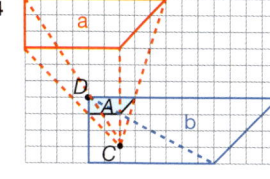

5

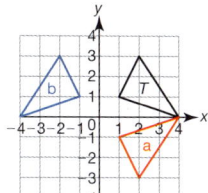

6

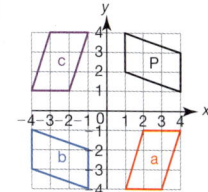

7 a Pentagon
  b

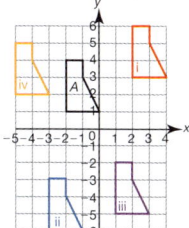

8 a Rhombus
  b

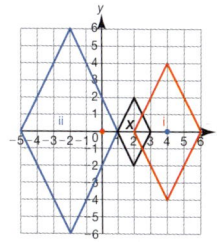

*9 a Kite
b–e

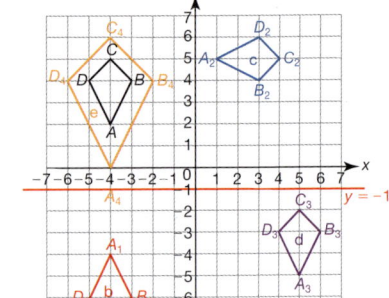

10 a Rotation around $(1, 1)$.   **b** Reflection in $y = 1$.

7.3A

1 a Rotation 90° CW (or 270° ACW) about $(0, 0)$.
b Rotation 90° ACW (or 270° CW) about $(0, 0)$.
c Rotation 180° CW (or ACW) about $(0, 0)$.
d Translation by vector $\binom{-4}{-5}$.
e Translation by vector $\binom{4}{5}$.

2 a Reflection in the y axis.   **b** Reflection in the y axis.
c Reflection in the x axis.   **d** Reflection in the x axis.
e Rotation 180° CW (or ACW) about $(0, 0)$.

3 a i Translation by vector $\binom{2}{-4}$.
ii Rotation 90° CW (or 270° ACW) about $(-4, 3)$.
iii Rotation 90° ACW (or 270° CW) about $(-1, 2)$.
iv Reflection in $y = 3$.   **v** Reflection in $x = 4$.
vi Rotation 180° CW (or ACW) about $(4, 3)$.
vii Enlargement, centre $(0, -2)$, scale factor 2.
viii Enlargement, centre $(-3, -6)$, scale factor 1.5
b i None   **ii** $(-4, 3)$ **iii** $(-1, 2)$ **iv** $y = 3$
v $x = 4$ **vi** $(4, 3)$ **vii** $(0, -2)$ **viii** $(-3, -6)$

4 a All co-ordinates are multiplied by 3.
b Enlargement, centre $(0, 0)$, scale factor 3.

5 a

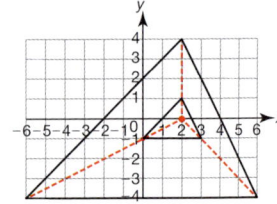

b Enlargement, centre $(0, 0)$, scale factor 4.

6 a i Rotation 180° CW about the centre of the clockface.
ii Rotation 60° CW about the centre of the clockface.
*b i Rotation 15° CW about the centre of the clockface.
ii Rotation 5° CW about the centre of the clockface.

*7 No, rotation 90° CW (or 270° ACW) about $(1.5, 4.5)$.

*8 Translation by vector $\binom{4}{0}$, Reflection in $x = 3$,
Rotation 180° CW (or ACW) about $(3, 3)$.

9 a i $(a, b) \rightarrow (a, -b)$ **ii** $(a, b) \rightarrow (-a, b)$
*iii $(a, b) \rightarrow (b, a)$ ***iv** $(a, b) \rightarrow (-b, -a)$
b i $(a, b) \rightarrow (b, -a)$ **ii** $(a, b) \rightarrow (-a, -b)$
iii $(a, b) \rightarrow (-b, a)$

7.4S

1

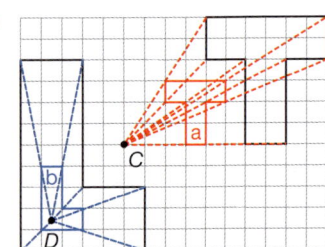

---

2 a, b

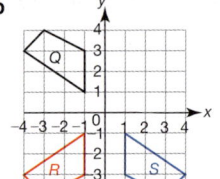

c Rotation 180° CW (or ACW) about $(0, 0)$.
d Rotation 90° ACW (or 270° CW) about $(2, 2)$.

4 a Trapezium
b, c

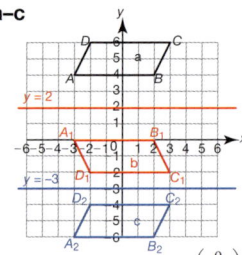

d Enlargement, centre $(6, 5)$, scale factor $\frac{1}{2}$.

3 a–c

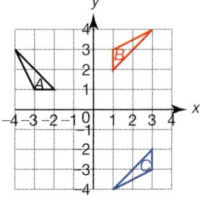

5 a–c

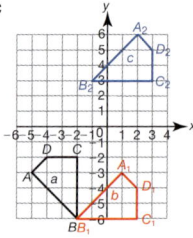

d Translation by vector $\binom{0}{-10}$.

6 a–c

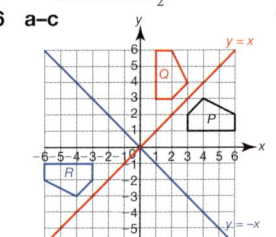

d Rotation of 180° CW (or ACW) about $(0, 0)$.

*7 a–c

d Rotation of 90° CW (or 270° ACW) about $(3, -2)$.
e Rotation of 90° ACW (or 270° CW) about $(3, -2)$.

7.4A

1 Translation by vector $\binom{1}{1}$.
2 Rotation 45° CW about the centre.
3 a i
    ii Rotation 180° CW (or ACW) about the point of intersection of the mirrors.
    b No, the final image is the same.

4 a i
    ii Translation by vector $\binom{4}{0}$.
    b Translation by vector $\binom{-4}{0}$.

5 Translation by vector $\binom{7}{3}$.
6 No, it is a clockwise rotation of 180°, about $(1, 2)$.
7 Students' answers, for example, Enlargement from the centre by scale factor 4, followed by enlargement from the centre scale factor $\frac{1}{2}$.
*8 a No, a rotation $x°$ CW followed by $x°$ ACW will produce a translation if the centre of rotation is moved.
b No, a reflection in $x = a$ followed by $y = b$ will produce a rotation 180° CW (or ACW) about (a, b).
9 a Enlargement centre O, scale factor $\frac{1}{4}$.
b Check students' diagrams.
10 Check students' investigations.

Review 7

1 a 55°   **b** 5.1 cm
2 Check $AC = 5.7$ cm, $\angle BAC = 73°$, $\angle BCA = 42°$.

3 a 28 m²   **b** 6 cm²   **c** 20 mm²   **d** 25 cm²
   e 55 m²

4 a, b

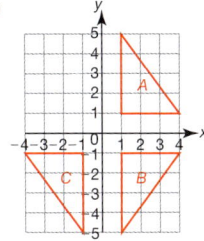

   c Reflection in the y-axis.

5 a

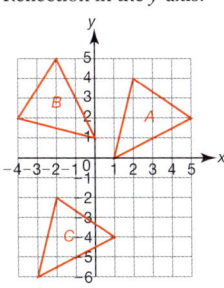

   b Rotation 90° CW (or 270° ACW), about (0, 0).

6 a

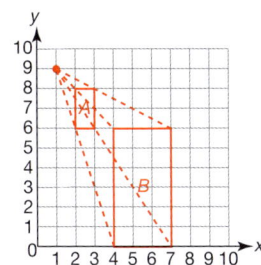

   b Enlargement scale factor $\frac{1}{3}$, centre of enlargement (1, 9).

Assessment 7

1 a Karl   **b** Neither, multiply by 1 000 000.

2 a Check student's drawing.
   b i 6.5 cm ± 3 mm   **ii** 60° ± 3°
     iii 135° ± 3°
   c Kite   **d** Rectangle

3 a Yes, $90° < a < 180°$.   **b** No, $b = 90°$, right-angle.
   c Yes, $c < 90°$.   **d** Yes, $d > 180°$.
   e Yes, $90° < e < 180°$.   **f** No, $f < 90°$, acute.
   g No, $g = 90°$, right-angle.   **h** No, $h < 90°$, acute.
   i Yes, $90° < i < 180°$.   **j** No, $j < 90°$, acute.
   k No, $k > 180°$, reflex.

4 a 195°   **b** 135°   **c** 015°   **d** 315°
   e 245°

5 a Yes   **b** No, 2.625 cm².
   c Yes, 294 cm².   **d** No, 21.12 cm².
   e No, 181.5 cm².   **f** No, 4.64 cm².
   g No, 22 cm.   **h** No, 22 cm.
   i No, 12 cm.   **j** No, 20 cm.

6 a 49
   b No, $2 \times 2 = 4$ and $3 \times 3 = 9$ are not factors of 49.
   c No, each side is 7 m so must include two 2×2 slabs and a 3×3 slab. There are three ways of arranging these slabs around the edges, all of which have gaps or overlaps.

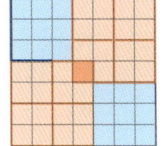 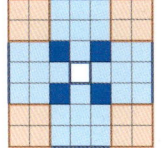 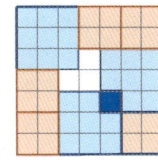

7 a 4   **b** 1   **c** 7   **d** 6
   e 5   **f** 3   **g** 2   **h** 10
   i 8

8 a, c

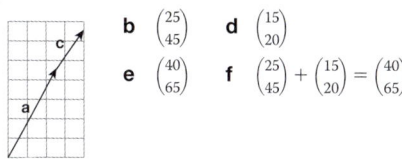

   b $\binom{25}{45}$ **d** $\binom{15}{20}$

   e $\binom{40}{65}$ **f** $\binom{25}{45} + \binom{15}{20} = \binom{40}{65}$

9 a Rotation 180° about the point (0, 0).
   b, c   **d** Reflection in the x-axis.

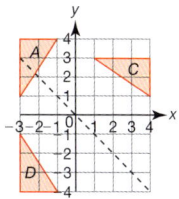

Chapter 8

Check in 8

1 a $\frac{2}{3}$ **b** $\frac{4}{5}$ **c** $\frac{1}{4}$ **d** $\frac{3}{5}$
   e 1

2 a 0.2, 0.25, 0.3   **b** 0.7, 0.75, 0.8
   c 0.8, 0.85, 1

3 a 1   **b** 1   **c** $\frac{1}{10}$ **d** $\frac{1}{5}$
   e $\frac{1}{4}$

4 a 0.8   **b** 0.3   **c** 0.1

8.1S

1 a $\frac{1}{2} = 0.5 = 50\%$
   b $\frac{2}{3} \approx 0.67 \approx 67\%$
   c $\frac{3}{10} = 0.3 = 30\%$
   d i $\frac{3}{5} = 0.6 = 60\%$ **ii** $\frac{2}{5} = 0.4 = 40\%$

2 a i $\frac{1}{5}$ **ii** 0.2   **iii** 20%
   b i $\frac{7}{40}$ **ii** 0.175   **iii** 17.5%
   c i $\frac{13}{40}$ **ii** 0.325   **iii** 32.5%
   d i $\frac{3}{10}$ **ii** 0.3   **iii** 30%

3 a Bag A, P(Red) $= \frac{1}{3} >$ Bag B, P(Red) $= \frac{1}{5}$.
   b Bag B, Bag A only has one red ball.

4 a 50 times
   b i $\frac{9}{50} = 0.18 = 18\%$ **ii** $\frac{14}{50} = \frac{7}{25} = 0.28 = 28\%$
     iii $\frac{27}{50} = 0.54 = 54\%$
   c Students' answers, for example, Red 2, Green 3, Blue 5.
   d Complete more trials.

5 a Check students' results $\approx$ 0.5, 0.25, 0.13, 0.06, 0.03, 0.02, 0.01, 0, ... (2 dp)
   b Students' answers, for example, P(Head on 1st toss) is the highest.

6 a Check students' results $\approx$ 0.67, 0.14, 0.12, 0.1, 0.08, 0.07, 0.06, 0.05, 0.04, 0.03, 0.03, 0.02, 0.02, 0.02, 0.01, 0.01, 0.01, 0.01, 0.01, 0.01 (2 dp)
   b Students' answers, for example, P(6 on 1st toss) is the highest.

8.1A

1 No, the relative frequency will only approach the theoretical probability with a high number of trials.

2 a His sample size is too small to give a reliable result.
   b No, relative frequency is an estimate.
   c i $\dfrac{\text{Total number of red balls}}{\text{Total number of trials}} = \dfrac{4+16+95}{5+20+100} = \dfrac{115}{125} = 0.92$
     ii Probably, this is only an estimate.

3 a Frequency 10, Relative frequency 0.4, the relative frequencies sum to 1.
b A and C
c Any spinner that has red, white and blue sections.
4 a **B** and **C**, **A** does not have any 4s.
b Correct net with at least one 1, 2, 3 and 4.
c **C**, it has more 1s and 3s.
5 No, the relative frequencies are very similar, Black = 0.34, White = 0.32, Green = 0.34
*6 Kieth needs a larger and more diverse sample as his is only made up of his friends who will all be close in age to him.

8.2S

1 a 5   **b** ≈ 2
2 a 15   **b** ≈ 5
3 a ≈ 7 **b** ≈ 13
4 a 25   **b** ≈ 13
5 ≈ 4 people
6 ≈ 4 people
7 ≈ 19 people
8 Score 1 = 25, Score 2 = 50, Score 3 $\approx$ 63, Score 4 $\approx$ 38, Score 5 = 25, Score 6 = 50
9 9 days
10 14 shares
11 a ≈ 44 **b** ≈ 33 **c** ≈ 44 **d** ≈ 67

8.2A

1 a 12 times   **b** 24 times
2 a 5 points
b Students' answers, for example, less because it is difficult for counters to land near the sides of the box.
3 60
4 2 games
5 15
6 Purple 15, Blue 10, Red 15, Brown 10, Green 15
*7 a 10 times   **b** More often
*8 a 6 times   **b** £6

8.3S

1 a $\frac{1}{6}$ **b** $\frac{1}{3}$ **c** $\frac{1}{2}$ **d** $\frac{1}{3}$
2 a Yes
b No, P(Point down) > P(Point up).
c No, planned births don't happen at the weekend.
d No, P(Head on 1st toss) = $\frac{1}{2}$ so the other outcomes cannot be equally likely.
e No, some letters are more common than others.
3 6
4 a Highest 9, Lowest 3   **b** 0.34 (2 dp)
5 a $\frac{1}{6}$ **b** $\frac{4}{9}$ **c** 1   **d** 0.2 (est)
e 1
6 a $\frac{5}{10} = 0.5$, $\frac{9}{20} = 0.45$, $\frac{14}{30} = 0.4\dot{6}$, $\frac{17}{40} = 0.425$, $\frac{19}{50} = 0.38$, $\frac{23}{60} = 0.38\dot{3}$, $\frac{29}{70} = 0.4\dot{1}4285\dot{7}$, $\frac{32}{80} = 0.4$, $\frac{35}{90} = 0.3\dot{8}$, $\frac{38}{100} = 0.38$, $\frac{43}{110} = 0.3\dot{9}0$, $\frac{47}{120} = 0.391\dot{6}$, $\frac{49}{130} = 0.3\dot{7}6923\dot{0}$, $\frac{53}{140} = 0.378571\dot{4}\dot{2}$
b 0.38 (2 dp)
7 0.24 (2 dp)

8.3A

1 a $\frac{7}{15}$ **b** $\frac{3}{10}$
2 Yes, relative frequency of 5 = 0.29 > $\frac{1}{6}$
3 $\frac{3}{8}$
4 0.79 (2 dp)
5 a $\frac{3}{5}$ **b** $\frac{13}{32}$
6 $\frac{1}{4}$
*7 Expected number of sectors, White $\approx$ 4, Black $\approx$ 7.6, Red $\approx$ 7.4, biased because Black and Red don't have a whole number of sectors.

8.4S

1 a No   **b** Yes   **c** No   **d** No
2 a No   **b** Yes   **c** Yes   **d** Yes
3 a 0   **b** $\frac{2}{3}$
4 a $\frac{1}{6}$ **b** $\frac{1}{2}$
5 a $\frac{1}{7}$ **b** $\frac{3}{7}$ **c** $\frac{11}{21}$
6 a 0.2   **b** 0.6   **c** 0.5
7 a $\frac{1}{3}$ **b** $\frac{1}{2}$
8 a 0.4   **b** 0.8   **c** 0.2

8.4A

1 a

	1	2	3	4	5	6
1	0	1	2	3	4	5
2	1	0	1	2	3	4
3	2	1	0	1	2	3
4	3	2	1	0	1	2
5	4	3	2	1	0	1
6	5	4	3	2	1	0

b i $\frac{5}{18}$ **ii** $\frac{1}{6}$ **iii** $\frac{5}{9}$

2 No, $\frac{1}{4} + \frac{1}{3} + \frac{1}{2} = \frac{3}{12} + \frac{4}{12} + \frac{6}{12} = \frac{13}{12} > 1$
3 a Yes, $\frac{1}{6} + \frac{1}{3} + \frac{1}{8} + \frac{3}{8} = \frac{4}{24} + \frac{8}{24} + \frac{3}{24} + \frac{9}{24} = \frac{24}{24} = 1$
b There are no other colours because the probabilities sum to 1.
4 $\frac{1}{7} + \frac{2}{7} + \frac{4}{7} = \frac{7}{7} = 1$, the probabilities sum to 1.
5 a $\frac{1}{2}$ **b** $\frac{3}{4}$ **c** $\frac{5}{12}$
Chocolate, fudge and caramel are not mutually exclusive.
6 a C and D   **b** C and D   **c** 2 is prime and even
7 a T You can't be both M and F.
b T You can only be M or F.
c F You can be M and D.
d F You can be F and R.
8 a

	1	2	3	4	5	6
1	0	−1	−2	−3	−4	−5
2	1	0	−1	−2	−3	−4
3	2	1	0	−1	−2	−3
4	3	2	1	0	−1	−2
5	4	3	2	1	0	−1
6	5	4	3	2	1	0

b i $\frac{1}{9}$ **ii** $\frac{1}{36}$ **iii** $\frac{35}{36}$

9 a 0.1   **b** 0.3   **c** 0.5

Review 8

1 a 0.4   **b** 120
2 a 0.2   **b** 24
3 a 0.1   **b** $\frac{1}{6}$ **c** $\approx \frac{1}{6}$
4 a $\frac{3}{10}$ **b** 0   **c** 1
5 0.25
6 $\frac{8}{11}$
7 a $\frac{2}{7}$ **b** $\frac{4}{7}$

Assessment 8

1 No, the probability of this event is greater than zero.
2 P(H) = 0.5 because each throw is an independent event.
3 $\frac{51}{100}$
4 a 200   **b** 2   **c** 4
5 a If $x = 0.2$ then total = 1.02, $x = 0.19$
b 52   **c** 128   **d** 118
6 a i 0.18   **ii** 0.17
b Ben's, because he sampled more packets.
7 a $\frac{1}{5}$ **b** $\frac{1}{2}$
8 9 packets
9 a Yes, P(Late) = 0.51 > 0.49   **b** 70
10 a i $\frac{3}{25}$ **ii** $\frac{8}{25}$ **iii** $\frac{7}{25}$ **iv** $\frac{11}{25}$
b P(Blue 4) = 0
11 a $\frac{2}{5}$ **b** $\frac{1}{5}$ **c** $\frac{1}{2}$ **d** $\frac{3}{5}$

12 a i No, 102 is a multiple of 3. **ii** No, 5 is prime.
iii Yes, no white numbers are multiples of 7
  b i No, 81 is a multiple of 3. **ii** No, 29 is prime.
  iii No, 84 is a multiple of 7.
13 No, sunny and snowing are not mutually exclusive events.
14 a P(Blue) = $\frac{72}{360}$ = $\frac{1}{5}$
  b Assume that all angles are equally likely to be chosen.

Chapter 9

Check in 9

1 a 400 **b** 14 000 **c** 31 **d** 1340
  e 6300 **f** 40 **g** 60 **h** 4.3
  i 0.64 **j** 0.31 **k** 3.4 **l** 0.078
2 −7°C
3 15 miles

9.1S

1 a i 8 **ii** 8
  b i 20 **ii** 19
  c i 40 **ii** 36
  d i 300 **ii** 279
  e i 1 **ii** 1
  f i 4000 **ii** 3895
  g i 0.008 **ii** 0
  h i 2000 **ii** 2400
  i i 9 **ii** 9
  j i 10 **ii** 14
  k i 1000 **ii** 1403
  l i 100 000 **ii** 140 306
2 a i 3490 **ii** 3500
  b i 611 000 **ii** 611 100
  c i 0.00372 **ii** 0
  d i 859 **ii** 900
  e i 1000 **ii** 1000
  f i 859 000 **ii** 859 200
3 a 10 **b** 5 **c** 7
  d 3 **e** 9 **f** 0.4
4 a 8 (7.95) **b** 560 (555) **c** 0.008 (0.00792)
  d 10 (15.06) **e** 3 (3.2045) **f** 50 (50.180 to 3 dp)
5 a 16 (15.8672) **b** 400 (399.6222)
  c 2 (1.826 to 3 dp) **d** 6 (6.172 to 3 dp)
6 a 100 (94.752 to 3 dp) **b** 1 (1.060 to 3 dp)
  c 81 (104.6529) **d** 60 (58.194 to 3 dp)
  e 7 (7.049 to 3 dp) **f** 7 (6.959 to 3 dp)
7 a 30 (30.4776) **b** 40 (39.8)
  c 2 (2.029 to 3 dp) **d** 25 (25.0183)
  e 55 (50.957 to 3 dp) **f** 100 (100.902 to 3 dp)
  g 8 (8.131 to 3 dp) **h** 10000 (10983.4592)
8 a 1.4 **b** 2.8 **c** 3.2 **d** 3.9
  e 4.5 **f** 5.1 **g** 5.7 **h** 6.7
  i 8.4 **j** 9.2
9 a 12.5 (12.35) **b** 64 (62.37)
  c 8 (7.745 to 3 dp) **d** 14 (14.422 to 3 dp)
10 a 2 (2.076 to 3 dp) **b** 144 (139.204 to 3 dp)
  c 2 (1.988 to 3 dp) **d** 1 (1.070 to 3 dp)
11 a 400 (431) **b** 80 (80)
12 £2.60 (£2.52)
13 50, 20, 10 or 45, 20, 9

9.1A

Your estimates may differ from the estimates given here.
Exact values are given in brackets.

1 a 100 cm (91.44) **b** 8.56 cm

2 a 100 m² (86.4)
  b Students' answer, for example, 4 because 3 tins cover
    90 m² < 100 m².
3 a i 700 (620) **ii** 2 (3.208) **iii** 120000 (13279.86)
    iv 40 (31.9)
  b Students' answers.
4 No, he is out by a factor of 10. (1036.105...)
5 a 4 (4) **b** 50 ml (16.99 ml)
6 a i 2 × 5 (10.10352) **ii** 20 × 500 (9530.23475)
    iii 12 ÷ 18 (0.692307)
  b Students' answers.
7 a 70 (75) **b** 210 (232)
8 a i £1900 (£1902.22) **ii** £450 (£435.30)
  b £3500 (£3525.55)
*9 a 35 (32.325...) **b** Less (29.391...)
*10 Students' answers. (≈250)

9.2S

1 a 21, 20.1 **b** 19, 18.6 **c** 13, 12.5 **d** 9, 9.8
  e 10, 11.1 **f** 3, 3.1 **g** 400, 412.4
2 a 71.1 **b** 29.624 **c** 2.07885304659
  d 186.408 **e** 0.1508856039 **f** 19.05
3 a 5.2 **b** 1.3 **c** 1.7 **d** 2.2
4 a 13 hrs 53 min 20 s **b** 1 day 3 hrs 46 min 40 s
  c 5 days 18 hrs 53 min 20 s **d** 11 days 13 hrs 46 min 40 s
  e 16 weeks 3 days 17 hrs 46 min 40 s
  f 49 weeks 4 days 5 hrs 20 min
5 a 5.6 **b** 73 **c** 0.085 **d** −35
  e −35 **f** 38
6 a 464.5923967 **b** 0.4536084142
7 a 178.4123835 **b** 0.1967089505
  c 3.210178253 **d** 3.350190476
  e 1.157007415 **f** 0.1356045007
*8 The number converges to 0.6180339887... every time.

9.2A

1 a Correct
  b Incorrect (36 ÷ 2.5) + 5.5 = 19.9
  c Incorrect 36 ÷ (2.5 + 5.5) = 4.5
  d Correct
2 a £23 **b** £13.20
3 a £120 **b** £128.28 **c** £71.37
  d No, she would need another box of style C.
4 a **D** (0) **b** **C** (56)
  c **B** (119.02) **d** **A** (−2.63)
5 a i £21.13 **ii** £16.37
  b Yes, it is cheaper both months.
*6 a 35.3 ft/s (3 sf) **b** 757 mph (3 sf)

9.3S

1 a Centimetre (cm) **b** Millilitre (ml)
  c Kilogram (kg) **d** Centimetre (cm)
  e Kilogram (kg) **f** Kilometre (km)
  g Millilitre (ml) **h** Litre (l)
  i Tonne (t) **j** Gram (g)
2 a 200 cm **b** 4 m **c** 4.5 m **d** 4 km
  e 5 mm **f** 4500 g **g** 6 kg **h** 6.5 kg
  i 2.5 tonnes **j** 3000 ml
3 a 2.5 mph **b** 32 km/h **c** 80 km/h **d** 8 km/h
  e 18 m/h **f** 10 mph
4 2 hrs 30 min
5 37.5 miles
6 a 480 kg/m³ **b** 41.7 g/cm³ (3 sf)
7 a 5 g/cm³ **b** 87.88 g
8 a 9.4575 kg **b** 6.15 l
9 a 5.75 ≤ d < 5.85 **b** 16.45 ≤ C < 16.55
  c 0.85 ≤ m < 0.95 **d** 6.25 ≤ f < 6.35

e $10.05 \leqslant t < 10.15$ **f** $104.65 \leqslant d < 104.75$
g $15.95 \leqslant d < 16.05$ **h** $9.25 \leqslant s < 9.35$
10 a $6.65 \leqslant d < 6.75$ **b** $7.735 \leqslant C < 7.745$
c $0.8125 \leqslant m < 0.8135$ **d** $5.5 \leqslant F < 6.5$
e $0.0005 \leqslant t < 0.0015$ **f** $2.535 \leqslant d < 2.545$
g $1.1615 \leqslant d < 1.1625$ **h** $14.5 \leqslant s < 15.5$
11 a $32.5 \leqslant d < 37.5$ **b** $37.5 \leqslant d < 42.5$
c $107.5 \leqslant d < 112.5$ **d** $4.25 \leqslant d < 4.75$

9.3A

1 a 1:45 pm **b** £1.35
2 a 5400 m/h **b** 90 m/min **c** 1.5 m/s
3 a 14 hrs 54 min **b** 25 mph
4 a 3408 ml **b** 4 litres
5 $44 \leqslant m < 46$
6 Yes, $948.75 \leqslant A < 1011.75$
7 14 boxes, $445 \div 30.5 = 14.59$ (2 dp)
8 a No, Tin, $6984 \leqslant D$ (4 sf) < 7590 **b** $772 \leqslant m$ (4 sf) < 838
*9 a 42 mph
b No, it should have done 450 miles for this to be true.

Review 9

1 a i 8750 **ii** 8800 **iii** 9000
b i 15.0 **ii** 15 **iii** 20
c i 0.0682 **ii** 0.068 **iii** 0.07
d i 0.509 **ii** 0.51 **iii** 0.5
2 a 200 (220.15) **b** 103 (118.842...)
c 5.5 (5.118...) **d** 4 (3.865...)
3 a 8, 8.114056225 **b** 80, 81.312
c 1, 1.052060738 **d** 300, 255.9035917
4 a cm **b** cm³ **c** litres **d** kg
e m²
5 a 3.5 m **b** 0.145 m **c** 200 m **d** 9.32 m
6 1 hr 52 min
7 a £72.50 **b** 9.5
8 a 5 s **b** 25 s
9 a 5 g/cm³ **b i** 27 cm³ **ii** 3 cm
10 $89.5 \leqslant x < 90.5$
11 Nearest kg

Assessment 9

1 a 3.23 (3 sf) **b** 29 (2 sf)
c 0.2 (1 sf) **d** 310 (nearest 10)
e 5700 (nearest 100) **f** 256000 (nearest 1000)
2 a B (37.7) **b** B (113.12)
c C (69.52) **d** A (5.076923̇)
e C (8.6813181̇8̇)
3 $6 \times 60 \times 25 = 9000$, Mia used 25 as an estimate for the number of hours in a day.
4 a 6 and 7 **b** $6^2 = 36$ and $7^2 = 49$
c 7, 46 is closer to 49.
5 a $2 \times 2^2 = 8$, 2 is a better estimate.
b 0.808629471, the difference is larger than his estimate.
6 40
7 a 94.18 m² **b** 11.7725 m³
c i 300 kg/m³ **ii** 3.53 tonnes
8 a $4.6 + (4.1 + 1.2) \times 2.6 = 18.38$
b $14.9 - 6.8 \div (3.7 - 1.2) = 12.18$
c $3.4 \times (2.4 - 0.8) + 5.9 - 2.8 = 8.54$
d $2.6 + (5.52 + 2.04) \div 1.8 - 0.72 = 6.08$
e $(12.3 - 5.2 \times 1.6 + 3.4) \times 2 = 14.76$
f $(5.9 + 2.2) \div 3.6 - 2.4 \times 0.3 = 1.53$
9 a $\approx 100-1000$ litres **b** ≈ 20 kilograms
c ≈ 2 grams **d** ≈ 150 metres
e $\approx 50-300$ tonnes **f** ≈ 30 millilitres
g ≈ 15 millimetres **h** ≈ 40 kilometres
i ≈ 15 centimetres **j** ≈ 300 millilitres

10 a £25.30 **b** 25 m 30 cm
c 25 kg 300 g **d** 25 cl 3 ml
e 25 hrs 18 min
11 a 66 km/h **b** 150 km/h (3 sf)
c 481.25 km **d** 3410 miles
e 17 min 52 s
12 a Yes, Density $= 0.81$ g/cm³ < 1
b 2.28 cm³ (3 sf) **c** 5400 tonnes
13 a 221 455, 221 465 **b** 85 cm, 95 cm
c 452.5 g, 457.5 g **d** 3 min 28.75 s, 3 min 28.85 s
e 238 bags, 242 bags **f** 27.5 tonnes, 28.5 tonnes
g 585.5 mm, 586.5 mm
14 a 390.1625 cm³, 261.4375 cm³
b 0.293 g/cm³ (3 sf), 0.194 g/cm³ (3 sf)

Chapter 10

Check in 10

1 a 3, 4 **b** 5, 7
2 a x **b** m **c** $3n$ **d** $2p$
3 a 6 **b** 11 **c** 9 **d** 6
4 a 7 **b** -1 **c** 12 **d** $2\frac{1}{4}$

10.1S

1 a $x = 14$ **b** $x = 53$ **c** $x = 13$ **d** $x = 70$
2 $x = 9$
3 $x = 14$
4 a $x = 8$
b   **c** They are the same.
5 a $x = 16$ **b** $x = 8$ **c** $x = 5$
d $x = 5$ **e** $x = 7$ **f** $x = 8$
6 a $y = -1$ **b** $y = -1$ **c** $y = 0$
d $y = 1$ **e** $y = -2$ **f** $y = 1$
7 a $m = 4$ **b** $m = 3$ **c** $p = 9$
d $p = 24$ **e** $f = 8$ **f** $h = 16$
8 a $b = -2$ **b** $a = -3$ **c** $x = 2$
d $y = 17$ **e** $w = -11$ **f** $x = 4$
9 a $x = 20$ **b** $m = 45$ **c** $p = 60$ **d** $m = 16$
10 a $x = 4.6 \pm 0.1$ **b** $x = -0.6 \pm 0.1$
c $x = 2$ **d** $x = 4.3 \pm 0.1$
11 a $x = 5$ **b** $x = 4.5$ **c** $x = 8$ **d** $x = 0$
*12 a
b i $x = 9$ **ii** $x = -9$ **iii** $x = 0$ ***iv** $x = 9$
*13 $x = 5$
14 3 cm

10.1A

1 a $5n$ **b** $5n = 40$ **c** $n = 8$
2 a $n - 12 = 11, 23$ **b** $\frac{n}{5} = 8, 40$
3 $6p + 12 = 16$
4 a 5 **b** 12
5 a 5 cm **b** 6 cm **c** 8 cm, regular hexagon
6 Students' answers with solutions $-1, 0.5, 4, 6$
7 Sarah 3, Josh 6, Millie 2
8 £12.50

9 14 cm, 1 cm

10 51°

10.2S

1 $x = 3$

2 $x = 5$

3 **a** $x = 7$ **b** $y = 5$ **c** $p = 2$
   d $f = -3$ **e** $x = 5$ **f** $x = -7$
   g $b = -2$ **h** $p = -0.5$

4 **a** $x = 2$ **b** $y = -5$ **c** $p = 7$
   d $f = -3$ **e** $f = 7$ **f** $p = \frac{1}{2}$

5 **a** $x = 3$ **b** $x = 1$ **c** $x = -2$
   d $x = 2$ **e** $p = 3$ **f** $h = 5$

6 $x \approx 1.3$

7 The graphs are parallel and so don't intersect.

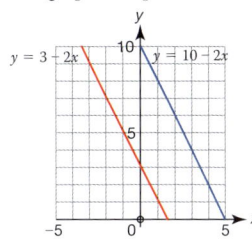

8 **a** $3x + 183 = 360$ **b** 44°

9 **a** $4x + 6 = 24$ **b** $x = 4.5$
   c Rectangle

11 **a** $12x - 18 = 6x - 6$ **b** $x = 2$
   c Their bases are equal.

*11 $b = -2$

12 1.5

*13 **a**, **b**   **c** $x = 0.3$

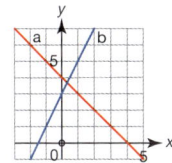

10.2A

1 6

2 15

3 $y = 1$

4 $z = 8$

5 $2(x + 3) = 3x - 2, x = 8$
$2(x + 3) = 4x + 1, x = 2.5$
$2(x + 3) = 8x - 3, x = 1.5$
$4(2x - 1) = 3x - 2, x = 0.4$
$4(2x - 1) = 4x + 1, x = 1.25$
$4(2x - 1) = 8x - 3,$ no solution
$3(4x + 1) = 3x - 2, x = -\frac{5}{9}$
$3(4x + 1) = 4x + 1, x = -0.25$
$3(4x + 1) = 8x - 3, x = -1.5$

6 $x = 5$

7 $x = 2$

8 8 buttons

9 **a** 10   **b** 6   **c** -4

10 $x = 9\,\text{cm}$

11 $x = 36\,\text{cm}$

10.3S

1 ± 4.5 (1 dp)

2 $0, -2$

3 **a** ± 2 **b** $0, -4$ **c** $0, 4$

4 **a** $-3, -2$ **b** $-4, -3$ **c** $-5, -3$
   d -4 **e** $-17, -1$ **f** $-13, -2$

5 **a** $-3, 2$ **b** $-4, 3$ **c** $-5, 3$
   d $-8, 2$ **e** $-6, 3$ **f** $-11, 2$

6 **a** $-2, 3$ **b** $-3, 4$ **c** $-3, 5$
   d $-2, 8$ **e** $-2, 15$ **f** $-4, 7$

7 **a** $-5, -4$ **b** $-12, -1$ **c** $-10, 2$ **d** $-9, 3$
   e $4, 6$ **f** $5, 7$ **g** $-5, -1$ **h** 3

8 **a** $-4, 5$ **b** $-12, 1$ **c** $-4, 2$
   d $-7, 3$ **e** $5, 9$ **f** $-20, 0$

9 **a** $-3, 0$ **b** $-3.6, 0.6$ **c** $-4, 1$ **d** $-2, -1$

10 The graph does not intersect the line $y = -5$.

*11 The graph of $y = x^2 + 20$ does not cross the x-axis ($y = 0$).

12 3 cm

10.3A

1 **a** The double brackets must equal zero.   **b** $-5, 3$

2 1, 5

3 **a** $-10, -6$ or $10, 6$ **b** $-6, 2$ or $-2, 6$

4 **a** $-8, 5$ **b** $-8, 10$ or $5, 16$

5 **a** $-10, 5$ **b** $t(t + 5) = 50, t^2 + 5t - 50 = 0$
   c $t = -10$ is impossible as a length cannot be negative.

6 10

7 3 cm

8 **a** $x^2 + (x + 1)^2 = 5^2, 2x^2 + 2x - 24 = 0, x^2 + x - 12 = 0$
   b 12 cm

9 4 cm

10 65, 66

10.4S

1 $x = 7, y = 3$

2 $x = 6, y = 2$

3 **a** Students' answers, for example, $x = 2, y = 7$.
   b Students' answers, for example, $x = 3, y = 9$.
   c $x = 4, y = 3$

4 **a** $x = 1, y = 3$ **b** $x = 5, y = 2$
   c $p = 2, q = 7$ **d** $a = 5, b = -1$

5 **a** $m = 4, n = 2$ **b** $x = 2, y = 5$
   c $x = 4, y = -1$ **d** $e = 1, f = 3$
   e $m = 4, n = 2$ **f** $x = 3, y = -2$

6 **a** $x = 6, y = 1$ **b** $x = 7, y = -2$
   c $p = 6, q = 1$ **d** $a = 7, b = 1$

7 **a** $x = 6, y = 1$ **b** $x = 7, y = 2$
   c $p = 2, q = -1$ **d** $a = 5, b = -3$

8 $x = -0.3, y = 1.4$ (1 dp)

*9 **a**   **b**

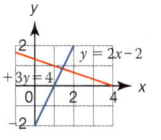 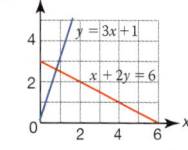

   $x = 1.4, y = 0.9$ (1 dp) $x = 0.6, y = 2.7$ (1 dp)

   c   **d**

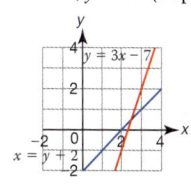

   $x = -0.5, y = -2.5$ (1 dp) $x = 2.5, y = 0.5$ (1 dp)

*10 $x = -1.9, y = 11.5$ or $x = 5.9, y = -19.5$

11 62, 73

10.4A

1 **A** and **B**

2 **a** $x = 12, y = -16$ **b** $x = 7, y = -1$
   c $a = 3, b = -2$ **d** $v = 3, w = 2$
   e $q = 3, p = 0$ **f** $x = 5, y = 2$

3 **AB** $x = -1, y = 14$; **AC** $x = 5, y = 2$; **BC** $x = -67, y = -52$

4 **a** £0.17   **b** 5.6 cm

5 **a** $x + y = 23, x - y = 5, x = 14, y = 9$
    b $x - y = 6, 2x + y = 6, x = 4, y = -2$

6 $s = 29, m = 17$

7 73 children

8 **a** $x = 2, y = 0$ **b** $x = 3, y = 1$

9 **a** The lines are parallel so they do not intersect, therefore no solution.
    b Yes, if either of the equations are non-linear.

10.5S

1 $-1, 0, 1, 2, 3, 4$

2 $1 \le x \le 4$

3 **a** $2, 3, 4, 5, 6$ **b** $3, 4$ **c** $2, 3, 4$

4 **a** $x > 6$ **b** $x < 8$ **c** $x < 7$ **d** $x \le 6$
    e $x \ge 3$ **f** $x \ge 4$ **g** $x \le 11$ **h** $x > -5$

5 **a** $5 > x > 1$ **b** $0 < x < 8$
    c $3 \le x < 7$ **d** $1 < x \le 6$
    e $-1 \le x \le 4$ **f** $-1 < x < 2$
    g $-5 < x < 8$ **h** $-5 < x < -4$

6 **a** $x > 12$ **b** $x < 8$ **c** $x < 14$ **d** $x \le 6$
    e $x \ge 5$ **f** $x \ge 4.5$ **g** $y \ge -1$ **h** $x > -\frac{2}{3}$

7 **a** $x > 8$ **b** $x < 19$ **c** $x > 6$ **d** $x \le 4$
    e $x > 3$ **f** $x < 3$

8 **a** $x > -3$ **b** $x \le -2$ **c** $x < -7$ **d** $x \le -4$
    e $x < 3$ **f** $x \ge 17$ **g** $x \le -9$ **h** $x < \frac{1}{3}$

9 **a** $x > 2$ **b** $x \le 4$ **c** $x < -1$ **d** $x < 5$
    e $x < -2$ **f** $x \le -29$ **g** $x \le \frac{1}{2}$ **h** $x < -2.5$

10 **a** Smallest 7   **b** Largest 7   **c** Largest 6   **d** Largest 6
     e Smallest 3   **f** Smallest 4   **g** Largest 11   **h** Smallest -4

11 **a** [number line: open circles at 2 and 10]   **b** [number line: closed circle at 1, open circle at 5]
       2 4 6 8 10   1 2 3 4 5

     c [number line: closed circles at 7 and 9]   **d** [number line: open circle at -3, arrow to 4]
       7 8 9 -3 -2 -1 0 1 2 3 4

12 **a** $4 \le x < 8$ **b** $3 < x \le 4$
      [number line: closed at 4, open at 8]   [number line: open at 3, closed at 4]
       4 6 8   3 4

     c $-2 \le x \le 0$ **d** $-1.5 < x < -0.25$
      [number line: closed at -2 and 0]   [number line: open at -1.5 and -0.25]
       -2 -1 0 -1.5 -1 -0.5 0

13 **a** Smallest 13   **b** Largest 7
     c Largest 13   **d** Largest 6
     e Smallest 5   **f** Smallest 5
     g Smallest -1 **h** Smallest 0

14 **a** $x > 2$ **b** $x \le 4$ **c** $x < 3$ **d** $x \le 1$
     e $x \le 3$ **f** $x \le 0$ **g** $x \ge \frac{1}{2}$ **h** $x < 4\frac{5}{7}$

10.5A

1 **a**, **b**, **c**, **e**

2 $x > \frac{4}{3}$ [number line: open circle between 1 and 2]
          1   2

3 **a** $6(x - 2) > 2x + 8, x > 5$ **b** 6

4 $-\frac{1}{3} < x \le 2$

*5 No, it is not possible for a linear inequality to have a solution with a value less than zero as well as greater than 2.

6 **a** $30 < x < 75$ **b** $45 < y < 81$

*7 $x > 3$ or $x < -3$

8 $x = 1, y = 1; x = 1, y = 2; x -= 2, y = 1; x = 2, y = 2;$
    $x = 3, y = 1$

9 **a** $-1 < x < 4$ **b** $-5 < x < 18$

10 It is not possible for $y \le 6$ and $y > 6$ at the same time.

11 **a** $x + 2y = 180, x = 180 - 2y$, so $180 - 2y < 30$
     b $y > 75$
     c $x > 0$, so $180 - 2y > 0, y > 90$

12 $2x + 2y = 360, y = 180 - x, 180 - x > 120, x < 60$

Review 10

1 **a** 48   **b** 5   **c** 7   **d** 16.5
    e 100   **f** -2

2 **a** 6   **b** 2   **c** -7

3 **a** $2x + 24 = 42$ **b** Vicky = 9, mum = 33

4 **a** $-3, -5$ **b** $1, 5$ **c** $-3, 2$ **d** $-8, 8$
    e $0, 12$

5 $-2, 10$

6 **a** $x = 3, y = 5$ **b** $a = 1, b = -2$
    c $x = 9, y = 3$ **d** $v = 2, w = -3$

7 **a** $2c + w = 3.75, 3c + 2w = 6$
    b $c = 1.5$ kg, $w = 0.75$ kg

8 $x = 1.5, y = 3.5$

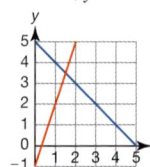

9 **a** $x > 5$ **b** $x \le 3$ **c** $x \ge 1$
      [5 6 7]   [1 2 3]   [1 2 3]

10 **a** [3 4 5]   **b** [4 5 6 7]   **c** [1 2 3 4 5 6]
     d [-2 -1 0]

Assessment 10

1 $10s = 49, s = 4.9$

2 11

3 2.75 kg

4 Albert 3, Oliver 9

5 154 g

6 **a** $x + \frac{x}{2} + 75 + x + 150 = 2000$ **b** 710
    c Brendan €140, Arsene €430, José €860

7 $a = -9$ or $5, a > 0$ because it is a length, so $a = 5$

8 **a** $5k - 2(30 - k) = 115, k = 25$ **b** 9 km

9 **a** $(w + 8)(w - 9)$ **b** $w = -8, 9$

10 $89^2 - 11^2 = (89 - 11)(89 + 11) = 78 \times 100 = 7800$,
     $6.89^2 - 3.11^2 = (6.89 - 3.11)(6.89 + 3.11) = 3.78 \times 10 = 37.8$

11 11 cm

12 7 and 9, 7 and -9

13 $n = 100$

14 1 s, the ball is going up and 3 s, the ball is coming down.

15 12 sides

16 Batman 28, Robin 18

17 **a** £6.50   **b** £4.50

18 50 g

19 Graham £21.50, Liz £19

20 **a** 72p   **b** 63p

21 DVD £11.95, CD £4.99

22 **a** £250   **b** £500

23 Maggot 4p, Worm 3p

24 [number line: open circle at -1, arrow right]
     -1 0 1 2 3

25 3, 2, 1 or 0

26 $0 < z < 84$

27 **a** $\frac{1}{a} < 1$ **b** No, Student's answers, $0 \le b \le 1$.

Lifeskills 2

1 a 1:10000

b, c

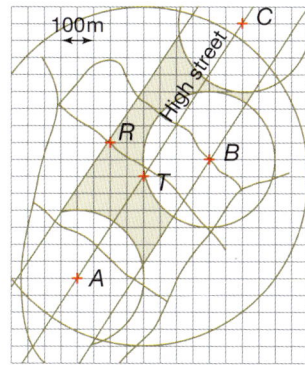
100 m
High street
C
R
B
T
A

2 a 5.8 cm² (1 dp)   **b** 36.0 cm² (1 dp)
  c 640

3 28

4 a $2 \times 0.41 + 0.58 = 1.4$ m
  b The octagon, as it completely contains the circle.
  c Circle = 1.59 m² (3 sf), Octagon = 1.62 m² (3 sf)
  d Students' answers, yes or no with a valid reason.

5 11

6 a Juliet   **b** 22
  c Students' answers, for example, assume the proportion of unanswered calls to answered calls will stay the same.

Chapter 11

Check in 11

1 a 42°   **b** 123°   **c** 230°

2 a 93 mm   **b** 9.3 cm

3 28 cm²

11.1S

1 a 49   **b** 156   **c** 9   **d** 20.9 (3 sf)
  e 15.7   **f** 50.3   **g** 6.60   **h** 5.01

2 a i 3 cm   **ii** 1.5 cm
  b i 9.42 cm   **ii** 7.07 cm²
  c Check students' technique.

3 a i 126 cm (3 sf)   **ii** 120
  b i 201 mm (3 sf)   **ii** 180
  c i 11.3 m (3 sf)   **ii** 12
  d i 1.32 km (3 sf)   **ii** 1.2

4 a i 1260 cm² (3 sf)   **ii** 1200
  b i 3220 mm² (3 sf)   **ii** 2700
  c i 10.2 m² (3 sf)   **ii** 12
  d i 0.139 km² (3 sf)   **ii** 0.12

5 28.0 cm (3 sf)

6 0.740 m (3 sf)

7 a 6 cm, 12 cm, 37.7 cm, 113 cm²
  b 2.3 m, 4.6 m, 44.5 m, 16.6 m²
  c 15.6 mm, 31.2 mm, 98 mm, 764 mm²
  d 2.20 m, 4.40 m, 13.8 m, 15.2 m²

8 a 9.25 m (3 sf)   **b** 5.09 m² (3 sf)

9 a i 7.88 m (3 sf)   **ii** 3.45 m² (3 sf)
  b i 28.6 cm (3 sf)   **ii** 49.1 cm² (3 sf)

10 1640 mm² (3 sf)

*11 622 mm² (3 sf)

12 a 3   **b** 9

11.1A

1 No, 106.8 (1 dp) >100

2 No, 900π cm² < 1125π cm²

3 a Yes, P = 11.8 m (3 sf)   **b** 10 times

4 a $P = 15\pi + 10\pi + 5\pi = 30\pi$ cm
  b 150π cm²

5 $30\pi - 90 = 4.25$ cm (3 sf)

6 No, P = 139 cm (3 sf) < $1\frac{1}{2}$ m, A = 1220 cm² (3 sf) > 0.1 m²

7 Yes, 531 (3 sf) > 500

*8 3850 cm² (3 sf) > $\frac{3}{4} \times 4900 = 3675$ cm²

*9 123 m (3 sf)

*10 a 600   **b** Yes, 21.5% (3 sf) > 20%

11.2S

1 a i $\frac{5}{12}$ **ii** $\frac{5}{9}$ **iii** $\frac{2}{5}$
  b i 8π **ii** 10π **iii** 8π **iv** 54π

2 a 12.6 cm (3 sf)   **b** 28.6 cm (3 sf)
  c 50.3 cm² (3 sf)

3 a 11.3 cm (3 sf)   **b** 2.26 m (3 sf)
  c 78.5 mm (3 sf)   **d** 39.3 cm (3 sf)

4 a 50.9 cm² (3 sf)   **b** 1.36 m² (3 sf)
  c 785 mm² (3 sf)   **d** 147 cm² (3 sf)

5 a 6.28 mm, 28.3 mm²   **b** 1.88 m, 0.754 m²
  c 7.70 cm, 6.73 cm²   **d** 1.63 km, 0.245 km²

6 a i $(\pi + 6)$ m   **ii** $\frac{3\pi}{2}$ m²
  b i $(14\pi + 24)$ cm   **ii** 84π cm²

7 143° (3 sf)

8 229° (3 sf)

*9 124 cm² (3 sf)

*10 50 cm²

11 a 7.09 m (3 sf)   **b** 2.62 m² (3 sf)

11.2A

1 No, 251 cm (3 sf) > 250 cm.

2 a 18.8 m² (3 sf)   **b** Yes, 18.3 m (3 sf) < 20 m.

3 Yes, 3312 m (3 sf) > 300 m.

4 $(24\pi + 96)$ inches

5 a 3   **b** 120°

*6 10.9 cm² (3 sf)

*7 a $P = \frac{90}{360} \times 2\pi \times 20 \times 2 = 20\pi$ cm
  b $A = 400 - 2(400 - 100\pi) = 200(\pi - 2)$ cm²

11.3S

1 a, b, c Check students' lines are bisected at 90°.

2

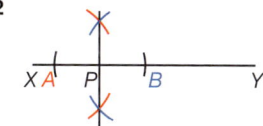
X A P B Y

3 a, b, c Students' own line.

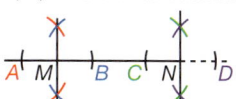
A M B C N D

  d Parallel

4 Students' own point P and line QR.

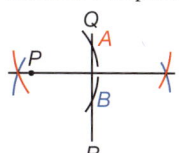
Q A P B R

5 a Students' own acute angle. **b** Students' own obtuse angle.

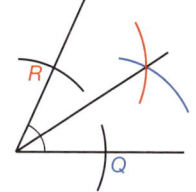
R Q R Q

493

6 a–c Students' own line.

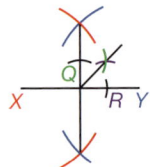

7 a, bi, iii   **ii** 120°

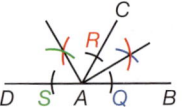

*8 a   **b**   ***c**

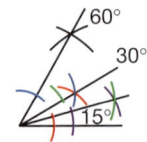   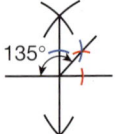   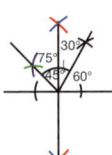

9 Check $\angle ACB = 75°$, $AC = 8.9$ cm and $BC = 7.3$ cm.

*10

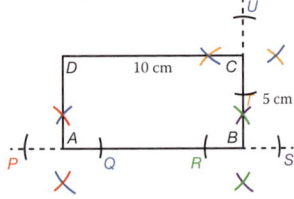

11.3A

1 a, b 120 m

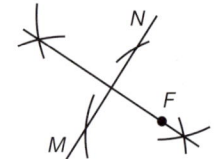

2 a

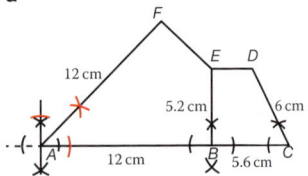

b $DE = 13$ m, $EF = 24$ m

3 $AB = r_{AB} = r_{AB}$, so ABC is an equilateral triangle with angles of 60°

4 Yes, $PR = PQ$, $RS = QS$, $PS = PS$, $\triangle PRS$ and $\triangle PQS$ are congruent (SSS).

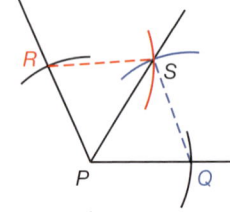

5 No, $AC = BC$, $r_A = AC$ and $r_B = BC$ so $r_A = r_B$.

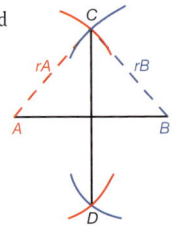

6 a i, ii iii (−1, 2)   **b i, ii iii** (−1, −1)

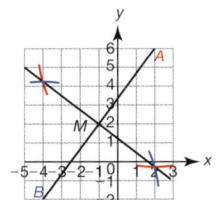   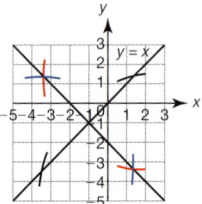

*7 a–d

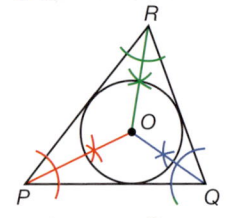

*8 a–d   **e**

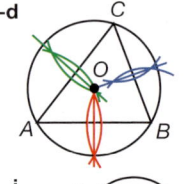   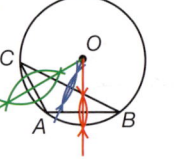

f i   **ii** It is a semi-circle.

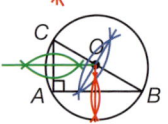

11.4S

1 a   **b**   **c**

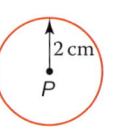

2 a

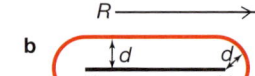

b

c

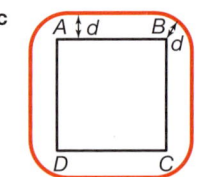

3 Check AB is bisected at 90° (3.9 cm).

4 Check $\angle PQR$ is bisected (38°).

5   **6**

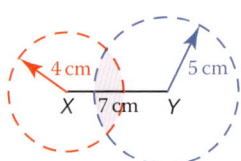

7   **8**

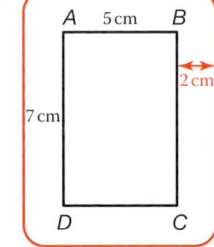   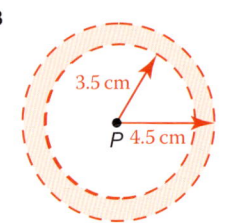

9

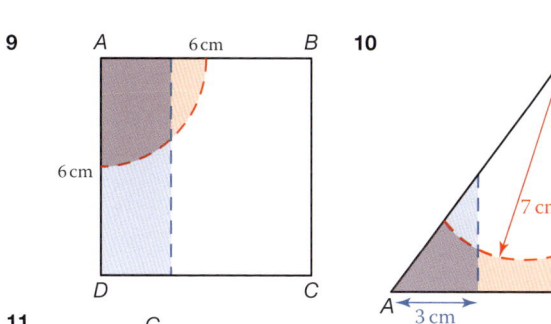

10

11

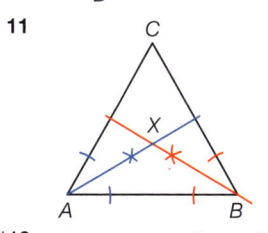

*12 a locus of P

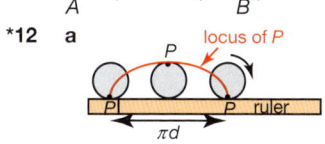

b locus of Q

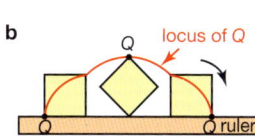

11.4A

1

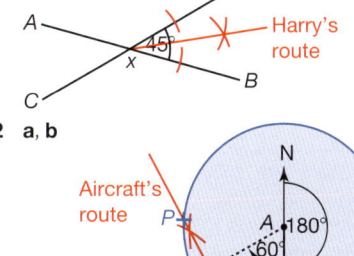

2 a, b

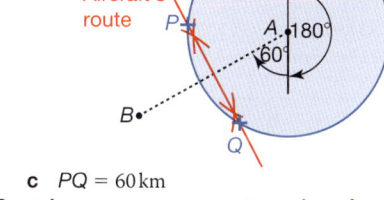

c $PQ = 60\,\text{km}$

3 a, b

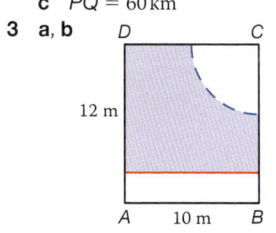

4 a, b

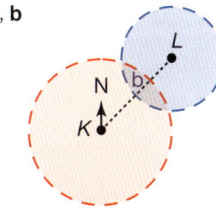

5 a

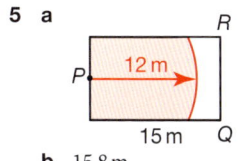

6 a

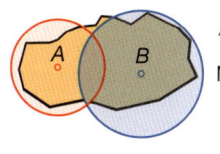

b $15.8\,\text{m}$

b Position mast B further NW.

*7 a

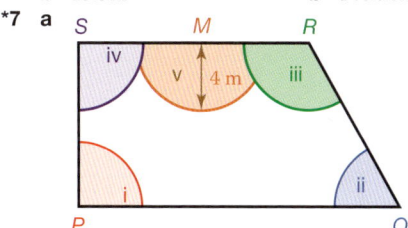

b i M
ii $P = 12.6\,\text{m}^2$, $Q = 8.38\,\text{m}^2$, $R = 16.8\,\text{m}^2$, $S = 12.6\,\text{m}^2$, $M = 25.1\,\text{m}^2$ (All answers 3 sf)

Review 11

1 a Radius **b** Chord
2 a $50.3\,\text{cm}^2$ **b** $154\,\text{cm}^2$
3 a $25.1\,\text{cm}$ **b** $44.0\,\text{cm}$
4 a $56.5\,\text{m}^2$ **b** $30.8\,\text{m}$
5 a $4.19\,\text{cm}$ **b** $8.38\,\text{cm}^2$
6 Check students' diagrams (nearest mm, degree).
    a $34°$, $44°$, $102°$ **b** $48°$, $6.4\,\text{cm}$, $62°$
    c $27°$, $6.7\,\text{cm}$, $63°$ **d** $3.4\,\text{cm}$, $120°$, $4.6\,\text{cm}$
7 a, b Check 7 cm with bisector (3.5 cm) at $90°$.
8 Check perpendicular ($90°$).
9 a, b Check both angles $= 35°$.
10 Check circle, radius 6 cm.

Assessment 11

1 a, b, c Check student's diagrams.
    d No, 11 cm > diameter of either circle.
2 a i **A** $x = 9.80\,\text{cm}$ (3 sf) **ii** **C** **iii** **B**
    b i **B** **ii** **A** **iii** **C** $y = 0.360\,\text{m}$ (3 sf)
3 A $21.5\,\text{cm}^2$ **B** $19.27\,\text{cm}^2$ **C** $51.1\,\text{cm}^2$
    a 29.3 is the incorrect answer.
4 63.7 times
5 a $14.3\,\text{cm}$ (3 sf) **b** $10.1\,\text{cm}^2$ (3 sf)
6 $\frac{\theta}{360} \times 2 \times \pi \times 10 = 10$, $\theta = \frac{360}{2\pi} = 57.3°$
7 a

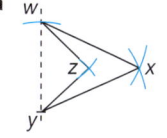

8

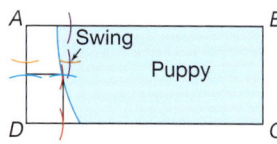

b $\angle WXY = 47°$ ($\pm 2°$),
    $\angle WZY = 276°$ ($\pm 2°$)
c Acute **d** $9.4\,\text{cm}$ ($\pm 2\,\text{mm}$)
d $4.6\,\text{cm}$ ($\pm 2\,\text{mm}$)

9

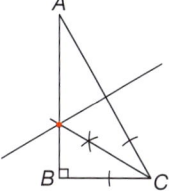

e $17.4\,\text{m}$ ($\pm 20\,\text{cm}$)
f Yes, the swing is within the locus of the puppy's lead.
10 a, b **c** 7.5 miles (± 0.5 miles)
    d 3 miles/s

Chapter 12

Check in 12

1 a 0.75 **b** 0.4 **c** 1.25
2 a $\frac{3}{5}$ **b** $\frac{1}{4}$ **c** $\frac{1}{5}$
3 $\frac{1}{3}$
4 £24

12.1S

1 a $\frac{1}{2}$ **b** $\frac{3}{4}$ **c** $\frac{1}{3}$
    d $\frac{2}{5}$ **e** $\frac{1}{3}$ **f** $\frac{1}{5}$

2 a $\frac{1}{3}$ **b** $\frac{1}{2}$ **c** $\frac{3}{10}$ **d** $\frac{1}{4}$

    e $\frac{2}{5}$ **f** 2 **g** $\frac{1}{8}$ **h** $\frac{1}{16}$

    i $\frac{1}{12}$ **j** $\frac{1}{4}$

3 a $\frac{26}{100}$ **b** $\frac{71}{100}$ **c** $\frac{2}{100}$ **d** $\frac{102}{100}$

4 a $\frac{19}{50}$, 0.38   **b** $\frac{23}{50}$, 0.46   **c** $\frac{41}{50}$, 0.82   **d** $\frac{7}{100}$, 0.07

5 a 80%   **b** 70%   **c** 75%   **d** 24%

    e 150%   **f** 104%

6 a i $\frac{3}{10}$ **ii** 30%   **b i** $\frac{5}{8}$ **ii** 62.5%

    c i $\frac{4}{9}$ **ii** 44.4%   **d i** $\frac{5}{12}$ **ii** 41.7%

7 a i $\frac{1}{2}$ **ii** 50%   **b i** $\frac{2}{7}$ **ii** 28.6%

    c i $\frac{2}{5}$ **ii** 40%   **d i** $\frac{3}{7}$ **ii** 42.9%

    e i $\frac{1}{3}$ **ii** 33.3%   **f i** $\frac{1}{2}$ **ii** 50%

    g i $\frac{1}{4}$ **ii** 25%   **h i** $\frac{4}{7}$ **ii** 57.1%

8 a i $\frac{7}{20}$ **ii** 35%   **b i** $\frac{2}{25}$ **ii** 8%

    c i $\frac{3}{10}$ **ii** 30%   **d i** $\frac{3}{40}$ **ii** 7.5%

    e i $\frac{9}{40}$ **ii** 22.5%   **f i** $\frac{3}{20}$ **ii** 15%

    g i $\frac{20}{3}$ **ii** 666.7%   **h i** $\frac{5}{2}$ **ii** 250%

    i i $\frac{103}{25}$ **ii** 412%   **j i** $\frac{32}{5}$ **ii** 640%

9 $\frac{1}{2}$, 0.5 or 50%

10 $\frac{4}{5}$, 0.8 or 80%

11 $\frac{5}{8}$, 0.625, 62.5%

12 A 30%   **B** 25.9%   **C** 23.3%   **D** 20.8%

12.1A

1 a 72%, 0.74, $\frac{3}{4}$ **b** 0.29, $\frac{3}{10}$, 31%

    c 0.78, 0.8, 83%, $\frac{7}{8}$ **d** 0.39, $\frac{2}{5}$, 41%, $\frac{3}{7}$

    e $\frac{3}{4}$, $\frac{8}{10}$, 83%, 0.84   **f** $\frac{3}{11}$, 28%, $\frac{7}{24}$, 0.3

2 a i Maths   **ii** English

    b Find the mean of the percentage scores.

3 50 cars

4 Yes, $\frac{165}{220} = \frac{3}{4}$

5 Yes, 0.124 g/ml (3 sf)

6 a i 40   **ii** 50

    b 50 employees.

7 No, both shapes are half shaded.

8 $\frac{5}{16}$

*9 a 64 cheeses   **b** 32%

12.2S

1 a 1:3   **b** 3:1   **c** 1:3   **d** 1:7

    e 1:10   **f** 5:1   **g** 3:1   **h** 1:15

2 a 2:1   **b** 3:1   **c** 9:7   **d** 5:3

3 a 2:1   **b** 1:3   **c** 3:2   **d** 9:8

    e 3:4   **f** 4:3   **g** 10:9   **h** 4:5

4 a £27:£63   **b** 287 kg:82 kg

    c 64.5 tonnes:38.7 tonnes   **d** 19.5 litres:15.6 litres

    e £6:£12:£18

5 a £40, £35   **b** £350, £650   **c** 260 days, 104 days

    d 142.86 g, 357.14 g   **e** 214.29 m, 385.71 m

6 a 2:5   **b** 11:16   **c** 5:2   **d** 5:3

    e 5:3   **f** 8:5

7 a £100:£250:£150   **b** 180°:45°:135°

    c 900 m:400 m:700 m

8 a 1:3   **b** 1:4   **c** 1:2   **d** 1:5

    e 1:2   **f** 1:3   **g** 1:5   **h** 1:3

    i 1:2   **j** 1:5   **k** 1:2   **l** 1:6

9 a 1000 m   **b** 4000 m   **c** 5000 m

    d 250 m   **e** 7250 m   **f** 5 m

10 a 325 m   **b** 0.6 cm

11 a 50 m   **b** 4 cm

12 a 200 m   **b** 12 cm

13 a 116 m   **b** 180 cm

12.2A

1 a 66.6%   **b** 360 cm

2 a 120%   **b** 102 kg

3 a 72 g   **b** Copper 115 g, Aluminium 69 g

4 $\frac{11}{50}$

5 a 300 m   **b** 120 m

6 75 cm by 200 cm

7 a 6 m, 9 m, 18 m   **b** 162 m²

8 0.7

9 a £280   **b** £80, £200

10 a 3:6:2   **b** $20

11 8 sides

12 P = 55°, Q = 125°

12.3S

1 a 0.5   **b** 0.6   **c** 0.25   **d** 0.085

    e 0.0015   **f** 0.0001

2 a £40   **b** 260 cm   **c** 3.2 kg   **d** 20 m

    e 190 p   **f** £35   **g** 3 kg   **h** £6.20

3 a 325.35 kg   **b** $120   **c** 10.35 kg   **d** 5.88 kg

    e £21.70   **f** 78.2 m

4 a £224   **b** £385.20   **c** €1458   **d** £77

    e €13.35   **f** £465.83

5 10% of £350 = £35 > £30

6 8% of £28 = £2 < £2.24

7 a £549.60   **b** £2519   **c** £842.72   **d** £1167.90

8 a £790   **b** £1109.25   **c** £54.60   **d** £132.43

9 a 1.2   **b** 1.3   **c** 1.45

10 a 0.6   **b** 0.4   **c** 0.65

11 a £495   **b** 672 kg   **c** £756   **d** 392 km

    e £658   **f** 256 m

12 a £275   **b** £2264   **c** £18060   **d** £2520

    e £4.23   **f** £2000

13 a £385   **b** 70.3 kg   **c** £550.20   **d** 491.4 km

    e 1128 kg   **f** £216

14 a £397.80   **b** 524.9 kg   **c** £1758.96   **d** 599.56 km

    e $3423.55   **f** 2154.75 m

12.3A

1 a $\frac{12}{20} = \frac{12 \times 5}{20 \times 5} = \frac{60}{100} = 60\%$ **b** $\frac{36}{75} = \frac{12}{25} = \frac{48}{100} = 48\%$

    c $\frac{24}{40} = \frac{6}{10} = \frac{60}{100} = 60\%$

2 a (19 ÷ 37) × 100% = 51.351...% = 51.4% (1 dp)

    b (42 ÷ 147) × 100% = 28.571...% = 28.6% (1 dp)

    c (8 ÷ 209) × 100% = 3.827...% = 3.8% (1 dp)

3 a 20%   **b** 20%

4 a 12.5%   **b** 20%

5 10.8%

6 £56 = 80% of the original price, the original price ≠ 120% of £56. The original price is £70.

7 £5

8 a 50   **b** 40   **c** 80   **d** 110

9 a £6   **b** £80

10 a Francesca £13.13, Frank £12.80

    b £78.03

11 a £50   **b** £30.94

12 £8450

13 13600

*14 a £321.63   **b** £1749.60

Review 12

1 a $\frac{2}{5}$ **b** 60%

2 Physics

3 a 6:7   **b** 3:5   **c** 1:400

4 a £7, £28   **b** £40, £50

5 a 1:3   **b** $\frac{1}{4}$

6 a Butter 240 g, Flour 600 g   **b** 750 g   **c** $\frac{5}{7}$

7 a 63 cm   **b** 3 cm

8 a 10 km   **b** 0.75 cm
9 a 57   **b** 106.5
10 £4060
11 a 2.5%   **b** 30%
12 a £1.20   **b** £44

Assessment 12

1 a $\frac{1}{3}$ **b** $\frac{1}{6}$ **c** $\frac{1}{8}$ **d** $\frac{1}{2}$
   e $\frac{3}{8}$ **f** $\frac{3}{4}$
2 a 0.48, 48%   **b** 108 000 rand
3 £13 110
4 a 12:5   **b** 2.4:1
5 55
6 a Chocolate 27, Plain 36   **b** 33
7 a 8:2:3   **b** Butter 45 g, Cheese 67.5 g
8 a 6:5   **b** $\frac{5}{3}$:1
9 a Girls 32, Boys 56
   b Girls 44, Boys 77, 44:77 = 4:7
10 a 200 m   **b** 1.17 km   **c** 20.75 cm
11 a 40.5 miles   **b** 25.6 in   **c** 236 miles
12 a 50p   **b** £728
13 a Gavin   **b** Gemma
   c Steven has calculated 30% of 90 kg, $90 \times 1.03 = 92.7$ kg
   d Shayda has calculated 45.4% of 550 ml, $550 \times (1 - 0.454) = 300.3$ ml
14 44.44%
15 30%
16 £2 712 500
17 90.82%
18 Yes, $(35 - 24) \div 24 = 45.83\% > 45\%$.
19 a 230   **b** 217
20 £2200
21 £2480
22 £236

Revision 2

1 9
2 72 cm²
3 a

   b 90°   **c** 30 km²
4 a 39 609 m²   **b** 3.96 hectares (3 sf)
5 a Enlargement, scale factor 3, centre (0,0).
   b, c

Translation by vector $\binom{-1}{-2}$.
6 a i B   **ii** B   **iii** D
   b 365.270, 0.0104
7 a 120 cm   **b** 40 kg
   c 110.2 cm, 41.8 kg
   d Length = 8.89%, Mass = 4.31%
   e Mass
8 a i 71.5 m², 119.5 m²   **ii** 39.375 m³, 86.625 m³
   b 3600 kg/m³   **c** 141 750 kg, 311 850 kg
9 4 cm × 25 cm

10 $r < 100$
11 10p coins 26, 50p coins 15
12 53
13 a 96   **b** 32
14 a 188 in (3 sf)   **b** 2802 in²
15 a 5 cm   **b** 19.6 cm² (3 sf)
   c 824 cm² (3 sf)   **d** 78.5% (3 sf)
16 a, b, c

No, the deckchair is not within the locus of the modem.

17 a 56 men, 48 women   **b** No, 70:60 = 7:6
   c 70 men, 60 women.   **d** 65 men, 65 women.
18 a 11.1 miles   **b** 7.3 inches
19 a 504 g   **b** 2.97% (3 sf)   **c** £10 669 (nearest £)

Chapter 13

Check in 13

1 a 49   **b** 8   **c** 1000   **d** 0.32
2 5
3 2, 3, 5, 7, 11, 13
4 1, 2, 3, 4, 6, 8, 12, 24

13.1S

1 a 1, 8; 2, 4   **b** 1, 16; 2, 8; 4, 4
   c 1, 23   **d** 1, 34; 2, 17
   e 1, 39; 3, 13   **f** 1, 44; 2, 22; 4, 11
   g 1, 42; 2, 21; 3, 14; 6, 7   **h** 1, 48; 2, 24; 3, 16; 4, 12; 6, 8
2 a 7, 14, 21   **b** 9, 18, 27
   c 12, 24, 36   **d** 15, 30, 45
   e 30, 60, 90   **f** 32, 64, 96
   g 45, 90, 135   **h** 50, 100, 150
3 a Yes   **b** Yes   **c** No   **d** Yes
   e No   **f** No   **g** Yes   **h** No
   i No   **j** No   **k** Yes   **l** Yes
   m Yes
4 a 1, 45; 3, 15; 5, 9
   b 1, 100; 2, 50; 4, 25; 5, 20; 10, 10
   c 1, 120; 2, 60; 3, 40; 4, 30; 4, 24; 6, 20; 8, 15; 10, 12
   d 1, 132; 2, 66; 3, 44; 4, 33; 6, 22; 11, 12
   e 1, 160; 2, 80; 4, 40; 8, 20; 10, 16; 5, 32
   f 1, 180; 2, 90; 3, 60; 4, 45; 5, 36; 6, 30; 9, 20; 10, 18; 12, 15
   g 1, 324; 2, 162; 3, 108; 4, 81; 6, 54; 9, 36; 12, 27; 18, 18
   h 1, 224; 2, 112; 4, 56; 8, 28; 7, 32; 14, 16
   i 1, 264; 2, 132; 3, 88; 4, 66; 6, 44; 8, 33; 11, 24; 12, 22
   j 1, 312; 2, 156; 3, 104; 4, 78; 6, 52; 8, 39; 12, 26; 13, 24
   k 1, 325; 5, 65; 13, 25
   l 1, 432; 2, 216; 3, 144; 4, 108; 6, 72; 8, 54; 9, 48; 12, 36; 16, 27; 18, 24
5 a 17, 34, 51   **b** 29, 58, 87   **c** 42, 84, 126
   d 25, 50, 75   **e** 47, 94, 141   **f** 35, 70, 105
   g 90, 180, 270   **h** 120, 240, 360   **i** 95, 190, 285
   j 208, 416, 624   **k** 144, 288, 432   **l** 111, 222, 333
6 a 2, 5, 10, 1   **b** 3, 1   **c** 5, 1   **d** 2, 4, 1
   e 7, 1   **f** 2, 3, 6, 1   **g** 2, 4, 1   **h** 3, 9, 1
7 a 30, 60   **b** 36, 72   **c** 24, 48   **d** 30, 60
   e 42, 84   **f** 60, 120
8 a 2   **b** 5   **c** 6   **d** 8
   e 15   **f** 18   **g** 25   **h** 12
   i 15   **j** 2

9 a 12   b 40   c 36   d 75
  e 54   f 150
10 2, 3, 5, 7, 11, 13, 17, 19, 23, 29, 31, 37, 41, 43, 47, 53, 59, 61, 67, 71, 73, 79, 83, 89, 97

13.1A

1 a Students' answers, for example, 220, 240, ...
  b Students' answers, for example, 105, 120, 135
  c Students' answers, for example, 72, 78, 84, 90, 96
2 Students' answers, for example, 24, 54, 84, ...
3 Students' answers, for example, 1, 2, 4, 5, ...
4 Students' answers, for example, 5, 11, 17, 23, ...
5 a Students' answers, for example, 9, 15, 21, 25, ...
  b 2
  c i Students' answers, for example, 6, 12, 18, 24, ...
    ii Students' answers, for example, 36, 66, 78, ...
6 56 s
7 20 cm
8 12
9 12 packets of burgers, 5 packets of burger buns
10 a Luca
   b Students' answers, for example, 72, 84, 90, 96
11 a 8 × 10 = 80, 2, 40; 12 × 18 = 216, 6, 36; 6 × 9 = 54, 3, 18; 15 × 20 = 300, 5, 60; 15 × 25 = 375, 5, 75
   b HCF × LCM = Product   c LCM = Product ÷ HCF
*12 a Deficient   b Abundant   c Deficient
   d Perfect   e Abundant   f Perfect

13.2S

1 a 1, 18; 2, 9; 6, 3   b 1, 14; 2, 7
  c 1, 30; 2, 15; 3, 10; 5, 6   d 1, 64; 2, 32; 4, 16; 8, 8
2 a 2   b 6   c 2   d 7
  e 6   f 8   g 12   h 9
3 a 112   b 40   c 48   d 175
  e 84   f 78   g 200   h 105
4 a 12   b 8   c 4   d 6
5 a $77 = 7 \times 11$ b $51 = 3 \times 17$ c $65 = 5 \times 13$
  d $91 = 7 \times 13$ e $119 = 7 \times 17$ f $221 = 13 \times 17$
6 9, 3, 3; $18 = 2 \times 3 \times 3$
7 a 8   b 75   c 40   d 63
  e 180   f 441
8 a 2×3^2 b $2^3 \times 3$ c $2^3 \times 5$ d 3×13
  e $2^4 \times 3$ f 2×41 g $2^2 \times 5^2$ h $2^4 \times 3^2$
  i $2^2 \times 3^2 \times 5$ j $3^2 \times 5 \times 7$
  k $2^2 \times 3 \times 37$ l $2 \times 3^3 \times 5^2$
9 a $2^2 \times 3^2$ b $2^3 \times 3 \times 5$ c 2×17 d 5^2
  e $3^2 \times 5$ f $2 \times 3^2 \times 5$ g 3^3 h 2^6
  i $2^3 \times 3^2$
10 a $2^2 \times 263$ b $2^9 \times 5$ c $2 \times 3^2 \times 5 \times 7$
   d $3 \times 5^2 \times 11$ e $5 \times 11 \times 13$ f $7 \times 11 \times 13$
   g 3×73 h 17^2 i $2^3 \times 5 \times 71$
   j $5 \times 7^2 \times 11$ k $7 \times 13 \times 19$ l $2 \times 3^2 \times 11 \times 17$
   m $2^2 \times 11 \times 13 \times 17$
   n $2^2 \times 23 \times 31$
   o $3^3 \times 13 \times 29$
11 a 5   b 16   c 3   d 5
   e 14   f 15
12 a 48   b 800   c 66   d 416
   e 280   f 5040
13 a 1260, 60   b 8085, 7   c 1680, 48   d 9216, 2
   e 314706, 2   f 82944, 16
14 a 420, 5   b 252, 2

13.2A

1 a $126 = 2 \times 3^2 \times 7$ b $210 = 2 \times 3 \times 5 \times 7$
2 a 2×3^2, $2 \times 3 \times 5^2$ b $2^3 \times 5$, $2 \times 3 \times 5^2$
  c $2 \times 3 \times 5^2$ d $2^3 \times 5$

3 a Students' answers, for example, 8, 12, 27, 28, 40, 42, 44, ...
  b Students' answers, for example, 16, 24, 36, 54, 81, 135, ...
  c Students' answers, for example, 108, 112, 120, 162, 168, 176, 180, 200, 208, 243, 252, 264, 270, 272, 280, 300.
  d Students' answers, for example, 64, 96.
4 $6 \times 5 \times 7$, $3 \times 7 \times 10$, $3 \times 5 \times 14$, $2 \times 7 \times 15$, $2 \times 5 \times 21$, $2 \times 3 \times 35$, $1 \times 6 \times 35$, $1 \times 10 \times 21$, $1 \times 14 \times 15$, $1 \times 1 \times 210$
5 a $1815 = 3 \times 5 \times 11^2$
  b $3 \times 5 \times 121$, $3 \times 55 \times 11$, $15 \times 11 \times 11$, $5 \times 33 \times 11$, $1 \times 15 \times 121$, $1 \times 33 \times 55$
6 120 s
7 18 cm
8 a LCM(6 and 9) = 18, LCM(12 and 18) = 36
  b Yes, both numbers have an extra factor of 2 in common which will become part of the LCM.
  c Yes, both numbers have an extra factor of 2 in common so the HCF doubles.
9 45
10 12

13.3S

1 a 1, 4, 9, 16, 25, 36, 49, 64, 81, 100, 121, 144, 169, 196, 225, 256, 289, 324, 361, 400
  b 1, 8, 27, 64, 125, 216, 343, 512, 729, 1000
2 a 1 + 8 = 9   b 25 + 100 = 125
3 a 36   b 121   c 196   d 529
  e 961   f 2209   g 64   h 216
  i 512   j 2197   k 5832   l 9261
4 a 13   b 16   c 28   d 80
  e 36   f 233   g 71   h 47
  i 509   j 1071
5 a 6.25   b 2401   c 32.77   d 23.04
  e 53.29   f 117.65   g 1.44   h 0.25
  i 98.01   j 970.30   k 25 cm²   l 64 m³
6 a 5   b 3   c 4   d 1
  e 2   f 8
7 a 6.32   b 7.81   c 13.42   d 15.78
  e 26   f 31.62
8 a 2   b 5   c −1   d 4
  e 10   f −3
9 a 3.42   b 8   c 15   d −2.15
  e 1.53   f −2.99
10 a 2   b 2   c 3   d 3
   e 3   f 3
11 a 81   b 1   c 128
   d 729   e 1000000   f 625
12 a 1728   b 46656   c 9261
   d 1048576   e 2197   f 256
13 a 2   b 3   c 2
   d 3   e 5   f 6
14 a 3723.88   b 387.42   c 1667.99
   d 217678.23   e 2460.38   f −1338278.22
15 a 0.01   b 0.001   c 0.0001
   d 0.25   e 0.0625   f $\frac{1}{27}$
16 a 25   b 40   c 259 947
   d 8000   e $\frac{5}{256}$ f $\frac{7}{6}$

13.3A

1 2, 5, 8, 10, 13, 17, 18, 20, 25, 26, 29, 32, 34, 37, 40, 41, 45
2 1F, 2C, 3E, 4A, 5B, 6G, 7H, 8D
3 144 cm
4 17
5 a 12 and 13   b 17 and 18   c 8 and 9
6 2.645751 is only an approximation of $\sqrt{7}$
7 56 and 57
8 15
9 12, 13, 14
10 a 27   b The only cube number less than 40 is 27.

11 Students' answers, for example, $(a, b) = (3, 5), (5, 3), (7, 13), \dots$

12 Students' answers, for example, $(p, q) = (4, 20), (3, 6), (6, 69), \dots$

13 a i Students' answers, for example, any number x where $0 \leq x \leq 1$.

    ii Students' answers, for example, any number x where $0 \leq x \leq 1$.

  b $0 < x < 1$ then $x^2 < x < \sqrt{x}$, $x = 0, 1$ then $x^2 = x = \sqrt{x}$, $x > 1$ then $x^2 > x > \sqrt{x}$, $x < 0$ then $x^2 > x$

14 a $2^2 \times 19^2$ **b** 38

Review 13

1 a 1, 2, 3, 4, 6, 12   **b** 1, 19

  c 1, 2, 3, 4, 6, 8, 12, 16, 24, 48

2 a 5, 10, 15, 20, 25   **b** 13, 26, 39, 52, 65

  c 18, 36, 54, 72, 90

3 a No, 5 is also a factor.   **b** Yes, exactly two factors.

  c No, 3 and 19 are factors.

4 a 2, 3   **b** 1, 2, 3, 9, 18   **c** 18, 54, 72

5 Check students' factor trees, $72 = 2 \times 2 \times 2 \times 3 \times 3$

6 a $2 \times 3 \times 3$ **b** $2 \times 2 \times 7$

7 a

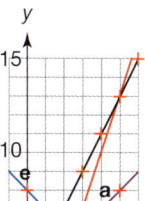

Prime factors of 60   Prime factors of 70

  b i 10   **ii** 420

8 a 4   **b** 27   **c** 1   **d** 1

9 a 12   **b** 24   **c** 99   **d** 143

10 a i 2×3^3 **ii** $2^3 \times 5^2$

  b LCM $= 2^3 \times 3^3 \times 5^2$, HCF $= 2$

11 a 4   **b** 9   **c** 2   **d** 10

12 a 9   **b** 49   **c** 125   **d** 32

13 a 3   **b** 3

Assessment 13

1 No, $1 \times 42, 2 \times 21, 3 \times 14, 6 \times 7$.

2 a 3 is not a factor.

  b i 100, 102, 104, 106, 108   **ii** 102, 105, 108

    iii 100, 104, 108   **iv** 100, 105

    v 102, 108   **vi** 105

    vii 104   **viii** 108

    ix 100

3 a No, the last two digits are not divisible by 4.

  b 5, 10, 11

4 No, $54 = 2 \times 3 \times 3 \times 3, 99 = 3 \times 3 \times 11$, HCF $= 3 \times 3 = 9$.

5 $36 = 2 \times 2 \times 3 \times 3, 60 = 2 \times 2 \times 3 \times 5$,

  LCM $= 2 \times 2 \times 3 \times 3 \times 5 = 180$

6 3rd

7 9 pm

8 $28 = 1 \times 2 \times 4 \times 7, 1 + 2 + 4 + 7 + 14 = 28$.

9 a 4   **b** 6   **c** 84

10 Isa, $2^3 \times 3^2 \times 5^2 \times 11$.

11 a Yes, $1024 > 625$.   **b** Yes, $1024 > 100$.

  c No, $16 = 16$.

12 a No, $33 = 3 \times 11$, factors $= 1, 3, 11, 33$.

  b No, all the numbers in column **A** are odd.

  c No, 2 is the only even prime number and column **B** is even.

  d 43 and 47

  e No, all the numbers in column **D** are multiples of 4.

13 a $25 = 2 + 23$ **b** $16 = 3 + 13 = 5 + 11$

  c $36 = 5 + 31 = 7 + 29 = 13 + 23 = 17 + 19$

14 a Glen, 3^5 **b** Giorgia, 14^6

15 a George (correct), no negative number can have a square root. Sam (partly correct), all positive numbers have 2 square roots.

  b No, a positive number has a positive cube root, a negative number has a negative cube root.

16 23

17 a $41 - 0 + 0^2 = 41, 41 - 3 + 3^2 = 47, 41 - 6 + 6^2 = 71$

  b $41, 41 - 41 + 41^2 = 41^2$

Chapter 14

Check in 14

1 a 8   **b** 5   **c** 8   **d** $3\frac{1}{2}$

2 a -1 **b** -5 **c** 1   **d** -7

3

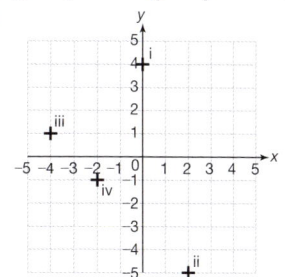

14.1S

1 $A (4, 1), B (-3, 6), C (0, -1), D (-2, 3),$

  $E (-3, -2), F (0, 7), G (5, 3), H (5, 0)$

2 a y-axis, $x = 0$

  b Yes, this line is labelled 'x'.

3 a $y = 4, 5, 6, 7, 8$ **b** $y = 3, 3, 3, 3, 3$

  c $y = 9, 5, 3, 7, 11$ **d** $y = -5, -2, 1, 4, 7$

  e $y = 6, 8, 3, 4, 7$

4

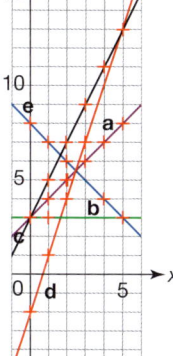

5 a $(0, 2), (1, 1)$

  b $(0, 9), (1, 8)$

  c $(0, 8) (1, 6)$

  d $(0, 4), (2, 3)$

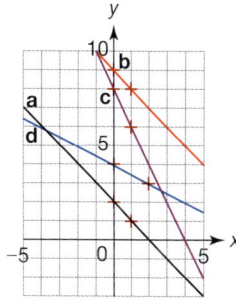

6 a–d

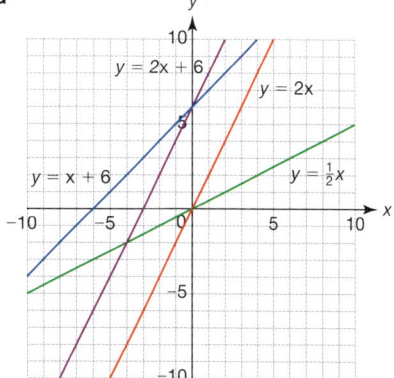

$y = 2x + 6$

$y = 2x$

$y = x + 6$

$y = \frac{1}{2}x$

*7

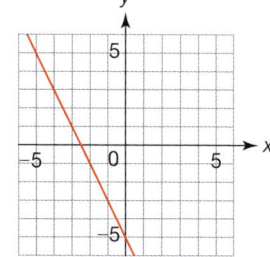

8 a $y = 5$ **b** $y = 4$ **c** $y = -1$ **d** $y = -3$
e $x = -5$ **f** $x = -2$ **g** $x = 1$ **h** $x = 3$

9

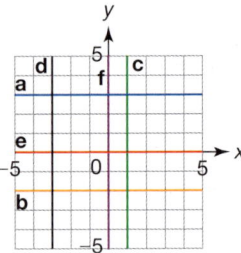

10 $(4, 3)$

14.1A

1 a

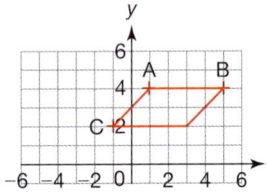

b, *c $(3, 2), (-5, 2), (7, 6)$

2

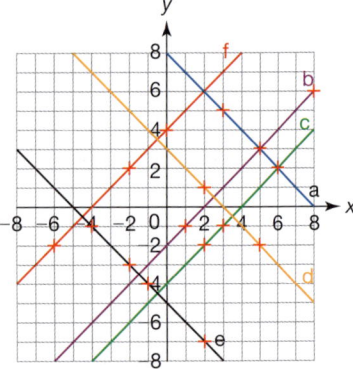

Students' answers, for example,
a $(3, 5)$ $3 + 5 = 8$ **b** $(1, -1)$ $1 - -1 = 2$
c $(3, -1)$ $3 - -1 = 4$ **d** $(2, 1)$ $2 + 1 = 3$
e $(-4, -1)$ $-4 + -1 = 5$ **f** $(0, 4)$ $0 - 4 = -4$

3

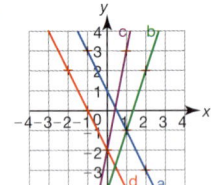

Students' answers, for example,
a $(2, -3)$ $2 \times 2 + -3 = 1$
b $(2, 2)$ $3 \times 2 - 2 = 4$
c $(1, 3)$ $5 \times 1 - 3 = 2$
d $(-2, 2)$ $2 \times -2 + 2 = -2$

4

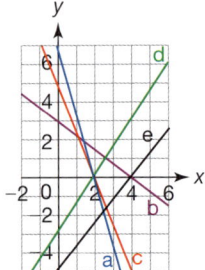

5 a $(0, 3)$ **b** $(0.5, 2.5)$ **c** $(2, -2)$ **d** $(-1, 2)$
6 $(3, 2)$
7 a A **b** E **c** C
 d D **e** B **f** F
*8 $x + y = 4$
9 a £280

b

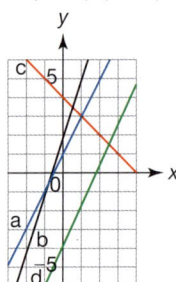

c Adam £225, Ben £240
d Adam 12, Ben 15
e Students' answers, for example, Adam because he is cheaper for this number of guests.

14.2S

1 a 2 **b** -2 **c** 0
 d 3 **e** 2 **f** 8
2 a 1 **b** 3 **c** 1 **d** $\frac{1}{2}$
 e -2 **f** -1
3 a $y = x + 2$ **b** $y = 3x - 2$
 c $y = x$ **d** $y = \frac{1}{2}x + 3$
 e $y = -2x + 2$ **f** $y = 8 - x$
4 a $y = x - 1$ **b** $y = \frac{3}{2}x$
 c $y = \frac{1}{2}x + 3$ **d** $y = 2 - \frac{1}{2}x$
5 a $y = x + 4$ **b** $y = 3x + 2$
 c $y = 2x + 5$ **d** $y = 5 - x$.
6 a i 7 **ii** 9 **b i** -3 **ii** $\frac{1}{2}$ **c i** 23 **ii** 1
 d i -2 **ii** 7 **e i** 1 **ii** 2 **f i** 10 **ii** -1
7 a $y = 3x + 1$ **b** $y = x + 7$
 c $y = -2x$ **d** $y = \frac{1}{2}x - 2$
 e $y = -\frac{1}{3}x - 3$ **f** $y = 4$
8 a i $(0, 1), (1, 3)$ **ii** 1 **iii** 2
 b i $(0, 2), (1, 5)$ **ii** 2 **iii** 3
 c i $(0, 4), (1, 3)$ **ii** 4 **iii** -1
 d i $(0, -5), (1, -3)$ **ii** -5 **iii** 2

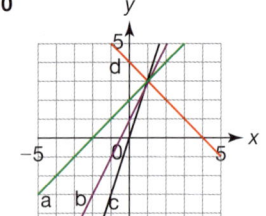

9 For any equation $y = mx + c$, $m =$ gradient and $c = y$-intercept.

10

11 **A**, **C**, **E** and **B**, **F** are parallel, they have the same gradient.
12 a E **b** F **c** B
 d D **e** C **f** A
13 a Students' answers, for all four functions m must be the same.
 b Students' answers, any four functions of the form $y = mx + 6$.
 *c Students' answers, any two functions whose gradients have a product of -1.

14.2A

1 a $y = 6x - 9$ **b** $y = -10x + 18$ **c** $y = \frac{1}{4}x + 1$
 d $y = 2x - 7$ **e** $y = 1 - 2x$ **f** $y = 5x - 7$
 g $y = 3x + 8$

2 a $y = 3x - 1$ **b** $y = x + 3$
  c $y = -3x + 6$ **d** $y = \frac{1}{2}x - 4$
3 a $y = x + 1$ **b** $y = 3x + 2$ **c** $y = \frac{1}{2}x - 1$
  d $y = -2x + 6$ **e** $y = -2x + 3$ **f** $y = -\frac{1}{3}x - 1$
4 a, b As a increases, the y-intercept increases.
  c, d As a increases, the gradient of the graph increases.
5 a E **b** D **c** F **d** B **e** A **f** C
6 $y = 3x - 16$
7 $p = 8$
*8 a $q = 23$ **b** $y = \frac{2}{3}x - 1$
9 a $y = \frac{20}{3}x + 50$
  b i Growth per year   **ii** Height at birth

14.3S

1 a i 20 km   **ii** 40 km
  b i 45 min   **ii** 2 hrs 15 min
2 a 10 : 45 am   **b** She stops for 30 min.   **c** 9 : 30 am
3 a 40 km   **b** 15 min   **c** 45 min
  d 4 pm   **e** 80 km/h
  f 14 : 00 to 14 : 30, this is the steepest section of the graph.
  g i 35 km   **ii** 55 km   **iii** 140 km
4 a 30 km/h   **b** 20 km/h   ***c** 12.5 km/h
5 a 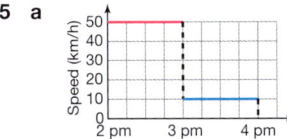   **b**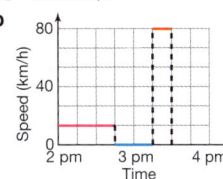

*c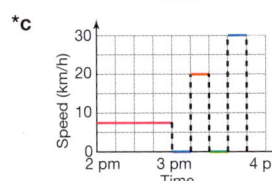

6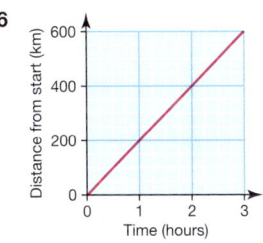

14.3A

1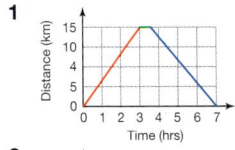

2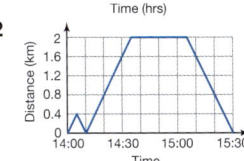

3 a 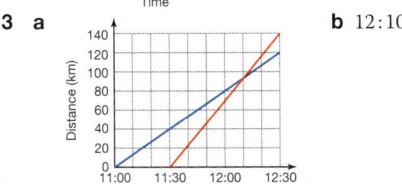   **b** 12 : 10

4 a Sian   **b** Asif overtook Sian.   **c** Asif
  d i 6 m/s   **ii** Sian, 2 s

5 a Time goes backwards in the 2nd section.
  b Distance is always positive.
  c The distance changes instantaneously in the 2nd section.
  d Time goes backwards in the 2nd and 4th section.
6 a C   **b** A   **c** B
7 a i 0.03 km/s^2 **ii** 0.05 km/s^2
  b 0.042 km/s^2 (2 sf)

Review 14

1 $A\,(2, 5), B\,(-3, 5), C\,(0, -4), D\,(3, 0), E\,(-2, 3)$
2 a $y = -6, -3, 0, 3, 6$ **b** $y = 3, 5, 7, 9, 11$
3 a 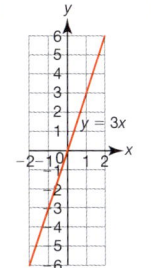   **b**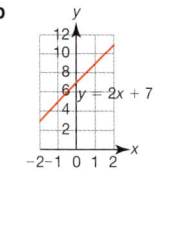

4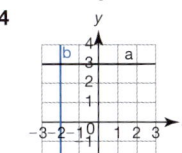

5 a 3   **b** $y = 3x + 1$
6 a $y = -4, -1, 2, 5, 8, 11$ **b** $y = 25, 20, 15, 10, 5, 0$
    $m = 3, c = -1$ $m = -5, c = 20$

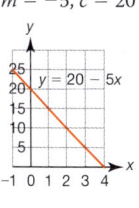

7 Students' answers, $y = 3x + c$, where c is any number.
8 $y = 7x - 4$
9 $y = 4x + 2$
10 a 4 km   **b** Speed   **c** 10 km/h

Assessment 14

1 a 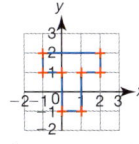   **b** T

2 a **A** $y = -1, 1, 3, 5, 7, 9$ **b**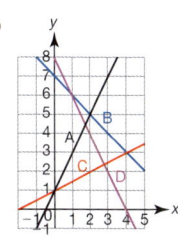
    B $y = 8, 7, 6, 5, 4, 3$
    C $y = \frac{1}{2}, 1, \frac{3}{2}, 2, \frac{5}{2}, 3$
    D $y = 10, 8, 6, 4, 2, 0$

  c i $(2, 5)$ **ii** $(0, 1)$ **iii** $(4, 3)$ **iv** $(1, 6)$
3 a–d $y = 6 - x$ **e** 4.5

4 No, point (0, 2) does not line on the line, $2 \times 0 + 3 \neq 2$.

5 a Yes   **b** $y = x - 1$

6 a $2, 7$ **b** $4, 9$ **c** $6, -11$ **d** $-4, 12$
    e $\frac{4}{7}, -2$ **f** $\frac{-15}{14}, \frac{35}{14}$

7 a $y = x + 1$ **b** $y = -x + 1$
    c $y = 2x + 6$ **d** $y = -4x + 13$

8 a i D to E   **ii** E to F   **iii** C to D
      iv B to C   **v** A to B
    b D to E, the gradient is steepest in this section.

9

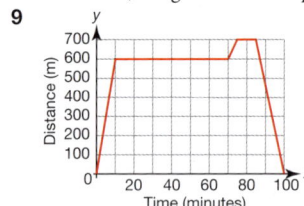

10 a 1 **b** -8 **c** -1 **d** $-\frac{1}{2}$

11 a $y = -5x - 61$ **b** $y = 3x - \frac{1}{2}$
    c $y = \frac{1}{3}x - 2$ **d** $y = \frac{1}{4}x - \frac{3}{4}$

12 a $y = 7x + 5$ **b** $y = -3x - 7\frac{1}{4}$
    c $y = -\frac{1}{3}x + \frac{8}{3}$

13 a A 2   B $\frac{3}{2}$ C $-\frac{5}{3}$ D $-\frac{2}{3}$
    b i C   **ii** B   **iii** A   **iv** D

Chapter 15

Check in 15

1 a $40\,\text{cm}^2$ **b** $30\,\text{cm}^2$

2 a Diameter   **b** Circumference   **c** Radius

15.1S

1 a 12, 8, 6; 9, 6, 5; 18, 12, 8; 6, 4, 4; 8, 5, 5   **b** $V + F = E + 2$

2 a   **b**    **c**

Cylinder   Cone   Sphere

3 a 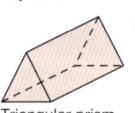   **b**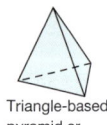

Triangular prism   Triangle-based pyramid or tetrahedron

4 a, b, c, d

5 Other nets are possible.

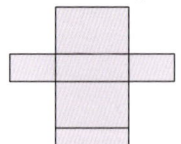

6 a 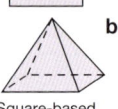   **b**    **c**

Square-based pyramid   Cylinder   Triangular prism

7 a   Plan   Front   Side

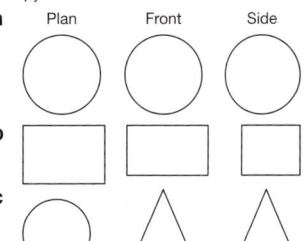

b

c

8 a 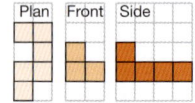   **b**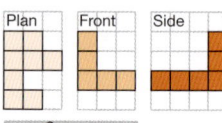

9 a 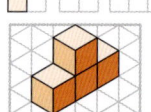   **b**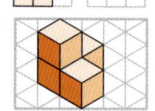

*__10__ Check students' 3D models.

15.1A

1 a The triangles need to be congruent.
    b

2 $a = 7.6\,\text{cm}$, $b = 11.5\,\text{cm}$, $c = 23.9\,\text{cm}$ (3 sf)

3 a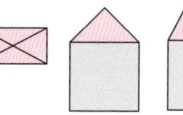
       Plan   Front   Side
    b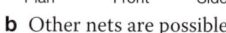
       Plan   Front   Side

4 a 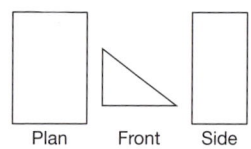   **b** Other nets are possible.
      Plan   Front   Side

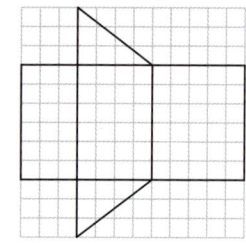

    c 12 cm by 12 cm

5

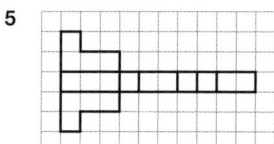

6 a i, ii Other nets are possible.

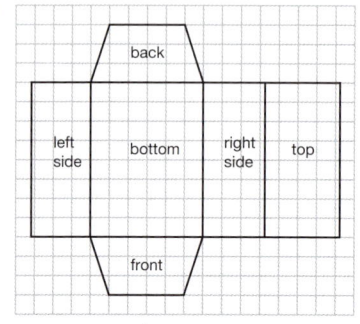

    b Students' answers, for example, the tray may not fit inside the sleeve, make the tray slightly smaller.

*__7__ **a**

    b No, $64.8\,\text{cm} > \frac{1}{2}\,\text{m}$.   **c** 8

*8 a Other nets are possible.

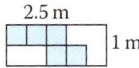

2.5 m
1 m

b 0.5 m

15.2S

1 a 160 cm³   **b** 3500 mm³
   c 94.5 m³   **d** 6615 cm³
2 a 283 cm³ (3 sf)   **b** 75.4 m³ (3 sf)
   c 503 cm³ (3 sf)   **d** 1570 mm³ (3 sf)
3 a 4800 mm³   **b** 270 cm³
4 126 litres (3 sf)
5 Volume 756 mm³, Height 3 cm, Length 8 m, Width 7 mm
6 a 420 cm³   **b** 4.41 kg
7 a 75 000 cm³   **b** 52.5 kg
*8 19 200 kg or 19.2 tonnes
*9 a 8.02 cm (3 sf)   **b** 7.48 cm (3 sf)
*10 66.24 litres

15.2A

1 a $V = 23.04$ litres < 24 litres   **b** 43.4 cm (3 sf)
2 5, assuming that the skip is filled exactly to the top.
3 a 160, assuming accurate dimensions and all of the metal is used.
   b 1.125 kg
4 5, assuming filled to the top. 6, assuming filled to 6 cm.
5 24 min (nearest min)
6 120
7 Sally, 3050 cm³ (3 sf) $>$ 2430 cm³ (3 sf)
*8 4.30 kg (3 sf)
*9 No, l = 15 m, h = 6 m, V = 482 m³ (3 sf) $<$ 500 m³.
*10 a 22.8 cm × 15.2 cm × 10.5 cm, assuming accucrate dimensions, the cans fit exactly and the thickness of the box can be ignored.
   b 78.5% (3 sf)

15.3S

1 All answers have been rounded to 1 dp.
   a i 452.4 cm³   **ii** 226.2 cm²   **iii** 326.7 cm²
   b i 589.0 cm³   **ii** 235.6 cm²   **iii** 392.7 cm²
   c i 6434.0 cm³   **ii** 804.2 cm²   **iii** 2412.7 cm²
   d i 10429.9 cm³   **ii** 1846.0 cm²   **iii** 2648.3 cm²
   e i 1760.3 cm³   **ii** 253.3 cm²   **iii** 1467.2 cm²
   f i 313.2 mm³   **ii** 96.4 mm²   **iii** 361.8 mm²
2 All answers have been rounded to 1 dp.
   a i 615.7 cm²   **ii** 1436.8 cm³
   b i 7238.2 mm²   **ii** 57905.8 mm³
   c i 2123.7 cm²   **ii** 9202.8 cm³
   d i 1520.5 cm²   **ii** 5575.3 cm³
   e i 7854.0 cm²   **ii** 65449.8 cm³
   f i 12076.3 cm²   **ii** 124788.2 cm³
   g i 3761.0 cm²   **ii** 21688.4 cm³
   h i 824.5 cm²   **ii** 2226.1 cm³
   i i 2222.9 mm²   **ii** 9854.7 mm³
   j i 20.0 m²   **ii** 8.4 m³
3 a i 36 cm²   **ii** 120 cm³
   b i 135 mm²   **ii** 900 mm³
4 a 10 368 mm³   **b** 3456 mm²
5 All answers have been rounded to 1 dp.
   a i 678.6 cm²   **ii** 1017.9 cm³
   b i 25.4 m²   **ii** 8.5 m³
   c i 628.3 cm²   **ii** 1005.3 cm³
   d i 217.9 cm²   **ii** 185.2 cm³
   e i 70.7 mm²   **ii** 39.3 mm³
   f i 10159.9 m²   **ii** 63862.3 m³
6 a i 144π cm²   **ii** 288π cm³
   b i 2304π mm²   **ii** 18 432π mm³

15.3A

1 a 66 113 cm³   **b** 352 m³ (3 sf)
2 a 8 cm   **b** 12 mm
3 3 cm
4 a $x = 20$ mm   **b** 30 mm
5 15 cm
6 $V_{cone} = \frac{1}{3}(\pi \times (2r)^2 \times r) = \frac{4}{3}\pi r^2 = V_{sphere}$
7 $\frac{2}{3}SA_{cylinder} = \frac{2}{3}(2\pi r^2 + \pi \times 2r \times 2r) = \frac{2}{3} \times 6\pi r^2 = 4\pi r^2 = A_{sphere}$
8 a 7.46 cm (3 sf)   **b** 3.91 cm (3 sf)
*9 4.92 cm (3 sf)
*10 a i 528π cm²   **ii** 1920π cm³
   b Radius = 9 cm, Height = 12 cm, Slant height = 15 cm
11 a 1560 cm² (3 sf)   **b** 5790 cm³ (3 sf)
12 a 384.8… ml (cm³) $<$ 400
   b i 10.4 cm (3 sf)   **ii** 306 cm² (3 sf)
*13 a Other nets are possible, 84 cm².

4 cm
6 cm
6 cm

   b 9.9 cm by 9.9 cm
   c 86% (nearest percent)

*14 a 508 cm² (3 sf)   **b** $66\frac{2}{3}$%

Review 15

1 a 6   **b** 8
2 a i Cone   **ii** 1 flat and 1 curved, 1, 1
   b i Triangular prism   **ii** 5, 9, 6
3 a   **b**

4

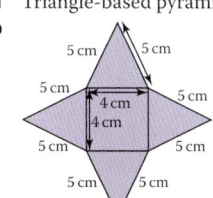

5 a 12 m³   **b** 34 m²   **c** 30 kg
6 a 7600 cm³   **b** 1380 cm²   **c** 2140 cm²
7 37.7 cm²
8 2150 cm³

Assessment 15

1 a Yes   **b** No, Tetrahedron   **c** No, Sphere
   d Yes   **e** No, Hexagonal-based prism
2 a Triangle-based pyramid
   b

5 cm   5 cm
5 cm   5 cm
4 cm
4 cm
5 cm   5 cm
5 cm   5 cm

3 a A   **b** B
4 0.372, 2.75, 5, 8, 9, 6.25
5 a 512   **b** 64   **c** 8
   d i 18   **ii** 26
6 1 × 1 × 60, 1 × 2 × 30, 1 × 3 × 20, 1 × 4 × 15, 1 × 5 × 12, 1 × 6 × 10, 2 × 2 × 15, 2 × 3 × 10, 2 × 5 × 6, 3 × 4 × 5
7 a True
   b False, 1 000 000 cubic cm
   c False, 1 000 000 000 cubic mm
8 a 432
   b No, large$_{SA}$: small$_{SA}$ = 666 : 13, large$_V$: small$_V$ = 432 : 1
9 a 22 200 cm³ (3 sf)   **b** 0.378 m²
10 a 50.4 in²   **b** 13.5 in³
11 9.93 cm³
12 a 47666.7 cm³ (1 dp)   **b** 91

13 a 4.71 cm³ (3 sf)   **b** 12.3 g (3 sf)
14 a Yes, 13.5 × 0.3 × 0.3 = 6 × 0.45 × 0.45 = 1.215 cm³.
  b Chunky Dunkies = 11.205 cm² < 16.38 cm²
15 a 94.2 cm³ (3 sf)   **b** 104 cm³ (3 sf)

Lifeskills 3

1   40
2 a

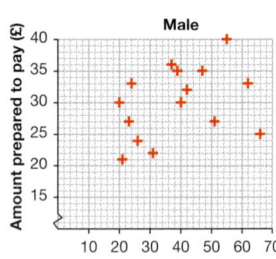
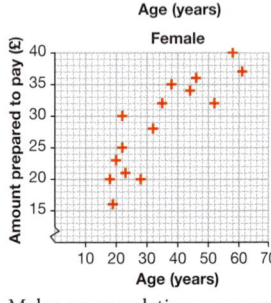

  b Males, no correlation, no connection between age and
     amount prepared to pay. Females, positive correlation, older
     females are prepared to pay more than younger females.
  c No, they should increase their sample size.
3 a

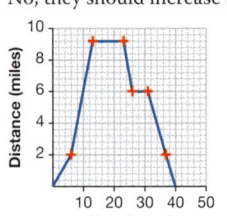

                           b 1:41 pm
4 a

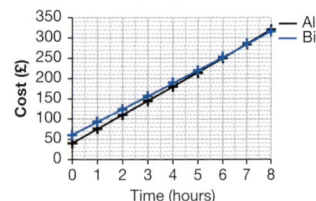

  b $t > 6$ hours 40 minutes
  c Students' answers, for example, Alan because he is cheaper.
5 a 17240.7 cm³ (1 dp)   **b** 1449000 cm³ = 1.449 m³
  c 35.7% (1 dp)   **d** 396.5 cm² (1 dp)
  e 43.0 cm² (1 dp)   **f** 10.6 cm (1 dp)
  g 18 cans   **h** Cartons, you can fit more in the cupboard.
6 a £866
  b

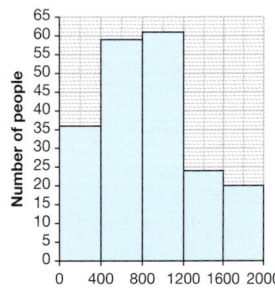

  c Yes, the median class is £800 - £1200.   **d** £216.50

Chapter 16

Check in 16

1 a 04:58   **b** 20:45
2 a 80, 80 ,80, 82, 83, 83, 84, 84, 84, 84
  b 45, 45, 46, 48, 48, 48, 48, 49, 49, 49

16.1S

1 a 4, 9, 16, 11, 10   **b** 50
2 a 8, 6, 7, 10, 3, 6   **b** 40
  c

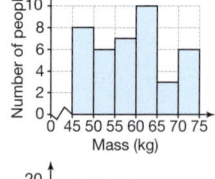

                              d 14, 17, 9
  e

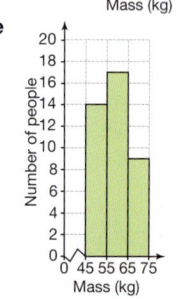

3 a 8, 8, 13, 6, 5
  b

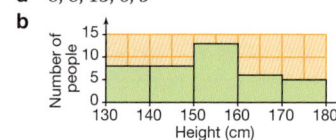

4

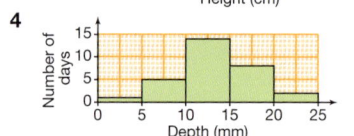

5

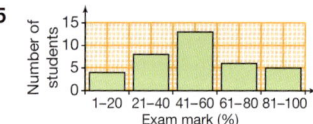

16.1A

1 a i 2   **ii** 1
  b 85 - 90 m   **c** 10
2 a i No   **ii** No
     iii Yes, smallest range = £200.01 − £39.99 = £160.02
  b 42   **c** $\frac{1}{3}$
3 $x = 5$, Frequency = 3, 5, 9, 13, 10, Check students' histogram.
4 7, 11, 9, 5, 4

16.2S

1 a 1, 6, 8, 3, 4, 3   **b** 70 to 74   **c** 70 to 74
2 a 3   **b** 23, 28, 33, 38
  c 23, 168, 66, 38; 29.5 mph
3 a i $10 < t \leq 15$ **ii** $15 < t \leq 20$ **iii** 16.1
  b i $10 < t \leq 20$ **ii** $20 < t \leq 30$ **iii** 23.5
  c i $5 < t \leq 10$ **ii** $10 < t \leq 15$ **iii** 14.7
  d i $15 < t \leq 25$ **ii** $25 < t \leq 35$ **iii** 30
4 a $165 \leq h < 170$ **b** 164.3 cm   **c** $165 \leq h < 170$

16.2A

1 a December 50 miles, January 59.7 miles
  b December, modal $40 < m \leq 60$, median $40 < m \leq 60$.
     January, modal $40 < m \leq 60$, median $40 < m \leq 60$

c The most common journey length is the same.
The mean is greater for January than December.

2 a Teachers 23.3 miles, Office workers 35.3 miles
b Teachers $20 < t \leqslant 30$, Office workers $30 < t \leqslant 40$
c On average, office workers take longer to travel home.

3 No, Mean 53.125 g, Range 40 g.

4 Yes, modal class and median class $24.5 \leqslant w < 25.5$, mean 25.28 g, all round to 25 g.

5 Machine A estimated mean = 0.3007 mm, median and modal class $0.30 \leqslant t < 0.31$ Machine B estimated mean = 0.2907, median and modal class $0.28 \leqslant t < 0.29$.

16.3S

1 a Negative correlation   **b** No correlation
c Positive correlation

2 a A Poor exam mark, lots of revision
B Very good exam mark, lots of revision
C Very good exam mark, little revision
D Poor exam mark, little revision
E Average exam mark, average amount of revision
b A Not much pocket money, equal eldest
B Lots of pocket money, equal eldest
C Lots of pocket money, equal youngest
D Not much pocket money, equal youngest
E Average pocket money, middle age
c A Low fitness level, lots of hours in gym
B Good fitness level, lots of hours in gym
C Good fitness level, few hours in gym
D Low fitness level, few hours in gym
E Medium fitness level, medium hours in gym

3 a No correlation   **b** Positive correlation
c Negative correlation

4 a
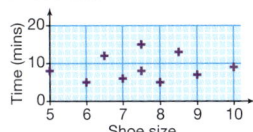
b Positive correlation
c i Increases   **ii** Decreases

5 a
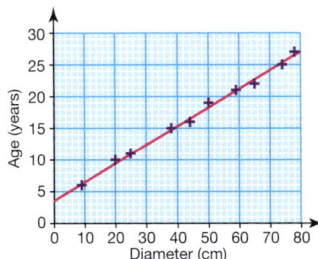
b No correlation
c No relationship

16.3A

1 a Positive correlation
b (24, 80), the student performed poorly in Paper 1 and well in Paper 2.
c 35
d 110%, no this is impossible.

2 a 16   **b** 24
c (20, 2), the student performed well on Paper 1 and poorly on Paper 2.

3 a, c
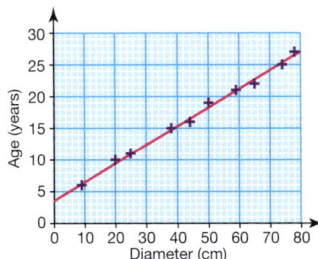

b Positive correlation   **d** 20 years
e ≈ -6 cm
f Predictions outside the range of data values can be unreliable.

4 a, c
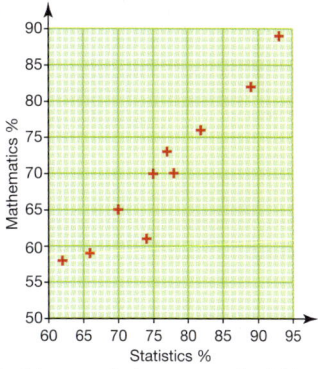

b Positive correlation   **d** 68%
e Predictions outside the range of data values can be unreliable.

16.4S

1
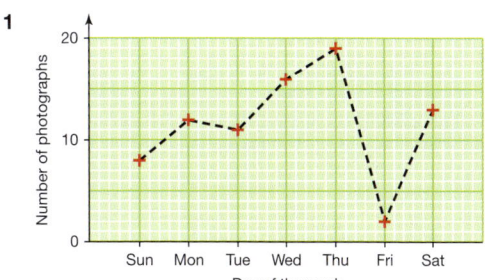

2
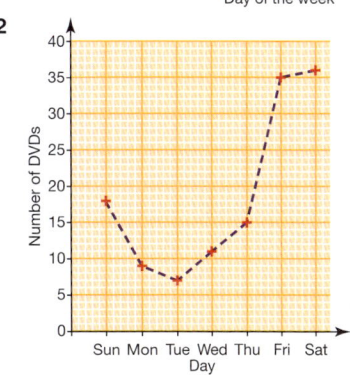

3
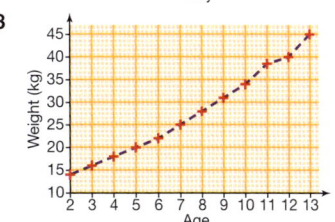

4
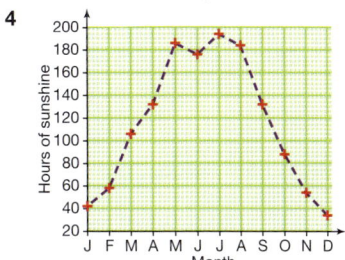

5

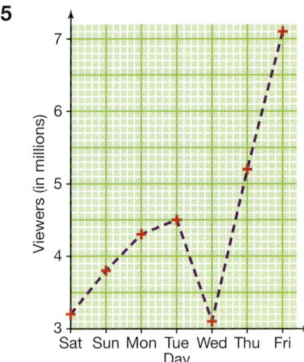

16.4A

1 a

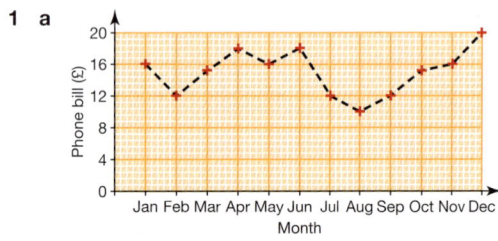

b Typical phone bills are about £15. These fall during the Summer months and peak in December

2 a

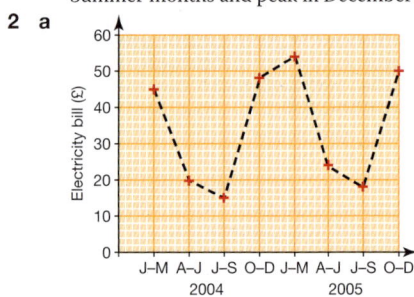

b Electricity bills are highest in the Winter months and lowest in the Summer months. This annual pattern repeats itself; there is a slight trend for bills to rise from year-to-year.

3 a

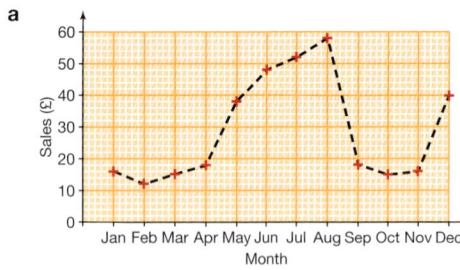

b Icecream sales grow steadily during Spring and Summer but drop sharply in the Autumn. Sales are low during Autumn and Winter except for a peak in December.

4 a

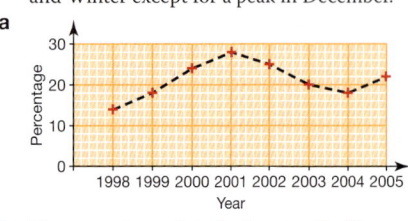

b The percentage of students using the library grows steadily from about 15% in 1998 to 28% in 2001. It has since fallen back to around 20%.

5 The average suggests Sell-a-lot, 116 is a valid range for both data sets.

Review 16

1 4, 5, 6, 1
2 a $12 \leqslant l < 14$ **b** $10 \leqslant l < 12$ **c** 11.5 cm
3 a Negative
   b No, the correlation may be a coincidence or y be affecting x.
4 a, c

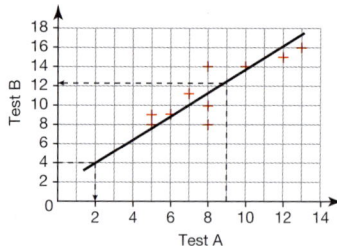

   b Positive   **d** 12−13
   e i 2−3
      ii Unreliable estimate because there is no data in this range.

5

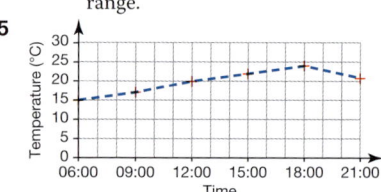

Assessment 16

1 a

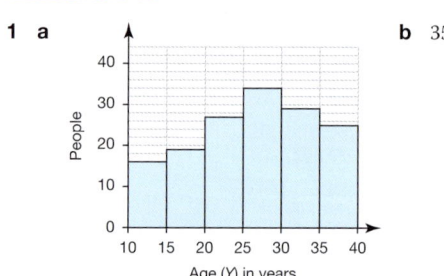

   b 35

2 Students' answers, for example, split all the classes into two equal parts. The police officer can now identify who is breaking the speed limit and record that no people were travelling at less than 10 mph.

3 a 20 − 30   **b** It is the highest bar.
   c 76   **d** 20 − 30

4 a

b 6.5
c d is always recorded as positive, so 395 tea bags and 405 tea bags would both be recorded as a difference of 5.

5 a

b Strong positive correlation
c i 20.2 cm
   ii 320 g
d No

Answers

6 a

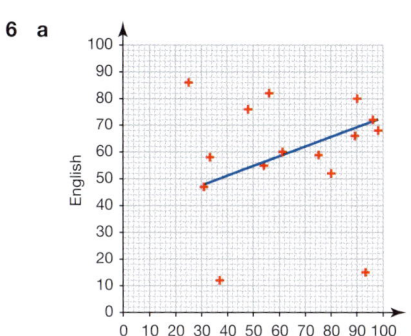

b Weak positive correlation

c (93Sc, 15E), (25Sc, 86E)

d i 42 (±2)
  ii 56 (±2)

7 a 23 (±1) **b** 18 (±1) **c** 25 (±1)

d The graph just shows a trend, there are no actual figures.

e 41

f It coincides with morning, lunchtime and afternoon/evening surgeries.

8 a

Sales (£1000s) plotted J F M A M J J A S O N D, Year 2 and Year 1

b Winter sales in Year 1 were lower than in Year 2. Summer sales in Year 1 were higher than in Year 2.

Chapter 17

Check in 17

1 a 45 **b** 100
2 a 0.23 m **b** 240 000 cm

17.1S

1 a 5 **b** 3 **c** 4 **d** 1
  e 2 **f** 6
2 a 6.32 **b** 7.81 **c** 13.42 **d** 15.78
  e 26 **f** 35.13
3 a 3 **b** 8 **c** 15 **d** 4.64
  e 29 **f** 25.27
4 a 25 **b** 8 **c** 27 **d** 64
  e 144 **f** 169
5 a 81 **b** 1 **c** 128 **d** 729
  e 1 000 000 **f** 256
6 a 1728 **b** 46 656 **c** 9261 **d** 1 048 576
  e 2197 **f** 7 529 536
7 a 5 **b** 6 **c** 1 **d** 1
  e 1 **f** 41 **g** 1 **h** 0
8 a 7^3 **b** 3^3 **c** 5^3 **d** 6^4
  e 5^2 **f** 8^5 **g** 9^6 **h** 8^8
9 a 6^5 **b** 4^9 **c** 2^{13} **d** 11^7
  e 1^{30} **f** 7^{12} **g** 3^9 **h** 9^{10}
10 a 7^2 **b** 8^4 **c** 3^1 **d** 9^3
  e 4^6 **f** 2^{-7} **g** 12^2 **h** 1
11 a 2^6 **b** 4^{10} **c** 7^4 **d** 5^{15}
  e 3^{16} **f** 6^4 **g** 5^{21} **h** 10^{16}
  i 8^{28}
12 a 3^1 **b** 5^{10} **c** 4^2 **d** 7^2
  e 8^{11} **f** 9^{10}
13 a 4^2 **b** 6^4 **c** 9^2 **d** 8^1
  *e 5^3 ***f** 6^5 ***g** 8^2 ***h** 1
*14 a 4^4 **b** 3^6 **c** 6^5 **d** 5^{12}

e 2^2 **f** 7^{12} **g** 3^8 **h** 9^4

17.1A

1 $3^2 + 3^2 + 3^2 = 9 + 9 + 9 = 27 \neq 9^2$
2 a $36 \times 216 = 6^2 \times 6^3 = 6^5 = 7776$ **b** 1296
3 a 6 **b** 6 **c** 10 **d** 5
  e 3 ***f** 3
4 p^3 has factors 1, p, p^2 and p^3.
5 a Incorrect, $2^3 \times 2^5 = 2^8$. **b** Correct
  c Correct **d** Incorrect, $(7^3)^4 = 7^{12}$.
  e Incorrect, $4^5 \times 4^3 = 4^8$. **f** Correct
6 0^8, 8^0, 1^7, 7^1, 6^2, 2^6, 5^3, 3^5, 4^4
7 a i and **iv** **b** Students' answers, for example, 6 and 8.
8 a 10^3 **b** 10^4 **c** Level 8
*9 a i 4^3 and 8^2 **ii** 2^9 and 2^{10}
  iii 10^2 and 1000 **iv** $5 \times 5 \times 5 \times 5$ and 125
  v 4^3 and $\frac{1}{4^3}$ **vi** 4^3 and $\frac{1}{4^3}$
  vii $5 \times 5 \times 5 \times 5$ and 125 **viii** 0.2 and 125
  b i $3^2 + 3^2$ **ii** $2^{10} (= 1024)$ **iii** $\frac{1}{4^3}$
*10 a 7 **b** −3
11 a 2 **b** 2

17.2S

1 a $\frac{8}{15}$ **b** $\frac{29}{35}$ **c** $\frac{5}{56}$ **d** $\frac{44}{45}$
  e $\frac{39}{8}$ **f** $\frac{59}{60}$ **g** $\frac{19}{40}$ **h** $\frac{341}{19}$
2 a $\frac{1}{2}$ **b** $\frac{5}{3}$ **c** $\frac{25}{16}$ **d** $\frac{8}{15}$
  e $\frac{14}{9}$ **f** $\frac{66}{10}$ **g** $\frac{147}{55}$ **h** $\frac{238}{9}$
3 Recurring **1 a, b, c, d, f, h**, and **2 b, d, e, g, h.**
  Terminating **1 e, g** and **2 a, c, f.**
4 a $\frac{17}{30}$ **b** $\frac{154}{87}$ **c** $\frac{1}{3}$ **d** $\frac{207}{1715}$
  e $\frac{505}{324}$ **f** $-\frac{8}{21}$
5 a 2π **b** 3π **c** $2 + \pi$ **d** 10π
  e 18π **f** 16π **g** $\frac{21\pi}{2}$ **h** 12π
6 a 32π **b** $7 + 2\pi$ **c** $4(7 + \sqrt{2})$ **d** 62π
  e 6π **f** 4π **g** $\frac{(2 + \pi)}{3}$ **h** $\frac{(3 - 2\pi)}{12}$
7 a $\frac{1}{4}$ ft wide by $\frac{3}{8}$ ft long **b** $\frac{5}{6}$ in tall **c** $\frac{5}{8}$ in wide
8 a 29 m **b** $17 + 18\pi$ m **c** 51 m² **d** $\frac{9\pi}{2}$ m²
  e $51 + 9\pi$ m²
9 a–d Yes, $\frac{5}{4}$

17.2A

1 a 2200π **b** **D** **c** $\frac{15000}{\pi}$
  d Joshua **e** Lex
2 a $6\frac{1}{6}$, $6\frac{3}{8}$, $6\frac{11}{24}$, $6\frac{1}{2}$ **b** 24π, 30π, $33\frac{1}{3}\pi$, 34π
3 a i $13\frac{1}{2}$ in × 17 in **ii** $6\frac{3}{4}$ in × $8\frac{1}{2}$ in **iii** $6\frac{3}{4}$ in × $4\frac{1}{4}$ in
  b i 34 in **ii** $13\frac{1}{2}$ in **iii** 459 in²
  c Yes, $8 \times 6\frac{3}{4} \times 4\frac{1}{4} = 13\frac{1}{2} \times 17$
4 a i $2(2 + \pi)$ **ii** $12 + \pi$ **iii** 4π **iv** $4 + 3\pi$
  *b $4\pi = \pi r^2$, $r = 2$ and $\sqrt{4} = 2$
5 a $2\frac{1}{4}\pi$ in² **b** 3π in **c** $12\frac{3}{4}\pi$ in²
  *d $\frac{15}{32}\pi$ in³ ***e** $9\frac{9}{16}\pi$ in³
6 a 320π
  b Yes, $225\pi > 220\pi$, she doesn't ice the base of the cake.
  *c No, 305π, you don't need to ice the bases of the two layers.
  *d 190π

17.3S

1 a 10^2 **b** 10^1 **c** 10^3 **d** 10^0
  e 10^4 **f** 10^6 **g** 10^5 **h** 10^8
2 a 10^{-2} **b** 10^{-1} **c** 10^{-3} **d** 10^{-5}

e 10^{-4} **f** 10^{-7} **g** 10^{-6} **h** 10^0

3 a 1000 **b** 1 000 000 **c** 100 000 **d** 1 000 000 000
   e 10 000 **f** 10 **g** 100 **h** 10 000 000

4 a 1 **b** 0.01 **c** 0.00001 **d** 0.001
   e 0.0000001 **f** 0.1 **g** 0.0001 **h** 0.000001

5 a 10^5 **b** 10^9 **c** 10^8 **d** 10^3
   e 10^4 **f** 10^4

6 a 10^{-2} **b** 10^{-4} **c** 10^{-8} **d** 10^{-8}
   e 10^{-8} **f** 10^6

7 a 2×10^2 **b** 8×10^2 **c** 9×10^3 **d** 6.5×10^2
   e 6.5×10^3 **f** 9.52×10^2 **g** 2.358×10^1 **h** 2.5585×10^2
   i 3×10^{-1} **j** 4.7×10^{-3} **k** 7.8×10^{-5} **l** 4.485×10^{-1}

8 a 500 **b** 3000 **c** 100 000 **d** 250
   e 4900 **f** 3 800 000 **g** 750 000 000 000
   h 8 100 000 000 000 000 000

9 a 6×10^2 **b** 4.5×10^4 **c** 6.5×10^0 **d** 5×10^6
   e 2.8×10^{-1} **f** 4×10^{-2} **g** 1.35×10^{-3} **h** 1.2×10^{-7}

10 a 4×10^5 **b** 9×10^7 **c** 2.5×10^8 **d** 2.4×10^{13}
   e 5×10^{-1} **f** 9.2×10^{-8}

11 a 2×10^2 **b** 2×10^4 **c** 5×10^1 **d** 7.5×10^2

12 a 7.74×10^{-3} **b** 9.63×10^5 **c** 4.38×10^{-5} **d** 2.55×10^2
   e 3.4×10^5 **f** 4.47×10^{-2}

17.3A

1 a 1×10^{-2} km **b** 2×10^{-3} g
   c 5×10^{-6} m **d** 1.1×10^{-2} l

2 a Correct **b** 3.28×10^{13}
   c 4.3×10^5 **d** Correct

3 10^3

4 No, 3.33×10^5 (3 sf).

5 1×10^{45}

6 5×10^2

7 a 3.33×10^{-7} s (3 sf) **b** 9.46×10^{15} m (3 sf)
   c 7.9×10^8 s $= 25.1$ years (3 sf)

8 2.0×10^{-26} kg

9 1.79×10^6

10 4.3×10^4 km

11 a Jupiter **b** 2.668612×10^{27}
   c No, Jupiter is about 3 times the mass of Saturn.
   d Jupiter **e** Jupiter and Saturn
   f 5.53×10^{-2}

Review 17

1 a 5^{10} **b** 3^4 **c** 7^{12}
   d 2^0 **e** 3^3 **f** 4^4

2 a 128 **b** 16 **c** 5 **d** 64
   e 1 **f** 27 **g** 6 **h** 2

3 a $\frac{3}{5}$ **b** $\frac{3}{14}$ **c** $\frac{3}{5}$ **d** $\frac{13}{24}$
   e $\frac{31}{24}$ **f** $2\frac{1}{20}$ **g** $2\frac{4}{9}$ **h** $\frac{11}{14}$
   i $\frac{2}{7}$ **j** $\frac{1}{6}$

4 a $1\frac{1}{5}$ m^2 **b** $\frac{5}{8}$ m^2

5 a $3 + 2\pi$ **b** 5π **c** $\frac{5\pi}{12}$ **d** $\frac{9\pi}{2}$

6 a i 64π **ii** $\frac{25}{2}\pi$
   b i 16π **ii** $5\pi + 10$

7 a 1.37×10^9 **b** 5.46×10^7
   c 6.97×10^{-2} **d** 6.25×10^{-5}

8 a 350 000 **b** 821 000 000
   c 0.0027 **d** 0.000 000 207

9 a 6×10^{10} **b** 7×10^4 **c** 4×10^5
   d 6×10^{-2} **e** 2.424×10^5 **f** 3.42×10^3

Assessment 17

1 a Soraya $0^2 = 0 \times 0 = 0$ **b** Peter $49^{\frac{1}{2}} = \sqrt{49} = 7$
   c Soraya $(-3)^2 = -3 \times -3 = 9$

2 a 9.261 **b** 2687 (3 sf) **c** 974 (3 sf)

d 29 **e** 8.77 (3 sf) **f** 8.68 (3 sf)

3 a No, 15^{12} **b** Yes, 3^{20} **c** Eliza, $(3^4)^0 = 3^0 = 1$
   d $7^{7 + 2 - 6} = 7^3$

4 12.167 cm^3

5 a 216 **b** 5

6 a 5 **b** 4 **c** 4

7 Students' answers, for example, $\frac{1}{4} + \frac{2}{8} = \frac{1}{2}$, $\frac{1}{3} + \frac{2}{12} = \frac{1}{2}$, etc.

8 a $\frac{1}{5}$ **b** 192

9 a 40 **b** 15

10 a $\frac{14}{15}$ **b** $\frac{1}{15}$ **c** 54

11 a Juliet, £350 **b** £9030

12 $\frac{7}{10}$

13 a English Channel 29 000 mi^2, Baltic Sea 1 46 000 mi^2, Bering Sea 876 000 mi^2, Caribbean Sea 1 060 000 mi^2, Malay Sea 3 140 000 mi^2, Indian Ocean 28 400 000 mi^2
   b 1×10^9, 1
   c i 4 **ii** 2 **iii** 6 **iv** -3
   d i 0.0008 joules **ii** 10 763 km
   e 4×10^{-6} km **f** 4 mm

14 a 8 710 **b** 199 times

15 a 1.332×10^{-6} **b** $\frac{1}{751\,000}$

16 a 226 km (3 sf) **b** 2.26×10^2 km

Chapter 18

Check in 18

1 a 4 **b** 5.061 **c** -26.368

2

3 $y = 1, 3, 5, 7, 9$

18.1S

1 a $y = 11, 6, 3, 2, 3, 6, 11$ **b** $y = 2, -1, -2, -1, 2, 7, 14$
   c $y = 6, 2, 0, 0, 2, 6, 12$ **d** $y = 4, 0, -2, -2, 0, 4, 10$
   e $y = -7, -2, 1, 2, 1, -2, -7$ **f** $y = 9, 4, 1, 0, 1, 4, 9$
   g $y = 18, 8, 2, 0, 2, 8, 18$ **h** $y = 6, 3, 1, 0, 0, 1, 3$

2

3 a i 2 **ii** $(0, 2)$ **iii** No roots
   b i -2 **ii** $(0, -2)$ **iii** ±1.4 (1 dp)
   c i 0 **ii** $(-0.5, -0.25)$ **iii** $-1, 0$
   d i 0 **ii** $(-1.5, -2.25)$ **iii** $-3, 0$
   e i 2 **ii** $(0, 2)$ **iii** ±1.4 (1 dp)
   f i 1 **ii** $(-1, 0)$ **iii** -1
   g i 0 **ii** $(0, 0)$ **iii** 0
   h i 0 **ii** $(0, 0)$ **iii** $-0.5, 0$

4 a $y = 2 - x^2$ **b** It has a $-x^2$ term.

5 a, b i

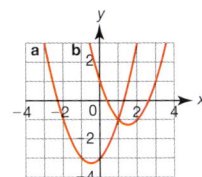

a ii $(-0.5, -3.25)$
   iii $-2.3, 1.3$ (1 dp)
b ii $(1.5, -1.25)$
   iii $0.4, 2.6$ (1 dp)

6 a i 4   **ii** $(0, 4)$ **iii** $-2, 2$

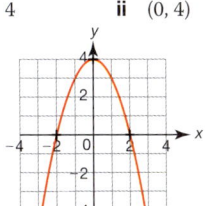

b i 2   **ii** $(1, 3)$ **iii** $-0.7, 2.7$ (1 dp)

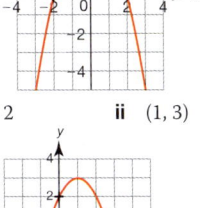

*7

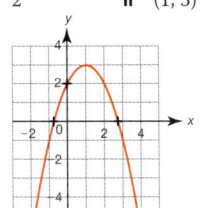

8 a $0, -7$ **b** $0, 7$ **c** $-2, -3$
   d $-3, -4$ **e** -2 **f** $-1, 6$

9 a

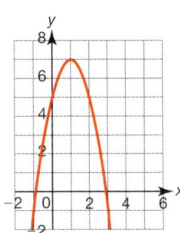

b i B, C and D   **ii** A and D

18.1A

1 a Yes, $f(x) = (x + 1)(x - 3), x = 3$ **b** -1
   c $(-1, 0), (3, 0), (0, -3)$ **d** $(1, -4)$
2 a $-3, -9$ **d**
   b $(-9, 0), (-3, 0), (0, 27)$
   c $(-6, -9)$

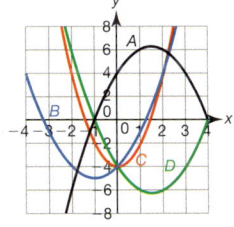

3 a Yes, $y = x(6 - x), x = 0$ **b** 6   **c** $(3, 9)$

4

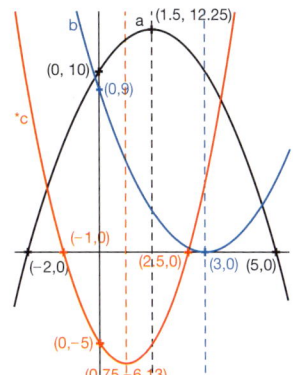

5

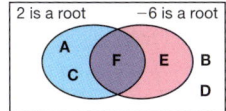

6 a

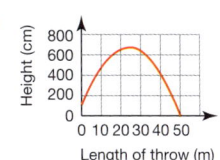

b 676 cm
c 50 m

7 a

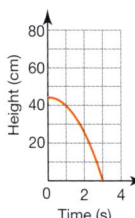

b 44.1 m   **c** 3 s

8 a 2   **b** 0   **c** 1   **d** 2
9 a Vertical translation by a units.
   b Horizontal translation by $-a$ units.
10 a $-4, -2$ **b** $-5, -1$ **c** $-7, 1$ **d** -3
   e $-4.4, -1.6$ (1 dp)

18.2S

1 a $y = -26, -7, 0, 1, 2, 9, 28$
   b $y = 27, 8, 1, 0, -1, -8, -27$
   c $y = -\frac{1}{4}, -\frac{1}{3}, -\frac{1}{2}, -1, 1, \frac{1}{2}, \frac{1}{3}, \frac{1}{4}$
   d $y = -1\frac{1}{4}, -1\frac{2}{3}, -2\frac{1}{2}, -5, 5, 2\frac{1}{2}, 1\frac{2}{3}, 1\frac{1}{4}$
   e $y = -1.875, 0, 0.375, 0, -0.375, 0, 1.875, 6$
   f $y = \frac{1}{2}, \frac{2}{3}, 1, 2, -2, -1, -\frac{2}{3}, -\frac{1}{2}$

2 a, b   **c, d, f**

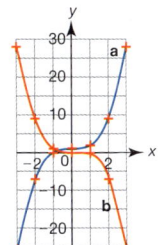   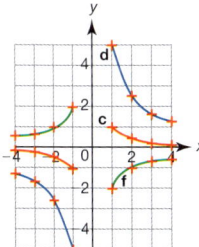

e

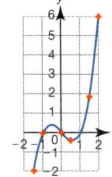

3   **4**

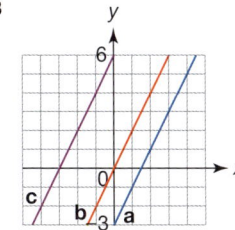   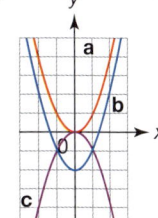

5

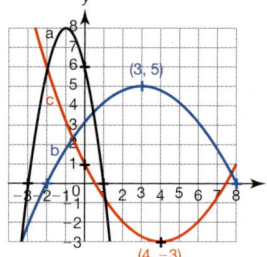

6

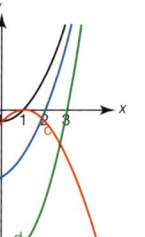

*7

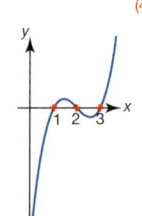

8

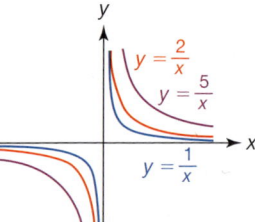

$y = \dfrac{2}{x}$

$y = \dfrac{5}{x}$

$y = \dfrac{1}{x}$

*9

$x = -1.9, 0.4, 1.5 \,(1\,\text{dp})$

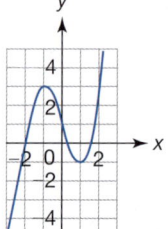

10 a, b The line rotates about $(-1, 0)$, $y = a(x + 1)$, so a varies the gradient and y-intercept but the line always passes through $(-1, 0)$.

11 a Vertical translation by a units.

   b i Horizontal translation by $-a$ units.
     ii Vertical stretch by factor a.
    iii Vertical translation by a units.
    iv Vertical stretch by factor a.

18.2A

1 a B **b** D **c** C **d** A

2 a i 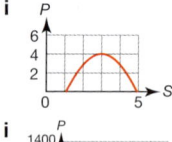   **ii** £1 and £5   **iii** £3   **iv** £4

   b i 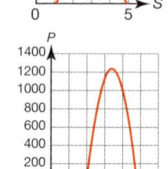   **ii** £50 and £120
                               iii £85
                               iv £1225

3 a £506.25
   b $P = £0$ when $S = £15, £60$, Max P when $S = £37.50$

4 a 400 cm   **b** 39 m

*5 a 0.5 m   **b** 0.625 m   **c** $y = -10x^2 + 5x$

6 a A, 3   **b** B, 2
   c D, 1   **d** C, 4

*7 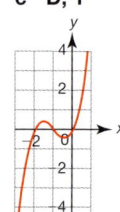   ***8**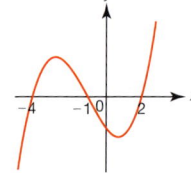

Answers

18.3S

1 a i £960   **ii** £1068   **iii** £954   **b** £833.34

2 a 30 km/h   **b** 60 km/h
   c The train slows down and stops.

3 a 1:45 pm   **b** 45 min
   c i 1 pm and 1:45 pm   **ii** 20 km/h

4 a 1400 m   **b** Ben   **c** 1 min 45 s   **d** Matt

5 a 40 kg   **b** 2.6 m/s   **c** Quadratic

6 a

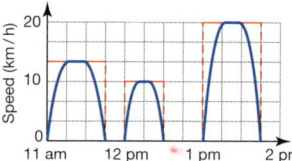

  **b No, speed cannot change instantly.

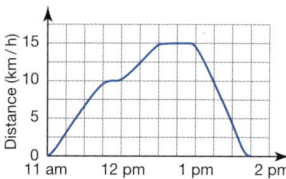

18.3A

1 a Quadratic   **b** Cubic (not exact)

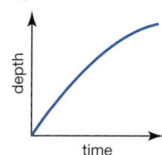   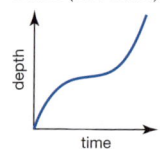

  c Linear   **d** No standard function (linear + quadratic)

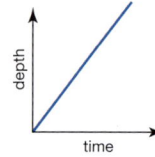   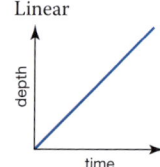

  e No standard function   **f** Linear

*2 a 15, 7.5, 5, 3.75, 3, 2.5
   b   **c** Reciprocal
                             d 85

3 a

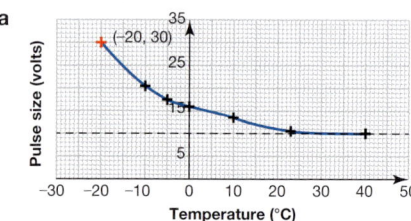

  b Reciprocal

c Yes, $(-20, 30)$ fits with the trend of the data and the lower limit of 11.8 V suggests there is an asymptote at $y = 11.8$

4 a Check students' descriptions. Bret starts at 11 am, $s = 13.3$ km/h, $d = 10$ km. 11:45 Bret rests for 15 min. 12 pm, $s = 10$ km/h, $d = 5$ km. 12:30 Bret rests for 30 min. 1 pm Bret cycles home, $s = 20$ km/h, $d = 15$ km.

b Check students' race commentaries. Ben 500 m/min, Matt 400 m/min. At 1 min Ben 200 m/min, Matt 400 m/min. At 1 min 45 s Matt overtakes Ben. At 2 min 30 s Ben 200 m/min, Matt 320 m/min. At 3 min Ben 400 m/min, Matt 320 m/min. At 3 min 40 s Matt wins. At 4 min Ben finishes the race.

c Check students' short stories. Ali starts at 18:00, $s = 1.34$ cm/min, $d = 20$ min. 18:15 Ali gets in and sits for 20 min, $d = 30$ cm. 18:35 Ali lets water out $s = 2$ cm/min, $d = 25$ cm, then adds water $s = 3$ cm/min, $d = 32.5$ cm. 18:40 Ali sits for 15 min. 18:55 Ali gets out $d = 22.5$ cm and lets water out, $s = 2.25$ cm/min until $d = 0$, 19:05.

*5 a, b

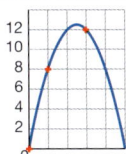

c $y = -0.5(x - 5)^2 + 12.5$ **d** $(10, 0)$
e Students' answers.

Review 18

1 a $(0, -5)$ **b** $(-1, 0), (5, 0)$ **c** -1 and 5 **d** $(2, -9)$
2 a $y = 16, 9, 4, 1, 0, 1, 4, 9, 16$

b

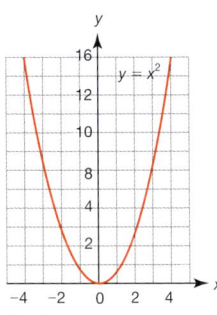

c $(0, 0)$

3 a

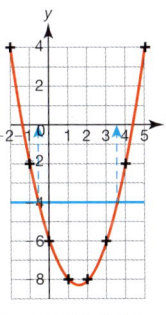

b i $-1.4, 4.4$ (1 dp)
ii $-0.6, 3.6$ (1 dp)

4

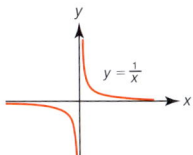

5 a Reciprocal, $y = \dfrac{2}{x}$ **b** Linear, $y = 2x + 3$
c Quadratic, $y = x^2 + 3x$ **d** Cubic, $y = x^3 - 4x$
6 a 15 km/h **b** 20 km/h
7 a

b Speed is decreasing.
c Constant speed 3 m/s
d 1 m/s²

Assessment 18

1 a

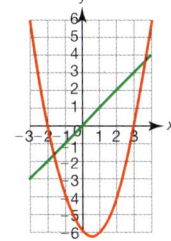

b $(-1.6, -1.6)$ and $(3.6, 3.6)$
c $(0.5, -6.25)$
d $x = -2, 3$
e $x = -2, 3$

2 a $d = 8.7, 9.3, 10, 10.7, 11.3, 12, 12.7, 13.3, 14, 14.7$
b

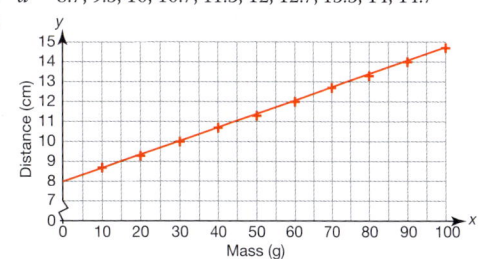

c i 13 cm **ii** 15 g **d** $\dfrac{1}{15}$ **e** 8 cm

3 a $d = 0, 6, 17, 33, 53, 79, 110, 146, 187, 233, 283$
b

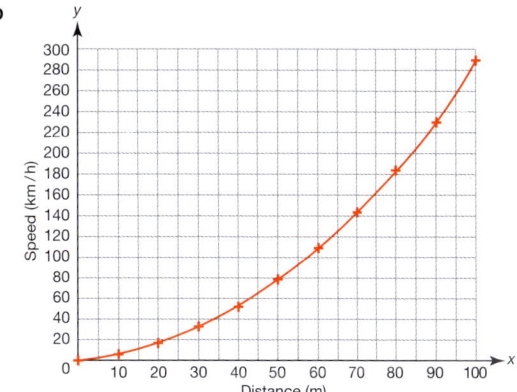

c i 24 m **ii** 58 m **iii** 174 m
4 a $y = -2000, 0.25, 3.25, 4, 4.25, 4, 3.25, 2, 0.25, -2$
b

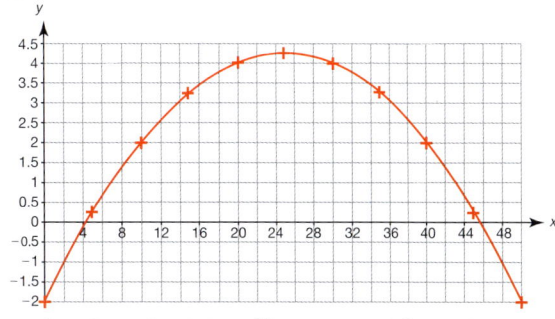

c i £1690 **ii** £2688 **iii** 1793, 3207 **iv** £4250
v 2500 **vi** 438 **vii** 4562 **viii** 1177 to 3823
5 i E **ii** C **iii** B **iv** A **v** D
6 a, b

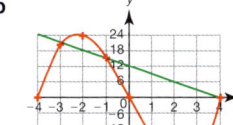

c $x = 4$
d $x = -4, 0, 4$
e $(-3, 21), (-1, 15), (4, 0)$
f $x^3 - 13x - 12 = 0$

7 a $y = \frac{36}{x}$ **b**

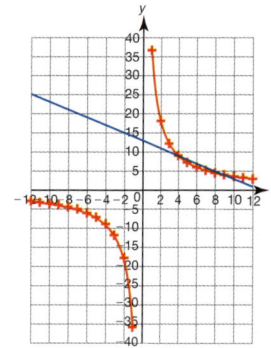

c 4 and 9, the two points of intersection between $y = 13 - x$ and $y = \frac{36}{x}$ are (4, 9) and (9, 4).

Revision 3

1 a 48 **b** HCF = 30, LCM = 1260
c $180 \times 5 = (2^2 \times 3^2 \times 5) \times 5 = 2^2 \times 3^2 \times 5^2$
2 a 121 **b** 6.76 **c** 21 **d** 1.5
e 1 **f** −27 **g** 32.768 **h** 3.6
i −8.4 **j** 6.1
3 a 8×10^3 **b** 7.5×10^4 **c i** 1.6^3 **ii** $4.096\,cm^3$
d $2^3 \times 2^{-3} = 2^{3\,+\,-3} = 2^0 = 1$
4 a *DE*, the can empties rapidly.
b *FG*, the level of water goes down slowly.
c *EF*, the graph is steepest in this section.
d *BC*, the graph is horizontal.
5 a $3 \times -2 + 4 \times 7 = -6 + 28 = 22$
b i **A** **ii** **C** **iii** **B** **iv** **D**
c i $y = -4x + 3$ **ii** $y = -4x - 7$
iii $y = -4x - 245$
6 No, A forms an open box, B is not a net and C forms a cube.
7 a $232\,cm^2$ (3 sf) **b** $333\,cm^3$ (3 sf)
8 $1900\,cm^3$
9 $4000 = \frac{4}{3} \times \pi \times r^3$ so $r = \sqrt[3]{\frac{3\,\times\,400}{4\pi}} = 9.85\,m$ (3 sf)
10 a $9450\,cm^3$ **b** $3150\,cm^2$
11 a i $2.1 \leqslant t < 2.6$ **ii** $2.1 \leqslant t < 2.6$
b 19 **c** 2.4 mm (1 dp)
12 a

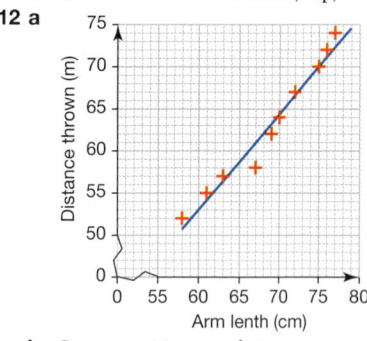

b Strong positive correlation
c i 57.5 m **ii** 71 m
13 a

b Unemployment rates fell much more rapidly in Oct 2013 − Sep 2014 compared to the previous year.
14 a 564 000 units **b** 1.87×10^{22}
c Yes, $-3 \times 10^2 < 3 \times 10^{-2}$ because it is negative.
d i 3 **ii** 6 **iii** 3

---

15 a

b 18.2 m (1 dp) **c** The boat can't go under the bridge.

Chapter 19

Check in 19

1 a 12 **b** 75 **c** 28.83 **d** 12
2 a 25 **b** 24 **c** 8.4

19.1S

1 1, 4, 9, 16, 25, 36, 49, 64, 81, 100, 121, 144, 169, 196, 225
2 a 10 **b** 8 **c** 9.9 (1 dp) **d** 2.8 (1 dp)
3 a $81\,cm^2$ **b** $144\,mm^2$ **c** $2.25\,m^2$ **d** $1225\,cm^2$
e $23.04\,m^2$ **f** $0.36\,km^2$
4 a 1 cm **b** 8 m **c** 4 mm **d** 16 cm
e 2.3 m **f** 0.5 km
5 a $21\,cm^2$ **b** $9\,mm^2$ **c** $14\,m^2$ **d** $16\,m^2$
6 a 13 cm **b** 25 m **c** 50 mm **d** 39 mm
e 2.9 m (1 dp) **f** 5.9 cm (1 dp)
7 a 9 cm **b** 60 mm **c** 24 m **d** 8.4 m
e 75.0 cm (1 dp) **f** 3.9 km (1 dp)
*8 $h = 72\,mm$
9 a 2.9 cm (1 dp) **b** 162.3 cm (1 dp)
10 34 cm

19.1A

1 a 35 mm **b** 7.2 cm **c** 29 m **d** 180 km
2 4.5 cm (1 dp)
3 a 14.1 cm (1 dp) ***b** 7.1 cm (1 dp)
4 4.8 m (1 dp)
5 1.2 m (1 dp)
6 a 5 **b** 5 **c** 3.2 (1 dp) **d** 4.2 (1 dp)
e 8.6 (1 dp) **f** 5 **g** 5.4 (1 dp) **h** 7.2 (1 dp)
7 29.4 m (1 dp)
8 a $60\,m^2$ **b** $240\,cm^2$
9 a $6^2 + 8^2 = 100 = 10^2$ **b** $8^2 + 10^2 = 164 \neq 12^2 (= 144)$
10 a $4^2 + 8^2 = 80 \neq 9^2 (= 81)$
*b Obtuse, $AC >$ hypotenuse of a right-angled triangle, so $\angle ABC > 90°$.
*11 a 6.0 cm (1 dp) **b** 14.1 cm (1 dp)
12 a 24
b Yes, the triangles will be similar (AAA).
*13 a There are 16 Primitive Pythagorean triples, for example, (3, 4, 5) (5, 12, 13), etc.
b Yes, for example, (6, 8, 10), (15, 36, 39), etc.

19.2S

> Answers given to 1 decimal place or nearest whole number when not exact.

1 a **b** **c**
2 a 2.5 **b** 8.2 **c** 3.0
d 92.7 **e** 17.8 **f** 92.9
3 a 7.5 cm **b** 53.8 mm **c** 4.1 m
4 a 8.6 m **b** 28.8 cm **c** 1.6 km
5 a $a = 6.7\,m$, $b = 2.2\,m$
b $c = 28.6\,cm$, $d = 22.2\,cm$
c $e = 124.6\,mm$, $f = 91.2\,mm$
d $g = 42.9\,m$, $h = 16.5\,m$
e $i = 132.5\,km$, $j = 120.1\,km$

6 a 19.4 mm   **b** 31.5 mm
7 a $PQ = 38.3\,\text{cm}, PR = 11.7\,\text{cm}$
   b $PR = 5.8\,\text{m}, QR = 7.6\,\text{m}$
   c $PQ = 73.1\,\text{km}, PR = 121.7\,\text{km}$
   d $QR = 91.2\,\text{mm}, PR = 40.0\,\text{mm}$
*8 a 12.3 cm   **b** 10.3 cm   **c** 28.3 cm   **d** 30.1 cm
*9 a $AC^2 = 1^2 + 1^2 = 2, AC = \sqrt{2}$
    i $\sin 45° = \frac{AB}{AC} = \frac{1}{\sqrt{2}}$ **ii** $\cos 45° = \frac{BC}{AC} = \frac{1}{\sqrt{2}}$
   b $\triangle PQR$ is equilateral, $QS = QR = 1, PS^2 = 2^2 - 1^2 = 3,$
    $PS = \sqrt{3}$
    i $\cos 60° = \frac{QS}{PQ} = \frac{1}{2}$ **ii** $\sin 60° = \frac{PS}{PQ} = \frac{\sqrt{3}}{2}$
   c $0°, 0, 1, 0; 30°, \frac{1}{2}, \frac{\sqrt{3}}{2}, \frac{1}{\sqrt{3}}; 45°, \frac{1}{\sqrt{2}}, \frac{1}{\sqrt{2}}, 1; 60°, \frac{\sqrt{3}}{2}, \frac{1}{2}, \sqrt{3};$
    $90°, 1, 0, \cdot$
10 a $h = 4.8\,\text{m}$ **b** $d = 1.3\,\text{m}$

19.2A

1 103 m (nearest m)
2 a 168 m (nearest m)
   b It is difficult to estimate distances and angles accurately.
3 a i 4.9 m (1 dp)   **ii** 2.1 m (1 dp)
    iii 1.0 m (1 dp)   **iv** 0.3 m (1 dp)
   b The shadow gets shorter as the sun rises in the sky.
4 No, 1.606... m > 1.5 m.
5 $144\sqrt{3}\,\text{cm}^2$
6 $12\sqrt{2}\,\text{cm}^2$
7 530 m (nearest m)
8 Yes, 11.134... m > 10 m.
*9 a $9\sqrt{3}\,\text{mm}$ **b** $81\sqrt{3}\,\text{mm}^3$
*10 9.0 m (1 dp)
*11 No, 2.41... m > 2 m.

19.3S

> Answers given to 1 decimal place or the nearest whole number when not exact.

1 a 60°   **b** 36.9°   **c** 23.3°
   d 68.3°   **e** 63.1°   **f** 58.7°
2 a 21.8°   **b** 47.7°   **c** 32.9°
   d 53.1°   **e** 55.3°   **f** 62.9°
3 a 66.4°   **b** 30°   **c** 56.3°
   d 19.5°   **e** 51.3°   **f** 29.4°
4 a $AC = 10\,\text{cm}, A = 36.9°, C = 53.1°$
   b $DF = 5\,\text{m}, E = 22.6°, F = 67.4°$
   c $GI = 32.3\,\text{m}, HI = 10.5\,\text{m}, G = 18°$
   d $KL = 15.1\,\text{mm}, JK = 23.5\,\text{mm}, K = 50°$
   e $MN = 116\,\text{km}, NO = 435\,\text{km}, M = 75°$
   f $PR = 71.6\,\text{cm}, P = 54.1°, R = 35.9°$
5 a $ST = 12.2\,\text{cm}, T = 35.0°, S = 55.0°$
   b $ST = 40.2\,\text{m}, T = 41.8°, R = 48.2°$
   c $RT = 3.08\,\text{km}, R = 28.5°, S = 61.5°$
6 a $QS = 8.1\,\text{cm}$ **b** $\angle PQS = 29.7°$ **c** $\angle PSQ = 60.3°$
7 $KN = KL = 19.2\,\text{cm}, NM = LM = 29.5\,\text{cm}, \angle LKN = 77.3°,$
   $\angle LMN = 47.9°, \angle KNM = \angle KLM = 117.4°$
8 a $\angle PRQ = \angle PQR = 73.4°, \angle RPQ = 33.2°$ **b** 6.7 cm
*9 a 116.4°, 63.6°, 116.4°, 63.6°   **b** 10.5 cm
10 a 11.3°   **b** 102 m

19.3A

1 125 km, 217° (nearest degree)
2 a 108 km (nearest km), 146° (nearest degree)
   b 130 km, 337° (nearest degree)
3 a $TS = 66.4\,\text{km}$ (1 dp), 211° (nearest degree)
   b 031° (nearest degree)
4 a 71.79...° < 75°
   b About 0.4 m nearer the wall.
5 a 3.3 m (1 dp)   **b** 64.2° (1 dp)
6 a 243 cm (nearest cm)   **b** No, 8.53...° > 4°.

7 a ii $\tan \theta = 2$ **b ii** $\tan \theta = 3$
   c ii $\tan \theta = \frac{1}{2}$
   d $\tan \theta = m$ in $y = mx$.

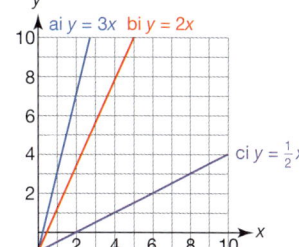

*8 a 60°   ***b** 120°
*9 a 31.0° (1 dp)   **b** 28.2 ft (1 dp)   **c** 591 ft² (nearest ft²)
*10 $\sin 40° = \frac{x}{H} = \cos 50°.$

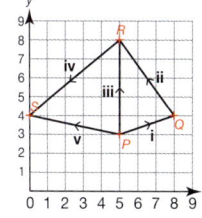

*11   **a** $H^2 = 4^2 + 3^2 = 16 + 9 = 25$ $H = 5$
                i $\sin \theta = \frac{O}{H} = \frac{3}{5}$ **ii** $\cos \theta = \frac{A}{H} = \frac{4}{5}$
                b i $\cos \theta = \frac{5}{13}$ **ii** $\tan \theta = \frac{12}{13}$

19.4S

1 $p = \begin{pmatrix} 3 \\ 2 \end{pmatrix}$ $q = \begin{pmatrix} 1 \\ -4 \end{pmatrix}$ $r = \begin{pmatrix} -1 \\ -3 \end{pmatrix}$ $s = \begin{pmatrix} -4 \\ 2 \end{pmatrix}$
2 a $\begin{pmatrix} 1 \\ 2 \end{pmatrix}$ **b** $\begin{pmatrix} -5 \\ -2 \end{pmatrix}$ **c** $\begin{pmatrix} 1 \\ -3 \end{pmatrix}$
   d $\begin{pmatrix} -1 \\ 3 \end{pmatrix}$ **e** $\begin{pmatrix} -4 \\ 0 \end{pmatrix}$ **f** $\begin{pmatrix} 0 \\ 5 \end{pmatrix}$
3 a i $\begin{pmatrix} 3 \\ 1 \end{pmatrix}$ **ii** $\begin{pmatrix} -3 \\ 4 \end{pmatrix}$ **b**
    iii $\begin{pmatrix} 0 \\ 5 \end{pmatrix}$ **iv** $\begin{pmatrix} -5 \\ -4 \end{pmatrix}$
    v $\begin{pmatrix} -5 \\ 1 \end{pmatrix}$ **vi** $\begin{pmatrix} 5 \\ -1 \end{pmatrix}$

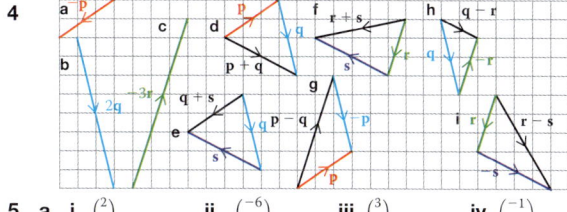

4

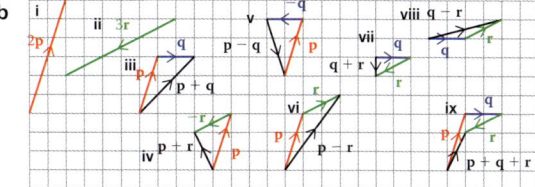

5 a i $\begin{pmatrix} 2 \\ 6 \end{pmatrix}$ **ii** $\begin{pmatrix} -6 \\ -3 \end{pmatrix}$ **iii** $\begin{pmatrix} 3 \\ 3 \end{pmatrix}$ **iv** $\begin{pmatrix} -1 \\ 2 \end{pmatrix}$
    v $\begin{pmatrix} -1 \\ 3 \end{pmatrix}$ **vi** $\begin{pmatrix} 3 \\ 4 \end{pmatrix}$ **vii** $\begin{pmatrix} 0 \\ -1 \end{pmatrix}$ **viii** $\begin{pmatrix} 4 \\ 1 \end{pmatrix}$
    ix $\begin{pmatrix} 1 \\ 2 \end{pmatrix}$

b

6 a $\begin{pmatrix} 1 \\ -4 \end{pmatrix}$ **b** $\begin{pmatrix} 4 \\ -4 \end{pmatrix}$
   c $\begin{pmatrix} 5 \\ 3 \end{pmatrix}$ **d** $\begin{pmatrix} -3 \\ 0 \end{pmatrix}$
   e $\begin{pmatrix} 4 \\ 7 \end{pmatrix}$ **f** $\begin{pmatrix} -1 \\ -7 \end{pmatrix}$
7 a $\begin{pmatrix} -8 \\ 4 \end{pmatrix}$ **b** $\begin{pmatrix} 3 \\ -4 \end{pmatrix}$
8 a i $\begin{pmatrix} 1 \\ 1 \end{pmatrix}$ **ii** $\begin{pmatrix} -1 \\ 7 \end{pmatrix}$ **iii** $\begin{pmatrix} -5 \\ 7 \end{pmatrix}$ **iv** $\begin{pmatrix} 3 \\ 5 \end{pmatrix}$
    v $\begin{pmatrix} 5 \\ 5 \end{pmatrix}$ **vi** $\begin{pmatrix} 8 \\ 0 \end{pmatrix}$

b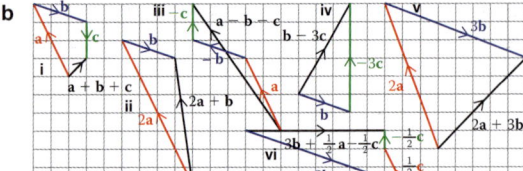

9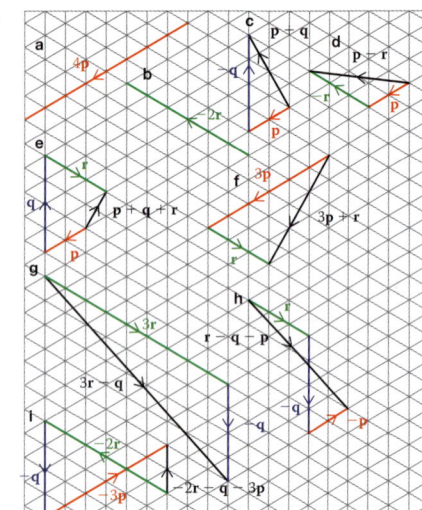

19.4A

1 $c = 4a, d = -3b, e = a + 3b, f = -3a + 2b, g = -3a - 2b,$
$h = 4a - b, i = 5a + 3b, j = -a + 4b$

2 $r = 2p, s = 3q, t = -p, u = p + q, v = q - p, w = 4p - 2q,$
$x = 4q + 2p, y = 2p - 2q, z = 4p - q$

3 a $a = 4$ **b** $b = 12$ **c** $c = -4$

4 $\binom{12}{8} = 4\binom{3}{2}$ and $\binom{-6}{-4} = -2\binom{3}{2}$

5 a The cave.
 b i $\binom{4}{0}$ **ii** $\binom{-1}{2}$ **iii** $\binom{2}{3} + \binom{3}{2} = \binom{5}{5}$
 c i, ii Students' answers, for example,
  $\binom{2}{-2} + \binom{4}{3} + \binom{0}{5} + \binom{-3}{-1} + \binom{-3}{0} + \binom{-2}{-2} + \binom{0}{-2} +$
  $\binom{-1}{-2} + \binom{3}{1} = \binom{0}{0}$, it starts and ends at the same place.

6 a $(6, 4)$ **b** $(7, 1)$ **c** $(-4, 10)$ **d** $(-1, -2)$

7 a $\binom{5}{-1}$ **b** $\binom{10}{9}$

8 a $\sqrt{x^2 + y^2}$ **b** $\tan^{-1}\frac{y}{x}$

*9 Speed = 131 km/h, bearing = 168°

*10 a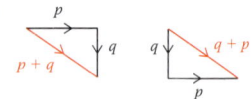

b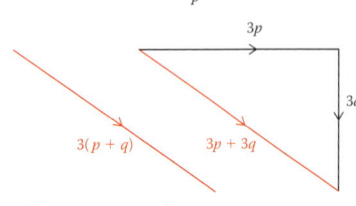

c

d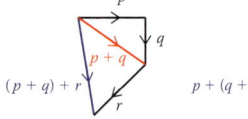

Answers

11 $x = 2, y = -1$

12 Translation by vector $\binom{1}{-6}$.

Review 19

1 $a = 21.6\,\text{cm}, \ b = 10.7\,\text{cm}$

2 a $x = 9\,\text{cm}, \ y = 5\,\text{cm}$ **b** Yes, scale factor 3

3 a i $\frac{12}{13}$ **ii** $\frac{5}{13}$ **iii** $\frac{12}{5}$
  b 67.4°

4 $a = 7.64\,\text{cm}, \ b = 4.77\,\text{cm}, \ c = 3.11\,\text{cm}$

5 a $\frac{\sqrt{3}}{2}$ **b** 1 **c** 1 **d** $\frac{\sqrt{3}}{2}$

6 a $u = \binom{2}{4}, v = \binom{-3}{-2}$
  b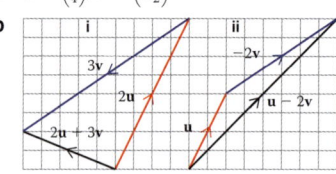

7 a $\binom{5}{9}$ **b** $\binom{6}{-6}$ **c** $\binom{15}{-5}$ **d** $\binom{1}{0}$

Assessment 19

1 a $a = \sqrt{181}\,\text{m} = 13.45\ (2\,\text{dp})$
  b $b \neq \sqrt{14.2^2 + 6.1^2}, b = \sqrt{14.2^2 - 6.1^2}\,\text{m} = 12.82\ (2\,\text{dp}).$

2 a $2^2 + 3^2 = 13$, not 13^2
  b $20^2 + 99^2 = 10\,201 = 101^2$, 101

3 9.85 (3 sf)

4 a $1 \times 28, 2 \times 14, 4 \times 7$
  b $1 \times 28, d = 28.02; 2 \times 14, d = 14.14; 4 \times 7, d = 8.062$

5 a Yes, $9^2 + 12^2 = 225 = 15^2$
  b No, $9^2 + 14^2 = 277 \neq 17^2$
  c Yes, $1.6^2 + 3.0^2 = 11.56 = 3.4^2$
  d No, $11^2 + 19^2 = 482 \neq 22^2$
  e Yes, $3.6^2 + 7.7^2 = 72.25 = 8.5^2$

6 a $\angle HBN = 118°$ (IA), $x = 90°$ (AP) **b** 4.38 km (3 sf)
  c 10.5 km (3 sf) **d** 297°

7 a i Sin, 41.8° **ii** Cos, 37.9° **iii** Tan, 40.6°
  b i 4.47 cm (3 sf) **ii** 11.7 cm (3 sf) **iii** 45.5 cm (3 sf)

8 a 8.09 m (3 sf) **b** 7.89 m (3 sf)

9 a Check students' drawings.
  b i $\binom{2}{13}$ **ii** $\binom{-6}{1}$ **iii** $\binom{6}{-1}$ **iv** $\binom{8}{12}$
   v $\binom{4}{-14}$ **vi** $\binom{14}{-9}$ Check students' drawings
  c i $\sqrt{173}$ **ii** $\sqrt{37}$ **iii** $\sqrt{37}$ **iv** $\sqrt{208}$
   v $\sqrt{212}$ **vi** $\sqrt{277}$
  d The vectors are parallel, and the same length.

10 a Correct **b** Incorrect, $-2x + 7y$
  c Incorrect, $-6y$ **d** Correct
  e Incorrect, $2x + 3y$ **f** Correct

11 a $\binom{35}{-10}$ **b** $\binom{16}{-10}$ **c** $\binom{-1}{3}$ **d** $\binom{3}{-3}$
   e $\binom{3}{4}$ **f** $\binom{-40}{18}$ **g** $\binom{56}{-15}$

Chapter 20

Check in 20

1 a $\frac{2}{3}$ **b** $\frac{1}{2}$ **c** $\frac{3}{4}$ **d** $\frac{1}{4}$
  e 1

2 a 1 **b** 1 **c** $\frac{3}{10}$ **d** $\frac{2}{5}$

3 a 0.9 **b** 0.4 **c** 0.85

4 a 2, 3, 5 or 7 **b** 1, 4 or 9 **c** 1, 3, 6 or 10
  d 4 or 8 **e** 1, 2, 5 or 10

20.1S

1 a P = {1, 4, 9, 16, 25, 36, 49, 64, 81, 100}
  b R = {Canada, Mexico, USA}
  c S = {2, 3, 5, 7, 11, 13, 17, 19, 23, 29}
  d T = {1, 2, 3, 4, 6, 9, 12, 18, 36}

2 a $P \cap T = \{1, 4, 9, 36\}$ **b** $S \cap T = \{2, 3\}$
  c $P \cap S = \varnothing$
  d $P \cup S = \{1, 2, 3, 4, 5, 7, 9, 11, 13, 16, 17, 19, 23, 25, 29, 36, 49, 64, 81, 100\}$

3 **a** Factors of 10   **b** Even numbers
  c Vowels
  d All the possible outcomes when you toss two coins.
  e Types of coins that you can get in pounds sterling.
  f The first 10 multiples of 3.

4 **a** $A = \{1, 2, 3, 4, 5, 6, 7, 8, 9, 10\}$
  b $\{1, 3, 5, 7, 9\}$
  c $\{2, 3, 5, 7\}$
  d $\{1, 4, 9\}$

5 **a** $B \cap C = \{2, 3, 5, 7\}$ **b** $B \cap D = \{1, 9\}$
  c $B \cup D = \{1, 3, 4, 5, 7, 9\}$ **d** $C \cup D = \{1, 2, 3, 4, 5, 7, 9\}$

6   A contains all the elements of B, C and D.

7 **a** Yes, 10.   **b** Yes, 2.   **c** Yes, 1, 4, 9, 36.
  d Yes, 9.   **e** No   **f** No
  g No   **h** Yes, 3.   **i** No

8 **a** **i** 8   **ii** 13   **iii** 12   **iv** 21
  b Students that don't play hockey or football.
  c $\frac{13}{25}$

9 **a** **i** $\frac{29}{50}$ **ii** $\frac{19}{50}$ **iii** $\frac{17}{50}$ **iv** $\frac{33}{50}$
  b 14

20.1A

1 **a** $P \cap Q = \{1, 3, 7, 21\}$ **b** 21
2 **a** 21   **b** 3   **c** 7
  d A pupil who is not a girl and does not have dark hair.
3 **a** Sketch of an isosceles right-angled triangle.
  b An equilateral triangle cannot have a right angle.
4 **a** $18 + (8 - x) + (14 + x) + 2 = 50$
  b **i** $\frac{11}{25}$ **ii** $\frac{2}{5}$ **iii** $\frac{48}{50}$ **iv** $\frac{2}{25}$
  c $x = 8$ **d** $A = B$
5   5
6   3
7 **a** 8
  b **i** $\frac{3}{10}$ **ii** $\frac{3}{10}$ **iii** $\frac{3}{5}$
8 **a** $\frac{1}{2}$ **b** $\frac{5}{16}$ **c** $\frac{13}{16}$

20.2S

1 **a** $\frac{1}{12}$ **b** $\frac{11}{12}$ **c** $\frac{7}{36}$ **d** $\frac{1}{6}$
  e $\{2, 3, 4, 5, 6, 7, 8, 9, 10, 11, 12\}$
2 **a** $\frac{1}{18}$ **b** $\frac{19}{12}$ **c** $\frac{1}{6}$ **d** $\frac{2}{9}$
  e $\{1, 2, 3, 4, 5, 6, 8, 9, 10, 12, 15, 16, 18, 20, 24, 25, 30, 36\}$
3 **a**

	1	2	3	4	5	6
1	0	1	2	3	4	5
2	1	0	1	2	3	4
3	2	1	0	1	2	3
4	3	2	1	0	1	2
5	4	3	2	1	0	1
6	5	4	3	2	1	0

  b $\{0, 1, 2, 3, 4, 5\}$
  c **i** $\frac{1}{6}$
    ii $\frac{1}{6}$
    iii 0
    iv $\frac{4}{9}$

4 **a** {HHH, HHT, HTH, HTT, THH, THT, TTH, TTT}
  b 3   **c** 2
5 **a**

·	1	3	5
1	2	4	6
2	3	5	7
4	5	7	9

  b $\{2, 3, 4, 5, 6, 7, 9\}$
  c **i** $\frac{1}{9}$
    ii $\frac{1}{9}$
    iii $\frac{1}{3}$

6 **a**

·	1	3	5
1	1	3	5
2	2	6	10
4	4	12	20

  b $\{1, 2, 3, 4, 5, 6, 10, 12, 20\}$
  c **i** $\frac{1}{9}$
    ii $\frac{1}{9}$
    iii $\frac{2}{3}$

7 **a** $\frac{1}{3}$ **b** $\frac{2}{3}$
8 **a** ≈ 8 times   **b** ≈ 6 times

  c **i** More often   **ii** About the same
9 **a** **i** $\frac{1}{6}$ **ii** $\frac{1}{6}$ **iii** $\frac{1}{12}$ **iv** $\frac{1}{12}$
  b 4 and 5

20.2A

1 **a** $\frac{1}{5}$ **b** $\frac{4}{5}$ **c** 1
2   $\frac{3}{4}$
3 **a**

		B1	R2	Y2	G3	B2	R4
Dice	**1**	1	2	0	3	2	4
	2	2	4	0	6	4	8
	3	3	6	0	9	6	12
	4	4	8	0	12	8	16
	5	5	10	0	15	10	20
	6	6	12	0	18	12	24

Spinner (column headers)

  b **i** $\frac{1}{9}$ **ii** $\frac{5}{12}$

4 **a** $\frac{9}{100}$ **b** $\frac{9}{10}$ **c** $\frac{19}{100}$
5 **a** **i** $\frac{1}{9}$ **ii** $\frac{1}{2}$ **b** $\frac{5}{18}$
*6 **a** $\frac{5}{12}$ **b** $\frac{1}{2}$
*7 **a** $P(A) = \frac{1}{6}$, $P(B) = \frac{1}{6}$, $P(C) = \frac{1}{2}$, $P(D) = \frac{7}{36}$, $P(E) = \frac{5}{36}$
  b **i** $\frac{1}{36}$ **ii** $\frac{1}{9}$
  c A and C, A and E, B and D, B and E.

20.3S

1 **a, b**

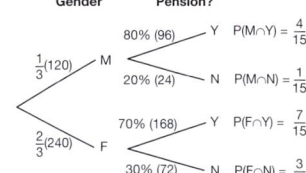

Gender   Pension?
$\frac{1}{3}$ (120) M — 80% (96) Y $P(M \cap Y) = \frac{4}{15}$
    — 20% (24) N $P(M \cap N) = \frac{1}{15}$
$\frac{2}{3}$ (240) F — 70% (168) Y $P(F \cap Y) = \frac{7}{15}$
    — 30% (72) N $P(F \cap N) = \frac{3}{15}$

2 **a, b**

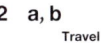
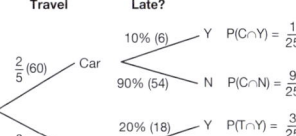

Travel   Late?
$\frac{2}{5}$ (60) Car — 10% (6) Y $P(C \cap Y) = \frac{1}{25}$
    — 90% (54) N $P(C \cap N) = \frac{9}{25}$
$\frac{3}{5}$ (90) Train — 20% (18) Y $P(T \cap Y) = \frac{3}{25}$
    — 80% (72) N $P(T \cap N) = \frac{12}{25}$

  c 24   **d** $\frac{4}{25}$

3 **a, b**

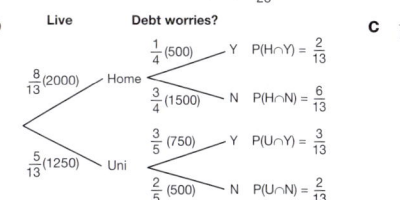

Live   Debt worries?
$\frac{8}{13}$ (2000) Home — $\frac{1}{4}$ (500) Y $P(H \cap Y) = \frac{2}{13}$
    — $\frac{3}{4}$ (1500) N $P(H \cap N) = \frac{6}{13}$
$\frac{5}{13}$ (1250) Uni — $\frac{3}{5}$ (750) Y $P(U \cap Y) = \frac{3}{13}$
    — $\frac{2}{5}$ (500) N $P(U \cap N) = \frac{2}{13}$

  c $\frac{5}{13}$

4 **a** The red dice is even and the blue dice is a multiple of 3.
  b $\frac{1}{6}$
  c $\frac{1}{6}$
  d Yes, $P(A \cap B) = P(A) \times P(B)$.

5 **a**

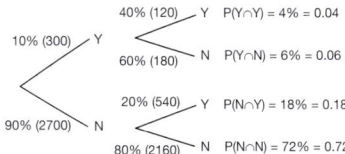

Brakes fail?   Lights fail?
10% (300) Y — 40% (120) Y $P(Y \cap Y) = 4\% = 0.04$
    — 60% (180) N $P(Y \cap N) = 6\% = 0.06$
90% (2700) N — 20% (540) Y $P(N \cap Y) = 18\% = 0.18$
    — 80% (2160) N $P(N \cap N) = 72\% = 0.72$

  b 840
  c Cars can fail for other reasons.

6 a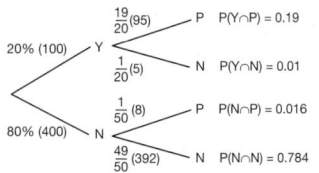

Taking drugs? Test

20% (100) Y
- $\frac{19}{20}$ (95) P $P(Y \cap P) = 0.19$
- $\frac{1}{20}$ (5) N $P(Y \cap N) = 0.01$

80% (400) N
- $\frac{1}{50}$ (8) P $P(N \cap P) = 0.016$
- $\frac{49}{50}$ (392) N $P(N \cap N) = 0.784$

b 103 **c** 8

20.3A

1 a 0.84 **b** $P(\text{Loss}) = 1 - (0.6 + 0.3) = 0.1$
c Students' answers, for example, no, the outcome of a match can affect the morale of the players during the next game.

2 a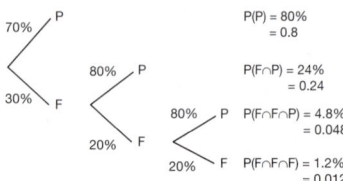

1st Attempt 2nd Attempt 3rd Attempt

70% P $P(P) = 80\% = 0.8$

30% F
- 80% P $P(F \cap P) = 24\% = 0.24$
- 20% F
  - 80% P $P(F \cap F \cap P) = 4.8\% = 0.048$
  - 20% F $P(F \cap F \cap F) = 1.2\% = 0.012$

b i 0.24 **ii** 0.012

3 a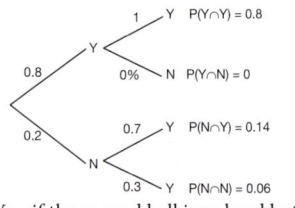

Bus connects? Train on time?

0.8 Y
- 1 Y $P(Y \cap Y) = 0.8$
- 0% N $P(Y \cap N) = 0$

0.2 N
- 0.7 Y $P(N \cap Y) = 0.14$
- 0.3 Y $P(N \cap N) = 0.06$

b 0.94

4 a Yes, if the second ball is red and both balls are the same colour then both balls are red.
b $P(X) = \frac{1}{2}$, $P(Y) = \frac{3}{7}$, $P(Z) = P(X \cap Y) = \frac{3}{14}$,
$P(X) \times P(Y) = \frac{1}{2} \times \frac{3}{7} = \frac{3}{14}$

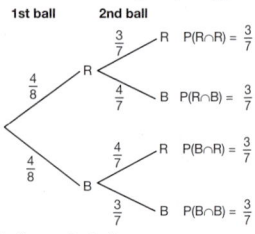

1st ball 2nd ball

$\frac{4}{8}$ R
- $\frac{3}{7}$ R $P(R \cap R) = \frac{3}{7}$
- $\frac{4}{7}$ B $P(R \cap B) = \frac{3}{7}$

$\frac{4}{8}$ B
- $\frac{4}{7}$ R $P(B \cap R) = \frac{3}{7}$
- $\frac{3}{7}$ B $P(B \cap B) = \frac{3}{7}$

5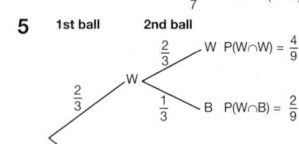

1st ball 2nd ball

$\frac{2}{3}$ W
- $\frac{2}{3}$ W $P(W \cap W) = \frac{4}{9}$
- $\frac{1}{3}$ B $P(W \cap B) = \frac{2}{9}$

$\frac{1}{3}$ B
- $\frac{2}{3}$ W $P(B \cap W) = \frac{2}{9}$
- $\frac{1}{3}$ B $P(B \cap B) = \frac{1}{9}$

No, $P(A) = \frac{4}{9}$, $P(B) = \frac{5}{9}$,
$P(A \cap B) = \frac{4}{9} \neq P(A) \times P(B)$
$= \frac{4}{9} \times \frac{5}{9} = \frac{20}{81}$.

*6 a $\frac{1}{9}$

b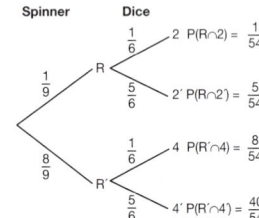

Spinner Dice

$\frac{1}{9}$ R
- $\frac{1}{6}$ 2 $P(R \cap 2) = \frac{1}{54}$
- $\frac{5}{6}$ 2' $P(R \cap 2') = \frac{5}{54}$

$\frac{8}{9}$ R'
- $\frac{1}{6}$ 4 $P(R' \cap 4) = \frac{8}{54}$
- $\frac{5}{6}$ 4' $P(R' \cap 4') = \frac{40}{54}$

c $\frac{1}{6}$
d It does not matter, the probabilities are the same.
e i The probability of winning would decrease, you cannot score an odd number using the spinner.
ii The probability of winning would increase, you cannot score an 8 unless you use the spinner.

Review 20

1 a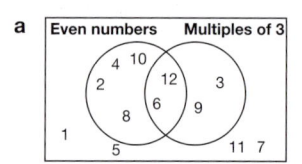

Even numbers Multiples of 3
- 4, 10, 2, 8 | 12, 6 | 3, 9
- 1, 5 | | 11, 7

b 2 **c** $\frac{1}{6}$

2 a

Cup \ Plate	Red	Blue	Green
Red	R, R	R, B	R, G
Blue	B, R	B, B	B, G
Green	G, R	G, B	G, G

b i $\frac{1}{9}$ **ii** $\frac{1}{3}$ **iii** $\frac{1}{3}$

3 a, b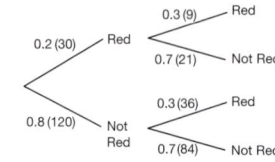

First Set Second Set

0.2 (30) Red
- 0.3 (9) Red
- 0.7 (21) Not Red

0.8 (120) Not Red
- 0.3 (36) Red
- 0.7 (84) Not Red

c i 0.06 **ii** 0.56

4 a 0.06 **b** 0.58

5 a $\frac{3}{10}$ **b** $\frac{1}{10}$ **c** $\frac{3}{5}$

Assessment 20

1 a i N = {2, 3, 5, 7, 11, 13, 17, 19, 23, 29}
ii M = {3, 6, 9, 12, 15, 18, 21, 24, 27, 30}
iii P = {a, d, e, i, l, p, r}
iv F = {F, a, h, r, e, n, i, t, 4, 5, 1}
b i {3}
ii {2, 3, 5, 7, 11, 13, 17, 19, 23, 29, 6, 9, 12, 15, 18, 21, 24, 27, 30}
iii {5} **iv** {a, r, e, i}
v {p, a, r, e, l, i, d, F, h, n, t, 4, 5, 1}
vi {5} **vii** ∅
viii {2, 3, 5, 7, 11, 13, 17, 19, 23, 29, 6, 9, 12, 15, 18, 21, 24, 27, 30, p, a, r, e, l, i, d, F, h, n, t, 4, 1}

2 a K | 18 (2) 13 | M, 1 **b** 1 **c** 31

3 a T | 24 (17) 19 | M, 20
b i $\frac{41}{80}$ **ii** $\frac{11}{20}$ **iii** $\frac{3}{4}$ **iv** $\frac{17}{80}$

4 a Heads = 8, 7, 9, 6, 7; Tails = 16, 12, 18, 9, 10
b $\frac{1}{5}$ **c** $\frac{3}{5}$ **d** $\frac{2}{5}$ **e** $\frac{1}{2}$

5 a

+	1	2	3	4	5	6	7	8
1	2	3	4	5	6	7	8	9
2	3	4	5	6	7	8	9	10
3	4	5	6	7	8	9	10	11
4	5	6	7	8	9	10	11	12
5	6	7	8	9	10	11	12	13
6	7	8	9	10	11	12	13	14
7	8	9	10	11	12	13	14	15
8	9	10	11	12	13	14	15	16

b i $\frac{1}{32}$ **ii** $\frac{7}{64}$ **iii** $\frac{5}{64}$
c 9
d All outcomes were assumed to be equally likely.

6 a T | 4 (2) 3 | F, 11 **b** $n = 16$

7 Tea 17, Coffee 15; Tea-Chocolate 12, Tea-Plain 5; Coffee-Chocolate 10, Coffee-Plain 5

8 a B $\frac{1}{4}$, A $\frac{3}{4}$; BB $\frac{1}{16}$, BA $\frac{3}{16}$; AB $\frac{3}{16}$, AA $\frac{9}{16}$
b $\frac{1}{16}$ **c** $\frac{7}{16}$

9 a 0.1

b

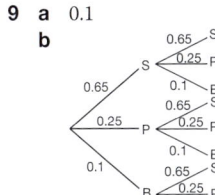

c i 0.025 **ii** 0.4225
iii 0.165 **iv** 0.19

Lifeskills 4

1 85

2 a Cuboid 288 m³, Cylinder 395.8 m³ (1 dp)
b Cuboid 216 m², Cylinder 245.0 m² (1 dp)
c Cuboid
d Total hire cost ÷ Area of base
e

3 a 33.7°
b True, $RT = 141$ m, $BT = 206$ m, $AT = 361$ m, $TC = 541$ m.
c 8 min

4 20, 50, 30, £12.80

5 810, 4761

Chapter 21

Check in 21

1 a 3 **b** 6 **c** 5 **d** 6
2 a −3 **b** −4 **c** −5 **d** −4
3 a 4, 8, 12, 16, 20, 24 **b** 3, 6, 9, 12, 15, 18
c 5, 10, 15, 20, 25, 30 **d** 6, 12, 18, 24, 30, 36
4 a 40 **b** 31 **c** 26 **d** 100

21.1S

1 a 2, 8, 14, 20 **b** 3, 8, 13, 18
c 25, 35, 45, 55 **d** 25, 28, 31, 34
e 2.5, 7.5, 12.5, 17.5 **f** $1\frac{3}{4}, 2\frac{1}{4}, 2\frac{3}{4}, 3\frac{1}{4}$
g 8, 10.5, 13, 15.5 **h** −10, −7, −4, −1
2 a 20, 18, 16, 14 **b** 100, 95, 90, 85
c 30, 26, 22, 18 **d** 12.5, 9.5, 6. 5, 3.5
e 5, 1, −3, −7 **f** −1, −5, −9, −13
3 a 2, 4, 8, 16 **b** 3, 6, 12, 24
c 5, 50, 500, 5000 **d** 10, 50, 250, 1250
e 0.5, 1, 2, 4 **f** −5, −10, −20, −40
4 a 64, 32, 16, 8 **b** 100, 50, 25, 12.5
c 27, 9, 3, 1 **d** 64, 16, 4, 1
e 26, 13, 6.5, 3.25 **f** 2, 4, 8, 16
5 a i Add 5 **ii** 13
b i Add 3 **ii** 10
c i Add 3 **ii** 23, 29
d i Add 4 **ii** 19, 27
e i Subtract 6 **ii** 18
f i Subtract 3 **ii** 4, −2
g i Multiply by 2 **ii** 12
h i Multiply by 10 5 **ii** 200
i i Multiply by 4 **ii** 16
j i Multiply by −2 **ii** −16
k i Divide by 2 **ii** 50, $12\frac{1}{2}$
l i Divide by 2 **ii** $\frac{1}{2}, \frac{1}{4}$
6 B, the rule is +4 instead of +3
7 a 6, 9, 12, 15, 18 **b** 1, 6, 11, 16, 21
8 a 2, 4, 6, 8, 10 **b** 3, 5, 7, 9, 11 **c** 5, 4, 3, 2, 1
d 15, 12, 9, 6, 3 **e** 1, 4, 9, 16, 25 **f** 1, 7, 17, 31, 49
g 1, 8, 27, 64, 125 **h** 1, 3, 6, 10, 25
i 32, 16, $10\frac{2}{3}$, 8, $6\frac{2}{5}$ **j** 2, 4, 8, 16, 32
9 8, 13, 21, 34

21.1A

1 a  **b**
c
d
2 a, b −2, 1, 4, 7, ... Add 3; 4, 12, 36, 108, ... Multiply by 3;
3, 5, 7, 9, 11, ... Add 2; 20, 17, 14, 11, ... Subtract 3;
11, 9, 7, 5, 3, ... Subtract 2
3 a 5, 13, 29, 61, 125, ... **b** 509
4 a No **b** Yes, 121 **c** No **d** No
5 7, 10, 13, 16, ... Add 3 $T(n) = 3n + 4$
4, 1, −2, −5, ... Subtract 3 $T(n) = 7 − 3n$
6, 10, 14, 18, ... Add 4 $T(n) = 4n + 2$
3, 10, 17, 24, ... Add 7 $T(n) = 7n − 4$
6, 2, −2, −6, ... Subtract 4 $T(n) = 10 − 4n$
6 Yes, $4n − 2 = 2(2n − 1)$.
7 a Yes, $n \to 4n \to 5n \to 9n \to 14n \to 23n \to 37n \to 60n$.
b Yes, check students' examples.
8 9
9 Check students' presentations.

21.2S

1 a 3, 6, 9, 12, 15 **b** 6, 10, 14, 18, 22
c 7, 9, 11, 13, 15 **d** 8, 13, 18, 23, 28
e 3, 4, 5, 6, 7 **f** 3, 7, 11, 15, 19
g 2, 7, 12, 17, 22 **h** −1, 1, 3, 5, 7
i −6, −5, −4, −3, −2 **j** −7, −4, −1, 2, 5
k 5.5, 6, 6.5, 7, 7.5 **l** 1.5, 4, 6.5, 9, 11.5
2 a −2, −4, −6, −8, −10 **b** 11, 7, 3, −1, −5
c 9, 8, 7, 6, 5 **d** 20, 15, 10, 5, 0
e 9, 6, 3, 0, −3 **f** 9, 3, −3, −9, −15
g 3.5, 3, 2.5, 2, 1.5 **h** −5, −12, −19, −26, −33
i $\frac{1}{2}, \frac{2}{3}, \frac{5}{6}, 1, 1\frac{1}{6}$
3 a 47 **b** 67 **c** 407 **d** 4007
4 a i 2 **ii** 10 **iii** 100 **iv** 200
b i −4 **ii** −20 **iii** −200 **iv** −400
c i 5 **ii** 17 **iii** 152 **iv** 302
d i 12 **ii** 32 **iii** 257 **iv** 507
e i 11 **ii** 7 **iii** −38 **iv** −88
f i 5 **ii** −3 **iii** −93 **iv** −193
g i 0.3 **ii** 1.5 **iii** 15 **iv** 30
h i 1.25 **ii** 2.25 **iii** 13.5 **iv** 26
i i 9.9 **ii** 9.5 **iii** 5 **iv** 0
j i 1 **ii** 3 **iii** 25.5 **iv** 50.5
5 a Linear **b** Non-linear **c** Non-linear
d Linear **e** Linear **f** Linear
g Non-linear **h** Non-linear **i** Linear
6 a +3
b $T(n) = 5, 8, 11, 14, 17$; $3n = 3, 6, 9, 12, 15$; Difference = +2
c $T(n) = 3n + 2$
7 a $2n + 1$ **b** $3n + 1$ **c** $3n + 2$
d $6n − 2$ **e** $10n − 3$ **f** $8n − 6$
g $1.5n + 0.5$ **h** $0.6n + 0.8$ **i** $0.5n + 1.5$
j $2.95 + 0.15n$
8 a $13 − 2n$ **b** $19 − 3n$ **c** $12 − 2n$
d $8 − 3n$ **e** $37 − 10n$ **f** $9.5 − 1.5n$
g $4\frac{1}{2} − \frac{1}{2}n$ **h** $6 − 5n$ **i** $6\frac{1}{2} − 1\frac{1}{2}n$
j $4.5 − 2.5n$
9 a Students' answers, for example, 36.
b $4n − 2$
c $4 \times 10 − 2 = 38$
10 No, $n = 15.6$, not an integer.

21.2A

1 No, $T(n) = 4n - 1$ so $T(10) = 39$.
2 No, $T(n) = 4n + 9$.
3 **a, b** 1, 3, 5, 7, ... $2n - 1$; 1, 4, 7, 10, ... $3n - 2$;
  3, 5, 7, 9, 11, ... $2n + 1$; 3, 7, 11, 15, ... $4n - 1$;
  3, 4, 5, 6, 7, ... $n + 2$
4 **a** 5, 7, 9 **b** $m = 2n + 1$ **c** 101 **d** 50th
5 57
6 Students' answers, for example,

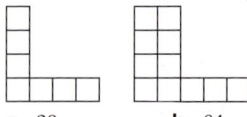

7 **a** 29 **b** 84
8 **a** 8, 10, 12, 14, ...; $2 \times 1 + 6 = 8$, $2 \times 2 + 6 = 10$,
  $2 \times 3 + 6 = 12$, $2 \times 4 + 6 = 14$, etc.
  b 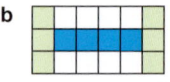 Three squares at either end gives '+6'.
  Two rows above and below the blue row
  gives '$2n$'.
9 Yes, $10n - 15 = 5(2n - 3)$.
10 **a** Yes, 15th term. **b** No, n is not an integer.
  c No, n is not an integer. **d** No, n is not an integer.
11 **a** $T(n) = 4n - 3$ **b** $T(n) = 12 - 3n$

21.3S

1 **a** 21 **b** 36 **c** 216
2

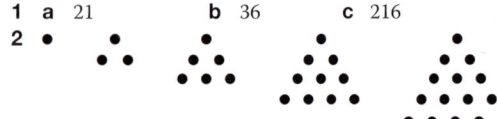

3 **a** Arithmetic **b** Arithmetic
  c Fibonacci-type **d** Quadratic
  e Geometric **f** Arithmetic
  g Fibonacci-type **h** Geometric
  i Quadratic **j** Arithmetic
  k Geometric **l** Geometric
4 **a** 2, 4, 6, 8, 10 **b** 2, 4, 8, 16, 32
  c 2, 4, 6, 10, 16 **d** 2, 4, 7, 11, 16
5 **a** 15 **b** 20 **c** 15 **d** 17
6 Both are correct.
7 Yes, $1^2, 2^2, 3^2, 4^2, ...$ are the square numbers.
8 No, this will generate the cube numbers.
9 Yes, 1, 3, 6, 10, ... are the triangular numbers.
10 **a** 3, 6, 12, 24 **b** 10, 50, 250, 1250
  c 3, 1.5, 0.75, 0.375 **d** 2, −6, 18, −54
  e $\frac{1}{2}, \frac{1}{4}, \frac{1}{8}, \frac{1}{16}$ **f** −3, 6, −12, 24
  g 4, $4\sqrt{3}$, 12, $12\sqrt{3}$
11 Students' answers, for example, $a = 1.5$, $b = 11.5$
12 5, 9, 17, 31
13 **a** 7.75 **b** 7.9921875
  c Students' answers, for example, the sum of the terms is
  getting closer to 8.
14 **a** $21 + 3 + 6$ **b** $28 + 3$ **c** $28 + 3 + 1$
15 **a** $9 + 4 + 1 + 1$ **b** $16 + 16$ **c** $36 + 16 + 4$

21.3A

1 Students' answers, check common difference is 4.
2 Students' answers, check common difference doubles.
3 **a** Students' answers, for example, 1, 64, 729, 4096, ...
  b i 1, 8, 27, 64, ... **ii** 1, 4, 9, 16, ...
4 Yes, but only if the terms in the triangular number sequence are
  consecutive, for example, $1 + 3 = 4$.
5 Yes, 1, $1 + 2$, $1 + 2 + 3$, ... = 1, 3, 6, ...
6 Students answers', for example,
  a i 4, 7, 10, 13, 16, ... **ii** 4, 12, 36, 108, 354, ...
  iii 4, 5, 9, 14, 23, ... **iv** 4, 6, 10, 16, 24, ...

b i Add 3 **ii** Multiply by 3
  iii 2nd term = 5, then add the previous two terms
  iv Square n, subtract n, add 4
7 **a** Check students' presentation.
  b i The diagonals of Pascal's Triangle sum to the Fibonacci
  numbers.
  ii The ratio of consecutive terms converge to the golden
  ratio.
8 Pentagonal numbers 1, 5, 12, 22, ...
  Hexagonal numbers 1, 6, 15, 28, ...
  Tetrahedral numbers 1, 4, 10, 20, ...
9 Check students' research.
10 Check students' research.
11 **a** Option 3 £2.7 million, Option 1 £500, Option 2 £6600
  b Option 3 £10.7 million, Option 1 £500, Option 2 £1000
12 **A** $T(n) = \frac{1}{2}n(n + 1)$ 1, 3, 6, 10, ... Triangular
  B $T(n) = 4n - 3$ 1, 5, 9, 13, ... Arithmetic
  C $T(n) = n^2 - n + 1$ 1, 3, 7, 13, ... Quadratic
  D $T(n) = n^2$ 1, 4, 9, 16, ... Square

Review 21

1 **a i** 21, 25, 29 **ii** 4, −6, −16 **iii** 2.7, 3.3, 3.9
  b i + 4 **ii** −10 **iii** + 0.6
2 **a** 29 **b** 61 **c** 3 **d** $8\frac{1}{2}$
3 **a** 27 **b** 146 **c** 10002
4 **a** $3n$ **b** $5n - 1$ **c** $6n + 2$ **d** $22 - 2n$
5 **a** [diagram of matchsticks]
  b Number of matchsticks = 4, 7, 10, 13, 16, 19
  c $m = 3s + 1$
6 **a** 24, 48, 96 **b** Geometric
7 **a i** 125, 216 **ii** 15, 21
  b i Cubic **ii** Triangular
8 **a** Geometric **b** Fibonnaci-type
  c Arithmetic **d** Quadratic

Assessment 21

1 **a i** −20, −30 **ii** 30 **iii** ↓ **iv** −10 **v** −10
  b i 188, 210 **ii** 154 **iii** ↑ **iv** +22 **v** +22
  c i 3, 6 **ii** −9, −3 **iii** ↑ **iii** +3 **iv** +3
  d i −10, −21 **ii** 12 **iii** ↓ **iv** −11 **v** −11
  e i 49, 54 **ii** 34, 39 **iii** ↑ **iv** +5 **v** +5
  f i 3.1, 3.5 **ii** 1.9 **iii** ↑ **iv** +0.4 **v** +0.4
  g i 84, 96 **ii** 24, 60 **iii** ↑ **iv** +12 **v** +12
  h i 1.07, 1.06 **ii** 1.09 **iii** ↓ **iv** −0.01 **v** −0.01
  i i −1.15, −1.20 **ii** −1, −1.1 **iii** ↓ **iv** −0.05 **v** −0.05
  j i 1 000 008, 1 000 011 **ii** 999 999, 1 000 002 **iii** ↑
  iv +3 **v** +3
  k i −0.75, −1.5 **ii** 0.75 **iii** ↓ **iv** −0.75 **v** −0.75
  l i −38, −45 **ii** −24 **iii** ↓ **iv** −7 **v** −7
  m i 6.625, 6.75 **ii** 6.125 **iii** ↑ **iv** +0.125
  v +0.125
2 **a** v **b** iv **c** vii **d** vi
  e i **f** iii **g** viii **h** ii
3 **a** +10 **b** $10n - 9$
4 **a** Jack [dots diagram] **b** $2n - 1$ **c** 19
5 **a** 16, 49, 144, 196 **b** Square numbers $1^2, 2^2, 3^2, 4^2, 5^2, ...$
6 **a** 9 red, 10 blue **b** 49 red, 50 blue
  c 99 red, 100 blue **d** $n - 1$ red, n blue
7 **a i** $2n + 8$ **ii** $2n + 2$ **b** 54 **c** 42
8 **a** $3n - 1$, $3 \times 1 - 1 = 2$, $3 \times 2 - 1 = 5$, $3 \times 3 - 1 = 8$,
  $3 \times 4 - 1 = 11$
  b 3, −3, each term is 6 less than the previous.
  c $33 - 6n$
  d −279

9 a Correct, $2 \times 10 + 7 = 27$.
  b Incorrect, $6 \times 1 - 5 = 1$, $6 \times 2 - 5 = 7$, $6 \times 3 - 5 = 13$.
  c Incorrect, $13 - 3 \times 100 = -287$.
  d Incorrect, $10^2 - 10 = 100 - 10 = 90$.
  e Correct. $15 - 3 \times 100^2 = 15 - 30000 = -29985$.
10 No, $1:2, 1:4, 1:6$

Chapter 22

Check in 22

1 £28.00
2 1.5
3 £68

22.1S

1 a 7.7 m/s **b** 7.1 m/s **c** 6.8 m/s **d** 5.1 m/s
2 a 160 km **b** 161 miles **c** 54 m **d** 288 miles
3 a 3 hours **b** 4 hours **c** 3 hours **d** 4 hours
4 1.425 g/cm³
5 a 9.4575 g **b** 6.15 litres
6 a 5704 kg **b** 3400 kg **c** 61 824 kg **d** 125 cm³
  e 24.7 cm³ (3 sf) **f** 0.002 m³
7 a 1.29 N/m² **b** 19.3 N/cm² **c** 0.46875 m² **d** 7424 N
  e 12.64 cm² **f** 68.88 N
8 £2.10/m
9 £11.95/hour
10 a 2.5 litres/s **b** 1.6 litres/s
11 1200 litres
12 a 2.4 units/hour **b** 57.6 units
13 a 500 km **b** 20 litres **c** 12.5 km/litre
14 a $1.5 per £ **b** $187.5 **c** £80

22.1A

1 89.25 mph
2 300 kg
3 8.5 g/cm³
4 32 mph
5 525, 2 hrs 12 min 30 s, 37.3 (1 dp), 37.1 (1 dp)
6 a 5 g/cm³ **b** 87.88 g
7 a 5.96 g/cm³ **b** 105 blocks
8 a 10.5 g/cm³ **b** 14.5 g/cm³ **c** 2.7 g/cm³
9 a £89.25 **b** 15 hours
10 £42.24
11 4 hrs 15 s
12 a 550 N/cm² **b** 4.32 N/cm²

22.2S

1 a £8.25 **b** £12.38 **c** £18.84 **d** £75.76
2 a £17.25 **b** £11.90 **c** £7.76 **d** £16.73
3 12 kg
4 75 kg
5 a £0.60 **b** £1.80
6 a £0.40 **b** £2.80
7 a £0.03 **b** 0.52 g
8 a £1.88 **b** £1.48 **c** £1.73 **d** £2.04
  e £1.86
9 £3.64
10 a Students' answers, for example, 2 cm = 20 mm, 3 cm = 30 mm, etc.
  b 100 mm
11 a 0 lbs, 22 lbs, 11 lbs
  b Straight line through (0, 0) and (10, 22).
  c i 4.5 kg **ii** 2.2–2.3 kg **iii** 6.6 lbs **iv** 5.5 lbs
12 a i 32 km **ii** 37 miles
  b Charlie **c** $y = \frac{5}{8}x$

22.2A

1 Pack A 1.2p per pin, Pack B 1.15p per pin
2 Regular 70p per sheet, Super 77p per sheet

3 a 640 g **b** 960 g **c** 1120 g **d** 2400 g
4 a 720 g **b** 2040 g **c** 2880 g **d** 12 000 g
5 a 6.5 tonnes **b** 51:14
6 No, his rate of pay for the four hours is £6/hour, the rate of pay for eight hours is £6.25/hour.
7 No, the formula is, number of radiators = days worked × 12
8 a Yes, the ratio is 1:1.4 for each size (1 dp).
  b Yes. Straight line passing through (148, 210) and (841, 1189)

22.3S

1 a Doubled **b** Halved
  c Multiplied by 6 **d** Divided by 10
  e Multiplied by 0.7
2 a Halved **b** Doubled
  c Divided by 6 **d** Multiplied by 10
  e Divided by 0.7
3 a 60 **b** 10 **c** 4 hours
4 a 1800 weeks **b** 6 weeks
  c 1.5 weeks
5 a 10 **b** 25 **c** 40 hours
6

Inverse proportion	Direct proportion	Neither
$y = \frac{20}{x}$	$y = 4x$	$y = 5x + 1$
$y = \frac{13.5}{x}$	$y = \frac{x}{10}$	$y = \frac{x+1}{10}$

7 a

x	$\frac{1}{4}$	1	2	4	8	16
y	64	16	8	4	2	1

  b Reciprocal curve passing through correct coordinates.
  c Yes
8 a 12 **b** Check students' explanations.

22.3A

1 2 hours
2 a 5.3 minutes (1 dp) **b** 3 minutes
3 a i 250 kg **ii** 250 kg
  b No, Safeload = 3200 kg < 3500 kg.
4 10 is left over.

y	0.5	1.5	2	3	4	4.8	7.5
z	48	16	12	8	6	5	3.2

5 a The product is always 48
  b

Length (cm)	Width (cm)	Perimeter (cm)
1	48	98
2	24	52
3	16	38
4	12	32
6	8	28
8	6	28
12	4	32
16	3	38
24	2	52
48	1	98

  The perimeter and length do not have a constant product.

22.4S

1 a 1.3 **b** 0.7 **c** 1.03 **d** 0.97
  e 1.25 **f** 0.75 **g** 1.025 **h** 0.975
2 a Population = 1, 2, 4, 8, 16, 32, 64, 128, 256, 512
  b i 3rd hour **ii** 4th hour **iii** 5th hour
  c i 180 **ii** 360
  d The population has doubled.
3 a Amount = 1000, 500, 250, 125, 62.5, 31.25, 15.625
  b i 4 **ii** 7 **iii** 11
4 a Price of boots (£) = 50, 40, 32, 25.60, 20.48, 16.38
  b 67.2% (3 sf)
5 a Number of trout = 800, 680, 578, 491, 417, 355, 302, 256, 218

*b

Trout in lake

Number of trout (y-axis, 0 to 800 in steps of 100) vs Time (years) (x-axis, 0 to 8)

c **i** $100\% - 15\% = 85\% = 0.85$, 800×0.85^n

    ii See part **a**.

6 **a** 400×1.35^n **b** In the 7th hour.

7 **a** Amount (£) = 2080, 2163.20, 2249.73, 2339.72, 2433.31, 2530.64, 2631.87, 2737.14, 2846.63, 2960.50

  b 48% (nearest %)

*c **i** $(100 + r)\% = \frac{100 + r}{100} = 1 + \frac{r}{100}$, so $A = P(1+\frac{r}{100})^n$

    ii £2960.49

22.4A

1 **a** **i** £1157.63   **ii** £157.63

  b **i** £316.33   **ii** £66.33

  c **i** £1075.27   **ii** £235.27

  d **i** £66 053.11   **ii** £21 053.11

2 **a** Lily has worked out simple interest, not compound interest.

  b £131.24

3 **a** £8352.10   **b** 9 years

4 **a** 56 859   **b** In the 10th year.

  c Immigration and emigration.

5 **a** 350

  b $0.96 = 1 - 0.04 = 1 - 4\%$, they expects the number of workers to go down by 4% each year.

6 **a** The number of vehicles per day will be 2400 when the road opens and the planner expects the number of vehicles per day to go up by 8% every month.

  b The number of vehicles on the first day may not be 2400. Road use may vary with the seasons. There will be a limit on the number of cars the road can take.

7 **a** Account A £7 058.31, Account B £7 057

  b Students' answers for example, Mark may not want to leave his money in the account for 3 years and Account B gives more interest in the first two years.

*8 **a** D   **b** C   **c** A   **d** B

*9 No, Easy saver's annual multiplier = 1.04, Half-yearly saver's annual multiplier = $(1.02)^2 = 1.0404$

Review 22

1 **a** 0.14 m   **b** 16 min, 40 s

  c 3000 cm³   **d** 18 km/h

2 2.5 N/m²

3 £0.24 per 100 g

4 18 000 cm²

5 **a** **i** 64 cm³   **ii** 1728 cm³

  b **i** 96 cm³   **ii** 864 cm³

6 **a** 32   **b** 11

7 **a** $\frac{2}{3}$ **b** 9

8 **a** $P = 3N$ **b** 3   **c** Price per magazine

9 **a** A   **b** C

10 **a** **i** £2040   **ii** £2122.42

  b $V = 2000 \times 1.02^t$

Assessment 22

1 **a** 30 min   **b** 30 mph   **c** 39 mph

2 8 min 20 s

3 **a** 39.37 in   **b** £9.75   **c** 14 min 56 s

4 The wombat, Bolt's speed = 36.73 km/h.

5 **a** 7.28 g/cm³ **b** 23.81 cm³ **c** 103 g

6 **a** 45 tins   **b** Yes, $100 \div (3 \div 2) = 67$ days.

  c No, $30 \div (9 \div 2) = 7$ days.

7 **a** 157.5 litres **b** 379 cm

8 **a** 69.12 kg   **b** $x = 24$ cm **c** $y = 10$ cm

9 **a** $\tan(x) = 0$, 0.0175, 0.0349, 0.0524, 0.0699, 0.0875, 0.1051, 0.1228, 0.1405, 0.1584, 0.1763

  b Graph of y (0 to 1.8) vs x (0° to 10°)

  c The graph is a straight line in this range.

10 **a** 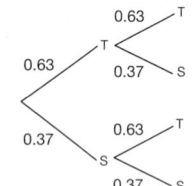 Graph of Monthly rent (£) (0 to 900) vs Distance (km) (0 to 7), labelled "Flat 5"

  b Flat 5 lies furthest from the curve.   **c** 3.5 km

11 **a** £2926   **b** £2858

12 **a** £27 000, 5%   **b** £16165.90 **c** £9545.80

Revision 4

1 6.71 km (3 sf)

2 **a** 7.14 km   **b** 161 cm   **c** 8.6°

3 **a** The diagonal from the vertical angle to the base bisects the base.

  b 14.5 cm (3 sf)

4 $RQ = \mathbf{p} - 2\mathbf{q}$, $NM = \frac{\mathbf{p}}{2} - \mathbf{q}$, $NM = \frac{1}{2}RQ$

5 $a = 40$

6 **a** $\binom{15}{-30}$ **b** $\binom{6}{-18}$ **c** $\binom{-9}{-9}$ **d** $\binom{6}{-15}$

  e $\binom{6}{-15}$ **f** $\binom{3}{6}$ **g** $\binom{0}{-6}$ **h** $\binom{-39}{69}$

  i $\binom{0.5}{0}$ **j** $\binom{0}{-6}$

7 **a** $1 - 0.63 = 0.37$

  b Tree diagram: 1st day, 2nd day
  0.63 → T; T → 0.63 T, 0.37 S; 0.37 → S; S → 0.63 T, 0.37 S

  c **i** 0.2331   **ii** 0.6031   **iii** 0.1369

8 **a** Yes, $1^2 + 1 = 2$, $2^2 + 2 = 6$, $3^2 + 3 = 12$, $4^2 + 4 = 20$.

  b 2550, 10 100   **c** $2n + 2$ **d** 102, 202

9 6

10 Liddi 27p, Addle 27.5p

11 **a** cm   **b** kg   **c** ml   **d** mm

  e l   **f** km   **g** g   **h** m

  i tonnes

12 **a**

3	4	5
6	10	15
3	6	10
9	16	25

  b **i** 1275   **ii** 5050   **c** r^2

  d $\frac{r(r-1)}{2}$ **e** $\frac{2(2-1)}{2} = 1$

  f 1770

13 **a** 1300 ml   **b** 0.076 kg

14 **a** 40 mph   **b** 1 hr 26 min   **c** 6 min

15 **a** 15.75 g/cm³ **b** 5.48 kg (3 sf) **c** 2.86 cm³ (3 sf)

16 **a** 10.5 m   **b** 2 hrs   **c** 4 hrs

  d Yes, Team size = $\frac{8}{3}$ Scaffolding.

  e No, Team size = $\frac{80}{3}$ Time.

Answers